INSTRUMENTAL THIN-LAYER CHROMATOGRAPHY

INSTRUMENTAL THIN-LAYER CHROMATOGRAPHY

Edited by

COLIN F. POOLE
Department of Chemistry, Wayne State University, Detroit, MI, USA

ELSEVIER

AMSTERDAM BOSTON HEIDELBERG LONDON NEW YORK OXFORD
PARIS SAN DIEGO SAN FRANCISCO SINGAPORE SYDNEY TOKYO

Elsevier
Radarweg 29, PO Box 211, 1000 AE Amsterdam, Netherlands
The Boulevard, Langford Lane, Kidlington, Oxford OX5 1GB, UK
225 Wyman Street, Waltham, MA 02451, USA

Notices
Knowledge and best practice in this field are constantly changing. As new research and experience broaden our understanding, changes in research methods, professional practices, or medical treatment may become necessary.

Practitioners and researchers must always rely on their own experience and knowledge in evaluating and using any information, methods, compounds, or experiments described herein. In using such information or methods they should be mindful of their own safety and the safety of others, including parties for whom they have a professional responsibility.

To the fullest extent of the law, neither the Publisher nor the authors, contributors, or editors, assume any liability for any injury and/or damage to persons or property as a matter of products liability, negligence or otherwise, or from any use or operation of any methods, products, instructions, or ideas contained in the material herein.

Library of Congress Cataloging-in-Publication Data
Application submitted

British Library Cataloguing-in-Publication Data
A catalogue record for this book is available from the British Library.

ISBN: 978-0-12-417223-4

For information on all Elsevier publications visit our web site at http://store.elsevier.com/

Printed and bound in Poland

Contents

Contributors

B. Avula National Center for Natural Products Research, School of Pharmacy, University of Mississippi, University, MS, USA

Pierre Bernard-Savary Chromacim SAS, France

Michael J. Brett Department of Electrical and Computer Engineering, University of Alberta, Edmonton, AB, Canada; NRC National Institute for Nanotechnology, Edmonton, AB, Canada

Vicente L. Cebolla ICB-CSIC, Zaragoza, Spain

Sychyi Cheng Department of Chemistry, National Sun Yat-Sen University, Kaohsiung, Taiwan

Irena Choma Chromatographic Method Department, M. Curie-Skłodowska University, Lublin, Poland

Łukasz Cieśla Department of Inorganic Chemistry, Medical University of Lublin, Lublin, Poland

Virginia Coman Raluca Ripan Institute for Research in Chemistry, Babeş-Bolyai University, Cluj-Napoca, Romania

Florina Copaciu Raluca Ripan Institute for Research in Chemistry, Babeş-Bolyai University, Cluj-Napoca, Romania

B. Dejaegher Department of Analytical Chemistry and Pharmaceutical Technology (FABI), Center for Pharmaceutical Research (CePhaR), Vrije Universiteit Brussel (VUB), Brussels, Belgium; Laboratory of Instrumental Analysis and Bioelectrochemistry, Institute of Pharmacy, Université Libre de Bruxelles (ULB), Brussels, Belgium

Sunil R. Dhaneshwar Department of Pharmaceutical Chemistry, RAK College of Pharmaceutical Sciences, Ras Al Khaimah Medical and Health Sciences University, Ras Al Khaimah, UAE; Poona College of Pharmacy, Bharati Vidyapeeth University, Erandwane, Pune, India

Tadeusz H. Dzido Department of Physical Chemistry, Medical University of Lublin, Lublin, Poland

Beate Fuchs Medical Department, University of Leipzig, Institute of Medical Physics and Biophysics, Härtelstr, Leipzig, Germany

Ahmed M. Galal National Center for Natural Products Research, School of Pharmacy, University of Mississippi, University, MS, USA

Rosa Garriga Department of Physical Chemistry, University of Zaragoza, Zaragoza, Spain

Vesna Glavnik Laboratory for Food Chemistry, National Institute of Chemistry, Ljubljana, Slovenia

Radosław Ł. Gwarda Department of Physical Chemistry, Medical University of Lublin, Lublin, Poland

Aneta Hałka-Grysińska Department of Physical Chemistry, Medical University of Lublin, Lublin, Poland

Mirosław A. Hawrył Department of Inorganic Chemistry, Medical University of Lublin, Lublin, Poland

Carmen Jarne ICB-CSIC, Zaragoza, Spain

Wioleta Jesionek Chromatographic Method Department, M. Curie-Skłodowska University, Lublin, Poland

Steven R. Jim Department of Electrical and Computer Engineering, University of Alberta, Edmonton, AB, Canada

Ikhlas A. Khan National Center for Natural Products Research, School of Pharmacy, University of Mississippi, University, MS, USA; Department of Pharmacognosy, School of Pharmacy, University of Mississippi, University, MS, USA; Cancer Institute, University of Mississippi Medical Center, Jackson, MS, USA

S. Krüger Institute of Nutritional Science, Justus Liebig University, Giessen, Germany

Luis Membrado ICB-CSIC, Zaragoza, Spain

Emil Mincsovics Department of Genetics and Plant Breeding, Faculty of Horticultural Sciences, Corvinus University of Budapest, Budapest, Hungary

Ágnes M. Móricz Centre for Agricultural Research, Plant Protection Institute, Hungarian Academy of Sciences, Budapest, Hungary

G. Morlock Institute of Nutritional Science, Justus Liebig University, Giessen, Germany

Colin F. Poole Department of Chemistry, Wayne State University, Detroit, MI, USA

Yulia Popkova Medical Department, University of Leipzig, Institute of Medical Physics and Biophysics, Härtelstr, Leipzig, Germany

Jürgen Schiller Medical Department, University of Leipzig, Institute of Medical Physics and Biophysics, Härtelstr, Leipzig, Germany

Wolfgang Schwack Institute of Food Chemistry, University of Hohenheim, Stuttgart, Germany

D.H. Shewiyo Tanzania Food and Drugs Authority, Dar es Salaam, Tanzania

Jentaie Shiea Department of Chemistry, National Sun Yat-Sen University, Kaohsiung, Taiwan; Department of Medicinal and Applied Chemistry, Kaohsiung Medical University, Kaohsiung, Taiwan

Bernd Spangenberg Department of Process Engineering, University of Applied Sciences Offenburg, Offenburg, Germany

Rosmarie Süß Medical Department, University of Leipzig, Institute of Medical Physics and Biophysics, Härtelstr, Leipzig, Germany

Tomasz Tuzimski Department of Physical Chemistry, Medical University in Lublin, Lublin, Poland

Ernő Tyihák Centre for Agricultural Research, Plant Protection Institute, Hungarian Academy of Sciences, Budapest, Hungary

Y. Vander Heyden Department of Analytical Chemistry and Pharmaceutical Technology (FABI), Center for Pharmaceutical Research (CePhaR), Vrije Universiteit Brussel (VUB), Brussels, Belgium

Jesús Vela Department of Analytical Chemistry, University of Zaragoza, Zaragoza, Spain

Irena Vovk Laboratory for Food Chemistry, National Institute of Chemistry, Ljubljana, Slovenia

Monika Waksmundzka-Hajnos Department of Inorganic Chemistry, Medical University of Lublin, Lublin, Poland

Paweł K. Zarzycki Section of Toxicology and Bioanalytics, Koszalin University of Technology, Koszalin, Poland

CHAPTER

1

Milestones, Core Concepts, and Contrasts

Colin F. Poole

Department of Chemistry, Wayne State University, Detroit, MI, USA

1.1 INTRODUCTION

Column chromatography and thin-layer chromatography are alternative formats for liquid chromatography [1]. Both formats exist as simple laboratory tools requiring little instrumentation and also as fully instrumental techniques. In both the cases, the stationary phase consists of a sorbent bed containing homogeneously packed particles or as a porous monolith. When movement of the mobile phase through the sorbent bed is controlled by capillary forces, the separation performance is suboptimal but requires little instrumentation affording a convenient and flexible arrangement for simple separations at the analytical or preparative scale. For faster separations, or separations with a higher peak capacity, a mechanism is required to enhance the mobile phase velocity. This requires instrumentation to pressurize the mobile phase and is the basis of high-pressure (or high-performance) liquid chromatography (HPLC) for columns and forced flow (or overpressured layer chromatography) for layers [2–4]. Although forced-flow instrumentation for thin-layer chromatography is commercially available, it is not in common use. Thus, whereas the practice of HPLC is a forced-flow technique, the practice of thin-layer chromatography is predominantly a capillary-controlled flow technique. In the latter case, instrumentation is required to optimize the various steps in the separation process and is referred to as high-performance thin-layer chromatography (HPTLC), or instrumental thin-layer chromatography, to distinguish the technique from conventional thin-layer chromatography (TLC) performed with much simpler equipment [5,6]. The general advantages of utilizing HPLC conditions versus conventional column chromatography are well known. The same argument cannot be made for

Instrumental Thin-Layer Chromatography
http://dx.doi.org/10.1016/B978-0-12-417223-4.00001-7

conventional TLC versus HPTLC, and the general migration of separations from conventional TLC practices to HPTLC has not been universal. In fact, one might say that conventional TLC thrives in the laboratory environment as a quick, inexpensive, flexible, and portable method for surveying the composition of simple mixtures while only a few laboratories are equipped to perform more complex and quantitative analyses by HPTLC.

1.2 MILESTONES

The origins of thin-layer chromatography can be traced to the experiments on drop chromatography performed by Izmailov and Shraiber in the late 1930s [7]. From this beginning, thin-layer chromatography evolved into a fast and more powerful tool than gravity flow column chromatography for analytical separations. Thin-layer chromatography, as we know it today, was established in the 1950s due in large part to the efforts of Stahl and Kirchner on different continents. Their main contribution was the development of standardized materials and procedures that led to improved performance and reproducibility, as well as popularizing the technique by contributing many new applications [8]. At about the same time, commercialization of materials and devices commenced making the technique accessible to all laboratories. This ushered in the golden era of thin-layer chromatography where it quickly displaced paper chromatography as the main analytical liquid chromatographic method. By the 1970s, high-pressure liquid chromatography was becoming firmly established as an alternative approach for liquid chromatography and eventually grew to eclipse thin-layer chromatography for analytical applications. Thin-layer chromatography did not disappear in subsequent years but became less well known to those working in analytical laboratories where its strengths were often under appreciated. Developments continued in thin-layer chromatography as indicated by the time line Figure 1.1 [6,9].

First the development of high-performance thin-layer chromatography in the late 1970s is discussed. Layers coated with smaller particles of a narrow size distribution required the development of instruments for their convenient use. This was achieved by the early 1980s and so began the second era of thin-layer chromatography, known as modern or instrumental thin-layer chromatography. The evolutionary changes during this second era are captured in a series of books, which if ordered chronologically, represent the state-of-the-art at different times during this period to the present [5,10–15]. The main characteristic features of modern thin-layer chromatography are the use of fine particle layers for fast and efficient separations; sorbents with a wide range of sorption

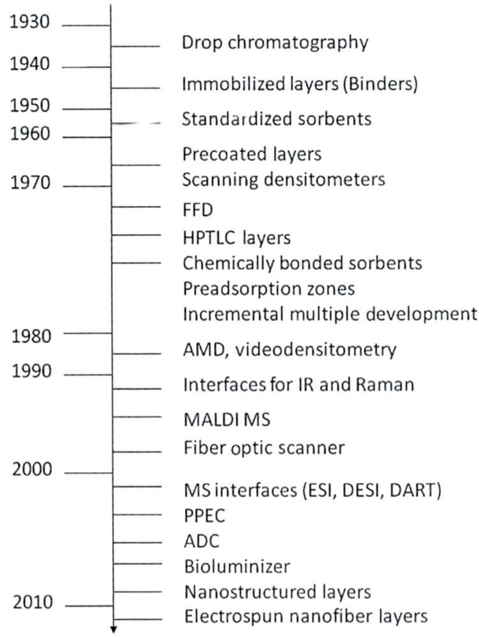

FIGURE 1.1 Time line depicting important developments in the evolution of modern thin-layer chromatography. FFD = forced-flow development in an overpressured development chamber; AMD = automated multiple development chamber; AMC = automatic development chamber; and PPEC = pressurized planar electrochromatography.

properties to optimize selectivity; the use of instrumentation for convenient and usually automated sample application, development and detection; and the accurate and precise in situ recording and quantification of chromatograms. Improvements in virtually all aspects of thin-layer chromatography continued over the next quarter century as indicated in Figure 1.1 and form the basis of subsequent chapters in this book. This period also marks the beginning of the philosophical division between conventional and high-performance thin-layer chromatography that has not been crossed by all those who use thin-layer chromatography. Expectations in terms of performance, ease of use, and quantitative information from the two approaches to thin-layer chromatography are truly opposite (see Section 1.1). As an example of expectations for a separation by modern thin-layer chromatography, see the chromatogram in Figure 1.2 for structurally similar ethyl estrogens (steroids used for birth control) [2]. Because of the small structural differences for these compounds, a high selectivity is required for their separation. Baseline separation is obtained with a short migration distance typical of fine particle layers and scanning densitometry provides a conventional record

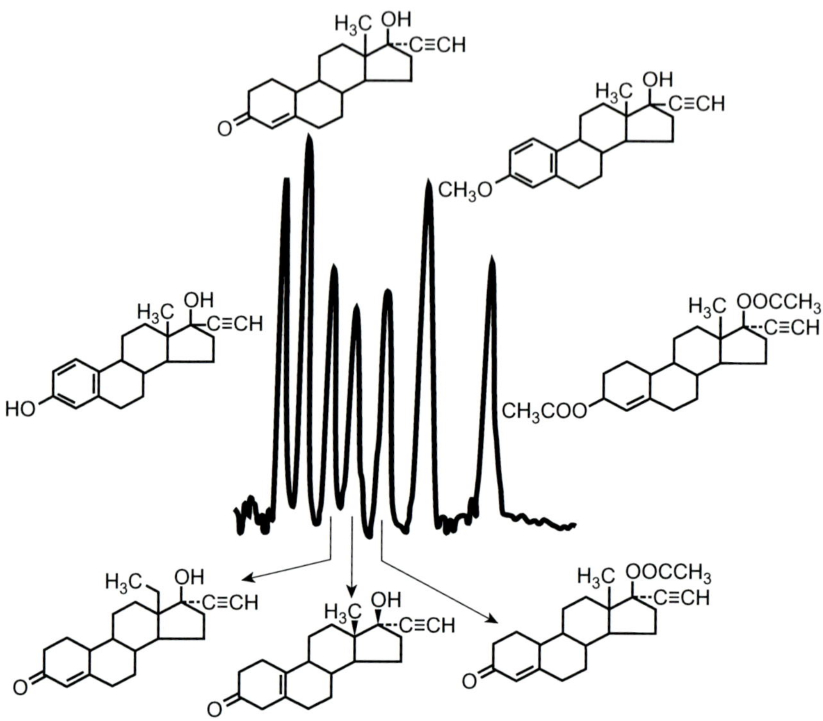

FIGURE 1.2 Separation of ethynyl steroids by modern thin-layer chromatography. Two 15 min developments with the mobile phase hexane–chloroform–carbon tetrachloride–ethanol (7:18:22:1) on a silica gel HPTLC plate. The chromatogram was recorded by scanning densitometry at 220 nm. *Reproduced with permission from Ref. [2].*

of the separation in the form of a chromatogram, as well as quantification of individual steroids after calibration. The quantitative results for tablet analysis are as accurate and precise as other chromatographic methods and the method is suitable for high-throughput routine tablet conformity analysis in which sample preparation requires no more than dissolution and filtration. Some specific reasons for choosing thin-layer chromatography for quantitative analysis are outlined in the next section.

1.3 ATTRIBUTES OF A PLANAR FORMAT

Columns afford a better arrangement for operation at high pressures and for variation of the separation conditions by the control of external parameters. The thin-layer format provides a better arrangement for high sample throughput, flexible detection strategies, and a greater tolerance of samples with a high-matrix burden [2,16]. The throughput advantage is a consequence of the possibility of separating multiple samples in parallel with each sample occupying a single lane (or track) on the layer and several samples assigned to different lanes for simultaneous separation. Column chromatography is inherently a sequential separation process in which the separation time for a group of samples is the product of the number of samples and the cycle time for an individual separation. A single 10 × 10 cm HPTLC plate can separate 18 samples and standards simultaneously if developed in one direction and twice that number if developed from two opposing edges to the center.

Separations in columns employ the elution mode in which all sample components experience the same separation distance defined by the column length, but because of the different nature of their interactions with the stationary phase, are separated in time. The typical experimental arrangement employs an injection device to insert the sample into the pressurized mobile phase close to the column entrance and an online detector at the column exit to record the separation [1]. For planar chromatography, it is more common to use a variation of the elution mode for the separation known as development. In this case, the mobile phase moves through the layer in a definite direction for a fixed distance, which is usually less than the bed length in the direction of mobile phase migration. In contrast to column chromatography individual sample components are separated in space (have different migration distances) achieved in an identical separation time. This has additional consequences for detection. Sample components are detected in the stationary phase, compared to the mobile phase in column chromatography, and at the completion of the separation the separated zones are stationary and can be interrogated free of time constraints. This simplifies the use of chemical and biological reagents for detection. A growing application of thin-layer chromatography is in effect-directed analysis where a biological response from, for example, luminescent bacteria or enzyme inhibition, is used to indicate the presence of substances with a specific toxic mechanism in contrast to structure-based detection strategies that are the basis of most common dynamic detection techniques [17,18]. In the development mode, the whole sample is contained in the sample lane and is available for detection. This is important for determining the integrity of a sample in contrast to column chromatography where the only sample components observed are those that are fully eluted from the column. Planar chromatography can be used to screen samples to predict their behavior in column chromatography and to assess the need for sample cleanup.

Typical applications in thin-layer chromatography are one-use applications after which the stationary phase (layer) is discarded. On account of their higher cost columns for HPLC are used to analyze multiple samples. Effective sample preparation improves the quality of information for target compounds in complex mixtures by both thin-layer and column chromatography, but a difference between the two techniques is

TABLE 1.1 Attributes of a Planar Format for Liquid Chromatography

Attribute	Application
Separation of samples in parallel	Low-cost analysis and high-throughput screening of samples requiring minimal sample preparation.
Disposable stationary phase	Analysis of crude samples (minimizing sample preparation requirements) analysis of a single or small number of samples when their composition and/or matrix properties are unknown. Analysis of samples containing components that remain sorbed to the separation medium or contain suspended microparticles.
Static detection	Samples requiring postchromatographic treatment for detection samples requiring sequential detection techniques (free of time constraints) for identification or confirmation separations can be evaluated in different locations or at different times.
Sample integrity	Total sample occupies the chromatogram not just that portion of the sample that elutes from the column.

the capability of thin-layer chromatography to handle samples with a heavy matrix burden without the need of potentially extensive remedial action. It is important that the sample contains no strongly retained components or suspended particles for column chromatography as these may alter the properties of the column and affect its ability to separate further samples. This is not a concern for thin-layer chromatography because reuse of the stationary phase is not planned, and therefore, in the initial phases of a study designed to gain an understanding of the properties of the sample matrix on the analysis, or to facilitate the analysis of samples with minimal sample preparation, then thin-layer chromatography is the method of choice [2,16].

The attributes of the planar format that are the basis for the continuing use of thin-layer chromatography in liquid chromatography are summarized in Table 1.1. The application chapters in this book provide definitive examples where these advantages have been realized in practice. In our own studies, we have preferred thin-layer chromatography for the screening step to identify samples requiring a detailed analysis (target analyte possibly present in a small number of samples), for samples with a heavy matrix burden or unknown matrix properties, for effect-directed analysis (target analytes unknown), for class fractionation (identification of analytes by group membership rather than individual identity), and for the standardization of plant materials by fingerprint analysis.

For some sample types, we generally prefer column chromatography, or use thin-layer chromatography for initial screening and column chromatography for the analysis. Samples that require a high peak capacity for their separation and identification are usually better handled by HPLC. The range of stationary phases available for HPTLC does not support the full range of separation mechanisms possible for column chromatography (e.g., size-exclusion chromatography, ion chromatography, etc.) [19]. In this case, format is no longer a matter of choice. Except for cellulose there are few high-performance layers suitable for the separation of biopolymers and HPTLC is little used in this field [20]. There are a wider range of detectors available for HPLC, such as electrochemical and light-scattering detectors, which offer more scope for selective or trace level detection. With the advent of convenient interfaces for thin-layer chromatography coupled to mass spectrometry, this is less of an issue than a few years ago [21,22]. The individual steps of sample application, development, and detection in thin-layer chromatography can be automated but complete automation of the whole process is not straightforward and manual intervention is required. When matrix effects are not conceived as a problem, HPLC can be set up for out-of-hours and unattended operation allowing staff to focus on other problems.

1.4 CONSEQUENCES OF CAPILLARY-CONTROLLED FLOW

The driving force for mobile phase migration in thin-layer chromatography is the decrease in free energy of the solvent as it enters the porous structure of the layer [12]. As it moves through the layer, its motion is opposed by frictional and gravitational forces. The mobile phase velocity declines as the solvent front moves further away from the solvent entry position, until at some position it falls to zero, and the solvent front becomes stationary. If the migration distance is not excessively long, the solvent front position as a function of time can be described by $Z_f^2 = \kappa t$ where Z_f is the distance of the solvent front from the solvent entry position; κ, the mobile phase velocity constant; and t, the time since the solvent commenced migration through the layer. At any position on the layer, the solvent front will be moving with a velocity u_f given by $u_f = \kappa/2Z_f$. The velocity constant is a system property, however, and depends on layer properties (permeability constant, particle size, thickness), mobile phase properties (viscosity, surface tension, contact angle), and environmental characteristics; in particular, the presence of a vapor phase and its saturation level [12,23,24]. Higher mobile phase velocities are obtained with layers of a relatively large particle size and mobile phases that fully wet the layer. Virtually all common solvents fully wet silica gel and polar chemically bonded phases but on hydrophobic layers with water-rich mobile phases this is not generally the case and separation times can be relatively long. This is addressed in part by increasing the particle size and reducing the degree of bonding of high-performance alkylsiloxane-bonded silica layers compared with silica gel and polar chemically bonded layers. Although larger particle sizes increase the mobile phase velocity, they also lead to an increase in band broadening and in practice a compromise must be made.

The presence of a vapor phase is a complication because it facilitates the exchange of mobile phase with the wet and dry portions of the layer that varies with the saturation level of the vapor phase. Mobile phase can be lost from the wet layer to the vapor phase by evaporation when in contact with an unsaturated vapor phase slowing down the migration velocity of the solvent front. In a saturated vapor phase, evaporative losses from the wet layer are minimized but the adsorption of mobile phase by the dry layer ahead of the solvent front is increased. This reduces the free volume of the layer that has to be filled up by the advancing solvent front and results in an increase in the mobile phase velocity. As a consequence, the mobile phase velocity constant is really a descriptive term that is defined by the experimental method and system properties, and is not a fundamental property. Compared to the conditions used in HPLC, there are three features of capillary-controlled flow systems to note: (1) there is no vapor phase in the column experiment; (2) in HPLC, the mobile phase velocity is independent of the column length; and (3) the mobile phase velocity is set by the system variables and cannot be independently optimized in thin-layer chromatography. These differences have consequences for the separation speed and performance of thin-layer chromatography compared with HPLC, and will be discussed below. The use of forced-flow conditions can mitigate the problems associated with capillary-controlled flow by operating the layer under conditions that approximate those of column chromatography. Some of the attributes of this technique are summarized in Table 1.2. Alternatively, multiple development strategies can provide a (partial) solution to some of the limitations of capillary-controlled flow with simpler instrumentation.

1.4.1 Retardation Factor

Substances are separated in space in the same time in thin-layer chromatography and the retardation factor is the standard measure of retention. The retention time, as used in column chromatography, has no meaning in regard to the identity of a substance when applied to thin-layer chromatography. The observed retardation factor is defined as the ratio of the distance migrated by the sample divided by the distance traveled by the solvent front, and for a linear development it is given by $R_F = Z_X/(Z_f - Z_0)$ where Z_X is the distance traveled by the sample from its origin, Z_f is the distance traveled by the solvent front measured from the mobile phase entry position (the solvent level at the start of the

TABLE 1.2 Advantages of Forced-Flow Development in Thin-Layer Chromatography

The solvent velocity can be optimized independent of other experimental variables.
A higher plate number can be achieved using smaller diameter particles and longer plate lengths than is possible for capillary-controlled flow systems.
A linear increase in the plate number with increasing migration distance is obtained. The limit is established by whether the pressure available can sustain the mobile phase velocity for the desired plate length.
Separation times are significantly shorter compared with capillary-controlled flow systems for a given plate number.
Mobile phases with poor sorbent wetting characteristics can be used.
Solvent-strength gradients (step or continuous) can be used for mixtures with a wide range of retention properties.

separation), and Z_0 is the distance from the sample origin to the mobile phase entry position. By definition, $1 \leq R_F \geq 0$ with R_F values indicated to two decimal places.

Systematic errors in determining R_F values result from the difficulty in locating the exact position of the solvent front [12,24]. As the solvent permeates the layer, the channels of narrower diameter are filled first leading to more rapid advancement of the mobile phase. Large pores below the solvent front fill more slowly, resulting in an increase in the thickness of the layer of mobile phase. The bulk mobile phase velocity representing saturated flow through the region occupied by the sample is moving at a slower velocity than the solvent front velocity (it is often approximated by $0.8u_f$). The solvent front corresponding to the saturated flow should be used to calculate the thermodynamic retardation factor and not the unsaturated solvent front observed as the border between the dry and wet regions of the layer. The saturated solvent front can be determined by using an unretained substance as a solvent front marker. The reproducibility of R_F values in unsaturated development chambers can be poor and concave solvent fronts formed resulting in higher R_F values for substances nearest to the edge of the layer. Fully automatic developing chambers with microprocessor control of all the steps in the development process provide the most reproducible conditions for R_F measurements [5].

Separations performed with mobile phases containing several solvents of different solvent strength may give rise to multiple solvent fronts. As the mobile phase moves through the layer, it becomes depleted in the solvent with the highest affinity for the stationary phase. Eventually, a secondary front is formed that separates the bulk mobile phase from the mobile phase now totally depleted in the solvent selectively adsorbed by the stationary phase. This process is referred to as solvent demixing and results in sample components moving in regions of different solvent strength and selectivity. Under these conditions, the retardation factor is no longer simply related to the bulk mobile phase composition. Solvent demixing is more common for separations in unsaturated developing chambers and layers with inadequate preconditioning. It is more common for some mobile phase compositions than others. Sample zones close to a secondary front are often focused into narrow bands with a different shape to neighboring zones, which allows demixing conditions to be identified.

Given the general difficulty of controlling absolute R_F values, it is common practice to separate standards and samples in the same system for identification purposes. For simple mixtures, the standards and analytes are usually the same. In surveillance programs, the simultaneous separation of selected reference standards is used to improve the certainty of identification by correcting observed R_F values to standard R_F values for automated library searches [25]. Using the mean list method, for example, all substances that migrate in an R_F window, which might be confused, are ranked and compared across a number of separation systems. Eventually only a single or handful of possible substances remain on the list as other candidates fall outside the window in one or more of the separation systems. This method is the basis of the systematic identification of toxic substances in forensic toxicology.

The R_F value is not related directly to the distribution properties of the separation system. The R_M value is used in studies that attempt to correlate migration properties with solute structure. The R_M value is equivalent to the ratio of the residence time of the solute in the stationary and mobile phases, and is formally equivalent to the retention factor ($\log k$) in column chromatography. It is calculated from the R_F value by using the formula $R_M = \log((1 - R_F)/R_F)$.

1.4.2 Plate Height

The plate height concept is not as useful in thin-layer chromatography as it is in column chromatography [3]. In column chromatography, the plate height assumes a constant value for compounds with similar diffusion coefficients and its variation with system properties affords an effective method for column design. For capillary-controlled flow in thin-layer chromatography system properties are interrelated and cannot be independently varied. The mobile phase velocity declines with the movement of the solvent front and solutes are forced to migrate through regions of different local efficiency. When they come to rest at the completion of the development, the observed plate height for each zone is an average of all the local plate heights the solute has encountered as it migrates behind the solvent front. The average plate height is thus dependent on the migration distance and the variation of the local plate heights for the regions below the zone center. It is a more complicated term than the plate height in column chromatography and more difficult to interpret in terms of layer properties [26]. For the purpose of standardization, an average value for a zone with an $R_F = 0.50$ or 1.00 is generally taken to compare separations on different layers. For typical mobile phases used in thin-layer chromatography, the bulk mobile phase velocity falls into the range 0.02–0.005 cm/s and never reaches the optimum value for high-performance layers of about 0.05 cm/s [27]. Capillary forces are too weak to provide optimum performance with modern thin-layer plates. There is also a dominant relationship between the solvent front migration distance, the average particle size, and the average plate height as shown in Figure 1.3 [28]. High-performance layers with an average particle size of about 5 μm afford compact zones of a small size if the solvent front migration distance does not exceed about 5–6 cm. For the more permeable conventional layers with an average particle size of about 10 μm, longer development distances up to 7–8 cm are feasible but the minimum in the average plate height is slightly larger than that for the high-performance layers. There is no point in using longer migration distances to improve separations in thin-layer chromatography since at higher values of the solvent front migration distance the mobile phase velocity is so low that zones broaden faster than their centers are

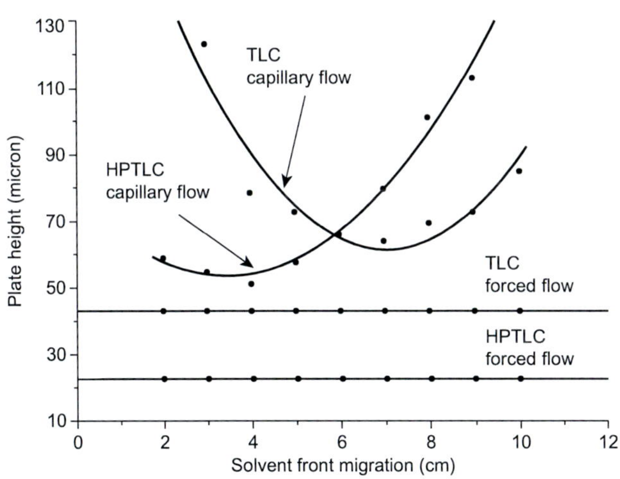

FIGURE 1.3 Variation of the average plate height as a function of the solvent front migration distance for conventional and high-performance layers using capillary-controlled flow and forced flow at the optimum mobile phase velocity. *Reproduced with permission from Ref. [28].*

separating. When the development length is optimized the separation performance of conventional layers is not very different from that of the high-performance layers; the primary virtue of the latter is that a shorter migration distance is required to achieve a given efficiency, resulting in faster separations and more compact zones that are easier to detect. The minimum in the average plate height under capillary-controlled flow conditions is always greater than the minimum observed for forced-flow conditions at the optimum mobile phase velocity for the same layer, and it can be concluded that for capillary-controlled flow conditions the potential performance of the layers is not fully realized.

The influence of the layer structure on the plate height can only be interpreted with the aid of suitable models. For thin-layer chromatography, general models have been adopted from column chromatography and then adapted to the prevailing conditions for thin-layer chromatography. The models developed by Guiochon and Siouffi [29,30] and Belenki et al. [31] are intuitively the most appealing and have been evaluated by others [14,26]. The general agreement with experimental data are only qualitative, and while suitable for providing insights into the likely contributing factors to zone broadening, are unsuitable for further optimizing the design of layers. For capillary-control flow, the interrelationship between variables and the difficulty of studying them independently by iteration of experimental conditions have diminished interest in theoretical treatments and studies in this area have fallen into neglect.

Under forced-flow conditions the minimum plate height is both higher and moved to a lower velocity compared to a fully optimized HPLC column, Figure 1.4 [14,19]. In addition, the optimum mobile phase

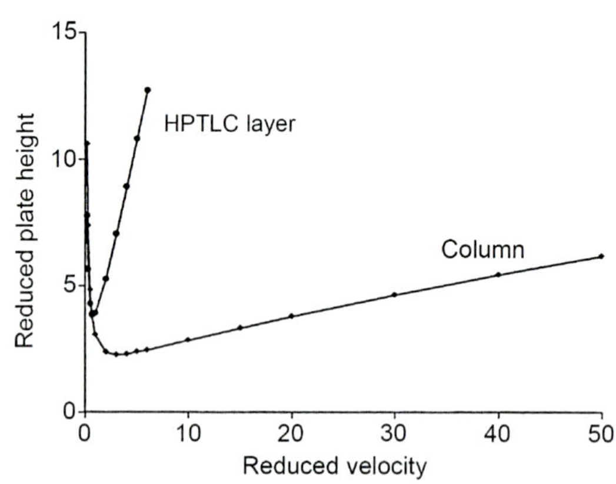

FIGURE 1.4 Variation of the reduced plate height as a function of the reduced mobile phase velocity (Knox plot) for a high-performance silica gel layer using forced-flow development and comparison to an ideal column for HPLC. *Reproduced with permission from Ref. [19].*

reduced velocity is shifted to a smaller value for the layer compared to the column. Also, at higher values of the mobile phase velocity, the plate height for the layer is significantly larger than that for the column. At high mobile phase velocities resistance to mass transfer is an order of magnitude higher for layers than for columns. This may be due to restricted diffusion within the porous particles or is a product of heterogeneous kinetic sorption on the sorbent and the binder added to layers to stabilize their structure. The consequences for forced-flow thin-layer chromatography are that separations will be slower than for columns and fast separations at high flow rates will be much less efficient than for columns. However, in terms of total efficiency and separation speed, the possibilities are more favorable for forced-flow than for capillary-controlled flow separations.

Some expectations for forced-flow separations are summarized in Table 1.3 [32]. For a typical development distance of 18 cm, a modest increase in performance (a maximum plate number of 8000) in a reasonable time of 9 min is achieved compared with the typical results for capillary-controlled flow (plate number <5000 in 25–45 min). Really significant increases in efficiency are achieved only by the use of smaller particles or longer development distances, in the latter case, at the expense of separation time. The optimum mobile phase velocity is sufficiently low that compared with columns the pressure drop remains modest for high-performance layers. Fast separations on high-performance layers result in low overall efficiency. If a relatively high plate number is required then small particles and high pressures are required, as is the case for columns. For layers of normal size, there are no useful benefits in separation time or performance to the use of conventional layers.

1.4.3 Multiple Development

Unidimensional multiple development techniques provide a complementary approach to forced flow for minimizing zone broadening in thin-layer chromatography. Multiple development is not an operation typically used in column chromatography and is more difficult to implement in column chromatography. All multiple development techniques employ successive repeated development of the layer in the same direction with the removal of mobile phase between developments. In consecutive developments, the experimental conditions can be conserved, the mobile phase composition varied, the entry position of the mobile phase increased, the solvent front migration distance increased, the number of developments varied, or a combination of these changes employed to optimize a separation [28,33–35]. Typical conditions employed to optimize multiple development in isocratic separations are summarized in Table 1.4 [19]. Each time the solvent front traverses the

TABLE 1.3 Simulation of Separation Performance with Forced-Flow Conditions

Development Length (cm)	Separation Time (min)	Pressure Drop (atm)	N_{max}
(A) HPTLC LAYER AND OPTIMUM CONDITIONS (H = 3.75, ν = 0.8, φ = 800, AND d$_P$ = 6 μM)			
8	4	2	3550
18	9	5	8000
50	25	13	22,200
100	50	26	44,400
(B) HPTLC LAYER FAST SEPARATION (H = 9 AND ν = 5)			
8	0.6	13	1490
18	1.4	29	3330
50	4.0	81	9250
100	8.0	161	18,500
(C) HPTLC LAYER AND OPTIMUM CONDITIONS (H = 3.75, ν = 0.8, φ = 800, AND d$_P$ = 3 μM)			
8	2.0	17	7610
18	4.5	37	17,100
50	12.5	103	47,600
100	25	207	95,200
(D) TLC LAYER (H = 4.5, ν = 0.8, φ = 600, AND d$_P$ = 9 μM)			
8	6	0.5	2000
18	13.5	1	4450
50	37.5	3	12,350
100	75	6	24,700

h = reduced plate height, ν = reduced mobile phase velocity, ϕ = flow resistance parameter, d_p = average particle size, and N_{max} = maximum plate number. See Ref. [24] for other conditions.

stationary sample zone, the zone is compressed in the development direction, Figure 1.5 The compression occurs because the mobile phase contacts the bottom edge of the zone first, and these molecules are transported into the immobilized molecules residing ahead of the solvent front. When the solvent front has reached beyond the zone, the focused zone migrates and is subjected to the normal zone-broadening mechanism. Experiment indicates that beyond a minimum number of developments, zone widths converge to a roughly constant value independent of the migration distance. Changing the solvent entry

TABLE 1.4 Optimum Separation Conditions for Isocratic Multiple Development

1. *Multiple chromatography* (experimental conditions conserved for each development)
   a. Resolution is controlled predominantly by the zone center separation since zone widths are (nearly) constant after the first few developments.
   b. The maximum zone center separation for two compounds with similar migration properties occurs when the zones have migrated 0.632 of the development distance.
   c. The maximum resolution is largely independent of the average R_F value. Although for separations with a small average R_F value, a large number of developments will be required to reach the region of maximum resolution.
   d. Difficult separations require solvents that result in a relatively low average R_F value (most selective mobile phase).
2. *Incremental multiple development*
   a. Increment the solvent entry position to maintain a fixed distance below the lowest zone.
      - Zone center separations increased due to the longer migration distances achieved.
      - Zone widths are minimized and reasonably constant throughout the chromatogram in later developments.
   b. Increment the solvent front migration distance.
      - Minimizes the build-up of compounds at the solvent front.
      - Enables resolution to be optimized throughout the chromatogram.
      - Resolution reaches a plateau value and is not reduced by increasing the number of developments.
      - The number of developments for a separation is determined by the properties.

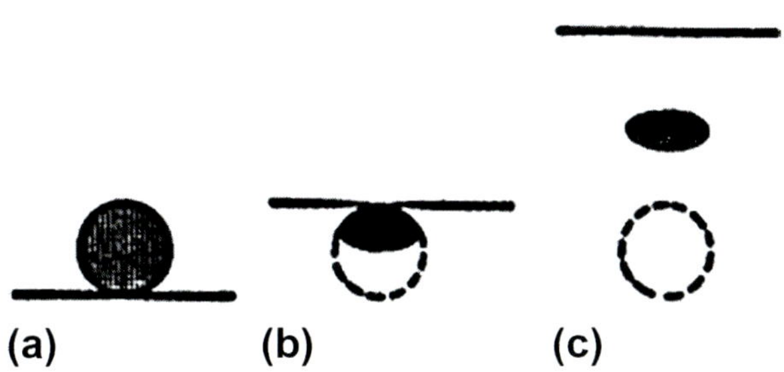

FIGURE 1.5 Illustration of the zone focusing mechanism in multiple development. (a) The solvent front approaches the stationary separated zone, (b) as the solvent front traverses the sample zone the lower region starts to migrate into the upper region of the zone, and (c) the focused zone in (b) migrates behind the solvent front and broadens in the usual way.

position for each or some of the development steps enables the separation in each development to be achieved with the highest available range of mobile phase velocities minimizing the separation time. With as few as 10 developments, it is relatively easy to achieve an apparent plate count of 10,000–25,000 with a migration distance of 6–11 cm for high-performance layers compared with a plate count of <5000 for a 5–6 cm single

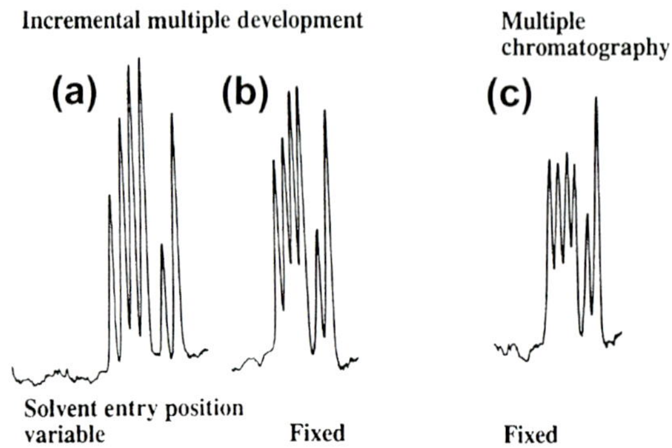

FIGURE 1.6 Separation of estrogens by incremental multiple development with a variable or fixed solvent entry position and with a fixed migration distance (multiple chromatography). A 9-step sequence with the mobile phase cyclohexane–ethyl acetate (3:1) was used for (a) and (b) and seven 7-cm developments for (c). The estrogens in order of migrations are: 17β-dihydroequilenin, 17α- dihydroequilenin, 17β-estradiol, 17α-estradiol, equilenin, and estrone. *Reproduced with permission from Ref. [35].*

development [35]. The simulation of zone broadening in multiple development is complex and limited as a predictive method or optimization tool [34]. When difficult separations are performed under capillary-control flow conditions, multiple development is the general strategy used to increase the spot capacity by simultaneously minimizing zone widths and increasing the useful migration distance available for the sample. In the absence of a suitable theoretical framework, separations are optimized by trial-and-error procedures. The general features of the different approaches to isocratic multiple development are illustrated by the separation of estrogens in Figure 1.6 For the use of multiple development in gradient methods see Section 1.5.

1.4.4 Resolution

Resolution is a geometrical function that describes how well two peaks are separated in a chromatogram. It has the same meaning and is calculated in the same way, for both column and thin-layer chromatography. For two symmetrical peaks, it is simply the ratio of the peak-to-peak separation divided by the average of the peak widths at the base. For two peaks of similar height, a resolution of 1.0 corresponds to a peak-to-peak valley separation of about 94%. For difficult separations, this value is often taken as the minimum value acceptable for the two most difficult components to separate in a mixture. Baseline separation corresponds to a resolution of 1.5. To ensure ruggedness in the separation of relatively simple mixtures a value of 2 or higher might be specified. Resolution is often used to set goals for method development in both column and thin-layer chromatography with its absolute value varied to accommodate the consequences of incomplete separation on the usefulness of the information derived from the chromatogram. Resolution also has another use based on its relationship to the adjustable experimental variables of the separation. In this regard, the results are quite different for column and thin-layer chromatography.

In column chromatography, the resolution can be described by three contributions (efficiency, selectivity, and retention) and to a first approximation these can be treated independently [1]. Resolution increases with the square root of the column plate number. This favors columns with a low plate height or long columns if the separation time is acceptable. The driving force in column technology is the design of columns with the highest possible intrinsic efficiency since this will make all separations easier. In practice, limits are imposed by system requirements and particle design. With the current state-of-the-art most separations are performed with a plate number less than 25,000. Lower plate numbers are suitable for fast separations of simple mixtures and higher values for complex mixtures. The separation factor, a measure of column selectivity, is the most important term for improving resolution. The general goal of method development is the optimization of this term by the selection of the most appropriate mobile and stationary phase combination for the separation of the sample components that are most difficult to separate. When the separation factor is small, only small increases in its value are required to facilitate a separation. Large values of the separation factor are virtually never needed, although with a large separation factor the separation is a trivial undertaking. The optimum resolution range for most separations occurs for retention factors between about 2 and 20. Higher values are associated with excessive retention time and small values with inadequate retention. For separations requiring a large peak capacity, higher values for the retention factor range may be inescapable and correspond to long separation times, but a useful feature of column separations is that resolution is roughly independent of the retention factor for high retention factors.

Resolution in thin-layer chromatography depends on the same three factors as for column chromatography but these are no longer independent for capillary-controlled flow conditions. For a single development resolution, R_S can be described by Eqn (1.1) [1,12,15]

$$R_S = \left[\sqrt{\left(N_f R_{F2}\right)}\right][(k_2/k_1) - 1][1 - R_{F2}] \tag{1.1}$$

where N_f is the maximum plate number corresponding to the solvent front position; k, a retention factor; R_F, a retardation factor; and the subscripts 1 and 2 refer to the migration order of the separated zones. The product term $N_f R_{F2}$ is used to accommodate the position dependence of the layer efficiency since each zone only experiences those regions of local efficiency that it migrates through. Resolution is expected to increase with the square root of the layer efficiency, as for a column, and to increase with the intrinsic efficiency or quality of the layer, but is dependent on the migration distance of the sample zone. Taken in isolation, it might be concluded that higher resolution is expected for compounds that migrate to the solvent front region, but this neglects the opposing contribution from the third term in Eqn (1.1). Taken together, resolution is expected to fall close to zero for small values of R_{F2} ($N_f R_{F2} \approx 0$) and for large values of R_{F2} ($1 - R_{F2} \approx 0$) with a maximum at intermediate values of R_{F2} as shown in Figure 1.7 [1]. The maximum resolution is obtained at an R_F value of about 0.3 but does not change much in the R_F range of 0.2–0.5. This then is the target R_F region for the components of a mixture than are most difficult to separate. The contribution of selectivity to the separation is indicated by the ratio of the retention factors (separation factor) and depends on the selection of both the layer and the mobile phase. Relatively small changes in selectivity have enormous impact on the ease of obtaining a given separation, since the plate number available for a separation is never very large (N_f is about 5000 for a high-performance layer). Separations in thin-layer

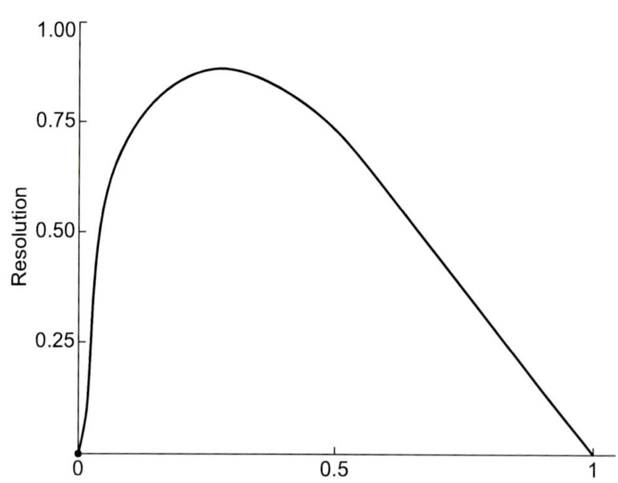

FIGURE 1.7 Variation of the resolution for two closely migrating zones as a function of the R_F value for the faster moving zone. *Reproduced with permission from Ref. [1].*

chromatography are fairly easy when $R_{F2} - R_{F1} > 0.1$ and very difficult or impossible for $R_{F2} - R_{F1} < 0.05$.

Resolution for forced-flow development is not restricted by the same factors observed for capillary-controlled flow. Resolution increases almost linearly with the solvent front migration distance and is highest for separations at the optimum mobile phase velocity. Resolution has no theoretical bounds for forced-flow development and is only limited by the practical constraints of plate length, inlet pressure, and separation time. For forced-flow separations in the elution mode, the resolution function and its dependence on system properties is the same as that for columns.

1.5 SOLVENT-STRENGTH GRADIENTS

For mixtures that span a wide retention range, isocratic methods are unsuitable for their separation. Gradient methods, which utilize a change in solvent strength during the separation, are required in this case. In column liquid chromatography, continuous mobile phase composition gradients are commonly used in reversed-phase chromatography and for normal-phase chromatography using polar chemically bonded stationary phases. Gradient methods are little used with inorganic oxide stationary phases in column chromatography. In this case, increasing solvent-strength gradients result in poor resolution across the chromatogram and often require long equilibration times to restore the column conditions to those at the beginning of the separation. In contrast, in thin-layer chromatography, normal-phase separations on inorganic oxides or polar chemically bonded phases dominate general practice with reversed-phase separations reserved for special applications. On account of the dominant use of inorganic oxide stationary phases in thin-layer chromatography for separations, continuous gradients are rarely used. Discontinuous solvent-strength gradients are almost always used in thin-layer chromatography. An approach consisting of a series of segments with the total mobile phase volume equal to the void volume of the layer was described in 2012 [36]. However, a more common approach is to couple a stepwise gradient with multiple development techniques (Section 1.4.3) to take advantage of the zone focusing mechanism that occurs at each step. The zone refocusing mechanism results in narrower separated zones compared with conventional development, a larger spot capacity, and easier-to-detect separated zones. If the mobile phase composition is varied for all, or some, of the development steps during multiple development, then solvent-strength gradients of different shapes can be produced.

Incremental multiple development with a discontinuous increasing solvent-strength gradient is used to fractionate complex mixtures by separating just a few components with similar properties at each, or some, of the development steps. Since separation of all components in a single chromatogram is not the general goal, it is usually necessary to scan the separation at several intermediate steps corresponding to the development at which different components of interest are separated. In later developments, these components may be merged again. This approach affords a higher spot capacity than can be achieved in a separation recorded as a single chromatogram [37,38]. Method development can be lengthy and this approach can be tedious when many components are of interest. It is also difficult to automate compared with incremental multiple development using a decreasing solvent-strength gradient.

In incremental multiple development with a decreasing solvent-strength gradient, the first development distance is the shortest and employs the strongest solvent composition, while subsequent developments are longer and employ mobile phase compositions of decreasing solvent strength [39]. The final development step is the longest and usually corresponds to the maximum useful development length for the layer and employs the weakest mobile phase. In this way, sample components migrate in each development until the strength of the mobile phase declines to a level at which some of the sample zones are immobile, while less retained zones continue to be separated in subsequent development steps, affording the separation of the mixture as a single chromatogram, Figure 1.8 [40]. Incremental multiple development with a decreasing solvent-strength gradient is easily automated and is the operational basis of the automated multiple development chamber (AMD). The two approaches for exploiting solvent-strength gradients in thin-layer chromatography are complementary, but in practice, the ease of automation and repeatability favor automated multiple development.

Optimized solvent-strength gradients for automated multiple development are usually arrived at by trial-and-error methods rather than based on the theory [28,39]. One approach for inorganic oxide layers is based on the use of a universal gradient which commences with methanol, ends with hexane, and uses dichloromethane or methyl t-butyl ether as the intermediate, or base solvent (typically employing 25 steps). By scaling and superimposing the chromatogram of the separation over the gradient profile those regions of the gradient affecting the chromatogram are easily identified and the gradient start, end and shape modified to better suit the needs of the separation. The universal gradient is not a linear solvent-strength gradient and the abrupt change in solvent strength occurring during the gradient can cause grouping of sample components resulting in poor resolution [41]. Alternatively, if the composition of the sample is known and standards are available, isocratic plots of the R_M

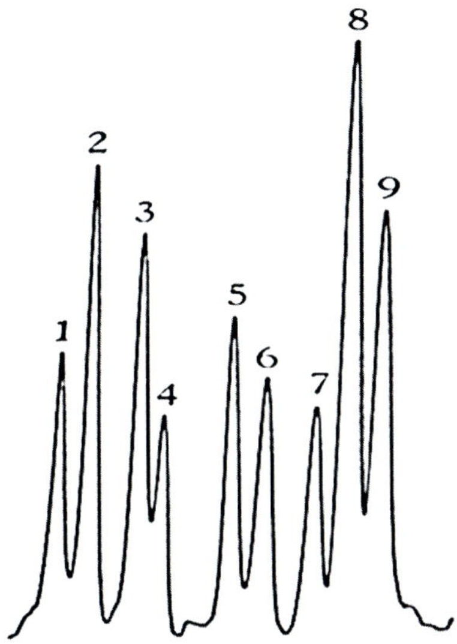

FIGURE 1.8 Separation of polar aromatic compounds found in vanilla extracts by automated multiple development on a silica gel high-performance layer. Identification: 1 = 3,4-dihydroxybenzaldehyde; 2 = 4-hydroxybenzoic acid; 3 = 4-hydroxybenzaldehyde; 4 = vanillic acid; 5 = anisic acid; 6 = vanillin; 7 = ethyl vanillin; 8 = coumarin; and 9 = piperonal.

Solvent Program (11 Steps)			
From step number	1	6	10
Ethyl acetate	1		
Chloroform	98	70	
Propan-1-ol	1		
Acetic acid	0.7		
Hexane		30	100

value against composition of binary solvent mixtures can be used to infer suitable gradient conditions [42].

1.6 MULTIDIMENSIONAL SEPARATIONS

Multidimensional separations take advantage of the higher separation power of coupled systems. In practice, only two-dimensional systems are of practical importance for both column and thin-layer chromatography [1]. Two general operations are considered as multidimensional separations. The first are heart-cut methods in which a part of the separation obtained in the first chromatographic system is passed to the second system to improve the separation of the components in the heart-cut fraction. This type of separation is commonly used for target compounds analysis in which a few compounds are designated important and need to be separated from each other and from the sample matrix, the separation of which is not considered necessary. Sometimes this approach is used in group-type separations, where instead of individual compounds, groups of compounds with a family resemblance is separated as fractions. Separations within a group are of secondary importance, and in fact, are often not desirable. The alternative method of multidimensional chromatography is known as comprehensive chromatography in which the complete separation from the first dimension is passed to the second dimension for separation with preservation of the separation obtained in the first dimension. These separations are associated with a high peak capacity and are used for fingerprint-type analyses and to characterize complex mixtures where the object is to separate as many components as possible for either classification of properties or identification.

In either case, the coupled chromatographic systems should employ retention mechanisms that are as different as possible for each dimension. For target compound analysis, the properties of the second dimension are tailored to provide the separation properties required for the small number of compounds of interest and the selectivity difference with respect to the first dimension can be relaxed to a suitable selectivity. Heart-cut separations are thus relatively straightforward and used routinely in column chromatography with only modest instrument requirements. This takes the form of a valve interface for online methods to isolate and inject fractions of eluent from the first column to the second column for separation. Online separations require compatibility of the mobile phase for the first separation with the mobile phase for the second column separation. Off-line methods usually employ a fraction collector to isolate portions of the first-dimension separation, and as the process is discontinuous, facilitate removal of the mobile phase for the first-dimension separation and its substitution with a mobile phase compatible with injection into the second dimension column. Each collected fraction is injected sequentially into the column for the second dimension separation. The off-line approach is flexible but labor intensive and time consuming. The analogous operation in thin-layer chromatography is a modification of conventional two-dimensional thin-layer chromatography [43]. The first-dimension separation along one edge of the layer is performed in the usual way. The layer is dried and the separated zones requiring further separation are identified and divided into lanes in the orthogonal direction by removing a thin strip of sorbent with a scribe parallel to the edge for the second dimension development. In sequence, each of the lanes for the second dimension separation are prepared in turn by scraping off a layer a few millimeters wide below those lanes to be developed in a different mobile phase to the lane selected for the first separation in the second dimension. The stationary phase removed below the layer determines which lanes will not be developed because the layer is not in contact with the mobile phase for that development. Repeating the layer preparation allows one lane at a time to be developed with an appropriate mobile phase to optimize the separation of the components that reside in that lane. If a suitable mobile phase exists for the separation of all components in the second dimension, or only compounds in a single lane are of interest, then no layer preparation after removal of the solvent from the first-dimension separation is required. The separation of multiple second dimension lanes with different mobile phase conditions is labor intensive and explains why this technique is reserved for special applications, which cannot be achieved by comprehensive thin-layer chromatography. In terms of automation, heart-cut separations are more straightforward in columns than when layers are used.

In comprehensive separations, significant differences and challenges exist for columns and layers [44,45]. The goals are the same but the processes are quite different. Comprehensive separations in column liquid chromatography occur in time and for thin-layer chromatography in space. In both cases, a high peak capacity is achieved when the retention mechanism in each dimension is orthogonal, likely an impossible goal, but at least they should be dissimilar for each dimension. It also presupposes that the sample components are varied in properties since highly similar compounds will occupy a narrow range of the selectivity space and their peak-to-peak separation is only weakly coupled to the dimensions of the selectivity space. For orthogonal separation systems, the maximum peak capacity is the product of the individual peak capacities for each dimension. For comprehensive column chromatography, the total peak capacity is limited by the short columns required for the second dimension, which is dictated by the need to sample each peak from the first-dimension separation a sufficient number of times to preserve the first-dimension retention information. Under-sampling in the first dimension and injection band broadening in the second dimension are additional sources of band broadening that reduce the peak capacity in comprehensive column chromatography. These contributions combined with peak dilution (loss of sensitivity) and instrumental constraints (difficult to design and operate suitable interfaces) have prevented comprehensive column liquid chromatography from establishing itself as a routine method for liquid chromatography. Considering experimentally verified results instead of theory, most online separations in comprehensive two-dimensional liquid chromatography in columns produce peak capacities between about 500 and 1000 in a time $\leq$2 h.

For maximum performance in comprehensive two-dimensional planar chromatography, it is assumed that layers can be operated as flat columns with a constant and optimal mobile phase velocity (forced-flow conditions). A sample is injected at one corner of the layer and eluted parallel to one axis of the layer. After a certain time, the flow of mobile phase is stopped when the first peak of interest approaches the opposite edge of the layer. A second mobile phase is then pumped in a direction orthogonal to the first-dimension separation resulting in migration of the sample components in the second dimension. A fundamental advantage of the spatial mode is that all second dimension separations are performed simultaneously allowing for much faster separations than for two-dimensional column separations. A practical advantage is that no complicated interface is required for the transfer of sample components from the first to the second dimension and transfer is achieved with minimal dilution. A disadvantage of the planar format is that part of the separation in the first dimension is lost in the second dimension because the separated zones broaden radially as they migrate and will be larger in the second dimension separation then they were at the end of the first-dimension separation. A peak capacity of about 780 should be possible for a 10×10 cm high-performance thin-layer plate operated under forced-flow conditions in only 2.7 min [44]. If layers with the properties of contemporary columns were used, the peak capacity rises to over a 1000 in about the same time but instrumentation for high-pressure operation above 50 bar is unavailable. Two-dimensional forced-flow thin-layer chromatography is attractive for fast separations, and competitive for extremely high-performance separations, but is largely underutilized in laboratory practice.

Two-dimensional thin-layer chromatography with capillary-controlled flow produces lower performance and slower separations than predicted for forced-flow methods but requires little in the way of instrumentation and is simple to implement. Separations with a reasonable increase in spot capacity still require that the separations in the orthogonal directions employ largely different retention mechanisms. If a single stationary phase is employed for the separation then the separation mechanisms differ only in the choice of mobile phase [28,37,46]. A common arrangement is the use of normal-phase and reversed-phase solvent systems on a polar chemically bonded stationary phase. Solvent-strength gradients can be employed in one or both dimensions to increase the spot capacity [47,48]. The use of reversed-phase conditions with a silica gel stationary phase or a bilayer plate with a strip of chemically bonded stationary phase butted against a larger strip of silica gel is complicated by alteration of the stationary phase properties by the mobile phase employed for the first development. A more practical solution is the use of two thin-layer plates with different stationary phases. The partial separation in a single lane on the first layer is transferred to a second layer with a different stationary phase by contacting the two layers in a special holder and using solvent elution to transfer the separation from the first layer to the second layer for subsequent development [47,49,50]. This process is known as graft thin-layer chromatography. Compared to forced-flow development, capillary-controlled flow results in additional band broadening in both separation directions due to the inadequate and variable mobile phase velocity, a limited migration distance for the solvent front due to the weak capillary forces driving the mobile phase, and low resolution for compounds close to the origin or solvent front (see Section 1.4.4). The spot capacity in two-dimensional capillary-controlled flow thin-layer chromatography is difficult to model because of the varying mobile phase velocity but it has been estimated, and in reasonable agreement with practice, that it should be relatively easy to generate between 100 and 250 separated spots in about 30 min with little prospect of exceeding 500, at least by single development separation for each dimension [51]. This is sufficient to handle mixtures of moderate complexity and suggests that two-dimensional thin-layer chromatography is underutilized in contemporary practice. One barrier, the recording of two-dimensional chromatograms, has largely been solved using videodensitometry, photodiode array detection, or mass spectrometric detection [15,52,53].

1.7 CONCLUSIONS

Capillary-controlled flow makes thin-layer chromatography the easiest of the chromatographic methods to implement. It provides the basis of its general use in laboratory practice for the qualitative analysis of simple mixtures. The general characteristics of capillary forces are not ideal for optimal performance in thin-layer chromatography. They are generally too weak to provide an optimum mobile phase velocity and vary with the migration distance of the solvent front. The variable mobile phase velocity results in longer separation times and an average plate height that is larger than that observed for forced flow with a constant and optimal mobile phase velocity. It is difficult to achieve a zone capacity of 20–25 for a single development with capillary-controlled flow. Typical values observed by experiment are on the order of 12–14 and largely independent of whether high-performance or conventional thin-layer plates are used [28]. For high-performance layers with forced-flow development, a spot capacity of 30–40 can be obtained with a theoretical limit of around 80–150, which depends on the available inlet pressure. To counteract the undesirable properties of capillary-controlled flow, multiple development techniques can improve layer performance by exploiting the zone focusing mechanism and raise the spot capacity to the same range as for forced-flow development at the expense of longer separation times. Automation, as in the automated multiple development chamber, makes this technique less tedious and facilitates the use of solvent-strength gradients for the separation of mixtures with a wide range of retention properties. Two-dimensional methods offer the highest spot capacity for both capillary-controlled and forced-flow development. In both cases, it is not difficult to reach a zone capacity of 200–300 in a reasonable time provided that the two orthogonal retention mechanisms are as different as possible with a potential to reach still higher values using forced-flow development if higher pressures than currently available with commercial instrumentation are used. This is still quite good performance and suggests that two-dimensional thin-layer chromatography is worthy of further investigation.

A comparison of the separation properties of thin-layer and column liquid chromatography should consider the attributes of a planar system as well as kinetic performance. Columns are generally prepared to a higher performance specification than layers [27] and forced-flow development is required to realize the full potential of available layers. The highest separation potential is available for comprehensive multidimensional chromatography where thin-layer chromatography has an advantage in that it is simpler to implement, but to reach its full potential, should exploit forced-flow development. This presents different challenges to column chromatography and instrumental difficulties limit both columns and layers in reaching their maximum potential at present. For capillary-controlled flow systems, the use of a disposable stationary phase, the capability of separating samples in parallel, the convenience of applying derivatization techniques, and the use of biological reagents to enhance detection and selectivity are important considerations when selecting the analytical method for a specific problem.

References

[1] Poole CF. The essence of chromatography. Amsterdam: Elsevier; 2003.
[2] Poole CF. Planar chromatography at the turn of the century. J Chromatogr A 1999;856: 399–427.

[3] Tyihak E, Mincsovics E. Forced-flow planar liquid chromatographic techniques (after twenty-two years). J Planar Chromatogr 2010;23:382–95.
[4] Nyiredy Sz. Progress in forced-flow planar chromatography. J Chromatogr A 2003; 1000:985–99.
[5] Reich E, Schibli A. High-performance thin-layer chromatography for the analysis of medicinal plants. New York: Thieme; 2007.
[6] Sherma J, Morlock GE. Chronology of thin-layer chromatography focusing on instrumental progress. J Planar Chromatogr 2008;21:471–7.
[7] Berezkin VG, Dzido TH. The discovery of thin-layer chromatography by N.A. Izmailov and M.S. Shraiber. J Planar Chromatogr 2008;21:399–403.
[8] Kreuzig F. The history of planar chromatography. J Planar Chromatogr 1998;11: 322–4.
[9] Siouffi AM. From paper to planar: 60 years of thin layer chromatography. Sep Purif Rev 2005;34:155–80.
[10] Zlatkis A, Kaiser RE, editors. HPTLC: high-performance thin-layer chromatography. Amsterdam: Elsevier; 1977.
[11] Bertsch W, Hara S, Kaiser RE, Zlatkis A, editors. Instrumental HPTLC. Heidelberg: Huthig; 1980.
[12] Geiss F. Fundamentals of thin-layer chromatography. Heidelberg: Huethig; 1987.
[13] Sherma J, Fried B, editors. Handbook of thin-layer chromatography. New York: Marcel Dekker; 1996.
[14] Nyiredy Sz, editor. Planar chromatography: a retrospective view of the third millennium. Budapest: Spriner; 2001.
[15] Spangenberg B, Poole CF, Weins C. Quantitative thin-layer chromatography. A practical survey. Berlin: Springer; 2011.
[16] Morlock GE, Schwach W. The contribution of planar chromatography to food analysis. J Planar Chromatogr 2007;20:399–406.
[17] Weins C. Overview of bioactivity based analysis by HPTLC bridging the gap between cause and effect – HPTLC detection of bioactive compounds in the environment and in food. J Planar Chromatogr 2008;21:405–10.
[18] Choma IA, Grzelak EM. Bioautography detection in thin-layer chromatography. J Chromatogr A 2011;1218:2684–91.
[19] Poole SK, Poole CF. High performance stationary phases for planar chromatography. J Chromatogr A 2011;1218:2648–60.
[20] Tuzimski T. Application of different modes of thin-layer chromatography and mass spectrometry for the separation and detection of large and small biomolecules. J Chromatogr A 2011;1218:8799–812.
[21] Cheng S-C, Huang M-Z, Shiea J. Thin-layer chromatography/mass spectrometry. J Chromatogr A 2011;1218:2700–11.
[22] Poole CF. Thin-layer chromatography: challenges and opportunities. J Chromatogr A 2003;1000:963–84.
[23] Poole SK, Ahmed HD, Belay MT, Fernando WPN, Poole CF. The influence of layer thickness on the chromatographic and optical properties of high performance silica gel thin layer chromatographic plates. J Planar Chromatogr 1990;3:133–40.
[24] Nyiredy S, Fater Z, Botz L, Sticher O. The role of chamber saturation in the optimization and transfer of the mobile phase. J Planar Chromatogr 1992;5:308–15.
[25] Poole CF. Thin-layer chromatography. In: Moffat AC, Osselton MD, Widdop B, Watts J, editors. Clarke's analysis of drugs and poisons. 4th ed., vol. 1. London: Pharmaceutical Press; 2011. pp. 600–35.
[26] Poole CF, Fernando WPN. Some musings on the measurement and interpretation of theoretical plate height in thin-layer chromatography. J Planar Chromatogr 1992;5: 323–33.

[27] Fernando WPN, Poole CF. Determination of kinetic parameters for precoated silica gel thin layer chromatography plates by forced flow development. J Planar Chromatogr 1991;4:278–87.
[28] Poole CF, Poole SK. Multidimensionality in planar chromatography. J Chromatogr A 1995;703:573–612.
[29] Siouffi AM, Bressolle F, Guiochon G. Optimization of thin-layer chromatography:some practical considerations. J Chromatogr 1981;209:129–47.
[30] Guiochon G, Siouffi AM. Studies of the performance of thin-layer chromatography: spot capacity in thin-layer chromatography. J Chromatogr 1982;245:1–20.
[31] Belenkii B, Kurenbin O, Litvinova L, Gankina E. A new approach to optimization in TLC. J Planar Chromatogr 1990;3:340–7.
[32] Poole CF, Wilson ID. Planar electrophoresis and electrochromatography: time to revisit these techniques. J Planar Chromatogr 1997;10:332–5.
[33] Poole CF, Poole SK, Belay MT. Insights into the mechanism and applications of unidimensional multiple development development in thin-layer chromatography. J Planar Chromatogr 1993;6:438–45.
[34] Szabady B, Ruszinko M, Nyiredy Sz. Prediction of retention data when using incremental multiple development techniques. Chromatographia 1997;45:369–72.
[35] Poole SK, Poole CF. The influence of the solvent entry position on resolution in unidimensional multiple development thin-layer chromatography. J Planar Chromatogr 1992;5:221–8.
[36] Markowski W, Wrobiewski K, Dzido TH. Stepwise gradient elution in RP HPTLC with a new horizontal development chamber. J Planar Chromatogr 2012;25:200–7.
[37] Poole CF, Poole SK, Fernando WPN, Dean TA, Ahmed HD, Berndt JA. Multidimensional and multimodal thin-layer chromatography: pathway to the future. J Planar Chromatogr 1989;2:336–45.
[38] Schuette SA, Poole CF. Unidimensional sequential separation of PTH-amino acids by high performance thin-layer chromatography. J Chromatogr 1982;239:251–7.
[39] Poole CF, Belay MT. Progress in automated multiple development. J Planar Chromatogr 1991;4:345–59.
[40] Belay MT, Poole CF. Determination of vanillin and related flavor compounds in natural vanilla extracts and vanilla-flavored foods by thin layer chromatography and automated multiple development. Chromatographia 1993;37:365–73.
[41] Colthup PV, Bell JA, Gadsdon DL. Use of linear solvent gradients in manual and automated multiple development for the separation and characterization of drugs and metabolites by HPTLC. J Planar Chromatogr 1993;6:386–93.
[42] Lodi G, Betti A, Brandolini V, Menziani E, Tosi B. Automated multiple development and HPTLC analysis of sugars on hydrophilic layers: 1. Amino layers. J Planar Chromatogr 1994;7:29–33.
[43] Tuzimski T. Strategy for the separation of complex mixtures by multidimensional planar chromatography. J Planar Chromatogr 2008;21:49–54.
[44] Vanhoutte DJD, Vivo-Truyols G, Schoenmakers PJ. Pareto-optimality study into the comparison of the separation potential of comprehensive two-dimensional liquid chromatography in the column and spatial modes. J Chromatogr A 2012;1235:39–48.
[45] Guiochon G, Marchetti N, Mriziq K, Shalliker RA. Implementations of two-dimensional liquid chromatography. J Chromatogr A 2008;1189:109–68.
[46] Zakaria M, Gonnord M-F, Guiochon G. Applications of two-dimensional thin-layer chromatography. J Chromatogr 1983;271:127–92.
[47] Ciesla L, Bogucka-Kocka A, Hajnos M, Petruczynik A, Waksmundska-Hajnos M. Two-dimensional thin-layer chromatography with adsorbent gradient as a method of chromatographic fingerprinting of furanocoumarins for distinguishing selected varieties and forms of *Heracleum Spp*. J Chromatogr A 2008;1207:160–8.

[48] Matysik G, Skaiska-kaminska A, Stefanczyk B, Wojciak-kosior M, Rapa D. Application of a new technique in two-dimensional TLC separation of multicomponent mixtures. J Planar Chromatogr 2008;21:233–6.

[49] Ciesla L, Petruczynik A, Hajnos M, Bogucka-kocka A, Waksmundzka-Hajnos M. Two-dimensional thin-layer chromatography of structural analogs. Part 1: graft TLC of selected coumarins. J Planar Chromatogr 2008;21:237–41.

[50] Tuzimski T. Separation of multicomponent mixtures of pesticides by graft thin-layer chromatography on connected silica and octadecyl silica layers. J Planar Chromatogr 2007;20:13–8.

[51] Guiochon G, Gonnord M-F, Siouffi A, Zakaria M. Study of the performance of thin-layer chromatography: VII spot capacity in two-dimensional thin-layer chromatography. J Chromatogr 1982;250:1–20.

[52] Komsta L, Ciesla I, Bogucka-kocha A, Jozetczyk A, Kryszen J. Waksmundzka-Hajnos. The start-to-end chemometric image processing of 2D thin-layer videoscans. J Chromatogr A 2011;1218:2820–5.

[53] Milz B, Karl-Friedrich K, Spangenberg B. Quantitative two-dimensional thin-layer chromatography using a diode-array detector. J Planar Chromatogr 2012;25:493–7.

CHAPTER

2

High-Performance Precoated Stationary Phases

Colin F. Poole

Department of Chemistry, Wayne State University, Detroit, MI, USA

2.1 INTRODUCTION

Most laboratories today use commercially available precoated plates for separations. Standardized procedures for layer preparation have been developed but this time consuming and skill-based task has largely been discontinued so that laboratory staff can concentrate on the analysis of samples. The preparation of layers with reproducible separation properties is easier to achieve in a manufacturing setting where in process quality control checks provide assurance of product performance and consistency.

The structure of layers is more complex than for columns. Columns for high-performance separations contain a homogeneous bed of densely packed sorbent particles of a narrow size distribution or a porous monolith of uniform structure. Layers are also composed of sorbent particle-beds but require other components to impart the necessary stability and durability to the layer, a consequence of its open structure. Detection in thin-layer chromatography occurs in the presence of the stationary phase, in contrast to column chromatography, and indicator compounds, such as magnesium-activated zinc silicate or zinc cadmium sulfide, of a similar particle size to the sorbent may be added to facilitate visualization. Binders used for the immobilization of high-performance layers are typically crosslinkable organic polymers, such as poly(methacrylate) and its salts, in amounts up to a few percent by weight. Other common binders for conventional thin-layer plates include gypsum and starch in amounts up to 20% (w/w). A reasonably consistent letter code is followed by most manufacturers to describe the layer properties [1–3].

Instrumental Thin-Layer Chromatography
http://dx.doi.org/10.1016/B978-0-12-417223-4.00002-9

2.2 INORGANIC OXIDE SORBENTS

Silica gel is the most important stationary phase for thin-layer chromatography and is also the main substrate for the synthesis of chemically bonded sorbents. Alumina is used to a lesser extent and kieselgur and Florisil rarely [3]. High-performance silica gel plates are prepared from irregular silica gel particles with a nominal particle size of 5 μm and a layer thickness of 0.1 or 0.2 mm. Thinner layers result in a 1.1–2.5-fold higher mobile phase velocity and a small increase in sample detectability of 1.1–1.5-fold by scanning densitometry [4]. However, the major use of thinner layers (0.1 mm) is in methods utilizing repeated developments, for example, in automated multiple development, where a significant time saving is possible, and to achieve increased sample detectability in atmospheric pressure ionization techniques for mass spectrometry. Spherical particles are expected to result in faster separations and higher performance compared with irregular shaped particles, but have not penetrated the market place for high-performance plates to a significant extent. Similar claims were made for smaller irregular particles of 3 μm silica gel but no significant advantages were observed for capillary flow controlled conditions [5]. Thus, the design characteristics of high-performance layers has not changed much over the past two decades while in the same time those for columns have changed dramatically [6]. The most credible reason for this is the limitations imposed by capillary-controlled flow which neutralizes many of the advantages of new particle formats. An exception might be superficially porous particles, which are yet to be explored for thin-layer chromatography.

There are two general types of silica gel used for the preparation of high-performance precoated layers. Silica gel 60 from Merck has a specific surface area of 550 m^2/g, a pore volume of 0.82 mL/g, and an average pore size of 6 nm and is typical of a number of other products [3,6]. These materials are optimized for the separation of small molecules with a molecular weight <750. Silica gel 50,000 from Merck has a specific surface area of 0.5 m^2/g, a pore volume of about 0.63 mL/g, and an average pore diameter of 5000 nm. Its low surface area makes it useful for the separation of polar compounds (e.g., phosphates, sulfonates, carbohydrates, etc.) that are excessively retained on high-surface area silica gels [7,8]. These materials are also used to prepare the concentration zones on bilayer plates where the low surface area silica gel in the form of a narrow strip along one edge of the plate acts as a sampling zone [3,9]. Samples burdened with suspended materials or partially insoluble materials when applied directly to the separation layer result in poor chromatography and improper quantification. Predevelopment of the sample on the low surface area silica gel facilitates extraction and concentration into a narrow band at the interface between the concentration zone and the higher surface area separation zone.

2.2.1 Kinetic Properties

The characteristic kinetic properties of high-performance silica gel layers determined by forced flow methods are summarized in Table 2.1 [10,11]. The interparticle porosity of layers is similar to that of homogeneously packed columns indicating that typical layers have a dense packing structure obtained without the application of high pressure. This is probably a result of the thinness of the layer. The total porosity and intraparticle porosity are relatively low compared to typical columns. This suggests that a significant amount of the binder used to stabilize the layers is located within the porous particles. The flow resistance parameter for layers is higher than for packed columns and is likely a result of the presence of a larger fraction of fine particles in the silica gel used for layer preparation. The observed apparent particle size for the high-performance layers is larger, and for conventional layers smaller, than the nominal particle size indicated by the layer manufacturers. Since this value is calculated from the flow resistance parameter, it is probably due to features of the particle size distribution, which are not captured by an average value. On the other hand, given the heterogeneous structure of layers and the variation observed in average particle size measurements by different techniques, there is reasonable agreement between the experimentally observed values and the nominal values indicated by the manufacturers.

The effect of the mobile phase velocity on zone broadening can be quantitatively evaluated by fitting the experimental plate height values to the van Deemter equation, or in terms of reduced parameters, the Knox equation (see Table 2.1). The minimum plate heights for the high-performance layers at 22–25 μm is significantly better than the values for conventional layers at 35–45 μm and occur at a similar optimum interparticle mobile phase velocity of about 0.5 mm/s. The minimum plate height values are about 1.5–3-fold larger than for an ideal column packed with porous particles and are shifted 2–4-fold to lower optimum mobile phase velocities with respect to an ideal column. The separation performance of layers in forced flow thin-layer chromatography is poorer than that can be obtained for columns in high-pressure liquid chromatography while considerably better than for separations on layers using capillary-controlled flow conditions. Interrogation of the coefficients of the Knox equation provides useful insight into the kinetic properties of layers. Values for the coefficients are given in Table 2.1 [6,12]. The *A* term characterizes the flow anisotropy within the streaming portion of the

TABLE 2.1 Characteristic Kinetic Properties of Precoated Silica Gel Layers Determined by Fully Online Forced Flow Development [7–9]

Parameter	Layers: High Performance	Layers: Conventional
POROSITY		
Total	0.65–0.70	0.65–0.75
Interparticle	0.35–0.45	0.35–0.45
Intraparticle	0.28	0.28
Flow resistance parameter (θ)	875–1500	600–1200
Apparent particle size (μm)	5–7	8–10
Minimum plate height (μm)	22–25	35–45
Optimum mobile phase velocity (mm/s)	0.3–0.5	0.2–0.5
Minimum reduced plate height	3.5–4.5	3.5–4.5
Optimum reduced mobile phase velocity	0.7–1.0	0.6–1.2
***VAN DEEMTER EQUATION* ($H = A + (B/U_E) + CU_E$) *COEFFICIENTS*[a]**		
A ($\times 10^4$) (cm)	2.5–4	15–27
B ($\times 10^6$) (cm^2/s)	60–80	40–80
C ($\times 10^3$) (s)	10–20	15–70
***KNOX EQUATION* ($H = A\nu^{1/3} + (B/\nu) + C\nu$) *COEFFICIENTS*[a]**		
Flow anisotropy (A)	0.4–0.8	1.7–2.8
Axial diffusion (B)	1.2–1.6	1.2–2.0
Resistance to mass transfer (C)	1.4–2.4	0.70–0.85
Separation impedance (E) $\times 10^3$	10–20	11–13

[a] *u_e = interparticle mobile phase velocity, ν = reduced mobile phase velocity, and h = reduced plate height defined in Ref. [10].*

mobile phase and is related to the density and homogeneity of the packing structure of the layer. The A coefficients for the high-performance layers are similar, or indeed smaller, than that of a typical porous particle column indicating that high-performance layers have a good packing structure. Conventional precoated silica gel layers are not as homogeneously packed, perhaps because of the greater difficulty of coating layers with larger particles, but more likely, because of the broader particle size range of conventional sorbents than for the sorbents used to prepare high-performance layers. The B term characterizes the contribution of longitudinal diffusion to the plate height and should be similar for layers of different types and columns (which is the case for the coefficients in

Table 2.1). The *C* term is a measure of the resistance to mass transfer between the solvated stationary phase and the streaming mobile phase. The poor values of the *C* term for layers (an order of magnitude larger than those for an ideal column packed with porous particles) may be a result of restricted diffusion within the porous particles or different rates of sorption for the silica gel surface and those portions of the silica gel surface coated with binder. For separations using forced flow conditions the relatively large *C* term is mainly responsible for the higher minimum reduced plate height and lower optimum reduced mobile phase velocity for layers compared with columns for high-pressure liquid chromatography. Thus, improvements in layer properties by reducing contributions to the *C* term would have a beneficial outcome for forced flow separations. However, this may not be important for capillary-controlled flow conditions since at the low range of flow rates available using capillary-controlled flow the plate height is dominated by the *B* term and the *C* term is usually negligible [13].

2.2.2 Retention Properties

Surface silanol groups are the dominant site for analyte interactions on silica gel and adsorption, the dominant retention mechanism. The surface silanol groups consist of different populations of isolated silanols, neighboring associated silanols (hydrogen bonded), germinal silanols, and a variable but low concentration, of dissociated silanols, Figure 2.1. The relative concentration of the different silanol groups depends on the method of preparation and the type and concentration of metal impurities. This variation in surface composition is the origin of selectivity differences for different silica gels while absolute retention is a strong function of the specific surface area of the silica gel. Since the silica network is rigid, the silanol groups are locked into fixed positions, adding a spatial recognition capability to the retention mechanism.

From a sample perspective the features of importance for retention are the type, number, and spatial location of functional groups. Figure 2.2

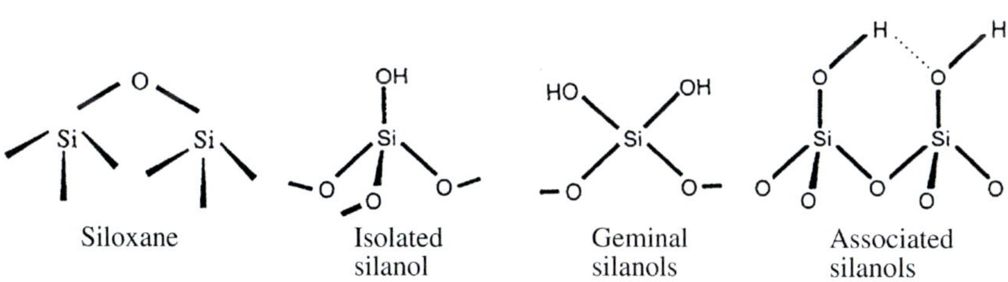

FIGURE 2.1 Types of functional groups present on a silica gel surface. *Reproduced with permission from Ref. [10].*

Silica gel
General adsorption scale

Difficult to separate because solvent strength is too high	Alkanes	**Weak**
	Aromatics	↓
	Halogenated compounds	
	Ethers	
	Nitro compounds	
	Nitriles	
	Carbonyl compounds	
	Alcohols	
	Phenols	
	Amines	
	Amides	
Difficult to separate because solvent selectivity is too low	Carboxylic acids	
	Sulfonic acids	**Strong**

FIGURE 2.2 General adsorption scale for unhindered functional groups on silica gel. *Reproduced with permission from Ref. [10].*

provides a general adsorption scale for unhindered functional groups on silica gel [10]. Compounds of low polarity have weak interactions with silanol groups and can be difficult to separate because most mobile phases are too strong and result in low retention caused by effective competition of solvent molecules for surface active sites. Compounds with highly-polar functional groups can be difficult to separate because their strong interactions with silanol groups limits selectivity as most common solvents compete poorly with the sample molecules for adsorption interactions at the silica surface. Compounds with functional groups found in the middle-range of those indicated in Figure 2.2 are those most easily separated by silica gel and allow for the widest selection of solvent types for method development. The significant difference in steric fitting of analyte functional groups with immobilized silanol groups results in a useful capability of silica gel to separate isomers and diastereoisomers. In addition, internal hydrogen bonding between functional groups and bulky alkyl groups adjacent to polar functional groups diminish retention and facilitate the separation of structurally related compounds. The competition model is used to formally explain retention on silica gel but its application is not straightforward in most cases and requires semi-empirical correction to account for site-competition delocalization, restricted-access delocalization, and solvent-type selectivity related to the relative solvent hydrogen-bond acidity/basicity [10,14].

The retention properties of alumina are not as well understood as silica gel but are generally complementary to those of silica gel. The alumina surface is more complex containing hydroxyl groups, aluminum cations, and oxide anions. The apparent surface pH can be controlled by treating alumina with buffer solutions and has a significant effect on retention and zone shape of acids and bases. Alumina is also more catalytically active than silica gel, and instances of sample decomposition are more prevalent.

2.2.3 Impregnated Inorganic Oxide Sorbents

The selectivity of inorganic oxide layers can be modified in several ways to suit the requirements for the separation of particular compound types. Impregnation with volatile reagents is usually performed through the gas phase by equilibrating the layer with an appropriate reagent solution in a developing chamber after sample application. Common reagent solutions are ammonia, hydrochloric acid, or formic acid. Preadsorption of acidic or basic vapors is used to minimize zone distortion and to adjust selectivity in the separation of acids or bases on silica gel layers. Solid or liquid additives are incorporated in the layer during coating by adding the reagent to the coating slurry, by dipping the layer into a solution of the additive followed by evaporation of the solvent, or by spraying the layer with a solution of the additive followed by evaporation of the solvent [2,10]. The development of chemically bonded sorbents for thin-layer chromatography has diminished interest in the use of physically adsorbed liquid phases such as hydrocarbon and silicone oils for reversed-phase separations. Solid additives that provide solvent-stable modified layers are more important today. Typical reagents include buffers for pH adjustment and ion pairing, silver nitrate for the separation of unsaturated compounds [15,16], boric acid for the separation of compounds with vicinal hydrogen-bonding functional groups [17,18], caffeine for the separation of polycyclic aromatic compounds [19], ethylenediaminetetraacetic acid (EDTA) to facilitate the separation of compounds with undesirable interactions with metal impurities in the layer, and ammonium sulfate for the separation of phospholipids [20]. Some of these modified layers are commercially available with additive concentrations adjusted for particular applications; all can be prepared in the laboratory, usually by dipping precoated silica gel plates into a solution of the additive. The dipping method allows the concentration of additive to be adjusted to an optimum value for a particular separation.

Silver nitrate layers are light and moisture sensitive with a limited shelf life. They are usually prepared in the laboratory shortly before use. A silver-thiolate silica sorbent was described in 2012 for the preparation of silver-impregnated silica gel layers with an extended shelf life and reduced sensitivity to laboratory light [21]. Silver ions are known to form weak reversible charge-transfer complexes with double bonds. Silver-impregnated silica gel layers posses high selectivity for the separation of lipids according to the number, configuration, and to some extent, the position of double bonds in the fatty acid side chains that would be difficult to achieve by other separation mechanisms. Reversible weak charge-transfer complexes are also thought to be responsible for the separation of polycyclic aromatic compounds by ring number on silica gel layers impregnated with caffeine.

2.2.4 Layer Pretreatment

Precoated inorganic oxide layers are invariably contaminated, or quickly become so when opened and handled in the laboratory. They contain residual contaminants from the manufacturing process and contaminants acquired by contact with packaging materials and by adsorption from the atmosphere. These contaminants can give rise to irregular and drifting baselines in scanning densitometry, ghost peaks mistaken for sample components, and a high general background with reduced sensitivity in postchromatographic derivatization [22]. Decontamination of inorganic oxide layers can be achieved by a combination of ascending development and immersion in a polar solvent, such as methanol or 2-propanol [22,23]. A single or double 5-min immersion in a polar solvent is generally superior to ascending development for removing contaminants. For trace analysis, both sequential ascending development followed by immersion may be required for acceptable results. Some manufacturers now offer especially clean precoated layers for trace analysis and mass spectrometric detection.

Reproducible retention on inorganic oxide layers requires careful control of the layer activity. This is achieved by the adsorption of reagents, most notably water, through the gas phase [22]. Physically adsorbed water can be removed at modest temperatures ($\approx$120 °C) but is quickly restored by exposure to the atmosphere after the plate is cooled to room temperature. Even if stored in a desiccator after drying, sample application is usually performed with the layer open to the atmosphere. Laboratories with good environmental control usually experience fewer problems with the control of layer activity. This aspect of the separation process, however, should not be overlooked when problems with reproducibility arise. Although heat activation is a common laboratory protocol, it remains a debatable question as to whether it is necessary or even useful, given the speed with which layer re-equilibration with atmospheric moisture occurs. Inorganic oxide layers can be adjusted to a defined activity level by exposure of the layer to a defined gas phase in a closed chamber. This is best performed after sample application in a developing chamber that allows both conditioning and development of the layer in the same chamber. Atmospheres of different constant relative humidities can be obtained by exposure to the vapor phase in equilibrium with solutions of concentrated sulfuric acid or saturated solutions of various salts [24]. The in situ absorption spectrum of Reichardt's betaine dye spotted on a silica gel layer can be used as a direct measure of layer activity [25]. There is an inverse linear relationship between the signal observed at 500 nm and the adsorbed water content.

2.3 CHEMICALLY BONDED SORBENTS

The silanol group on silica gel can be reacted with organosilane reagents to form surface-bonded organosiloxane groups that modify the sorptive properties of the silica gel. This general route facilitates the creation of a family of stationary phases retaining the favorable structural morphology of silica gel but with complementary separation properties. Since the siloxane bonds formed at the surface are reasonably durable, layers prepared from these sorbents are stable to the typical separation conditions used in thin-layer chromatography. Silica-based, chemically bonded layers with dimethyl-, ethyl-, octyl-, octadecyl-, 3-aminopropyl-, 3-cyanopropylsiloxane-bonded groups and a siloxane-bonded spacer bonded propanediol group are commercially available [3,7,26–30]. Some characteristic properties are summarized in Table 2.2 [30,31].

Those layers prepared with bifunctional reagents most likely have a monomeric structure with a single ligand bonded to one or two surface silanol groups. Those prepared with trifunctional reagents are probably surface bonded to one or two silanol groups (similar to the bifunctional reagents) but, in addition, internally bonded to their neighbors forming a polymer-like network extending over the surface. For steric reasons, not all surface silanol groups can be reacted. In addition, unreacted reagent functional groups bound to the silica surface regenerate further silanol groups during workup. Thus, chemically bonded sorbents retain a significant concentration of silanol groups, some of which are accessible to the sample and participate in the retention process.

Alkylsiloxane-bonded layers with a high degree of surface bonding are incompatible with mobile phases containing a significant amount of water. This is because aqueous mobile phases poorly wet the largely hydrophobic surface and the capillary forces responsible for establishing the mobile phase velocity are rendered inadequate [26,29,30]. This problem is overcome by using a slightly larger particle size for the silica substrate, a reproducible although lower degree of surface bonding than the maximum value, and a different binder to enhance compatibility with water. These layers are referred to as water wettable and indicated by the suffix W after the general abbreviation for the surface treatment, Table 2.2. They are used for all reversed-phase separations. The development time for a water-wettable layer is an intrinsic property of the mobile phase composition and properties (viscosity, surface tension) and properties of the layer (permeability, contact angle), Figure 2.3 [32]. Solvents like

TABLE 2.2 Characteristic Properties of Precoated Chemically Bonded Layers

Manufacturer	Derivatizing Reagent	Percent Silanol Groups Reacted	Average Particle Size (μm)
MERCK			
RP-2	Bifunctional	50	5–7
RP-8 W	Bifunctional	25	11–13
RP-8	Bifunctional	37	5–7
RP-18 W	Bifunctional	22	11–13
RP-18	Bifunctional	35	5–7
3-Aminopropyl	Trifunctional	50	5–7
3-Cyanopropyl	Bifunctional	27	5–7
WHATMAN			
KC-2	Monofunctional		10–14, or 20
KC-8	Monofunctional		10–14
KC-18	Trifunctional	16	10–14
Diphenyl	Difunctional		10–14
MACHEREY–NAGEL			
Sil C18-100	Trifunctional	45	5–10
Sil C18-50	Trifunctional	30	5–10

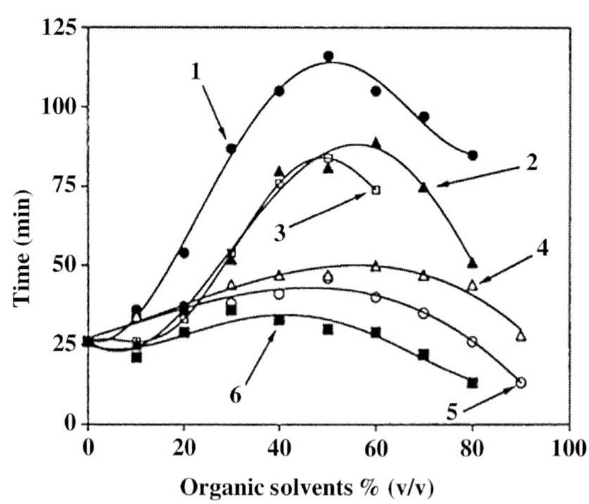

FIGURE 2.3 Plot of the time required for the solvent front to migrate 5 cm as a function of the volume composition of aqueous solutions of (1) 2-propanol; (2) N,N-dimethylformaide; (3) 2,2,2-trifluoroethanol; (4) methanol; (5) acetone; and (6) acetonitrile on a water-wettable, octadecylsiloxane-bonded silica gel layer (Merck HPTLC RP-18 WF_{254s}).

2-propanol that form aqueous mixtures with an unfavorable surface tension to viscosity ratio are not recommended. For fast separations, solvents such as acetonitrile, acetone, and methanol are preferred. Alkylsiloxane-bonded sorbents with a high phase loading and small particle size are suitable for the separation of compounds of low polarity employing nonaqueous mobile phases. Polar chemically bonded phases are compatible with water in all proportions and are suitable for use in both normal- and reversed-phase separations.

2.3.1 Retention Properties

Chemically bonded phases provide a useful alternative for mixtures that cannot be separated on silica gel because of excessive retention or limited selectivity [30]. Some general guidelines for stationary phase selection are summarized in Table 2.3 [10]. With nonaqueous solvents, retention on chemically bonded phases is dominated by adsorption interactions involving unreacted and accessible silanol groups and with the polar functional groups bonded to the silica surface. This includes dipole-type interactions with 3-cyanopropylsiloxane-bonded groups and hydrogen-bonding interactions with 3-aminopropylsiloxane-bonded and spacer bonded propanediol groups. The retention mechanism for alkylsiloxane-bonded sorbents with nonaqueous mobile phases is unclear but probably involves partition into the solvated stationary phase and adsorption interactions with accessible silanol groups. In the reversed-phase mode, separations are dominated by the properties of water, as discussed in the following paragraph. In the reversed-phase mode, separations employing ion suppression and ion pairing are possible for ionizable or ionized compounds difficult to handle on silica gel [10,30,33,34]. The 3-aminopropylsiloxane-bonded layers can be used as a weak anion exchanger with a partially aqueous acidic (buffered) mobile phase [10,30,35]. A still incompletely understood feature of 3-aminopropylsiloxane-bonded layers is the capability to visualize a wide range of polar compounds by inducing fluorescence simply by heating the layer [36].

An interphase model has been proposed to describe the stationary phase contribution to retention for chemically bonded phases in the reversed-phase mode [32,37,38]. In this approach, retention occurs by distribution of a solute between the bulk mobile phase and an interphase region that represents the active volume of the solvated stationary phase. The interphase region is bound on one side by the silica surface and on the other by an imaginary dividing plane separating the bulk mobile phase from the interphase region. The interphase region contains the silica surface groups, the bond ligands of the stationary phase, and mobile phase components selectively sorbed from the bulk mobile phase. The

TABLE 2.3 Retention Properties of Silica-based Chemically Bonded Layers

Type of Modification	Functional Group	Application[a]
Alkylsiloxane	$Si{-}CH_3$ $Si{-}C_2H_5$ $Si{-}C_8H_{17}$ $Si{-}C_{18}H_{37}$	For reversed-phase separations generally but not exclusively Separation of water soluble polar organic compounds (RPC) Weak acids and bases after ion suppression (RPC) Strong acids and bases by ion-pair mechanism (RPC) Homologs and oligomers (RPC) Hydrocarbon-like and polycyclic aromatic compounds (RPC & NPC)
Cyanopropylsiloxane	$Si{-}(CH_2)_3CN$	Useful for both RPC and NPC In NPC it exhibits properties similar to a low capacity silica gel In RPC it exhibits properties similar to short-chain alkylsiloxane-bonded layers (it has no selectivity for dipole-type interactions)
Aminopropylsiloxane	$Si{-}(CH_2)_3NH_2$	Used mainly in NPC & IEC. Limited retention in RPC Selectively retains compounds by hydrogen-bond interactions in NPC. Separations unlike those obtained on silica gel. Functions as a weak anion exchanger in acidic mobile phases (IEC)
Spacer bonded propanediol	$Si\text{-}(CH_2)_3OCH_2CH(OH)CH_2OH$	Used in NPC and RPC but more useful for NPC because of low retention in RPC Reasonable retention of polar compounds by hydrogen bond and dipole-type interactions in NPC More hydrogen-bond acidic and less hydrogen-bond basic then aminopropylsiloxane-bonded layers in NPC. More retentive than aminopropylsilsoxane-bonded layers in RPC Similar retention to short-chain alkylsiloxane-bonded layers but different selectivity for hydrogen-bonding compounds

[a] *NPC = normal-phase chromatography; RPC = reversed-phase chromatography; and IEC = ion-exchange chromatography.*

interphase region is structurally heterogeneous resulting from the presence of immobile components. In addition, the solvent composition of the interphase region is dynamic and varies in both volume and composition with changes in the bulk mobile phase composition. The interphase model suggests that retention is a result of differences in cavity formation in each phase, the capability of each phase for intermolecular interactions, and to changes in the phase ratio that accompany variation in the bulk mobile phase composition.

A characteristic feature of reversed-phase separations is the dominant role of water in the retention mechanism. Water is the most cohesive and hydrogen-bond acidic of the solvents normally encountered in reversed-phase chromatography. Although water is expected to be present in the interphase region, its relative concentration at equilibrium is much lower than in the bulk mobile phase, at least for water-rich mobile phases. The strong intermolecular interactions between water molecules tend to favor self-association over interactions with dissimilar solvent molecules. This preference for water to reform water—water interactions assists in the transfer of solutes to the interphase region and is the main driving force for retention in reversed-phase chromatography. The main process that opposes transfer to the interphase region and is responsible for lower retention is the selective distribution of hydrogen-bond basic solutes to the water-rich mobile phase. Dipole-type interactions and solute hydrogen-bond acidity have a minor role in the transfer process but may affect selectivity.

The combined contribution of the solvated stationary phase and bulk mobile phase to retention is described by a system map, which provides a quantitative assessment of the relative contributions of cavity formation and intermolecular interactions to the retention mechanism extracted by the solvation parameter model [10,37—39]. A system map for a 3-cyanopropylsiloxane-bonded silica gel layer with methanol—water mobile phase compositions is shown in Figure 2.4. Those contributions indicated by positive coefficients contribute to higher retention, and conversely, those with negative coefficients favor the mobile phase and lower retention. The importance of water is clear from the relative contributions to retention in the water-rich and water-poor regions of the plot. The cavity formation term (the difference in cavity formation in the bulk mobile phase and interphase region as well as residual dispersion interactions that are not fully canceled in the transfer from the bulk mobile phase to the interphase region) and hydrogen-bond acidity (difference in hydrogen-bond acidity of the bulk mobile phase and interphase region) dominate across the water-rich region. This is a direct result of the high cohesive energy and hydrogen-bond acidity of water. Adding organic solvent to water dilutes the effect of water on the retention mechanism; and in the water-poor region of the system map, the retention mechanism is no longer determined by the same factors as in the water-rich region.

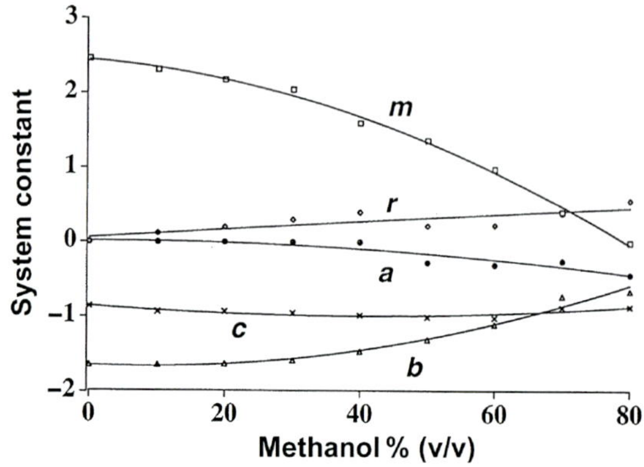

FIGURE 2.4 A system map for separations of neutral organic compounds on a 3-cyanopropylsiloxane-bonded layer with methanol–water as a mobile phase. A system map is a plot of the system constants of the solvation parameter model against the volume fraction of organic solvent in the bulk mobile phase. A guide to the construction of system maps is given in Refs [10,39]. Identification: m = cavity formation and dispersion interactions, r = contribution from electron lone pair interactions, a = hydrogen-bond basicity, c = the model intercept and is strongly correlated with the phase ratio, and b = hydrogen bond acidity. The system constants represent the difference for the interaction in the interphase region and bulk mobile phase.

The cavity formation term for polar chemically bonded phases is positive for water-rich mobile phases and the main factor that promotes retention. This term, however, is considerably smaller than for alkylsiloxane-bonded silica layers indicating that the interphase region for the polar chemically bonded layers is more cohesive than for the alkylsiloxane-bonded silica layers. This is the main reason why retention is significantly lower on 3-aminopropylsiloxane-bonded and spacer bonded propanediol layers compared with alkylsiloxane-bonded layers for all mobile phase composition. The contribution of hydrogen-bond acidity to retention is also significantly smaller for the polar chemically bonded phases than alkylsiloxane-bonded layers. The solvated 3-aminopropylsiloxane-bonded and spacer bonded propanediol interphase regions are more competitive with the bulk mobile phase and show higher retention of hydrogen-bond bases compared with both the alkylsiloxane-bonded and 3-cyanopropylsiloxane-bonded layers. The role of hydrogen-bond acidity in the retention mechanism is also significantly dependent on the identity of the organic solvent selected as modifier. Both the difference in cohesion and hydrogen-bond acidity of the alkylsiloxane-bonded and polar chemically bonded silica layers are probably a reflection of the attraction of a larger fraction of water into the interphase region for the polar chemically bonded layers. Selectivity differences among the polar chemically bonded phases are then largely the result of their ability to attract water into the interphase region rather than specific interactions of the solute with the polar functional groups themselves. It is likely that accessible silanol groups also play an important role in attracting water into the interphase region.

The contribution of dipole-type interactions to retention is again different for the solvated alkylsiloxane-bonded and polar chemically bonded layers. The alkylsiloxane-bonded layers exhibit moderate and negative contributions to retention from dipole-type interactions while for the polar chemically bonded phases these interactions are generally similar for the interphase region and bulk mobile phase, and therefore, self-canceling. Thus, dipole-type interactions are not important for retention on the 3-cyanopropylsiloxane-bonded layers because they are nearly equal in the interphase region and bulk mobile phase region. Dipole-type interactions are important for the alkylsiloxane-bonded layers because the interphase region is not competitive with the bulk mobile phase for these interactions and dipole-type interactions result in lower retention. This may seem counter intuitive if only the properties of the nonsolvated layers are considered but is easily understood if solvation of the interphase region is taken into account.

The solvated polar chemically bonded phases are generally more hydrogen-bond basic than the alkylsiloxane-bonded layers but for both layer types the interphase region either has similar properties to the bulk mobile phase or is less hydrogen-bond basic than the bulk mobile phase. These differences are small and depend on the identity of the organic modifier, which is probably more important than the identity of the stationary phase. This is an indication that differences in hydrogen-bond basicity are more a mobile phase property than a property of the solvated stationary phase.

As well as the capacity for intermolecular interactions, changes in the phase ratio with mobile phase composition make an important contribution to the observed retention for chemically bonded layers. Because of the differences in the phase ratio, a mobile phase composition providing suitable retention on an alkylsiloxane-bonded layer will often fail to provide useful retention on a polar chemically bonded layer. For method development, mobile phase optimization will almost always result in different solvent compositions for the two general types of chemically bonded layers.

2.4 CELLULOSE

Native and microcrystalline cellulose layers are used for the separation of polar compounds (e.g., carbohydrates, amino acids, peptides, nucleic acid derivatives, phosphates, etc.) with aqueous mobile phases [3,7]. Cellulose is a linear homopolymer of β-glucosepyranose. Native cellulose consists of 2–20 μm fibers with 400–500 monomer units. Microcrystalline cellulose is prepared by the acid hydrolysis of native cellulose with a narrower monomer distribution (in the range 49–200 monomer units) and a rod-shaped structure. Both forms of cellulose have surface areas of about 2 m^2/g. Ion-exchange celluloses are prepared by surface modification to introduce aminoethyl (AE), carboxymethyl (CM), diethylaminoethyl (DEAE), and phosphate (P) groups. Acylation of cellulose results in layers suitable for the separation of compounds of low polarity with nonaqueous mobile phases. Impregnation of cellulose with liquid poly(ethylenimine) or poly(phosphate) is also used to prepare modified layers for ion-exchange separations [40]. The reaction of cellulose with compounds containing amino groups results in the formation of conjugated imines. This reaction is the origin of discoloration and weak fluorescence observed with aging or thermal treatment of impregnated or modified cellulose layers. It can lead to baseline interference in scanning densitometry. Applications using cellulose layers are not that common for general analysis, particularly outside of the life sciences. Their recent use for two-dimensional separations of peptide digests and favorable properties for atmospheric pressure ionization mass spectrometry has promoted interest in these layers [41,42].

2.5 CHIRAL SORBENTS

The separation of enantiomers has more specific requirements than other compound types and utilizes application-tailored materials containing a chiral selector. There are comparatively few chiral sorbents used for thin-layer chromatography compared with column chromatography, which is more widely used in practice. There are several reasons for this. Chiral sorbents are relatively expensive and the single use characteristic of thin-layer chromatography is a disadvantage in this case. The higher efficiency of column separations allows baseline resolution with significantly smaller separation factors. This facilitates faster method development and increases the probability of success. Since detection in thin-layer chromatography usually occurs in the presence of the stationary phase, chiral selectors with weak ultraviolet–visible absorption properties are required. This reduces the choice of useful chiral selectors that might be employed. Scale up to preparative separations is straightforward for column liquid and supercritical fluid chromatography compared with preparative thin-layer chromatography. There are, however, a large number of successful enantiomer separations by thin-layer chromatography [43,44]. These methods predominantly employ chemical modification of the enantiomers to form diastereomeric derivatives or utilize mobile phase chiral additives and achiral stationary phases for the separation.

The most widely used chiral stationary phase for the separation of enantiomers is cellulose and its ester derivatives [43–46]. Fibrous cellulose contains linear chains of D-glucopyranose folded into a sheet-like crystalline structure maintained by intramolecular hydrogen bonding. The presence of low order (amorphous) regions, alternating with crystalline regions, in the fiber structure diminishes the potential of fibrous cellulose for the separation of enantiomers. These amorphous regions are largely absent from microcrystalline cellulose, which is prepared from fibrous cellulose by acid hydrolysis. Microcrystalline cellulose is generally preferred for the separation of enantiomers. Enantiomer recognition typically involves the formation of inclusion complexes with the helical cavity of the supramolecular crystal structure and hydrogen-bonding and dipole-type interactions with monomer hydroxyl groups. Cellulose, however, is limited to the separation of polar compounds such as amino acids and dipeptides using nonaqueous solvents.

Cellulose esters, particularly the triacetate, tribenzoate, and triphenylcarbamates, are more versatile chiral sorbents used mainly with aqueous-organic solvent mobile phases [43–46]. They have been used to separate an impressive number of enantiomers, typically aromatic compounds with hydroxyl or ketone groups close to the stereogenic center, heterocyclic compounds, and dinitrophenyl amino acid derivatives Their separation capability was shown to depend on the degree of esterification (not always correctly indicated as some materials may be only partially esterified), the crystallinity of the esterified sorbent, and the choice and amount of binder. The majority of applications are described for laboratory prepared sorbents that facilitate optimization of sorbent properties and layer composition that may have different separation characteristics to the limited number of precoated layers. Several attributes are believed to contribute to the chiral recognition mechanism [45–48]. The ability of the enantiomers to fit into the chiral cavity of the crystalline polymer is generally considered an important feature. Since cellulose esters swell to different extents in different solvents, changes in the dimensions and the shape of the chiral cavity with changes in the mobile phase composition can affect enantiomer separations. Hydrogen-bonding and dipole-type interactions between the polar functional groups of the enantiomer and the carbonyl group of the cellulose ester usually enhance enantiomer recognition. Also, enantiomer recognition is generally enhanced when the polar functional groups of the enantiomer that interact with cellulose ester are close to the stereogenic center. In the case of benzoate and phenylcarbamate esters, π–π interactions between the aromatic rings of the enantiomer and those of the cellulose ester enhance enantiomer recognition. The presence of the stereogenic center in a rigid molecular structure also serves to enhance enantiomer recognition.

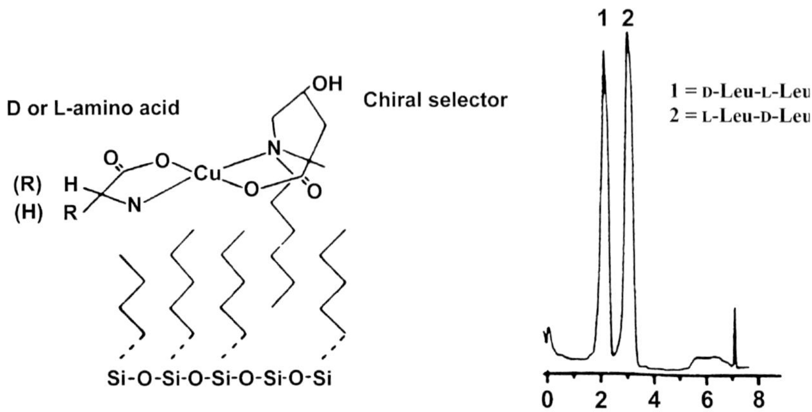

FIGURE 2.5 Schematic representation of the interaction between an amino acid and a physically immobilized chiral selector (copper(II) complex) in ligand exchange chromatography (Chiralplate). Separation of D-leu–L-leu and L-Leu–D-Leu on a Chiralplate with methanol–propan-1-ol–water (5:1:4 v/v/v) as mobile phase with absorbance detection at 410 nm. *Reproduced with permission from Ref. [49].*

The separation of enantiomers based on ligand exchange chromatography employs precoated reversed-phase layers impregnated with a copper(II) complex of (2S,4R, 2′RS)-4-hydroxyl-1-(2-hydroxydodecyl) proline in optimized amounts, Figure 2.5 [44,48–50].

Enantiomers are separated based on the differences in stability of the diasteromeric complexes formed between the sample, copper, and the proline selector. A prime requirement for separation is that the sample must be capable of forming complexes with copper. Suitable samples include amino acids and their derivatives, peptides, α-hydroxycarboxylic acids, β-blockers (drugs), and some lactones.

In addition, a large number of separations on silica gel or reversed-phase layers impregnated with different chiral selector have been described but with little subsequent use. An extensive bibliography is available in Ref. [43] and useful selective reviews in Refs [44,45].

2.6 CONCLUSIONS

The range of retention mechanisms employed in thin-layer chromatography is only a fraction of those employed for column chromatography. Those in significant use employ precoated high-performance layers based on silica gel or chemically modified silica gel. These products have changed little for 20 years or more, reflecting an acceptance of a certain performance standard. One reason for the persistence of this standard is that higher performance standards would require improvements in instrumentation so that the gains made are not lost in the laboratory practice. The performance specifications of instruments have tended to advance slowly over the last 20 years and were focused largely on quality assurance issues rather than probing the boundaries of speed and zone capacity. These are interesting times since major changes have occurred in column design and instrument performance in column liquid chromatography in the last five years that could provide the catalyst for change in thin-layer chromatography as well. This requires an end to complacency and a new expectation of what should be considered the standard of acceptance for thin-layer chromatography.

References

[1] Hahn-Deinstrop E. Stationary phases, sorbents. J Planar Chromatogr 1992;5:57–61.

[2] Hahn-Deinstrop E. Applied thin-layer chromatography. Weisbaden: Wiley-VCH; 2006.

[3] Rabel F, Sherma J. Stationary phases for modern thin-layer chromatography. LCGC North Am 2012;30:458–72.

[4] Poole SK, Ahmed HD, Belay MT, Fernando WPN, Poole CF. The influence of layer thickness on the chromatographic and optical properties of high performance silica gel thin-layer chromatographic plates. J Planar Chromatogr 1990;3:133–40.

[5] Poole SK, Ahmed HD, Poole CF. Evaluation of high performance thin layer chromatographic plates prepared from 3 μm silica gel. J Planar Chromatogr 1990;3:277–9.

[6] Poole SK, Poole CF. High performance stationary phases for planar chromatography. J Chromatogr A 2011;1218:2648–60.

[7] Hauck HE, Halpaap H. HPTLC on inert pre-coated layers of cellulose and silica 50000. Chromatographia 1980;13:538–47.

[8] Doner LW. Determining sugar composition of food gum polysaccharides by HPTLC. Chromatographia 2001;53:579–81.

[9] Halpappp H, Krebs K-F. Thin-layer chromatographic and high-performance thin-layer chromatographic ready-for-use preparations with concentrating zones. J Chromatogr 1977;142:823–53.

[10] Poole CF. The essence of chromatography. Amsterdam: Elsevier; 2003.

[11] Poole CF, Fernando WPN. Comparison of the kinetic properties of commercially available precoated silica gel plates. J Planar Chromatogr 1993;6:357–61.

[12] Fernando WPN, Poole CF. Determination of kinetic parameters for precoated silica gel thin layer chromatography plates by forced flow development. J Planar Chromatogr 1991;4:278–87.

[13] Poole CF, Fernando WPN. Some musings on the measurement and interpretation of theoretical plate height in thin-layer chromatography. J Planar Chromatogr 1992;5:323–33.

[14] Snyder LR. Solvent selectivity in normal phase TLC. J Planar Chromatogr 2008;21:315–23.

[15] Momchilova S, Nikolova-Damyanova B. Stationary phases for silver-ion chromatography of lipids: preparation and properties. J Sep Sci 2003;26:261–70.

[16] Nikolova-Damyanova B. Retention of lipids in silver ion high performance liquid chromatography: facts and assumptions. J Chromatogr A 2009;1216:1815–24.

[17] Ando Y, Satake M, Takahashi Y. Reinvestigation of positional distribution of fatty acids in docosahexaenoic acid-rich fish oil triacyl-sn-glycerols. Lipids 2000;35:579–82.

[18] Panalu HA, Homayon A, Rahimi A, Mori E, Izadi A, Parvin MM. HPTLC separation and quantitative analysis of asprin, salicylic acid and sulfosalicyclic acid. J Planar Chromatogr 2010;23:137–40.

[19] Jarne C, Cebolla VL, Membradio L, Le Maplham K, Giusti P. High-performance thin-layer chromatography using automated multiple development for the separation of heavy petroleum products according to their number of aromatic rings. Energy Fuels 2011;25:4586–94.

[20] Abidi SJ. Separation procedures for phosphatidylserines. J Chromatogr B 1998;717: 279–93.

[21] Dillon JT, Aponle JC, Tsai Y-J, Huang Y. Thin-layer chromatography in the separation of unsaturated organic compounds using silver-thiolate charomatographic material. J Chromatogr A 2012;1251:240–3.

[22] Poole CF, Poole SK, Dean TA, Chirco NM. Sample requirements for quantitation in thin-layer chromatography. J Planar Chromatogr 1989;2:180–9.

[23] Maxwell RJ, Lightfield AR. An improved method for the efficient predevelopment washing of HPTLC plates. J Planar Chromatogr 1999;12:109–13.

[24] Geiss F. Fundamentals of thin-layer chromatography. Heidelberg: Huethig; 1987.

[25] Spangenberg B, Kaiser RE. The water content of stationary phases. J Planar Chromatogr 2007;20:307–8.

[26] Halpaap H, Krebs K-F, Hauck HE. Thin-layer chromatographic and high-performance thin-layer chromatographic ready-for-use layers with lipophilic modifications. J High Resolut Chromatogr 1980;3:215–40.

[27] Jost W, Hauck HE. High-performance thin-layer chromatographic pre-coated plates with amino modification and some applications. J Chromatogr 1983;261:235–44.

[28] Jost W, Hauck HE, Fischer W. Cyano-modified pre-coated plates – a medium polar modification for high performance thin-layer chromatography. Chromatographia 1986;21:375–8.

[29] Heilwell E, Rabel FM. Reversed-phase thin-layer chromatography on C2, C8, C18, and diphenyl bonded phases. J Chromatogr Sci 1985;23:101–5.

[30] Jost W, Hauck HE. The use of modified silica gels in TLC and HPTLC. Adv Chromatogr 1987;27:129–65.

[31] de Haan JW, van de Ven LJM, de Vries G, Brinkman UATh. Characterization of chemically bonded phases on precoated thin-layer plates by ^{29}Si and ^{13}C cross-polarization and magic angle spinning NMR. Chromatographia 1986;21:687–92.

[32] Kiridena W, Poole CF. Structure-driven retention model for method development in reversed-phase thin-layer chromatography on octadecylsiloxane-bonded layers. J Planar Chromatogr 1999;12:13–25.

[33] Petruezynik A. Effect of chromatographic conditions on sorption and system efficiency in HPTLC of selected quinoline standards on cyanopropyl stationary phases. J Planar Chromatogr 2010;23:56–64.

[34] Jost W, Hauck HE. Reversed-phase thin-layer chromatography of nitrogen bases using alkyl sulphonates as ion-pair reagents. J Chromatogr 1983;264:91–8.

[35] Jost W, Hauck HE. A modified high-performance thin-layer plate for the separation of purines and pyrimidines. Anal Biochem 1983;135:120–7.

[36] Klaus R, Fischer W, Hauck HE. Use of a new adsorbent in the separation and detection of glucose and fructose by HPTLC. Chromatographia 1989;28:364–6.

[37] Kiridena W, Poole CF. Structure-driven retention model for solvent selection and optimization in reversed-phase thin-layer chromatography. J Chromatogr A 1998;802: 335–47.

[38] Ali Z, Poole CF. Insights into the retention mechanism of neutral organic compounds on polar chemically bonded stationary phases in reversed-phase liquid chromatography. J Chromatogr A 2004;1052:199–204.

[39] Poole CF, Poole SK. Column selectivity from the perspective of the solvation parameter model. J Chromatogr A 2002;965:263–99.

[40] Urreaga JM, de la Orden MU. Modification of cellulose with amino compounds. A fluorescence study. Carbohydr Polym 2007;69:14–9.

[41] Pasilis SP, Kerbesz V, van Berkel GJ, Schulz M, Scharcht S. HPTLC/DESI-MS imaging of tryptic protein digests separated in two dimensions. J Mass Spectrom 2008;43:1627–35.

[42] Smrke S, Vovk I. Comprehensive thin-layer chromatography mass spectrometry of flavanols from *Juniperus communis* L. and *Punica granatum* L. J Chromatogr A 2013;1289:119–26.

[43] Kowalska T, Sherma J, editors. Thin layer chromatography in chiral separations and analysis. Boca Raton: CRC Press; 2007.

[44] Lepri L. Enantiomer separations by TLC. J Planar Chromatogr 1997;10:820–31.

[45] Del Bubba M, Checchini A, Lepri L. Thin-layer chromatography eneantioseparations on chiral stationary phases: a review. Anal Bioanal Chem 2013;405:533–54.

[46] Del Bubba M, Cincinelli A, Lepri L. Chiral separations and quantitative analysis of optical isomers on cellulose tribenzoate plates. J Chromatogr A 2013;1218:2737–44.

[47] Lepri L, Cincinelli A, Del Bubba M. Structure and substituent effects on retention and chiral resolution of ketones and alcohols on microcrystalline cellulose triacetate plates. Chromatographia 2010;71:685–94.

[48] Siouffi AM, Piras P, Roussel C. Some aspects of chiral separations in planar chromatography compared with HPLC. J Planar Chromatogr 2005;18:5–12.

[49] Mack M, Hauck HE, Herbert H. Enantiomeric separation in TLC with the new HPTLC pre-coated plate CHIR with concentrating zone. J Planar Chromatogr 1988;1:304–8.

[50] Bhushan R, Tanwar S. Different approaches of impregnation for resolution of enantiomers of atenolol, propanolol and salbutamol using Cu(II)-L-amino acid complexes for ligand exchange on commercial thin-layer chromatographic plates. J Chromatogr A 2010;1217:1395–8.

CHAPTER

3

Ultrathin and Nanostructured Stationary Phases

Steven R. Jim [1], Michael J. Brett [1,2]

[1] Department of Electrical and Computer Engineering, University of Alberta, Edmonton, AB, Canada; [2] NRC National Institute for Nanotechnology, Edmonton, AB, Canada

3.1 INTRODUCTION

Many analytical separation techniques become more powerful upon miniaturization. The invention of monolithic silica gel ultrathin-layer chromatography (UTLC) sorbents in 2001 marked the beginning of a trend toward higher performing miniature planar chromatography plates with ultrathin and nanostructured stationary phases [1–4]. While traditional thin-layer chromatography (TLC) and high-performance thin-layer chromatography (HPTLC) layers are composed of silica gel particles fixed to a substrate using a binder (*Chapter 2: High Performance Stationary Phases*), these isotropic monolithic layers generally have a single homogenous chromatographic surface [1]. Such materials therefore exhibit consistent surface chemistry and chromatographic interactions throughout, since no competing analyte–binder interactions occur. Thinner UTLC sorbents with carefully selected morphologies achieve faster separations over shorter distances with higher sensitivity and lower reagent requirements [2]. These potential advantages motivate continued UTLC stationary phase innovation.

This chapter describes leading approaches to fabricating advanced ultrathin and nanostructured stationary phases. The chromatographic merits of monolithic silica gels, monolithic polymers, electrospun nanofibrous mats, nanostructured columnar thin films, and carbon nanotube-templated silica are surveyed. The properties of these materials are summarized in Table 3.1. Standard TLC and HPTLC sample application,

Instrumental Thin-Layer Chromatography
http://dx.doi.org/10.1016/B978-0-12-417223-4.00003-0

TABLE 3.1 Comparison of Ultrathin and Nanostructured Stationary Phase Properties and Performance

	Monolithic Silica Gel[a]	Monolithic Polymer	Electrospun Polymer	Nanostructured GLAD Thin Films	Carbon Nanotube Templated Silica
Stationary phase material	Silica sol-gel [1–3,5–7]	Photopolymerized poly(butyl methacrylate-*co*-ethylene dimethacrylate) [8,9]; Poly(glycidyl methacrylate-*co*-ethylene dimethacrylate) [10]	Polyacrylonitrile [11–14]; Glassy carbon [12]; Polyvinyl alcohol[15];	SiO_2 [16–23]; Al_2O_3 [21,22]; TiO_2 and ZrO_2 [21]; SiO_2 functionalized with C18 [24]; SiO_2 functionalized with ALD Al_2O_3, ZrO_2, and ZnO (<10 nm thick) [22]; Ag [25]	SiO_2 nanowire hedges [26,27]; (also functionalized with 3-aminopropyltriethoxysilane (APTES) [28,29])
Thickness	10–12 μm [1–3,6]; ~100 μm [5]	50–200 μm [8–10]	5–25 μm [11–15]	2.6–9 μm [16–24]; 200 nm Ag thin film [25]	~50 μm [26–29]
Morphology[b]	Isotropic monolith; macroporous and mesoporous [2,5,7]; SSA ~ 350 m^2/g [2]	Isotropic monolith; macroporous and mesoporous [8–10]	Isotropic [11–15] or Anisotropic [13]; Mat composed of nanofibers (width 150–520 nm [11–15]); SSA 21.7 m^2/g [11]	Isotropic or anisotropic; ~100–400 nm thick columns or blades separated by ~100–800 nm: Hexagonal helices [16,17]; Chevrons [16,18]; Vertical posts [18,20,22–24]; Blade-like [18,19,21,23];	Anisotropic; herring-bone (zig-zag) pattern [26–29]; composed of 3–4 μm wide hedges spaced 4–6 μm apart [27–29]; with 100–120 nm diameter SiO_2 nanowires [29]; Specific surface area 19–27 m^2/g for oxidized SiCNTs [27]

				Inclined posts [25] (also reactive ion etched columns [20]); SSA $\sim 300\ m^2/g$ (SiO_2 vertical posts) and $\sim 250–340\ m^2/g$ (SiO_2 coated with ALD Al_2O_3) [22]	
Substrate	Glass [1–3,5–7]	Glass [8–10]	Al foil [11,14,15]; Stainless steel [12,13]	Glass [16–25]	Si wafer [26–29]
Substrate size	25–40 mm × 30–100 mm [1–3,5–7]	26–30 mm × 33–76 mm [8–10]	20–30 mm × 40–70 mm [11–15]	20–25 mm × 25–100 mm [16–24]	Limited by 100 mm circular wafer [26–29]
Elution distance and time	10–30 mm in 1–8 min [1–3,5,7,30]	30–60 mm in <6 min [8,10]	25–35 mm in 3–12 min [11–15]; 25 mm in 1–1.5 min for aligned nanofibers [13]	5–12 mm in <6 min [16–21,23,25]	25–45 mm in 1–4.5 min [26–29]
Tested analytes	Dyes [1,5,7]; Pharmaceutically active ingredients [2,3,30–32]; Phenols and plasticizers [2,3]; Amino acids and pesticides [2]; Steroids [3]; Triazoles [30]	Proteins and peptides [8–10]; Dyes [8]	UV fluorescent laser dyes [11–13]; Steroids [11,13]; Amino acids [12,15]; β-blockers [13]; Beverage preservatives [14]; Aspartame hydrolysis products (aspartic acid, phenylalanine) [15]	Dyes [16–22,24,25]; Sugars and chocolate samples [23]; Melamine and Trans-1,2-bis(4-pyridyl)ethylene [25]	Dyes [26–29]; Analgesics [27]

[a]*The monolithic silica gel properties listed here include those of Merck UTLC plates and others. Table 3.2 includes only Merck UTLC plate data.*

[b]*The reported specific surface areas (SSA) were measured using the Brunauer, Emmett, and Teller (BET) method [33], where available.*

development, and documentation instruments are generally inappropriate for UTLC layers. The chapter therefore concludes with a discussion about recent progress toward specialized equipment designed to help these advanced chromatographic media approach their full analytical potential.

3.2 MONOLITHIC SILICA GELS

Porous monolithic stationary phases exhibit a bimodal distribution in pore size [5] that affects both surface activity and migration velocities. Abundant small mesopores (2–50 nm [34]) ensure sufficient surface area over which analytes may interact with the sorbent while larger macropores (>50 nm [34]) enable rapid mobile phase migration.

The porosity and morphology of monolithic normal phase silica layers are entirely dictated by the chemical reactions occurring during fabrication. Silica sol-gel monoliths are generally prepared from a mixture of silane precursors such as methyltrimethoxysilane (MTMOS) [6], tetraethoxysilane (TEOS) [5,7], tetramethoxysilane (TMOS) [5,7], or tetra-alkoxysilane [1–3]. As in the synthesis of silica gel particles for traditional TLC and HPTLC layers, condensation of silicic acids in the mixture produces the mesoporous sol-gel. However, porogens (such as polyethylene glycol, PEG) added to this mixture serve as sacrificial templates during monolith fabrication [5,7]. Removing these porogens produces the macropores in the silica gel monoliths. Additives ensure a favorable pH during the condensation reaction and reduce cracking when drying the layer. An early study of 10–60 μm thick monoliths investigated the suitability of different silica gel mixtures in planar chromatography [6].

The first normal phase UTLC plates were prepared using similar methods optimized for exceptionally thin 10 μm monolithic layers [1–3]. These media were invented by Hauck et al. (Merck KGaA, Darmstadt, Germany) in 2001 [1] and were commercially available from Merck Darmstadt until 2011. (Factors contributing to their discontinuation are discussed below.) These binder-free UTLC layers were prepared on 36×60 mm glass slides using tetra-alkoxysilane precursors [1]. The resultant layers exhibit fine pores and a sponge-like appearance very different from that of traditional media composed of silica gel granules (Figure 3.1). Table 3.2 compares the physical properties and representative separation performance of TLC, HPTLC, and UTLC layers. UTLC layers enable faster separations over shorter distances with higher sample detectability [2].

Monolithic silica UTLC plates possess sorbent layers that are thinner and have lower specific surface area than TLC and HPTLC plates. A consequence of the lower resultant total surface area is reduced sample

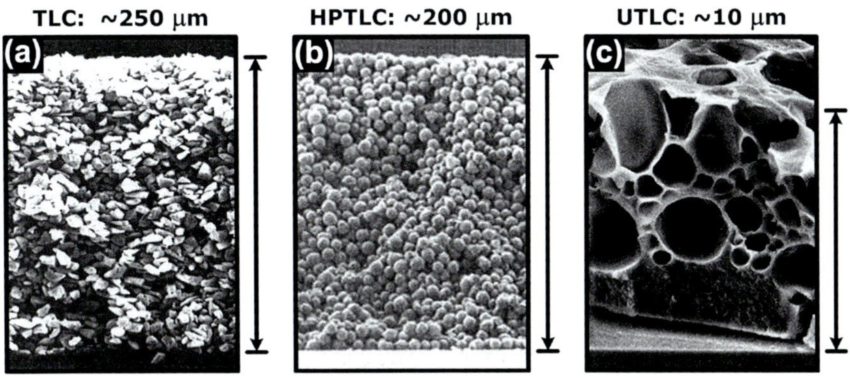

FIGURE 3.1 Cross-sectional scanning electron micrographs of (a) thin-layer chromatography (TLC), (b) high-performance TLC (HPTLC) [35], and (c) ultrathin-layer chromatography (UTLC) silica gel stationary phases. Representative sorbent thicknesses provided. *Parts (a) and (b) modified with permission from Ref. [35], copyright (2003) Marcel Dekker (Taylor and Francis). Part (c) modified with permission from Ref. [17], copyright (2010) American Chemical Society.*

TABLE 3.2 Comparison of TLC, HPTLC, and Merck Monolithic UTLC Stationary Phase Properties and Performance

Stationary Phase Layer Properties	**TLC [36,37] (particles, silica gel 60)**	**HPTLC [36,37] (particles, silica gel 60)**	**UTLC [2,36,37] (monolithic, silica gel)**
Typical thickness (μm)	100–250	100–200	10
Characteristic size (μm)	10–12 (mean particle size)	4–6 (mean particle size)	1–2 (macropore size)
Pore size (Å)	60	60	30–40
Specific surface area (m^2/g)	520	480–540	~350
Representative Elutions	**TLC** [2,36,37]	**HPTLC** [36,37]	**UTLC** [2,36,37]
Development time (s)	>900	>180	60–360
Development distance (mm)	70–150	30–70	10–30
Limit of detection (ng, absorption)	1–5	0.1–0.5	0.5
Theoretical plate height (μm)	30–75	12–25	80

capacity. Increasing the thickness of the monolithic silica gel to ~100 μm offers a method of increasing sample capacity but introduces other challenges [5,7]. Adhesion with the underlying substrate material and cracking within the silica gel can reduce the mechanical stability and influence mobile phase migration in these films. Careful optimization of glass substrate surface treatments and silica gel additives is required.

3.3 MONOLITHIC POLYMERS

Porous polymer monoliths have also been used in reversed-phase planar chromatography. Superhydrophobic ultraviolet (UV) photopolymerized layers (50–200 μm thick) are better suited than silica gel monoliths in separations of peptides and other large molecules [8–10]. During layer preparation, complex mixtures of methacrylate monomers, styrene monomers, alcohols, and other additives are cast into a glass mold. These molds are constructed from a glass substrate, a Teflon gasket (to determine layer thickness), and a glass top plate. UV irradiation cures the mixture within the mold to create thin porous polymer slabs. The porosity and morphology of these hydrophobic monoliths (Figure 3.2) are influenced by the amount and type of alcohol porogens in the polymer mixture [9]. Polymer mixture composition must be optimized to produce a material dense enough to be mechanically stable but porous enough to allow rapid mobile phase migration. Surface treatments are again required to improve adhesion between the monolith and the glass substrate.

The hydrophobicity of the polymer monolith can also be selectively patterned to introduce interesting 2D separation behaviors. A narrow hydrophilic line optically patterned onto the hydrophobic monolith can

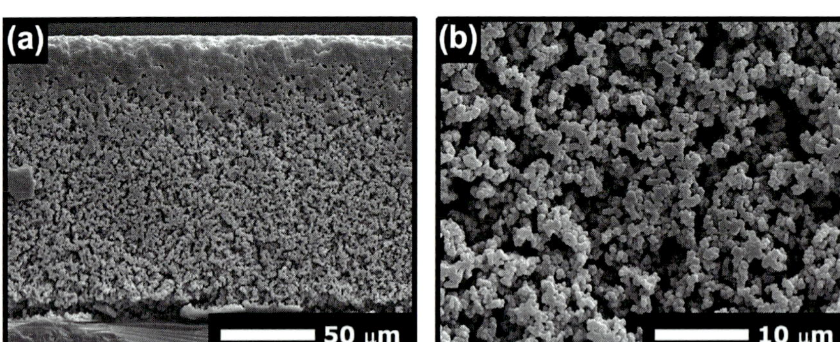

FIGURE 3.2 (a) Low and (b) high magnification cross-sectional scanning electron micrographs of a 150 μm thick poly(butyl acrylate-*co*-ethylene dimethacrylate) monolith. *Figure modified with permission from Ref. [8]. Copyright (2007) American Chemical Society.*

act as a virtual channel [9]. Performing the first elution along this channel (ion exchange chromatography) and the second elution perpendicular to this channel (reversed-phase chromatography) improved separations of a peptide sample. Polymer monoliths with optically patterned hydrophobicity gradients can also improve 2D separation quality [10]. Similarly prepared polymer monoliths 125-μm thick have also been used in planar electrophoresis and pressurized planar electrochromatography [38] (*Chapter 7: Pressurized Planar Electrochromatography*). Preliminary work on thermally initiated polymer monoliths has also been recently reported [39].

3.4 ELECTROSPUN NANOFIBERS

Electrospinning is a powerful method of microstructuring polymers [40]. A relatively simple apparatus (Figure 3.3) enables rapid fabrication of nanofibrous stationary phase mats. Solutions containing dissolved polymers such as polyacrylonitrile (PAN) [11–14], polyvinyl alcohol (PVA) [15], and SU-8 [12] are injected into the setup using a syringe pump. A high voltage DC power supply generates a potential difference (for example, 9–20 kV [11–15]) between the syringe tip and the metal substrate (such as aluminum foil [11,14,15] or stainless steel [12,13]). The resultant electric field produces a polymer jet that arrives at the substrate as a solid nanofiber. This nanofiber winds randomly across the substrate and can be meters in length. The nanofiber width (150–520 nm [11–15]) and mat thickness (5–25 μm [11–15]) are controlled by adjusting the polymer type, syringe pump flow rate, solution viscosity, power supply

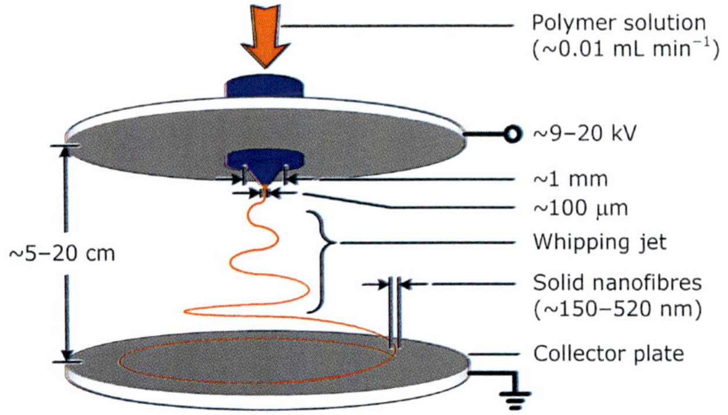

FIGURE 3.3 Schematic illustration of electrospinning apparatus. Injected polymer solution is drawn into a thin whipping jet by an applied electric field. The polymer jet solidifies into a thin nanofiber as it reaches the collector plate. Typical operating parameters and dimensions are shown.

voltage, tip-substrate separation distance, electrospinning time, and other parameters. An initial report of nanofibrous PAN mats studied the effects of electrospinning time on morphology (Figure 3.4) and UTLC separation performance [11]. Several recent advancements have increased the capabilities of this technique.

UTLC stationary phases with aligned nanofibers may be fabricated by moving the substrates on a rotating drum during electrospinning [13]. Careful selection of rotation speed increases nanofiber ordering and results in an anisotropic medium. Chromatographic separations performed along the aligned nanofibers occurred faster and more efficiently than on unordered materials. Incorporating cross-linking additives made PVA stationary phases insoluble in water and allowed reversed-phase separations of amino acids [15]. High temperature pyrolysis of electrospun SU-8 layers produced glassy carbon UTLC sorbents appropriate for reversed-phase chromatography [12]. However, the dark color

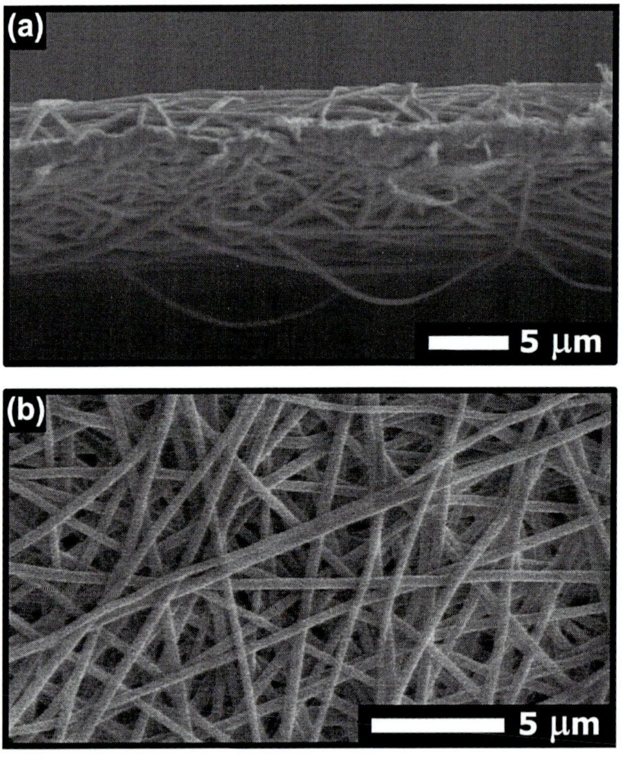

FIGURE 3.4 (a) Edge and (b) top scanning electron micrographs of electrospun nanofibrous PAN planar chromatography stationary phases. *Figure modified with permission from Ref. [11]. Copyright (2009) American Chemical Society.*

of the resultant plates restricted their use to UV fluorescent laser dyes. Adding a photoluminescence indicator to the electrospun PAN nanofibers enabled detection of UV-active compounds [14]. Eluted beverage preservative analytes absorbed UV and appeared as dark spots on the UTLC plates. The robustness of the electrospun nanofibrous UTLC plates also makes them appealing in electrospray ionization–mass spectroscopy (ESI-MS) [14]. The elution head used to extract separated analytes can be pressed firmly against the layer without causing substantial damage.

3.5 NANOSTRUCTURED THIN FILMS

Nanostructured columnar thin films may be fabricated using glancing angle deposition (GLAD) [41,42]. This single-step physical vapor deposition technique takes place within high vacuum chambers (Figure 3.5) and has been used to engineer planar films in various morphologies from several materials. Source material evaporated within the chamber is incident on the substrate at a given deposition angle (α, measured with respect to the substrate normal) and forms small nuclei early into the

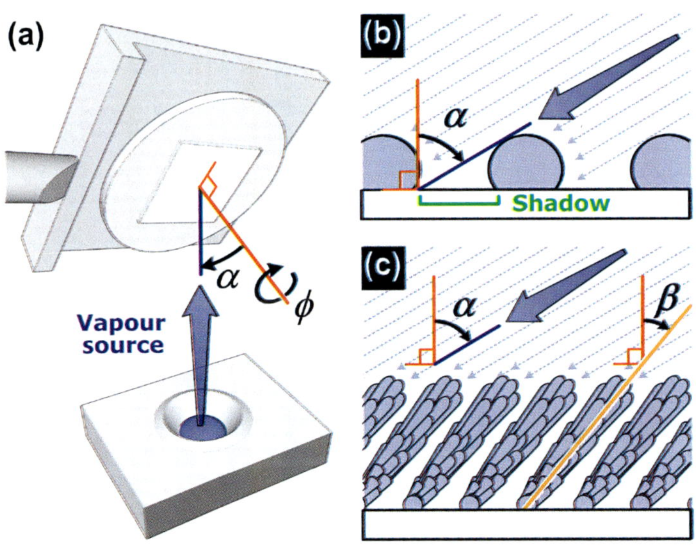

FIGURE 3.5 Schematic representation of the GLAD thin film physical vapor deposition technique and growth mechanisms. (a) Directional source material evaporant reaches the substrate at a given oblique angle α and azimuthal angle ϕ. (b) When evaporant is incident at large α, growing nuclei cast shadows across the substrate into which additional material cannot be deposited. (c) These nuclei grow into columns inclined toward the apparent vapor source by an angle β.

deposition. At large α, these nuclei cast microscopic shadows across the substrate into which additional material cannot be deposited. The nuclei grow into columns inclined (by an angle β) toward the evaporant source over the course of the deposition. Increasing α (toward a maximum value of 90°) increases the film porosity. The columns may be sculpted into different architectures by changing the azimuthal angle of this evaporant (ϕ). Motors enable dynamic manipulation of substrate orientation throughout the deposition.

Several of these morphologies have been studied in UTLC. Early investigations considered ~5 μm thick normal phase SiO_2 stationary phases with helical, vertical post, chevron, and blade-like features [16–19]. Anisotropic GLAD UTLC microstructures (Figure 3.6) exhibited channel features that strongly influenced chromatographic behaviors; elutions along these channels proceeded faster. GLAD UTLC morphologies and separation performance may also be tuned using reactive ion etching

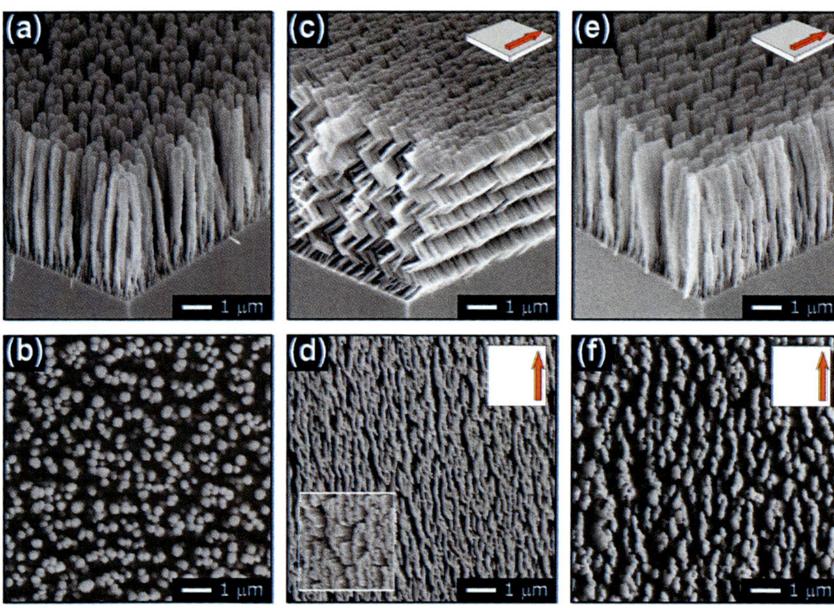

FIGURE 3.6 SEM micrographs of macroporous GLAD thin film separation media. Micrographs of (a, b) isotropic vertical posts as well as anisotropic (c, d) chevron and (e, f) blade-like media viewed from an oblique angle and above, respectively. Red arrow indicates the along-channel vector in anisotropic films. Top images of the (b) vertical post and (f) blade-like films were imaged normal to the substrate surface. Pores within the top layer of the chevron film are apparent only when the film is viewed along the angled segments in this layer (~53° to the substrate normal) (d). The inset in (d) shows the film when viewed normal to the substrate. *Figure reproduced with permission from Ref. [18]. Copyright (2010) American Chemical Society.*

(RIE) [20]. This dry fluorocarbon plasma etching process increases the porosity of SiO_2 GLAD films (Figure 3.7). Selectively etching specific regions of the GLAD layers can be used to pattern highly porous concentration zones adjacent to denser separation zones [20].

Methods of achieving different GLAD UTLC surface chemistries have also been investigated. The merits and unique properties of GLAD Al_2O_3, TiO_2, and ZrO_2 stationary phases have been explored using food dye mixture separations [21]. TiO_2 and ZrO_2 media are particularly interesting because analyte retention can be changed through thermal oxidation and UV irradiation. Layers with tunable reversed-phase functionalization were produced by immersion in octadecyltrichlorosilane and subsequent oxygen plasma exposures [24]. This processing enabled variation in analyte retention and aqueous mobile phase wettability impossible on commercial reversed-phase HPTLC plates. Another report investigated GLAD UTLC layers coated with ~5 nm thick atomic layer deposition (ALD) metal oxides (Al_2O_3, ZrO_2, ZnO) (Figure 3.8) [22]. As analyte retention was almost entirely dictated by outermost surface, the combined GLAD ALD approach potentially broadens the selection of

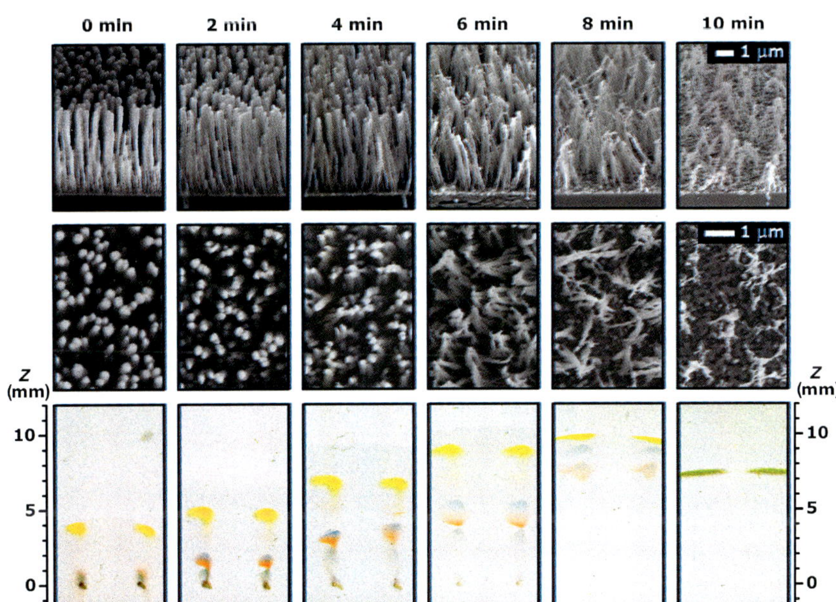

FIGURE 3.7 Reactive ion etching (0–10 min etch durations) to modify GLAD SiO_2 film morphology and UTLC performance. Oblique (first row) and top (second row) scanning electron micrographs are provided. Corresponding dye separation images (third row) are enhanced for presentation. Etching increased GLAD film porosity, reduced analyte retention, and increased migration velocities. *Figure modified with permission from Ref. [20]. Copyright (2011) Elsevier.*

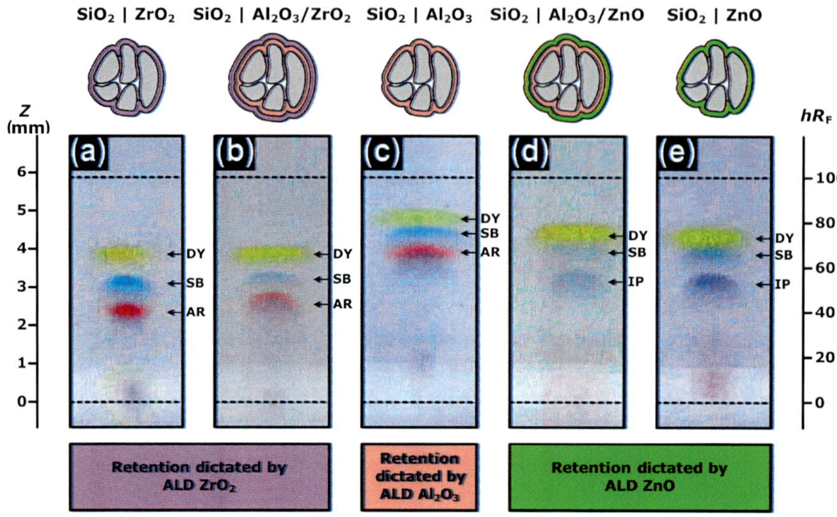

FIGURE 3.8 Dye separations on GLAD SiO_2 media functionalized with single (a, c, e) and double (b, d) ALD metal oxide coatings (5 nm per coating). The similarity between (a, b) and (d, e) indicates that chromatographic behaviors were dominated by the outermost metal oxide surface rather than the GLAD column core. Dye separation images enhanced for clarity. The typical elution order (descending retention factor) was Dimethyl Yellow (DY), Sudan Blue II (SB), and Ariabel Red (AR), and Indophenol (IP). *Figure modified with permission from Ref. [22]. Copyright (2013) Elsevier.*

chromatography materials to include all those being researched in the highly active ALD material science field.

Although much of this work has been performed using test dye models, GLAD UTLC was recently applied to sugar separations [23]. This study incorporated the benefits of inkjet-printed samples, derivatizing agent spraying, and coupling with ESI-MS. Another related contribution involved UTLC separations performed on oblique angle deposition silver nanorods [25]. This stationary phase enabled subsequent detection of separated dyes using spatially resolved surface enhanced Raman spectroscopy (SERS).

3.6 CARBON NANOTUBE-TEMPLATED LAYERS

Chromatography media have also been microfabricated using processes employed in the microelectronics industry. A series of several thin film deposition, photolithography, and other steps enable fabrication of complex media exhibiting herring-bone (zig-zag) shaped channel features comprised of SiO_2 templated onto carbon nanotubes (CNTs) [26–29] (Figure 3.9 and Figure 3.10). The initial report of these

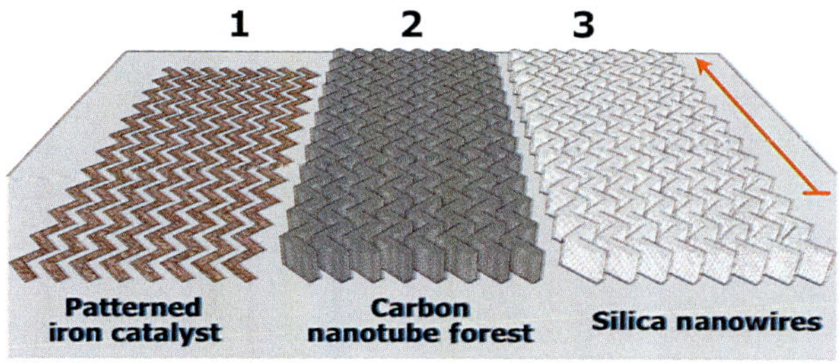

FIGURE 3.9 Overview of carbon nanotube-templated layer microfabrication. (1) Photolithography is used to pattern an Al_2O_3 diffusion barrier layer and Fe catalyst layer into herring-bone structures. (2) The Fe catalyzes chemical vapor deposition growth of carbon nanotube forests. (3) Subsequent processing introduces SiO_2 into the forests and burns away the carbon nanotubes. Chromatographic elutions are performed along macropores that separate the anisotropic porous herring-bone hedges (red arrow).

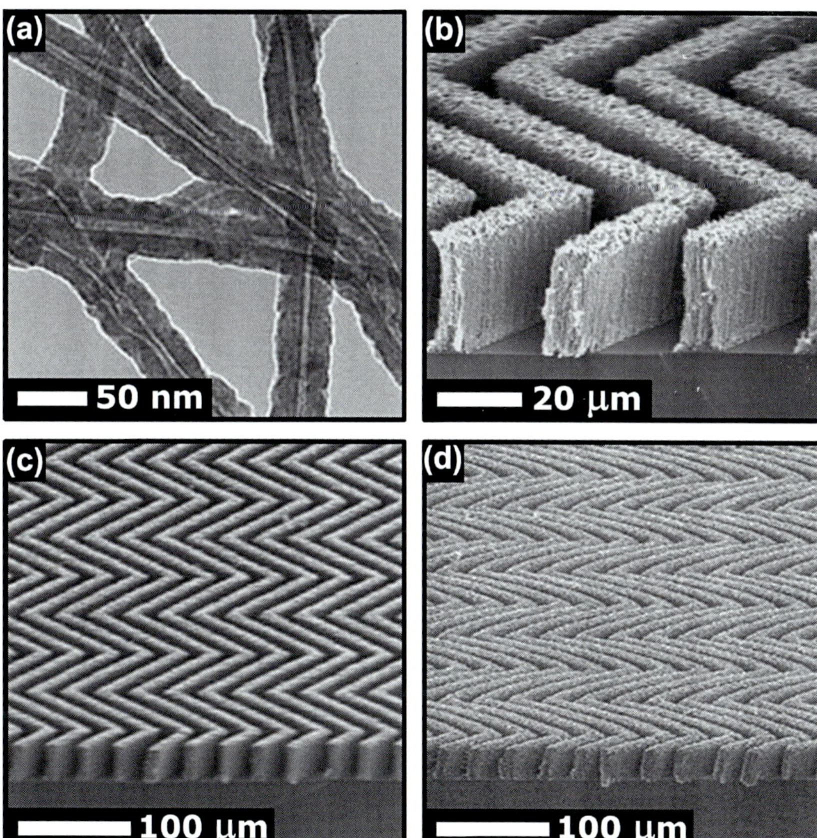

FIGURE 3.10 Microfabricated carbon nanotube-templated planar chromatography media. (a) Transmission electron micrograph showing carbon nanotubes coated with Si using low-pressure chemical vapor deposition. (b, c) High and low magnification scanning electron micrographs of patterned Si coated carbon nanotube hedges. (d) Scanning electron micrograph of structures after oxidizing Si to SiO_2 and burning out the carbon nanotubes. *Figure modified with permission from Ref. [26]. Copyright (2011) Wiley.*

media [26] employed an iron thin film deposited onto an Al_2O_3 diffusion barrier that was patterned into herring-bone shapes by photolithography (lift-off). The iron film was used to catalyze carbon nanotube growth by chemical vapor deposition (CVD). The resultant carbon nanotube "hedges" (4 μm wide, separated by 7 μm spaces) were then coated with low-pressure CVD (LPCVD) silicon. Thermal processing burned away the CNTs and oxidized the silicon into 50 μm tall SiO_2 nanowires <200 nm in diameter. Mesopores within the hedges enabled access to high surface areas while macroporous channels between hedges permitted rapid mobile phase migration. The herring-bone pattern of the hedges had better mechanical stability than straight line hedges [43].

Subsequent work optimized the thickness of the iron thin film catalyst layer [27] and addressed pattern distortion problems associated with the thermal processing step [28,29]. The volume expansion that occurs upon oxidation distorted the herring-bone structures. This annealing step was eliminated by rapid direct deposition of an Al-doped SiO_2 film and amino-functionalization to cover the Al impurities [28,29]. The high temperature steps presently limit this process to Si wafer substrates.

3.7 ADVANCED INSTRUMENTATION AND TECHNIQUES

The size and unconventional elution characteristics of UTLC plates introduce unique challenges and capabilities. The small format (the shortest dimension can be ~25 mm) and exceptionally thin layers on these miniaturized planar chromatography plates can make them difficult to handle and easy to damage. Conventional HPTLC sample application systems, development chambers, and documentation tools are often incompatible with these plates. Modified or entirely new instruments are required if UTLC media are to reach their ultimate performance. However, ultrathin layers can also improve hyphenation with post-separation characterization methods such as mass spectroscopy. These factors must be considered carefully when applying miniaturized media to analytical problems.

3.7.1 Sample Application

Sample application significantly affects the overall separation performance in all forms of planar chromatography (*Chapter 4: Instrumentation for Sample Application and Development*). This dependence is pronounced on miniaturized plates. Large applied spot sizes can negate improvements in separation efficiency and analyte resolution. Reduced thickness and lower specific surface areas also make UTLC stationary phases more susceptible to sample overloading than TLC or HPTLC layers. In their first report, Hauck et al. (2001) recognized this limitation and recommended sample volumes <20 nL to maximize UTLC's benefits [1].

Narrow gauge syringes (such as Hamilton model 7000.5 with total volume 0.5 μL) enable hand-application of sample volumes as low as 10 nL [1,3]; however, precise manual placement of repeatable spots may require customized apparatus [22]. The CAMAG Automatic TLC Sampler 4 is the state-of-the-art tool in HPTLC spot application. Although an exceptionally powerful spotting robot, modifications to its hardware and firmware settings are required to achieve <25 nL volumes [14,18]. The Office Chromatography concept advanced by Morlock et al. adapts inexpensive high-performance consumer electronics as planar chromatography instrumentation [17,23]. In this approach, inkjet printers perform rapid and precise application of sharp ~10 nL aqueous sample bands [17,23]. However, the plastic components in current inkjet printers preclude samples dissolved in organic solvents.

3.7.2 Development Chambers

Successful ascending (upright) UTLC separations have been achieved within custom sealed glass containers [11–14,16] and existing HPTLC chambers (for example, CAMAG Twin Trough chamber designed for 10 × 10 cm plates [26]). Other separations have been performed in small, modified horizontal development chambers such as the 50 × 50 mm Desaga H-chamber [18]. Nonetheless, these chambers are generally grossly oversized for miniaturized UTLC plates. In addition to increased reagent requirements, large reservoirs and dead volumes adjacent to the stationary phases can affect mobile phase composition, flow uniformity, and reproducibility. Details about these effects are discussed in Chapter 4 (*Instrumentation for Sample Application and Development*).

3.7.3 Documentation

Adequate recording of the short separation patterns on miniature planar chromatography plates requires higher imaging resolution than most conventional TLC and HPTLC instruments can produce (*Chapter 8: Theory and Instrumentation for In situ Detection*). The Office Chromatography concept employs consumer reflection-mode flatbed scanners to digitize developed UTLC plates for subsequent chromatogram extraction [17]. Meanwhile, flatbed film scanners can be used to image transparent

UTLC media in transmission mode [18]. Consumer digital cameras capable of high image resolutions are often effective in capturing separation patterns illuminated by visible [23] and UV light [13,28]. Time-resolved UTLC exploits the high-definition video recording capabilities of modern digital cameras to record elutions of visible analytes [44]. Extracted video frames are processed to quantify separation performance at different times into the development.

3.7.4 Hyphenation with Mass Spectroscopy

Ultrathin stationary phases can increase the sensitivity of mass spectroscopy characterization performed on separated analytes. Benefits have been observed for several sampling mechanisms including matrix assisted laser desorption ionization (MALDI) [10,45], atmospheric pressure MALDI (AP-MALDI) [30,31], desorption electrospray ionization (DESI) [32], and electrospray ionization (ESI) [14,23,46] (*Chapter 11: Advanced Spectroscopic Detectors for Identification and Quantification: Mass Spectrometry*). The significant sensitivity improvements (as high as 10–100 times better than for HPTLC plates [30]) are attributed to the thinness of UTLC stationary phases. Reduced layer thickness produces higher analyte surface densities (molecules per unit area) and increases extraction from the layer interior (ions are closer to the surface) in AP-MALDI performed on UTLC layers [30]. These monolithic layers are also especially compatible with DESI since incident sprayed solvent tends to erode them less than the granular HPTLC layers composed of silica gel particles [32].

3.8 CONCLUSIONS

Ultrathin and nanostructured planar chromatography stationary phases have progressed significantly since UTLC's invention in 2001 by Hauck et al. [1] and its commercialization by Merck Darmstadt in 2002 [47]. Relatively simple chemical reactions can produce monolithic silica gels and porous polymer sorbents of varied thickness, sometimes as thick as traditional TLC and HPTLC layers. In new approaches, nanofibrous mats rapidly prepared using electrospinning enable highly efficient separations. GLAD is a powerful platform for engineering nanostructured thin films of varied material, porosity, and architecture that are well suited for UTLC. Planar chromatography media templated onto patterned carbon nanotube forests produce tunable geometries by leveraging established microfabrication processes. Initial reports of new UTLC media produced from layers of coated <1 μm silica particles [45], brush-gel polymers [48], and microfabricated pillar arrays [49] were not discussed in this chapter, but illustrate that new approaches to producing alternative sorbents continue to emerge and evolve.

Although each ultrathin and nanostructured stationary phase has its own merits, there are several similarities. Miniaturized plates utilize these layers to achieve faster separations over shorter distances with higher sensitivity and reduced reagent requirements [2]. However, their smaller size, thinness, and reduced specific surface areas prevent direct transfer of HPTLC methods to UTLC [2,23]. Smaller total surface areas often result in lower analyte retention and higher susceptibility to overloading. Shorter development distances can reduce sample capacity and resolution [36]. Their miniature format can make manual handling awkward and HPTLC instrument limitations can stifle overall separation performance [36]. These obstacles slowed widespread adoption of monolithic UTLC plates by chromatographers and contributed to Merck Darmstadt's decision to remove them from product catalogs after 2011 [50]. Improved devices for precise small sample application, development, and documentation (as per Office Chromatography, for example) are required before UTLC's full separation power can be realized. As a whole, the UTLC (and broader planar chromatography) community recognizes that success is contingent upon creation of this infrastructure. Nonetheless, the significant potential performance benefits motivate continued parallel development of ultrathin stationary phases and instrumentation optimized for these layers.

References

[1] Hauck HE, Bund O, Fischer W, Schulz M. Ultra-thin layer chromatography (UTLC)—a new dimension in thin-layer chromatography. J Planar Chromatogr 2001;14:234–6.

[2] Hauck HE, Schulz M. Ultrathin-layer chromatography. J Chromatogr Sci 2002;40: 550–2.

[3] Hauck HE, Schulz M. Ultra thin-layer chromatography. Chromatographia 2003;57: S313–5.

[4] Patel RB, Gopani MC, Patel MR. UTLC: an advanced technique in planar chromatography. Chromatographia; 2013.

[5] Frolova AM, Konovalova OY, Loginova LP, Bulgakova AV, Boichenko AP. Thin-layer chromatographic plates with monolithic layer of silica: production, physical-chemical characteristics, separation capabilities. J Sep Sci 2011;34:2352–61.

[6] Tsionsky M, Vanger A, Lev O. Macroporous thin films for planar chromatography. J Sol-gel Sci Technol 1994;2:595–9.

[7] Frolova AM, Chukhlieb MA, Drobot AV, Kryshtal AP, Loginova LP, Boichenko AP. Producing of monolithic layers of silica for thin-layer chromatography by sol-gel synthesis. Open Surf Sci J 2009;1:40–5.

[8] Bakry R, Bonn GK, Mair D, Svec F. Monolithic porous polymer layer for the separation of peptides and proteins using thin-layer chromatography coupled with MALDI-TOF-MS. Anal Chem 2007;79:486–93.

[9] Han Y, Levkin P, Abarientos I, Liu H, Svec F, Fréchet JMJ. Monolithic superhydrophobic polymer layer with photopatterned virtual channel for the separation of peptides using

two-dimensional thin layer chromatography-desorption electrospray ionization mass spectrometry. Anal Chem 2010;82:2520–8.
[10] Urbanova I, Svec F. Monolithic polymer layer with gradient of hydrophobicity for separation of peptides using two-dimensional thin layer chromatography and MALDI-TOF-MS detection. J Sep Sci 2011;34:2345–51.
[11] Clark JE, Olesik SV. Technique for ultrathin layer chromatography using an electrospun, nanofibrous stationary phase. Anal Chem 2009;81:4121–9.
[12] Clark JE, Olesik SV. Electrospun glassy carbon ultra-thin layer chromatography devices. J Chromatogr A 2010;1217:4655–62.
[13] Beilke MC, Zewe JW, Clark JE, Olesik SV. Aligned electrospun nanofibers for ultra-thin layer chromatography. Anal Chim Acta 2013;761:201–8.
[14] Kampalanonwat P, Supaphol P, Morlock GE. Electrospun nanofiber layers with incorporated photoluminescence indicator for chromatography and detection of ultraviolet-active compounds. J Chromatogr A 2013;1299:110–7.
[15] Lu T, Olesik SV. Electrospun polyvinyl alcohol ultra-thin layer chromatography of amino acids. J Chromatogr B 2013;912:98–104.
[16] Bezuidenhout LW, Brett MJ. Ultrathin layer chromatography on nanostructured thin films. J Chromatogr A 2008;1183:179–85.
[17] Morlock GE, Oellig C, Bezuidenhout LW, Brett MJ, Schwack W. Miniaturized planar chromatography using office peripherals. Anal Chem 2010;82:2940–6.
[18] Jim SR, Taschuk MT, Morlock GE, Bezuidenhout LW, Schwack W, Brett MJ. Engineered anisotropic microstructures for ultrathin-layer chromatography. Anal Chem 2010;82:5349–56.
[19] Oko AJ, Jim SR, Taschuk MT, Brett MJ. Analyte migration in anisotropic nanostructured ultrathin-layer chromatography media. J Chromatogr A 2011;1218:2661–7.
[20] Jim SR, Oko AJ, Taschuk MT, Brett MJ. Morphological modification of nanostructured ultrathin-layer chromatography stationary phases. J Chromatogr A 2011;1218:7203–10.
[21] Wannenmacher J, Jim SR, Taschuk MT, Brett MJ, Morlock GE. Ultrathin-layer chromatography on SiO2, Al2O3, TiO2, and ZrO2 nanostructured thin films. J Chromatogr A 2013;1318:234–43.
[22] Jim SR, Foroughi-Abari A, Krause KM, Li P, Kupsta MR, Taschuk MT, et al. Ultrathin-layer chromatography nanostructures modified by atomic layer deposition. J Chromatogr A 2013;1299:118–25.
[23] Kirchert S, Wang Z, Taschuk MT, Jim SR, Brett MJ, Morlock GE. Inkjet application, chromatography, and mass spectrometry of sugars on nanostructured thin films. Anal Bioanal Chem; 2013.
[24] Hall JZ, Taschuk MT, Brett MJ. Polarity-adjustable reversed phase ultrathin-layer chromatography. J Chromatogr A 2012;1266:168–74.
[25] Chen J, Abgrall P, Huang Y-W, Zhao Y-P. On-chip ultra-thin layer chromatography and surface enhanced Raman spectroscopy. Lab Chip 2012;12:3096–102.
[26] Song J, Jensen DS, Hutchison DN, Turner B, Wood T, Dadson AE, et al. Carbon-nanotube-templated microfabrication of porous silicon-carbon materials with application to chemical separations. Adv Funct Mater 2011;21:1132–9.
[27] Kanyal SS, Jensen DS, Miles AJ, Dadson AE, Vail MA, Olsen R, et al. Effects of catalyst thickness on the fabrication and performance of carbon nanotube-templated thin layer chromatography plates. J Vac Sci Technol B 2013;31:031203.
[28] Jensen DS, Kanyal SS, Gupta V, Vail MA, Dadson AE, Engelhard M, et al. Stable, microfabricated thin layer chromatography plates without volume distortion on patterned, carbon and Al(2)O(3)-primed carbon nanotube forests. J Chromatogr A 2012;1257:195–203.
[29] Jensen DS, Kanyal SS, Madaan N, Miles AJ, Davis RC, Vanfleet R, et al. Ozone priming of patterned carbon nanotube forests for subsequent atomic layer deposition-like

deposition of SiO_2 for the preparation of microfabricated thin layer chromatography plates. J Vac Sci Technol B 2013;31:031803.

[30] Salo PK, Salomies H, Harju K, Ketola RA, Kotiaho T, Yli-Kauhaluoma J, et al. Analysis of small molecules by ultra thin-layer chromatography-atmospheric pressure matrix-assisted laser desorption/ionization mass spectrometry. J Am Soc Mass Spectrom 2005;16:906–15.

[31] Salo PK, Vilmunen S, Salomies H, Ketola RA, Kostiainen R. Two-dimensional ultra-thin-layer chromatography and atmospheric pressure matrix-assisted laser desorption/ionization mass spectrometry in bioanalysis. Anal Chem 2007;79:2101–8.

[32] Kauppila TJ, Talaty N, Salo PK, Kotiaho T, Kostiainen R, Cooks RG. New surfaces for desorption electrospray ionization mass spectrometry: porous silicon and ultra-thin layer chromatography plates. Rapid Commun Mass Spectrom 2006;20: 2143–50.

[33] Brunauer S, Emmett P, Teller E. Adsorption of gases in multimolecular layers. J Am Chem Soc 1938;60:309–19.

[34] Rouquerol J, Avnir D, Fairbridge CW, Everett DH, Haynes JM, Pernicone N, et al. Recommendations for the characterization of porous solids (Technical report). Pure Appl Chem 1994;66:1739–58.

[35] Rabel FM. Sorbents and precoated layers in thin-layer chromatography. In: Sherma J, Fried B, editors. Handbook of thin layer chromatography. Marcel Dekker [Taylor and Francis]; 2003. pp. 129–76.

[36] Poole SK, Poole CF. High performance stationary phases for planar chromatography. J Chromatogr A 2011;1218:2648–60.

[37] Merck KGaA ChromBook 2008|09; 2008.

[38] Woodward SD, Urbanova I, Nurok D, Svec F. Separation of peptides and oligonucleotides using a monolithic polymer layer and pressurized planar electrophoresis and electrochromatography. Anal Chem 2010;82:3445–8.

[39] Lv Y, Lin Z, Tan T, Svec F. Preparation of porous styrenics-based monolithic layers for thin layer chromatography coupled with matrix-assisted laser-desorption/ionization time-of-flight mass spectrometric detection. J Chromatogr A 2013;1316:154–9.

[40] Ma M, Rutledge GC. 7.10 Nanostructured electrospun fibers. In: Matyjaszewski K, Möller M, editors. Polym. Sci. A Compr. Ref. Amsterdam: Elsevier; 2012. pp. 187–210.

[41] Hawkeye MM, Brett MJ. Glancing angle deposition: fabrication, properties, and applications of micro- and nanostructured thin films. J Vac Sci Technol A 2007;25: 1317–35.

[42] Taschuk MT, Hawkeye MM, Brett MJ. Glancing angle deposition. In: Martin P, editor. Handbook of deposition technologies for films and coatings: science, applications and technology. Oxford (UK): William Andrew (Elsevier); 2010. pp. 621–78.

[43] Linford MR. Fabrication and chromatographic separations on binder-free, carbon nanotube-fabricated TLC plates (July 6–8), Basel, Switzerland Int Symp Thin-Layer Chromatogr; 2011.

[44] Oko AJ, Jim SR, Taschuk MT, Brett MJ. Time resolved chromatograms in ultra-thin layer chromatography. J Chromatogr A 2012;1249:226–32.

[45] Zhang Z, Ratnayaka SN, Wirth MJ. Protein UTLC-MALDI-MS using thin films of submicrometer silica particles. J Chromatogr A 2011;1218:7196–202.

[46] Vovk I, Popović G, Simonovska B, Albreht A, Agbaba D. Ultra-thin-layer chromatography mass spectrometry and thin-layer chromatography mass spectrometry of single peptides of angiotensin-converting enzyme inhibitors. J Chromatogr A 2011;1218: 3089–94.

[47] Sherma J, Morlock GE. Chronology of thin-layer chromatography focusing on instrumental progress. J Planar Chromatogr 2008;21:471–7.

[48] Costantini F, Domenici F, Mura F, Scipinotti R, Sennato S, Manetti C, et al. A new nano-structured stationary phase for ultra-thin layer chromatography: a brush-gel polymer film. Nanosci Nanotechnol Lett 2013;5:1155–63.
[49] Kirchner TB, Hatab NA, Lavrik NV, Sepaniak MJ. Highly ordered silicon pillar arrays as platforms for planar chromatography. Anal Chem; 2013.
[50] Merck KGaA ChromBook 2012; 2012.

CHAPTER

4

Automated Multiple Development

Luis Membrado [1], *Vicente L. Cebolla* [1], *Carmen Jarne* [1], *Rosa Garriga* [2], *Pierre Bernard-Savary* [3], *Jesús Vela* [4]

[1] **ICB-CSIC, Zaragoza, Spain;** [2] **Department of Physical Chemistry, University of Zaragoza, Zaragoza, Spain;** [3] **Chromacim SAS, France;** [4] **Department of Analytical Chemistry, University of Zaragoza, Zaragoza, Spain**

4.1 INTRODUCTION

Thin-layer chromatography (TLC) began and still is, at least in part, a manual and modular technique. In column chromatographic techniques, a sample is injected into some sort of apparatus at the head of a column and the chromatographic separation is registered at another part of the apparatus at the opposite end of the column. In TLC, the sample is put on a plate that has to be moved to another apparatus to be developed, and finally to a further apparatus to record the chromatogram.

Application, development, and even detection in the case of colored compounds can be performed with minimal equipment requirements and TLC is, therefore, at least in part, a low-cost technique. Nevertheless, even in these simple cases, TLC results can be quite good.

It is not surprising, then, that efforts to improve TLC have focused on a particular step in the process, so we now have different methods of sample application and development systems, and detectors, that can be combined depending on the analytical needs. A completely automated TLC system consists of several instruments that afford almost unattended operation, apart from the need to move the plate from one instrument to another. It is also capable of providing increased reproducibility,

Instrumental Thin-Layer Chromatography
http://dx.doi.org/10.1016/B978-0-12-417223-4.00004-2

separation capability, sensitivity, and many other benefits, but at the same time with an increase in cost.

As it is still possible to obtain quite good TLC results at minimal cost, the success of the different instruments developed to improve TLC operation have always been dependent on a not so easy balance between cost and performance. Many good ideas have not survived, or are even awaiting a commercial opportunity to become of general use.

One of the most successful additions to TLC in the last decades is an improved development system that balances cost with a series of important benefits in some aspects of the technique. Since Poole [1] wrote a comprehensive review in 1991, automated multiple development (AMD) is no longer a promising technique but a reality that seems to be here to stay, and so it deserves some additional attention.

We will try to show its origins, current applications, and possible future.

4.1.1 Manual Multiple Development (MMD)

Multiple development in TLC [1] consists of the repeated development of a plate after the previous mobile phase has been removed. No special equipment is required and many different schemes are possible. The main drawback is a total development time that is the sum of the many development steps and the drying time between successive steps, which can become quite long. The three main modes of multiple development are as follows:

Unidimensional multiple chromatography (UMC) consists of repeated development of a plate for the same length and with the same mobile phase. In incremental multiple development (IMD), the same mobile phase is used, but the plate is progressively developed for longer distances at each step. The development length is usually incremented by a constant amount at each step. Gradient multiple development (GMD) can be performed in a unidimensional or incremental mode, and employs a change in the mobile phase composition between some or all of the developments.

Multiple development TLC benefits from a spot reconcentration mechanism that reduces peak widths in the chromatogram. Their final width, in UMC, is expected to be $I_W(1 - R_f)^n$ where n is the number of developments and I_W the initial zone width. In practice, this value is a theoretical minimum because of zone broadening during each development, but at least this reconcentration effect compensates for some of the natural zone broadening with migration, and in favorable cases, results in an almost constant zone width across the total length of the chromatogram. See Ref. [1] for additional details. The equation shows that width will depend on R_f and number of developments, as can be seen in Figures 4.1 and 4.2.

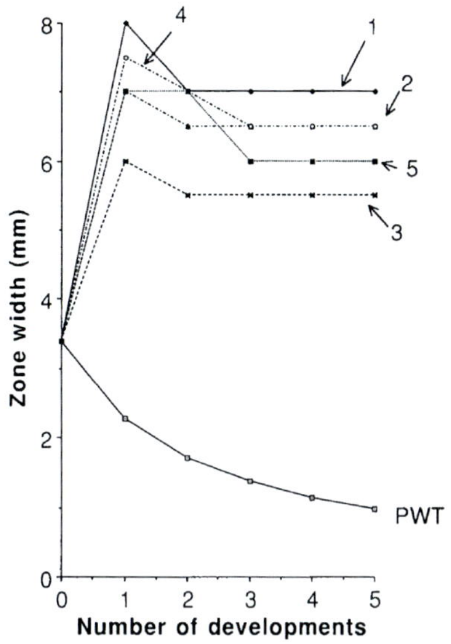

FIGURE 4.1 Change in the zone width in the direction of development during UMC. The results were obtained using five 7 cm developments with hexane on silica gel 60 high-performance thin-layer chromatography (HPTLC) plates. The sample composition (and R_f value in the first development) where: 1, nitronaphtalene (0.14); 2, benzo[e]pyrene (0.25); 3, fluoranthene (0.31); 4, biphenyl (0.58); and 5, decafluorobiphenyl (0.70). PWT represents the theoretical zone width calculated by assuming $W_n = W_I(1-R_f)^n$ where W_n is the zone width after n identical developments, W_I the initial zone width, and the R_f used for the calculation was 0.33. *Reproduced with permission from Ref. [1]. Copyright Research Institute for Medicinal Plants.*

The final result is that narrower zones improve chromatogram resolution and sample detectability, as their components are more concentrated.

IMD effects on zone width or resolution are more complicated to analyze and they depend on the increase in development length as well. Best results are obtained for small increases in the development length that make total development time much longer than in UMC (see Figure 4.3).

As GMD can be used in UMC or IMD modes, peak width will depend on using unidimensional or incremental modes, and changing the mobile phase composition makes possible the separation of samples that are a mixture of different polarity compounds. Decreasing polarity profiles use a strong solvent to get an initial separation of the sample. Next steps increase development length and use weaker mobile phases to complete the development. As the more polar compounds will not be greatly affected by the final steps, much richer chromatograms can be obtained.

Another possibility consists of increasing solvent strength of the mobile phase to provide a progressive separation of the sample components.

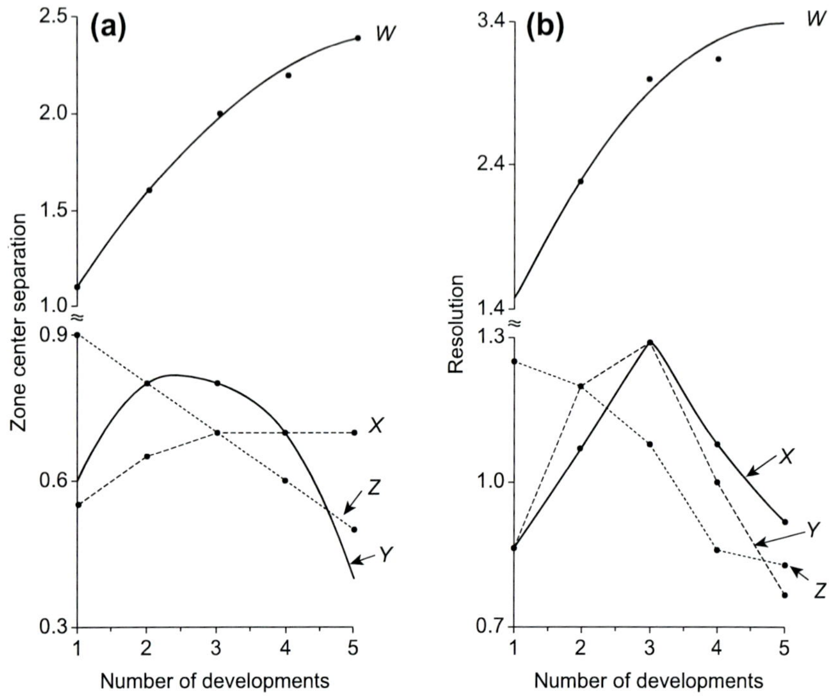

FIGURE 4.2 Variation of the zone center separation (a) and resolution (b) as a function of the number of developments in unidimensional multiple chromatography (UMC). Separation conditions as given in Figure 4.1. *W*, zones with R_f values 0.14 and 0.25; *X*, zones with R_f values 0.25 and 0.31; *Y*, zones with R_f values 0.31 and 0.56; *Z*, zones with R_f values 0.56 and 0.70. *Reproduced with permission from Ref. [1]. Copyright Research Institute for Medicinal Plants.*

By scanning intermediate steps it is possible to detect interesting compounds at any time in the process.

In any case, using GMD in IMD mode makes possible a unique chromatogram where a wide polarity mixture is separated along the full development length. Coupling changes in mobile phase composition and progressive development in many steps we could reach a good separation of a wide polarity samples with narrow zones. The cost is a rather long total development time with much more complicated handling that makes it impractical in MMD.

4.1.2 Programmed Multiple Development

In summary, with the possible exception of UMC, profiting from and even exploring the possibilities of multiple development in TLC was difficult and time consuming in a manual mode. Moreover, a complex

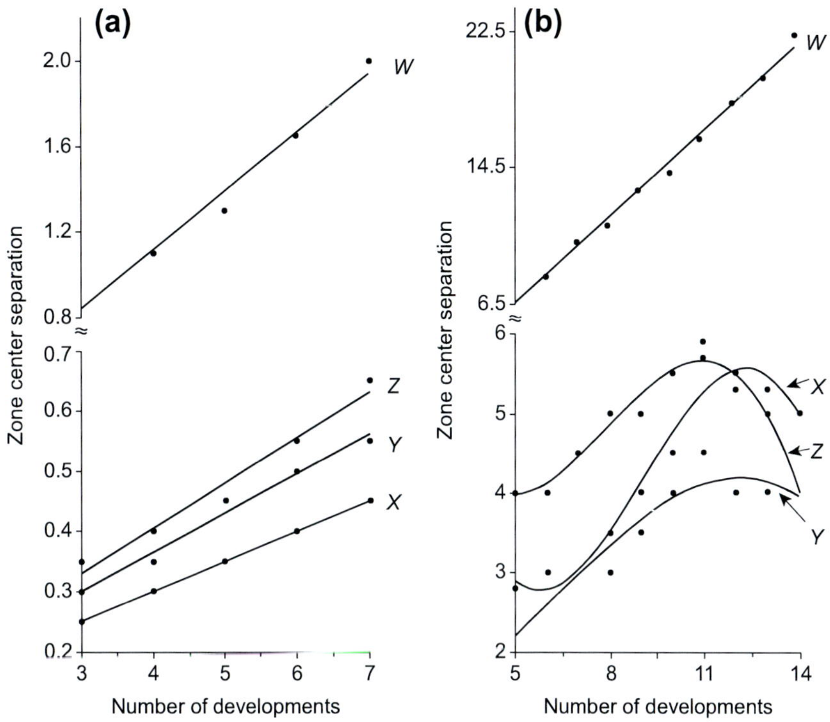

FIGURE 4.3 Variation of the zone center separation as a function of the number of developments in incremental multiple development (IMD). The total development distance was 7 cm and the incremental distance 1 cm in (a) and 0.5 cm in (b). Other conditions as for Figure 4.2. *Reproduced with permission from Ref. [1]. Copyright Research Institute for Medicinal Plants.*

development could lead to repeatability concerns, which became worse as the number of steps increased. There was a clear need to automate the process that led Jupille, Perry, Curtice and Regis Company of Illinois, USA, [1–3] to develop an instrumental version of the technique that was named Programmed Multiple Development (PMD). Some home made alternatives were also developed (see references 23,24 in Ref. [1]). The apparatus was sold by the Regis Co. for several years and basically consisted of a custom sandwich developing chamber and associated infrared electrical heater and a controller module.

PMD made IMD using isocratic development with a volatile solvent practical, and so benefits from zone refocalization could be exploited. In modern terms, automation was rather primitive and was limited to heating the plate according to three time-programs provided by the controller. Heating and/or a nitrogen flow dried the plate, although it was still in contact with the mobile phase reservoir. The number of steps and time-increase between them could be set as "economical," "optimal," and "medium" operating modes that used empirical parameters to adjust the IMD development sequence.

Maximizing the spot reconcentration mechanism, PMD improved chromatograms at a time where sample application and plate quality followed 1970 standards. But "automated isocratic IMD" was still only a partial solution that did not cover the complete possibilities of multiple development in TLC.

4.1.3 Automated Multiple Development

The final goal in building a completely automated instrument for multiple development working in the GMD–IMD mode was realized by Burger, Jaenchen, and Wilson in 1985–88 [4–6] based on earlier ideas represented by a custom-made apparatus developed by Burger that he applied to the analysis of pesticides in water. Automated Multiple Development (AMD) was born. Figure 4.4 shows a schematic diagram of a Burger AMD and Figure 4.5 a photograph of his custom machine.

The main difference with the PMD is the use of a vacuum system to dry the plate and remove excess mobile phase at each development step and the addition of a stepwise gradient generation system allowing plate conditioning prior to each development step. And last, but not least, increased automation of the operation of the apparatus.

Modifications removed concern from component degradation caused by too much heating; drying was no longer simultaneous with development; IMD provided zone width enhancements; and GMD facilitated the

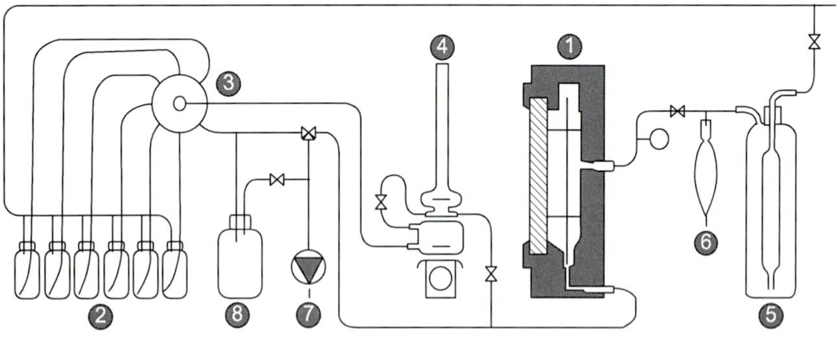

FIGURE 4.4 Schematic diagram of an apparatus for automated development: 1, enclosed developing chamber; 2, solvent reservoir bottles; 3, switching valve for selecting solvent composition; 4, gradient mixing chamber; 5, wash bottle to prepare gas phase; 6, reservoir for gas phase; 7, vacuum pump; 8, solvent water bottle. *Reproduced with permission from Ref. [1]. Copyright Research Institute for Medicinal Plants.*

FIGURE 4.5 Photograph of Burger's original automated multiple development (AMD) apparatus as shown in Figure 4.4. *Reproduced with permission from Ref. [4]. Copyright Springer-Verlag.*

separation of samples with components spanning a wide polarity range. In addition, AMD separations demonstrated a useful interesting matrix independence. The only remaining limitation was that the mobile phase components must be volatile enough to be removed efficiently by the vacuum system. It is capable, however, of handling limited amounts of water, for example, methanol–water mixtures containing up to 30% (v/v) water [7] and pH gradients are possible.

Around 1990 an AMD1 instrument was manufactured by Camag and the system became widely available. Later on, around 2000, Camag improved on the original design and introduced the AMD2 version that monitored the development length using a CCD sensor. The better general automation provided, among other benefits, more flexible gradient generation and plate drying systems [8]. Camag AMD2, shown in Figure 4.6, remains the reference system today.

AMD can be described as providing fully automated multiple development capability including stepwise gradients. In the Camag AMD2 instrument, up to five different solvents can be mixed to prepare the mobile phase for any development step and plates can be preconditioned with a defined gas phase. As a summary, AMD provides focalization to constant narrow zones for samples of wide polarity and matrix independent results with a separation number of around 40–80 mm (migration) with performance roughly equivalent to HPLC. Automation and the classical TLC advantage of simultaneous separation of samples in parallel (20 or even more) more than compensate development times of about 2 h for 20–25 development steps.

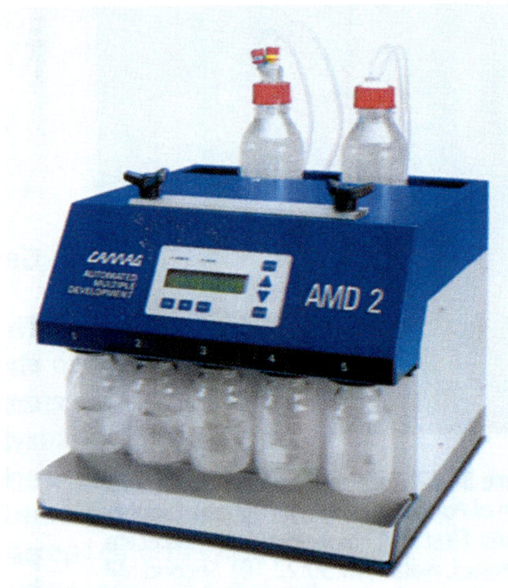

FIGURE 4.6 An image of the Camag AMD2 instrument. *Freely available at: http://www.camag.com.*

A classical objection to AMD [1] was that real mobile phase compositions in a stepwise gradient could be quite different to that defined by its theoretical profile (see Figures 4.7 and 4.8), and that the lack of a model relating solvent mixing, dispensing, front segregation effects and many other aspects of gradient development made the correlation of migration characteristics to solvent properties an almost impossible task. Method development in AMD lacked a theory to support it.

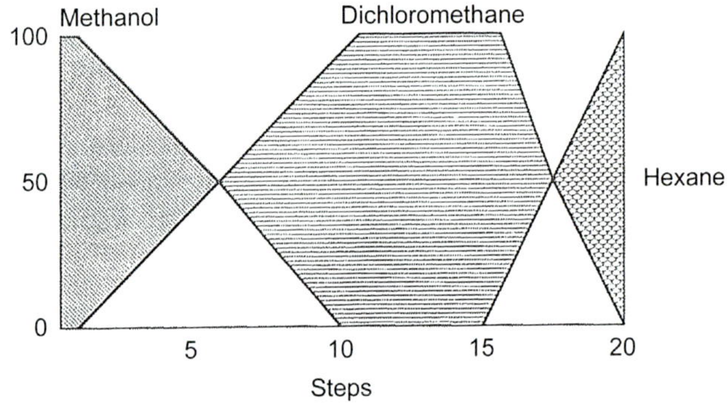

FIGURE 4.7 Diagramatic representation of a theoretical universal gradient for automated multiple development (AMD). *Reproduced with permission from Ref. [1]. Copyright Research Institute for Medicinal Plants.*

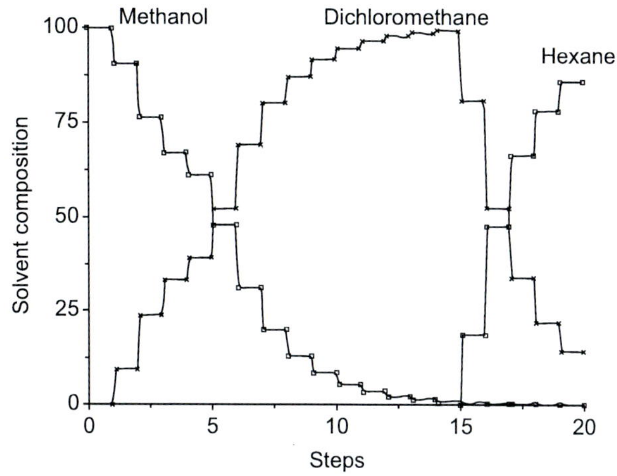

FIGURE 4.8 Actual mobile phase composition profile for the idealized universal gradient plotted in Figure 4.7. *Reproduced with permission from Ref. [1]. Copyright Research Institute for Medicinal Plants.*

In practice, AMD proved to be reproducible and provided good results, and empirical procedures for method development were adopted. The most common being the comparison of gradient profiles and chromatograms to guide method optimization. The starting point was a so called "universal gradient" of decreasing mobile phase polarity including three migration zones with methanol, dichloromethane and hexane as reference components that already appeared in Burger's first publication on the subject [4].

Some early attempts at method optimization were performed as described in Ref. [9]. Later on, Summanen [10] carried out a systematic study of the experimental optimization of AMD operating parameters (development time, number of steps, drying time, proportion of high and medium polarity solvents, and proportion of acid modifier) on the separation of a mixture of six phenolic compounds using an AMD1 apparatus. Many nonlinear effects for peak widths were observed, while peak separation showed a much simpler evolution. The optimized method consisted of a 220 min development, with 20 development steps, 15 s conditioning of plates using nitrogen bubbled through a solution of 10% formic acid, and a drying time of 2 min after the first step followed by 3 min for the remaining 19 steps.

But CCD development length control in the AMD2 apparatus removed the need to define development methods in terms of development time, and permitted a more evident use of development lengths that made comparison of gradient profiles and chromatograms easier as they could be directly superimposed to identify those parts of the method requiring improvement [1,11].

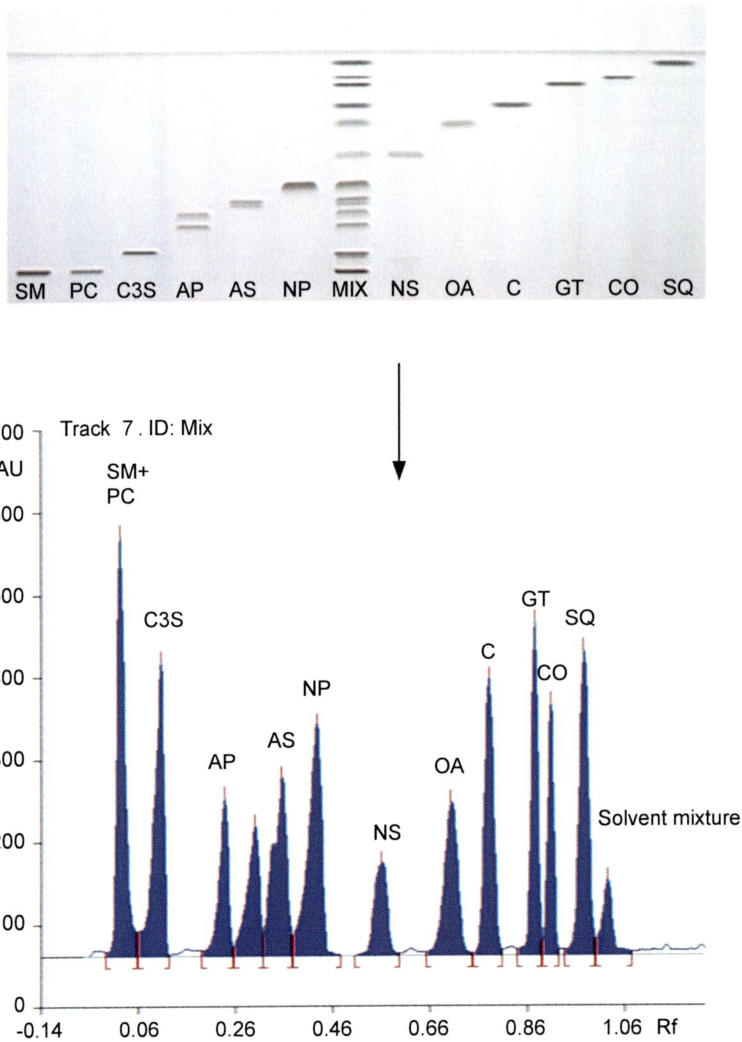

FIGURE 4.9 Top: AMD2 chromatogram of a standard mixture of lipids. Bottom: Corresponding densitogram of the standard mixture. SM-Sphingomyeline; PC-Phosphatidylcholine; C3S-Cholesteryl-3-Sulphate; AP-Ceramide AP; AS-Ceramide AS; NP-Ceramide NP; MIX-Standard mixture; NS-Ceramide NS; OA-Oleic acid; C-Cholesterol; GT-Glyceryl trioleate; CO-Cholestryloleate; SQ-Squalene. *Reproduced with permission from Ref. [32]. Author's Copyright.*

Finally, work on gradient theory for AMD has been scarce and it can be said that AMD might well be a good example of a general tendency in instrumental TLC (or just technical instrumentation) where a successful enough empirical approach obfuscates the development and application of a complicated (or just difficult) theory.

4.2 APPLICATIONS

Currently, AMD offers high reproducibility, separation performance similar to conventional HPLC for samples spanning a wide polarity range, results which are generally independent of the matrix, and unattended operation for long development methods that can be carried out overnight. Working mainly on silica plates, it is the development technique of choice for difficult samples.

We will try to provide a sensible summary of its current applications in the following sections.

4.2.1 Lipids

Several types of complex polar lipids of clinical interest have been analyzed by AMD [12–36]. Numerous references deal with stratum corneum extracts and their dermatological effects, mainly focused on ceramides. This complex mixture of lipids includes cholesterol and its derivatives, ceramides, oleates, squalene, sphingomyeline and other compounds, and have been analyzed using different development methods. Bonte [14] used an initial isocratic step to separate serum lipids followed by a 25-step development with a methanol–water to hexane gradient, a method that was modified later by Farwanah et al. [23,24,30].

Some optimizations have been proposed in 2010 and 2011 [32,35], and using Schellenberg work as an example, an eight step gradient of methanol–chloroform–hexane/5% toluene with 4M acetic acid preconditioning before step 6 and a drying time of 2 min was able to separate a mixture of 13 reference compounds in 1.5 h consuming 60 mL of solvent (see Figure 4.9).

Other lipid groups such as phospholipids [12], gangliosides [13,15,31], alkylphosphocholines [21], medium and long chain triglycerides [22], and ceramides or cholesterol and other products with a different origin than the stratum corneum have been described [25,28,29,33]. Plates containing a fluorescence indicator were frequently used to facilitate detection, although a modification of this technique using berberine impregnated plates for fluorescence detection by intensity changes (FDIC) [33] gave good sensitivity for cholesterol and some of its derivatives.

4.2.2 Pesticides/Environmental Analysis/Hazardous Compounds

AMD application began with Burger analysis of pesticides in water, so it is not surprising that the use of AMD in the field of environmental analysis is quite common. Water regulations in Germany have promoted interest in the development of improved TLC development system, and increasingly rigorous environmental regulations clearly contribute to its current use in the field. Since the previous AMD review by Poole [1] many papers deal with the application of current AMD techniques to products with environmental impact [37–68]. Many of these methods employ universal gradients as the development method and solid-phase extraction as the preconcentration step. We will focus in the first place on water and soil analysis as most references in the field deal with this application.

Zietz [37,38] described methods to monitor ground and drinking water contamination using AMD and tested their capability to meet water regulations. Phenylureas, methylcarbamates, modified ureas, and N-heterocyclic compounds were determined with a detection limit below 0.1 mg/L. De la Vigne et al. [39,40] were able to detect trace amounts of pesticides, like 28 ppt of atrazine, using a gradient based on dichloromethane. Jork [41] worked on method limitations and recommended a prewashing step for plates by immersion in 2-propanol for 30 min and the use of thin layers. Detection limits for triazines were around 50 ng. A review by Sherma [42] discussed 13 references on the subject. Pfaab et al. [43] studied solid-phase extraction of samples prior to AMD analysis. Wolff et al. [44] focused on the analysis of editic acid (EDTA). Limits of detection and quantification were 250 and 450 ng. Butz et al. [45,46] described a method for screening 265 pesticides by AMD using a universal gradient based on dichloromethane. Special gradients allowed confirmation during a second development of individual pesticide classes. Later on, the number of pesticides was increased to 283. In addition, a data base of migration distances and UV spectra were developed to facilitate pesticide identification. Burger et al. proposed a similar approach [47].

Other extraction techniques prior to AMD separation were evaluated by. Koeber et al. [49]. Good results were obtained using supercritical fluid extraction with carbon dioxide and methanol as a modifier. A good summary of current methods for the analysis of contaminants in water using the AMD is provided by Morlock et al. [52]. Figure 4.10 illustrates a 30 step universal gradient for pesticides, and Figure 4.11 for a pesticide sample. A later review by Albert [59] is also available.

The general detection of pesticides in water could be considered a solved problem by 1995 and focus of research changed to identification of more specific or unknown contaminants in water. Mass spectrometric detection after AMD separation became the preferred tool for this task. Plass et al. [61] described the determination of amitrol in water using an AMD2 apparatus. Weber et al. [62,64–68] identified chloridazon, triazole and rhodamine B derivatives. Figures 4.12 and 4.13 illustrate an AMD2 gradient for screening of water samples and a chromatogram in which triazoles were detected.

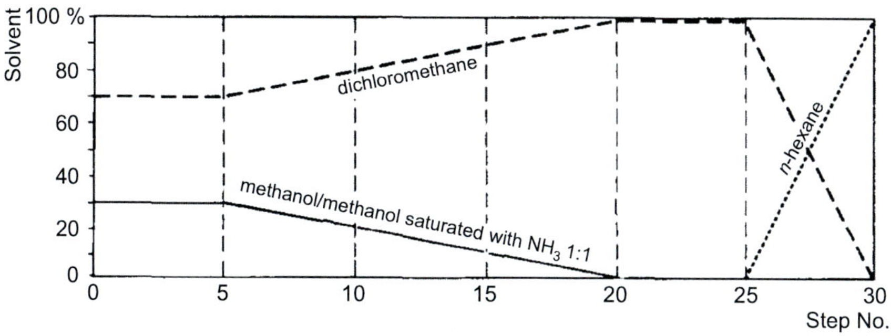

FIGURE 4.10 Thirty step universal gradient from methanol/dichloromethane to *n*-hexane using dichloromethane as "host" solvent. *Reproduced with permission from Ref. [52]. Copyright Elsevier.*

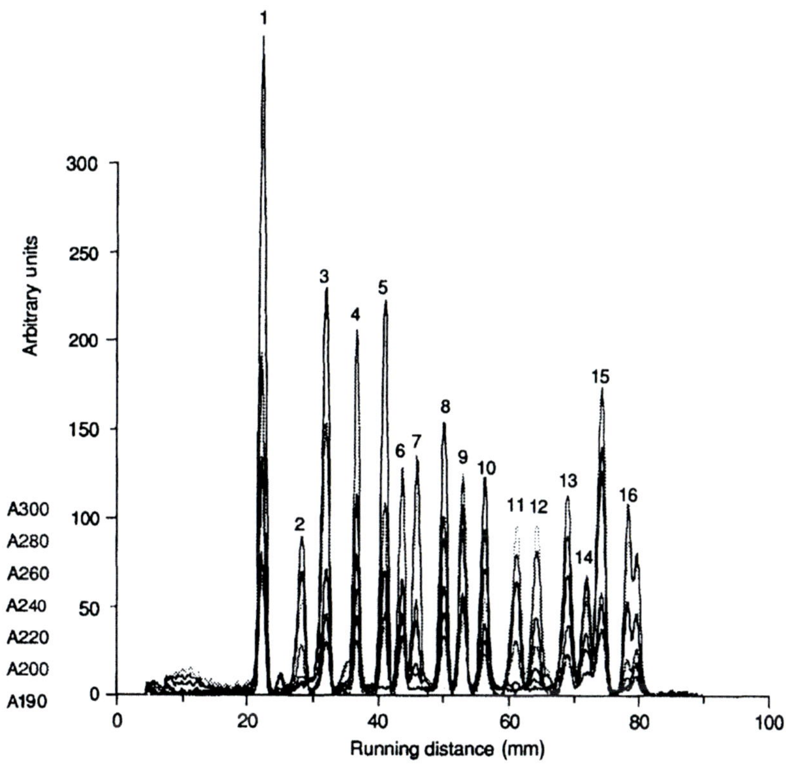

FIGURE 4.11 Multiwavelength scan of various pesticides (100 ng each) at seven different wavelengths from 190 to 300 nm after separation by the gradient illustrated in Figure 4.10. (1 = naphthalene-1-sulfonic acid, 2 = procloraz, 3 = triazoxid, 4 = ethidimuron, 5 = simazine, 6 = bromazil, 7 = carbofuran, 8 = metribuzin, 9 = azinphos methyl, 10 = coumaphos, 11 = prosulfocarb, 12 = dichlofuanid, 13 = parathion, 14 = fenthion, 15 = dinoseb, 16 = prothiofos). *Reproduced with permission from Ref. [52]. Copyright Elsevier.*

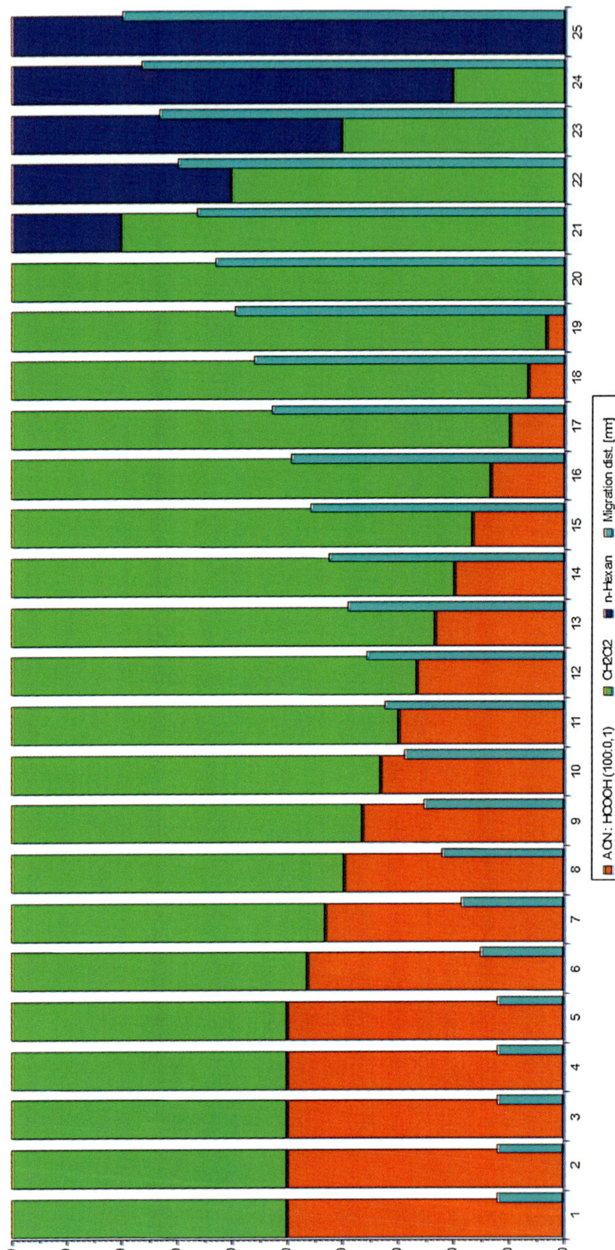

FIGURE 4.12 A 25-step AMD2 gradient for the screening of pesticides in water after extraction, from 50% dichloromethane –50% acetonitrile–formic acid (100–0.1) to 100% dichloromethane (step 20) to 100% methanol. Thin bars represent migration distances in mm. *Reproduced with permission from Ref. [66]. Author's Copyright.*

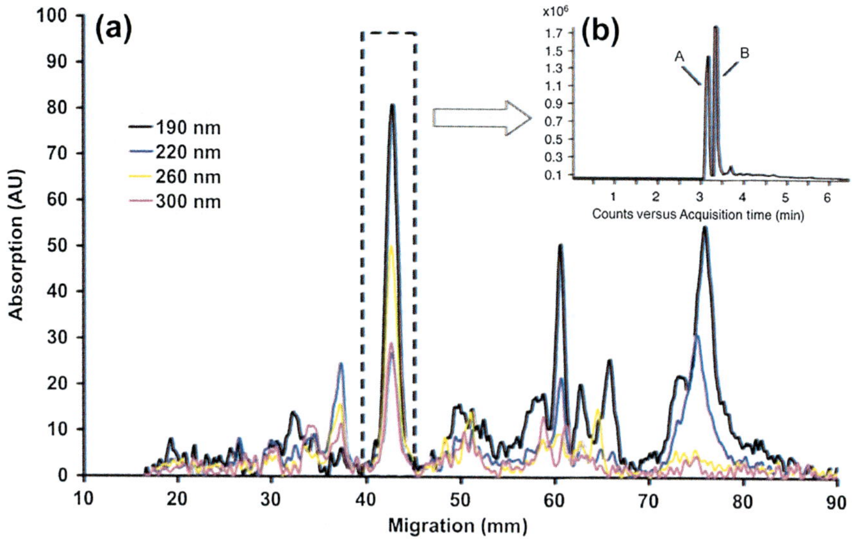

FIGURE 4.13 (a) High-performance thin-layer chromotography (HPTLC)/Automated multiple development (AMD) multiwavelength scan of an extract from secondary effluents. (b), Nano LC/QTOF-MS chromatogram of the extracted zone, peak A: 1H-benzotriazole; peak B: tolyltriazole. *Reproduced with permission from Ref. [66]. Author's Copyright.*

Mill effluents were a further source of contaminants studied using AMD. Dethlefs et al. [48] used a methanol–dichloromethane gradient to analyze lignin, phenols, aromatic acids, aldehydes, alcohols and 2-resin acids in effluents from the alkali-oxygen-peroxide (EOP) bleaching stage of a pulp mill. Reemtsma et al. [54,55] described a method using AMD and toxicity detection by *Vibrio fischeri* for the analysis of industrial wastewater. Topfer et al. [58] characterized several methods with the same aim. Kralicek [50] studied the detection of azo dyes in textiles and Lautie et al. [51] studied phenylurea herbicides in plant materials. Stan et al. [53] determined pesticide residues in plant food products by AMD and Yue et al. [63] determined organophosphorous pesticide residues in tea. Some works on bioluminescence detection of toxicity are of interest with the methods described in this section [56,57,60].

4.2.3 Lubricants, Additives, and Petroleum Products

These products seem well suited to AMD analysis but little work has been performed so far. It may be a new field for AMD applications in the near future. AMD was used for the fast identification and quantification of lubricant additives in mineral oil [69] and to characterize groups of heavy petroleum products according to their number of aromatic rings [70]. Caffeine impregnated plates and a dichloromethane–heptane gradient in 34 development steps were used in the later case.

4.2.4 Pharmaceutical and Clinical Applications

We discussed previously that new developments in TLC depend greatly on their particular balance of price versus performance to become successful. Analysis of pharmaceutical and clinical products may be one of the fields where this balance in the AMD case was more clearly favorable. Therefore, papers on AMD analysis of pharmaceutical products are abundant [71–93], as also is its use in closely related fields, such as phytochemical extracts. These studies contain useful information but are not easily summarized as they cover a wide range of structures and properties, linked together only by their clinical importance. It is not surprising, therefore, and this is only an example, that Belay and Poole [75] applied AMD to the analysis of amino acids from protein sequencing in 1993, but it does not have much resemblance with the work of Colthup et al. [76] in the same year on the separation of drug metabolites with linear solvent gradients (from methanol to hexane).

We will use then a mainly chronological enumeration of papers using AMD, with the exception of the need to identify and quantify possible drugs of abuse, where reports often refer to similar samples or matrices. Wolff et al. [77] made use of AMD coupled to FTIR to obtain favorable detection limits for hexobarbital (55 ng) and other compounds. Kovar et al. [78] used a similar procedure to identify 38 of the most important drugs of abuse divided into several groups (barbiturates, monosubstituted benzenes, morphine derivatives, amphetamines, and indole derivatives). Later on Roberts [80] analyzed drugs in human urine. A mixture of 25 benzodiazepines was also analyzed by Werner et al. [83] using AMD with a methanol–diisopropyl ether gradient in nine steps. The method was indicated as a cost efficient and rapid screening tool. Pothier and Garland [88] improved the classical TLC analysis of opiate alkaloids using AMD and reported a clear separation for each of the three groups (poppy alkaloids, semisynthetic and synthetic derivatives, and narcotic compounds). Best results were obtained with a universal gradient of methanol–dichloromethane–hexane for opium alkaloids and 5% of 28% (v/v) ammonia in methanol–acetone–dichloromethane for semisynthetic and synthetic derivatives. Detection limits using Dragendorff and iodoplatinate reagents were 1 ng for ethylmorphine, thebaine, papaverine, codeine, and 2 ng for morphine, noscapine and other alkaloids. Bowles et al. [81] used the AMD to study protein biosensors kinetics. Erythromycin in pharmaceutical preparations was determined with a methanol gradient in AMD by Liawruangrath et al. [82]. Jamshidi [84–86] studied progesterone and

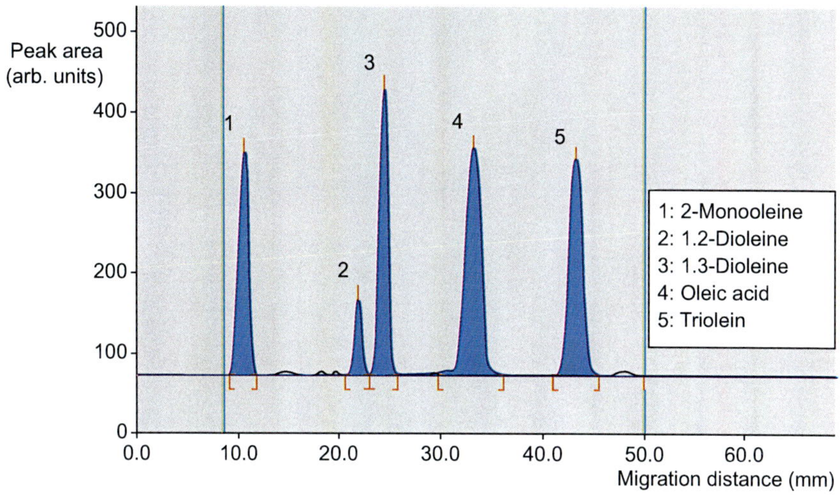

FIGURE 4.14 Densitogram of a lipid standard mixture. *Reproduced with permission from Ref. [87]. Author's Copyright.*

levonorgestrel in drug release media using toluene-2-propanol (90–10) as mobile phase with limits of detection of 5 ng for progesterone. In vitro lipolysis products were analyzed by Mäder et al. [87] using an 11 step *n*-hexane (70–30)-*n*-hexane (100) gradient and 20 s dipping into an aqueous copper sulfate solution heated during 30 min at 150 °C for detection of lipolysis hydrolysis products. Figure 4.14 illustrates results for a standard mixture.

Jamshidi developed a stability indicating assay for leuprolide acetate using a 5-step IMD development with ethyl acetate–methanol–25% ammonia in water (60–30–10) as mobile phase with a limit of detection of 25 ng [89] and also reported the analysis of tamoxifen citrate in drug-release media [90]. A review of the separation of coumarins is also available [91]. Impurity profiling provides a similar example of the use of TLC techniques for the characterization of pharmaceutical products [92]. Finally, Rothenhoefer et al. [93] studied the identification of the oligomers of hyaluronic acid on amino-modified silica plates using High-performance thin-layer chromatography (HPTLC)–MS.

4.2.5 Organics

This is a somewhat miscellaneous group of compounds difficult to fit under other section headings. Kralecik et al. [94] used AMD for the determination of amines derived from textile azo dyes. Forskolin, a diterpene extracted from Indian koleus and 11 related compounds, were separated and quantified by Bonte et al. [95] using a 25-step gradient of methylene chloride–methanol–*n*-hexane. Raverkar [96] used AMD to

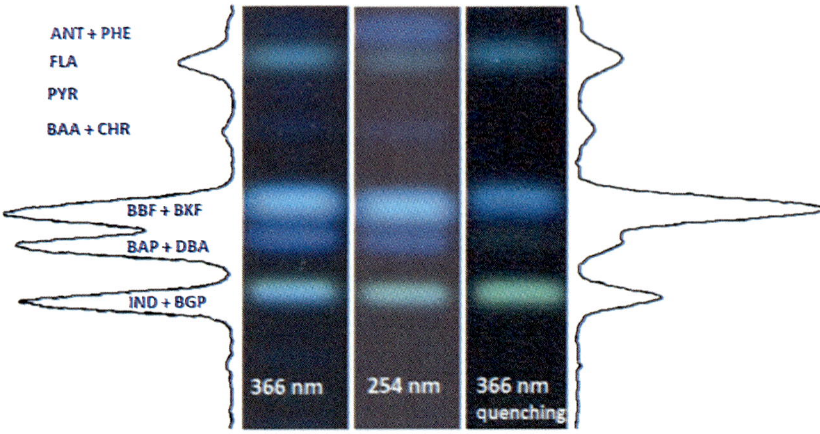

FIGURE 4.15 Plate images of a separated standard mixture of PAHs (10 ng/zone) illuminated under different wavelengths and after fluorescence quenching by nitromethane; fluorescence scans at 366 nm/>400 nm. *Reproduced with permission from Ref. [99]. Author's Copyright.*

monitor nitrobenzene process streams during benzene nitration. Using a binary chloroform–*n*-hexane gradient mobile phase in a program of 22 steps, eight reaction byproducts were separated and quantified. Sunscreen photodegradation products of potential toxicity were analyzed [97,98] on HPTLC LiChrospher plates using AMD with a *tert*-butyl-methyl ether and *n*-hexane gradient. Detection was by UV and bioluminescence with *V. fischeri*. A quite peculiar use of AMD is described by Behringer and Schwack for the detection of polycyclic aromatic hydrocarbons in toys [99]. The AMD separation used a three step (45, 55, 65 mm) IMD separation with an acetonitrile–water 9:1 mobile phase, nitrogen atmosphere, and a drying time of 2 min. AMD provided light and oxygen protection during development and drying. Fourteen polycyclic aromatic hydrocarbons were separated into eight sharp zones. Figure 4.15 illustrate some results of PAH detection in toys.

4.2.6 Plant Extracts and Bioactive Metabolites

This is a classical application of TLC, connected with the identification of new active pharmaceutical entities. AMD was soon applied to improve the separation of extracts. Numerous reports can be found in Refs [7,100–117]. Most applications employed silica gel plates with a wide range of development conditions. The general objective is to obtain a fingerprint record for the plant or the identification of interesting compounds. Fingerprints and quantification of individual components are useful to identify plant materials or compounds present in plant products and to determine the authenticity of phytopharmaceuticals. The matrix independence of AMD separations is a useful feature for this application.

Gocan et al. [103] studied the alcoholic extracts of 10 plants (*Cinchona succirubra, Aesculus hippocastanum, Berberis vulgaris, Artemisia abrotanum, Carduus marianus, Thuja occidentalis, Baptisia tinctoria, Paullinia cupana, Lycopus europaeus,* and *Echinacea angustifolia*). Olah et al. [104] studied alcoholic extracts of several Artemisa plants (*A. abrotanum, Artemisia absinthium, Artemisa vulgaris,* and *Artemisia cina*) and compared the fingerprint results of their separation by normal TLC and two-dimensional AMD. Scholl et al. [105] studied calystegines, a class of nortropane alkaloids. A combination of TLC methods allows the separation of the individual calystegines and some of their possible biosynthesis precursors. A review by Galand et al. [106] compares the separation performance of overpressured layer chromatography (OPLC) and AMD for plant extracts and synthetic products. Di et al. [107] investigated fingerprints for the products from hydrolyzed polysaccharides from Lingzhi mushroom. These fingerprints allowed detection of different varieties of mushrooms in commercial products. Galand et al. [108] studied the separation of cannabis components by TLC, AMD, and OPTLC. AMD provided higher resolution due to lower spot diffusion. Gocan and Cimpan [7] prepared a review of the use of those techniques for the analysis of medicinal plants.

Markowski [109] studied the screening of gingenosides with methanol-chloroform mobile phases. Pozharitskaya [110] worked on the separation of terpenoids from *Boswellia serrata* and quantified the four boswellic acids by means of direct densitometry or after derivatization with anisaldehyde- sulfuric acid reagent. Klöppel [111,112] studied metabolites in sponges using bioluminescence and MS as detectors and AMD with a 15-steps universal gradient of methanol-dichloromethane-*n*-hexane for the separation of methanol extracts with mass spectrometric detection (see Figure 4.16).

Schulte et al. [113] used AMD for the characterization of pollen carotenoids. Simoes-Pires et al. [114] developed a method to detect α- and β-glucosidase inhibitors in plant extracts. Yan et al. [115] studied fingerprints from triterpenoic acids in Ganoderma (Lingzhi) mushrooms to authenticate Lingzhi preparations. Cui et al. [116] analyzed flavonoids in leaves from eight species of *Indocalamus*. Misra et al. [117] performed a general screening of several Indian plants (*Hippophae rhamnoides, Valeriana wallichii, Triticum aestivum,* and *Cordyceps sinensis* fungus) seeking to identify products with pharmacological activity.

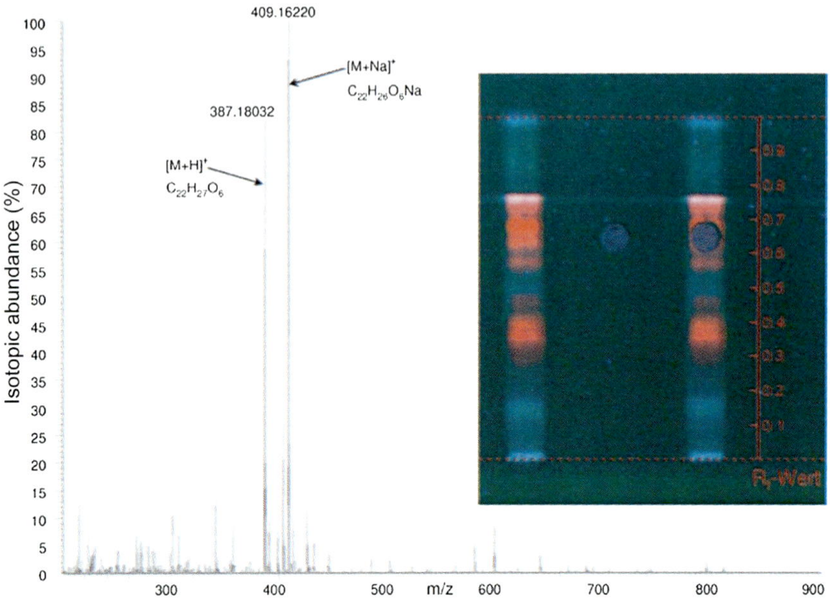

FIGURE 4.16 Left: HPTLC-ESI-MS-spectrum (positive-ion mode) of a bioactive zone from the freshwater sponge *Ephydatia fluviatilis* leading to the monoisotopic mass of the protonated molecule with m/z 387.18032 $[M+H]^+$ and a mass divergence of 0.7 ppm. Right: Image detail of a developed high-performance thin-layer chromatography (HPTLC) plate documented at UV 366 nm: the bioactive substance (hRF 65) was extracted via the extraction interface (4 mm diameter) (left track: before extraction, middle: blank value, right track: after extraction). *Reproduced with permission from Ref. [112]. Author's Copyright.*

4.2.7 Food and Food Additives

Apart from carbohydrates, discussed in Section 4.2.8, AMD has been used to analyze other food components. Wippo and Stan [118] used reversed-phase TLC and AMD for the determination of ioprodine in vegetables. Ioprodine was identified by its UV spectrum and quantified with a detection limit of 30 mg/kg 2,3-Butanediol and acetoin were used to characterize four different species of wine yeasts (*Kloeckera apiculata, Saccharomyces cerevisiae, Saccharomycodes ludwigii, Zygosaccharomyces bailii*) [119]. Glycerol was used as a target compound with the same purpose using AMD with an acetonitrile–acetone–hexane gradient on silica gel layers [120].

Morlock et al. [121] used a diethyl ether–methanol–chloroform gradient for the separation of heterocyclic aromatic amines (HAA) on silica gel 60 plates with fluorescent indicator. Scanning densitometry with multiwavelength scanning at 252, 262, 316 nm in the absorption mode and 366/>400 nm in the fluorescence mode was used for quantification.

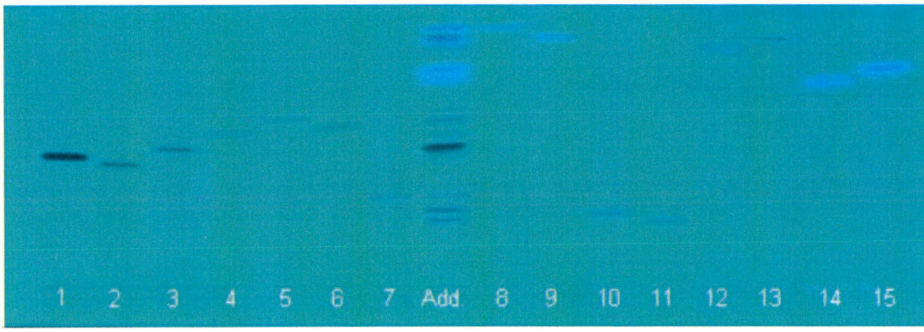

FIGURE 4.17 Chromatogram of 15 heterocyclic aromatic amines (HAA) at UV 254 nm (reflectance), automated multiple development (AMD) separation on high-performance thin-layer chromotography (HPTLC) plates silica gel 60 WRF254s, track 1–15 standard substances, middle track standard mixture. *Reproduced with permission from Ref. [121]. Author's Copyright.*

The method provides a fast and cost-efficient alternative to quantitative determination of HAA by HPLC. The AMD method is faster than HPLC by a factor of seven and solvent consumption is reduced by about 90%. A typical HAA chromatogram is shown in Figure 4.17. Unknown, but separated, compounds were identified by HPTLC-MS.

4.2.8 Carbohydrates

Buffered amino and diol layers are more appropriate than bare silica gel for the separation of sugars and oligosaccharides by AMD [122–126]. Samples analyzed include beverages with no clean-up, molasses, fructo-oligosaccharides (FOS), and inulin mixtures. Acetonitrile–acetone–water gradients were mainly used but in some cases sample-specific gradients were required. Silica gel layers have been used in a few cases [127] A Si-Wu-Tang extract was analyzed showing four major zones identified off-line by electrospray ionization–mass spectrometry (ESI–MS) as monosaccharides, disaccharides, trisaccharides, and tetra-saccharides, respectively. The main monosaccharides were fructose and glucose.

4.2.9 Explosives

Steuckart et al. [128] used AMD for the analysis of explosives in soil near ammunition plants in Germany. A gradient of methanol, ethyl acetate, and petroleum ether allows the separation of hexyl, picric acid, RDX, HMX, tetryl, 2,4- and 2,6-dinitrotoluene, and 2,4,6-trinitrotoluene, as well as some byproducts, such as 1,3-dinitrobenzene,2-amino-4,6-dinitrotoluene, and 4-amino-2,6-dinitrotoluene. Detection limits by

UV–visible absorption were between five and 20 ng. As it was possible to analyze 20 samples simultaneously, the technique was a useful screening tool for contaminated sites.

4.3 FUTURE TRENDS

As demonstrated by its many applications, AMD is nowadays a successful and proven technology. Being an established and mature technology does not mean that it can not be improved or put to new use. Even if we assume that the current commercial AMD apparatus will not change significantly in the near future, technologies change in an attempt to remove their limitations. AMD is still mainly used with silica plates. Gradient development may need a comprehensive theory to support new applications. Coupling AMD with other separation or detection techniques remains an interesting possibility. Some logical evolution lines can already be seen and are summarized below.

4.3.1 Stationary Phases Other than Silica Gel

In most applications, AMD has been used with silica gel plates. There are, of course, other possibilities. In the first place [129], significant improvements in the total development time (up to 50%) can be obtained by using silica plates coated with spherical particles, such as LiChrospher® or by using "extra thin" (100 μm) layers requiring shorter development increments and shorter drying times. In the second place, some work has already been done using polar chemically bonded layers, largely for the separation of sugars (Section 4.2.8).

4.3.2 New AMD Instruments

The AMD2 instrument is not inexpensive and a wider use of AMD would be catalyzed by a more affordable instrument. Therefore, it is not surprising that the development of new AMD-like systems has received some attention. They mainly try to add gradient development capability to TLC development. Markowski et al. [130] described a horizontal chamber with gradient capabilities (DS-II and DS-M from Chromdes) which was the basis for a manual gradient development system (see Figure 4.18), including a sensible summary of current gradient theory. The system was tested with a mixture of 10 dyes using a reversed-phase gradient separation with methanol and water.

Another solution is to preserve automation to the extent possible while employing readily available or custom components. Wang et al. [131] followed this path to build what can be described as an AMD1 clone using

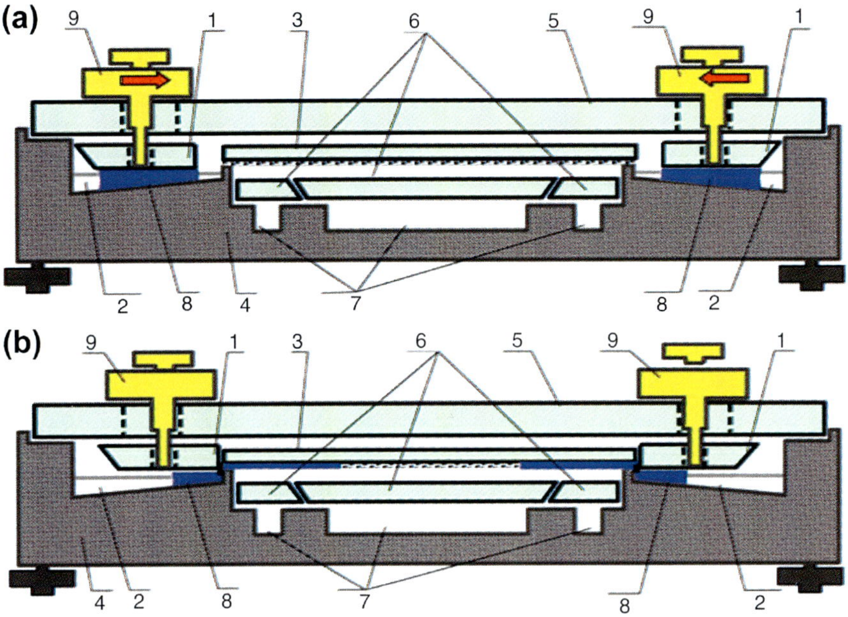

FIGURE 4.18 Horizontal DS-II chamber (Chromdes): (a) before development, (b) during development; 1, cover plate of the mobile-phase reservoir; 2, mobile phase reservoir; 3, chromatographic plate with layer face down; 4, body of the chamber; 5, main cover plate; 6, cover plates (removable) of the troughs for vapor saturation; 7, troughs for saturation solvent; 8, mobile phase; 9, mobile-phase distributor/injector. Yellow: Inert plastic mobile pieces; Light blue: Glass; Dark blue: Mobile phase; Grey: Inert plastic chamber body; Black: Regulable feet. *Freely available at: http://www.chromdes.com.*

a horizontal development chamber for less than 100 USD. They also focused on economical operation through solvent savings. The features of the custom horizontal development chamber are shown in Figures 4.19 and 4.20. Chamber design included a fan to dry the plate between development steps and the system was tested with a mixture of 13 dyes with good results.

New and probably surprising AMD developments will appear in the future. A potential example is a new AMD apparatus based on office peripherals, as discussed by Morlock et al. [132] for use with ultra-thin-layer plates.

4.3.3 Gradient Properties

There are few fundamental studies of gradient properties in AMD. General progress has been achieved using empirical solutions (Section 4.1.3). Gradient properties have been studied to some extent as a basis for manual gradient development chambers (Section 4.3.2). Gradient

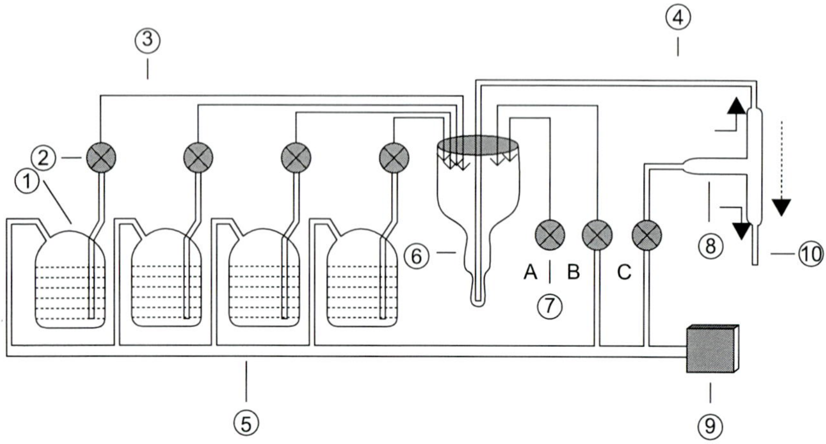

FIGURE 4.19 Schematic diagram of the solvent-supply system (not drawn to scale). 1, solvent reservoir bottles; 2, modified solenoid valves (labeled from valve 1 to valve 4); 3,4, polythene tubing; 5, nylon tubing; 6, gradient mixer; 7, conventional solenoid valves (labeled from valve A to valve C); 8, three-way tube; 9, air pump; 10, solvent pumping-out pipe. The arrowhead on the solid lines shows the direction of flow of stirring air; and that on the dotted line shows the direction of flow of the solvent. Note: Valve A is used to equalize the air pressure of the gradient mixer, so is connected not to the pump but directly to the atmosphere. *Reproduced with permission from Ref. [131]. Copyright Research Institute for Medicinal Plants.*

optimization was studied by Lodi [9] and other theories, and even computer-aided optimization methods, exist [133]. Further studies would be welcome to shorten the learning curve when AMD methods are first explored.

4.3.4 Coupling with Column Methods and Hyphenation

Hyphenation techniques with an AMD base are quite common [113] and provide good results in fields like pesticide residue analysis (Section 4.2.2). An explorative use of AMD-MS is found in Morlock's work [121] and the interface became commercially available shortly afterward. This interface may provide a general method of coupling with other detection techniques, such as NMR [134]. Further examples of AMD-MS coupling can be found in Refs [135,136]. A general review of coupling methods in TLC was authored by Tuzimski [137]. Kertesz et al. [138] described the automation of a TLC-electrospray interface for mass spectrometry. Coupling AMD to FTIR detection was also described [139]. Coupling of two separation techniques is also unknown, for example HPLC/AMD [140,141].

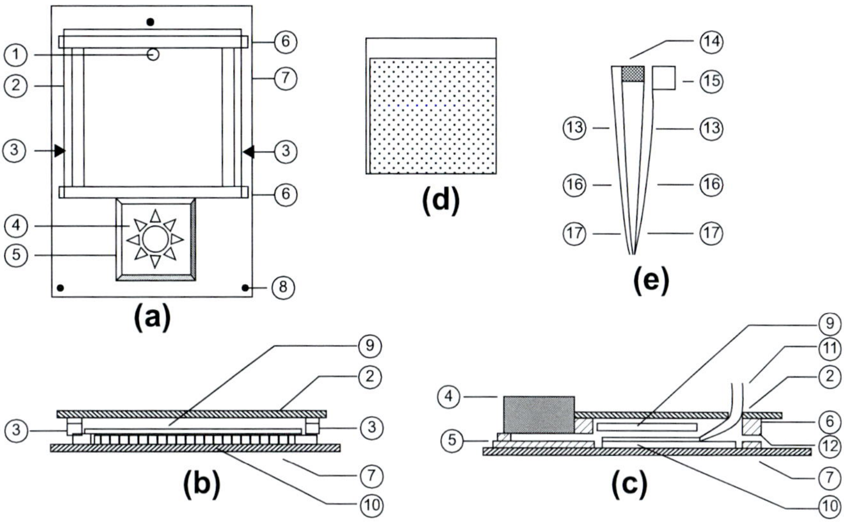

FIGURE 4.20 Schematic diagram of the laboratory-made horizontal sandwich chamber and its components (not drawn to scale). (a) Top view of the development chamber (the thin-layer plate and the inside cover plate are omitted); (b) front cut-away view; (c) side cut-away view; (d) thin-layer plate (20 cm × 20 cm; the layer of silica gel is 17.0 cm × 19.4 cm); (e) side view of the funnel distributor (for front and side perspective view see Figure 4.6(b)). 1, hole; 2, outside cover plate; 3, spacing plate; 4, fan; 5, glass frame (supporting the fan); 6, spacing frame (exhaust path); 7, base plate; 8, screw; 9, inside cover plate; 10, thin-layer chromatography (TLC) plate; 11, guide tube; 12, Teflon tape (as an automatic door for exhaust path); 13, specially treated superlong slide (19.6 cm × 2.6 cm × 0.07 cm); 14, glue; 15, small piece of glass; 16, surface formed by erosion; 17, slope formed by polishing. *Reproduced with permission from Ref. [131]. Copyright Research Institute for Medicinal Plants.*

4.4 CONCLUSIONS

Multiple development, gradient development, and automation, together, solve many difficult problems in the analysis of complex samples. There is no doubt that AMD is here to stay and that it will be more popular in the future. The current need to use volatile solvents and its price are the main problems that limit its widespread use. Easily available or even public domain hardware and software alternatives exist nowadays to make a low cost AMD2 clone machine possible, but reliability, system integration, and market size must be considered before it could become a reality. A complete TLC system also needs well established communication protocols between different modules (application, development, detection) that support a flawless and easy handling of the chromatographic plate. Open communication protocols and data interchange standards should be developed and adopted. Most of these concerns have no immediate solution, so the most probable prediction is that

AMD will remain mainly as it is today in the near future. However, it might be improved to enter new fields of application, such as reversed-phase chromatography, that would require low solvent-consuming horizontal developing chambers or enhanced drying systems. Ultrathin plates might also be considered, among other possibilities.

References

[1] Poole CF, Belay MT. Progress in automated multiple development. J Planar Chromatogr 1991;4:345–59.

[2] Jupille TH, Perry JA. Programmed multiple development - spot behavior during solvent advance – a mathematical treatment. J Chromatogr 1974;99:231–42.

[3] Jupille TH, Curtice AB. Reproducibility of spot position in programmed multiple development (PMD). Chromatographia 1975;8:193–4.

[4] Burger K. DC-PMD, Dünnschicht-Chromatographie mit Gradienten-Elution im Vergleich zur Säulenflüssigkeits-Chromatographie (TLC-PMD, thin-layer chromatography with Gradient-Elution in Comparison to HPLC). Fresenius' Z Anal Chem 1984; 318:228–33.

[5] Jaenchen DE. Instrumental multiple development of chromatograms in HPTLC. LaborPraxis 1985;9:1160–2, 1164, 1166, 1173.

[6] Wilson ID, Lewis S. Separation of ecdysteroids by high-performance thin-layer chromatography using automated multiple development. J Chromatogr 1987;408:445–8.

[7] Gocan S, Cimpan G. Review of the analysis of medicinal plants by TLC: modern approaches. J Liq Chromatogr Relat Technol 2004;27:1377–411.

[8] Sherma J, Fried B, editors. Sherma-handbook of thin-layer chromatography. 3rd ed. New York: Marcel Dekker Inc; 2003. p. 144.

[9] Lodi G, Betti A, Menziani E, Brandolini V, Tosi B. Some aspects and examples of automated multiple development (AMD) gradient optimization. J Planar Chromatogr 1991;4:106–10.

[10] Summanen J, Hiltunen R, Vuorela H. The choice of parameters in the optimization of automated multiple development. J Planar Chromatogr 1998;11:16–24.

[11] Wippo U. Conversion of a gradient from an AMD1 to an AMD2 system. Camag Bibliogr Serv 2004;92:10–2.

[12] De la Vigne U. Selective analysis of phospholipids by multiple-development high-performance TLC. LaborPraxis 1991;15:232, 234, 236, 238.

[13] Müthing J. Improved thin-layer chromatographic separation of gangliosides by automated multiple development. J Chromatogr B 1994;657:75–81.

[14] Bonte F, Pinguet P, Chevalier JM, Meybeck A. Analysis of all stratum corneum lipids by automated multiple development high-performance thin-layer chromatography. J Chromatogr B 1995;664:311–6.

[15] Müthing J, Ziehr H. Enhanced thin-layer chromatographic separation of GM1b-type gangliosides by automated multiple development. J Chromatogr B 1996;687:357–62.

[16] Bonte F, Saunois A, Pinguet P, Meybeck A. Existence of a lipid gradient in the upper stratum corneum and its possible biological significance. Arch Dermatol Res 1997;289: 78–82.

[17] Cansell M, Gouygou JP, Jozefonvicz J, Letourneur D. Phospholipids AMD-primuline quantitative scanning 366 nm. Lipids 1997;32:39–44.

[18] Gildenast T, Lasch J. Isolation of ceramide fractions from human stratum corneum lipid extracts by high-performance liquid chromatography. Biochim Biophys Acta, Lipids Lipid Metab 1997;1346:69–74.

[19] Zellmer S, Lasch J. Individual variation of human plantar stratum corneum lipids, determined by automated multiple development of high-performance thin-layer chromatography plates. J Chromatogr B 1997;691:321–9.

[20] Raith K, Zellmer S, Lasch J, Neubert RHH. Profiling of human stratum corneum ceramides by liquid chromatography-electrospray mass spectrometry. Anal Chim Acta 2000;418:167–73.

[21] Ratz H, Schnell H, Rischer M, Eibl HJ. Separation and quantitation of alkylphosphocholines and analogues of different liposome formulations by HPTLC. J AOAC Int 2001;84:1277–82.

[22] Sek L, Porter CJH, Charman WN. Characterization and quantification of medium chain and long chain triglycerides and their in vitro digestion products, by HPTLC coupled with in situ densitometric analysis. J Pharm Biomed Anal 2001;25:651–61.

[23] Farwanah H, Neubert RHH, Zellmer S, Raith K. Improved procedure for the separation of major stratum corneum lipids by means of automated multiple development thin-layer chromatography. J Chromatogr B 2002;780:443–50.

[24] Farwanah H, Neubert RHH, Zellmer S, Raith K. Improved procedure for the separation of major stratum corneum lipids by means of automated multiple development thin-layer chromatography. J Chromatogr B 2002;780:443–50.

[25] Figueroa CE, Vega MH, Saelzer RF, Rios GG, Rodriguez H. Analysis of egg yolk lipid fraction by AMD-HPTLC. Bol Soc Chil Quim 2002;47:61–6.

[26] Neubert RHH, Farwanah H, Raith K, Zellmer S. Improved analysis of skin lipids by AMD. Camag Bibliogr Serv 2003;90:2–4.

[27] Raith K, Farwanah H, Neubert RHH. HPTLC and LC/MS in ceramide analysis. Trends Clin Exp Dermatol 2003;1:106–17.

[28] Grether-Beck S, Timmer A, Brammertz D. Quantification of signaling ceramides in primary keratinocytes. Camag Bibliogr Serv 2004;93:2–4.

[29] Diaz-Del Consuelo I, Jacques Y, Pizzolato GP, Guy RH, Falson F. Comparison of the lipid composition of porcine buccal and esophageal permeability barriers. Arch Oral Biol 2005;50:981–7.

[30] Farwanah H, Raith K, Neubert RHH, Wohlrab J. Stratum corneum sphingolipids profiles. Arch Dermatol Res 2005;296:514–21.

[31] Dreisewerd K, Müthing J. Structural characterization of gangliosides by HPTLC/IR-MALDI-o-TOF. Camag Bibliogr Serv 2006;97:2–5.

[32] Schellenberg I, Kabrodt K. Optimization of an AMD 2 method for determination of stratum corneum lipids. Camag Bibliogr Serv 2010;105:10–2.

[33] Cebolla VL, Jarne C, Domingo P, Domínguez A, Delgado-Camón A, Garriga R, et al. Fluorescence detection by intensity changes for high-performance thin-layer chromatography separation of lipids using automated multiple development. J Chromatogr A 2011;1218:2668–75.

[34] Kabrodt K, Lüttich J, Dittler I, Schellenberg I. Improved HPTLC separation of lipids by using automated multiple development (AMD) and identification with the TLC-MS-interface. In: Proceedings of the HPTLC international symposium, July 2011, Basel, Switzerland; 2011. p. 63.

[35] Opitz A, Wirtz M, Melchior D, Mehling A, Kling HW, Neubert RHH. Improved method for stratum corneum lipid analysis by automated multiple development HPTLC. Chromatographia 2011;73:559–65.

[36] Ochalek M, Heissler S, Wohlrab J, Neubert RHH. Characterization of lipid model membranes designed for studying impact of ceramide species on drug diffusion and penetration. Eur J Pharm Biopharm 2012;81:113–20.

[37] Zietz E, Ricker I. Methods to monitor pesticides in ground and drinking water by HPTLC/AMD according to drinking water regulations. J Planar Chromatogr 1989;2:262–7.

[38] Zietz E, Ricker I, Arendt G. Possibilities of checking of the highest permissible concentrations of pesticides by high-performance thin-layer chromatography in terms of the new drinking water ordinance. Gewaesserschutz, Wasser, Abwasser 1989;106:136–59.

[39] De la Vigne U, Jaenchen D. Determination of pesticides in water by HPTLC using automated multiple development (AMD). J Planar Chromatogr 1990;3:6–9.

[40] De la Vigne U, Jaenchen DE, Weber WH. Application of high-performance thin-layer chromatography and automated multiple development for the identification and determination of pesticides in water. J Chromatogr 1991;553:489–96.

[41] Jork H, Keller G, Kocher U. Applications of AMD to the determination of crop-protection agents in drinking water. Part II: limitations. J Planar Chromatogr 1992;5:246–50.

[42] Sherma J. Modern thin-layer chromatographic pesticide analysis using multiple development. J AOAC Int 1992;75:15–7.

[43] Pfaab G, Jork H. Application of AMD [automated multiple development] for the determination of pesticides in drinking water. Part 3. Solid phase extraction and influencing factors. Acta Hydrochim Hydrobiol 1994;22:216–23.

[44] Wolff SC, Kovar KA. Determination of edetic acid (EDTA) in water by online coupling of HPTLC and FTIR. J Planar Chromatogr 1994;7:286–90.

[45] Butz S, Stan HJ. Screening of 265 pesticides in water by thin-layer chromatography with automated multiple development. Anal Chem 1995;67:620–30.

[46] Stan HJ, Butz S. Multimethod applying AMD-TLC analysis to drinking water. Chem Plant Prot 1995;12:197–216.

[47] Burger K, Jork H, Koehler J. Application of AMD to the determination of crop protection agents in drinking water. Part 5. Fundamentals of a confirmatory test. Acta Hydrochim Hydrobiol 1996;24:6–15.

[48] Dethlefs F, Stan HJ. Analysis of pulp mill EOP bleaching process effluent by thin-layer chromatography with automated multiple development (AMD-HPTLC). Acta Hydrochim Hydrobiol 1996;24:77–84.

[49] Koeber R, Niessner R. Screening of pesticide-contaminated soil by supercritical fluid extraction (SFE) and high-performance thin-layer chromatography with automated multiple development (HPTLC/AMD). Fresenius' J Anal Chem 1996;354:464–9.

[50] Kralicek P, Reifler F. New TLC/AMD method for testing of textiles for azo dyes. Melliand Textilber 1996;77:140.

[51] Lautie JP, Stankovic V. Automated multiple development TLC of phenylurea herbicides in plants. J Planar Chromatogr 1996;9:113–5.

[52] Morlock GE. Analysis of pesticide residues in drinking water by planar chromatography. J Chromatogr A 1996;754:423–30.

[53] Stan HJ, Wippo U. Pesticide residue analysis in plant food products by TLC with automated multiple development. GIT Fachz fuer das Lab 1996;40:855–8.

[54] Reemtsma T, Putschew A, Jekel M. Industrial wastewater analysis: a toxicity-directed approach. Waste Manage (Oxford, UK) 1999;19:181–8.

[55] Reemtsma T, Putschew A, Jekel M. Application of toxicity directed analysis to industrial wastewaters. Vom Wasser 1999;92:243–55.

[56] Kreiss W, Eberz G, Weisemann C. Bioluminescence detection for planar chromatography. Camag Bibliogr Serv 2002;88:12–3.

[57] Kastelan-Macan M, Babic S. Pesticides. Chromatographic science series. In: Handbook of thin-layer chromatography. 3rd ed; 2003. pp. 767–805.

[58] Topfer A, Stan HJ. Characterization of pulp mill effluents. Vom Wasser 2000;95:1–24.

[59] Albert R, Weber W. HPTLC/AMD in water analysis. Determination of hydroxylated triazine decomposition products and acidic pesticides. CLB Chemie Labor und Biotech 2004;55:172–5.

[60] Weber W, Seitz W, Aichinger A, Albert R. Luminographic detection of toxicity with *Vibrio fischeri* (luminescent bacteria). Camag Bibliogr Serv 2005;94:2–4.

[61] Plass E. Determination of Amitrol in water by AMD. Camag Bibliogr Serv 2006;96:2–5.
[62] Weber WH, Seitz W, Schulz W, Aichinger A. Evidence of the metabolites desphenyl-chloridazon and methyl-desphenyl-chloridazon in surface, ground and drinking water. Camag Bibliogr Serv 2007;98:12–5.
[63] Yue Y, Zhang R, Fan W, Tang F. High-performance thin-layer chromatographic analysis of selected organophosphorous pesticide residues in tea. J AOAC Int 2008;91: 1210–7.
[64] Weber WH, Müller A, Weiss SC, Seitz W, Schulz W. 1H-benzotriazole and tolyltriazoles in the aquatic environment. Occurrence in ground, surface and wastewater. Vom Wasser 2009;107:16–24.
[65] Müller A, Weiss SC, Schulz W, Seitz W, Albert R, Ruck W, et al. Combination of different liquid chromatography/mass spectrometry technologies for the identification of transformation products of rhodamine B in groundwater. Rapid Commun Mass Spectrom 2010;24:659–66.
[66] Weber WH, Seitz W, Weiss SC, Müller A, Schulz W. 1H-Benzotriazole and tolyltriazole in the aquatic environment. Camag Bibliogr Serv 2010;105:7–9.
[67] Weber WH, Müller A, Weiss SC, Seitz W, Schulz W. Benzotriazoles in the aquatic environment. Multidimensional screening using HPTLC/AMD and nano-HPLC/QTOF-MS. GIT Spez Sep 2010;30:26–8.
[68] Schulz W, Weber WH, Müller A, Weiss SC. Drinking water treatment – identification of reaction by-products of 4- and 5-methyl-1H-benzotriazole formed during ozonation. Camag Bibliogr Serv 2012;109:13–5.
[69] Dytkiewitz E, Morlock GE. Analytical strategy for rapid identification and quantification of lubricant additives in mineral oil by high performance thin-layer chromatography with UV absorption and fluorescence detection combined with mass spectrometry and infrared spectroscopy. J AOAC Int 2008;91:1237–43.
[70] Jarne C, Cebolla VL, Membrado L, Le Mapihan K, Giusti P. High-performance thin-layer chromatography using automated multiple development for the separation of heavy petroleum products according to their number of aromatic rings. Energy Fuels 2011;25:4586–94.
[71] Ebel S, Voelkl S. Thin-layer chromatographic analysis of pharmacognostic products. Improvement with automated multiple development. Dtsch Apoth Ztg 1990;130:2162–9.
[72] Reddy M, Vijayaraj B, Gary R. Phosphorus-32-postlabeling assay for carcinogen-DNA adducts: description of beta shielding apparatus and semiautomatic spotting and washing devices that facilitate the handling of multiple samples. Carcinogenesis 1990;11:683–7.
[73] Schuetz H, Erdmann F, Funk W. TLC-scanners in toxicological analysis. GIT Suppl 1990;1:17–8, 20–1, 24–6, 28–9.
[74] Belay M. Method development and solvent optimization for the separation of pharmaceutically and biologically important compounds by high-performance planar chromatography and automated multiple development. SciFinder; 1993. 221 pp.
[75] Belay MT, Poole CF. Separation of PTH amino acid derivatives by thin-layer chromatography and automated multiple development. J Planar Chromatogr 1993; 6:43–50.
[76] Colthup PV, Bell JA, Gadsdon DL. Use of linear solvent gradients in manual and automated multiple development for the separation and characterization of drugs and metabolites by HPTLC. J Planar Chromatogr 1993;6:386–93.
[77] Wolff SC, Kovar KA. Direct HPTLC-FTIR measurement in combination with AMD. J Planar Chromatogr 1994;7:344–8.
[78] Kovar KA, Dinkelacker J, Pfeifer AM, Pisternick W, Woessner A. Identification of drugs by HPTLC-ultraviolet spectrometry/FTIR on line coupling. GIT Spez Chromatogr 1995;15:19–20, 22–24.

[79] Kralicek P. Detection of carcinogenic amines arising from azo dyes. Chimia 1995;49: 222–5.
[80] Roberts RQ. Thin-layer chromatographic (TLC) analysis of drugs in multiple human urines with parallel automated sample extraction and direct TLC extract application. Clin Chem 1996;42:1570–1.
[81] Bowles MR, Hall DR, Pond SM, Winzor DJ. Studies of protein interactions by biosensor technology: an alternative high-performance TLC plates has been obtained by automated multiple development approach to the analysis of sensorgrams deviating from pseudo-first-order kinetic behavior. Anal Biochem 1997;244: 133–43.
[82] Liawruangrath B, Liawruangrath S. High-performance thin-layer chromatographic determination of erythromycin in pharmaceutical preparations. Chromatographia 2001;54:405–8.
[83] Werner R, Arndt C, Demme U. Improved separation of benzodiazepines by AMD. Camag Bibliogr Serv 2003;91:5–7.
[84] Jamshidi A. Determination of progesterone in drug release media. Camag Bibliogr Serv 2004;93:10–2.
[85] Jamshidi A. A convenient and high throughput HPTLC method for determination of progesterone in release media of siliconbased controlled-release drug-delivery systems. J Planar Chromatogr 2004;17:229–32.
[86] Khakpour M, Jamshidi A, Entezami AA, Mirzadeh H. HPTLC procedure for determination of levonorgestrel in the drug-release media of an in-situ forming delivery system. J Planar Chromatogr 2005;18:326–9.
[87] Mäder K, Rübe A, Klein S. Quantitation of in vitro lipolysis products with HPTLC. Camag Bibliogr Serv 2005;95:14–5.
[88] Pothier J, Galand N. Automated multiple development thin-layer chromatography for separation of opiate alkaloids and derivatives. J Chromatogr A 2005;1080:186–91.
[89] Jamshidi A, Mobedi H, Ahmad-Khanbeigi F. Stability-indicating HPTLC assay for leuprolide acetate. J Planar Chromatogr 2006;19:223–7.
[90] Jamshidi A, Sharifi A. HPTLC analysis of tamoxifen citrate in drug-release media during development of an in-situ-cross-linking delivery system. J Planar Chromatogr 2009;22:187–9.
[91] Glowniak K, Widelski J. Coumarins: TLC analysis. In: Cazes J, editor. Encyclopedia of chromatography. 3rd ed. vol. 1; 2010. pp. 511–3.
[92] Ferenczi-Fodor K, Végh Z, Renger B. Impurity profiling of pharmaceuticals by thin-layer chromatography. J Chromatogr A 2011;1218:2722–31.
[93] Rothenhoefer M, Scheruebl R, Bernhardt G, Heilmann J, Buschauer A. Qualitative and quantitative analysis of hyaluronan oligosaccharides with high performance thin layer chromatography using reagent-free derivatization on amino-modified silica and electrospray ionization-quadrupole time-of-flight mass spectrometry coupling on normal phase. J Chromatogr A 2012;1248:169–77.
[94] Kralecik P. A new TLC/AMD (automated multiple development) method for the testing of textiles for carcinogenic amines arising from azo dyes. GIT Spez Chromatogr 1996;16:126–9.
[95] Bonte F, Pinguet P, Saunois A, Chevalier JM, Meybeck A. Separation of forskolin derivatives by automated multiple development high-performance thin-layer chromatography. J Chromatogr A 1997;791:231–5.
[96] Raverkar SR. High performance thin layer chromatographic analysis of nitrobenzene process streams. Anal Sci 2001;17(Suppl.):a287–9.
[97] Hauri U, Baumgartner V, Hohl C. Bio-activity based analysis of irradiated sunscreens using HPTLC and in situ detection with *Vibrio fischeri*. Camag Bibliogr Serv 2008;100: 2–5.

[98] Baumgartner V, Hohl C, Hauri U. Bioactivity-based analysis of sunscreens using the luminescent bacteria *Vibrio fischeri*. J Planar Chromatogr 2009;22:19–23.

[99] Behringer S, Schwack W. Determination of PAHs in toys by HPTLC. Camag Bibliogr Serv 2012;108:12–5.

[100] Trypsteen MFM, Van Severen RGE, De Spiegeleer BM. Planar chromatography of Echinacea species extracts with automated multiple development. Analyst 1989;114: 1021–4.

[101] Menziani E, Tosi B, Bonora A, Reschiglian P, Lodi G. Automated multiple development high-performance thin-layer chromatographic analysis of natural phenolic compounds. J Chromatogr 1990;511:396–401.

[102] Read H, Wilson ID, Lafont R. A note on overpressure thin-layer chromatography of ecdysteroids. In: Chromatogr Isol. Insect Horm. Pheromones, Proc. Int. Symp. 1st; 1990. pp. 127–30.

[103] Gocan S, Cimpan G, Muresan L. Automated multiple development thin layer chromatography of some plant extracts. J Pharm Biomed Anal 1996;14:1221–7.

[104] Olah NK, Muresan L, Cimpan G, Gocan S. Normal-phase high-performance thin-layer chromatography and automated multiple development of hydroalcoholic extracts of *Artemisia abrotanum*, *Artemisia absinthium*, *Artemisa vulgaris*, and *Artemisia cina*. J Planar Chromatogr 1998;11:361–4.

[105] Scholl Y, Asano N, Drager B. Automated multiple development thin layer chromatography for calystegines and their biosynthetic precursors. J Chromatogr A 2001;928: 217–24.

[106] Galand N, Pothier J, Dollet J, Viel C. OPLC and AMD, recent techniques of planar chromatography: their interest for separation and characterization of extractive and synthetic compounds. Fitoterapia 2002;73:121–34.

[107] Di X, Chan KKC, Leung HW, Huie CW. Fingerprint profiling of acid hydrolyzates of polysaccharides extracted from the fruiting bodies and spores of Lingzhi by high-performance thin-layer chromatography. J Chromatogr A 2003;1018:85–95.

[108] Galand N, Ernouf D, Montigny F, Dollet J, Pothier J. Separation and identification of cannabis components by different planar chromatography techniques (TLC, AMD, OPLC). J Chromatogr Sci 2004;42:130–4.

[109] Markowski W, Ludwiczuk A, Wolski T. Analysis of Ginsenosides from *Panax quinquefolium L.* by automated multiple development. J Planar Chromatogr 2006;19:115–7.

[110] Pozharitskaya ON, Ivanova SA, Shikov AN, Makarov V. Separation and quantification of terpenoids of *Boswellia serrata* Roxb. extract by planar chromatography techniques (TLC and AMD). J Sep Sci 2006;29:2245–50.

[111] Klöppel A, Grasse W, Brümmer F, Morlock GE. HPTLC coupled with bioluminescence and mass spectrometry for bioactivity-based analysis of secondary metabolites in marine sponges. J Planar Chromatogr 2008;21:431–6.

[112] Klöppel A, Kolm A, Brümmer F, Morlock GE. Screening for bioactive natural products in sponges. Camag Bibliogr Serv 2009;102:4–7.

[113] Schulte F, Maeder J, Kroh LW, Panne U, Kneipp J. Characterization of pollen carotenoids with in situ and high-performance thin-layer chromatography supported resonant Raman spectroscopy. Anal Chem 2009;81:8426–33.

[114] Simoes-Pires C, Hmicha B, Marston A, Hostettmann K. A TLC bioautographic method for the detection of alpha- and beta-glucosidase inhibitors in plant extracts. Phytochem Anal 2009;20:511–5.

[115] Yan YZ, Xie PS, Lam WK, Chui E, Yu QX. Study on triterpenoic acids distribution in ganoderma mushrooms by automatic multiple development high performance thin layer chromatographic fingerprint analysis. J AOAC Int 2010;93:1384–9.

[116] Cui J, Vue Y, Tang F, Wang J. HPTLC analysis of the flavonoids in eight species of Indocalamus leaves. J Planar Chromatogr 2011;24:394–9.

[117] Misra K, Tulsawani R, Shyam R, Meena DK, Morlock GE. Hyphenated high-performance thin-Layer chromatography for profiling of some indian natural efficiency enhancers. J Liq Chromatogr Relat Technol 2012;35:1364–87.
[118] Wippo U, Stan HJ. Determination of iprodione residues in vegetable food samples using online coupling of RP-HPLC followed by AMDTLC. Dtsch Lebensm -Rundsch 1997;93:144–8.
[119] Romano P, Brandolini V, Ansaloni C, Menziani E. The production of 2,3-butanediol as a differentiating character in wine yeasts. World J Microbiol Biotechnol 1998;14:649–53.
[120] Brandolini V, Salzano G, Maietti A, Caruso M, Tedeschi P, Mazzotta D, et al. Automated multiple development method for determination of glycerol produced by wine yeasts. World J Microbiol Biotechnol 2002;18:481–5.
[121] Morlock GE, Häberle S, Jautz U, Schwack W. New HPTLC-MS method for determination of heterocyclic aromatic amines. Camag Bibliogr Serv 2004;93:14–5.
[122] Lodi G, Betti A, Brandolini V, Menziani E, Tosi B. Automated multiple development HPTLC analysis of sugars on hydrophilic layers: I. Amino layers. J Planar Chromatogr 1994;7:29–33.
[123] Brandolini V, Menziani E, Mazzotta D, Cabras P, Tosi B, Lodi G. Use of AMD-HPTLC for carbohydrate monitoring in beers. J Food Compos Anal 1995;8:336–43.
[124] Lodi G, Bighi C, Brandolini V, Menziani E, Tosi B. Automated multiple development HPTLC analysis of sugars on hydrophilic layers: II. Diol layers. J Planar Chromatogr 1997;10:31–7.
[125] Vaccari G, Lodi G, Tamburini E, Bernardi T, Tosi S. Detection of oligosaccharides in sugar products using planar chromatography. Food Chem 2001;74:99–110.
[126] Bernardi T, Tamburini E, Vaccari G. Separation of complex fructo-oligosaccharides (FOS) and inulin mixtures by HPTLC–AMD. J Planar Chromatogr 2005;18:23–7.
[127] Liang Q, Li W, Wang H, Wang S. Saccharide analysis of si-wu-tang by automated multiple development and off-line coupling thin layer chromatography electrospray ionization mass spectrometry techniques. China J Chin Mater Med 2003;28:632–5.
[128] Steuckart C, Berger-Prelss E, Levsen K. Determination of explosives and their biodegradation products in contaminated soil and water from former ammunition plants by automated multiple development high-performance thin-layer chromatography. Anal Chem 1994;66:2570–7.
[129] Camag AMD brochure. http://www.camag.com; 2014.
[130] Markowski W, Wróblewski K, Dzido TH. Stepwise gradient Elution in RP HPTLC with a new horizontal developing chamber. J Planar Chromatogr 2012;25:200–7.
[131] Wang Y, Wang D, Wang J, Xiong Z, Zhang H, She G, et al. A new instrument for automated multiple development in thin-layer chromatography. J Planar Chromatogr 2004;17:290–6.
[132] Morlock GE, Oellig C, Bezuidenhout LW, Brett MJ, Schwack W. Miniaturized planar chromatography using office peripherals. Anal Chem 2010;82:2940–6.
[133] Markowski W, Soczewinski E. Computer-aided optimization of stepwise gradient and multiple-development thin-layer chromatography. Chromatographia 1993;36:330–6.
[134] Adhami HR, Cherer U, Kaehlig H, Hettich T, Schlotterbeck G, Reich E, et al. Combination of bioautography with HPTLC-MS/NMR: a fast identification of acetylcholinesterase inhibitors from galbanum. Phytochem Anal 2013;24:395–400.
[135] Morlock GE, Schwack W. Coupling of planar chromatography to mass spectrometry. TrAC 2010;29:1157–71.
[136] Morlock GE, Schwack W. Hyphenations in planar chromatography. J Chromatogr A 2010;1217:6600–9.
[137] Tuzimski T. Basic principles of planar chromatography and its potential for hyphenated techniques. In: Srivastava MM, editor. High-performance thin-layer chromatography (HPTLC). Berlin Heidelberg: Springer-Verlag; 2011. pp. 247–310 [Chapter 14].

[138] Kertesz V, Ford MJ, Van Berkel GJ. Automation of a surface sampling Probe/Electrospray mass spectrometry system. Anal Chem 2005;77:7183–9.

[139] Kovar KA, Ensslin HK, Frey OR, Rienas S, Wolff SC. Applications of online coupling of thin layer chromatography and FTIR spectroscopy. J Planar Chromatogr 1991;4:246–50.

[140] Burger K. On-line coupling HPLC/AMD-TLC (automated multiple development-thin-layer chromatography). Analusis 1990;18:i13–6.

[141] Stan HJ, Schwarzer F. Online coupling of liquid chromatography with thin-layer chromatography. J Chromatogr A 1998;819:35–44.

CHAPTER

5

Forced-Flow Development in Overpressured Layer Chromatography

Emil Mincsovics [1], *Ágnes M. Móricz* [2], *Ernő Tyihák* [2]

[1] **Department of Genetics and Plant Breeding, Faculty of Horticultural Sciences, Corvinus University of Budapest, Budapest, Hungary;** [2] **Centre for Agricultural Research, Plant Protection Institute, Hungarian Academy of Sciences, Budapest, Hungary**

5.1 INTRODUCTION

Overpressured layer chromatography (OPLC) is a forced-flow, planar layer, liquid chromatographic technique, which was developed and introduced by Hungarian scientists in 1979 [1]. The ancestor of this technique is the ultramicro (UM) chamber [2] in which the adsorbent layer is covered by a counterplate eliminating the vapor phase above the adsorbent layer and the separation is fulfilled by capillary action using a conventional normal chamber. The pressurized UM chamber was designed for OPLC development, as the first realization of a separation chamber, where an external pressure was used to close the adsorbent layer surface allowing the mobile phase to be forced into or through the adsorbent bed by an overpressure generated by a pump [3,4]. The essence of OPLC is that the adsorbent layer is closed and pressurized during the development/elution; the mobile phase flow rate can be set to the optimum linear velocity yielding the highest efficiency; moreover, contrary to high-performance liquid chromatography (HPLC) columns, the adsorbent bed can be open at any development distance or any time for evaluating or finishing the separation process as in TLC. As a consequence, OPLC integrates the advantages of TLC/HPTLC and HPLC.

OPLC is a collective term for techniques using a pressurized UM chamber, a liquid delivery system for mobile phase control, a chamber

Instrumental Thin-Layer Chromatography
http://dx.doi.org/10.1016/B978-0-12-417223-4.00005-4

isolation of the selected bands can be fulfilled off-line according to conventional preparative layer chromatography by scraping off the actual band from the layer together with the adsorbent from which the material can be eluted by a strong solvent. Of course the combined online/off-line detection and isolation is also possible [9].

5.2.2 Transfusion, Infusion, Infusion–Transfusion OPLC Operations

The conventional OPLC systems with one open inlet and one outlet are suitable for transfusion operations enabling outflow of both air and mobile phase (Figure 5.1). Starting the development with a dry adsorbent layer as in the original fully off-line OPLC resulted in spot and band deformation in the region of the front of total wetness yielding a zigzag appearance for the spots and/or bands [10]. This disturbing effect can be dramatically reduced by performing a prerun with weak and noneluting mobile phase prior to the admission of the mobile phase [13] or by first pumping mobile phase into the adsorbent layer from the outlet side until it reaches the region adjacent to the point of sample application before introduction of the mobile phase for the separation through the inlet, thus the bulk of the air is eliminated prior to separation [14]. Another solution, the infusion and infusion–transfusion process was also developed to solve this problem [15,16]. In infusion operations there is either no outlet or the outlet valve is closed during the development and the adsorbent layer is sealed on all four sides. This operation allows off-line OPLC development only, starting the separation with a dry adsorbent layer. The use of overrun and online separation and detection are impossible. This was solved by the infusion–transfusion method where the outlet tube of the chamber was equipped with a clip or valve (Figure 5.1, No. 3) and the first period of development is infusion (the outlet is closed), which is followed by the transfusion operation.

5.2.3 Flowing Eluent Wall Processes and Aliquot Online Detection

An OPLC procedure has been developed for single and multichannel OPLC separation using a nonsegmented adsorbent bed and a flowing eluent wall (FEW) for segmentation [17–19]. This arrangements can be applied at the inlet and/or outlet sides as well, accordingly FEW-I, FEW-O, and FEW-I/O variations are feasible. The conceptual view can be seen in Figure 5.2.

In FEW-I and FEW-I/O the adsorbent bed is divided into active and nonactive regions during the separation. Only eluent is introduced to the nonactive part, while sample and mobile phase can be admitted to the

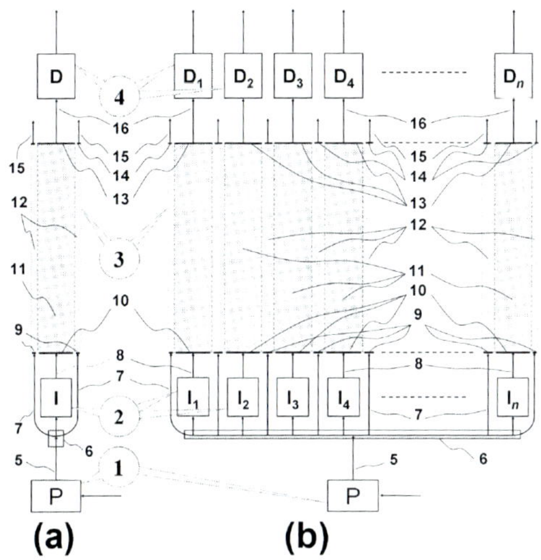

FIGURE 5.2 Scheme of one-channel (a) and multichannel (b) configurations of partially and fully online forced-flow separations on a nonsegmented adsorbent bed using a flowing eluent wall (FEW) for segmentation. 1, Mobile phase delivery system (pump, P); 2, injector/multichannel injector (I); 3, nonsegmented adsorbent bed; 4, detector/multichannel detector (D); 5, mobile phase stream on the inlet side; 6, mobile phase distributor; 7, mobile phase lines for FEW; 8, mobile phase line for sample application by injection; 9, distributor space of mobile phase for FEW; 10, distributor space of sample application; 11, separation part of adsorbent bed (lane); 12, FEW part of adsorbent bed; 13, sample and/or mobile phase collector space at outlet; 14, waste collector space at outlet; 15, stream of waste at outlet; 16, stream of sample and/or mobile phase at outlet side. *Reproduced on the basis of Figure 9 from Ref. [19].*

active parts by means of injector(s) for single or parallel separations. This configuration at the inlet eliminates the edge effects in the case of single sample injection and also the sample mixing effect between neighboring lanes in multichannel separations. These settings can be used at the inlet or both inlet and outlet sides allowing separately online sample injection, separation, detection, and fraction collection for the given lanes. In addition, aliquot online detection can be implemented at outlet side irrespective of FEW or non-FEW sample injection modes, where the central part of the eluted sample components is directed to a flow through detector and the side parts are collected in the waste containing the eluent and part of the sample; moreover, the partially mixed samples can be excluded from online detection in the case of parallel separations and non-FEW injections. The detection and waste ratio at the outlets can be adjusted by the flow resistance of the internal tube dimensions [20]. A typical chromatogram illustrates the effect of aliquot detection in Figure 5.3. As can be seen, the sensitivity may vary by the detection-to-waste ratios.

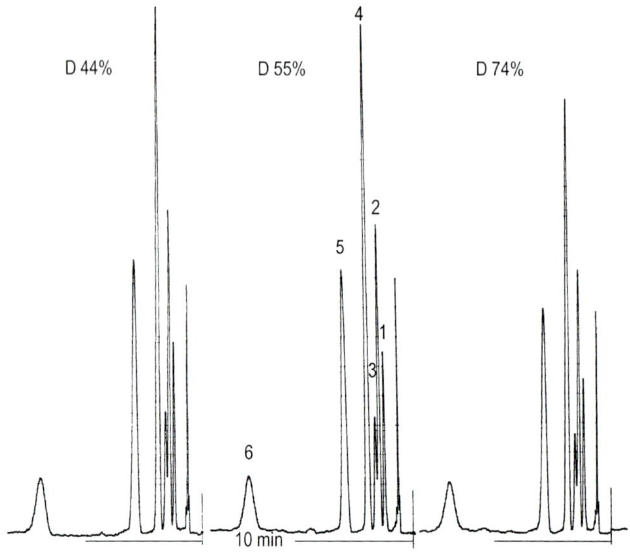

FIGURE 5.3 One-channel fully online FEW-OPLC separation of some phenylthiohydantoin (PTH)-amino acids on 5 × 20 cm TLC silica gel 60 layer (Merck Co.) sealed using the experimental OPLC Separation Unit 100 at 10 MPa external pressure and different detection-to-waste ratios. Mobile phase, chloroform–ethyl acetate (9:1, v/v) at a flow rate of 0.8 ml/min; injection volume, 0.5 μL; cell volume, 1 μL; detection, 275 nm (AUFS 0.05/0.5 V). 1, PTH-proline, 0.15 mg/ml; 2, PTH-leucine, 0.18 mg/ml; 3, PTH-isoleucine, 0.11 mg/ml; 4, PTH-valine, 0.36 mg/ml; 5, PTH-methionine, 0.52 mg/ml; 6, PTH-glycine, 0.51 mg/ml. FEW-OPLC, flowing eluent wall–overpressured layer chromatography; TLC, thin-layer chromatography. *(Reproduced from Ref. [20], with permission.)*

A similar approach was applied for the reduction (elimination) of the so-called wall effects in HPLC columns and the process was termed active flow management covering the combination of conventional and curtain flow injection with the detection of the central part of the parallel segmented outlet flow [21,22].

5.3 INSTRUMENTS AND LAYERS FOR OPLC

5.3.1 Conventional Cushion and Modern Cassette Systems for OPLC

The first experimental pressurized UM chambers made of plexiglass were constructed for circular and linear OPLC development using 20 × 20 cm adsorbent layers [1,3,4]. A cushion system with flexible membrane made of polyethylene foil was applied to form a homogeneous external pressure seal with the adsorbent layer surface containing one connection for the eluent inlet. In the case of circular development, the eluent is introduced at the center of the adsorbent layer, however, in the linear development mode the mobile phase first enters a distributing trough at the side of the adsorbent bed for linear front formation. This can be achieved by forming a narrow channel in the adsorbent layer; furthermore, when the separation starts with a dry adsorbent layer at the start of the separation, a rapid introduction of eluent is also necessary. Based on the results of work with these experimental systems, Labor Instrument Works (Budapest, Hungary) developed the first commercial OPLC instruments having a pressurized UM chamber, a pump for the water cushion, and a second pump for the mobile phase, called the Chrompres 10 and Chrompres 25, permitting 1 and 2.5 MPa external pressures for layers 40 or 20 cm long and 20, 10, or 5 cm wide, respectively. To protect the cushion system, a polytetrafluoroethylene (PTFE) insert sheet with eluent directing troughs and holes for eluent conduction was applied for linear development. The Chrompres 10 was equipped with one eluent inlet initially [5] and then later two mobile phase connections were provided, one for the inlet and the other for the outlet similar to the Chrompres 25 system, thus both systems were suitable for fully off-line and fully online separations and their combinations as well [10].

To minimize the handling difficulties of the Chrompres systems, a user-friendly automated personal OPLC 50 basic system (OPLC-NIT Ltd., Budapest, Hungary) was developed, which included a hydraulic separation chamber and liquid delivery system [23]. The separation chamber equipped with two eluent connections could be used at a maximum 5 MPa external pressure and consisted of a holding unit, hydraulic unit with an attached drain valve, and a tray-like layer cassette. The mobile phase distribution troughs were formed in the PTFE cover sheet of the cassette. The microprocessor controlled integral liquid delivery system includes a hydraulic liquid delivery pump and an eluent delivery pump. The following parameters can be selected and also be stored in the software of the liquid delivery unit for automatic isocratic or stepwise gradient (maximum three steps) developments: external pressure, eluent flow rate, and mobile phase volume for rapid eluent admission and for development. To generate straight front in linear separations starting with a dry layer a rapid eluent admission is required at first, exceeding the flow by capillary action, which is followed by lower flow rates for the separation. During the development, the external pressure and the inlet pressure of the mobile phase are measured by pressure sensors. The inlet pressure is limited to 80% of the external pressure, which is continuously checked by the software.

A stand alone chamber was also developed, called the OPLC Separation Unit 50. It is equipped with one inlet and one outlet and has a built-in hydraulic pump to form 5 MPa external pressure. The cassette enveloping the adsorbent layer can be inserted into the slit of the chamber for the separation similar to the personal OPLC 50 system. For mobile phase delivery, an HPLC pump can be connected to the chamber inlet [17].

Both OPLC systems with cassette solutions are suitable for off-line and online development using analytical adsorbent layers (20 cm long and 20, 10, or 5 cm wide) or preparative adsorbent layers (20 × 20 cm) with the appropriate cassette.

Kaiser and Rieder developed a circular OPLC instrument permitting about 4.0 MPa pressure on the adsorbent layer surface and termed this HPPLC [24].

An instrument was also developed for linear OPLC separation by Witkiewicz et al. in which gas pressure was used for the external pressure source with the eluent admitted to the adsorbent layer (20 cm long and 10 cm wide) by a syringe type pump [25].

To evaluate some theoretical concepts concerning OPLC, a high-pressure experimental instrument was constructed by Flodberg and Roerade using a metal cover and a hydraulic press capable of applying up to 50 MPa on the adsorbent layer surface [26].

Recently, a one dimensional apparatus was utilized by Guiochon et al. for online detection with an array of microelectrochemical detectors. The construction of a pressurized wide and flat column (100 × 100 × 1 mm) could be operated with an inlet pressure up to 5 MPa [27].

5.3.2 OPLC Systems with More Inlets and Outlets

The OPLC system equipped with one inlet and one outlet allows sequential sample processing as for HPLC. For parallel sample injection and/or detection and fraction collection additional inlets and outlets are required. For FEW arrangements, the number of separation channels should be tailored to the number of FEW-s.

One, four, and eight channel FEW-I (I means inlet side) and FEW-I/O (I/O means inlet and outlet sides) models were developed by modifying the hydraulic unit of the personal OPLC 50 and OPLC Separation Unit 50 systems. Of course for different FEW models, the appropriate cassettes have been constructed, too. For FEW-I and FEW-I/O 2, 5, 9 inlets and 1, 4, 8 outlets are available. For the FEW-I/O solution, one of the inlets is used for the formation of the FEW-s; the other inlet connections are for parallel sample injection. At the outlet side, one outlet collects the FEW lines and the other outlets collect the separation lines. The FEW-O can be carried out with FEW-I arrangement, but in the reverse direction [17–19].

Recently, a stand-alone experimental OPLC Separation Unit 100 was developed allowing 10 MPa external pressure for use with 20 × 5 cm analytical layers. This instrument has three inlets and three outlets, and apart from off-line processes, it can be used for single- or double-channel FEW separations with the appropriate cassettes [20].

5.3.3 Adsorbent Layers for Different OPLC Separations

Commercially available ready-to-use analytical and preparative adsorbent layers made from irregular and spherical particles with different particle size ranges and various modifications on glass, plastic, or aluminum supports can be used for circular and linear OPLC separations. Adsorbent layers without a support can also be used, but are inconvenient compared with the widely used mechanically stable and abrasion resistance adsorbent layers containing gypsum or organic polymer binder. Cellulose layers are less useful because of their compressibility, but even these can be used at low external pressure. For circular development, where the eluent is introduced to the center of the adsorbent bed precoated layers can be used without modification, except for methods employing overrun. For linear development, it is necessary to prevent the mobile phase from flowing off the layer, and the edges of the adsorbent layer should be sealed along the sides.

For two-directional transfusion development, the opposite sides of the adsorbent layer are sealed, however, for infusion development all four sides should be sealed. In both cases the eluent is introduced by a trough positioned in the middle of the bed. For unidirectional linear development, the adsorbent layer is sealed on three or four sides and the mobile phase is admitted to the trough close to the sealed edge. The layer sealed on three sides can only be used for transfusion development; however, the layer sealed on all four sides is suitable for infusion and transfusion developments, as well as for their combinations [1,5,16,23]. For two-dimensional development, the adsorbent layer should be sealed along all four sides [5,28]. The all-sides-sealed layer is universal and can be used for all off-line and online OPLC processes [11,23].

Two or more layers can simultaneously be developed with superimposed adsorbent layers using overpressured-multilayer chromatography (OPMLC). For eluent conduction to the next layer slit-like perforations are applied to the upper foil backed adsorbent layers for parallel linear development [29], however, for circular development only the central holes need to be connected [30]. With OPMLC the sample throughput can be increased dramatically and 50–100 or more samples can be separated in parallel. Long-distance OPLC was described in the linear development mode by Botz et al. where the layers superimposed can be connected in series with narrow slit-like perforations to increase the separation distance and the plate number [31]. The potential of this solution can further be increased by using different phases for a single run [32].

5.4 MAIN CHARACTERISTICS OF OPLC

5.4.1 Mobile Phase Movement

In conventional thin-layer chromatography where the mobile phase moves by capillary action, the solvent front position can be described by the quadratic equation ($Z^2 = kt$; Z, the distance of the alpha front; k, a velocity constant; and t, the development time). Accordingly the solvent front velocity decreases continuously during the development [33].

In OPLC the velocity of the alpha front (u) depends on the chosen flow rate (F) of the pump and the free cross-sectional area (A) and total porosity (ε) of the adsorbent bed ($u = F/A\varepsilon$). Only linear development yields a constant linear velocity along the bed.

Figure 5.4 illustrates the variation of the solvent front distance, external and inlet pressures during linear transfusion (a), infusion (b), and infusion–transfusion (c) OPLC operating modes using a personal OPLC 50 basic system [16]. In all three cases rapid eluent admission is used initially followed by a lower flow rate for the separation. This high-velocity step yields straight front lines by quick filling of the area of the adsorbent layer at the eluent inlet trough and remains straight during the separation. In all

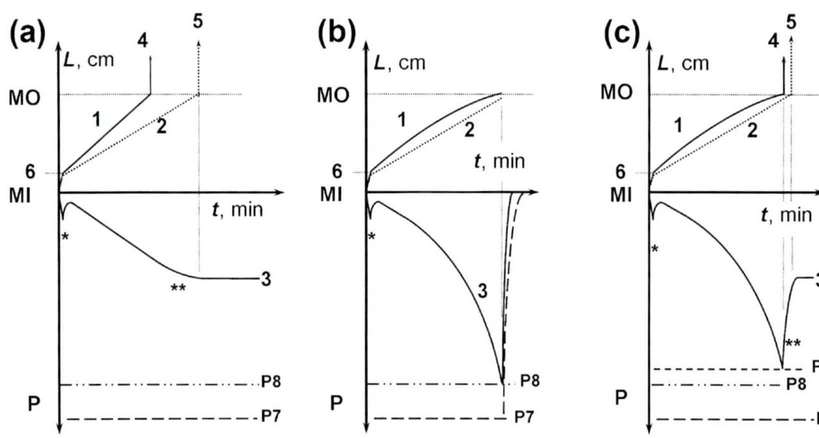

FIGURE 5.4 Variation of front position and pressures during OPLC development using transfusion (a), infusion (b), and infusion–transfusion (c) operating modes. 1, alpha front; 2, front of total wetness; 3, mobile phase inlet pressure; 4, leaving alpha front; 5, leaving front of total wetness; 6, point of off-line sample application; L, distance on the adsorbent layer; P, pressure; t, time; MI, mobile phase inlet position; MO, mobile phase outlet position; P7, external pressure; P8, limit of mobile phase inlet pressure; P9, limit of mobile phase inlet pressure for outlet valve activation; * transition period of rapid mobile phase admission; **transition period of leaving the partially wetted zone. OPLC, overpressured layer chromatography. *(Reproduced from Ref. [16], with permission.)*

three operating modes the off-line sample application should be placed above point 6, where the eluent is already moving with a lower constant speed than in the initial period. Starting the development with dry adsorbent layer in the zone under the alpha front (1), the adsorbent layer is filled partially with eluent and air. The next zone toward the inlet trough is totally wetted, which is completely filled by the eluent. The border between these zones is the front of total wetness (2), which migrates proportionally to the alpha front (1) at constant flow rate and its waviness increases with migration distance in the transfusion mode. The inlet pressure (3) increases gradually with time until the mobile phase begins to flow from the chamber (4) and after this transition period during which the front of total wetness reaches the outlet (5), the inlet pressure becomes constant. It should be mentioned that during this transition period online detection is difficult owing to the air content of the effluent (bubble effect). Contrary to the transfusion mode, the infusion OPLC process (b) results in a continuously decreasing alpha front velocity (1) and a dramatically decrease in the unevenness for the front of total wetness (2) and sample band shape, because of the air originally contained in the bed, is continuously compressed at the outlet side, which helps in the pore filling of particles. During this process the inlet pressure increases (3) more quickly, and when the inlet pressure reaches the pressure limit (P8), the eluent admission is terminated automatically. After a waiting time the external pressure (P7) is released. In the infusion–transfusion mode, the compressed air is flushed out quickly at the first infusion period by opening the outlet valve at lower inlet pressure (P9) rather than at the pressure limit (P8). This solution provides better conditions for online detection by a dramatic reduction of the bubble effect. It should be noted, to prevent the mobile phase movement anomaly between the adsorbent layer surface and the cover sheet during OPLC development, the inlet pressure should be kept lower than the external pressure [34].

5.4.2 Plate Height and Plate Number

In the case of off-line sample application, separation, and detection, the plate height (HETP, H) of a given component (i) can be calculated from the densitogram according to Guiochon and Siouffi [35]:

$$H = \sigma_i^2 \Big/ L_i = \sigma_i^2 \Big/ L_f R_F$$

where σ_i is the band variance, L_i is the migration distance, L_f is the front distance measured from the sample application position, and R_F is the retardation factor.

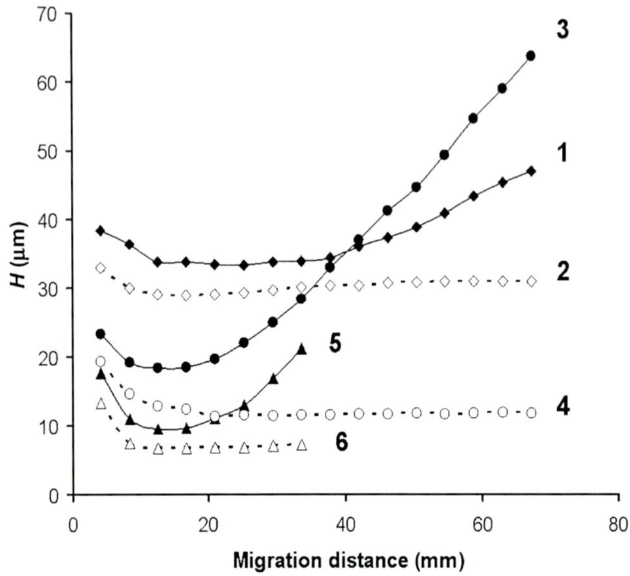

FIGURE 5.5 Effect of migration distance (L_i) on the plate height (H) using PTH-methionine (R_F, 0.42), dichloromethane–ethyl acetate (92:8, v/v) and different silica gel layers developed in a conventional unsaturated chamber and experimental OPLC 100 system at 10 MPa. 1 and 2, TLC layer (10–12 µm; thickness, 0.215 mm; 1.1 µg/5 mm band); 3 and 4, HPTLC layer (5–6 µm; thickness, 0.17 mm; 1 µg/5 mm band); 5 and 6, spherical Raman sheet (3–4 µm; thickness, 0.1 mm; 0.5 µg/5 mm band). Mobile phase linear velocity (cm/min) of OPLC separation: 2, TLC, 1.32; 4, HPTLC, 2.63; 6, Raman, 3.88. OPLC, overpressured layer chromatography; TLC, normal particle adsorbent layer for thin-layer chromatography; HPTLC, fine particle adsorbent layer for high performance thin-layer chromatography.

Contrary to conventional TLC, in transfusion and infusion fully off-line OPLC the plate height is practically constant along the adsorbent layer if small spots or narrow bands are applied [4,23], however, the starting spot size effects the efficiency and larger diameter starting spots increase the HETP value. This is valid for components having a low R_F value and especially for short migration distances [11]. The correlation between the plate height and the migration distance for phenythiohydantoin (PTH)-methionine can be seen in Figure 5.5 for silica gel layers of different particle sizes developed in a conventional unsaturated chamber and in an OPLC 100 system at 10 MPa external pressure. Figure 5.5 shows the essential feature of OPLC and the dissimilarity from conventional TLC development regarding the observed plate height. The HETP value decreases with particle size in both techniques, however, for OPLC this results in more efficient separations than for conventional development at all migration distances using the same adsorbent layer. The HETP difference between TLC and OPLC is especially dramatic at long development or migration distances.

In the case of online OPLC separation and detection with off-line or online sample application the HETP value can be calculated from the chromatogram according to HPLC practice [36] as follows:

$$H = L(\sigma/t_R)^2$$

where L is the bed length used (in fully online process, the distance between the inlet and outlet troughs, and in the case of off-line sample application, the distance between the sample application point and the outlet trough), σ is the peak variance expressed in units of time, and t_R is the retention time of the peak. The total peak variance measured in fully online OPLC is the sum of the chromatographic variance and the "extra-column" variance of the instrument (injector, connecting tubes, troughs, detector, and electronics).

The HETP is influenced by the linear velocity in OPLC as it is in HPLC [36], however, it depends on the combination of the off-line and online operating steps applied. The HETP versus linear velocity result in similar curves, however, the fully off-line process yields the lowest and the fully online the highest HETP values, and between them are two curves of partial off-line/online and partial online/off-line operation. The reason for the difference between the plots arises from the effects of "extra-column" band broadening [11].

The external pressure also affects the HETP value. In fully off-line OPLC, the HETP values can significantly be decreased by increasing external pressure contrary to fully online process, where this effect is negligible [37].

Figure 5.6 shows the effect of the linear velocity on the HETP using a fine particle irregular silica gel 60 adsorbent layer with different external pressures. As can be seen the curves are similar and have a minimum HETP at a series of optimum velocities. In the case of fully off-line transfusion development the high external pressure results in lower HETP values and a broader range of optimum linear velocity, however, the difference between 7.5 and 10 MPa is slight. Owing to the effect of "extra-column" band broadening, fully online OPLC results in about a 2.5 times higher HETP value than fully off-line operation at the optimum linear velocity (3 cm/min).

A representative fully online OPLC chromatograms of some PTH-amino acids is shown in Figure 5.7 using the experimental OPLC Separation Unit 100 with a two-channel FEW arrangement for both the inlet and outlet sides and 10 MPa external pressure. The sample was injected into one channel only and both channels were monitored online. Figure 5.7 shows that no signals appear in the sample free channel and quite symmetrical peaks are detected in the other channel.

Fully online OPLC at 2.5 MPa external pressure was used by Fernando and Poole to study the kinetic parameters of conventional and

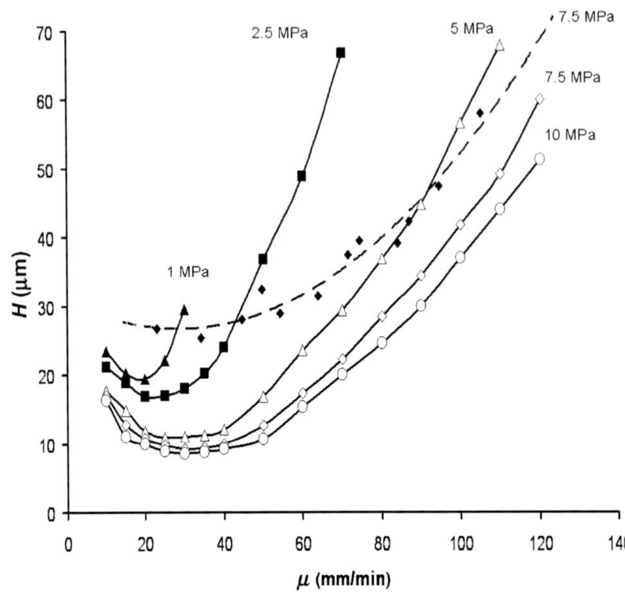

FIGURE 5.6 Effect of linear velocity (μ) on the plate height (H) of PTH-leucine with different external pressures on an HPTLC silica gel 60 layer (5 × 20 cm, made of irregular particles, Merck Co.) using an experimental OPLC Separation Unit 100. Mobile phase, dichloromethane–ethyl acetate, 92:8 (v/v); fully off-line separation (continuous lines): development distance, 170 mm; 1 μg/5 mm band dissolved in the eluent; fully online separation (dashed line), bed length, 170 mm, injection, 0.1 μg/0.5 μL; cell volume, 1 μL. HPTLC, fine particle adsorbent layer for high performance thin-layer chromatography.

high-performance precoated silica gel layers from different manufacturers using polycyclic aromatic hydrocarbons as samples and hexane as the mobile phase [38]. They obtained a minimum HETP value of 24 μm at 2 cm/min optimum linear velocity for HPTLC silica gel 60. This HETP is about twofold larger and the optimum linear velocity two- to fourfold lower than for an ideal column of totally porous particles with same size as the layer. The differences between the layer- and column-shaped beds come from the distinct packing technologies applied and from the presence of binder in the layer which is absent in the column.

The theoretical plate number (N) can be calculated in OPLC with the following equation as for HPLC [36]:

$$N = L/H$$

where L is the bed length used for the separation and H is the plate height (calculated as described above). Since the HETP is practically constant along the adsorbent layer in OPLC, the plate number increases linearly with development distance, contrary to conventional development. Botz

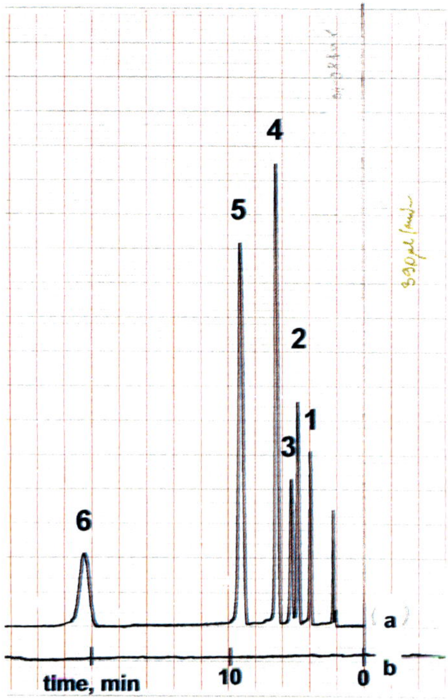

FIGURE 5.7 Fully online OPLC separation of some PTH-amino acids by experimental OPLC Separation Unit 100 with a two-channel FEW arrangement at the inlet and outlet sides with sample injection in one channel only (a) and detection in both channels (a and b). External pressure, 10 MPa; adsorbent layer, 5 × 20 cm HPTLC silica gel 60 layer (Merck Co.); mobile phase, chloroform–ethyl acetate, 9:1 (v/v); flow rate, 0.39 ml/min. Injection volume, 0.5 μL; cell volume, 1 μL; detection, 275 nm (AUFS 0.05/0.5 V). 1, PTH-proline, 0.15 mg/ml; 2, PTH-leucine, 0.18 mg/ml; 3, PTH-isoleucine, 0.11 mg/ml; 4, PTH-valine, 0.36 mg/ml; 5, PTH-methionine, 0.52 mg/ml; 6, PTH-glycine, 0.51 mg/ml. OPLC, overpressured layer chromatography.

et al. achieved a plate number of about 70,000 for a 70 cm fully off-line long distance development [39].

5.4.3 Resolution

Assuming a Gaussian peak profile, the resolution (R_s) of adjacent spots or bands can be described by the following equation [40], which can also be used in fully off-line OPLC:

$$R_s = 0.25(R_F N)^{1/2}\left(1 - R_F\right)\left[\left(K_1/K_2\right) - 1\right]$$

where R_F is the average retardation factor of the peak pair, N is the plate number and K_1, K_2 are the distribution coefficients for the neighboring two substances. The equation shows that resolution is determined by three factors, the layer quality (first term), the average R_F value (second term) and the selectivity factor (third term).

A plot of resolution against R_F shows a curve with a maximum and an optimum range for resolution in conventional development between R_F 0.3–0.4. For fully off-line OPLC the optimum occurs at R_F 0.5–0.6 [41]. Hauck and Jost found that the resolution increases nonlinearly with increasing solvent front distance in agreement with the resolution equation [42].

5.4.4 Spot and Peak Capacity

The spot capacity, given as the number of spots resolved with a resolution of unity, is limited for conventional TLC (HPTLC) development in a single run, as shown by Guiochon and Siouffi. It is about 20 and cannot exceed 30 even for the most favorable case [43].

For a thin-column and fully off-line OPLC, the maximum value of the spot capacity (n_M) can be calculated with the following equation [8]:

$$n_M = 0.5(L/H)^{1/2}$$

where L is the distance between the start and front and H is the average HETP of the compounds.

At isocratic separation conditions in HPLC and also in OPLC using online separation and detection process with off-line or online sample application the peak capacity (n) is:

$$n = 1 + 0.25\,N^{1/2}\ln\left(1 + k\right)$$

where N is the plate number and k is the retention factor of the most retained compound [8].

Higher peak and/or spot capacities can be achieved in OPLC using combined online and off-line detection than with single online or single off-line separation and detection processes because the individual peak and spot capacity are additive [9].

5.5 ANALYTICAL AND PREPARATIVE APPLICATIONS

5.5.1 Analytical Applications, Hyphenations with Other Separation and Detection Techniques

Fully off-line infusion and transfusion OPLC is suitable for quick analytical separations with parallel sample separation. For linear unidirectional development 17–18 spots or 12–14 bands can be applied (interval and band width, 6–7 mm) to a 20 × 20 cm analytical TLC or HPTLC layer, however, 68–70 spots with 5 mm intervals or 38 bands (interval and band width, 4 mm) can be deposited for bidirectional development, on account of the lower lateral band broadening for shorter development distances. This solution is proposed if the resolution is acceptable for shorter development distances [23] e.g., for cleaning-validation in a steroid manufacturing plant using a single isocratic development [44]. To increase the development distance, multiple development can be used with the same or different development distances and mobile phase compositions. For impurity profile and purity of levonorgestrel/norgestrel bulk drug substances, OPLC procedures were evaluated using the same mobile phase and increasing consecutive development distances providing higher separation performance [45]. To determine biogenic amines in cauliflower and broccoli as their dansyl derivatives a stepwise gradient separation was employed [46]. Using two-dimensional separation, the spot capacity can be dramatically increased in fully off-line OPLC. This approach was used for the separation of 16 closely related coumarins from the genus *Angelica* [28].

In circular development the diameter of the circle around the center of the plate limits the number of samples that can be applied. D,L-alanine was separated on a chiralplate® with acetone–methanol–water (10:2:2) with 48 samples spotted on 5 cm, or 78 samples on 10.5 cm, diameter circle [47]. For high throughput screening, circular OPMLC was used for the semi-quantitative determination of the alkaloid content of poppy samples [30]. One hundred and forty-two samples and two standard mixtures containing the alkaloids at the limiting concentration were separated in a single run.

Off-line OPLC separation combined with digital autoradiography has been used for the study of ^{3}H- and ^{14}C-labeled deramciclane and its major and minor metabolites in pooled dog plasma, where, after separation they can be selectively detected and quantified in situ on the adsorbent layer [48]. In addition, the separated components localized on the adsorbent have been transferred to fast atom bombardment–mass spectrometry for structure elucidation [49]. An indirectly coupled OPLC-GC-MS system was used for the investigation of acetylenic thiophene derivatives in extracts of *Tagetes patula*. The fractions separated by off-line sample application and online separation and fraction collection were analyzed by GC-MS [50].

A fully online stepwise gradient FEW-OPLC separation and peak collection of red wine extracts was described [51]. OPLC and MS were directly connected online and the components eluted directed into the MS. This hyphenated system was used for the separation and electrospray mass spectrometric detection (ESI-MS) of glycolipids [51].

ESI-MS was hyphenated with fully online FEW-OPLC for the separation of xanthine compounds using the experimental OPLC 100 separation unit [52].

Direct coupling of OPLC to HPLC was used for the separation and identification of components in a tea leaf extract. The OPLC instrument was used as an interface for the combined system. Only components of interest were transferred to the HPLC system, while matrix compounds remained on the OPLC layer [53].

5.5.2 Analytical Separation and Detection with Direct Bioautography and BioArena

Planar chromatography (PLC) can be coupled with various effect-directed detection strategies for the relatively inexpensive and fast screening of many samples in parallel. The main advantage of this method is the possibility for the analysis of single components in various matrices. PLC was combined with several in situ tests for biodetection, for example, antimicrobial (direct bioautography, DB) [54], antioxidant (free radical or superoxide scavenging test), estrogenic (p-YES test) and different enzyme inhibition (e.g., anticholinesterase, antiglucosidase) activity as well as with immunostaining [55–57].

OPLC coupled with effect-directed detection is especially beneficial with forced-flow technique due to better separation and more compact spots that afford more selective and sensitive detection of biologically active compounds. OPLC-2,2-diphenyl-picrylhydrazyl (DPPH*) and OPLC-DB with various strains of microorganism were applied for the visualization of bioactivity. The purple DPPH* is a stable, free radical that is decolorized in the presence of antioxidant by electron or hydrogen atom transfer. So the radical trapping components on the developed layer dipped into methanol solution of DPPH* appear as bright spots against a purple background (Figure 5.8).

Bacteria, fungi, and yeast were introduced as test microorganisms in the OPLC-DB system. In the case of plant pathogenic *Colletotrichum* species, the antifungal components of *Scaligeria tripartite* essential oil and *Diospyros virginiana* root separated by OPLC were indicated by clear zones with little fungal growth [58,59]. The visualization of alive *Saccharomyces cerevisiae* [60] was performed by vital dye reagent (3-(4,5-dimethylthiazol-2-yl)-2,5-diphenyltetrazolium bromide, MTT). The yellow MTT is reduced to a bluish formazan by metabolically active cells; thereby the killing/inhibiting activity of copper ions was observed as bright zones against the bluish background. Generally the latter method is used for the visualization of the bioautogram prepared with bacteria, like *Bacillus subtilis, Pseudomonas savastanoi* pv. *phaseolicola,* and *Erwinia amylovora*. Aflatoxins, wine ingredients as well as

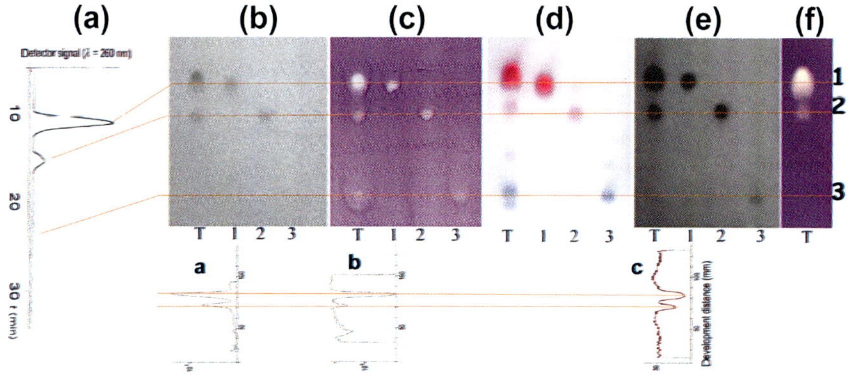

FIGURE 5.8 Isolation of antibacterial components of thyme essential oil (T) by OPLC with off-line sample application and online separation/detection and fractionation using TLC silica gel 60 and 1 ml/min flow rate of chloroform. Fractions 1–3 were collected according to UV signal (a) or prediction from R_F. The OPLC developed fractions were visualized under UV 254 nm (b) and after the use of vanillin–sulfuric acid reagent (d). Their antibacterial effect was tested against *Bacillus subtilis* (c) and luminescence gene-tagged *Pseudomonas syringae* pv. *maculicola* (d). (f) shows the result of the OPLC-DPPH* test. The isolated three components were identified by GC-MS as thymol (1), carvacrol (2), and linalool (3) and a–c represent the densitometric evaluation of the track T from plate (b), (c), and (e), respectively. OPLC, overpressured layer chromatography; TLC, normal particle adsorbent layer for thin-layer chromatography.

ascorbigen and methyl-ascorbigen separated by OPLC were tested against *P. savastanoi* pv. *phaseolicola* [61–63] and *E. amylovora* [64]; as well as the separated essential oil components of thyme and oregano against *B. subtilis* [65–67]. In the last decade, a new detection method was introduced in the DB exploiting bioluminescence. Using luminescent bacteria, the bioautogram is visualized by taking a photograph with a long exposure time. The antibacterial chamomile components separated by infusion OPLC were detected by naturally luminescent marine bacterium *Aliivibrio fischeri* [68], while thyme components were tested against luminescent gene-tagged plant pathogen *Pseudomonas syringae* pv. *maculicola* [67,69].

OPLC-DB is suitable for quantitative analyses as well. For this purpose, the inhibition zones of the stained bioautogram are evaluated with a densitometer (e.g., at $\lambda = 590$ nm after MTT dyeing) [70], or in the case of bioluminescence, by the intensity of the light emitted from the tracks [65].

The BioArena system [70–72] is an extension of DB that is appropriate for investigating the action mechanism of different endogenous and/or exogenous substances dissolved in a cell suspension for the antimicrobial effect of the separated components. This effect can be characterized with the tools applied for quantitative analyses previously described. BioArena also includes all qualitative technical solutions (spectroscopy and spectrometry), which make possible the determination of the chemical changes of the analytes in the presence of a microorganism. Based on OPLC-BioArena, it seems that formaldehyde and its reactive reaction products play a potential role in the antibacterial activity of resveratrol [72,73], aflatoxins [74,75], ascorbigens [63,64], as well as thyme and *Origanum* essential oil components [65,66].

5.5.3 Preparative Applications

The preparative OPLC isolation processes can be carried out in different modes. In the fully off-line mode, the isolation of separated zones can be achieved by scraping bands from the adsorbent layer followed by elution of the isolated materials, similar to conventional preparative layer chromatography. Phorbol diester constituents of croton oil were isolated by off-line OPLC separation followed by extraction and chemical ionization mass spectrometry for identification [76].

The online separation and fraction collection can be accomplished with or without online detection and the sample can be applied to the dry adsorbent layer as a band or it can be injected into the mobile phase. These solutions are more effective because time consuming scraping and elution can be eliminated. The loading for OPLC isolation can be modeled by fully off-line OPLC separation using different sample amounts (e.g., extract) and 10 mm band widths for sample applications. The optimum-loading capacity is selected from the chromatogram and can be estimated for the preparative adsorbent layers by taking into account the band size and layer thickness [16,23]. Isolation of hemp constituents was accomplished by off-line sample application with online OPLC elution and UV detection. The cannabinoid acid fractions were analyzed without further purification [77].

Off-line sample application combined with online OPLC separation on normal-phase layers with radioactivity detection (RD) was used for the isolation of urinary metabolites of a ^{14}C-labeled drug candidate [78]. The fractions collected by the RD signal were purified further by reversed phase HPLC.

Off-line sample application and online separation and fraction collection was used for the isolation of components from *Glehnia littoralis* roots using a 0.5 mm silica gel adsorbent layer and stepwise gradient elution at 5 MPa external pressure. Seven components were isolated (4–10 mg of each) in five runs for identification by ^{1}H and ^{13}C NMR spectroscopy and MS measurements [79]. Biologically active compounds of the plants (*Rhamnus frangula, Heracleum sphondylium, Gentiana purpurea*) were separated by online OPLC using 2-mm-thick layers and 50–500 mg loadings [12]. Alkaloids and glycosides were isolated from *Simaba multiflora,*

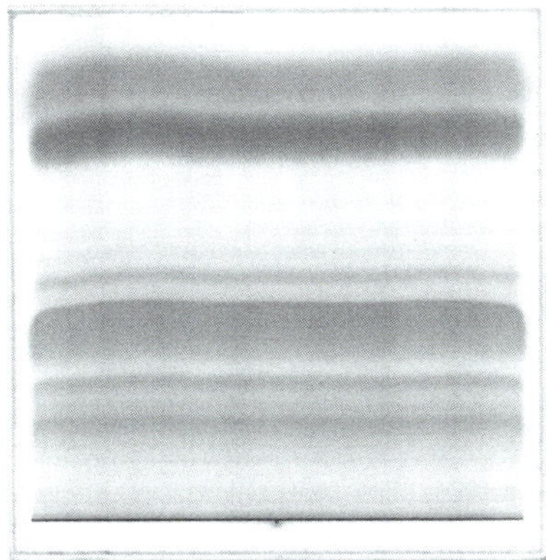

FIGURE 5.9 Chromatogram obtained from the separation of chamomile oil using an OPLC purification unit with one-channel FEW configuration at the inlet side, HPTLC silica gel 60 adsorbent layer (20 × 20 cm, Merck Co.), hexane–ethyl acetate (90:10, v/v; 1 ml/min) eluent and 22.5 mg/250 μL injection. Visualization by vanillin–sulfuric acid reagent. FEW, flowing eluent wall; OPLC, overpressured layer chromatography. *(Reproduced from Ref. [18], with permission.)*

Crossopteryx febrifuga, Streblus asper by means of fully online OPLC [80] and the effectiveness of long distance OPLC demonstrated for a raw extract from the roots of *Peucedanum palustre* [39].

A fully online FEW-OPLC system with one-channel configuration was used for the isolation of components of *Matricaria chamomilla* L. oil using an analytical silica gel layer and a 22.5 mg loading [18]. Figure 5.9 shows the chromatogram for isocratic development without overrun and visualized by vanillin–sulfuric acid at this high loading. As can be seen, the bands do not affect the sealed layer edges owing to the use of FEW at the injection side. Figure 5.10 shows the chromatogram detected online at 254 nm using a stepwise gradient and the same loading.

5.5.4 Activity-Guided Isolation

After off-line OPLC separation the adsorbent layer containing active compounds can be scraped off and eluted, as in the case of antibacterial chamomile components tested against *A. fischeri* by OPLC-DB [68]. The active ingredients were identified by solid phase microextraction (SPME)-GC-MS as *cis-* and *trans-*spiroethers, alpha-bisabolol, and herniarin. Chamomile flower extract was also fractionated by overrun stepwise gradient OPLC with online detection, where the fractions were collected according to their UV absorption [81,82]. Each fraction showed antibacterial effects against *B. subtilis* and/or *P. syringae* pv. *maculicola* and/or *A. fischeri* tested by TLC-DB and were further analyzed by SPME-GC-MS

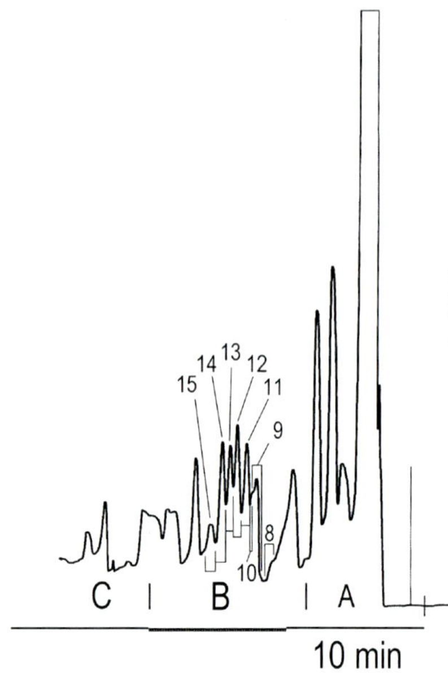

FIGURE 5.10 Fully online stepwise gradient isolation of chamomile oil on HPTLC silica gel 60 layer (20 × 20 cm, Merck Co.) sealed at the periphery by means of the OPLC purification unit with one-channel FEW configuration at the inlet. Mobile phase, hexane–ethyl acetate, A, 98:2; B, 92.5:7.5; C, 85:15; flow rate, 2 ml/min; injection, 22.5 mg/250 μL and detection at 254 nm. FEW, flowing eluent wall; OPLC, overpressured layer chromatography; HPTLC, fine particle adsorbent layer for high performance thin-layer chromatography. *(Reproduced from Ref. [18], with permission.)*

and/or LC-MS/MS. To scale-up the OPLC separation, a preparative layer with a 0.5 mm thickness was used [83]. In these studies chamomile oil components, coumarins, flavonoids, and flavonoid glucosides were characterized as antibacterial substances.

The isolation of antibacterial thyme essential oil components with activity against *B. subtilis* and *P. syringae* pv. *maculicola* was performed by OPLC on analytical TLC layers in an overloaded condition [67] (Figure 5.8). The UV active (at $\lambda = 260$ nm) thymol and carvacrol were collected based on the detector signal, while the retention time of the non-UV active linalool was calculated from its retention factor obtained from off-line OPLC.

The further OPLC purification of antifungal fractions obtained by column chromatography of acetone extract of *Diospyros virginiana* root was achieved on an analytical layer, resulting in the isolation of two new and nine known natural products possessing antifungal properties [59]. OPLC/TLC-DB was the biomonitoring system using the plant pathogenic *Colletotrichum* species.

5.6 CONCLUSION

OPLC, as one of the forced-flow techniques, provides a high flexibility for off-line and online connections with operating steps, such as sample application, separation, detection, and fraction collection. The fully off-line OPLC process is similar to the TLC/HPTLC technique and can be used for parallel separations. The fully online mode is analogous to HPLC. Contrary to conventional development in fully off-line OPLC, the plate height is practically constant along the bed even if viscous eluents are used and results in a lower plate height than TLC/HPTLC. Moreover, the differences are dramatic at longer development distances, especially for layers made from fine particles. The process of online separation, detection, and fraction collection combined with off-line or online sample application is suitable for rapid isolation of compounds from different origins. OPLC is suitable for direct and indirect coupling to other separation and/or spectroscopic techniques e.g., OPLC-DAD, OPLC-RD, OPLC-MS, OPLC-HPLC. The combination of an OPLC separation with DB helps to identify active components against microbes; in addition, the BioArena system aids the study of the interaction of microbes with endogenous and/or exogenous cofactor molecules in the adsorbent layer bed.

References

[1] Tyihák E, Mincsovics E, Kalász H. New planar liquid chromatographic technique: overpressured thin-layer chromatography. J Chromatogr 1979;174:75–81.

[2] Tyihák E, Held G. Thin-layer chromatography in pharmacognosy. In: Niederwieser A, Pataki G, editors. Progress in TLC and related techniques, vol. II. Ann Arbor (MI): Ann Arbor Science Publishers; 1971. pp. 183–234.

[3] Kalász H, Nagy J, Mincsovics E, Tyihák E. Circular development with overpressured thin-layer chromatography. J Liq Chromatogr 1980;3:845–90.

[4] Mincsovics E, Tyihák E, Kalász H. Resolution and retention behaviour of some dyes in overpressured thin-layer chromatography. J Chromatogr 1980;191:293–300.

[5] Tyihák E, Mincsovics E, Körmendi F. Overpressured layer chromatography I. Basic principles, instruments and developments. Hung Sci Instrum 1983;55:33–40.

[6] Tyihák E, Mincsovics E, Móricz ÁM. Overpressured layer chromatography: from the pressurized ultramicro chamber to BioArena system. J Chromatogr A 2012;1232:3–18.

[7] Beaver L, Guiochon G. System and apparatus for multi-dimensional real-time chromatography. US Patent 4,469,601.

[8] Guiochon G, Beaver L, Gonnord M, Siouffi A, Zakaria M. Theoretical investigation of the potentialities of the use of multidimensional column in chromatography. J Chromatogr 1983;255:415–37.

[9] Mincsovics E, Tyihák E. Combination of off-line and on-line operating steps in OPLC. J Planar Chromatogr 1988;1:309–12.

[10] Mincsovics E, Tyihák E, Siouffi AM. Characteristics of the one-dimensional on-line separation and detection in a modified CHROMPRES chamber. In: Tyihák E, editor. Proceedings of international symposium on TLC with special emphasis on overpressured layer chromatography (OPLC). Budapest: Labor MIM; 1986. pp. 251–64.

[11] Mincsovics E, Tyihák E, Siouffi AM. Comparison of off-line and on-line overpressured layer chromatography (OPLC). J Planar Chromatogr 1988;1:141–5.
[12] Nyiredy Sz, Erdelmeier CAJ, Dallenbach-Toelke K, Nyiredy-Mikita K, Sticher O. Preparative on-line overpressure layer chromatography (OPLC): a new separation technique for natural products. J Nat Prod 1986;49:885–91.
[13] Nyiredy Sz, Mészáros S, Dallenbach-Toelke K, Nyiredy-Mikita K, Sticher O. The "disturbing zone" in overpressure layer chromatography (OPLC). J High Resol Chromatogr Chromatogr Commun 1987;10:352–6.
[14] Mincsovics E, Tyihák E. Analytical and preparative overpressured layer chromatography. In: Dallas, editor. Recent advances in thin-layer chromatography. New York: Plenum Press; 1988. pp. 57–65.
[15] Mincsovics E. Efficiency characteristics of off-line infusion and conventional transfusion OPLC. In: Nyiredy Sz, Kakuk Á, editors. Proceedings of planar chromatography 2000. Budakalász (Hungary): Research Institute for Medicinal Plants; 2000. pp. 109–19.
[16] Mincsovics E, Sárdi É, Velich I, Kátay Gy, Tyihák E. Micro-preparative OPLC—rapid isolation by transfusion and infusion-transfusion processes. J Planar Chromatogr 2002;15:280–5.
[17] Mincsovics E, Manach M, Kecskés L, Tapa B, Papillard D, Tyihák E. Single- and multichannel OPLC separation on non-segmented sorbent bed using flowing eluent wall for operating segmentation. J Liq Chromatogr RT 2003;26:2611–27.
[18] Mincsovics E. Flowing eluent wall processes in OPLC: using segmentation of non-segmented adsorbent layer for single and parallel separations. J Planar Chromatogr 2004;17:411–9.
[19] Mincsovics E, Manach M, Kecskés L, Tapa B, Tyihák E. Apparatus for separating sample components by liquid chromatography under pressure. US Pub. No. 20040020834.
[20] Mincsovics E. Potential of the flowing eluent wall and aliquot on-line detection in the OPLC 50 and 100 Systems. J Planar Chromatogr 2008;21:97–102.
[21] Camenzuli M, Ritchie HJ, Ladine JR, Shalliker RA. The design of a new concept chromatography column. Analyst 2011;136:5127–30.
[22] Camenzuli M, Ritchie HJ, Ladine JR, Shalliker RA. Enhanced separation performance using a new column technology: parallel segmented outlet flow. J Chromatogr 2012; 1232:47–51.
[23] Mincsovics E, Garami M, Kecskés L, Tapa B, Végh Z, Kátay Gy, et al. Personal overpressured-layer chromatography (OPLC) basic system 50, flexible tool in analytical and semipreparative work. J AOAC Int 1999;82:587–98.
[24] Kaiser RE. Einführung in die HPPLC. Heidelberg: Hüthig; 1987.
[25] Witkiewicz Z, Mazurek M, Bladek J. A new instrument for overpressured layer chromatography. J Planar Chromatogr 1993;6:407–12.
[26] Flodberg G, Roeraade J. A device for high pressure thin layer chromatography. J Planar Chromatogr 1995;8:10–3.
[27] Abia JA, Putnam J, Mriziq K, Guiochon GA. Design and implementation of an array of micro-electrochemical detectors for two-dimensional liquid chromatography—Proof of principle. J Chromatogr A 2010;1217:1695–700.
[28] Härmälä P, Botz L, Sticher O, Hiltunen R. Two-dimensional planar chromatographic separation of a complex mixture of closely related coumarins from the genus *Angelica*. J Planar Chromatogr 1990;3:515–20.
[29] Tyihák E, Mincsovics E, Székely TJ. Overpressured multi-layer chromatography. J Chromatogr 1989;471:375–87.
[30] Szűcs Z, Szabady B, Szatmáry M, Cimpan G, Nyiredy Sz. High-throughput analytical strategy with combined planar and column liquid chromatography for improvement of the poppy (*Papaver somniferum* L.) with a high alkaloid content. Chromatographia 2002; 56(Suppl.):49–54.

[31] Botz L, Nyiredy S, Sticher O. The principles of long distance OPLC, a new multi-layer development technique. J Planar Chromatogr 1990;3:352–4.
[32] Nyiredy Sz. The bridge between TLC and HPLC: overpressured layer chromatography (OPLC). Trends Anal Chem 2001;20:91–101.
[33] Guiochon G, Siouffi A-M. Study of the performances of thin-layer chromatography. III. Flow velocity of the mobile phase. J Chromatogr Sci 1978;16:598–609.
[34] Tyihák E, Mincsovics E, Kalász H, Nagy J. Optimization of operating parameters in overpressured thin-layer chromatography. J Chromatogr 1981;211:45–51.
[35] Guiochon G, Siouffi AM. Study of the performances of thin layer chromatography II. Band broadening and plate height equation. J Chromatogr Sci 1978;16:470–81.
[36] Snyder LR, Kirkland JJ. Introduction to modern liquid chromatography. 2nd ed. New York: John Wiley; 1979.
[37] Tyihák E, Mincsovics E, Siouffi AM. Effect of the external pressure on the efficiency of off-line and on-line OPLC. J Planar Chromatogr 1990;3:121–5.
[38] Fernando WPN, Poole CF. Determination of kinetic parameters for precoated silica gel thin layer chromatography plates by forced flow development. J Planar Chromatogr 1991;4:278–87.
[39] Botz L, Nyiredy S, Sticher O. Applicability of long distance overpressured layer chromatography. J Planar Chromatogr 1991;4:115–22.
[40] Geiss F. Fundamentals of thin-layer chromatography, planar chromatography. Heidelberg: Hüthig; 1987.
[41] Tyihák E, Petro-Turza M, Kardos K, Mincsovics E. Behavior of the dinitrophenylhydrazones of saturated aldehydes and ketones in normal and reversed phase TLC and OPLC. J Planar Chromatogr 1992;5:376–82.
[42] Hauck HE, Jost W. Investigations and results obtained with overpressured thin-layer chromatography. J Chromatogr 1983;262:113–20.
[43] Guiochon G, Siouffi AM. Study of the performances of thin-layer chromatography. Spot capacity in thin-layer chromatography. J Chromatogr 1982;245:1–20.
[44] Katona Z, Vincze L, Végh Z, Trompler A, Ferenczi-Fodor K. Cleaning validation procedure eased by using overpressured layer chromatography. J Pharm Biomed Anal 2000; 22:349–53.
[45] Ferenczi-Fodor K, Mahó S, Pap-Sziklay S, Török I, Borka L. Purity testing of levonorgestrel/norgestrel bulk drug substances by personal OPLC. Pharmeuropa 1997;9:736–42.
[46] Kovács Á, Simon-Sarkadi L, Mincsovics E. Stepwise gradient separation and quantification of dansylated biogenic amines in vegetables using personal OPLC instrument. J Planar Chromatogr 1998;11:43–6.
[47] Nyiredy Sz, Dallenbach-Toelke K, Sticher O. Applicability of forced-flow planar chromatographic methods for the separation of enantiomers on chiralplate®. J Chromatogr A 1988;450:241–52.
[48] Szúnyog J, Mincsovics E, Hazai I, Klebovich I. A new tool in planar chromatography: combination of OPLC and DAR for fast separation and detection of metabolites in biological samples. Planar Chromatogr 1998;11:25–9.
[49] Ludányi K, Vékey K, Szúnyog J, Mincsovics E, Karancsi T, Újszászy K, et al. Application of overpressured layer chromatography combined with digital autoradiography and mass spectrometry in the study of deramciclane metabolism. J AOAC Int 1999;82: 231–8.
[50] Betti A, Lodi G, Bighi C, Chiorboli G, Coppi S. Use of overpressured layer chromatography and coupled OPLC-GC-MS for the analysis of acetylenic thiophene derivatives in extracts of *Tagetes patula*. J Planar Chromatogr 1994;7:301–4.
[51] Chai W, Leteux C, Lawson AM, Stoll MS. On-line overpressure thin-layer chromatographic separation and electrospray mass spectrometric detection of glycolipids. Anal Chem 2003;75:118–25.

[52] Mincsovics E, Pápai K, Ludányi K, Dávid ÁZ, Budai M, Antal I, et al. Fully on-line hyphenation of an experimental OPLC separation unit with diode-array detection and mass-spectrometry (OPLC–DAD–MS) for analysis of xanthine compounds. Planar Chromatogr 2008;21:361–6.
[53] Mincsovics E, Garami M, Tyihák E. Direct coupling of OPLC with HPLC: clean-up and separation. Planar Chromatogr 1991;4:299–303.
[54] Goodall RR, Levi AA. A micro-chromatographic method for the detection and approximate determination of the different penicillin in the mixture. Nature 1946; 158:675–6.
[55] Morlock G, Schwack W. Hyphenations in planar chromatography. J Chromatogr A 2010;1217:6600–9.
[56] Cieśla Ł, Staszek D, Hajnos M, Kowalska T, Waksmundzka-Hajnos M. Development of chromatographic and free radical scavenging activity fingerprints by thin-layer chromatography for selected *Salvia* species. Phytochem Anal 2010;22:59–65.
[57] Choma IM, Grzelak EM. Bioautography detection in thin-layer chromatography. J Chromatogr A 2011;1218:2684–91.
[58] Tabanca N, Demirci B, Baser KH, Mincsovics E, Khan SI, Jacob MR, et al. Characterization of volatile constituents of *Scaligeria tripartita* and studies on the antifungal activity against phytopathogenic fungi. J Chromatogr B 2007;850:221–9.
[59] Wang X, Habib E, León F, Radwan MM, Tabanca N, Gao J, et al. Antifungal metabolites from the roots of *Diospyros virginiana* by overpressure layer chromatography. Chem Biodivers 2011;8:2331–40.
[60] Tyihák E, Móricz MÁ, Ott PG. Biodetection and determination of biological activity of natural products. In: Waksmundzka-Hajnos, editor. Thin layer chromatography in phytochemistry. Boca Raton: CRC Press; 2008. pp. 193–214.
[61] Móricz Á, Otta KH, Ott P, Tyihák E. Separation and detection of aflatoxins using overpressured-layer chromatography and bioautography. J Planar Chromatogr 2003; 16:417–20.
[62] Király-Véghely Z, Móricz ÁM, Ott PG, Kátay Gy, Bélai I, Tyihák E. Comparison of components from red and white wines for antimicrobial activity by biodetection after OPLC separation. J Liq Chromatogr RT 2009;32:1259–72.
[63] Kátay Gy, Ott PG, Kátay E, Magyar D, Tyihák E. Potential role of formaldehyde in the mechanism of action of ascorbigens on the basis of BioArena studies. Biomed Chromatogr 2009;23:412–8.
[64] Hevesi M, Móricz ÁM, Király-Véghely Z, Tóth M, Kátay G, Tyihák E. Effect of trans-resveratrol and ascorbigens on the fire blight pathogen *Erwinia amylovora* in the BioArena system. J Planar Chromatogr 2010;23:411–4.
[65] Móricz ÁM, Horváth G, Molnár P, Kocsis B, Böszörményi A, Lemberkovics É, et al. Investigation of thyme (*Thymus vulgaris* L.) essential oil in BioArena system. J Planar Chromatogr 2010;23:406–10.
[66] Mincsovics E, Tabanca N, Móricz ÁM, Wedge DE, Tyihák E. Preliminary investigation of *Origanum onites* essential oil by overpressured layer chromatography and BioArena. J Planar Chromatogr 2010;23:225–6.
[67] Móricz AM, Ott PG, Böszörményi A, Lemberkovics E, Mincsovics E, Tyihák E. Bioassay-guided isolation and identification of antimicrobial compounds from thyme essential oil by means of overpressured layer chromatography, bioautography and GC-MS. Chromatographia 2012;75:1–9.
[68] Móricz ÁM, Szarka S, Ott P, Kertesy D, Héthelyi ÉB, Szőke É, et al. Application of direct bioautography and SPME-GC-MS for the study of antibacterial chamomile ingredients. J Planar Chromatogr 2012;25:220–4.

[69] Móricz ÁM, Tyihák E, Ott PG. Usefulness of transgenic luminescent bacteria in direct bioautographic investigation of chamomile extracts. J Planar Chromatogr 2010;23: 180–3.
[70] Móricz ÁM, Adányi N, Horváth E, Ott PG, Tyihák E. Applicability of the BioArena system to investigation of the mechanisms of biological effects. J Planar Chromatogr 2008; 21:417–22.
[71] Tyihák E, Botz L, Ott P, Nagy S, Kocsis B, Kiraly-Veghely Z, et al. The combination of the overpressured layer chromatography and bioautography and its applications to the analysis of molecules influencing cell proliferation. Chem Anal Wars 2003;48:543–53.
[72] Tyihák E, Ott P, Móricz Á, Kátay Gy, Király-Véghely Zs. Antibiosis, antibiotics, and the formaldehyde cycle: unique importance of planar chromatographic techniques to progress in this direction. J Planar Chromatogr 2004;17:84–8.
[73] Tyihák E, Móricz ÁM, Ott PG, Kátay Gy, Király-Véghely Zs. The potential of BioArena in the study of the formaldehydome. J Planar Chromatogr 2005;18:66–71.
[74] Móricz ÁM, Ott PG, Billes F, Otta KH, Tyihák E. The influence of L-ascorbic acid on the antibacterial – toxic activity of aflatoxins on adsorbent layer. J Appl Microbiol 2007;103: 2525–32.
[75] Móricz AM, Ott PG, Otta KH, Tyihák E. Investigation of the mechanism of antibacterial effect of aflatoxins in BioArena system. Nat Prod Commun 2011;6:657–60.
[76] Erdelmeier CA, Van Leeuwen PA, Kinghorn AD. Phorbol diester constituents of croton oil: separation by two-dimensional TLC and rapid purification utilizing reversed-phase overpressure layer chromatography. Planta Medica 1988;54:71–5.
[77] Oroszlán P, Verzár-Petri G, Mincsovics E, Székely T. Separation, quantitation and isolation of cannabinoids from *Cannabis sativa* L. by overpressured layer chromatography. J Chromatogr 1987;388:217–24.
[78] Mincsovics E, Kiss BD, Morovjan G, Klebovich I. A new tool in metabolism research—combination of OPLC on-line radioactivity detection with HPLC-RD technique. J Planar Chromatogr 2001;14:312–7.
[79] Su X, Li X, Tao H, Zhou J, Wu T, Chou G, et al. Simultaneous isolation of seven compounds from *Glehnia littoralis* roots by off-line overpressured layer chromatography guided by a TLC antioxidant autographic assay. J Sep Sci 2013;36:3644–50.
[80] Erdelmeier CA, Kinghorn AD, Farnsworth NR. On-plate injection in the preparative separation of alkaloids and glycosides. J Chromatogr 1987;389:345–9.
[81] Móricz ÁM, Szarka S, Ott PG, Héthelyi ÉB, Szőke É, Tyihák E. Separation and identification of antibacterial chamomile components using OPLC, bioautography and GC-MS. Med Chem 2012;8:85–94.
[82] Mincsovics E, Ott PG, Alberti Á, Böszörményi A, Héthelyi ÉB, Szőke É, et al. In-situ clean-up and OPLC fractionation of chamomile flower extract searching active components by bioautography. J Planar Chromatogr 2013;26:172–9.
[83] Móricz ÁM, Ott PG, Alberti Á, Böszörményi A, Lemberkovics É, Szőke É, et al. Applicability of preparative overpressured layer chromatography and direct bioautography in search of antibacterial chamomile compounds. J AOAC Int 2013;96:1214–21.

CHAPTER

6

Pressurized Planar Electrochromatography

Aneta Hałka-Grysińska, Radosław Ł. Gwarda, Tadeusz H. Dzido

Department of Physical Chemistry, Medical University of Lublin, Lublin, Poland

6.1 INTRODUCTION

Pressurized planar electrochromatography (PPEC) is a forced-flow layer liquid chromatography technique in which an electric field is used to drive the flow of the mobile phase through a porous media (electroosmotic flow). Forcing the flow of mobile phase by electric field provides a number of important advantages of PPEC over conventional thin-layer chromatography (TLC). Among these advantages, we can distinguish high efficiency, short separation times, and different separation selectivities.

TLC with electroosmotic flow generated by an electric field was first described by Pretorius et al., in 1974. Due to the significantly shorter time to develop a chromatogram, they called the technique high-speed thin-layer chromatography (HSTLC) [1]. The current name planar electrochromatography (PEC) was introduced by Pukl et al., 20 years later [2]. To obtain more reproducible results, the PEC technique was improved by Nurok et al., in 2004 [3]. They applied external pressure to the adsorbent layer via a special foil and the technique was renamed as PPEC. Since then PPEC has been developed intensively. A recent review on this relatively new separation technique was published by our group in 2011 [4]. The theoretical background, development, applications, constructional details, and principle of action of PPEC devices were described therein. In this chapter, our earlier review is broadened and updated. We describe also a new achievement, reported during the last three years, e.g., online

Instrumental Thin-Layer Chromatography
http://dx.doi.org/10.1016/B978-0-12-417223-4.00006-6

sample application and orthogonal simultaneous electrophoresis/electrochromatography, and chromatography (orthogonal pressurized planar electrochromatography, OPPEC).

6.2 THEORETICAL BACKGROUND

PPEC is a separation technique similar to capillary electrochromatography (CEC) but performed in a thin layer of adsorbent, which is covered with a special foil under pressure, and in this way provides a closed separation system. It is a combination of two independent techniques: electrophoresis and chromatography. It should be mentioned, that the flow of mobile phase relative to a stationary phase, is controlled by an electric field and originates from the electroosmotic effect.

Electroosmosis occurs due to an electrical double layer present in the boundary between the surface of the stationary phase and the mobile phase. Movement of ions in the layer drags solvent molecules against the pneumatic resistance of the stationary phase and generates the mobile phase flow. The phenomenon of electroosmosis can be easily explained by using a fused-silica capillary filled with an aqueous solution as a model. At pH higher than 1.5, the silica surface is negatively charged due to the dissociation of silanol groups:

$$\equiv \mathrm{SiOH} \leftrightarrow \equiv \mathrm{SiO}^- + \mathrm{H}^+$$

Negative charge on silica is balanced by positive ions present in solution near the solid surface, and both charged zones form an electrical double layer. The electrical double layer consists of two parts according to a model suggested by Stern (Figure 6.1). The first part of the layer

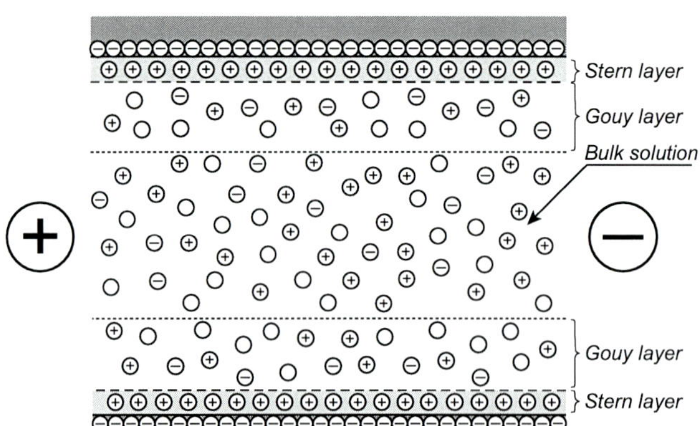

FIGURE 6.1 Scheme of electrical double layer in silica capillary. *Reproduced with permission from Ref. [4].*

(Stern layer), located adjacent to the surface of the solid phase, has a dimension approximately equal to the diameter of single ion and contains a high concentration of positive ions. The second layer (Gouy layer) extended into the bulk solution, and consists of the diffusion layer in which the concentration of positive ions decreases with the distance from the silica surface. The potential drops significantly in the first part of the double layer and then decreases monotonously in the second layer until the bulk solution is reached, where the uniform distribution of cations and anions take place (Figure 6.2). When an electric field is applied, the excess of cations in the diffusion layer moves toward the cathode, causing flow of the bulk solution present in the capillary. Such flow is characterized by a flat profile, resulting from the similar value of stream flow across the entire diameter of the capillary (Figure 6.3(a)). This is contrary to laminar flow (resulting from hydrodynamic force) characterized by a parabolic flow profile intrinsic to packed column liquid chromatography (Figure 6.3(b)). The flat electroosmotic flow profile is advantageous for limiting band broadening and therefore for the separation efficiency. This results in a higher plate number for electrochromatography systems compared with column/capillary liquid chromatography.

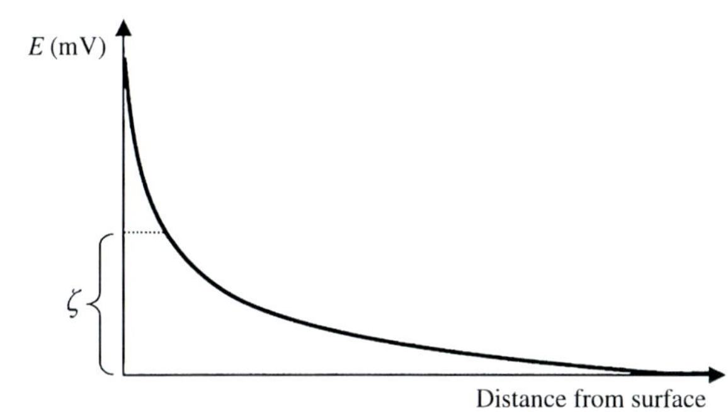

FIGURE 6.2 Potential drop in stationary–mobile phase interface. *Reproduced with permission from Ref. [4].*

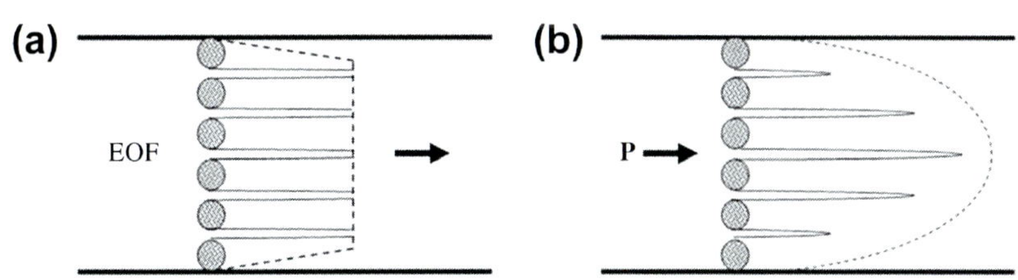

FIGURE 6.3 Flow profiles. (a) Electroosmotic flow and (b) laminar (hydrodynamic) flow. *Reproduced with permission from Ref. [4].*

One of the main parameters characterizing electroosmotic flow is the zeta potential, ζ. It reflects the potential drop in the diffusion part of the electrical double layer, and influences the linear electroosmotic flow velocity of the mobile phase, ν_{eo}, expressed by the Smoluchowski equation:

$$v_{eo} = \frac{\varepsilon \zeta E}{\eta} \tag{6.1}$$

According to this equation, the electroosmotic flow velocity depends on the mobile phase dielectric constant, ε; the applied electric field, E; and the viscosity of the solvent, η. It is independent of capillary parameters, such as its diameter, length, and the diameter of stationary phase particles. It is constant over the entire length of the capillary. Moreover, no pump is necessary to force the flow of the mobile phase—only a potential gradient.

Equation (6.1) is reliable when the capillary diameter is significantly greater than the thickness of the electrical double layer. In practice, however, stationary phase particles used in electrochromatography form channels of various diameters. This can lead to overlap of the electrical double layer in narrow capillaries, and therefore, to disturbance of the flat electroosmotic flow profile, reduction of the flow, and a decrease of the separation efficiency. It is particularly important for stationary phases with a broad particle size distribution [5,6]. The adverse effect of a broad particle size distribution was confirmed by Nurok et al. for PEC [7].

On the other hand, in conventional TLC the flow velocity, ν_{tlc}, decreases with the development distance, according to the following relation:

$$v_{tlc} = 0.5\kappa (Z_f)^{-1} \tag{6.2}$$

Other parameters in Eqn (6.2) are Z_f, the distance of the solvent front migration from the solvent entry position and κ, the velocity constant. The decrease of the mobile phase flow velocity during development is one of the main disadvantages of TLC, because the flow cannot be optimized and strongly depends on the distance of solvent front migration. Use of electroosmosis to drive the mobile phase flow often resolves this vital issue.

Contrary to high-performance liquid chromatography (HPLC) and overpressured layer chromatography (OPLC), there is no pressure drop when the mobile phase flow is driven by an electric field. In a system with hydrodynamic flow, the mobile phase velocity, v_c, is expressed by:

$$v_c = \frac{d_p^2 \Delta P}{\theta \eta L} \tag{6.3}$$

Parameters in Eqn (6.3) are as follows: θ, flow resistance parameter; ΔP, inlet pressure; d_p, average particle size; η, mobile phase viscosity, and L, the column length.

As discussed above, the movement of charged analyte zones results from not only the electroosmotic flow of the mobile phase but also from its electrophoretic mobility. The electrophoretic flow velocity, v_{ep}, may be expressed by an equation similar to that of the electroosmotic flow velocity:

$$v_{ep} = \frac{\varepsilon \zeta E a}{\eta} \tag{6.4}$$

The parameter a represents the size and shape of an analyte and the rest of the parameters are as of Eqn (6.1). For an easier comparison of data obtained with different electric fields, E, the electroosmotic mobility, μ_{eo} and electrophoretic mobility, μ_{ep}, are introduced:

$$\mu_{eo} = \frac{v_{eo}}{E} \tag{6.5}$$

and

$$\mu_{ep} = \frac{v_{ep}}{E} \tag{6.6}$$

Usually all species of the solution—cation, anion, and neutral solutes—migrate toward the cathode, due to the larger electroosmotic mobility compared with the electrophoretic velocity. The velocity of each type of solute is expressed by the following equations:

$$\text{For neutral solutes}: v = \mu_{eo} E \tag{6.7}$$

$$\text{For cations}: v_{+} = \left(\mu_{eo} + \mu_{ep+}\right) E \tag{6.8}$$

$$\text{For anions}: v_{-} = \left(\mu_{eo} - \mu_{ep-}\right) E \tag{6.9}$$

Thus, the overall mobility of cations is the highest, the mobility of neutral solutes is intermediate, and the mobility of anions is the lowest. However, in electrochromatography the order of solute migration may be significantly changed due to molecular interactions with the stationary and mobile phase components that provide an additional contribution to separation selectivity (solute partition between mobile and stationary phases). Therefore, in comparison to chromatography and electrophoresis, electrochromatographic systems offer new, additional mechanisms of separation, which can be exploited and optimized for particular separations.

The new type of PPEC is OPPEC, which is a combination of PPEC and OPLC techniques [8]. In this case, both processes (electrophoresis/electrochromatography and chromatography) are performed simultaneously and orthogonally (Figure 6.4). It should be mentioned that OPPEC may be used in two ways: analytical and preparative modes. The analytical mode

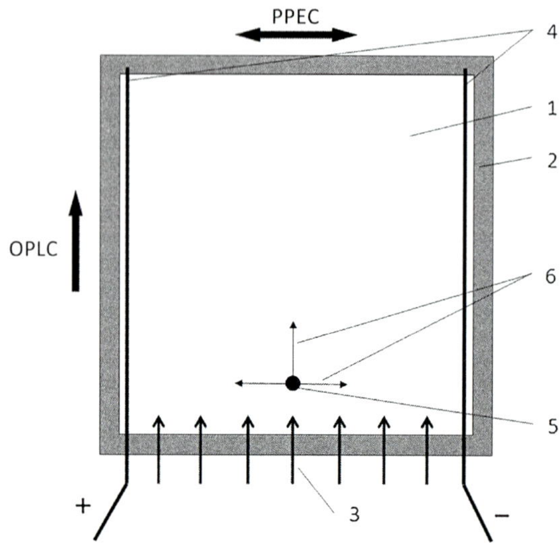

FIGURE 6.4 Orthogonal pressurized planar electrochromatography(OPPEC) as orthogonal combination of overpressured layer chromatography (OPLC) and pressurized planar electrochromatography (PPEC) into single continuous process; (1) chromatographic plate, (2) sealed margins of adsorbent layer, (3) mobile phase pumping, (4) electrodes, (5) position of sample application, (6) directions of solutes' migration.

provides additional separation potential due to the possibility of simultaneous two-dimensional separations based on the two independent techniques. This approach allows faster separations because the application of two consecutive PPEC and OPLC techniques in conventional two-dimensional separation are replaced with one in which PPEC and OPLC proceed orthogonally and simultaneously. In the preparative mode, electrophoresis and partition (involved in PPEC) and partition (involved in OPLC) mechanisms are responsible for the separation of samples. Hydrodynamic mobile phase flow of OPLC elutes the bands from the adsorbent layer enabling continuous recovery of solutes from the separation system.

6.3 DEVELOPMENT OF EQUIPMENT FOR PPEC

The first PPEC device was presented by Nurok et al., in 2004 [3], and soon after that our group presented an alternative device [9]. Since then, the research on the development of prototype PPEC devices has been carried out by three independent groups [3,8–17].

The device by Nurok et al. enables PPEC to be performed in the vertical mode. It contains a prewetted chromatographic plate with samples applied to it mounted vertically in a pressurized chamber using a polyacetal frame. The sorbent layer of the chromatographic plate is covered with a Teflon membrane and aluminum nitride ceramic sheet. These elements are pressurized between two die blocks using a hydraulic press. One edge of the chromatographic plate and one of two electrodes are dipped in the solvent mixture in the mobile phase reservoir. The second electrode is attached to the upper edge of the sorbent layer together with the paper wick, which accumulates excess liquid migrating to the top of the plate during the separation process (after application of the polarization voltage). An exploded view of the device is shown in Figure 6.5. In a subsequent report, Nurok et al. introduced a cooling system for the PPEC device, which comprised the die blocks with channels for circulating water [10]. This improvement is advantageous due to the possibility of temperature control of the separating system. This is important in the case of excessive Joule heat generation. The device is shown in Figure 6.6.

Another PPEC device was presented by Dorsey's group [11,12]. It was designed to facilitate investigations of the equilibration of the separating system and to monitor the electric potential drop across the chromatographic plate in the horizontal mode. The device consists of a plastic block (Kel-F) equipped with a reader electrode grid (32 platinum electrodes), covering the sorbent layer of the chromatographic plate and uses a hydraulic press to seal the layer from the atmosphere, Figure 6.7.

Our group adapted a commercially available horizontal DS (Dzido, Soczewiński) chamber for TLC to PPEC for the first time [9]. In this device, the sorbent layer of the chromatographic plate is covered with a Teflon foil

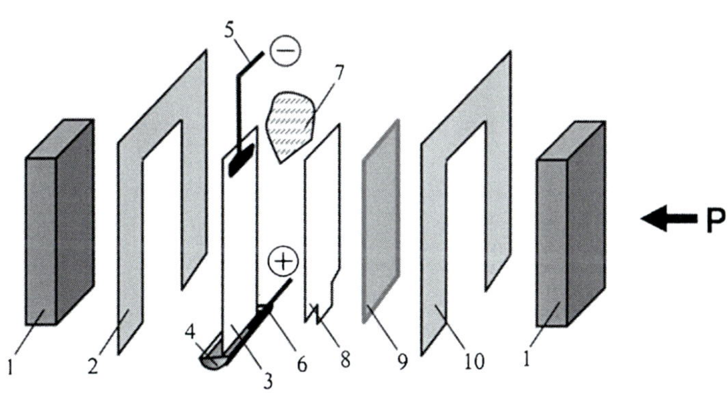

FIGURE 6.5 Exploded view of the elements of pressurized planar electrochromatography (PPEC) chamber by Nurok et al.: (1) metal die block, (2) frame (first part) for the chromatographic plate, (3) chromatographic plate, (4) mobile phase, (5) cathode, (6) anode, (7) paper wick, (8) Teflon foil, (9) ceramic sheet, and (10) frame (second part) for the chromatographic plate. *Adapted from Ref. [3] and reproduced with permission from Ref. [4].*

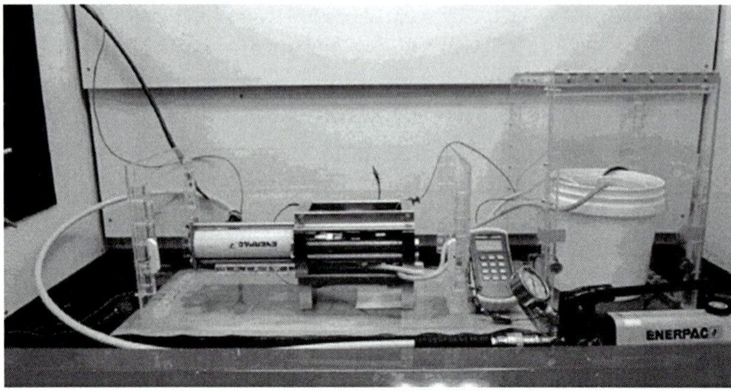

FIGURE 6.6 Photograph of the pressurized planar electrochromatography (PPEC) device by Nurok et al. *Reproduced with permission from Ref. [10].*

FIGURE 6.7 Photograph of the pressurized planar electrochromatography (PPEC) device by Tate and Dorsey. *Reproduced with permission from Ref. [12].*

maintained in contact with the layer by a bolted cover. However, some part of the chromatographic plate extends beyond the cover. This may allow some evaporation of the mobile phase from the plate. Therefore, in a subsequent device [13], the chromatographic plate was completely covered and pressurized. Moreover, this arrangement enables simple and precise monitoring of the electroosmotic flow velocity of the mobile phase through the sorbent layer, which is advantageous for investigating the relationship between retention and mobile phase composition and between mobile phase velocity and kinetic performance. This device was further modified [14] by construction of new electrode cells. This afforded a solution to the problem of air bubble formation in the electrode compartments during electrochromatography. Additionally, bolts used to pressurize the chromatographic plate were replaced by a hydraulic press. A schematic view of this device is presented in Figure 6.8.

A major disadvantage of all the devices described so far is the necessity for the manual transfer of the prewetted chromatographic plate from a prewetting vessel to a PPEC chamber/device. This operation may result in evaporation of the mobile phase, and therefore, in the disturbance of the separation system equilibrium. Lower repeatability of the results (especially migration distance) is a consequence of this operation. To solve this problem, our group has developed another device [15] which supports prewetting of the adsorbent layer and development of the chromatogram in a single chamber. The samples are applied to the dry plate

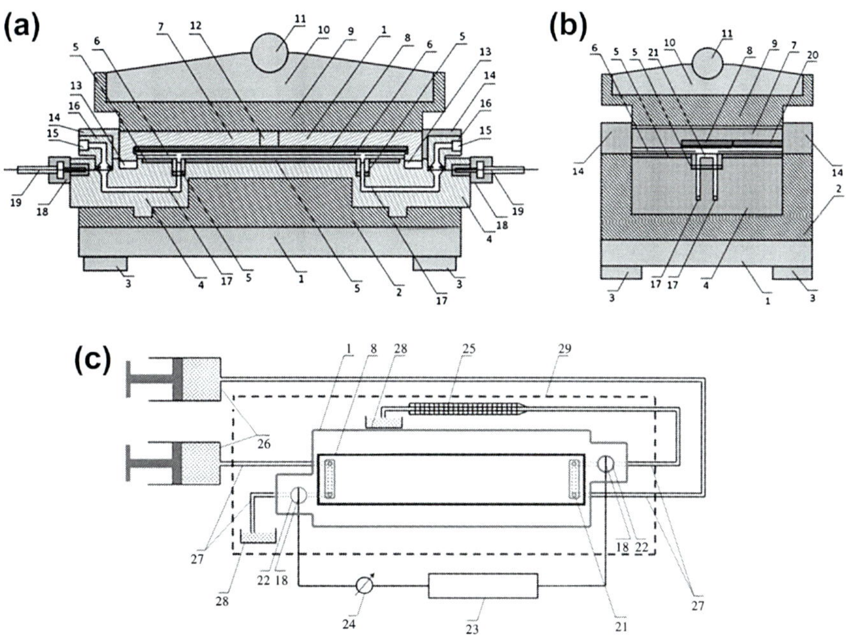

FIGURE 6.8 Schematic view of the device for pressurized planar electrochromatography (PPEC): (a) longitudinal section of the chamber for PPEC, (b) cross-section of the chamber for PPEC, and (c) conceptual view of the complete device for PPEC. (1) Steel base plate, (2) Tarnoform body of the chamber, (3) wheel legs, (4) Teflon block, (5) silicone gasket, (6) Teflon foil, (7) the adaptor of the chromatographic plate, (8) chromatographic plate, (9) polyacetal body of the lid, (10) steel body of the lid, (11) steel ball, (12) the hole for thermocouple, (13) the recess for the adaptor of the chromatographic plate, (14) electrode cell plug, (15) mobile phase in/out, (16) channel unit, (17) channel for the mobile phase solution, (18) electrode housing, (19) electrode, (20) supplement glass plate, (21) trough, (22) electrode cell, (23) high-voltage power supply DC, (24) ammeter, (25) micro-pipette, (26) reservoir, (27) Teflon tube, (28) waste, and (29) the dashed rectangle represents a Plexiglas cabinet, which stands as a safety interlock. *Reproduced with permission from Ref. [14].*

and covered with a partition block during the prewetting process of the adsorbent layer. During the prewetting process, two solvent fronts migrating in opposite directions under the partition block meet at the sample application area. This results in sharper zones (just in the middle lower part of the partition block). The device has one disadvantage—before covering the adsorbent layer with the cover, the prewetting solution has to be removed from the three compartments formed by the partitions and the PPEC chamber body walls. This weak point has been overcome, with the design of a single partition block (Figure 6.9) [16]. During the movement of the cover down the chromatographic plate, the mobile phase solution is automatically transferred to the electrode compartment/reservoir.

Despite the improvements discussed above, traditional methods of sample application in PPEC result in some disadvantages and inconveniences, from the necessity of prewetting the sorbent layer. This mainly concerns zone dispersion and tedious manual operations. Moreover, these operations are usually time consuming. Therefore, a new PPEC modification was introduced based on the last prototype described above [16,17]. The new device includes an online sample application system. In this case, the partition of the PPEC chamber is equipped with six Teflon tubes connected to mobile phase reservoirs and syringes. The sorbent layer of the chromatographic plate contains special grooves, exactly located in the places where online application lines of the partition contact the plate. After the prewetting process, the cover is pressed against the sorbent layer, syringes combined with a stepping motor aspirate a mobile phase from the reservoirs through the tubes, filling the grooves under the partition. Next, an electric potential is applied to the electrodes and the electroosmosis flow starts up. Then, sample solutions are injected via the tubes by the microsyringes. Injected sample zones migrate through the sorbent grooves perpendicular to the electroosmotic flow of the mobile phase, and some portions of the samples are introduced by electroosmosis and electrophoresis into the sorbent layer. The remaining sample solutions are withdrawn by the syringes and the sorbent layer groves filled with the mobile phase. The separation process then proceeds as usual. The device enables simultaneous application of six samples during an electroosmosis run. Its main advantages are shortening of the separation procedure and elimination of excessive dispersion of the sample starting zone. Moreover, sample application and separation proceed in a fully equilibrated system. Additionally, the device is equipped with a cooling/heating system, which enables precise temperature control during the separation process. A schematic view of the device is shown in Figure 6.10.

The first efficient device for OPPEC was described in 2014 [8]. In this device, electrodes are placed along the direction of mobile phase flow

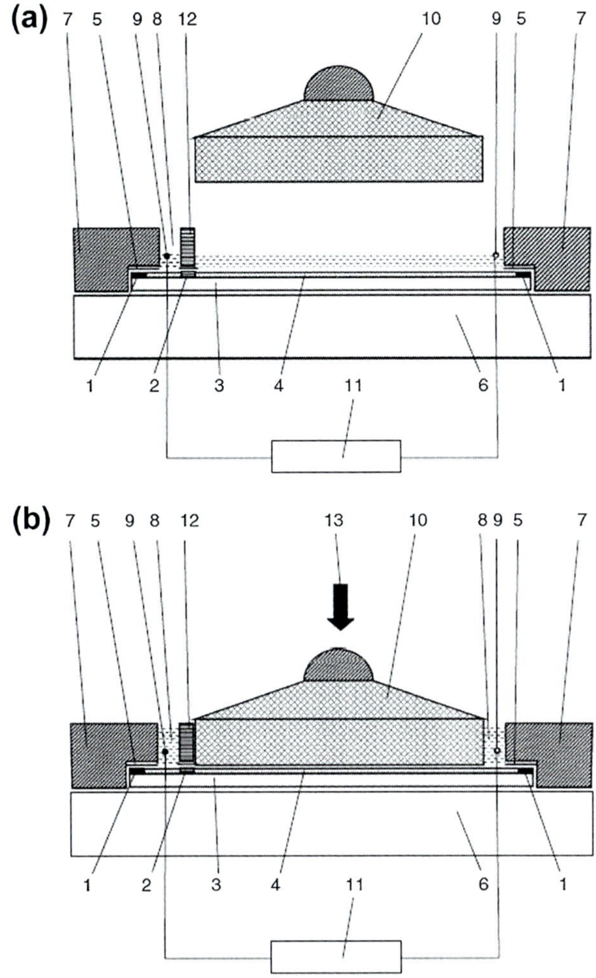

FIGURE 6.9 Schematic view of the device for pressurized planar electrochromatography (PPEC): (a) schematic view of the PPEC chamber during prewetting of an adsorbent layer; (b) schematic view of the PPEC chamber during the separation process. (1) Sealant margin on the adsorbent layer, (2) substance zone or groove in the case of online sample application/injection, (3) chromatographic plate, (4) adsorbent layer, (5) Teflon foil, (6) base, (7) frame of the chamber, (8) electrode compartment with the mobile phase solution, (9) electrodes, (10) cover, (11) high-voltage power supply, (12) partition, (13) external hydraulic press. *Reproduced with permission from Ref. [16].*

(OPLC direction), driven by hydrodynamic pressure originating from a syringe pump. The direction of the electric field (PPEC direction) is orthogonal relative to the mobile phase flow direction. This means that two different mechanisms of separation act in the same time, but in orthogonal directions. To start the separation process the sorbent layer

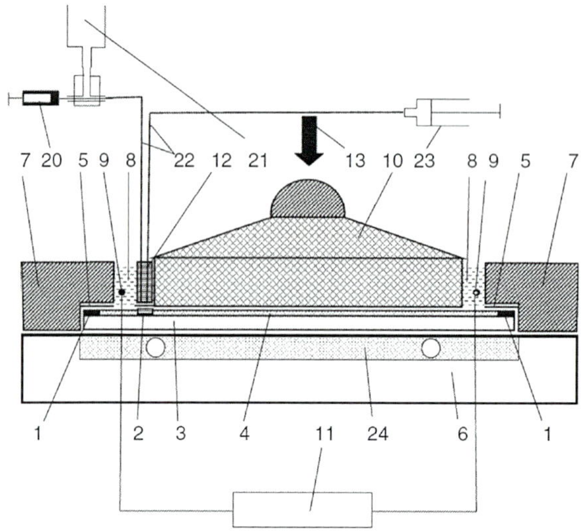

FIGURE 6.10 Schematic view of the pressurized planar electrochromatography (PPEC) device with online sample application system: (1) sealant margin on the adsorbent layer, (2) substance zone or groove in the case of online sample application, (3) chromatographic plate, (4) adsorbent layer, (5) Teflon foil, (6) base, (7) frame of the chamber, (8) electrode compartment with the mobile phase, (9) electrodes, (10) cover, (11) high-voltage power supply, (12) partition, (13) external hydraulic press, (20) microsyringe, (21) reservoir, (22) Teflon tube, (23) syringe, (24) cooling plate. *Reproduced with permission from Ref. [16].*

must be prewetted and the electrodes dipped into the mobile phase. Consequently, at the beginning all electrode compartments have to be filled with mobile phase and the sorbent layer prewetted with mobile phase supplied by a syringe pump. After that, the power supply is turned on. Just then sample is applied/injected into the separation system. The device is equipped with an online sample application system using a microsyringe, microinjection valve, or syringe pump. In addition, it is equipped with a temperature control system and pressure bladder filled with hydraulic liquid, which provides a uniform distribution of pressure over the entire sorbent layer of the chromatographic plate. A schematic view of the cross-section of the OPPEC chamber is shown in Figure 6.11 and a photograph of the device is shown in Figure 6.12 [8].

The OPPEC device may work in two separation modes: analytical and preparative (continuous) [8]. In the analytical mode, a sample solution is introduced into the OPPEC system as a specific volume and the separation process is terminated after a defined volume of a mobile phase flows through the sorbent layer. Eluate from the chamber is collected as a waste. This mode enables fast two-dimensional separations with quite different selectivities (resulting from the different separation mechanisms) in each direction. A weak point of this mode is the small sample volume which

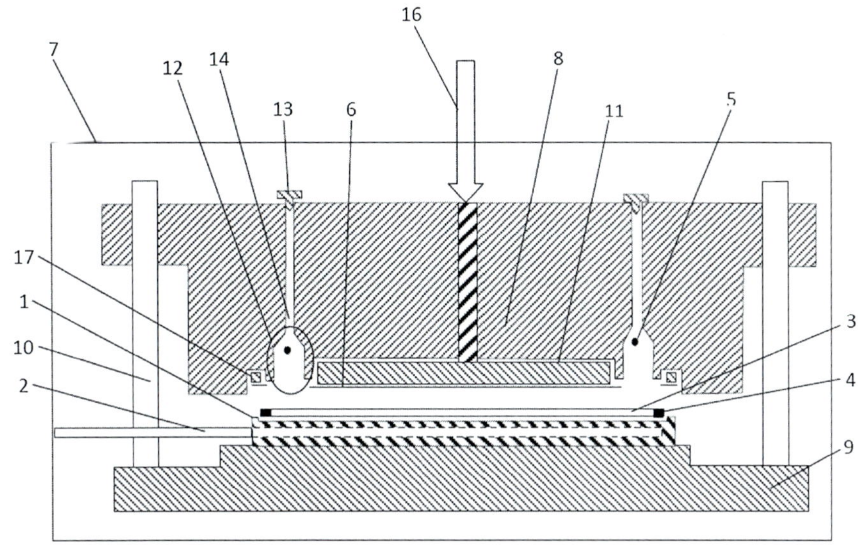

FIGURE 6.11 Schematic view of cross-section of the orthogonal pressurized planar electrochromatography (OPPEC) chamber: (1) removable cooling/heating frame, (2) channels for cooling/heating liquid, (3) chromatographic plate, (4) special sealing margin, (5) electrode, (6) Teflon foil, (7) security cabined, (8) cover, (9) base, (10) screw, (11) silicone bladder, (12) electrode compartment, (13) shut-off valve, (14) channels, (16) hydraulic press, (17) silicone rubber. *Reproduced with permission from Ref. [8].*

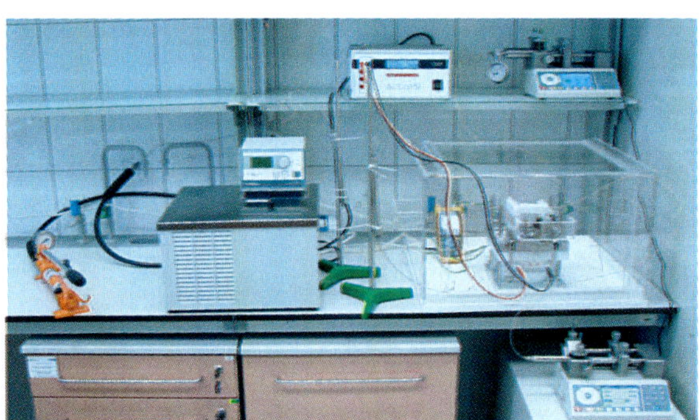

FIGURE 6.12 Picture of the complete device for orthogonal pressurized planar electrochromatography (OPPEC) with full equipment. *Reproduced with permission from Ref. [8].*

can be applied. In the preparative mode, the separation system is continuously fed with sample solution (e.g., by syringe pump). Eluate from the chamber is collected at the end of the chromatographic plate by microtubes and delivered to the fraction collector. A schematic view of a preparative separation is shown in Figure 6.13. The preparative mode

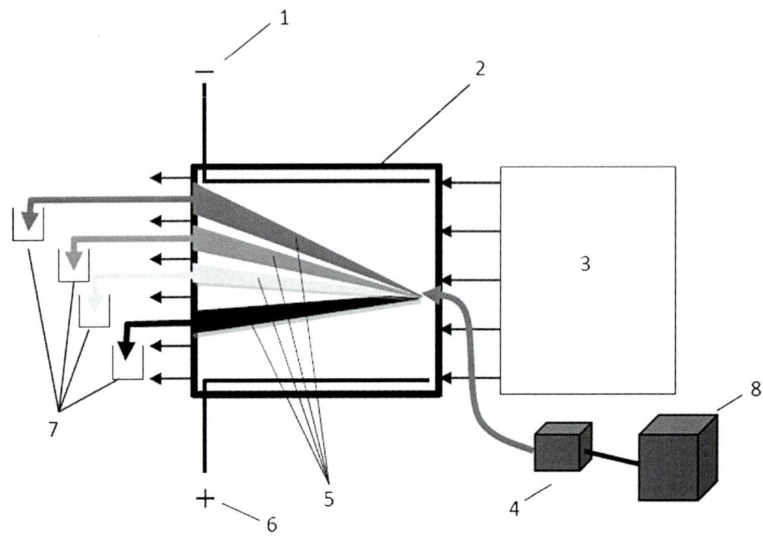

FIGURE 6.13 Conceptual view of the (orthogonal pressurized planar electrochromatography) OPPEC device by Dzido et al. for application to preparative separation: (1) cathode, (2) chromatographic plate, (3) mobile phase pump, (4) sample injector, (5) solute bands, (6) anode, (7) reservoirs of a fraction collector, (8) sample syringe pump. *Reproduced with permission from Ref. [8].*

allows two-dimensional separations with continuous collection of the eluate at the outlet of the OPPEC device.

6.4 ADVANTAGES OF PPEC

6.4.1 Kinetic Performance of PPEC Systems

The kinetic performance of a chromatographic system is usually characterized by its plate number and/or plate height. Both parameters reflect the ability of a chromatographic system to produce narrow solute zones. The performance of a planar chromatographic system strongly depends on a few variables, e.g., mobile phase flow velocity, diameter of starting solute band, the average diameter of the sorbent particles, temperature, etc. It was established that the narrow zones, i.e., the lowest plate height, can be obtained in capillary flow controlled systems when the mobile phase front has migrated approximately 4 and 7 cm in high-performance thin-layer chromatography (HPTLC) and TLC systems, respectively. This restriction does not apply to PPEC. In PPEC, the mobile phase velocity does not depend on the migration distance of the mobile phase.

The effect of the mobile phase flow velocity on plate height in PPEC was reported by our group. Such investigation with our early equipment was possible because it was possible to measure the plate height continuously during the separation process [13]. In Figures 6.14 and 6.15, the relationships between the plate height and mobile phase velocity, and the plate height and migration distance of the mobile phase front are plotted for PPEC and conventional HPTLC for systems with HPTLC RP-18W plates (Merck, Darmstadt), respectively [18]. The minimum plate height value of the HPTLC system is about 27 μm and for PPEC 10 μm. The performance increase in the PPEC system compared with HPTLC is between two- and threefold. The minimum plate height of the PPEC system is approximately twice the average diameter of the stationary phase particles. This is similar to the observed values for well packed HPLC columns. In addition, from Figure 6.14 it can be seen that at a mobile phase velocity equal to 80 mm/min the plate height of the PPEC system increases about 30% relative to its minimum value, observed at a mobile phase velocity of 50 mm/min.

Other groups have also reported the performance of PPEC systems [10,12]. Nurok et al. obtained a minimum plate height of 10.6 μm in an HPTLC system with 2-nitroaniline as the test solute [10]. Tate and Dorsey obtained a plate number greater than 10,000 for Rhodamine B on a 10 cm HPTLC-RPS chromatographic plate (8.5 cm migration distance) with a methanol—aqueous buffer mobile phase as the test solute [12].

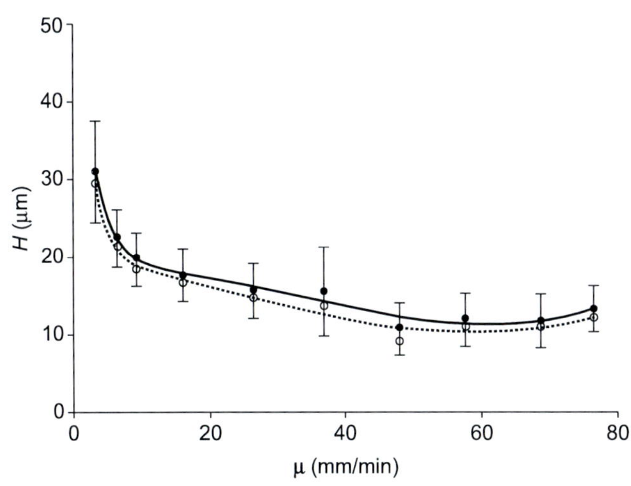

FIGURE 6.14 Plate height versus flow velocity of the mobile phase for pressurized planar electrochromatography (PPEC) system. (•) HPTLC RP-18W plate from Merck and (○) after subtracting variance concerned with starting spot width. The test solute is prednisolone succinate. Mobile phase 80% acetonitrile in buffer (1.96 mM citric acid, 4.08 mM disodium hydrogen phosphate, pH = 5.0). *Reproduced with permission from Ref. [18].*

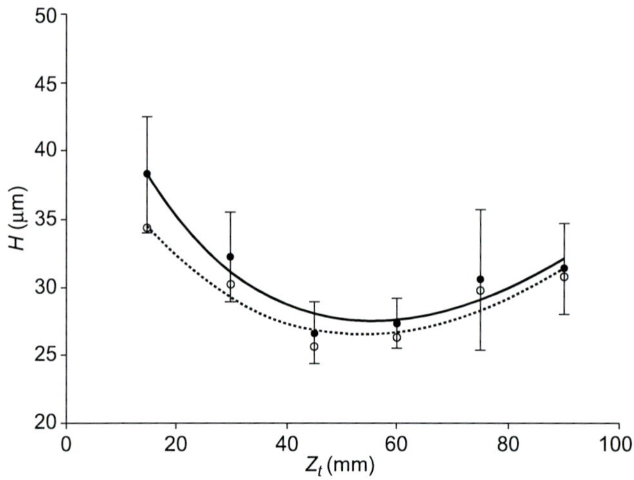

FIGURE 6.15 Plate height versus migration distance of solvent front for HPTLC system. The thin-layer chromotograpghy (TLC) chamber (DS-II-5 × 10 from Chromdes) with (•) chromatographic plate HPTLC RP-18W from Merck and (○) after subtracting variance concerned with starting spot width. The test solute is prednisolone succinate. Mobile phase 80% acetonitrile in buffer (1.96 mM citric acid, 4.08 mM disodium hydrogen phosphate, pH = 5.0). *Reproduced with permission from Ref. [18].*

In TLC and PPEC, all solute bands remain on the layer after the separation process. Some solute bands migrate a short distance and the contribution of the starting sample width to zone broadening is especially important. Nurok and coworkers investigated this problem and obtained the plate height data against migration distance for the test solute 2-nitroaniline shown in Figure 6.16 for starting zones of 1.0 mm [10]. The plate height of the PPEC system diminishes with an increase in the migration distance before reaching a plateau for migration distance longer than 40 mm. Similar results were obtained by our group when investigated influence of sample application mode on performance of the PPEC system [19]. In this chapter, we also demonstrated that sample application with a commercially available aerosol applicator leads to the highest efficiency for the PPEC system in comparison with sample zones applied with a hand operated 5 μL syringe or using a scrap-out (approximately 1.0 mm in diameter) of the sorbent layer in which the sample solution was deposited.

Our group reported online sample application for the PPEC system [16,17]. Optimized sample application on an HPTLC RP-18W plate enabled plate heights of about 39 and 25 μm for acetanilide and acetaminophen as the test solutes, respectively. The performance of the PPEC system is still too low in comparison to that discussed above. However, solute bands in electrochromatograms with short migration distance are

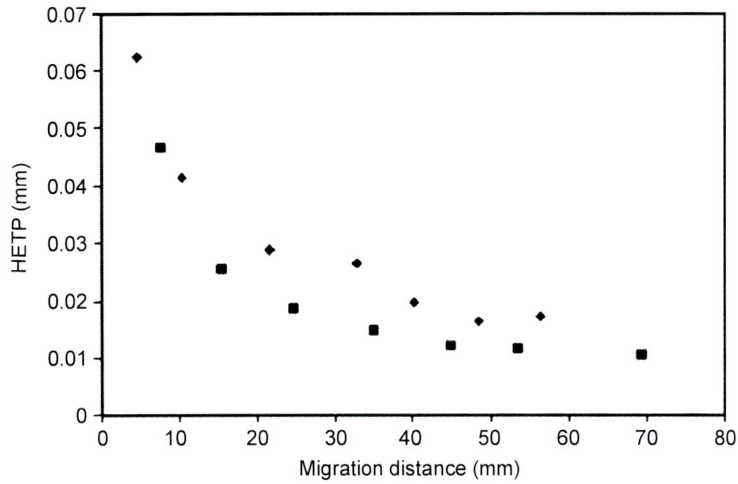

FIGURE 6.16 Plate height (HETP) for 2-nitroaniline versus migration distance. The high-performance plates conditioned at 160 °C for 20 min. Pressurized planar electrochromatography (PPEC) was under conditions voltage 6 kV, pressure 63 atm, mobile phase 55% acetonitrile in buffer (pH 4.7, 5 mM acetate buffer) for between 1.0 and 7.5 min, depending on distance migrated. The upper plot (◆) is for the regular plates and the lower plot (■) is for the high-performance plates. *Reproduced with permission from Ref. [10].*

symmetrical and promising for further development of this sample application mode, Figure 6.17 [17].

Nurok and coworkers investigated the influence of temperature on performance for a PPEC system [10]. They observed that the best kinetic performance was obtained for the separation system at 26 °C. Diminution of the system performance below 26 °C was attributed to the predominance of mass transfer effects over molecular diffusion and at higher temperatures, the opposite behavior, Figure 6.18 [10].

In the PPEC systems, the sorbent layer is pressurized during the separation process. It is reasonable that a too high pressure can lead to damage of the layer structure reflected in lowering of the separation system performance. Tate and Dorsey reported that the optimum pressure exerted on the sorbent layer should be in the range 1000–2000 psi (68–136 atm) [12]. Nurok and coworkers showed a decrease in the separation quality of a test mixture in the PPEC system at a pressure of 118 atm compared with two lower pressures (11.8 and 19.7 atm), Figure 6.19 [3].

6.4.2 Separation Time and Development Distance in PPEC

The mobile phase velocity in the PPEC system is independent of the average particle diameter of the stationary phase and is a constant during

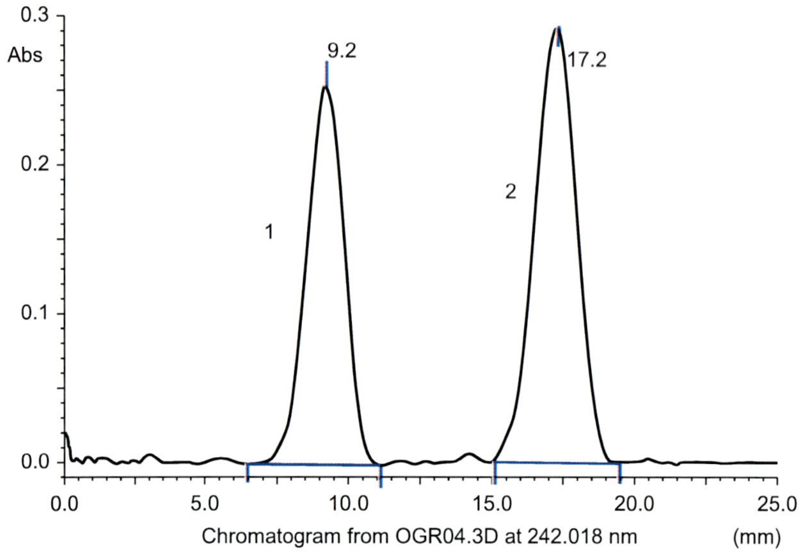

FIGURE 6.17 Chromatogram of acetanilide (1) and acetaminophen (2) obtained with pressurized planar electrochromatography (PPEC) technique with online sample application and with partition of sample solution and mobile phase solution in the injection tubing with air bubbles; flow velocity 60 mL/h external hydraulic press 15 bar, polarization voltage 1.5 kV, time of experiment 1 min. *Reproduced with permission from Ref. [17].*

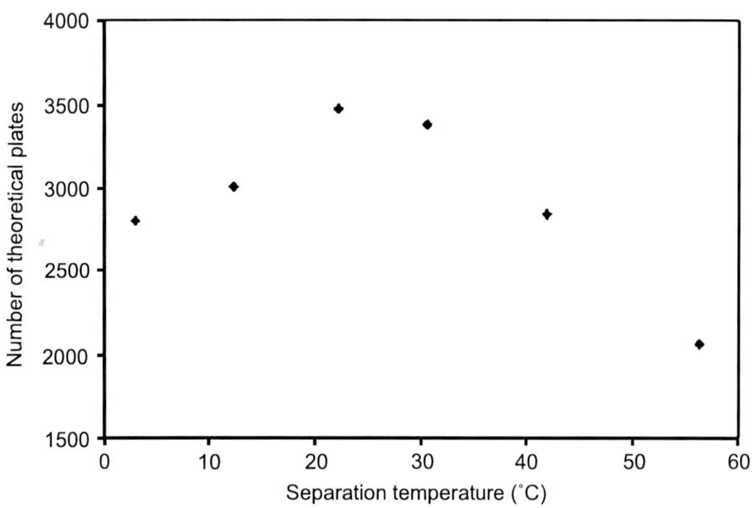

FIGURE 6.18 Number of theoretical plates for 2-nitroaniline versus separation temperature on HPTLC plates conditioned at 160 °C for 20 min. Pressurized planar electrochromatography (PPEC) conditions: voltage 6 kV, pressure 63 atm, and mobile phase 55% acetonitrile in buffer (pH 4.7, 5 mM acetate buffer). Separation time was adjusted from 8 min at 3.0 °C to 3.0 min at 56 °C such that migration distance at each temperature was close to 49 mm. *Reproduced with permission from Ref. [10].*

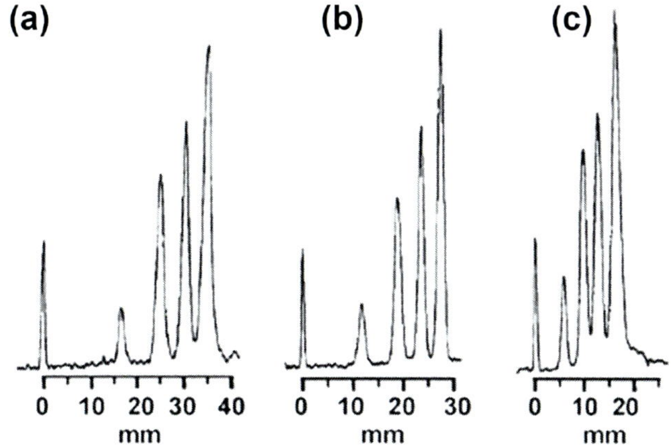

FIGURE 6.19 A 1 min separation on a LiChrospher C18 plate at 9 kV and a pressure of (a) 11.8 atm, (b) 19.7 atm, and (c) 118 atm. The mobile phase 55% acetonitrile in buffer (pH 4.7; 25 mM acetate buffer). Solute mixture: 4-cholesten-3-one, 17-acetoxyprogesterone, 2-acetonaphthone, and benzanilide. *Reproduced with permission from Ref. [3].*

the separation process (contrary to conventional TLC/HPTLC). The flow velocity can be easily changed by varying the electric field (polarization voltage) applied to the prewetted sorbent layer. The linear relationship observed is shown in Figure 6.20 for the range 0–3.5 kV [13].

The plot shows an increase in slope above 3.5 kV. It can be explained by the decrease in viscosity of the mobile phase at higher voltage attributed to stronger Joule heat generation. This relationship illustrates that fast

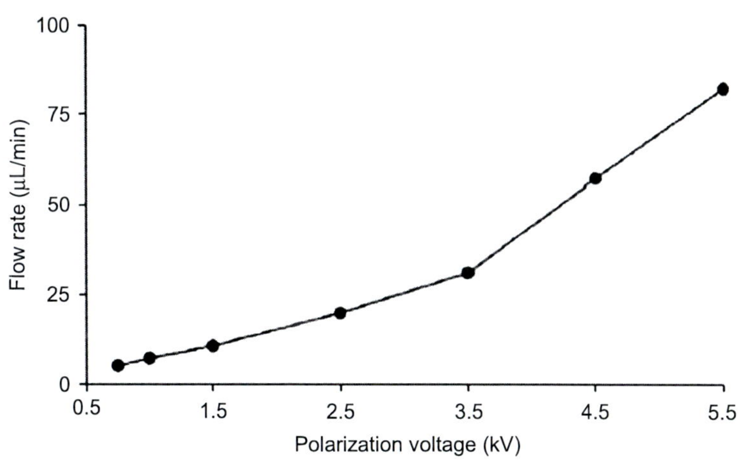

FIGURE 6.20 Plot of mobile phase flow rate versus polarization voltage. thin-layer chromotography (TLC) RP-18 F254 plate (Merck), mobile phase: 80% acetonitrile in buffer (pH 6.0; 3.74 mM citric acid, 12.52 mM disodium hydrogen phosphate). *Reproduced with permission from Ref. [13].*

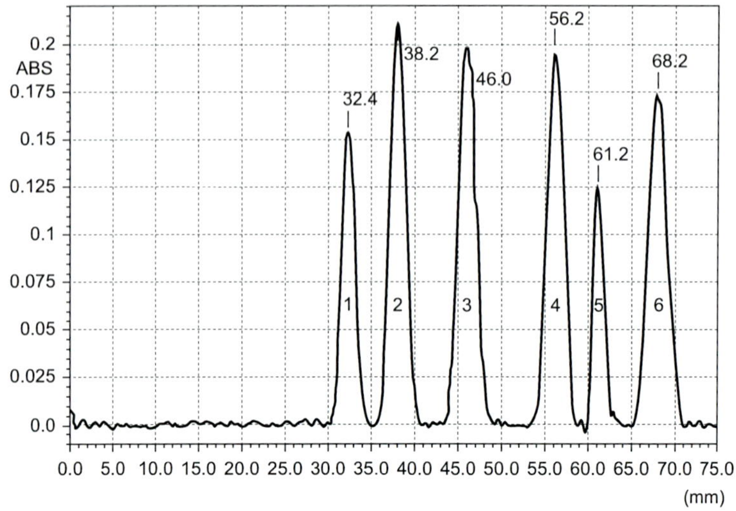

FIGURE 6.21 Electrochromatogram pressurized planar electrochromatography (PPEC) of the test mixture: RP-18W HPTLC plate from Merck, applied polarization voltage 2.5 kV, separation time 5 min, the mobile phase 80% acetonitrile in buffer (1.96 mM citric acid, 4.08 mM disodium hydrogen phosphate, pH = 5.0), the order of mixture components according to the increase of migration distance: (1) testosterone isobutyrate, (2) testosterone acetate, (3) methandienone, (4) hydrocortisone acetate, (5) 16-dehydropregnenolone acetate, and (6) prednisolone succinate. *Reproduced with permission from Ref. [18].*

separations are achievable with the PPEC technique. Several groups have reported fast separations by PPEC [3,13]. Very fast separations by PPEC are a promising future direction.

In TLC/HPTLC, especially in reverse-phase systems, chromatogram development cannot be performed for a long distance due to an inadequate mobile phase velocity. For PPEC systems, this does not take place. The mobile phase migrates with a constant velocity, which is dependent on the electric field strength, but not on the plate length.

In conventional HPTLC techniques, the migration of sample zones is determined by the solvent front migration. The maximum value of the solvent front migration is equal to the plate length. However, in PPEC the mobile phase can migrate continuously as in column chromatography. This feature allows the maximum migration distance for the sample zones. In Figures 6.21 and 6.22, a six-component steroid mixture is separated by PPEC and conventional HPTLC, respectively, using the same mobile and stationary phase system [18]. The chromatographic plate was 10 cm in length. The resolution of the mixture by PPEC is considerably higher than that by HPTLC. The solvent front migration distance for the HPTLC separation was 50 mm, time for development 6.5 min, and the least retained steroid had a retardation factor of 0.63. The PPEC

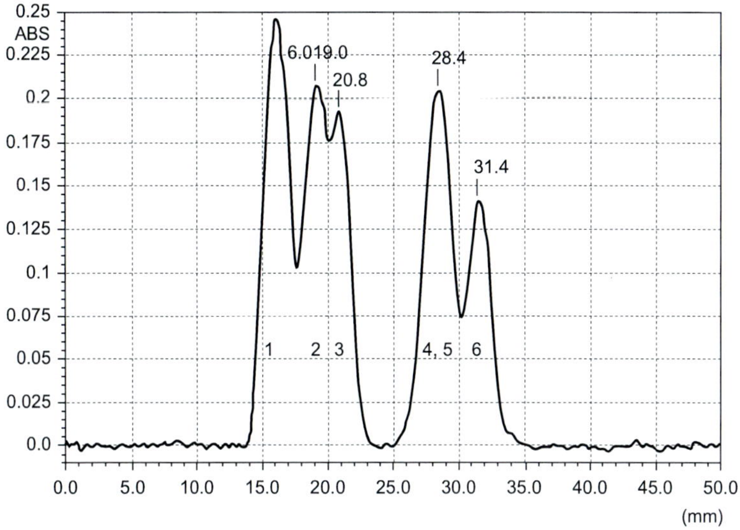

FIGURE 6.22 Separation of the test mixture by conventional high-performance planar chromatography with high-performance thin-layer chromatography (HPTLC) RP-18W plate (Merck): separation time 6.5 min, the mobile phase and the test solute mixture as in Figure 6.20. *Reproduced with permission from Ref. [18].*

chromatogram was obtained after 5.0 min by applying a voltage 2.5 kV with a mobile phase velocity of 24.6 mm/min. The corresponding virtual solvent front migration distance was approximately 120 mm (more than double the distance for the HPTLC system).

6.4.3 Separation Selectivity

The distribution of solutes between the stationary and mobile phases, and electrophoresis are involved in the PPEC separation mechanism. The electrophoretic effect is responsible for alterations of the separation selectivity in PPEC relative to liquid chromatography for partially ionized compounds. For neutral solutes, the separation selectivity in liquid chromatography and electrochromatography should be practically the same.

In Figure 6.23(a and b), the separation of acetylsalicylic acid, caffeine, and acetaminophen is shown for HPTLC and PPEC, respectively, with the same mobile and stationary phase system [20]. The separation selectivity in both the systems is quite different. In the PPEC system, acetylsalicylic acid migrates in the opposite direction to caffeine and acetaminophen. This feature of the PPEC mode is advantageous compared with CEC. All solute bands that migrate in opposite directions can be registered on the

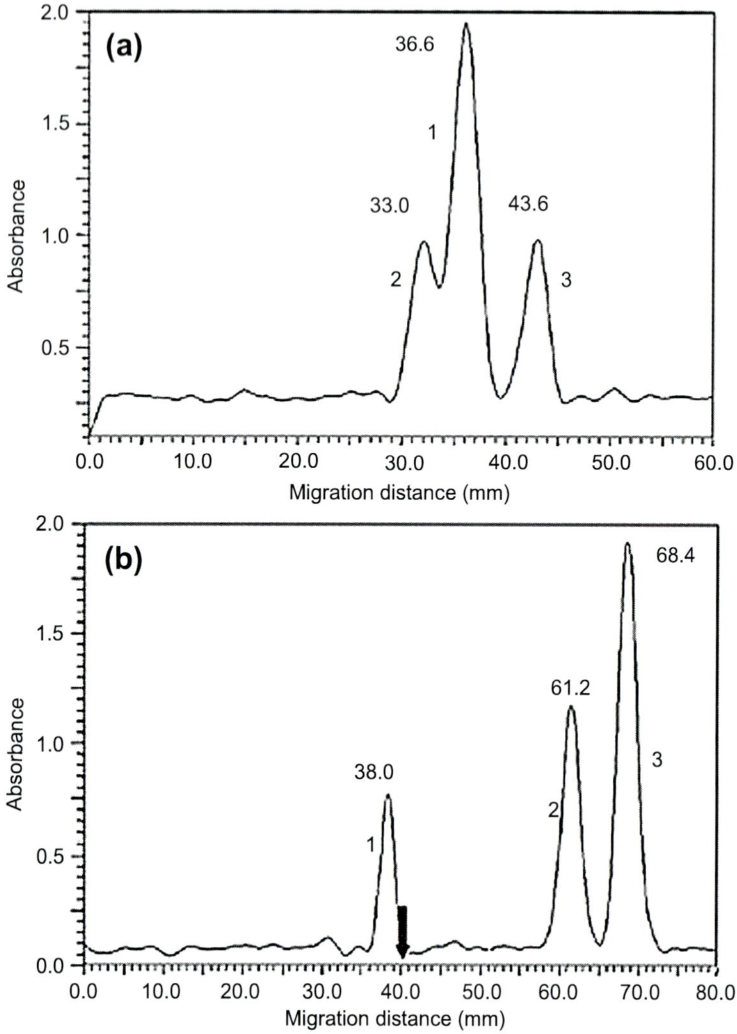

FIGURE 6.23 (a) Thin-layer chromatography (TLC) and (b) Pressurized planar electrochromatography (PPEC) chromatograms of acetylsalicylic acid (1), caffeine (2), and acetaminophen (3) on reversed-phase thin-layer chromatography (RPTLC) plate. Mobile phase 60% acetonitrile in buffer (pH 3.8). PPEC conditions: polarization potential 2.0 kV (the black arrow indicates the point of application). *Reproduced with permission from Ref. [20].*

chromatographic plate. On the other hand, solute bands migrating only to one electrode can be registered in a single run by CEC. Large differences in separation selectivity for electrochromatography and liquid chromatography systems were utilized to demonstrate the advantages of two-dimensional separations with PPEC [14,21].

6.5 APPLICATIONS

Pressurized planar chromatography is relatively a new separation technique characterized by high separation efficiency, fast separations, and changes in separation selectivity in comparison to liquid chromatography, especially TLC. However, instrumental and methodological challenges of PPEC are obstacles to its rapid development and broad application in laboratory practice.

6.5.1 Quantitative Analysis

PPEC is well suited to qualitative analysis except for the requirement to prewet the sorbent layer with mobile phase before development and before or after sample application. For PPEC, the different modes of sample application include direct application of the sample solution by microsyringe to the chromatographic plate [3,9,10,13,19], deposition of sample solution in a scrap-out of the sorbent layer [19], and application of the sample solution with an aerosol applicator [19]. All of these methods are suitable for qualitative analysis. For quantification in TLC/HPTLC, the application of relatively large sample volumes in the form of bands is recommended [22]. For this application, an aerosol applicator is the preferred choice. However, how to perform sample application on the prewetted chromatographic plate with aerosol technique?

Attempts at quantitative analysis using PPEC have only been carried out by our research team so far. The first report of the quantitation of active components in a pharmaceutical preparation (cefalgin) by PPEC was published in 2012 [23]. The following procedure was used: the plate was spotted with sample solution with an aerosol applicator and then prewetted with mobile phase using two prewetting reservoirs equipped with a special partition. The partition prevented the unwanted broadening of the starting sample zones [14,19,23]. In the first reservoir, a cuboic shaped partition tightly covered the sample on the plate during the prewetting procedure (PPEC 1). In the second reservoir, the sample spot on the layer was covered with an inverted U-shaped partition so that the sample spot was not in direct contact with the lower surface of the partition (PPEC 2) [23]. After the prewetting procedure, the 2×5 cm chromatographic plate was immediately introduced into the PPEC chamber. It should be noted that calibration curves were prepared in a simplified manner—different volumes of standard solution were applied with the aerosol applicator. The highest limit of detection (LOD) and limit of quantitation (LOQ) were obtained when the reservoir with a cubic partition was used compared with the values obtained with HPTLC or PPEC and the inverted U-shaped partition (LOQ values with HPTLC mode and PPEC one are comparable). It seems that some quantity of the sample could remain on the Teflon foil (attached to the lower surface of the partition) when the chromatographic plate was transferred from the prewetting reservoir to the PPEC chamber. It is noticeable that there is a larger spread for the results obtained for PPEC 1 mode. In this case, relative standard deviation (RSD) ranges from 5.15% to 11.91% when calculations are based on peak area, and from 1.35% to 9.45% when based on peak height. On the other hand, for the PPEC 2 mode, the RSD values ranged from 0.04% to 5.92%, based on the peak area, and from 2.22% to 4.69% based on the peak height. Analogous RSD values for HPTLC mode ranged from 2.54% to 5.1% and from 0.85% to 4.53% for peak area and peak height, respectively. These data reveal that the PPEC 2 mode is more suitable for quantitative analysis than PPEC 1. However, there are some disadvantages (e.g., evaporation of solvent from the sorbent layer during the transfer of the chromatographic plate from the prewetting reservoir to the PPEC chamber). Furthermore, despite the short time of development, the total duration of the PPEC analysis of many samples was longer than for HPTLC due to the single separation of samples by PPEC on each chromatographic plate [16,23].

In a later chapter, quantitative analysis of the active components in cefalgin preparations was carried out using an improved device for PPEC that allowed the use of wider plates, 10 cm × 5 cm (PPEC 3). This device allowed prewetting of the sorbent layer after sample application [16]. These modifications resulted in a significant reduction in analysis time and partial elimination of manual operations (see above). The procedure for quantitative analysis and validation of the selected parameters was in accordance with the International Conference on Harmonisation of Technical Requirements for Registration of Pharmaceuticals for Human Use (ICH) guidelines. Calibration curves were best fitted by polynomial equations. The repeatability of the method was considerably improved. RSD values ranged from 3.10% to 3.84%, while recovery values, which characterize the accuracy, ranged from 95.60% to 107.36%.

Recently our group reported preliminary results with online sample application during the PPEC process [16]. In the prototype device, the sample application, prewetting of the sorbent layer, and the separation process were performed in a single unit. It should be emphasized that there was no possibility to inject reproducible sample volumes for each track/path on the chromatographic plate in the online application mode. Therefore, the internal standard addition method was applied to overcome this difficulty. Acetanilide was selected as an internal standard for the assay of acetaminophen in a pharmaceutical formulation. In this case, the procedure for quantitative analysis and validation (selected parameters) was also in accordance with the ICH guidelines. Calibration curves were linear. Relatively good intraday precision was obtained. The dispersion of the results for PPEC with online sample application is about two times lower than that for PPEC 3, using the aerosol sample application mode (RSD = 1.54% and RSD = 3.10%, respectively). The recovery ranged from 97.80% to 103.10%. The main advantage of online sample application in PPEC is the shorter analysis time with a fully equilibrated separation system and reduced dispersion of sample zones during the prewetting process. The developed prototype device is promising for automation and future applications in laboratory practice.

6.5.2 Enantiomer Separations

The separation of chiral compounds is a difficult issue. Due to the similar properties of enantiomers, the use of special direct or indirect chromatographic methods is required. In the direct mode, a chiral selector, bonded to the stationary phase or present in the mobile phase, forms diastereomeric complexes with sample components. These diastereomeric complexes are separated during the chromatographic process. In the indirect method, chiral derivatizing reagent (CDR) reacts with solutes, leading to the formation of a mixture of diastereomers. The components of this mixture can be separated during the chromatographic process. In separation mechanism of PPEC, two processes (partition and electrophoresis) are involved. These could be used for the separation of enantiomers. Polak et al. have published a few papers describing the application of both PPEC and TLC techniques for the separation of amino acid enantiomers [15,24,25]. The separation of enantiomers of amino acids was performed with commercially available chiral plates. The effects of different variables such as voltage, buffer pH and its concentration, and the concentration of the organic component of the mobile phase on migration distance were investigated. In Figure 6.24(a and b) PPEC and TLC separations of the enantiomers of tryptophan are presented, respectively [15]. The separation efficiency obtained with PPEC is substantially better than that with TLC.

Variables such as buffer pH, buffer concentration, and concentration of organic modifier in the mobile phase can be easily optimized for enantiomer separations using PPEC. Using the indirect method for the separation of some amino acid enantiomers (leucine, isoleucine, valine, asparagine, cysteine, tryptophan), a comparison of HPTLC and PPEC was made [25]. In both the cases, the separations were obtained on HPTLC RP-18 plates. The influence of various factors on solute migration was investigated. However, in this case the effect of buffer type was also examined. The results showed that variables such as buffer pH or acetonitrile concentration affected the solute migration distances for both the techniques. The electrophoretic effect is responsible for the different selectivity observed for PPEC compared with HPTLC. The PPEC system turns out to be more sensitive to buffer type and pH than HPTLC. The separation time for PPEC is considerably shorter than that for HPTLC.

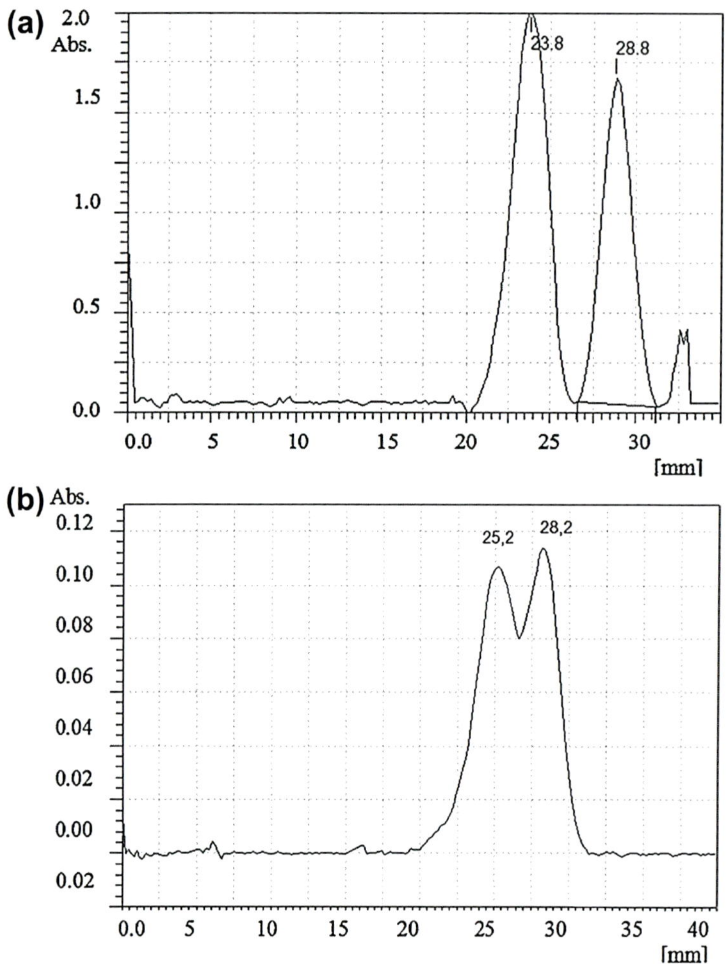

FIGURE 6.24 (a) Pressurized planar electrochromatography (PPEC) separation of the enantiomers of tryptophan. The mobile phase was 70% acetonitrile, 10% methanol, 10% bidistilled water, 10% buffer (acetic acid + sodium acetate, buffer concentration 5 mM, pH 5.57), and the polarization potential was 1.7 kV. The separation time was 7.5 min and detection was at 556 nm. (b) thin-layer chromatography (TLC) separation of the enantiomers of tryptophan. The mobile phase was as for (a), the separation time was 4.5 min, and detection was at 556 nm. *Reproduced with permission from Ref. [15].*

6.5.3 Two-Dimensional Separations

Despite enormous progress in the field of chromatography, the separation of complex samples such as plant extracts and biological samples is still a challenge. In most cases, it is difficult to separate and identify all compounds by means of one-dimensional liquid chromatography. In this context, multidimensional chromatography offers several advantages for solving many of the most challenging analytical problems [26]. PPEC separations with regard to HPTLC are characterized by higher performance, shorter duration of the chromatogram development, but most of all, offer changes in separation selectivity. For the first time, the use of PEC and conventional planar chromatography in a two-dimensional process was described by Panchagnula et al., in 2007 for the separation of phosphopeptides [27].

The preliminary separations indicate that 2D HPTLC-PPEC is a promising tool for further applications [14,21]. It should be stressed that in the experiments the 5 cm × 20 cm HPTLC RP-18W plates were used and electrochromatogram developments were performed along the longer edge (20 cm) of the chromatographic plate [14]. It is well known that conventional chromatogram developments on such long plates are not realized in laboratory practice.

6.5.4 Orthogonal pressurized planar electrochromatography

The newest separation technique, OPPEC, is especially attractive for continuous preparative (or micropreparative) separations although analytical applications are also possible [8]. Due to the recent development of OPPEC, there are only a few examples of its application so far. In Figure 6.25, a two-dimensional electrochromatogram with the OPPEC device in the analytical mode is presented. As can be seen the sample zones (food dye mixture) are spread over the whole area of the sorbent layer.

In Figure 6.26, the orthogonal electrochromatogram after continuous separation with the OPPEC device is presented. In this experiment, the separation system was fed with a mixture of two dyes (brilliant black and patent blue) as the sample solution using an additional syringe pump. The main feature of the device construction is the pattern of the sealing margin on the edges of the sorbent layer. In a closed system comprised of two phases (liquid and solid) under pressure this enables the two orthogonal processes of liquid chromatography and electrophoresis/electrochromatography to be performed simultaneously with a removable chromatographic plate. It is worthy of note that in the past year another interesting example regarding a similar operating principle was reported [28]. Here simultaneous chromatography and electrophoresis (SCE) were evaluated for two-dimensional separations. However, in this technique the separation process occurred in an open system consisting of three phases: gas, liquid, and solid.

6.5.5 Combination of mass spectrometry (MS) and PPEC

The introduction of few ion sources and interface devices, which easily transfer solute zones from any chromatographic plate to a mass

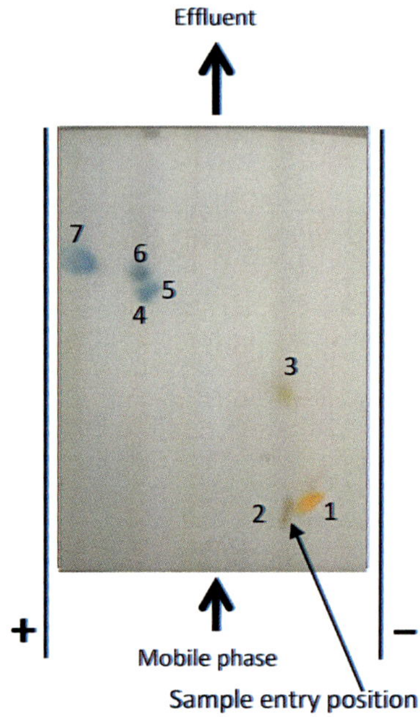

FIGURE 6.25 The picture of the chromatographic plate after orthogonal pressurized planar electrochromatography (OPPEC) separation of a six food dye mixture (the plate is cut, with no sealing margin). Sample volume (1 μL) introduced with microsyringe; polarization voltage 1.0 kV, the mobile phase (70% acetonitrile, ACN) flow 75 μL/min: (1) not identified dye, (2) rhodamine G, (3) 4-(2-pirydylazo)resorcinol (PAR), (4) green S, (5) patent blue, (6) brilliant blue, (7) brilliant black. *Reproduced with permission from Ref. [8].*

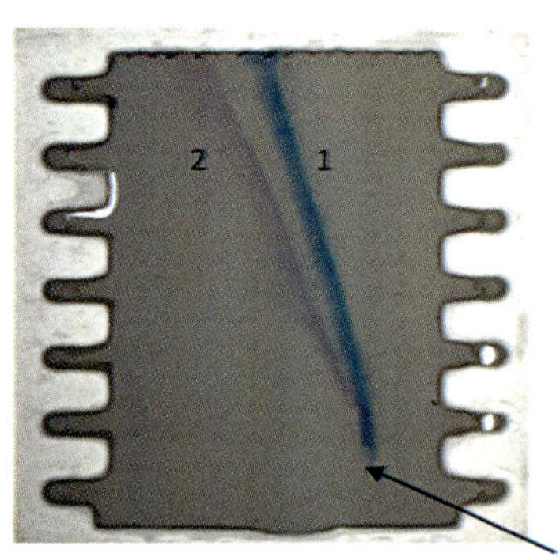

FIGURE 6.26 The picture of the chromatographic plates after micropreparative orthogonal pressurized planar electrochromatography (OPPEC) separation of two dye mixture: (1) patent blue and (2) brilliant black, the mobile phase 30% ACN, mobile phase flow velocity 100 μL/min, sample solution flow velocity 1.0 μL/min, polarization voltage 1.0 kV. *Reproduced with permission from Ref. [8].*

spectrometer has increased significantly the interest and use of TLC-MS approaches [29–32]. However, there are only two publications on the combination of mass spectrometry and PEC. Patton et al. [27] described a two-dimensional separation of phosphopeptides by PEC and TLC with detection of phospholipids by matrix-assisted laser desorption/ionization time-of-flight mass spectrometry. Nurok et al. described a preliminary application of the desorption electrospray ionization (DESI) source to transfer solute zones from the chromatographic plate to a mass spectrometer after separation by PPEC [32].

6.6 CHALLENGES AND CONCLUSIONS

The features and advantages of PPEC make it attractive for applications in laboratory practice. However, in spite of 10-year development of PPEC, this technique has not been commercialized so far. The main challenges of PPEC are concerned with the development of equipment and stationary phases. Many manual operations are necessary to perform a complete separation procedure. At this stage of development, the PPEC devices described so far are not commercially attractive. The further automation of PPEC equipment could provide the stimulus for commercial development and customer acceptance. A new company, InChromatics, has been formed with a view to develop instrumentation for PPEC [33].

The other weak point of the PPEC technique is the stationary phase. Commercially available chromatographic plates with silanized silica gel have been used in nearly all applications of PPEC so far. These plates are not fully suitable for PPEC. They are too thick with respect to the problem of controlling Joule heat generation. In addition, these sorbent layers have a relatively low capacity to generate electroosmotic flow. The production of sorbent layers specifically for application in PPEC is an important challenge for the further development of PPEC. Nurok and coworkers employed monolithic polymer layers for the separation of six peptides in 1 min by PPEC [34]. Although the results are preliminary, they are promising for the fast separations of compounds of biological interest by PPEC.

Special attention should be paid to the further development of OPPEC, which is a new technique especially attractive to preparative separations.

References

[1] Pretorius V, Hopkins BJ, Schieke JD. Electro-osmosis—a new concept for high-speed liquid chromatography. J Chromatogr A 1974;99:23–30.

[2] Pukl M, Prosek R, Kaiser E. Planar electrochromatography. Part. 1. Planar electrochromatography on non-wetted thin-layers. Chromatographia 1994;38:83–7.

[3] Nurok D, Koers JM, Novotny AL, Carmichael MA, Kosiba JJ, Santini RE, et al. Apparatus and initial results for pressurized planar electrochromatography. Anal Chem 2004;76:1690–5.
[4] Dzido TH, Płocharz PW, Chomicki A, Hałka-Grysińska A, Polak B. Pressurized planar electrochromatography. J Chromatogr A 2011;1218:2636–47.
[5] Wan QH. Capillary electrochromatography: effect of electrolyte concentration on electroosmotic flow and column efficiency. J Chromatogr A 1997;782:181–9.
[6] Wan QH. Effect of electrical double-layer overlap on the electroosmotic flow in packed-capillary columns. Anal Chem 1997;69:361–3.
[7] Nurok D, Koers JM, Carmichael MA. Role of buffer concentration and applied voltage in obtaining a good separation in planar electrochromatography. J Chromatogr A 2003; 983:247–53.
[8] Dzido TH, Łopaciuk E, Płocharz PW, Chomicki A, Zembrzycka M, Frank H. Equipment and preliminary results for orthogonal pressurized planar electrochromatography. J Chromatogr A 2014;1334:149–55.
[9] Dzido TH, Mróz J, Jóźwiak GW. Application of a horizontal DS chamber to planar electrochromatography. J Planar Chromatogr 2004;17:404–10.
[10] Novotny AL, Nurok D, Replogle RW, Hawkins GL, Santini RE. Results with an apparatus for pressurized planar electrochromatography. Anal Chem 2006;78:2823–31.
[11] Tate PA, Dorsey JG. Characterization of flow and voltage profiles in planar electrochromatography. J Chromatogr A 2005;1079:317–27.
[12] Tate PA, Dorsey JG. Linear voltage profiles and flow homogeneity in pressurized planar electrochromatography. J Chromatogr A 2006;1103:150–7.
[13] Dzido TH, Płocharz PW, Ślązak P. Apparatus for pressurized planar electrochromatography in a completely closed system. Anal Chem 2006;78:4713–21.
[14] Chomicki A, Ślązak P, Dzido TH. Preliminary results for 2-D separation with high-performance thin-layer chromatography and pressurized planar electrochromatography. Electrophoresis 2009;30:3718–25.
[15] Polak B, Hałka A, Dzido TH. Pressurized planar electrochromatographic separation of the enantiomers of tryptophan and valine. J Planar Chromatogr 2008;21:33–7.
[16] Hałka-Grysińska A, Ślązak P, Torbicz A, Sajewicz M, Dzido TH. A modified device for pressurized planar electrochromatography and preliminary results with on-line sample application. Chromatographia 2013;76:1271–9.
[17] Hałka-Grysińska A, Płocharz PW, Szczap R, Dzido TH. Optimization of some variables of on-line injection in pressurized planar electrochromatography. J Liq Chromatogr Relat Technol 2013;36:2512–23.
[18] Płocharz PW, Klimek-Turek A, Dzido TH. Pressurized planar electrochromatography, high-performance thin-layer chromatography and high-performance liquid chromatography – comparison of performance. J Chromatogr A 2010;1217:4868–72.
[19] Płocharz PW, Dzido TH, Ślązak P, Jóźwiak GW, Torbicz A. Influence of sample application mode on performance of pressurized planar electrochromatography in completely closed system. J Chromatogr A 2007;1170:91–100.
[20] Hałka-Grysińska A, Płocharz PW, Torbicz A, Dzido TH. Reversed-phase pressurized planar electrochromatography and planar chromatography of acetylsalicilic acid, caffeine and acetaminophen. J Planar Chromatogr 2010;23:420–5.
[21] Chomicki A, Kloc K, Dzido TH. Two-dimensional separtaion of some amino acids by HPTLC and pressurized planar electrochromatography. J Planar Chromatogr 2011;24:6–9.
[22] Hahn-Deinstrop E. Applied thin-layer chromatography. Best ractice and avoidance of mistakes. 2nd ed. Weinheim: Wiley-vch; 2007.
[23] Hałka-Grysińska A, Ślązak P, Zaręba G, Markowski W, Klimek-Turek A, Dzido TH. Simultaneous determination of acetaminophen, propyphenazone and caffeine in cefalgin preparation by pressurized planar electrochromatography and high-performance thin-layer chromatography. Anal Methods 2012;2:973–82.

[24] Polak B, Wojtanowski KK, Ślązak P, Dzido TH. Separation of some aromatic amino acid enantiomers with pressurized planar electrochromatography and TLC. Chromatographia 2011;73:339–45.
[25] Polak B, Balasa K, Dzido TH. Separation of amino acid 2,4-dinitrophenyl-5-L-valine amide diastereomeric derivatives with high-performance planar chromatography and pressurized planar electrochromatography. J Planar Chromatogr 2013;26:180–9.
[26] Cieśla Ł, Waksmundzka-Hajnos M. Two-dimensional thin-layer chromatography in the analysis of secondary plant metabolites. J Chromatogr A 2009;1216:1035–52.
[27] Panchagnula V, Mikulskis A, Song L, Wang Y, Wang M, Knubovets T, et al. Phosphopeptide analysis by directly coupling two-dimensional planar electrochromatography/thin-layer chromatography with matrix-assisted laser desorption/ionization time-of-flight mass spectrometry. J Chromatogr A 2007;1155:112–23.
[28] Stevenson PR, Dunlap BE, Powell PS, Petersen BV, Hatch CJ, Chan H, et al. Simultaneous chromatography and electrophoresis: two-dimensional planar separations. Anal Bioanal Chem 2013;405:3085–9.
[29] Gusev AI, Vasseur OJ, Proctor A, Sharkey AG, Hercules DM. Imaging of thin-layer chromatograms using matrix/assisted laser desorption/ionization mass spectrometry. Anal Chem 1995;67:4565–70.
[30] Van Berkel GJ, Ford MJ, Debel MA. Thin-layer chromatography and mass spectrometry coupled using desorption electrospray ionization. Anal Chem 2005;77:1207–15.
[31] Morlock G, Ueda Y. New coupling of planar electrochromatography with direct analysis in real time mass spectrometry. J Chromatogr A 2007;1143:243–51.
[32] Janecki DJ, Novotny AL, Woodward SD, Wiseman JM, Nurok D. A preliminary study of the coupling of desorption electrospray ionization-mass spectrometry with pressurized planar electrochromatography. J Planar Chromatogr 2008;21:11–4.
[33] www.inchromatics.com.
[34] Woodward SD, Urbanova I, Nurok D, Svec F. Separation of peptides and oligonucleotides using a monolithic polymer layer and pressurized planar electrophoresis and electrochromatography. Anal Chem 2010;82:3445–8.

CHAPTER

7

Theory and Instrumentation for In situ Detection

Bernd Spangenberg

Department of Process Engineering, University of Applied Sciences Offenburg, Offenburg, Germany

7.1 INTRODUCTION

Thin-layer chromatography (TLC) and high-performance thin-layer chromatography (HPTLC) are versatile methods for the separation and quantification of compounds in mixtures of different complexities. Separations occur in parallel facilitating the simultaneous separation of the samples. A major advantage of TLC methods compared with other separation techniques is that they use disposable layers. The sample pretreatment procedure, therefore, can often be reduced to just a sample dissolution step. This helps to reduce the uncertainty in the assay. In this chapter, we will not distinguish between TLC and HPTLC. HPTLC is the modern version of TLC and the use of high-performance layers for separations is strongly recommended.

One of the most important steps in TLC analysis is the measuring process. The pattern of the separated compounds is transformed into digital data, which can then be used for further calculations. At the present time there are differently working systems on the market, all of which allow quantification of separated zones without removal from the layer: mono-wavelengths scanners, diode-array scanners, charge-coupled device (CCD) cameras, and flatbed scanners. The first two types are commonly called scanners. The last two types are classified as video-densitometric devices. We call these methods densitometry because the optical density of light (the spectra-dependent light intensity) is measured. We, therefore, measure a densitogram and not a chromatogram. According to *M. Tswett* (the inventor of chromatography), a chromatogram is the space-separated distribution of compounds in the stationary phase and not its measurement curve [1].

Instrumental Thin-Layer Chromatography
http://dx.doi.org/10.1016/B978-0-12-417223-4.00007-8

Systems for direct radioactivity measurements are also commercially available. Beside this, infrared and Raman spectrometry, as well as mass-spectrometric systems, are available for measurements in TLC. These systems are not designed exclusively for TLC. All these systems allow measuring an intensity distribution directly on a TLC plate. The main focus of this chapter is densitometric methods because these are the commonly used techniques in TLC. This chapter is divided into two main sections, a theoretical part and a practical part.

7.2 THEORY FOR IN SITU DENSITOMETRIC DETECTION

All the above-mentioned densitometric systems use light-to-detect separated sample zones by illuminating the TLC plate from above with a light of known intensity. If the illuminating light shows a higher intensity than the reflected light, a fraction of the light must have been absorbed by the sample (the analyte) and/or the layer. Increasing sample amounts will induce decreasing light reflection. A transformation algorithm is recommended to turn the decreasing light intensity into an increasing signal. Ideally there should be a linear relationship between the transformed measurement and the analyte amount [2]. To make densitometric measurements, the surface of the TLC plate is illuminated with white or monochromatic light of intensity I_0. The clean plate surface should reflect light with an intensity J_0 and the sample spot with an intensity J. Both values are combined in the expression of reflectance R in Eqn (7.1).

$$R = \frac{J}{J_0} \tag{7.1}$$

where,

$R(\lambda)$, is the relative reflectance
$J(\lambda)$, intensity distribution of the analyte zone
$J_0(\lambda)$, intensity distribution of the clean layer
(All intensity abbreviations in the text are used without a wavelength-indication. This is done to ease readings although the intensity values are obviously wavelength dependent.)

In planar chromatography the absorption of light by the layer is constant, but the sample light absorption depends on the sample mass. The sample distribution within a sample zone has an unknown shape. This unknown sample distribution affects the intensity of the reflection signal. The deeper a sample is located within the layer, the weaker is its absorption signal! For quantitative TLC, a constant sample concentration within the layer is not recommended (and cannot be verified experimentally). A constant sample distribution in each sample zone is essential. This makes the mobile phase evaporation before scanning so important. Uneven drying of the layer results in different sample distributions for the same compound in different tracks [1].

Historically, the *Kubelka–Munk* theory presented the first Eqn (7.2) in 1931, which described a relationship between the measured remission and the sample amount in the layer [2].

$$KM = \frac{(1 - R_\infty)^2}{2R_\infty} = \frac{a}{s} = 2.3\frac{\varepsilon}{s}c = 2.3\frac{\varepsilon}{s}\frac{n}{dA} \tag{7.2}$$

where,

R_∞, is the absolute reflectance of an infinitely thick layer
a, the absorptivity (absorption coefficient)
ε, the molar absorption coefficient
s, the scattering coefficient
c, the TLC spot concentration
n, molar analyte amount
A, the spot area
d, the layer thickness

The *Kubelka–Munk* theory is based on the assumption that half of the scattered light flux is directed forward and half backward. Both light fluxes show the same intensity. According to this theory, scattering in every layer will illuminate the next layer above and below it with half of its nonabsorbed light intensity. The main result of the *Kubelka–Munk* theory is that only the quotient of the absorption coefficient and the scattering coefficient is linear with respect to the analyte mass!

7.2.1 Absorption and Scattering Coefficient

In the case of a scattering material like TLC plates, part of the scattered light is emitted as diffuse reflectance J when viewed perpendicular to the plate surface. (Here the vector I represents the light flux in the direction of the incident light and the vector J describes the light intensity in the antiparallel direction.) If J is the scattered light in the direction of the layer surface and I_{abs} is the abbreviation for the light absorbed by the layer and sample, the incident light intensity I_0 splits into scattered and absorbed light. The abbreviation a stands for the absorption coefficient and s stands for the scattering coefficient.

$$I_0 = J + I_{abs} = sI_0 + aI_0 = I_0(s + a) \tag{7.3}$$

where,

$I_0(\lambda)$, is the incident light intensity
$J(\lambda)$, the diffuse reflectance at the plate surface ($J = sI_0$)
s, the scatter coefficient

$I_{abs}(\lambda)$, the light absorption ($I_{abs} = aI_0$)
a, the absorption coefficient

In Eqn (7.3), the absorbed share of the incident light I_0 is described as aI_0 and the fraction of scattered light is sI_0. If all the incident light is either absorbed or scattered, and if there is no other light loss, then the sum of both the shares must be the original incident light intensity. Thus, a simple connection between the scatter coefficient and the absorption coefficient can be derived [1].

$$s = (1 - a)$$

This equation explains the observation that opaque TLC plates shine brightly when incident light is reflected from the surface at a low angle. In this case, the light does not penetrate the stationary phase, and there is no light absorption, hence $a = 0$ and all the light is scattered from the layer surface.

Assuming, that the incident light beam with intensity I_0 is scattered in all directions by particles inside the layer, some light may be absorbed by either the sample or the layer itself. Beside this loss of incident light by absorption, light is also lost by transmission from the back and edges of the layer. We can also expect fluorescence from some analytes. All in all, different processes occur within the layer but only the diffuse reflected light J from the top of the plate can be easily measured. To simplify Eqn (7.3), all losses of light in the absence of sample are described with the expression $I_{abs,u}$. The expression $I_{abs,s}$ describes the light absorbed by the sample. The scattered light J in the direction of the layer surface is:

$$J = I_0 - I_{abs} = I_0 - (I_{abs,u} + I_{abs,s}) \tag{7.4}$$

where,

$J(\lambda)$, the reflectance
$I_0(\lambda)$, the incident light intensity
$I_{abs}(\lambda)$, the light absorbed by the plate and the sample
$I_{abs,u}(\lambda)$, the light absorbed by the plate material
$I_{abs,s}(\lambda)$, the light absorbed by the sample

Equation (7.4) describes two different types of absorption which occur simultaneously. A sample "recognizes" the light intensity I_0 of the illumination lamp minus the light intensity $I_{abs,u}$ absorbed by the layer. The sample absorption coefficient as is now defined as a_s (Eqn (7.5a)). The same is true for the absorption of light by the layer, which is illuminated by I_0 minus the light absorbed by the sample. The plate absorption coefficient is therefore defined as a_u (Eqn (7.5b)).

$$a_s \equiv \frac{I_{abs,s}}{I_0 - I_{abs,u}} \tag{7.5a}$$

$$a_u \equiv \frac{I_{abs,u}}{I_0 - I_{abs,s}} \tag{7.5b}$$

Substituting $I_{abs,u}$ from Eqn (7.5a) in Eqn (7.5b) and $I_{abs,s}$ vice versa results in:

$$\frac{I_{abs,s}}{I_0} = \frac{a_s - a_s a_u}{1 - a_s a_u} \quad \text{and} \quad \frac{I_{abs,u}}{I_0} = \frac{a_u - a_s a_u}{1 - a_s a_u}$$

The absorption coefficient of the layer and sample then becomes the sum of both the factors [1,3].

$$a = \frac{I_{abs,s}}{I_0} + \frac{I_{abs,u}}{I_0} = \frac{a_s + a_u - 2a_s a_u}{1 - a_s a_u} \tag{7.6}$$

It can be shown that only the quotient of a and the scattering factor $(1 - a)$ is linear with respect to the sample mass m. That is exactly what the *Kubelka–Munk* equation states.

$$\frac{a}{1-a} = \frac{a_s + a_u - 2a_s a_u}{(1 - a_s a_u) - (a_s + a_u - 2a_s a_u)} = \frac{a_s(1 - 2a_u) + a_u}{(1 - a_s)(1 - a_u)} \tag{7.7}$$

The light intensity, absorbed by the sample mass m depends on the mass absorption coefficient a_m and must be written as $I_{abs,s} = m a_m I_0$. The rest of the light $I_{abs,u} = (1 - m) a_m I_0$ either leaves the layer or is absorbed by the stationary phase. It then follows that the sample absorption coefficient a_s becomes:

$$a_s = \frac{m a_m}{1 - a_m(1 - m)}$$

Substituting a_s in Eqn (7.7) results in the following expression [1–3]:

$$\frac{a}{(1-a)} = m\frac{a_m}{(1 - a_m)} + \frac{a_u}{(1 - a_u)} \tag{7.8}$$

This is the proof that there is a linear connection between the mass m of a compound and the quotient a/s. The intercept in Eqn (7.8) contains the constant a_u, and therefore, describes the layer absorption [1–3].

7.2.2 The Connection between Measured Data and Analyte Mass

The quotient $a/(1 - a)$ is proportional to the amount of analyte in the TLC layer. The absorption coefficient a comprises the absorption of the TLC layer and the analyte absorption. The absorption coefficient describes the fraction of incident light aI_0, which is absorbed. The non-absorbed radiation will be scattered inside the layer and only this light can be emitted as J_1 from the layer surface (which we also call the first virtual layer).

$$J_1 = I_0 - I_{abs} = I_0 - aI_0 = I_0(1 - a) \tag{7.9}$$

Opaque TLC plates shine brightly when light from the surface is reflected at a low angle. This observation from all TLC plates led to the conclusion that remission is not only influenced by the incident light intensity but also by the angle of the incident light. Thus, the extent of reflection is a function of the incident light intensity and the incident light angle. This angle changes not only according to the position of the incident light source placed over the plate but also with the crystal structure of the stationary phase material and its orientation within the layer. If we assume that every layer will illuminate the next layer above and below with different light intensities, we need to introduce a backscattering factor k. The factor k describes the fraction of remaining light in a virtual layer, which is scattered in the direction of the plate surface. Logically the fraction $(1 - k)$ describes the remaining light in a virtual layer, which is scattered in the direction of the plate bottom, which is also the direction of the incident light. To make the situation clear, the abbreviations $q = k(1 - a)$ represents the fraction of non-absorbed light toward the plate surface and $p = (1 - k)(1 - a)$ stands for the fraction of light in the direction of the incident light. Both expressions together describe how much light was scattered in this process. Therefore, we can conclude that:

$$s = p + q = (1 - a) \tag{7.10}$$

The schematic diagram in Figure 7.1 illustrates the interactions of the incident light (I_0) with at least five virtual layers.

For the light intensity of the first layer $J_1 = J$, the layer-model is expressed as:

$$\begin{aligned}\frac{J_1}{I_0} &= 2^0 q + \frac{1}{1*2}2^1 pq^2 + \frac{1*3}{1*2*3}2^2 p^2 q^3 + \frac{1*3*5}{1*2*3*4}2^3 p^3 q^4 \\ &+ \frac{1*3*5*7}{1*2*3*4*5}2^4 p^4 q^5 + \frac{1*3*5*7*9}{1*2*3*4*5*6}2^5 p^5 q^6 + \ldots +\end{aligned} \tag{7.11}$$

The sequence of factors (1, 2, 5, 14, 42, 132, 429,...,) is well known as the Catalan numbers (C_n) [4]. Equation (7.11) can be written as the sum of the Catalan numbers:

$$\begin{aligned}R_0 \equiv \frac{J_1}{I_0} &= q + 1pq^2 + 2p^2q^3 + 5p^3q^4 + 14p^4q^5 + 42p^5q^6 + 132p^6q^7 \\ &+ 429p^7q^8 + \ldots + = \sum_{n=0}^{\infty} C_n q(pq)^n pR_0^2 + q = R_0\end{aligned} \tag{7.12}$$

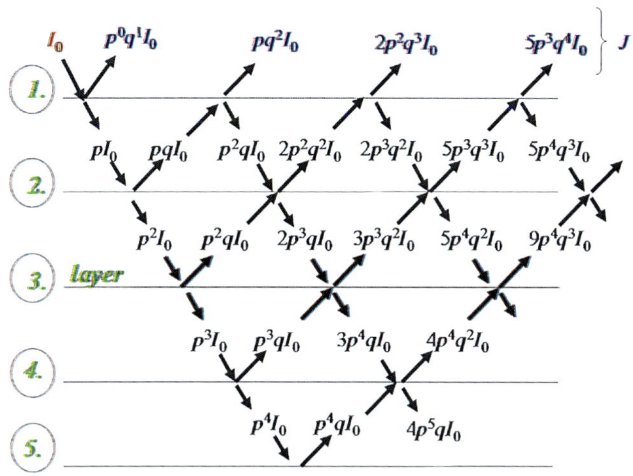

FIGURE 7.1 Illustration of the interactions of the incident light (I_0) with at least five virtual layers. The fraction q of I_0 is reflected to the top and the fraction p is scattered in the incident light direction. The expression for reflectance from the top five virtual layers is $J = qI_0 + pq^2I_0 + 2p^2q^3I_0 + 5p^3q^4I_0 + 14p^4q^5I_0 + \ldots +$.

Equation (7.12) results if the function that contains all of the Catalan numbers is multiplied by itself [4]. Equation (7.12) can be resolved to Eqn (7.13).

$$R_0 = \frac{1}{2p} - \frac{1}{2p}\sqrt{1 - \frac{4p^2q}{p}} = \frac{1}{2p}\left(1 - \sqrt{1 - 4pq}\right) \tag{7.13}$$

This equation can be rearranged to the quotient $a/(1-a)$ because only this expression is proportional to the analyte mass.

$$1 - 4pq = (1 - 2pR_0)^2 = 1 - 4pR_0 + 4p^2R_0^2$$

After rearranging and with the help of Eqn (7.10), we can write:

$$q\left(\frac{1}{R_0} - 1\right) + p(R_0 - 1) = a$$

If this equation is resolved to an expression as a function of the total absorption coefficient a and the backscattering factor k, we obtain Eqn (7.14), the modified *Kubelka–Munk* equation [1,2].

$$\mathrm{KM}(R_0, k \geq 0, k \leq 1) = k\left(\frac{1}{R_0} - 1\right) + (1-k)(R_0 - 1) = \frac{a}{(1-a)} \tag{7.14}$$

Equation (7.14) in conjunction with Eqn (7.8) demonstrate linearity between the transformed intensity data and the quotient $a/(1-a)$. We obtain direct linearity between the transformed data and the analyte mass. With the transformed intensity data according to Eqn (7.8), the fraction of light absorbed by the analyte can be separated from the light absorbed by the layer. The diffuse reflected light intensity J_0 can be determined at layer positions free of absorbing compounds. If this signal is used in Eqn (7.14) instead of the lamp intensity I_0 to calculate the relative reflection R, we assume that the TLC plate does not show any loss of light instead of sample absorption. Thus the corrected relative reflection R can be written as:

$$R = \frac{J}{I_0 - I_{\text{abs,u}}} = \frac{J}{J_0} \tag{7.15}$$

As a result the plate absorption intensity $I_{\text{abs,u}}$ becomes zero but in fact the light intensity of the lamp I_0 is only replaced by J_0. Mathematically, the original light flux of the lamp I_0 is reduced to J_0 by all loss of light at the plate surface, and hence, we define $I_{\text{abs,u}}$ to be zero. This turns a_{u} to zero as well. With Eqns (7.8), (7.14), and (7.15), we obtain the fundamental Eqn (7.16) for quantitative planar chromatography.

$$k\left(\frac{1}{R} - R\right) + (R - 1) = m\frac{a_{\text{m}}}{(1 - a_{\text{m}})} \tag{7.16}$$

where,

$R(\lambda)$, the relative reflectance ($R = J/J_0$)
k, backscattering factor ($k \geq 0$ and $k \leq 1$)
m, analyte mass
a_{m}, analyte absorption coefficient

For absorptions in scattering media, Eqn (7.16) is the equivalent expression to the *Lambert–Beer law* in clear solutions (also known as the *Beer–Lambert–Bouguer* law). The first term describes the scattering and the second term the fluorescence of the measured zone. In Eqn (7.16), the factor k indicates the fraction of remaining light in a virtual layer, which is scattered in the direction of the layer surface. In the case of $k = 1$, no scattered light illuminates the top of the second virtual layer. In this case, all light is absorbed in the first virtual layer or is scattered from the top of it. No light reaches the second virtual layer. This expression is useful in trace analysis and is indicated as reflectance formula $(1/R - 1)$. The second extreme case of Eqn (7.16) is $k = 1/2$. In this case, Eqn (7.16) renders the well-known *Kubelka–Munk* formula (Eqn (7.2)) with $R = R_\infty$. This equation is suitable for high analyte concentrations. The third extreme case of the extended *Kubelka–Munk* equation is $k = 0$, where no incident light is scattered toward the layer surface. This case indicates that the light emitted from the plate surface must be fluorescence.

7.2.3 Fluorescence Measurements

Sample molecules located in light-scattering media exhibit fluorescence. If we indicate this fluorescence as J_F, the absorbed light is transformed into fluorescent emission according to Eqn (7.17). The extent of this transformation is described by the quantum yield factor q_F. The light intensity absorbed by the sample can be calculated from the light intensity reflected from the clean plate surface J_0, minus the light intensity J reflected from the sample.

$$J_F = q_F(J_0 - J) = q_F J_0(1 - R) \tag{7.17}$$

where,

$J_F(\lambda)$, the emitted fluorescence intensity
$q_F(\lambda)$, the fluorescence quantum yield factor
$J_0(\lambda)$, the light intensity reflected from a clean plate surface
$J(\lambda)$, the light intensity reflected from the sample
$R(\lambda)$, the relative reflectance ($R = J/J_0$)

In case of trace analysis (which means we obtain k-values near $k \sim 1$), we observe linearity according to the extended *Kubelka–Munk* Eqn (7.16) for the reflectance formula $(1/R) - 1$. Multiplying Eqn (7.17) by $1/R$ and taking Eqns (7.10) and (7.16) into account, we observe:

$$\frac{J_F}{J_0 R} = q_F \frac{(1 - R)}{R} = q_F m \frac{a_m}{s}$$

If R is substituted by $R = s/(a + s)$, derived from expression $(1 - R)/R = a/s$ and taking Eqn (7.4) into account, we obtain:

$$J_F = m J_0 q_F \frac{a_m}{s} \frac{s}{(a + s)} = m J_0 q_F a_m \tag{7.18}$$

The reflected sample light intensity J is the sum of light–scattering J_{0F} and fluorescence at the fluorescence wavelength ($J_F = J - J_{0F}$). Equation (7.18) can therefore be rewritten as:

$$\frac{J - J_{0F}}{J_{0F}} = R_F - 1 = \frac{J_0}{J_{0F}} m q_F a_m \tag{7.19}$$

where,

$J(\lambda)$, the scattered light intensity of the sample (at the wavelength of the fluorescence)
$J_{0F}(\lambda)$, the intensity of the reference at the wavelength of the fluorescence
$J_0(\lambda)$, the intensity of the reference at the wavelength of absorption
m, the sample mass

$q_F(\lambda)$, the fluorescence quantum yield factor
a_m, the mass absorption coefficient

The fluorescence intensity is directly proportional to the sample amount in the layer at low analyte concentrations.

7.3 INSTRUMENTATION FOR IN SITU DENSITOMETRIC DETECTION

The extended *Kubelka–Munk* expression is valid for in situ infrared, UV–visible and fluorescence measurements. Linear calibration functions can be obtained for more than three orders of magnitude. Modern scanners use this theory for raw data evaluations.

7.3.1 Densitrometric TLC Scanner

In 1964, *H. Jork* described a scanner, which was able to record a TLC separation in the reflectance mode using a source beam of nearly monochromatic light [1]. For reflectance measurements, the lamp and detector are positioned above the TLC plate. This measurement principle has not changed in modern times. In mono-wavelength scanners, a monochromatic light beam of low intensity illuminates the plate surface from above. The detector at an angle of 45° measures the scattered light leaving the plate surface in all directions. The incidental light spot is scattered within the layer, to an extent of more than 150 μm. Therefore, the spatial resolution cannot be better than 300 μm. The wavelength dispersion device of the scanner illuminates the plate in a range ±2.5 nm of the chosen wavelength. The wavelength range generally lies in the range of 200 nm up to 700 or 800 nm. Measurements at different wavelengths require additional track scanning. For recording spectra the light beam must be spatially fixed at the spot center or on the sides, and the monochromatic device illuminates the spot with light of different wavelengths. The intensity of the reflected light is measured and plotted against the illumination wavelengths producing a conventional spectrum. This is a time-consuming process. In contrast, scanning a single track at a given wavelength can be completed within seconds. Various companies sell these more-or-less old-fashioned items. The market leader is the company CAMAG, Muttenz, Switzerland.

The first evaluation of a TLC separation using the diode-array technique was published in 1989 [5]. The basic idea of using a diode array for evaluating TLC separations was first included in a thesis written by *W. Wuthe* [1]. There are two systems on the market: The TIDAS TLC 2010 system from J&M company, Aalen, Germany, and (from summer 2015 on) there will be a system from Biostep Company, Jena, Germany. The J&M system consists of 50 identical optical fibers, each with a 100 μm diameter. These fibers transport light of different wavelengths from a deuterium lamp to the TLC plate and back to the diode array detector. In contrast to mono-wavelength scanners, the plate is illuminated by white light and the remission is analyzed at different wavelengths. This fiber-array arrangement produces overlapping areas in the illuminating and detecting fiber spots. The result is a small scanning slit, smaller than 100 μm, which is mainly dependent on the fiber-array distance above the layer. Establishing the optimum distance between the fiber arrangement and the TLC plate is the only adjustment needed. Measurements at different wavelengths can be simultaneously performed. The measured data must be transformed according to Eqn (7.16) to obtain so-called contour plots which comprise all measured densitograms and all spectra in a single plot.

To demonstrate the performance of the diode-array scanner, a dequalinium chloride containing ointment was analyzed (Figures 7.2 and 7.3). Dequalinium chloride [decamethylene-bis-(4-aminoquinaldinium)-chloride] is an antibacterial and antifungal agent, active against many Gram-positive and Gram-negative bacteria. For quantification, 1.25 g of the ointment was dissolved in methanol to a final concentration of

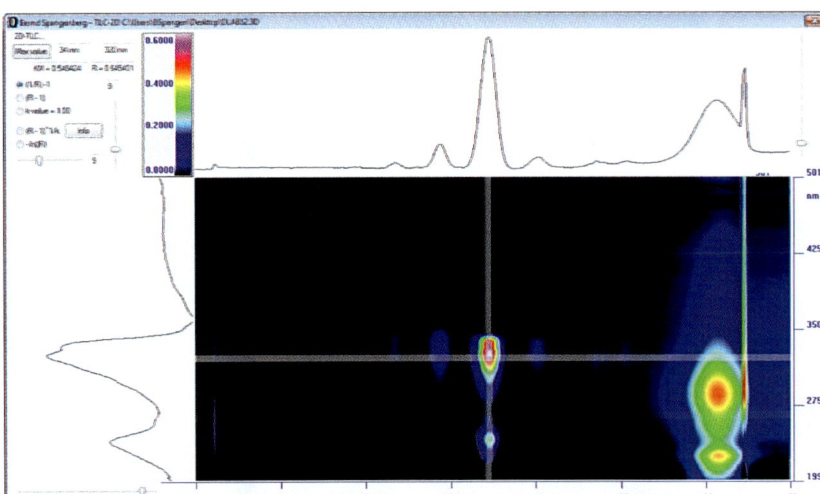

FIGURE 7.2 Contour-plot of a separation of an ointment with absorption evaluated according to Eqn (7.16) and a *k*-value of 1 (reflectance formula). The signal for dequalinium chloride can be seen at 34 mm separation distance together with some side-products. Plotted above is the densitogram and left the appropriate spectrum. Sample application was carried out at 3 mm and the mobile phase moved to 65 mm distance. The ointment matrix signal can be seen at 63 mm separation distance.

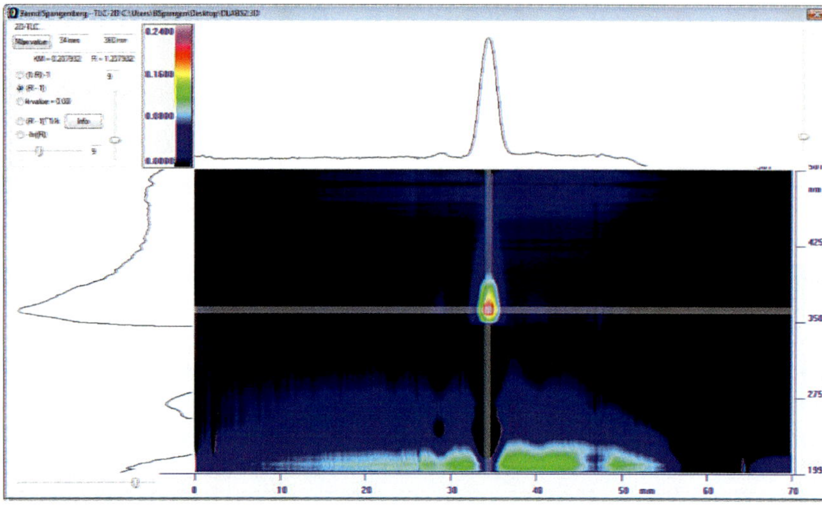

FIGURE 7.3 Fluorescence contour-plot of an ointment sample using Eqn (7.16) for evaluation with $k = 0$. The fluorescence signal of dequalinium chloride can be seen at 34 mm separation distance. Sample application was done at 3 mm and the mobile phase moved to 65 mm.

c. 4 mg/10 mL dequalinium chloride. One microliter of this solution was applied band-wise (7 mm) to the TLC plate. The silica gel HPTLC plates were developed in a vertical developing chamber without vapor saturation to a distance of 65 mm from the starting point (immersion line), using methanol and 1 M ammonium acetate in water (17 + 3, v/v) as the mobile phase. A sample pretreatment step can be avoided because the matrix remains on the disposable HPTLC plate. This allows relative standard deviations of the total analytical chain of less than 3%.

A contour plot, evaluated by the reflectance and the fluorescence formulae, instantly shows absorptions and reveals compounds showing fluorescence. The symmetric signals in the contour plots of Figures 7.2 and 7.3 indicate peak purity. Diode-array spectrometry allows checking peak purity automatically, which is required for quantitative TLC. Diode bundling reduces the spectral noise, which can be seen in Figures 7.2 and 7.3. In this case, nine diodes were combined into a single densitogram and a single spectrum thus reducing the noise by a factor of three. Using the diode-array technique even 2D-separations can be scanned [6].

7.3.2 Video-Densitometric Devices

In the beginning, TLC and paper chromatography relied on visual detection, thus it was measured in the wavelength range from 400 to 800 nm.

Nonabsorbing analytes were stained by using more-or-less specific reaction mixtures. Even today, this kind of analysis is widespread and used as a simple and inexpensive detection method. Today this simple TLC approach is gaining in importance duc to CCD-arrays. The term video-densitometry was introduced to describe these systems. Video-densitometric devices are small in size and weight, which make these camera systems portable [7]. The use of a CCD-device provides ultraviolet response and a larger dynamic range than a photomultiplier. The evaluation of 2D-separations is possible, which is difficult with slit-scanners. The principle of CCD-scanning is not restricted to cameras. A flatbed scanner can also be used for this purpose [7]. Commonly used flatbed scanners illuminate the plate with white light and can scan colored zones. Even fluorescence can be measured if the flatbed scanner is equipped with a UV-lamp [7]. The cheapest CCD-device for TLC evaluations is to use a hand-scanner [1,7]. The disadvantage of a video-densitometer is that spectral information is not available. This makes spectral peak identification and spectral peak-purity tests impossible. Otherwise, most substances show no light absorption or fluorescence in the visible range. To make separations more specific, we recommend staining, which often makes spectral identification and peak-purity testing unnecessary.

What features should be taken into consideration when using a CCD camera for TLC measurement purposes? Quantitative video-densitometric measurements require a detector, which can linearly digitalize light intensities. A double-fold light intensity must result in a doubled signal value, which can be checked by changing the measurement time. Therefore, a doubled measurement time must result in doubled measured values. The digital resolution of commonly used cameras is 8 bit. A signal is rendered in $2^8 = 256$ different increments (gray levels), which is not sufficient for quantification purposes. At least a minimum resolution of 16 bits is recommended, rendering $2^{16} = 65{,}536$ gray scales. These cameras must store data as TIFF images, since only the TIF-Format (Tagged Image File Format) supports 16-bit data storage.

Although inexpensive flatbed scanners and cheap cameras are not linearly calibrated, it is nevertheless possible to quantify planar chromatographic separations. In Figure 7.4 the calibration of 17-ethinylestradiol is measured by a yeast estrogen screen (YES) method. YES is a reporter gene assay used for measuring the activation of an estrogen receptor. The estrogen receptor activation is measured by inducting the reporter gene lacZ which encodes the enzyme β-galactosidase. Finally, the β-galactosidase activity—as a measure for the samples estrogenic potential—is determined by using the fluorescent substrate MUG (4-methylumbellifreyl β-D-galactopyranoside). This compound is cleaved to the fluorescent product 4-methylumbelliferone, which will be detected. The test strain according to *MCDonnell* (1991) is derived from *Saccharomyces cerevisiae* BJ3505 [8].

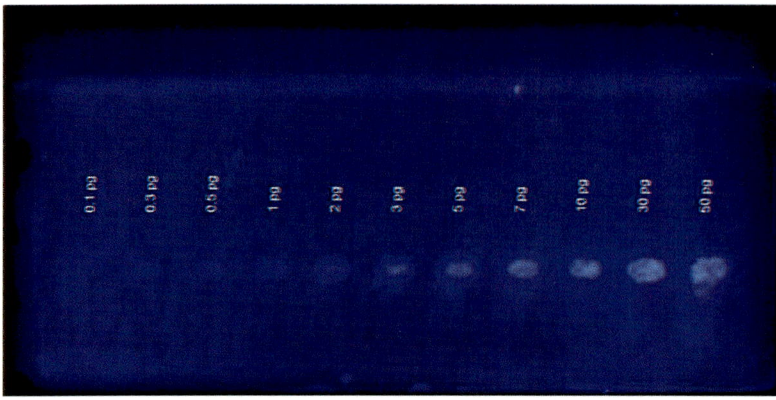

FIGURE 7.4 Fluorescence of 4-methylumbelliferone, induced by different 17-ethinylestradiol amounts (0.1–50 pg/spot). The mobile phase for separation on a silica gel 60 plate was hexane–ethyl acetate (8 + 3, v/v). The plate was measured in fluorescence mode using a simple 8-bit CCD camera.

Figure 7.4 shows the fluorescence of 4-methylumbelliferone spots, induced by different 17-ethinylestradiol amounts (0.1–50 pg/spot). The calibration shows a nonlinear curve because the plate was measured using an 8-bit CCD camera (Figure 7.5). Nevertheless, quantification is possible.

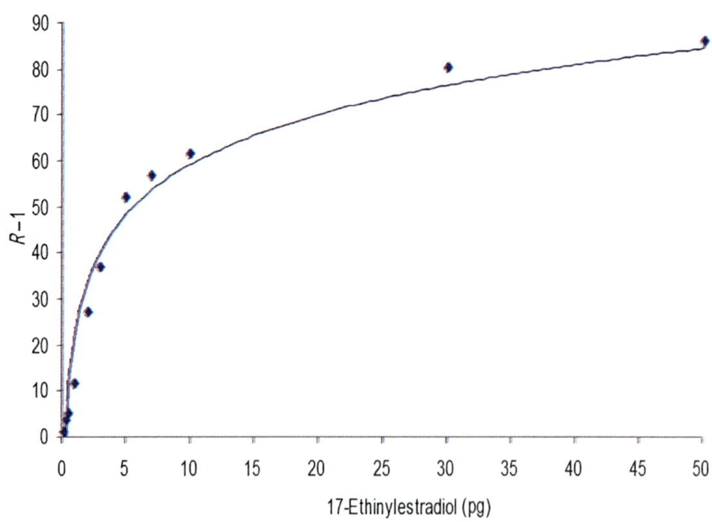

FIGURE 7.5 Logarithmic calibration curve induced by different 17-ethinylestradiol amounts (0.1–50 pg/spot), calculated from the fluorescent spots shown in Figure 7.4. A simple 8-bit CCD camera with a γ-factor of $\gamma = 0.4$ was used.

CCD cameras and flatbed scanners measure colored images. During the measurement, the TLC plate is illuminated by green, red, and blue light. Thus, a red channel (600–750 nm), a green channel (520–565 nm), and a blue channel (425–450 nm) arc available for evaluation. In Figure 7.4, only the blue channel was evaluated. All CCD cameras that are not designed for analytical measurements change their measured light intensity on purpose to adapt the result to human eyes. A so-called γ-factor is used for this. The gray level is calculated according to Eqn (7.20).

$$\text{Grey level} = (\text{reflected Light intensity})^{\gamma} \qquad (7.20)$$

$$\gamma(\text{human eye}) = 0.4$$

Using such a camera for TLC measurements will likely result in nonlinear, mostly logarithmic, calibration curves (Figure 7.5). If linear calibration curves are desired, the reflected light intensities should be transformed according to Eqn (7.16) and measured using a γ-factor of $\gamma = 1$. Cameras where the γ-factor can be altered should be preferably used for analytical measurements.

In Figure 7.6, a lozenge preparation was separated on an HPTLC silica gel cyanopropyl plate. The plates were developed in a vertical developing chamber without vapor saturation to a distance of 50 mm from the starting point, using water–acetonitrile–dioxane–ethanol–aqueous ammonia (25%) (8 + 2 + 1 + 1 + 0.05, v/v) as mobile phase. The

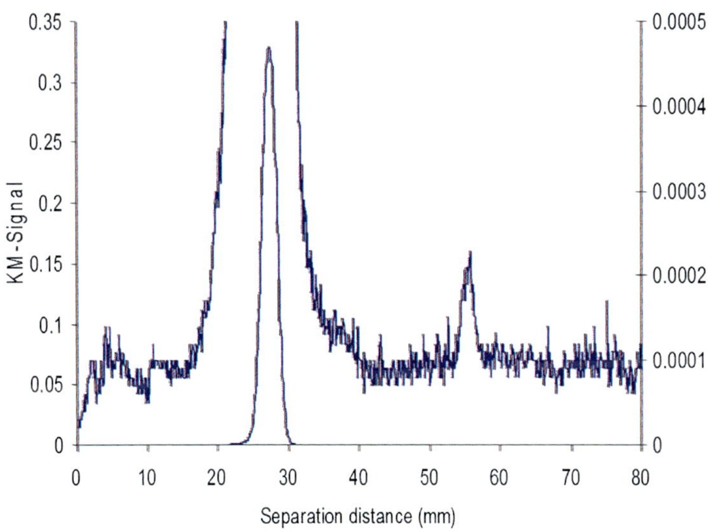

FIGURE 7.6 The densitogram of 1000 ng benzocaine and 0.1% aminobenzoeic acid (at 56 mm separation distance) on a cyanopropyl HPTLC plate measured with a 16-bit flatbed scanner with a γ-factor of 1.0.

benzocaine zones were stained using *Ehrlich's* reagent. The plate was dipped into the staining solution for 1 s, which instantly produces yellow spots on a white background. The plate was evaluated using the blue channel [9].

Figure 7.6 illustrates the densitogram for 1000 ng benzocaine (at 28 mm separation distance) and the amount of 0.1% aminobenzoeic acid (at 56 mm separation distance) measured with a 16-bit flatbed scanner. The performance of the 16-bit flatbed scanner can be seen in the enhanced plot of the same densitogram. The calibration plot, Figure 7.7, is linear in the working range from 10 to 1000 ng of benzocaine because a γ-factor of $\gamma = 1.0$ was used for the measurements in conjunction with Eqn (7.16). Equation (7.16) was applied with a k-factor of $k = 1/2$ which renders the original *Kubelka–Munk* expression (Eqn (7.2)). This formula is preferably used for the evaluation of spots with high analyte concentration.

7.3.3 Video-Densitometric Luminescence Measurements

The drawback of reflectance and fluorescence measurements by CCD cameras is the illumination problem. It is nearly impossible to evenly illuminate a 10 × 10 cm HPTLC plate. Bio- and chemiluminescence can be easily induced by *Vibrio fishery* bacteria or by oxidation of diaryl ethanedioates [10]. Dipping a TLC plate in a suspension of photobacteria is an

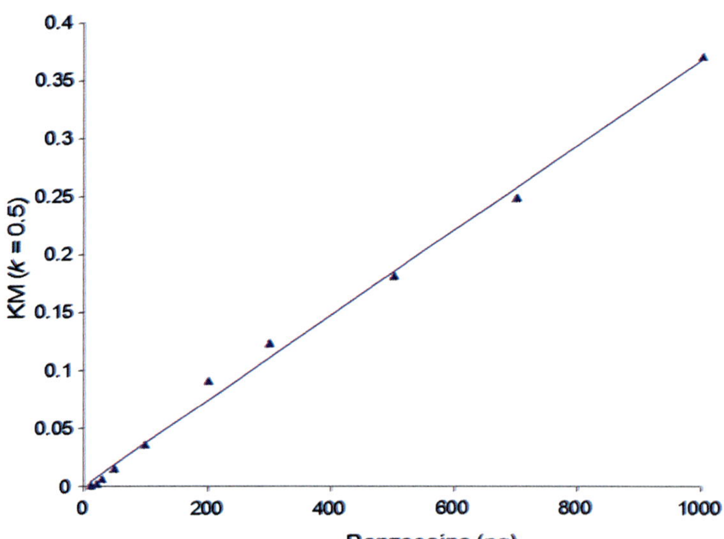

FIGURE 7.7 The calibration plot for different amounts of benzocaine separated on cyanopropyl HPTLC plate. The method is linear in the working range from 10 to 1000 ng because a γ-factor of 1.0 was used in conjunction with Eqn (7.16) with a k-factor of 1/2.

elegant method to detect in situ toxic substances on the plate. Photobacteria emit light via luciferase enzymes using long-chained aldehydes as a substrate. Their bio-luminescence depends directly on the metabolic status of the cell. If there is a certain concentration of these cells in the medium, this sets an auto-inducer free, (*N*-(β-hydroxybutyryl)homoserine lactone), which triggers the bacteria to luminescence. This is the case when the (slightly shaken) bacteria suspension has been kept incubated at room temperature for 25–30 h at ambient temperature. The luminous emission lasts for a maximum of about 24 h. The culture will be ready for use after 25–30 h shaking [10]. The TLC plate is dipped into the bacteria suspension for 3 s and then the suspension is gently wiped off from the plate surface. The plate luminescence is measured for 1–10 min with a light-sensitive 16-bit camera.

In Figure 7.8, the result of a *Vibrio fisheri* assay for two shampoo samples (the right two tracks) is shown. The four paraben standards for comparison (from left to right: methyl-, ethyl-, propyl-, and butylparaben) in amounts of 100 ng, each are also shown. The plate was measured within 10 min, using a light-sensitive 16-bit camera.

The compounds bis(2,4-dinitrophenyl)oxalate (DNPO) and bis(2,4,6-trichlorophenyl)oxalate (TCPO) are often used for inducing chemiluminescence. Both compounds are quickly oxidized by H_2O_2. In the presence of fluorescent compounds the reaction energy is transferred to the fluorescent compound which emits light while relaxing from its excited state. A solution of 250 mg DNPO in 40 mL n-butyl acetate was

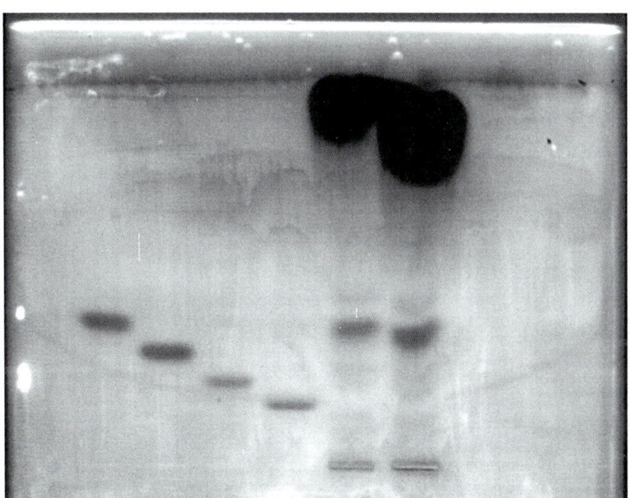

FIGURE 7.8 Two tracks of a shampoo sample measured with the *Vibrio fisheri* assay (the two tracks on the right) and the appropriate standard tracks for four parabens (from left to right: methyl-, ethyl-, propyl- and butylparaben, 100 ng each).

vigorously shaken for 20 min with 0.5 mL H_2O_2 (35%). The mixture is suitable for chemiluminescence measurements within a period of 3 h. Best results are observed when the plate is dipped for 1 s in this solution. The wet plate is dried until no light reflection can be seen on the surface. Then the TLC plate is covered by a glass plate and measured for 30 min using a light-sensitive CCD camera. The chemiluminescent reaction takes place on RP-18, RP-8, or RP-2 plates without the addition of a catalyst. The reaction is much faster on silica gel and this drastically reduces the measurement time. All other stationary phases show little or no light emission. Benzo[a]pyrene can be measured directly on plate down to a detection limit of 50 pg [11].

7.3.4 In situ Scintillation Measurements

A sample zone can emit light if a radioactive substance is present. For this, the X-ray light must be transformed into visible light. The two scintillation substances 1,4-bis-(5-phenyloxazole-2-yl)-benzene (POPOP) and 2,5-diphenyloxazole (PPO) can be used, dissolved in toluene, xylol, or cumol. POP transforms X-ray photons into UV light and POPOP transforms UV light into visible light at about 410 nm. A mixture of 5 g PPO and 0.3 g POPOP, dissolved in 1 L toluene can be used for in situ X-ray measurements on TLC plates [12].

7.3.5 Infrared and Raman Measurements

So far the most important range for TLC measurements is from 200 to 800 nm. This range can be extended to beyond 800 nm by the use of the so-called DRIFT-technique. DRIFT is the abbreviation for "Diffuse Reflectance Infrared Fourier Transformation" for measurements in the IR wave number range from 400 to 4000/cm (2.5–25 μm). Intense silica gel absorption bands from 1000 to 1350/cm and above 3550/cm restrict the available wave number range for in situ measurements to this range. For DRIFT measurements, a commercially available Fourier-transform infrared spectrometer is connected via mirrors and lenses to the surface of the TLC plate [1]. As stationary phase, glass-supported layers formed from 10 μm particles with a narrow size distribution are suitable. Better is to use a stationary phase mixture of 1:1 silica gel and magnesium tungstenate particles. For measurements the infrared light beam is moved over the plate surface and interacts with the stationary phase. The wave number dependent intensity of the reflected light is calculated by use of a Michelson interferometer. The spectra measured by the DRIFT-technique show more structured bands than those measured by UV–visible spectroscopy. The method is useful for substance identification [1]. Its disadvantages lay in its high cost and poor detection limits, which are 50-fold higher than typical for UV–visible measurements. The *Kubelka–Munk* equation can be used for evaluation.

Raman spectroscopy is a technique for studying molecular vibrations named after *C.V. Raman*, who invented it in 1928. It relies on the nonelastic scattering of monochromatic light when it interacts with matter. The detection limit is reported to be in the lower μg range (>2 μg/spot) [1]. Raman measurements show a rather high background, which must be subtracted from the sample spectra. An alternative to the FT-Raman method is Surface-Enhanced Raman Scattering Spectroscopy (SERS) [1]. Spraying an HPTLC plate with a silver colloidal solution produces a remarkable increase in Raman intensity. Raman scattering from molecules adsorbed on metal surfaces are enhanced by a factor of 1000. After separation, the dried plate is immersed in a silver colloid solution prepared from one volume of 1 mM aqueous silver nitrate solution added dropwise to three volumes of a 2 mM sodium borohydride solution in water. SERS spectra can be measured from 400 up to 1600/cm using a Raman spectrometer. The method is suitable for measurements in the pg range, but little work has been done in this field [1].

7.3.6 Measurement Error in Densitometry

Scanners consist of different parts contributing statistically defined noise to the measurement result. The light source is most important but the detector, signal amplifier, and A/D-converter also contribute to the noise. Equation (7.1) is calculated from two error-dependent values (J and J_0) which project their uncertainties into the reflectance R. In other words, both measurements contribute to the total error in R. The larger the value the better it can be measured. The brighter a clean plate surface reflects the incident light as J_0, the better we will measure J. If we assume that a sample spot reflects nearly no light ($J = 0$) the uncertainty in measuring this sample spot will be unpredictably high. From these facts, we can draw a simple conclusion to keep the measurement uncertainty low: neither J nor J_0 should have a value which is too low. On the other hand, we measure the amount of light loss, which is calculated as the difference between J and J_0. If this difference is too small, the relative calculation error will also dramatically increase. If we take both the effects into account, a minimum in the total measurement error is given for a situation in which both measurement values have the same size, thus $R = 1/2$! From this result, we can conclude that it is a fairy tale that the best results are achieved for measurements at the peak maximum [1]. This simple conclusion allows us to predict that the lowest error will be achieved if we measure around an R-value of $R \approx 1/2$! To achieve a minimum uncertainty, the measurement conditions must be prepared in such a way that all R-values are close to 1/2 [1].

7.4 IN SITU MASS-SPECTROMETRY

The combination of TLC and mass spectroscopy has often been described [1,13]. For desorption from the plate and for ionization purposes, several techniques, such as matrix-assisted laser desorption/ionization mass spectrometry (MALDI techniques using a UV/IR-Laser), by atmospheric-pressure chemical ionization (APCI), single ion beam (SIMS), or desorption electrospray ionization (DESI) have been described. The general aim of recent research is to liberate the mass analysis from the constraints of placing the surface to be sampled into a vacuum system. The number and type of ambient or atmospheric-pressure (AP) techniques suitable for HPTLC sampling is rapidly expanding. In principle, all in situ TLC-MS methods can be divided into two categories: fast atom and ion bombardment and laser irradiation. All TLC-MS systems offer structural information for the analyte and extend the scope of the TLC technique.

7.4.1 Fast Atom and Ion Bombardment Ionization

The most commonly used ion bombardment technique is EASI-MS (Easy Ambient Sonic Spray Ionization Mass Spectrometry) which offers a simple ionization interface. The method is capable of directly characterizing analytes on a solid surface [13]. Coupling TLC with ambient ionization techniques allows an atmospheric-pressure measurement. The TLC plate is placed on an XY-stage for precise movements of the plate. Desorption and ionization from the plate surface are performed by the charged droplets produced by a sonic spray using a 1:1 acidic (0.01% formic acid) water/methanol solution. The spatial resolution on the plate is 330 μm.

DESI employs high voltages and a methanol and/or formic acid spiked nitrogen gas for nebulization. The evaluation of 2D-separations is possible but time consuming. As DESI emitter a 7–13 cm long stainless steel tube (0.84 mm i.d., 1.27 mm o.d.) is moved over the TLC plate surface and press fitted into a standard heated atmospheric-pressure sampling capillary maintained at 200 °C. A high voltage of 4 kV is applied to the stainless steel body. The tip of the DESI spray emitter is mounted 2–3 mm above the surface of the layer at a ~55° angle to the surface.

In the APGD-method (Atmospheric Pressure Glow Discharge Mass Spectrometry), a simple ion source using plasma under atmospheric pressure desorbs and ionizes the analyte. The spatial resolution of this technique is better than 2.5 mm.

A new method for in situ TLC measurements is DART (Direct Analysis in Real Time). The DART-system works with an excited helium gas stream, forming protonated water clusters from the surrounding air.

These clusters transfer their energy to the analyte forming molecular cations. The spatial resolution of this device is better than 7 mm and the detection limit is in the lower ng range. All these methods are able to measure the molecular mass and characteristic fragments in situ from TLC plates. This is very important information to identify—in combination with UV–vis spectra—the composition of TLC-zones. Nevertheless, quantification is possible only by using an internal standard, because desorption processes are difficult to reproduce.

7.4.2 Laser Irradiation Ionization

All the atmospheric-pressure methods are in some way complementary to MALDI-MS, since MALDI is primarily suited to the detection of large molecules (>500 Da), such as peptides, proteins, and phospholipids. EASI, DESI, APGD, and DART are suitable for the detection of smaller molecules (<500 Da). For MALDI-MS, the plate must be prepared prior to analysis [13]. The plate is dipped in a coating solution (e.g., 2,5-dihydroxybenzoic acid), which will absorb the laser energy for ablation. The suitably impregnated plate is then placed under high vacuum. The mass-spectrometric information is subsequently read out automatically. The spatial resolution is better than 200 μm [1]. A new method AP-MALDI is able to operate under ambient pressure using a pulsed laser beam to irradiate cocrystal spots containing the analyte and the matrix. A high spatial resolution on plate is possible thus allowing the acquisition of molecular images of analytes, separated in one- or two-dimensional TLC [13]. The disadvantages of MALDI and AP-MALDI are interferences from matrix ions in the low mass range and lateral diffusion of analytes along the TLC track [13].

7.5 IN SITU RADIOISOTOPE DETECTION

Radioactive isotopes have been used for decades to elucidate biochemical pathways and TLC plays an important role in this approach. Radionuclides are used as tracers in pharmaceutical and biological research. Appropriate compounds are labeled with ^{3}H, ^{14}C, ^{35}S, ^{32}P, ^{18}F, ^{99}Tc, or ^{125}I and then separated by TLC. The β- or γ-radiation from these compounds is subsequently measured with a special scanner, using techniques such as gas counting, β-positron-*Geiger-Müller* counter, γ-scintillation counter, or storage phosphor screen imaging scanner [1,14].

For in situ gas counting of all β- and γ-emitting isotopes, a counting gas is ionized by interaction with the radioactive TLC zone. This produces a pulse of electrons that is proportional to the amount of radioactive compound in the separated zone. ^{3}H-labeled compounds can be measured only by use of an open-window gas flow-through counter. The open-window configuration can lead to difficulties for routine operations. For example, the high voltage counting wire can attract dust. Closed-window gas counting avoids this. Isotopes with medium-energy emission such as ^{14}C can be detected because the emitted radiation is able to pass through a very thin Mylar foil, which closes the ionization chamber. For high-energy β-emissions, the simplest detection method is to use a *Geiger-Müller* counter. During a scan of a TLC track, the radioactive material may decay due to its short half-life. To avoid loss of radioactivity by decay the *Geiger-Müller* scanner moves relatively quickly over the track many times. A γ-scintillation scanner can be used for the detection of γ-emitting isotopes. All instruments show high sensitivity, linearity over 4–5 decades of activity, and a spatial resolution on the layer of 0.5–3 mm. A TLC track can be measured within 1–10 min, depending on the isotope. The instrumentation for these methods is described in the review [14].

Storage phosphor screen imaging is an advantageous method for detecting radioactivity of compounds separated by TLC. After separation of the radionuclide labeled compounds the TLC plate is exposed to a phosphor imaging plate. The phosphor imaging plate provides a flexible and reusable image sensor in which crystals of a photo-stimulatable phosphor are coated as a film. The exposure time can be quite long, even overnight. The imaging plate accumulates and stores the radioactive energy as a latent image. After this exposure, a laser beam scans the phosphor imaging plate. The resulting photo-stimulated luminescence of exposed zones on the phosphor imaging plate (photo-stimulated luminescence, PSL) is measured by a photomultiplier and plotted as a densitogram. For TLC separations, the linearity, sensitivity, and resolution are excellent [1]. Bio-imaging analyzers are sold by different companies. Phosphor image detection results in better spatial resolution than for digital autoradiography and is able to scan 2D-separations [14].

References

[1] Spangenberg B, Poole CF, Weins Ch. Quantitative thin-layer chromatography. A practical survey. Heidelberg: Springer; 2011.

[2] Spangenberg B. Does the Kubelka–Munk theory describe TLC evaluations correctly? J Planar Chromatogr 2006;19:332–41.

[3] Spangenberg B, Post P, Ebel S. Fibre optical scanning in TLC by use of a diode-array detector – linearization models for absorption and fluorescence evaluations. J Planar Chromatogr 2006;15:88–93.

[4] Koshy Th. Catalan numbers with applications. Oxford University Press; 2008.

[5] Bayerbach S, Gauglitz G. Spectral detection in thin-layer chromatography by linear photodiode array spectrometry. Fresenius' Z Anal Chem 1898;335:370–4.

[6] Milz B, Klein KF, Spangenberg B. Quantitative two-dimensional thin-layer chromatography using a diode-array detector. J Planar Chromatogr 2012;25:493–7.
[7] Broszat M, Ernst H, Spangenberg B. A simple method for quantifying triazine herbicides using thin-layer chromatography and a CCD camera. J Liq Chromatogr Relat Technol 2010;33:948–56.
[8] Spira D, Reifferscheid G, Buchinger S. Combination of high-performance thin-layer chromatography with a specific bioassay – a tool for effect-directed analysis. J Planar Chromatogr 2013;26:395–401.
[9] Milz B, Spangenberg B. A validated quantification of benzocaine in lozenges using TLC and a flatbed scanner. Chromatographia 2013;76:1307–13.
[10] Choma IM, Grzelak EM. Bioautography detection in thin-layer chromatography. J Chromatogr A 2011;1218:2684–91.
[11] Spangenberg B. A new way of using chemiluminescence in thin-layer chromatography. J Planar Chromatogr 2011;24:357–9.
[12] Miwa M, Matsumoto M, Tezuka M, Okada S, Ohsuka S, Fujiwake H. Quantitative fluorographic detection of 3H and 14C on two-dimensional thin-Layer chromatographic sheets by an ultra-high sensitive TV camera system. Anal Biochem 1986;152:391–5.
[13] Cheng SC, Huang MZ, Shiea J. Thin layer chromatography-mass spectrometry. J Chromatogr A 2011;1218:2700–11.
[14] Sherma J, Larkin JD, Larkin FH. A field guide to instrumentation. J AOAC Int 2009;92: 29A–35A.

CHAPTER

8

Staining and Derivatization Techniques for Visualization in Planar Chromatography

Paweł K. Zarzycki

Section of Toxicology and Bioanalytics, Koszalin University of Technology, Koszalin, Poland

8.1 PROBLEM OVERVIEW

Detection and quantification of target compounds, particularly trace amounts of micropollutants or biomarkers in complex samples, is a key step in chromatographic analysis. The most important feature of planar chromatography is that after development and removal (evaporation) of the mobile phase, the separated compounds remain immobilized on the sorbent layer. Such a situation, which is specific to thin-layer chromatography (TLC), substantially extends the number of detection procedures and approaches that can be applied for quantification of target compounds. For direct detection in TLC the mobile phase components need not be transparent to the detector because they are removed before the detection step, in contrast to column chromatography. In case of nondirect detection using visualization reagents, different solvent systems after the separation step may be used as the carrier for reagent application. Generally, dry plates make an excellent support for postrun visualization via physicochemical interactions or chemical reactions. Planar chromatography allows robust quantitative visualization due to the fact that each analytical run can be performed on a new stationary phase. Moreover, most modern stationary phases are chemically stable, which significantly extends the capability of TLC for detection of target compounds at elevated temperatures.

Instrumental Thin-Layer Chromatography
http://dx.doi.org/10.1016/B978-0-12-417223-4.00008-X

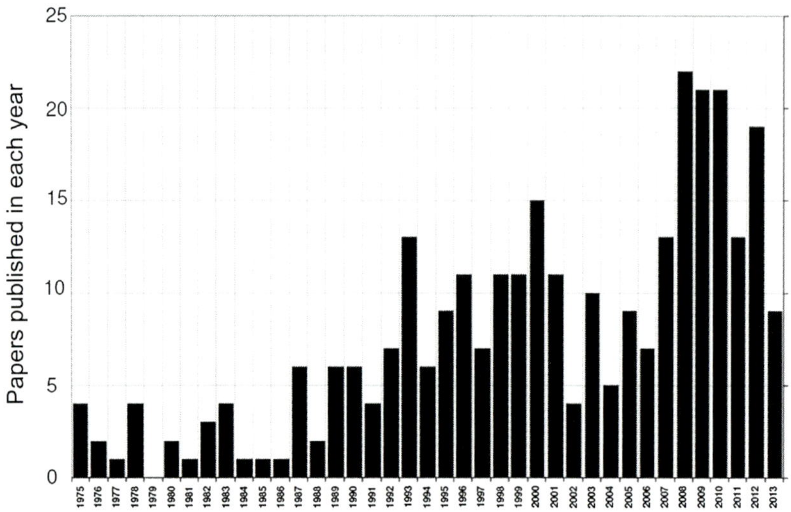

FIGURE 8.1 Publications in planar chromatography associated with topic phrase "*TLC derivatization*"; Comparison of the manuscripts number within years 1975–2013 according to Web of Knowledge database.

In principle, any qualitative or quantitative method requires a suitable signal to noise ratio for the detected compounds with an appropriate contrast. In planar chromatography and related techniques (e.g., microfluidic-paper-based devices) this can be achieved by online or post-separation reactions of compounds on the plate with a wide range of visualization reagents. This topic has been extensively described in the scientific literature (Figure 8.1), guide-books, and brochures [1–14]. Physicochemical visualization and/or chemical derivatization allow effective detection of components of interest that in their native form are transparent to a given detector for specific data acquisition conditions. Typically, such analytical signals result from physicochemical interactions or chemical reactions. Visualized zones can be acquired by a detector, usually operating within the UV–Visible–IR region, in the absorbance or fluorescence mode, by reflectance or transmission measurements using digital cameras, scanners, or video densitometers. In planar chromatography involving derivatization, the key issue is advanced postacquisition signal processing, as for thermal waves measurements, that was also demonstrated for TLC [15–18]. Presently, new types of efficient algorithms involving wavelet transforms, correlation functions and sensitivity analysis based on thermography that can be adapted to postrun TLC image processing has been developed and extensively investigated [19–23].

8.2 REAGENT APPLICATION, EQUIPMENT, AND PROTOCOLS

In general, there are three techniques used to apply a derivatizing reagent to a TLC plate after development: (i) dipping, (ii) spraying, and (iii) vapor phase contact. Dipping and spraying is appropriate for solutions and liquid reagents and vapor phase methods for gas phase reagents. In laboratory practice all are extensively used for postrun detection of sample zones on developed plates. Depending on the reagent type and reaction protocol, sample zones are visualized immediately, within seconds or minutes, or after solvent evaporation and plate heating.

The main advantage of dipping is the homogeneous distribution of the reagent across the plate surface combined with full saturation of the stationary phase. Uniform coverage is essential for quantitative densitometric evaluation of sample zones. The dipping method usually reduces the background noise and strongly diminishes large scale tide marks in the resulting chromatographic profile. Manual dipping is commonly performed using wide diameter plastic or glass jars filled with relatively small amounts of an appropriate liquid derivatizing reagent (1–3 cm above the bottom level). In the manual dipping procedure, the developed plate is horizontally placed under the surface of the liquid reagent for a short time. This step must be performed with the plate positioned stationary phase down and with a small plate-liquid angle to avoid further blurring of the separation. Optimal dipping time may vary depending on the temperature and the liquid viscosity. Usually, the plate is immersed for a short time (1–3 sec), which is sufficient to penetrate the stationary phase and to obtain a satisfactory uniform layer of derivatizing reagent. After the dipping step the plate must be immediately removed from the jar and positioned vertically, with its edge on a filter paper or porous material (typically touching the edge opposite to the start line) to allow excess reagent to drain from the impregnated layer. Moreover, it minimizes the possibility of distortion of the chromatogram as a result of reagent application postseparation. Considering the price of reagents and chemical-waste disposal, manual dipping is preferred for reactions on small TLC plates e.g., 10 cm × 10 cm or less.

The dipping process can be automated and several devices are commercially available for this purpose (Figure 8.2). In these devices the TLC plate is positioned vertically and then moved to a narrow solvent container filled with reagent. The advantage of the automated approach is that the dipping rate and total immersion time can be strictly controlled. Both parameters are important for reproducible quantitative analysis. For large plates (e.g., 10 cm × 20 cm or 20 cm × 20 cm) automated methods

FIGURE 8.2 Commercially available immersing device for automatic dipping of developed TLC plates. *Reprinted with permission from CAMAG Catalog Instrumental Thin-Layer Chromatography 2013.*

have the advantages of lower cost and produce less hazardous waste, because narrow tanks require minimal amount of the reagent solution.

In recent times, the dipping method has become more common, because this procedure significantly minimizes laboratory staff exposure to poisonous or irritating aerosols composed of reactive chemicals, which are produced by spraying methods. A general disadvantage of both manual and automatic plate dipping is that some derivatizing reagents are highly unstable and may quickly decompose after direct contact with air, the layer surface, and/or separated sample components. The consequence of this is that for satisfactory quantitative analysis some reagent solutions must be regularly replaced, even after each dipping procedure.

The spraying technique is frequently applied as a universal method for transferring a uniform film of derivatizing reagent to the developed TLC plate. Importantly, this process requires only small amount of reagent solution, while unused reagent solution remains in the sealed bottle compartment and out of contact with air, the layer surface, or separated sample components, which as mentioned above, may result in

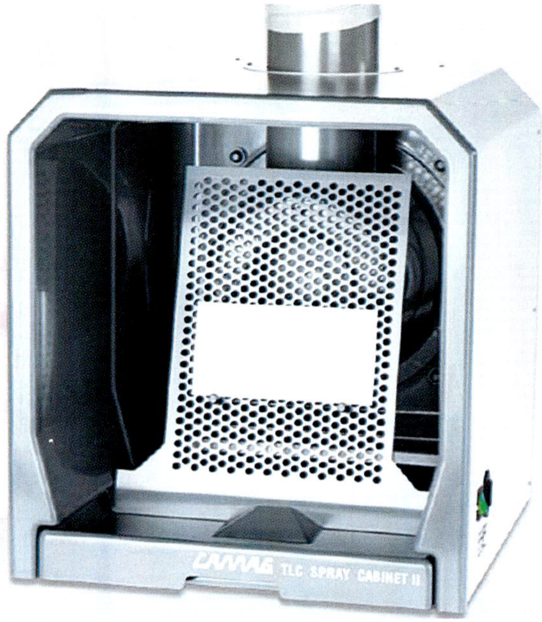

FIGURE 8.3 Air ventilated spray cabinet. *Reprinted with permission from CAMAG Catalog Instrumental Thin-Layer Chromatography 2013.*

deterioration of the reagent solution. Detection of target compounds via spraying can be completed in a short time. However, hazardous reagents must be sprayed in well ventilated hoods or cabinets (Figure 8.3). This insures that laboratory staff are not exposed to corrosive vapors by inhalation or through contact with eyes and skin. Spay containment-devices also protect laboratory equipment from exposure to corrosive agents like sulfuric acid. On the other hand, the setup of equipment for spraying in some cases can be tedious and time consuming.

A spray consists of small droplets of the reagent solution mixed with gas and is propelled from the spray head by gas or liquid pressure. An aerosol jet can be formed using different types of spray heads operated with air, inert gas, or reagent solution only. Typical sprayers, or atomizers, may run with gas pressure generated manually (e.g., rubber bulb) or electro-pneumatically as well as from an external compressed gas source, such as air pumps or gas cylinders. For application of relatively nonreactive reagent solutions, spray devices made from polyethylene units are adequate. Nevertheless, the reagent container and spray head must be resistant to organic solvents. Glass, stainless steel, or PTFE containers are used with corrosive reagent solutions. Some examples of commercially available and home-made spray devices are shown in Figure 8.4.

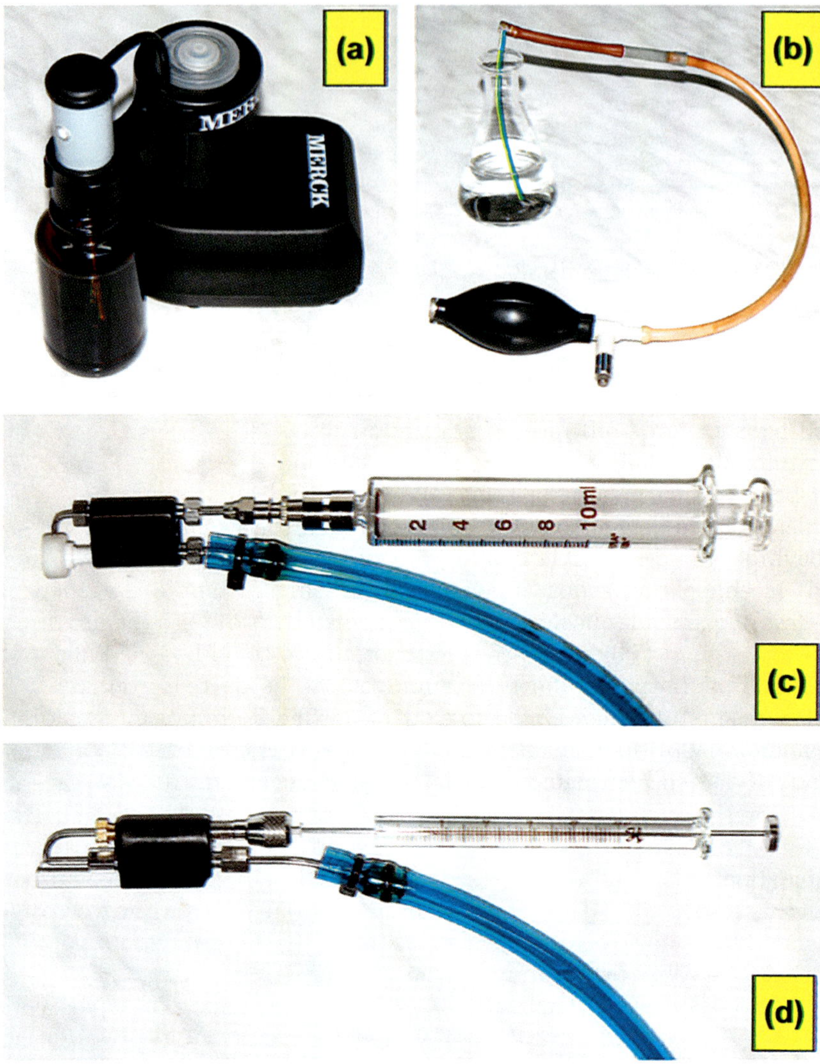

FIGURE 8.4 Examples of compressed air operated sprayers. Low pressure battery (a) and manual (b) driven devices. Compressed gas (0.5 bar) home-made units for handling large (milliliters range; (c)) and low (microliters range; (d)) volumes of visualization reagents [24].

From practical point of view the viscosity of the reagent solution may strongly influence the spraying results. Insufficient gas pressure and/or small diameter capillaries inside the spray head and reagent container may cause the formation of large droplets and a too narrow aerosol jet cone. Both phenomena can strongly affect the homogeneity of the reagent solution deposited on the layer. This can be observed at both small and large scale, affecting quantification by creating a nonuniform baseline as well as increasing background noise. To avoid this problem, equipment with suitable capillary dimensions and an optimized working pressure for each solution must be employed. A wide range of spray devices are currently available from different manufacturers (e.g., Alltech, Camag, Desaga, Macherey–Nagel, Merck, Supelco, Sigma–Aldrich and more) with a wide range of sophistication, including fully automated, computer controlled, robotic systems with selectable operating modes (for example Desaga Chroma Jet DS20).

In our laboratory we constructed simple spray heads that can work with a wide range of solution viscosities (Figure 8.4 (c and d)) employing an adjustable pressure provided from an external gas line or cylinder. Typically, a jet of reagent solution is generated at a pressure of 1–2 bar supplied by an adjustable compressed air line. Construction details for spray heads are presented in Figure 8.5 [24,25]. These compressed air-operated spray devices can handle small or large volumes (from microliters to milliliters) of reagent solutions. They provide a fine spray at the jet exit nozzle even for difficult solutions, for example starch solution, concentrated phosphoric or sulfuric acids. Both heads are made from

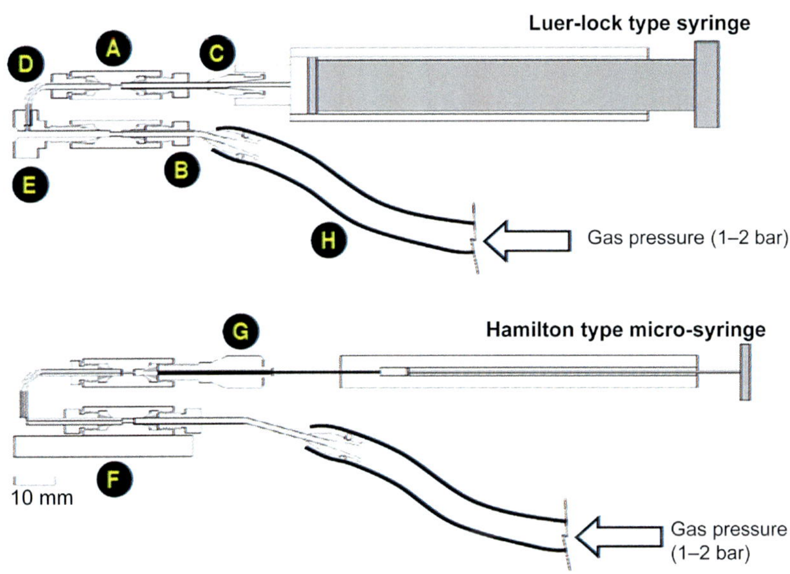

FIGURE 8.5 Sectional drawings and construction details of homemade compressed gas sprayers (displayed in Figure 8.4(c and d)). Top and bottom modules (working with Luer-lock syringe and Hamilton micro-syringe, respectively) are assembled from the chemically resistant HPLC fitting accessories including: stainless-steel HPLC union (A), hexagonal head fitting with ferrule (B), metal Luer needle (C), stainless steel HPLC tubing (pipe internal diameter = 0.5 mm); (D), PTFE finger-tight fitting (E), sample splash-protection bar (F), HPLC injection port for Hamilton type micro-syringe (G), flexible pipe (working gas pressure 1–2 bar; H). *(Reproduced from Ref. [25]with permission).*

chemically resistant HPLC accessories, including standard PTFE or stainless-steel fittings and tubing, and can tolerate different types of corrosive and nonstable liquids. Figure 8.6 illustrates the result of depositing different reagent solutions on TLC plates. It is apparent that despite the viscosity of the reagent solution they can be uniformly spread on the TLC plate. It is worth noting that within the sprayed zone the background noise does not increase. This indicates that small solution droplets are generated by the spray head. The sprayer can be used to apply parallel zones of different reagent solutions as well as dispensing small sample volumes (starting from 1 μL) to a TLC plate by the sprayed-on band technique. In practice, homogenous zones and bands at the origin can be easily formed using aluminum or plastic shields, perforated with band-shaped holes or different patterns (Figure 8.7).

The Automated spray-on technique is commonly used in TLC for sample application in the form of spots or bands without damaging the stationary phase surface. This process can be partially or fully automated and several instruments are commercially available (see Chapter 4). Such

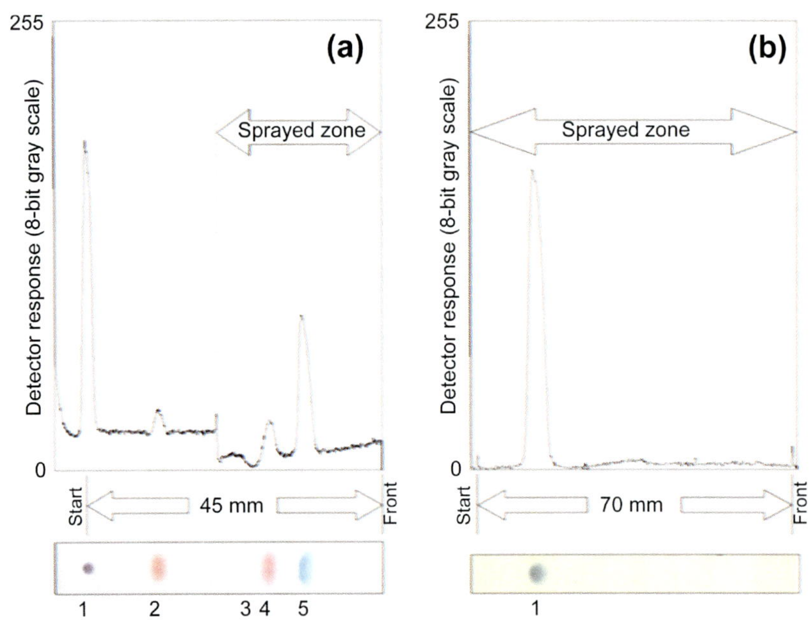

FIGURE 8.6 Application of low volumes zones of visualization reagents using home-made compressed air units (Figures 8.4(c and d), and 4) dispensing of an aqueous solution of sodium carbonate (a) and a methanolic solution of phosphomolybdic acid (b). Samples ID: crystal violet (1), methyl red (2), fluorescein (3), phenolphthalein (4) and bromocresol green (5). Visualization conditions—PMA: temperature 40 °C, heating time 10 min; sodium carbonate: saturated aqueous solution at room temperature. *(Reproduced from Ref. [25]with permission).*

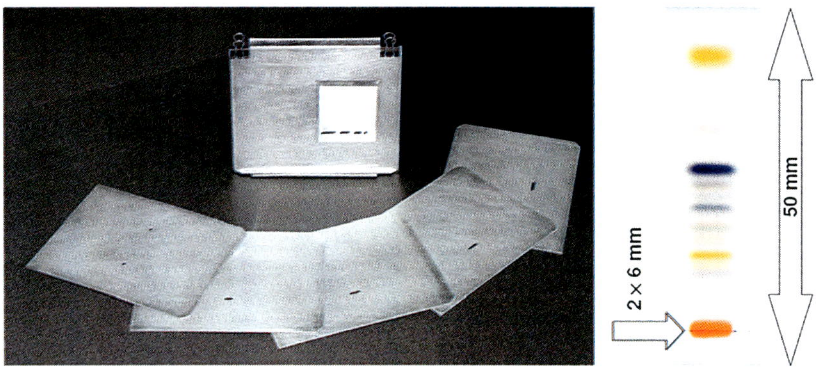

FIGURE 8.7 Thin-layer chromatography (TLC) plate support and aluminum shields with perforated holes and bands as well as the example of planar chromatogram with the sample (*Spirulina maxima* extract) dispensed by means of the microsprayer module, which was equipped with Hamilton type syringe (Figure 8.4(d)).

devices can be applied for accurate dispensing of reagent solutions, allowing selective postdevelopment visualization of sample components. An example employing a small amount of quantum dots fluorescent solution is shown in Figure 8.8.

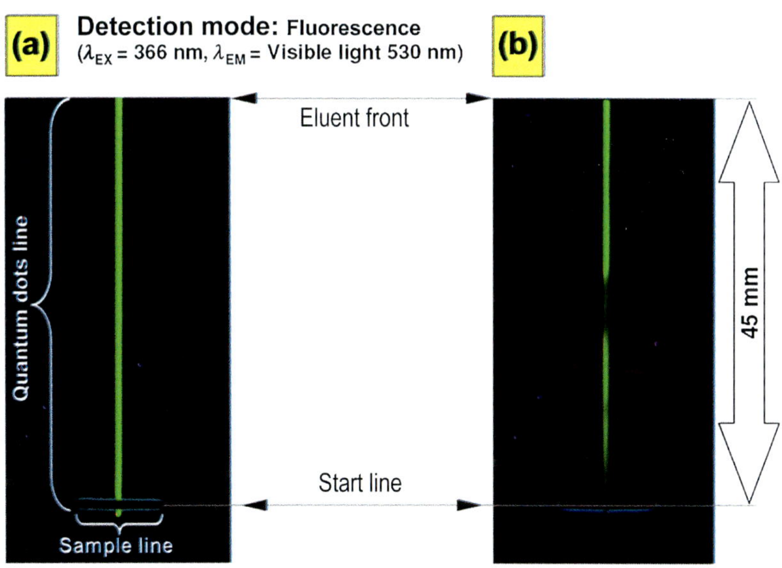

FIGURE 8.8 Sample and detection lanes (methanolic extract of *Spirulina maxima* desiccated cells and quantum dot nanoparticles: toluene solution of Lumidot™CdSe/ZnS 530 nm, respectively) formed using a semiautomated spray-on device (Linomat 5, Camag). Micro-TLC plate (HPTLC RP18W) before (a) and after (b) development with mobile phase composed of acetone: *n*-hexane,(3:7, v/v) and temperature 30 °C.

The use of volatile reagents enables a simple and nonexpensive detection of target compounds on developed TLC plates. A homogeneous distribution of reagent is easily obtained using sealed tanks (usually glass desiccators) containing a small amount of solid or liquid reagent forming a chemically active gas phase. As discussed for spraying, for safety reasons, the exposure of the layer to the reagent vapors must be performed in well ventilated hoods or cabinets. Reaction chambers should be fully presaturated with reagent vapor, allowing a rapid penetration of the stationary phase layer by the reagent vapor. It should be noted that vapor equilibration is rather time consuming (the saturation process ranging from minutes to hours), but full saturation of both chamber and stationary phase volume is necessary for quantitative analysis. Unfortunately, this useful protocol can be used at room temperature and normal pressure conditions with only a few volatile reagents (e.g., iodine, ammonia, nitric or hydrochloric acids). As can be seen from the chromatograms in Figure 8.9, the simple exposure of developed microplates to iodine vapors at room temperature, significantly improves the detection of a wide range of visible light transparent substances [26]. Under such conditions around 20

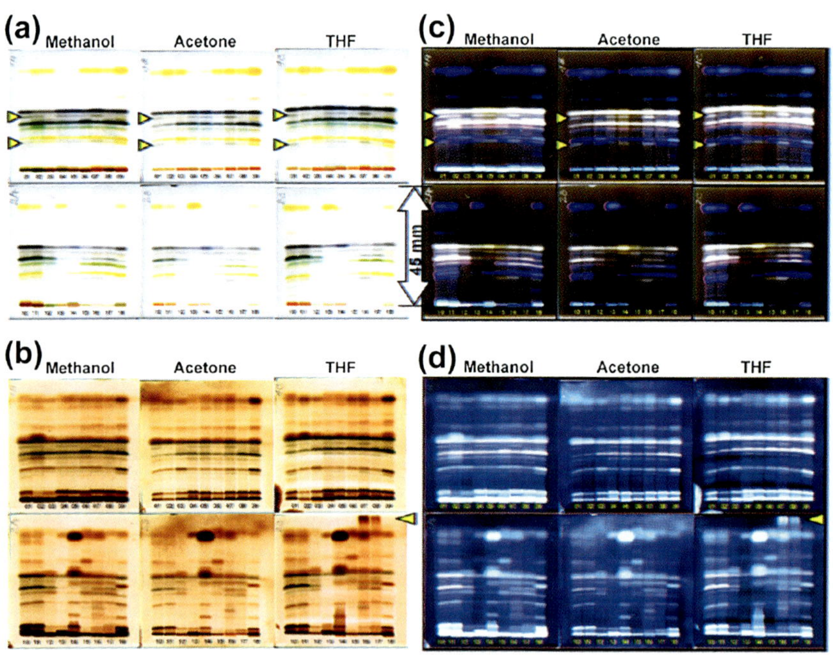

FIGURE 8.9 Micro-TLC separations of low-molecular-mass metabolites extracted from dry cyanobacteria cells and herb leaves before (top) and after (bottom) staining with iodine vapor ((a), (b)—raw scans; (c), (d)—inverted colors). *(Reproduced from Ref.[26]with permission).*

individual bands were detected on the micro-TLC plates. This visualization method combined with chemometric data mining allowed a distinction to be made between genuine spirulina and nonspirulina samples as well as fresh from expired commercial products. In addition, identification of some biodegradation compounds in the micro-TLC chromatograms was possible.

8.3 TECHNIQUES FOR HEATING OR DRYING LAYERS AFTER DEVELOPMENT

After exposure of the immobilized sample on the layer to the reagent solution or vapors some reactions may occur immediately resulting in colored or fluorescent bands that can be visually inspected and/or analyzed with a densitometer. However, the reagent solvent must usually be removed from the plate. How this is achieved will depend on the solvent's volatility and can be performed at room or elevated temperature in a well ventilated cabinet under a stream of air or inert gas. Plate heating increase the solvent evaporation rate and in most cases facilitates efficient reaction between the sample and reagent to complete the derivatization reaction within a reasonable time. In laboratory practice, simple devices called plate heaters or hotplates are used to complete solvent evaporation and to perform the staining reaction on the TLC plate Figure 8.10. The temperature of the top surface of the hotplate can be accurately controlled and programed, with temperature operating ranges from 25 to 200 °C. Such devices are usually protected from overheating. Hotplates provide uniform temperature distribution of the TLC plate surface and are sufficient for qualitative analysis. However, it should be noted that using a hot plate to heat a TLC plate may result in an unknown temperature at stationary phase surface, since one side of the plate, usually with the adsorbent layer, is in direct contact with room temperature air. This may cause uncertainty for the real temperature of the

FIGURE 8.10 Electronically controlled hot plate produced. *Reprinted with permission from CAMAG Catalog Instrumental Thin-Layer Chromatography 2013.*

adsorbent layer and further problems with accurate quantification of sample components.

Alternatively, the heating process can be performed inside a convection oven or custom-made heating compartments as shown in Figure 8.11. Such devices enable isothermal conditions for the derivatization process in comparison to the plate heaters described above. In particular, this approach provides a uniform temperature distribution over the whole volume of the TLC plate, which significantly increases the accuracy of quantitative analysis. Furthermore, it allows precise optimization of temperature and time for the derivatization process [27–30]. The main issue is that the TLC plate is thermally inhomogenous because it consists of two layers. A relatively thick plate support

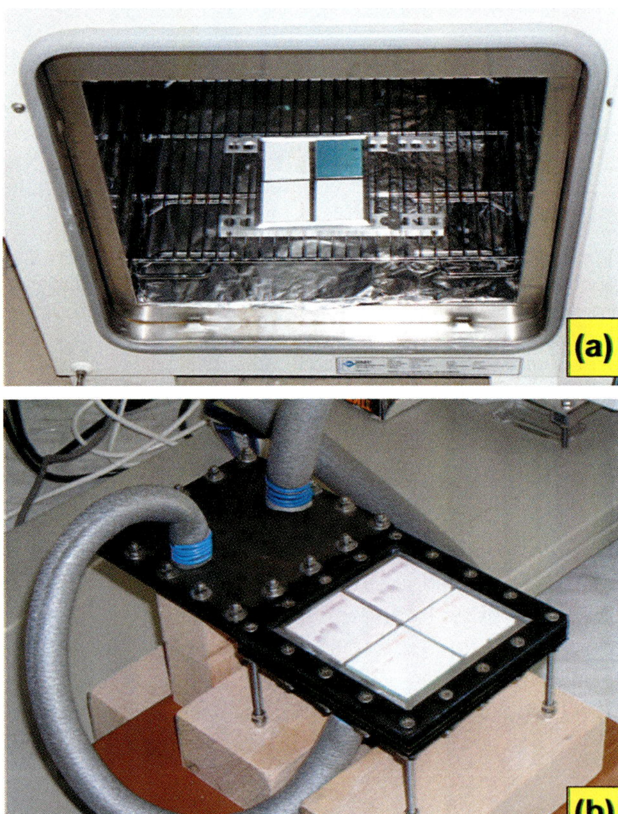

FIGURE 8.11 Thin-layer chromatography (TLC) plates heated in an air-convection oven (a) and a home-made hot plate with thermostated cover connected to an external liquid circulating thermostat (b).

(typically glass, aluminum, or plastic) and a thin stationary phase (e.g., porous silica, octadecylsiloxane-bonded silica, aluminum oxide, cellulose, or polyamide) are characterized by different values of their specific heat capacity and thermal conductivity. Therefore, the real heating rate and temperature at a given time, which are the key parameters for quantitative derivatization, are in fact difficult to predict and control, especially under nonisothermal conditions provided by simple hot plate heaters. It has been determined that during the first few minutes of heating the temperature of the plate changes very quickly. Such unsteady, nonlinear, and, in fact, uncontrolled conditions can be a significant source of poor precision in quantitative TLC [27–29]. Moreover, the average temperature of the stationary phase surface for glass-based TLC plates within the first 20 min of heating is substantially different from the destination derivatization temperature, which is usually set at 100 °C or more. This problem is also an important source of uncertainty for quantification because it significantly affects the reproducibility of the detector response, particularly when the heating time is relatively short, e.g., within 3–5 min. For that reasons, temperature and heating time optimization is necessary to minimize uncertainty and maximize reproducibility in detection and quantification of compounds based on postrun derivatization. As an example, for the robust quantification of target components (e.g., steroids) visualized by reaction with phosphomolybdic acid (PMA), the derivatization process should be performed at relatively low temperatures (close to 60 °C) and for longer heating times (more than 15 min). For the optimized reaction conditions the background noise can be minimized and the signal maximized, which allows application of this derivatization protocol for quantitative determination of analytes in complex samples (Figure 8.12) [31].

It should be mentioned that in some particular cases the optimization of the derivatization protocol (particularly temperature and pH) to maximize the signal cannot be justified and may provide misleading or difficult to interpret quantitative data. Significant problems may occur for signals recorded from compounds which exhibit specific detection properties, for example in the determination of the antioxidant properties (within cells, tissues, or living organisms) for which the determined values are used in further investigations. Detection at elevated temperature (where the best sensitivity and signal/noise ratio is obtained) may substantially change the signal ratio between separated compounds in comparison to the ratio measured at a different temperature (e.g., important for antioxidants interactions occurring within living cells). Recently, this problem was indicated and discussed for several reagents including DPPH (2,2-diphenyl-1-picrylhydrazyl), ABTS (2,20-azinobis-(3-ethylbenzthiazoline-6-sulfonic acid)), and FCR (Folin–Ciocalteu's

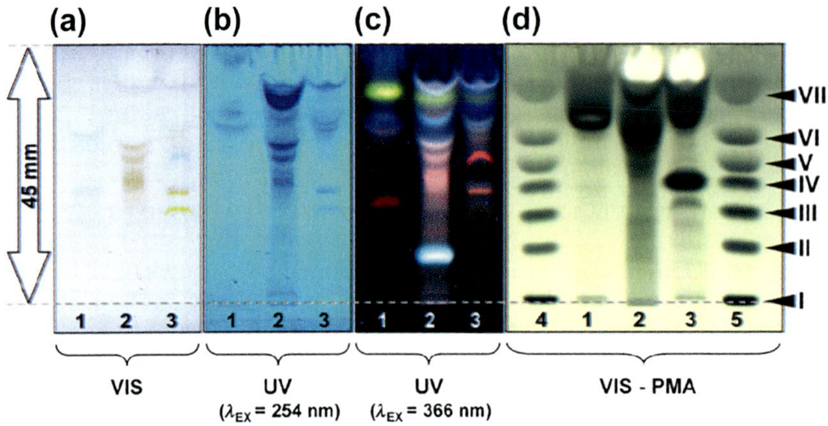

FIGURE 8.12 Detection of raw and chemically decomposed components of sea trout bile with different exposure to UV–Visible light and optimized staining with phosphomolybdic acid. Direct visible light scan (a), fluorescence quenching mode (b), fluorescence (c), phosphomolybdic acid staining using 10% PMA at 60 °C for 25 min (d). Lane no. 1 corresponds to the sample of deproteinized bile methanolic solution, lanes nos. 2 and 3 reflect separation of the bile extract after acid and base hydrolysis, respectively. Lanes nos. 4 and 5 correspond to cholesterol (I) and litocholic (II), deoxycholic (III), cholic (IV), glycodeoxycholic (V), glycocholic (VI), and the sodium salt of taurodeoxycholic acid (VII) standards mixture (quantity 1 and 2 μg/spot, respectively) [31].

phenol reagent), which are commonly used for the visualization of antioxidants in food products as well as active substances in pharmaceutical formulations [32]. In principle, the detection of antioxidant profiles must not significantly affect the antioxidant activity of target compounds. In the case of animal and human foods, reaction parameters such as pH and temperature should be as close as possible to the real physiological values.

8.4 COMMON DETECTION PROTOCOLS FOR TARGET COMPOUNDS

A unique feature of planar chromatography is that a number of low-molecular-mass compounds can be detected and quantified in the presence of interfering matrix, without the use of sophisticated detectors, simply by application of staining reagents and plate heating. Moreover, visualization protocols can be performed in a stepwise manner for the entire plate or for selected vertical lanes as well as horizontal zones or even irregular areas on developed plates. Modern stationary phases are both chemically and mechanically stable and can tolerate numerous washing steps after reagent application to avoid cross-contamination and unwanted interference in a subsequent reaction. The detection of derivatization products, particularly involving fluorescence, can be sensitive to small amounts of analytes starting from low nanogram levels per spot. Visualization methods can be driven by physical interactions (e.g., van der Waals forces), thermal decomposition, and a number of chemical reactions resulting in the formation of stable covalent bonds between specific functional groups of the compound or its decomposition products and the reagent. Visualization process can be destructive (e.g., spraying with acids and plate heating to degrade organic compounds to carbon) or nondestructive and reversible (e.g., spraying with water to detect nonpolar zones on a polar stationary phase or exposure to iodine vapor). Some target compounds can increase fluorescence after plate heating by interaction with the stationary phase (e.g., aminopropyl groups bonded to silica gel), which is referred to as thermochemical activation. Visualization reagents may give color products after reaction with particular chemical groups, and therefore, they are called group specific reagents. Such reactions are usually based on hydrazone formation, dansylation, diazotization, metal complexation, supramolecular complexes formation, Schiff's base reactions, etc. A number of visualization reagents employ less specific reactions and are considered as universal chemical reagents, for example: Dragendorf and iodoplatinate reagents or nitric acid vapor. There are also a number of redox reactions that are frequently involved in visualization process on TLC plates.

There are number of research papers and guide-books that provide details for the preparation of reagents and protocols for specific applications [1–14,33]. Table 8.1 summarizes the most commonly applied staining and derivatization reactions for specific target compounds, which can be detected on TLC plates.

It should be noted that in organic chemistry and pharmacy TLC is frequently applied for fast screening of reaction products and sensitive detection of UV transparent impurities and active compounds from complex mixtures (e.g., pharmaceutical formulations and herbal extracts) without sample purification. Remarkably, nearly 50% of the separation protocols that are listed in General Monographs of the European Pharmacopoeia are still based on planar chromatography, using silica gel as the main stationary phase (Figure 8.13) [24,42]. TLC separation procedures strongly contribute to quantification, purity confirmation and substance identification protocols in EP monographs compared with the remaining chromatographic techniques. According to data presented in Table 8.2 qualitative and quantitative planar chromatographic analysis involve more than 50 types of visualization reagents for detection of transparent active chemicals and drugs from complex pharmaceutical products.

TABLE 8.1 Derivatization, Staining, and Visualization Improvement Chemicals Commonly Used for Postseparation Detection in Planar Chromatography

No/Reagent Type	Analytes Detected	References
1. ABTS (2,2′-azino-di-(3-ethyl benzthiazoline-6-sulfonic acid)) **(redox indicator)**	Antioxidants, reducing compounds	[34]
2. AHTT (4-amino-5-hydrazino-1,2,4-triazole-3-thiol)	Aldehydes	[13]
3. Ammonia vapor **(universal reagent for spots/background contrast improvement)**	Organic acids with pH indicators	[8]
4. Ammoniacal silver nitrate	Halogen derivatives	[6]
5. Anisaldehyde	Terpenes	[1]
6. *p*-Anisidine	Carbohydrates, mono-saccharides, uronic acid, waxes	[3]
7. Antimony trichloride	3-beta-hydroxy steroids, resins, terpenes	[5]
8. Benzocyanide benzyltrimethylammonium hydroxide	Nitro derivatives	[6]
9. Berberine **(universal fluorescent reagent)**	Fatty acids, lipids, saturated compounds, steroids	[8]
10. Bial's orcinol reagent (orcin, hydrochloric acid, and ferric chloride)	Gangliosides, glycolipids, sugars, sulfolipids	[11]
11. Biodetection using living cells, usually microorganisms	Detection of biologically active plant substances	[11]
12. Blood–gelatin mixture	Saponins	[10]
13. Blue tetrazolinum reagent	Carbohydrates, corticosteroids	[8]
14. Bromocresol green	Organic acids	[13]
15. N-Bromosuccinimide reagent	Amino acid, hydroxyflavoness, hydroxyquinones	[8]
16. Bromothymol blue	Acid lipids	[13]
17. 7-Chloro-4-nitrobenzo-2-oxa-1,3-diazole	Phenols, thiols	[6]
18. Ceric ammonium sulfate in phosphoric acid	Estrogens	[35]
19. Cupric acetate	Phospholipids, polar lipids, prostaglandins.	[13]

TABLE 8.1 Derivatization, Staining, and Visualization Improvement Chemicals Commonly Used for Postseparation Detection in Planar Chromatography—cont'd

No/Reagent Type	Analytes Detected	References
20. Copper sulfate/o-phosphoric acid/methanol	Steroids	[36]
21. 2,7- Dichlorofluorescein **(universal fluorescent reagent)**	Saturated and unsaturated lipids	[8]
22. Dodecane (50% w/v in hexane) **(fluorescence intensifier)**	Polycyclic aromatic hydrocarbons	[8]
23. Dicobalt octacarbonyl	Acetylene compounds	[6]
24. 2,4-Dinitrophenylhydrazine	Aldehydes, ketones	[6]
25. Diphenylamine/aniline/ phosphoric acid in acetone	Fructooligosaccharides, polysaccharides	[37]
26. DPPH (2,2-diphenyl-1-picrylhydrazyl radical) **(redox indicator)**	Antioxidants, reducing compounds	[38]
27. Dragendorff's reagent (potassium iodobismuthate solution)	Alkaloids, cholines and other nitrogen-containing organic compounds	[13]
28. Ehrlich's reagent (4-dimethylaminobenzaldehyde in hydrochloric acid/ethanol solvent)	Amines, antibiotics, amino-glycosides, indoles and pyrroles hydrazines, primary and aromatic amines, indoles, sulfonamides, terpenes, sesquiterpene esters, urea and thiourea derivatives	[13]
29. Ferric chloride solution	Cholesteryl esters, enols, ergot alkaloids, hydroxamic acids, inorganic anions, phenolic compounds	[8]
30. Fluorescein **(universal fluorescent reagent)**	Barbituranes, chlorinated hydrocarbons, lipids, purines, pyrimidines, unsaturated compounds	[8]
31. Folin and Ciocalteu's reagent (phosphomolybdate and phosphotungstate mixture) **(redox indicator)**	Antioxidants, reducing compounds	[8]
32. Gibb's reagent (dichloroquinonechloroimide)	Barbituranes, indoles, phenols, thiols	[9]
33. Iodine solution/iodine vapor **(universal nondestructive visualization reagent)**	Carbohydrates, detergents, lipids, mycotoxins, pesticides, phopholipids, polymers, steroids	[13]

(*Continued*)

TABLE 8.1 Derivatization, Staining, and Visualization Improvement Chemicals Commonly Used for Postseparation Detection in Planar Chromatography—cont'd

No/Reagent Type	Analytes Detected	References
34. Iodoplatinate	Alkaloids, amines, organic nitrogen compounds	[14]
35. Jenson's reagent (chloramines T/trichloroacetic acid in chloroform/methanol/water solution)	Digitalis glycosydes	[8]
36. Lead acetate dichlorofluorescein	Alcohols	[6]
37. Molybdenum blue	Phospholipids, phosphate esters	[13]
38. α-Naphthol	Cyclodextrins	[39]
39. Ninhydrin	Amino-acids, amines, antibiotics, glycosides, peptides, sugars	[13]
40. Nitrous vapor	Phenols	[6]
41. Paraffin liquid (33% v/v in hexane) **(fluorescence intensifier)**	Aflatoxins, dansyl amines, gentamycins, steroids	[8]
42. Perchloric acid **(universal charring reagent)**	Calixarenes, polycyclic aromatic hydrocarbons	[40]
43. Phosphomolybdic acid	Alkaloids, bile acids, essential oils, lipids, prostaglandins, reducing compounds, steroids	[12]
44. o-Phathaldehyde/sulfuric acid reagent	Beta-blokers, ergot alkaloids, histidyl peptides, indole derivatives	[8]
45. Pinacryptol yellow **(universal fluorescent reagent)**	Anionic and nonionic surfactants, sweeteners, organic anions	[8]
46. Potassium permanganate	Organic compounds, reducing compounds	[40]
47. Potassium hydroxide solution	Antraquinones, anthrones, cumarins, flavonoides	[11]
48. Polyethylene glycol **(fluorescence intensifier)**	Flavonoids, compounds with −OH groups	[9]
49. Potassium dichromate/sulfuric acid **(universal charring reagent)**	Organic compounds	[13]
50. Rhodamine B **(universal fluorescent reagent)**	Detergents, fatty acids and methyl esters, flavonoids, gangliosides, glycerides, phenols, polyphenols	[8]

TABLE 8.1 Derivatization, Staining, and Visualization Improvement Chemicals Commonly Used for Postseparation Detection in Planar Chromatography—cont'd

No/Reagent Type	Analytes Detected	References
51. Rhodamine 6G **(universal fluorescent reagent)**	Gangliosides, fatty acids and methyl esters, phospholipids, triterpene alcohols, steroids, ubiquinones	[8]
52. Sodium nitroprusside/ acetaldehyde	Amines	[7]
53. Sulfuric acid (diluted with water or alcohols) **(universal charring reagent)**	Organic compounds	[30]
54. Tillman's reagent (2,6-dichlorophenolindophenol)	Organic acids including vitamin C	[8]
55. 2-Trichloromethylbenzimidazole (highly selective chromogenic reagent)	Azines, azoles, benzazoles, heteroaromatic nitrogen containing compounds	[41]
56. Vanillin/sulfuric acid	Alcohols, carbaminians, cardenolides, steroids, fatty oils	[2]
57. Water	Nonpolar compounds	[8]

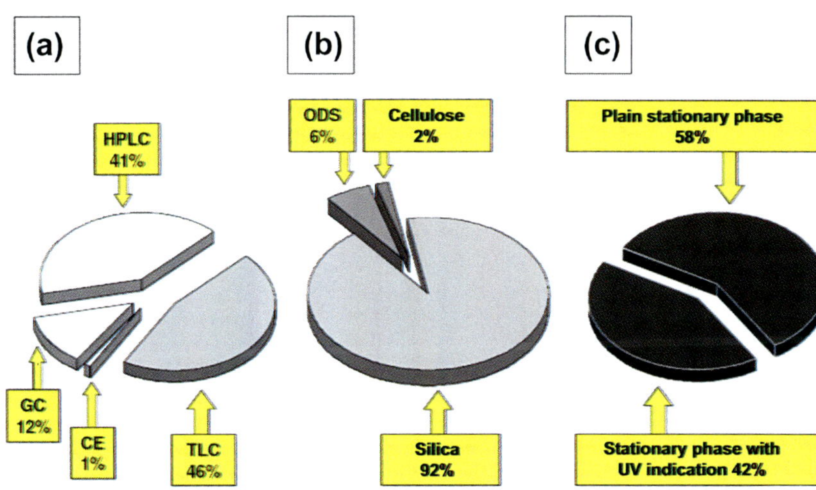

FIGURE 8.13 Contribution of separation techniques for quantification, purity confirmation and substance identification protocols used in the European Pharmacopoeia General Monographs (a) as well as types of stationary phases applied for thin-layer chromatography (TLC) separations (b, c) [24,42].

TABLE 8.2 Typical Visualization Chemicals and Protocols for Detection of Target Components on thin-layer chromatography (TLC) Plates Reported by the European Pharmacopoeia General Monographs [42]

No/Reagent Type and Composition	Derivatization and Visualization Conditions	Spots Inspection	Stationary Phase Type	Application Examples
1. 4-acetylbenzoate reagent	Spraying with solution composed of 0.25 g of methyl 4-acetylbenzoate in a mixture of 5 mL of sulfuric acid and 85 mL of cooled methanol. Temperature 100–105 °C for 10 min.	Daylight (hot plate within 5 min)	Silica gel GF254R	Anise oil; star anise oil.
2. 4-aminobenzoic acid solution	Spraying with solution composed of 1 g of 4-aminobenzoic acid in a mixture of 18 mL of anhydrous acetic acid, 20 mL of water and 1 mL of phosphoric acid. Immediately before use, mix 2 volumes of the solution with 3 volumes of acetone R. Temperature 100 °C for 15 min., allow to cool and spray with sodium periodate (2 g/L solution). After drying, heat the plate at temperature 100 °C for 15 min.	Daylight	Silica gel GR	Lactitol monohydrate; maltitol; mannitol; sorbitol; xylitol.
3. Aminohippuric acid reagent	Spraying with solution composed of 0.3 g of aminohippuric acid and 3 g of phthalic acid in ethanol (96%) diluted to 100 mL with the same solvent. Temperature 120 °C for 5 min.	Daylight	Silica gel	Guar; ispaghula husk; guar galactomannan.
4. Ammonia vapor	Developed plate sprayed with an appropriate visualization reagent and then ammonia vapor treating until the background is white or colored spots appears.	Daylight	Silica gel HR; TLC octadecylsilyl silica gel plate R	Dithranol; glycyrrhetic acid; capsicum; hop strobile; tormentil; turmeric Javanese; (20β)-3β-hydroxy-11-oxo-18a-olean-12-en-29-oic acid; glycyrrhetic acid.

5. Ammonium and cerium nitrate	Visualization reagent concentration 0.1 M. Room temperature.	Daylight	Silica gel R	Alcuronium chloride.
6. Ammonium vanadate solution	Spraying with solution composed of 1.2 g ammonium vanadate in 95 mL of water and dilute to 100 mL with sulfuric acid. Temperature 100 °C for 10 min.	Daylight	Silica gel F254	Hydroxyethyl salicylate.
7. Anisaldehyde solution	Spraying with solution composed of 10 mL anisaldehyde, 90 mL ethanol (96%) and 10 mL of sulfuric acid. Temperature 110 °C for 10 min/100–105 °C for 10 min.	Daylight (hot plate within 5 min)	Silica gel R	Acacia; aescin; *Anamirta cocculus*; anise oil; benzoin Sumatra; *Bistort rhizome*; borneol; bornyl acetate; busulfan; caraway fruit; cassia oil; centaury; centella; cineole; cinnamon bark oil; citronella oil; clove; colophony; coriander; dwarf pine oil; *Eleutherococcus*; enoxolone; erythromycin; eugenol; eucalyptus leaf; garlic; ginseng; glycyrrhetic acid; hexylresorcinol; iceland moss; juniper; lavender flower; lemon verbena leaf; levomenthol; liquorice dry extract; matricaria flower; melissa leaf; menthol; mint oil partly dementholised; myrrh; neroli oil; notoginseng root; olive leaf

(*Continued*)

TABLE 8.2 Typical Visualization Chemicals and Protocols for Detection of Target Components on thin-layer chromatography (TLC) Plates Reported by the European Pharmacopoeia General Monographs [42]—cont'd

No/Reagent Type and Composition	Derivatization and Visualization Conditions	Spots Inspection	Stationary Phase Type	Application Examples
				dry extract; oregano; oxprenolol hydrochloride; peppermint leaf; propranolol hydrochloride; primula root; restharrow root; sSaffron; senega root; shellac; spike lavender oil; spiramycin; star anise oil; sweet orange oil; tea tree oil; thyme; tragacanth; turpentine oil *Pinus pinaster* type; valerian dry aqueous extract; verbena herb; wild thyme; yarrow.
8. Antimony trichloride solution	Spraying with solution prepared using 30 g of antimony trichloride with two quantities, each of 15 mL, of ethanol-free chloroform, resulting crystals must be dissolved immediately in 100 mL of ethanol-free chloroform at slightly elevated temperature. Spots visualization at room temperature.	Daylight (3–4 min after spraying)	Silica gel GR	Cholesterol; 5α-cholesta-7, 24-dien-3β-ol; desmosterol; ergocalciferol; lathosterol.
9. Bromocresol green solution	Spraying with solution composed of 50 mg of bromocresol green in 0.72 mL of 0.1 M sodium hydroxide and 20 mL of ethanol (96%) and diluted to 100 mL with water.	Daylight	Silica gel	Valproic acid.

10. Bromocresol purple in methanol	Spraying with a 1 g/L solution of bromocresol purple in methanol, adjusted to a violet-red color with 0.05 mL of diluted ammonia, then the plate drying in a current of hot air at 100 °C. Spraying with a 0.4 g/L solution of bromocresol purple in ethanol (50% v/v), adjusted to pH 10 using 1 M sodium hydroxide.	Daylight	Silica gel R	Clonidine hydrochloride; codergocrine mesilate; dihydroergocristine mesilate; pefloxacin mesilate dehydrate.
11. Chloramine and trichloroacetic acid in alcohol	Detection mixture: 2 volumes of 10 g/L solution of chloramine and 8 volumes of a 250 g/L solution of trichloroacetic acid in ethanol (96%), after spraying plate heating at 100–105 °C for 10 min.	UV 365 nm	Silica gel GR	Digitalis leaf; digoxin.
12. Chloroplatinic acid in hydrochloric acid and hydriodic acid	Spraying with a 0.1 g/L solution of chloroplatinic acid in hydrochloric acid containing 0.4% v/v of hydriodic acid, visualization at room temperature.	Daylight	Silica gel GR	Trihexyphenidyl hydrochloride.
13. Dichlorofluorescein solution	Spraying with a solution containing 0.2 g/L of dichlorofluorescein and 0.1 g/L of rhodamine B in ethanol (96%) at room temperature. Spraying with a solution containing 0.2 g/L of dichlorofluorescein and 0.1 g/L of rhodamine B in ethanol (96%). Maintain the plate for a few seconds in the vapor above a water-bath.	UV 254 nm; UV 365 nm	ODS gel high-performance thin-layer chromotography (HPTLC)/ SilicaGel HR/ SilicaGelGF254R	Refined almond oil; chloramphenicol palmitate; erythromycin stearate; olive.
14. Dichlorophenolindo-phenol, sodium salt in alcohol	Spraying with a 2 g/L solution of dichlorophenolindophenol sodium salt in anhydrous ethanol and heating in an oven at 120 °C for a few minutes to intensify the spots.	Daylight	Silica gel F254 R	Dog rose; polyolefines; polyethylene without additives for containers.

(*Continued*)

TABLE 8.2 Typical Visualization Chemicals and Protocols for Detection of Target Components on thin-layer chromatography (TLC) Plates Reported by the European Pharmacopoeia General Monographs [42]—cont'd

No/Reagent Type and Composition	Derivatization and Visualization Conditions	Spots Inspection	Stationary Phase Type	Application Examples
15. Dichloroquinone-chlorimide and sodium carbonate solutions	Spraying with a 50 g/L solution of sodium carbonate in a mixture of 30 volumes of ethanol (96%) and 70 volumes of water; after drying in a current of air, spraying with a 1 g/L solution of dichloroquinonechlorimide in ethanol (96%); room temperature.	Daylight	Silica gel GR	Bearberry leaf; capsicum tincture standardized; pyridoxine hydrochloride; turmeric javanese.
16. Diethylamine and silver nitrate solutions	spraying with diethylamine and heat at 125 °C for 10 min. After cooling plate spray with silver nitrate solution (17 g/L). Brown to black spots appear after exposure to ultraviolet light at 365 nm.	UV 365 nm/ Vis	Silica gel	Carmustine.
17. 1,3-Dihydroxy-naphthalene in alcohol and solution of sulfuric acid	Spraying with a 1.0 g/L solution of 1,3-dihydroxynaphthalene in a mixture of 10 volumes of sulfuric acid and 90 volumes of methanol. Spraying with a mixture of equal volumes of a 2 g/L solution of dihydroxynaphthalene in alcohol and a 460 g/L solution of sulfuric acid; temperature 110–150 °C for 5–10 min.	Daylight	Silica gel HR/ Carbomer R	Dihydrostreptomycin sulfate; kanamycin monosulfate; lactulose; tobramycin.
18. Dimethylamino-benzaldehyde solution	Spraying with dimethylaminobenzaldehyde solution (1 g in 50 mL of hydrochloric acid and 50 mL of ethanol 96%)) and drying in a current of hot air for 2 min.	Daylight/ UV 254 nm	Silica gel 254F/ Cellulose R	Allantoin; dihydroergocristine mesilate; ergometrine maleate; Ffuorouracil;

	Spraying with solution composed of 0.2 g dimethylaminobenzaldehyde without heating, in a mixture of 4.5 mL of water and 5.5 mL of hydrochloric acid. Drying the plate at 100 °C–105 °C for 10 min and spraying with a 1 g/L solution of dimethylaminobenzaldehyde in alcohol containing 1% v/v of hydrochloric acid. Spraying with a 10 g/L solution of dimethylaminobenzaldehyde in ethanol 96%. Spraying with a 10 g/L solution of dimethylaminobenzaldehyde in 1 M hydrochloric acid. Spraying with a mixture of 200 mL of a 10 g/L solution of dimethylaminobenzaldehyde in anhydrous ethanol and 20 mL of hydrochloric acid; plate drying in an oven at 80 °C for 3–4 min.			hydroxycarbamide; metoclopramide hydrochloride; oxybuprocaine hydrochloride; propacetamol hydrochloride; sulfathiazole; tyrothricin.
19. 4-Dimethyl-aminocinnamaldehyde solution	Spraying with 4-dimethylaminocinnamaldehyde solution (2 g in a mixture of 100 mL of hydrochloric acid and 100 mL of anhydrous ethanol. Immediately before use the mixture should be diluted to four times of its volume with anhydrous ethanol). Spraying with 4-dimethylaminocinnamaldehyde solution and heat at 60 °C for 10 min.	Daylight	Silica gel R	Biotin; niflumic acid.

(Continued)

TABLE 8.2 Typical Visualization Chemicals and Protocols for Detection of Target Components on thin-layer chromatography (TLC) Plates Reported by the European Pharmacopoeia General Monographs [42]—cont'd

No/Reagent Type and Composition	Derivatization and Visualization Conditions	Spots Inspection	Stationary Phase Type	Application Examples
20. Dinitrophenyl-hydrazine–aceto-hydrochloric solution	Spraying with dinitrophenylhydrazine-aceto-hydrochloric solution (0.2 g of dinitrophenylhydrazine in 20 mL of methanol and 80 mL of a mixture of equal volumes of acetic acid and hydrochloric acid).	Daylight	Silica gel F254	Vanillin.
21. Diphenylboric acid aminoethyl ester solution and macrogol 4000 solution	Spraying with a 10 g/L solution of diphenylboric acid aminoethyl ester in methanol and. subsequently with a 50 g/L solution of macrogol 400 in methanol. Spots examining at room temperature after drying in air for about 30 min. After spraying plate heating at 100–105 °C for 5 min.	UV 365 nm	Silica gel R	Agrimony; alchemilla; arnica flower; belladonna leaf dry extract standardized; birch leaf; black horehound; buckwheat herb; calendula flower; chamomile flower roman; dandelion herb with root; ginkgo dry extract refined and quantified; goldenrod; hawthorn berries; *Hypericum*; knotgrass; lime flower; loosestrife; marshmallow leaf; matricaria liquid extract; milk thistle dry extract refined and standardized; oregano; passion flower; oriental cashew; saffron; St. John's wort; star anise; wild pansy flowering aerial parts.

22. Diphenylcarbazone mercuric reagent and potassium hydroxide solution	Spraying with diphenylcarbazone mercuric reagent (prepared by mixing of equal volumes of 0.1 g diphenylcarbazone in 50 mL of anhydrous ethanol and 1 g of mercuric chloride in 50 mL of anhydrous ethanol). After drying plate spraying with freshly prepared alcoholic potassium hydroxide solution diluted 1 in 5 with aldehyde free alcohol. Then plate heating at 100 °C–105 °C for 5 min and examining immediately.	UV 254 nm and daylight	Silica gel GF254R	Amobarbital; barbital; pentobarbital.
23. Etylenediamine and potassium ferricyanide	Temperature 60 °C for 10 min.	UV 254/365 nm	Silica gel GR	Noradrenaline hydrochloride.
24. Fast blue B salt/0.1 M sodium hydroxide	Spraying with a 5 g/L solution of fast blue B salt R and after drying in air plate spraying with 0.1 M ethanolic sodium hydroxide. Spraying with a 100 g/L solution of potassium hydroxide in methanol R and then with a freshly prepared 2 g/L solution of fast blue B salt in a mixture of 50 volumes of anhydrous ethanol and 50 volumes of water. Spraying with a freshly prepared 5 g/L solution of fast blue B salt. Spraying with a freshly prepared 5 g/L solution of fast blue B salt R; plate heating to 60 °C for 1 min.	Daylight	Silica gel GF254R	Gentian root; methylphenidate hydrochloride; rhatany root; tormentil.

(Continued)

TABLE 8.2 Typical Visualization Chemicals and Protocols for Detection of Target Components on thin-layer chromatography (TLC) Plates Reported by the European Pharmacopoeia General Monographs [42]—cont'd

No/Reagent Type and Composition	Derivatization and Visualization Conditions	Spots Inspection	Stationary Phase Type	Application Examples
	acetic acid. Add 80 mL of distilled water, 100 mL of a 400 g/L solution of potassium iodide, 200 mL of glacial acetic acid and dilute to 1000 mL with distilled water. Mix 2 volumes of this solution with 1 volume of a 200 g/L solution of barium chloride. Spraying with potassium iodobismuthate solution prepared as follows: To 0.85 g of bismuth subnitrate add 10 mL of glacial acetic acid and gently heat until completely dissolved. Add 40 mL of water and allow to cool. To 5 mL of this solution, add 5 mL of a 400 g/L solution of potassium iodide, 20 mL of glacial acetic acid and 70 mL of water.			
36. Iodoplatinate reagent	Spraying with reagent prepared as follows: To 3 mL of a 100 g/L solution of chloroplatinic acid add 97 mL of water and 100 mL of a 60 g/L solution of potassium iodide. Spraying with a mixture prepared immediately before use of 1 volume of sulfuric acid and 40 volumes of iodoplatinate reagent then with hydrogen peroxide solution (30%).	Daylight; UV 254 nm	Silica gel R	Cinchona bark; mefloquine hydrochloride; quinidine sulfate; quinine sulfate; trimetazidine dihydrochloride.

37. Methylbenzo-thiazolone hydrazone hydrochloride followed by potassium ferricyanide	Sequence spraying with a 1 g/L solution of methylbenzothiazolone hydrazone hydrochloride in a 90% v/v solution of methanol, followed by a 20 g/L solution of potassium ferricyanide in a mixture of 1 volume of concentrated ammonia and 3 volumes of water, followed by a further spraying with a 1 g/L solution of methylbenzothioazolone hydrazone hydrochloride in 90% v/v methanol.	Daylight	Silica gel R	Salbutamol.
38. Naphthylethylene-diaminedihydrochloride.	Spraying with a 10 g/L solution of sodium nitrite in 1 M hydrochloric acid and plate diping after 10 min in a 4 g/L solution of naphthylethylenediaminedihydrochloride in methanol.	Daylight	SilicaGel F254R	Clenbuterol hydrochloride; dipotassium clorazepate.
39. Ninhydrin solution	Spraying with a 2 g/L solution of ninhydrin in anhydrous ethanol and plate heating at 60 °C for 10 min. Spraying with a 2 g/L solution of ninhydrin in ethanol 96%and plate heating at 110 °C for 10 min. Spraying with a 2 g/L solution of ninhydrin in a mixture of 5 volumes of glacial acetic acid and 95 volumes of butanol and plate heating at 105–110 °C for 5–10 min.	Daylight	Silica gel R/octadecylsilyl TLC silica gel plate for chiral separations	Alanine; amikacin; aminocaproic acid; amisulpride; aprotinin; arginine aspartate; asparagine monohydrate; aspartame; aspartic acid; bacampicillin hydrochloride; bacitracin; baclofen; alcium pantothenate; carbocisteine; colistimethate sodium;

(*Continued*)

TABLE 8.2 Typical Visualization Chemicals and Protocols for Detection of Target Components on thin-layer chromatography (TLC) Plates Reported by the European Pharmacopoeia General Monographs [42]—cont'd

No/Reagent Type and Composition	Derivatization and Visualization Conditions	Spots Inspection	Stationary Phase Type	Application Examples
	Spraying with a 2 g/L solution of ninhydrin in a mixture of 5 volumes of glacial acetic acid and 95 volumes of butanol and plate heating at 100–105 °C for 15 min. Spraying with a solution of 0.1 g of ninhydrin in a mixture of 6 mL of a 10 g/L solution of cupric chloride, 21 mL of glacial acetic acid and 70 mL of anhydrous ethanol. Plate drying at 60 °C. Spraying with ninhydrin solution (4 g/L in a mixture of 5 volumes of anhydrous acetic acid and 95 volumes of butanol) until the plate is slightly wet. Spots visualization: In an oven maintained at 100 °C for 10 min. Spraying with reagent prepared as follows: Dissolve 0.2 g of ninhydrin in 4 mL of hot water, add 5 mL of a 1.6 g/L solution of stannous chloride, allow to stand for 30 min, then filter and store at a temperature of 2–8 °C. Immediately before use dilute 2.5 mL of the solution with 5 mL of water and 45 mL of 2-propanol. Heating the plate at 110 °C for 15 min. Spraying with a 1 g/L solution of ninhydrin and heating in an oven at 110 °C for about 5 min.			colistin sulfate; cysteine hydrochloride monohydrate; common stinging nettle; dexpanthenol; ephedra herb; ephedrine anhydrous; ethambutol hydrochloride; flecainide acetate; fluorodopa; garlic powder; gentamicin sulfate; glutamic acid; glycine; histamine dihydrochloride; histidine; honey bee; isoleucine; kanamycin acid sulfate; leucine; lymecycline; lysine acetate; magnesium aspartate dehydrate; magnesium pidolate; methionine; mexiletine hydrochloride; pamidronate disodium pentahydrate; phenylalanine; phenylpropanolamine hydrochloride; pivampicillin; polymyxin B sulfate; potassium hydrogen aspartate hemihydrate; proline; pseudoephedrine

				hydrochloride; dl-methionine; serine; sodium alendronate; somatostatin; sulpiride; threonine; tiapride hydrochloride; tryptophan; tyrosine; valine; zinc acexamate.
40. Nitric acid	Spraying with a 20% v/v solution of nitric acid and heating at 120 °C for 10 min. After cooling plate spraying with a 50 g/L solution of potassium hydroxide in alcohol (50% v/v) until the zones appear.	Daylight	TLC silica gel	Senna leaf.
41. Nitric acid vapor	After plate drying from the mobile phase residues in an oven at 160 °C for 15 minhot plate positioning in a closed tank containing about 20 mL of fuming nitric acid for 30 min. Then plate removing and heating again at 160 °C for 15 min. Spots examination under UV light.	UV 254 nm	ODS gel	Diphenoxylate hydrochloride.
42. Phenylhydrazine hydrochloride solution	Spraying with phenylhydrazine hydrochloride solution prepared as follows: 0.9 g of Phenylhydrazine hydrochloride in 50 mL of water. Decolorize with activated charcoal and filter. To the solution add 30 mL of hydrochloric acid and dilute to 250 mL with water. Room temperature/heating the plate at 100 °C to 105 °C for a 10 min.	Daylight	Silica gel GR	Nitrofural; nitrofurantoin.

(Continued)

TABLE 8.2 Typical Visualization Chemicals and Protocols for Detection of Target Components on thin-layer chromatography (TLC) Plates Reported by the European Pharmacopoeia General Monographs [42]—cont'd

No/Reagent Type and Composition	Derivatization and Visualization Conditions	Spots Inspection	Stationary Phase Type	Application Examples
43. Phosphomolybdic acid solution	Spraying with a 100 g/L solution of phosphomolybdic acid in ethanol 96%. Plate heating at 120 °C for about 3 min. Spraying with a 40 g/L solution of phosphomolybdic acid in anhydrous ethanol. Plate heating in an oven at 120 °C until spots appear. Spraying with a freshly prepared 200 g/L solution of phosphomolybdic acid in ethanol 96% and plate heating at 120 °C for 5–15 min. Spraying with solvent consisting of 4 g phosphomolybdic acid in 40 mL of water and with 60 mL of sulfuric acid (cautiously cooling while mixing). Spraying with a solution prepared as follows: Dissolve 5 g of phosphomolybdic acid in a mixture of 50 mL of glacial acetic acid and 10 mL of sulfuric acid, and dilute to 100 mL with glacial acetic acid. Plate heating at 100–105 °C. Spraying with a 50 g/L solution of phosphomolybdic acid in ethanol 96%; plate heating for about 3 h at 120 °C until spots appear. Spraying with a 80 g/L solution of phosphomolybdic acid in 2-propanol and plate heating at 120 °C for 1–2 min.	Daylight	Silica gel; ODS gel HPTLC	Refined almond oil; aniseed; atty oils; bitter-fennel fruit oil; carisoprodol; carnauba wax; cetostearyl alcohol type A emulsifying; cinnamon tincture; macrogol 15 hydroxystearate; macrogolglycerol hydroxystearate; mandarin oil; norgestrel; Peru balsam; polyolefines; polyethylene without additives for containers; polypropylene for containers and closures; polyethylene-vinyl acetate for containers and tubing; phytomenadione; sage leaf three-lobed; sodium cetostearyl sulfate; spanish sage oil.

	Spraying with a 100 g/L solution of phosphomolybdic acid in alcohol, plate heating at 100–105 °C for 15 min.			
44. Phosphomolybdo-tungstic reagent and ammonia vapor	Spraying with solvent prepared as follows: Dissolve 100 gof sodium tungstate and 25 g of sodium molybdate in 700 mL of water. Add 100 mL of HCl and 50 mL of phosphoric acid. Heat the mixture under a reflux condenser in a glass apparatus for 10 h. Add 150 g of lithium sulfate, 50 mL of water and a few drops of bromine. Boil to remove the excess of bromine (15 min), allow to cool, dilute to 1000 mL with water and filter. Dilute 1 volume of above reagent with 2 volumes of water. After spraying plate expose to ammonia vapor.	Daylight	Silica gel F254	Hop strobile.
45. Potassium dichromate	Spraying with a 50 g/L solution of potassium dichromate in a 40% m/m solution of sulfuric acid. Results after 5 min at room temperature. Spraying with a 5 g/L solution of potassium dichromate in a 20% v/v solution of sulfuric acid. Examine immediately at room temperature. Spraying with a 5 g/L solution of potassium dichromate in a mixture of 1 volume of sulfuric acid and 4 volumes of water. Examine immediately at room temperature.	Daylight	Silica 3gel GR	Calcium gluconate; chlorhexidine digluconate; clomipramine hydrochloride; desipramine hydrochloride; methyltestosterone; trimipramine maleate.

(Continued)

TABLE 8.2 Typical Visualization Chemicals and Protocols for Detection of Target Components on thin-layer chromatography (TLC) Plates Reported by the European Pharmacopoeia General Monographs [42]—cont'd

No/Reagent Type and Composition	Derivatization and Visualization Conditions	Spots Inspection	Stationary Phase Type	Application Examples
	Spraying with a saturated solution of potassium dichromate in a mixture of 30 volumes of water and 70 volumes of sulfuric acid. Examine immediately at room temperature.			
46. Potassium hydroxide solution	Spraying with a 100 g/L solution of potassium hydroxide in methanol. Spot visualization at room temperature under UV light. Spraying with about 10 mL of a 50 g/L solution of potassium hydroxide in ethanol 50% v/v. Plate heating at 100–105 °C for 15 min and spots examining in ultravioled light and/or daylight. Spraying with 2 M alcoholic potassium hydroxide and examining in ultraviolet light at 365 nm. Spraying with a 20% v/v solution of nitric acid and heating at 120 °C for 10 min; then allow to cool and spray with a 50 g/L solution of potassium hydroxide in ethanol 50% v/v until the zones appear.	UV 365 nm/ daylight	Silica gel GR	Aloes dry extract standardized; cascara; frangula bark; gentian root; melilot; pelargonium root; rhubarb; Alexandrian senna pods.

47. Potassium permanganate	Spraying with a mixture of equal volumes of a 10 g/L solution of potassium permanganate and a 50 g/L solution of sodium carbonate, prepared immediately before use. Then after 30 min and spots examining in daylight. Spraying with 0.02 M potassium permanganate at room temperature. Spraying with a 30 g/L solution of potassium permanganate and plate drying in a current of warm air for about 5 min. Spraying with a 16 g/L solution of potassium permanganate. Spraying with a 1 g/L solution of potassium permanganate. Spraying with a 5 g/L solution of potassium permanganate in 1 M sodium hydroxide and plate heating at 110 °C for 5 min. Spraying with a 10 g/L solution of potassium permanganate and plate examining in daylight after 15 min. Spraying lightly with a 10 g/L solution of potassium permanganate in dilute sulfuric acid and plate heating at 120 °C for 20 min; visualization under ultraviolet light at 365 nm.	UV 365 nm and daylight	Silica gel F254R	Adenosine; altizide; calcium glucoheptonate; calcium levulinate dehydrate; clemastine fumarate; clindamycin hydrochloride; dobutamine hydrochloride; erythromycin lactobionate; etilefrine hydrochloride; fenoterol hydrobromide; ibuprofen; isoxsuprine hydrochloride; orciprenaline sulfate; trometamol.
48. Rhodamine B	Spraying with a 0.1 g/L solution of rhodamine B in ethanol 96% and plate examining under ultraviolet light.	UV 365 nm	Silica gel	Caprylocaproylmacrogol-glycerides; chloramphenicol palmitate; erythromycin stearate; glycerol monolinoleate; lauroyl macrogolglycerides; linoleoyl macrogolglycerides; oleoyl macrogolglycerides; stearoyl macrogolglycerides.

(Continued)

TABLE 8.2 Typical Visualization Chemicals and Protocols for Detection of Target Components on thin-layer chromatography (TLC) Plates Reported by the European Pharmacopoeia General Monographs [42]—cont'd

No/Reagent Type and Composition	Derivatization and Visualization Conditions	Spots Inspection	Stationary Phase Type	Application Examples
49. 4-(4-Nitrobenzyl) pyridine	Spraying with a 50 g/L solution of 4-(4-nitrobenzyl) pyridine in acetone and plate heating at 120 °C for 15 min. Then, spraying the still-warm plate with a 100 g/L solution of tetraethylenepentamine in acetone.	Daylight	Silica gel R	Metrifonate.
50. Sodium nitroprusside and potassium ferricyanide	Spraying with a mixture of equal volumes of a 100 g/L solution of sodium nitroprusside, a 100 g/L solution of potassium ferricyanide and a 100 g/L solution of sodium hydroxide, prepared 20 min before use.	Daylight	Silica gel GR	Metformin hydrochloride.
51. Starch solution.	Spraying with potassium iodide and starch solution prepared as follows: Dissolve 0.75 g of potassium iodide in 100 mL of water. Heat to boiling and add while stirring a solution of 0.5 g of soluble starch in 35 mL of water. Boil for 2 min. Spraying with freshly prepared potassium iodide and starch solution; exposing to ultraviolet light at 254 nm for 15 min and examining in daylight. Spraying with strong sodium hypochlorite solution. After 10 min spraying abundantly with glacial acetic acid R. After next 10 min	Daylight	Silica gel; cellulose R1	Calcitonin; chlorpropamide; cyclophosphamide; flucytosine; glyceryl trinitrate solution; gonadorelin acetate; isosorbide dinitrate diluted; lomustine; magnesium pidolate; pentaerythrityl tetranitrate diluted; rifabutin; sodium cyclamate; xylometazoline hydrochloride.

	drying at 100–105 °C for 2 min. Spraying with potassium iodide and starch solution until spots appear. Spraying with freshly prepared potassium iodide and starch solution, exposing to ultraviolet light at 254 nm for 15 min and examining in daylight. Developed plate drying and exposing to iodine vapor for about 5 min, then spraying with potassium iodide and starch solution and plate examining after 5 min.			
52. Sulfuric acid solution	Spraying with a 10% v/v solution of sulfuric acid in alcohol. Spraying with alcoholic solution of sulfuric acid (20 mL of sulfuric acid diluted to 100 mL with ethanol 96%) and plate heating at 140 °C for 5–10 min. Spraying with a 100 g/L solution of sulfuric acid in methanol and plate heating in an oven at 120 °C for 7 min; spots examining in daylight. Spraying with a mixture of 1 volume of sulfuric acid and 9 volumes of acetic anhydride. Plate examining in ultraviolet light at 365 nm. Spraying with a mixture of 5 volumes of sulfuric acid and 95 volumes of methanol. Plate heating at 100–105 °C for 5 min and examining in daylight.	Daylight and UV 365/254 nm	Silica gel 254	Beclometasonedipropionate; betamethasone; budesonide; cholecalciferol concentrate oily form; cortisone acetate; deslanoside; cyclopentolate hydrochloride; desoxycortone acetate; dexamethasone; diethylstilbestrol; digitoxin; estriol; etoposide; fennel bitter; fludrocortisone acetate; Fluocortolone pivalate; flumetasone pivalate; flupentixol dihydrochloride; hedera helix; hydrocortisone hydrogen succinate; josamycin propionate; lactulose; methylprednisolone;

(*Continued*)

TABLE 8.2 Typical Visualization Chemicals and Protocols for Detection of Target Components on thin-layer chromatography (TLC) Plates Reported by the European Pharmacopoeia General Monographs [42]—cont'd

No/Reagent Type and Composition	Derivatization and Visualization Conditions	Spots Inspection	Stationary Phase Type	Application Examples
	Spraying with alcoholic solution of sulfuric acid. Plate heating at 120 °C for 10 min or until the spots appear. After cooling plate examining in daylight and in ultraviolet light at 365 nm. Spraying with a mixture of 5 volumes of sulfuric acid and 95 volumes of alcohol and plate heating at 140 °C for 15 min. Spraying with alcoholic solution of sulfuric acid, plate heating at 120 °C for 10 min or until the spots appear, and after cooling plate examining in daylight and in ultraviolet light at 365 nm. Spraying with alcoholic solution of sulfuric acid. Plate heating at 100 °C for 10 min. Spraying with a 15% v/v solution of sulfuric acid in methanol and plate heating at 125 °C for 10 min.			metixene hydrochloride; mometasone furoate; ouabain; phenothiazine; prednicarbate; prednisolone acetate; progesterone; schisandra fruit; sucralose; turmeric Javanese; willow bark dry extract; wormwood.
53. Thymol solution	Spraying with a solution containing 0.5 g of thymol in a mixture of 5 mL of sulfuric acid and 95 mL of ethanol 96%. Plate heating at 110 °C for 20 min or until the spots appear. Spraying with a solution of 0.5 g of thymol in a mixture of 5 mL of sulfuric acid and 95 mL of ethanol 96%. Plate heating at 130 °C for 10 min.	Daylight	Silica gel GR	Anticoagulant and preservative solutions for human blood; cladribine; fructose; galactose; glucose anhydrous; lactose monohydrate; sucralose; sucrose; xylose.

54. Toluenesulfonic acid	Spraying with a 200 g/L solution of toluenesulfonic acid in alcohol and plate heating for 10 min at 120 °C. Spraying with a 200 g/L solution of toluenesulfonic acid in ethanol and plate heating at 105 °C for 10 min. Spots examining in ultraviolet light at 365 nm.	Daylight;UV 365 nm	Silica gel R	Mesterolone; medroxy-progesterone acetate; testosterone; testosterone enantate.
55. Vanillin reagent	Spraying with vanillin reagent prepared as follows: Carefully add, dropwise, 2 mL of sulfuric acid to 100 mL of a 10 g/L solution of vanillin in ethanol 96%. Plate heating in an oven at 100–105 °C for 10 min. Spraying with a 5 g/L solution of vanillin in a mixture of 20 volumes of anhydrous ethanol and 80 volumes of sulfuric acid. Plates examining in daylight after 5 min. Spraying with a 10 g/L solution of vanillin in sulfuric acid and plate examining in daylight while heating at 100–105 °C for 10 min. Spraying with a 5 g/L solution of vanillin in a mixture of 20 volumes of ethanol 96% and 80 volumes of sulfuric acid. Plate examining in daylight immediately after heating at 130 °C for 5–10 min. Spraying with a 50 g/L solution of vanillin in a mixture of 5 volumes of sulfuric acid, 10 volumes of glacial acetic acid and 85 volumes of methanol. Plate heating at 100–105 °C.	Daylight	Silica gel R	Betaxolol hydrochloride; bogbean leaf; Butcher's broom; clary sage oil; feverfew; ginger; macrogol cetostearyl ether; mastic; nutmeg oil; meprobamate; octyldodecanol; olive leaf; pygeum africanum bark; stanozolol; tolu balsam; white horehound; zidovudine.

(Continued)

TABLE 8.2 Typical Visualization Chemicals and Protocols for Detection of Target Components on thin-layer chromatography (TLC) Plates Reported by the European Pharmacopoeia General Monographs [42]—cont'd

No/Reagent Type and Composition	Derivatization and Visualization Conditions	Spots Inspection	Stationary Phase Type	Application Examples
	Spraying with vanillin-sulfuric acid reagent prepared as follows: Dissolve 0.5 g of vanillin in 50 mL of ethanol 96% and dilute to 100 mL with sulfuric acid; allow to dry in air. Plate heating at about 130 °C for 15 min and allow to cool in air. Spraying with a solution of 0.25 g of vanillin in a cooled mixture of 10 mL of alcohol and 40 mL of sulfuric acid. Plate heating at 100 −105 °C for 30 min. Spraying with about 7 mL of a mixture of 1 volume of a 25 g/L solution of vanillin in ethanol 96% and 4 volumes of sulfuric acid. Plate heating at 130 °C for 5–10 min.			
56. Phosphoric vanillin solution.	Spraying with the solution prepared as follows: Dissolve 1.0 g of vanillin in 25 mL of ethanol 96%. Add 25 mL of water and 35 mL of phosphoric acid. Plate heating at 120 °C for about 10 min.	Daylight	Silica gel R	Acesulfame potassium.
57. Zinc chloride.	Spraying with reagent prepared as follows: Dissolve 20 g of zinc chloride in 30 mL of glacial acetic acid, add 3 mL of phosphoric acid and 0.8 mL of hydrogen peroxide solution (30%) and dilute to 60 mL with water. Plate heating at 120 °C for 15 min and examine under UV light at 365 nm.	UV 365 nm	Silica gel F 254	Doxepin hydrochloride.

8.5 CONCLUSIONS

Physicochemical visualization of separated spots on TLC plates allows effective and selective detection of target compounds that in their native form are transparent to a given detector. Commonly applied staining and derivatization reagents combined with postreaction data acquisition and image digital processing allows qualitative inspection as well as full quantitative analysis of complex mixtures. Staining process can be fully automated via robotic stations for the routine analysis of a plethora of samples. However manual spraying protocols are still extensively used, due to their low cost and adaptability. In most cases quantitative analysis involving staining or derivatization requires proper optimization of the key experimental conditions (reagent concentration, solvent selection, pH, plate heating time and temperature), because visualization results are specific for a given stationary phase, sample and matrix composition as well as target compound properties.

References

[1] Stahl E. Chromatographische und mikroskopische Analyse von Drogen. Stuttgart: Gustav Fischer Verlag; 1970.

[2] Borkowski B. Chromatografia cienkowarstwowa w analizie farmaceutycznej. Warszawa: Państwowy Zakład Wydawnictw Lekarskich; 1973.

[3] Jork H, Funk W, Fischer W, Wimmer H. Dünnschicht-Chromatographie, Reagenzien und Nachweismethoden, vol. 1a. Weinheim: VCH; 1989.

[4] Jork H, Funk W, Fischer W, Wimmer H. Dünnschicht-Chromatographie, Reagenzien und Nachweismethoden, vol. 1b. Weinheim: VCH; 1994.

[5] Striegel MF, Holl J. Thin-layer chromatography for binding media analysis. Los Angeles: The Getty Conservation Institute; 1996.

[6] Sherma J, Fried B. Handbook of thin-layer chromatography. New York: Marcel Dekker; 2003.

[7] Wilson ID, Cooke M, Poole CF. Encyclopedia of separation science. San Diego Calif: Academic Press; 2000.

[8] Wall PE. Thin-layer chromatography. A modern practical approach. Cambridge: The Royal Society of Chemistry; 2005.

[9] Reich E, Schibli A. High-performance thin-layer chromatography for the analysis of medicinal plants. New York: Thieme; 2006.

[10] Hahn-Deinstrop E. Applied thin-layer chromatography. Darmstadt: Wiley; 2007.

[11] Waksmundzka-Hajnos M, Sherma J, Kowalska T. Thin layer chromatography in phytochemistry. New York: Taylor & Francis Group; 2008.

[12] Srivastava M. High-performance thin-layer chromatography (HPTLC). Berlin: Springer-Verlag; 2011.

[13] Data Sheet 18200U, Alltech; 2005.

[14] Chromatography products for analysis and purification. St. Louis: Sigma-Aldrich-Supelco; 2013.

[15] Rosencwaig A, Hall SS. Thin-layer chromatography and photoacoustic spectrometry. Anal Chem 1975;47:548–9.

CHAPTER

9

Advanced Spectroscopic Detectors for Identification and Quantification: UV–Visible, Fluorescence, and Infrared Spectroscopy

Tomasz Tuzimski

Department of Physical Chemistry, Medical University in Lublin, Lublin, Poland

9.1 INTRODUCTION

Various chemical, physical, and biochemical methods have been used for the detection and identification of compounds after separation by thin-layer chromatography (TLC). For the detection of UV-absorbing compounds by dye, TLC plates are prepared with fluorescence indicators (e.g., manganese-activated zinc silicate). This dye absorbs light at 254 nm showing a green fluorescence at approximately 520 nm; the sample inhibits the absorption of light by the fluorescence indicator and the sample zones appear as dark spots or zones on a bright fluorescent background. A commonly used method is to expose the chromatographic plate to iodine vapor in a closed chamber containing some iodine crystals. The sample zones are visualized as dark brown zones with a yellow–brown background. The use of biological methods in planar chromatography is justified because they are highly specific and the detection limits are lower than those in other methods. The possibility of using reagents (especially color reagents) is one of the main advantages of planar chromatography [1].

However, for the qualified identification and quantification of compounds after separation more sophisticated spectroscopic methods are

Instrumental Thin-Layer Chromatography
http://dx.doi.org/10.1016/B978-0-12-417223-4.00009-1

often required. At the present time, there are four complementary systems which allow the in situ quantification of samples on thin-layer plates based on scanners (multiwavelength and diode-array scanners) and video-densitometers (charge-coupled device cameras and flatbed scanners).

Depending on the mode used, spectrophotometric detection techniques can be referred to as TLC-UV/Vis detection performed in the absorption mode at single or multiple wavelengths with diode array detector scanning (TLC-DAD), mass spectrometry (TLC-MS), nuclear magnetic resonance (TLC-NMR), and Fourier transform infrared spectroscopy (TLC-FTIR).

9.2 ADVANCED SPECTROSCOPIC DETECTORS FOR IDENTIFICATION AND QUANTIFICATION

The spectral properties of the substances to be measured in the chromatogram and the properties of the adsorbent layer have an important effect on the scanning mode. In most popular physical detection modes, absorption and fluorescence properties of the analyzed compounds are used. To identify separated zones on a thin-layer plate, the surface is illuminated by white or monochromatic light of intensity I_0. The clean plate surface reflects the light intensity J_0 and the sample spot the light intensity J [2]. Both values are combined in the expression of reflectance R (Eqn (9.1)) [2]:

$$R = \frac{J}{J_0} \tag{9.1}$$

where:

$R(\lambda)$ is the reflectance;
$J(\lambda)$ is the intensity distribution of the analyte zone; and
$J_0(\lambda)$ is the intensity distribution of the clean layer.

(All intensity abbreviations in the text are used without a wavelength indication. This is done to ease reading although the intensity values are obviously wavelength-dependent.)

Mostly, this is fulfilled using Expression (9.2) for the transformation of absorption measurements [2]:

$$KM = \frac{1}{R} - 1 = m\frac{a_m}{1 - a_m} \tag{9.2}$$

and for fluorescence evaluations by Expression (9.3):

$$Fl = R - 1 = m\frac{a_m}{1 - a_m} \tag{9.3}$$

where:

$R(\lambda)$ is the relative reflectance;
a_m is the total mass absorption coefficient; and
m is the analyte mass.

In TLC, light absorption of the layer is constant, but absorption by the sample zone depends on the sample mass. The sample distribution within a zone has a distribution of unknown shape.

The mathematical relationship between the concentration of a substance on the TLC/high performance thin-layer chromatography (HPTLC) plate and signal intensity when measuring concentrations in reflectance (absorbance) mode can be described by the Kubelka–Munk equation [2]. In their theory, *Kubelka* and *Munk* demonstrated that the quotient of absorption and scatter is independent of the layer thickness. For quantitative measurements, this leads to the fundamental expressions given by Eqn (9.4) [2]:

$$\begin{aligned} k\left(\frac{1}{R}-1\right)+(1-k)(R-1)&=m\frac{a_m}{1-a_m} \\ k\left(\frac{1}{R}-R\right)+(R-1)&=m\frac{a_m}{1-a_m} \end{aligned} \tag{9.4}$$

where:

$R(\lambda)$ is the relative reflectance ($R = J/J_0$); and
k is the backscattering factor ($k \geq 0$ and $k \leq 1$).

9.2.1 TLC-UV/Vis and TLC-DAD

In TLC, light is used to detect separated sample zones by illuminating the layer from the top with light of known intensity (I_0). If the illuminating light shows higher intensity than the reflected light (J), a fraction of light must have been absorbed by the sample and/or layer. The definition of the total absorption coefficient a, which describes the plate and sample light absorptions, is [1–4]:

$$I_{abs} = I_0 - J = aI_0 \tag{9.5}$$

Theoretical considerations lead to the following equation for transformation purposes, showing linearity between the transformed measurement data (TDM) and the absorption coefficient [1–4].

$$TDM(\lambda) = k\left(\frac{I_0}{J}-\frac{J}{I_0}\right)+\left(\frac{J}{I_0}-1\right)=\frac{a}{1}-a \tag{9.6}$$

Equation (9.6), where k is the backscattering factor, which has values in the range 0–1, I_0 the illuminating light intensity at different wavelengths,

J the intensity of reflected light at different wavelengths, and a the total absorption coefficient.

In the case of $k = 0$, no incident light is reflected from the plate surface [1–4]. All light is scattered "forward," in the direction of the incident light, and in this situation, the light leaving the layer from the top surface must be fluorescent light.

$$TDM(\lambda, k = 0) = (R - 1) \tag{9.7}$$

In the case of $k = 1/2$, the Kubelka–Munk expression results from [1,5,6] (Eqn (9.6)):

$$TDM\left(\lambda, k = \frac{1}{2}\right) = \frac{(1 - R)^2}{2R} = \frac{a}{s} \tag{9.8}$$

In the case of $k = 1$ Eqn (9.9), can be derived from (Eqn (9.6)):

$$TDM(\lambda, k = 1) = \left(\frac{1}{R} - 1\right) = \frac{a}{s} \tag{9.9}$$

where R is the reflectance ($R = J/I_0$), and s the scattering coefficient ($s = 1 - a$).

With $k = 1$, Eqn (9.6) describes the situation where all light is reflected from the plate surface [1,4]. In general, fluorescence measurements are more specific than absorption measurements, due to the restricted number of compounds showing fluorescence [1,4].

TLC combined with modern scanning densitometry provides the possibility of quantitative analysis [1,3–10]. The spectral data of a substance measured directly on a TLC/HPTLC plate can be processed postrun for a variety of purposes, such as to determine the optimum wavelength(s) for quantitative scanning; to identify individual fractions by comparison with the spectra of standards cochromatographed on the same plate and/or stored in a spectrum library; to check identity by superimposing the spectra of all fractions within the same R_F window; to check the purity of fractions by superimposing the spectra from different positions within a zone. Modern fiber optic TLC scanners with a DAD have several advantages [1,3–13], for example, the scanner can evaluate separations using different wavelengths simultaneously without destroying the plate surface and permits parallel recording of chromatograms and in situ UV spectra in the range 191–1033 nm; therefore, it is possible to obtain doubly credible correct identification of the compounds on a chromatogram. The TLC scanner permits analysis of each compound at its optimum wavelength, thus offering optimum sensitivity for detection of each component. The TLC-DAD scanner permits measurement of a three-dimensional (3D) chromatogram, $A = f(\lambda, t)$, with absorbance as a function of wavelength and distance. The TLC-DAD

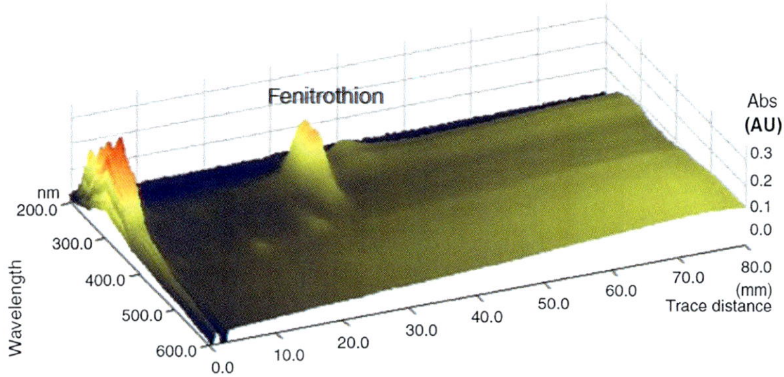

FIGURE 9.1 Three-dimensional plot obtained from an apple extract containing fenitrothion. *Adopted from Tuzimski with permission [13].*

scanner can compare parallel UV spectra of an unknown compound and a standard from a spectral library. Figure 9.1 shows an example of the 3D plot (scanning range × migration distance × absorbancy) obtained from an apple extract [13].

Software is available that allows the user access to all common parameters used in HPLC-DAD: peak purity, resolution, identification via spectral library match, etc. The TLC-DAD scanner is especially useful for correct identification of components of difficult, complicated mixtures, such as in plant extract and toxicological analysis [13]. Figure 9.2 shows a UV spectra obtained from fenitrothion standards at eight concentrations (100–1000 mg/mL) and the UV spectrum obtained from fenitrothion in an extract from freshly squeezed apple juice [1,13].

The peak purity index (P) is a numerical index for the quality of the coincidence between two data sets. It is given by the least squares fit coefficient calculated for all intensity pairs in the two data sets under consideration [1,13,14]. The least squares fit value of the spectrum from a fortified sample of water and the spectrum from a clofentezine standard by way of example is presented in Figure 9.3. The following equation is applied [1,13]:

$$P = \frac{\sum_i (s_i - \bar{s})(r_i - \bar{r})}{\sqrt{\sum_i (s_i - \bar{s})^2 (r_i - \bar{r})^2}} \tag{9.10}$$

where s_i and r_i are the respective intensities for the same abscissa value, i is the number of data points, and $\bar{s}$ and $\bar{r}$ are the average intensities of the first and second data set.

A peak purity index has values in the range from 0 to 1. A peak purity index of 1 indicates that the compared spectra are identical. Spectral comparisons should be performed at similar concentration

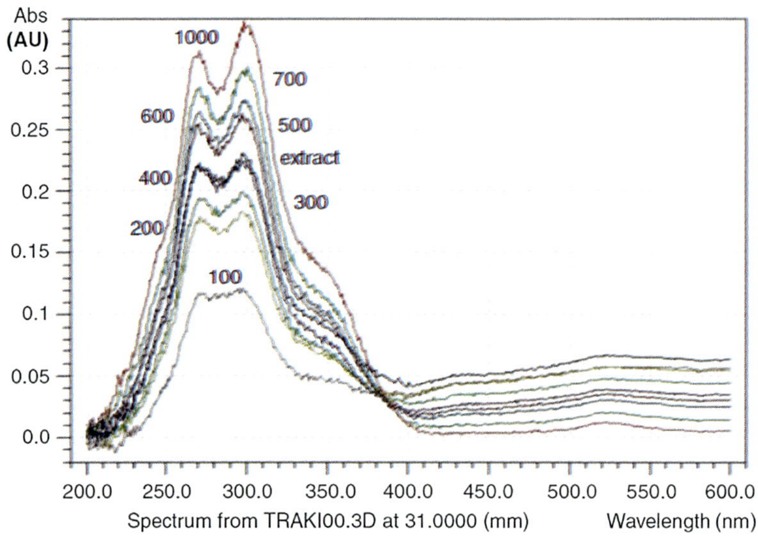

FIGURE 9.2 UV spectra obtained from fenitrothion standards at eight concentrations (100−1000 mg/mL) and from an extract of freshly squeezed apple juice containing fenitrothion. *Adopted from Tuzimski with permission [13].*

levels (standard versus compound) for reliable identification by the peak purity index.

9.2.2 Thin-Layer Chromatography with Fluorescence Detection

Fluorescent compounds absorb light at one wavelength and emit light at a longer wavelength. The emitted photons have less energy than the absorbed photons. Sample molecules located in light-scattering media exhibit fluorescence J_F, if the absorbed light is transformed into fluorescence emission. The extent of this transformation is described by the quantum yield factor q_F. The light intensity absorbed by the sample can be calculated from the light intensity reflected from the clean plate surface (J_0) minus the light intensity (J) reflected from the sample [2]:

$$J_F = q_F(J_0 - J) = q_F J_0(1 - R) \tag{9.11}$$

where:

$J_F(\lambda)$ is the emitted fluorescence intensity;
q_F (λ) is the fluorescence quantum yield factor;
$J_0(\lambda)$ is the light intensity reflected from the clean plate surface;
$J(\lambda)$ is the light intensity reflected from the sample; and
$R(\lambda)$ is the relative reflectance ($R = J/I_0$).

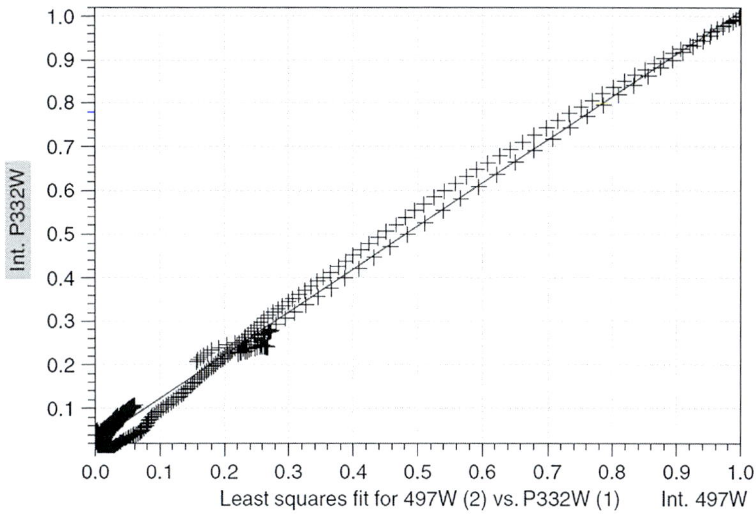

FIGURE 9.3 Least squares fit value (obtained by cross-correlation) of spectrum from fortified sample of water and spectrum from clofentezine standard. Purity index (Pearson's r) $P = 0.9959$. *Adopted from Tuzimski with permission [14].*

A general expression for fluorescence in scattering media can be described by Eqn (9.12) [2]:

$$J_F^{1/k} = mJ_0^{1/k} q_F^{1/k} a_m \tag{9.12}$$

This equation facilitates quantitative TLC and can be used for analysis [2,15].

9.2.3 TLC-FTIR and TLC-NMR

The comparison of R_F values and UV–Vis or fluorescence spectra, obtained for analytes and references—especially in complex multicomponent mixtures with difficult matrices—in many cases is insufficient to correctly confirm the identity of a compound. For direct structural characterization of analytes, several coupled techniques have been proposed, for example, TLC-FTIR and TLC-NMR. TLC was first coupled with FTIR 25 years ago, in 1989 [1,16]. Even the compounds that do not absorb UV have an infrared spectrum, so that the TLC-FTIR technique is universal. The quality of TLC-FTIR spectra is sufficient for identification of unknown substances and discrimination between closely related substances [1,17]. Because infrared absorption is observed for nearly all chemical compounds, it is possible to detect and quantify non-UV-absorbing substances on TLC plates, which makes this hyphenated technique universally applicable [1,18]. In comparison with the potassium bromide salt pellet technique, fewer absorbance bands are observed and—on commercially available layers—only the region between 3550 cm^{-1} and 1370 cm^{-1} can be evaluated [1,18]. Diffuse-reflectance infrared Fourier transform spectra (DRIFT) can be recorded from HPTLC/TLC plates [1,17]. The development of an optimized sorbent for direct HPTLC-FTIR online coupling was described [1,19]. A short introduction to the theory, instrumentation, possibilities, and limitations for direct online coupling is given in [1,18]. A report on 10 years of coupling HPTLC-FTIR online is presented in [20]. Qualitative and quantitative analysis by FTIR coupled with HPTLC is described in [1,21]. Figure 9.4 shows a schematic diagram of the DRIFT unit constructed for online measurement.

The spectrometer is connected to the external unit by means of the interface mirror, which diverts the infrared beam through a one-sided mirror with an aperture to the parabolic mirror, which in turn directs the infrared beam onto the plate surface at an exact angle of 90°. The parabolic mirror collects photons reflected from the surface of the layer and directs them through the aperture back into the spectrometer, so that most of the Fresnel reflectance is not detected (because it is eliminated in the 3600–1350 cm^{-1} region). The diffuse reflectance is gathered at the edges of the parabolic mirror and reflected by the one-sided mirror to the elliptical mirror, which again collects the diffuse reflectance and directs the converged beam to the mercury, cadmium, telluride detector [1,18,20]. By the use of commercially available GC-IR software, it is possible to obtain a chromatogram from the HPTLC plate. There are several different ways of presenting the huge amount of data acquired: as the Gram–Schmidt trace or as frequency dependent chromatograms at any wavelength of interest. In addition to the 3D plot of spectra, a contour plot can be derived from 3D stacked plots by acquiring a series of horizontal sections at different altitudes. This method enables rapid checking of peak purity. The DRIFT spectra of the substances can be calculated for reflectance and for quasi absorbance, by division of the single-beam spectra of reference and analyte or in Kubelka–Munk units [1,18,20].

The characterization of substances separated by TLC can also be obtained by NMR spectroscopy (TLC-NMR). TLC combined with high-resolution solid-state NMR for compound identification was described in [1,22]. The result presented show that it is possible to obtain interpretable NMR spectra for analytes separated on reversed-phase TLC layers without recourse to exhaustive isolation procedures [1,22].

More efficient identification of substances separated by planar chromatography is possible by modern scanning densitometry and the best characterization of analytes is possible by TLC coupled with modern MS [1].

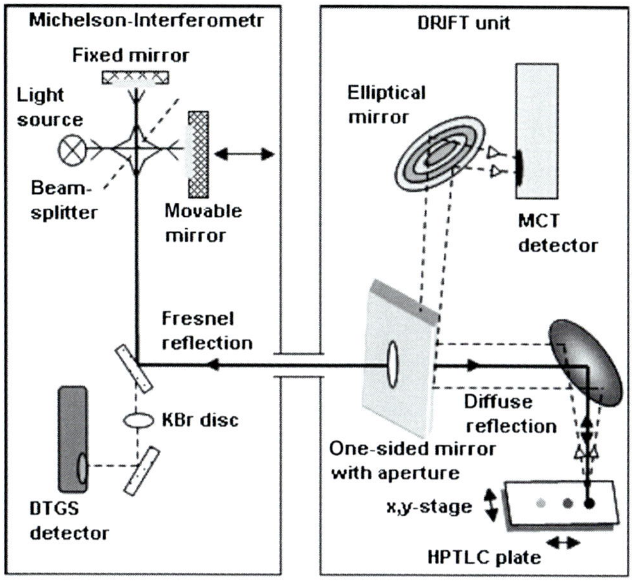

FIGURE 9.4 Schematic diagram of diffuse-reflectance infrared Fourier transform unit constructed for online measurement. HPTLC, high performance thin-layer chromatography; KBr, pottasium bromide; MCT, mercury, cadmium, telluride; DTGS, deuterated triglycine sulfate. *Adopted from Rager and Kovar [18].*

9.3 CONCLUSION

In combination with the modern spectroscopic detectors, TLC is well suited to the demands of analyte identification and quantitative analysis. New TLC methods with fast scanning detectors should enable us to detect and determine many more analytes in complex samples.

References

[1] Tuzimski T. Basic principles of planar chromatography and its potential for hyphenated techniques [chapter 14]. In: Srivastava MM, editor. High-performance thin-layer chromatography (HPTLC). Heidelberg: Springer-Verlag Berlin; 2011. pp. 247–310.

[2] Spangenberg B. Quantitative detection of drugs by densitometry and video scanning [chapter 10]. In: Komsta Ł, Waksmundzka-Hajnos M, Sherma J, editors. Thin-layer chromatography in drug analysis. Boca Raton: CRC Press, Taylor & Francis Group; 2014. pp. 169–86.

[3] Spangenberg B. Does the Kubelka–Munk theory describe TLC evaluations correctly? J Planar Chromatogr – Mod TLC 2006;19(111):332–41.

[4] Hiegel K, Spangenberg B. New method for the quantification of dequalinium cations in pharmaceutical samples by absorption and fluorescence diode array thin-layer chromatography. J Chromatogr A 2009;1216:5052–6.

[5] Sz Nyiredy. Planar chromatography, a retrospective view for the third millennium. Budapest: Springer Scientific Publisher; 2001.

[6] Hahn-Deinstrop E. Applied thin-layer chromatography. Weinheim: Wiley-VCH; 2000.

[7] Spangenberg B, Klein K-F. Fibre optical scanning with high resolution in thin-layer chromatography. J Chromatogr A 2000;898:265–9.

[8] Spangenberg B, Klein K-F. New evaluation algorithm in diode-array thin-layer chromatography. J Planar Chromatogr – Mod TLC 2001;14(4):260–5.

[9] Spangenberg B, Lorenz K, Nasterlack S. Fluorescence enhancement of pyrene measured by thin layer chromatography with diode-array detection. J Planar Chromatogr – Mod TLC 2003;16(5):331–7.

[10] Ahrens B, Blankenhorn D, Spangenberg B. Advanced fiber optical scanning in thin-layer chromatography for drug identification. J Chromatogr B 2002;772:11–8.

[11] Tuzimski T. Use of planar chromatography in pesticide residue analysis. In: Rathore H, editor. Handbook of pesticides: methods of pesticides residues analysis, Nollet LML. Boca Raton: CRC Press, Taylor and Francis Group; 2010. pp. 187–264 [chapter 9].

[12] Tuzimski T. Determination of pesticides in complex samples by one dimensional (1D-), two-dimensional (2D-) and multidimensional chromatography [chapter 12]. In: Stoytcheva M, editor. Pesticides in the modern world – trends in pesticides analysis. Intech Open Access Publisher; 2011. pp. 281–318.

[13] Tuzimski T. Use of thin-layer chromatography in combination with diode array scanning densitometry for identification of fenitrothion in apples. J Planar Chromatogr – Mod TLC 2005;18(106):419–22.

[14] Tuzimski T. Determination of pesticides in water samples from the Wieprz–Krzna Canal in the Łeczyńsko-Włodawskie Lake District of Southeastern Poland by thin-layer chromatography with diode array detection. J AOAC Int 2008;91(5):1203–9.

[15] Spangenberg B, Weyandt-Spangenberg M. Quantitative thin-layer chromatography using absorption and fluorescence spectroscopy. J Planar Chromatogr – Mod TLC 2004;17:164–8.

[16] Glauninger G, Kovar K-A, Hoffmann V. Untersuchungen zum online DC-IR transfer. In: Gauglitz G, editor. Software Entwicklung in der Chemie 3, Proc 3rd workshops comput chem. Berlin: Springer; 1989. pp. 171–80.

[17] Kovar K-A, Enblin HK, Frey OR, Rienas S, Wolff SC. Applications of on-line coupling of thin layer chromatography and FTIR spectroscopy. J Planar Chromatogr – Mod TLC 1991;4(3):246–50.

[18] Rager IOC, Kovar K-A. Planar chromatography and IR. In: Nyiredy Sz, editor. Planar chromatography, a retrospective view for the third millennium. Budapest: Springer Scientific Publisher; 2001. pp. 247–60.

[19] Bauer GK, Pfeifer AM, Hauck HE, Kovar K-A. Development of an optimized sorbent for direct HPTLC-FTIR on-line coupling. J Planar Chromatogr – Mod TLC 1998; 11(2):84–9.

[20] Stahlmann SA. Ten-years report on HPTLC-FTIR online coupling. J Planar Chromatogr – Mod TLC 1999;12(1):5–12.

[21] Cimpoiu C. Qualitative and quantitative analysis by hyphenated (HP)TLC-FTIR Technique. J Liq Chromatogr Relat Technol 2005;28(7–8):1203–13.

[22] Wilson ID, Spraul M, Humpfer E. Thin-layer chromatography combined with high-resolution solid-state NMR for compound identification without substance elution: preliminary results. J Planar Chromatogr – Mod TLC 1997;10(3):217–9.

CHAPTER

10

Advanced Spectroscopic Detectors for Identification and Quantification: Mass Spectrometry

Sychyi Cheng [1], *Jentaie Shiea* [1,2]

[1] Department of Chemistry, National Sun Yat-Sen University, Kaohsiung, Taiwan; [2] Department of Medicinal and Applied Chemistry, Kaohsiung Medical University, Kaohsiung, Taiwan

10.1 INTRODUCTION

The detection of compounds separated by thin-layer chromatography (TLC) is usually achieved through conventional chemical methods, such as optical examination and biochemical reactions. Although fast and simple, the aforementioned detection methods do not provide comprehensive information on molecular weight and structure of the analytes. It is generally accepted that a chemical compound can be identified by mass spectrometry (MS) based on the mass-to-charge (m/z) value of the molecular ion and its fragment ion pattern [1,2]. Since analytes are ionized via gain or loss of electrons forming a cation or anion before MS interrogation, the success of TLC-MS methods depends on whether analytes adsorbed on the TLC sorbent matrix are efficiently ionized. To successfully ionize analytes on a TLC plate, traditional methods such as extraction, elution, and desorption are employed [3–6]. Analytes on the TLC plate can be analyzed directly by MS without separation from the TLC sorbent matrix as well. Different interfaces for TLC-MS have been developed; enabling online, direct, and rapid detection for high-throughput analysis. This chapter describes different approaches for identifying compounds separated by TLC with MS.

Instrumental Thin-Layer Chromatography
http://dx.doi.org/10.1016/B978-0-12-417223-4.00010-8

10.2 CLASSIFICATION OF TLC-MS TECHNIQUES

Many TLC-MS techniques have been developed to characterize compounds on TLC plates. These techniques can be classified based on differences in the analytical processes (i.e., online or off-line coupling), detection environment (i.e., ambient or vacuum conditions), ionization methods, or sampling approaches (i.e., indirect or direct sampling). In this chapter, the existing TLC-MS techniques are classified as indirect and direct sampling TLC-MS. For indirect sampling TLC-MS, analytes removed from the layer prior to MS analysis are subjected to solvent extraction, concentration, desorption, and ionization. For direct sampling TLC-MS, analytes on TLC plate surfaces are directly characterized without additional sample pretreatment. Figure 10.1 is a tree diagram that separates TLC-MS techniques based on the sampling approaches.

10.3 INDIRECT SAMPLING TLC-MS

Indirect sampling TLC-MS techniques require sample preparation prior to MS analysis. Using indirect sampling approaches to characterize the analytes by MS, optical methods, such as absorption of light or derivatization by chemical and biological reagents must be used to find the zones of interest on the layer first. The analytes are then recovered through a series of processes, such as scraping, solvent extraction and preconcentration, or are transferred to other surfaces so that the isolated analytes are ready for MS analysis.

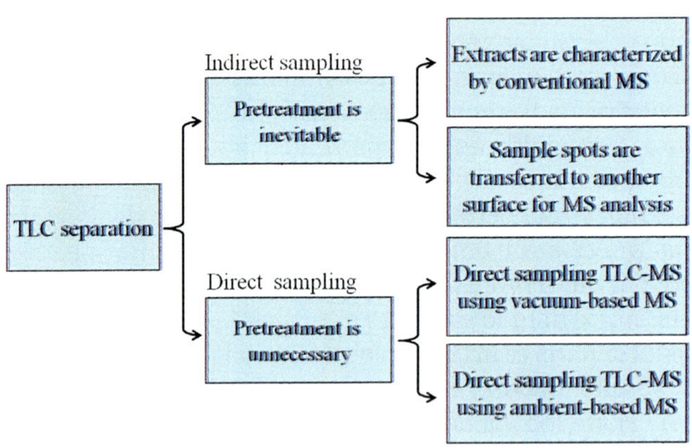

FIGURE 10.1 Classification of TLC-MS based on sampling methods. TLC, thin-layer chromatography; MS, mass spectrometry.

10.3.1 Solvent Extraction

Solvent extraction is a traditional method for recovering analytes from the TLC sorbent matrix. The zones of interest are scraped from the TLC plate and the analytes adsorbed on the gel particles are subsequently extracted by organic solvents in an ultrasonicator. After centrifugation, the suspensions are recovered and concentrated for MS analysis. A direct inlet system coupled with electron impact/chemical ionization (EI/CI) was used to characterize drug extracts from silica-based TLC plates [7]. However, the need to evaporate extracts under high temperatures (e.g., 250 °C) in an EI/CI source requires analytes to be thermally stable and volatile.

Since analytes may not be totally resolved by TLC separation, additional chromatographic techniques combined with MS, such as gas chromatography/mass spectrometry (GC-MS) and liquid chromatography/mass spectrometry (LC-MS) are used to further analyze analytes in the extracts [8–12]. The resolution of GC separation, the volatility and thermal stability of the analyte, and the detection sensitivity of the MS can be increased via analyte derivatization with hydrophobic groups. Several chemical reagents have been used to derivatize different analytes in sample extracts. For example, isomers of cholesteryl ester hydroperoxides were characterized by GC-MS after derivatization to introduce a trimethylsilyl groups, while the extracts of small organic compounds, such as phytosterol oxidation products, acylglycerols, and peroxidized cholesterol were characterized by GC-MS after derivatization with N,O-bis(trimethylsilyl)trifluoroacetamide or N,O-bis(trimethylsilyl)acetamide [10–13].

In contrast to the use of GC-MS to analyze thermally stable and volatile analytes, thermally unstable and nonvolatile analytes in the extracts are characterized by LC-MS without derivatization. An LC-MS system equipped with electrospray ionization (ESI) or atmospheric pressure chemical ionization (APCI) source is commonly used to characterize the analytes extracted from TLC layers [14].

Desorption/ionization (DI) techniques, such as fast atom bombardment (FAB), liquid secondary ion mass spectrometry (LSIMS), laser desorption (LD), and matrix-assisted laser desorption/ionization (MALDI) desorb and ionize analytes by bombarding or irradiating the extracted analytes in the matrix with primary fast ions (i.e., Ar^+ and Cs^+, 2–10 kV), fast atoms (i.e., 5–10 kV), or a pulsed laser beam (UV, 337 nm, 5 ns, 200 μJ) in vacuum [2]. Since protonated molecular ions (secondary ions) of fragile and polar compounds are produced by FAB, LSIMS, or MALDI; these techniques are suitable for characterizing thermally labile and low volatility compounds.

For FAB and LSIMS, a stream of fast atoms or ions, such as argon (Ar), cesium (Cs^+), or gallium ($^{69}Ga^+$) are accelerated through a high electric field (5–20 kV) to impact sample spots, generating neutral and charged analyte ions from the sample surface. The analyte ions are subsequently detected by sector, quadrupole, ion trap, or time-of-flight (TOF) mass analyzers. The functions of the viscous matrix used in this approach include: (1) absorbing and transferring the energy brought by bombardment of the TLC surface with fast atoms or ions to the analytes, (2) self-cleaning the bombarded area, (3) providing protons for ionization of analytes, and (4) extending the duration of the ion signals for MS/MS analysis. There are still some limitations on the detection of analytes via FAB and LSIMS: (1) to have a high efficiency of DI, the analytes must be soluble or mix well with the liquid matrix; this limit the analytes to polar compounds; (2) the proton affinities of analytes need to be higher than that of the matrix so the analytes can be ionized; and (3) the molecular weight of the detectable analytes should be less than 2000.

Among these DI techniques, MALDI is most commonly used for characterizing biological compounds, such as peptides, proteins, and lipids. The TLC extracts are mixed with a matrix solution (e.g., saturated solution of sinapinic acid, 2,5-dihydrobenzoic acid, or α-cyano-4-hydroxycinnamic acid) on a stainless steel target to form matrix/analyte cocrystals after drying [15,16]. The cocrystals are then irradiated with a pulsed laser beam (UV or IR), thereby desorbing and ionizing the analytes. The functions of the matrix are: (1) separating the analyte molecules from each other, (2) adsorbing the laser energy and transferring the energy to the analytes for desorption, and (3) serving as a proton donor or acceptor to ionize the desorbed analytes in the positive or negative ion modes. Nevertheless, the presence of matrix ions at low mass ranges may interfere with the detection of low mass analyte ions. With the use of MALDI/MS compounds such as phospholipids in brain and broncho alveolar lavage fluid, synthetic polymers (e.g., PEG, DPPG, TPPG, PTMG), and lipopolysaccharides were successfully characterized [17–23]. Although there have been fewer studies on the use of other DI techniques, such as FAB and LSIMS for characterization of TLC extracts, they are useful in characterizing hydrophobic organic compounds [24].

TLC extracts can also be analyzed using solution-based mass spectrometric ionization techniques, such as electrospray ionization mass spectrometry (ESI/MS), atmospheric pressure chemical ionization mass spectrometry (APCI/MS), and atmospheric pressure photoionization mass spectrometry (APPI/MS). ESI/MS has been used to characterize lipids, drugs, and alkaloids in sample solutions which were prepared by dissolving TLC extracts in organic solvents [23–28].

10.3.2 Transfer-based Indirect Sampling TLC-MS

The transfer of analytes from sample zones adsorbed in the layer to the surface of another material for further mass spectrometric analysis is an approach to separate the analytes from the sorbent matrix. Analytes can be transferred through side elution or press-to-transfer to a double-sided tape or polyvinylidene difluoride (PVDF) membrane [29–34].

Side elution is a technique developed to transfer analytes from a TLC plate to a porous surface, which preserves the chromatographic integrity of the TLC separation. To form the porous surface on one side of a TLC plate, an aqueous solution of mercury (II) chloride is applied to an aluminum-backed silica gel plate along one side of the thin layer [29,30]. The silica gel is removed and the aluminum sheet is oxidized to form Al_2O_3. Impurities from modification are removed by washing with distilled water, so that a porous surface is formed on the TLC plate. The sample is separated on the chromatographic TLC bed on another side, after which a lateral solvent development is performed to elute the analytes from the separation layer to the porous surface. Since analytes are distributed on the porous surface, they can be characterized by the aforementioned DI mass spectrometric techniques. LSIMS and FAB have been employed for the analysis of side-eluted compounds, such as methyl red indicator, terpinolen, giberrelic acid, and morphine [29–31].

The analytes in the sample spots can be transferred to a direct inlet probe without the need of extraction and lateral elution as well. A strip of double-sided tape attached to the tip of a FAB probe was used to sample the analytes in the TLC gel bed by their adhesion to the tape. A FAB matrix, such as glycerol or thioglycerol, was applied to the tape, after which the sample was introduced into the ion source to acquire the FAB mass spectra of lasalocid, septamycin, and monensin [32].

Another transfer method uses a PVDF membrane to transfer analytes [33,34]. This technique is coupled with MALDI/MS for the analysis of lipids extracted from a human brain with Alzheimer's disease. The transfer of lipids from a TLC plate to a PVDF membrane includes several steps: (1) visualizing the sample spots under UV light after spraying the plate with primuline reagent; (2) immersing the plate in a blotting solvent (e.g., iso-propanol/$CaCl_2$/MeOH) for 10 s; (3) mounting the plate with a PVDF membrane, Teflon membrane, and glass fiber filter paper; and (4) pressing the assembly at 180°C for 30 s. The lipids are successfully transferred to the PVDF membrane for further MS analysis. A solution of 10 pmol ganglioside was detected by this TLC-blot-MALDI-TOF MS approach [33].

The advantages of using indirect sampling TLC-MS for chemical analysis are that trace analytes removed from the TLC gel bed can be concentrated and characterized by the MS equipped with different ionization sources. Since TLC sorbent particles are removed during sample pretreatment, the approach greatly reduces the risk of damage to the vacuum system (especially for turbo molecular pump) by introducing a clean sample into the mass spectrometer without particles.

Although the sample pretreatment is simple, easy, and can be performed in most chemical and biochemical laboratories; the sample preparation processes are usually labor intensive and time consuming. Therefore, the application of indirect TLC-MS techniques for high-throughput analyses is impractical. Furthermore, the aforementioned approaches may fail to provide chromatographic information if analytes cannot be visualized via staining. Direct TLC-MS approaches that characterize TLC tracks without sample preparation would overcome the above shortcomings.

10.4 DIRECT SAMPLING TLC-MS

To directly characterize the analytes on TLC plate with MS, the TLC plate is delivered to the ionization source without removal of the analytes from the layer and those analytes that cannot be visualized by traditional detection methods are analyzed directly on the TLC plate. Since tedious sample preparation is avoided, direct sampling TLC-MS is potentially useful for automatic and high-throughput analysis. The techniques developed for simultaneous sampling and ionizing analyte on TLC plates include elution and desorption approaches. Based on the differences in operating environments, direct sampling TLC-MS techniques can also be grouped into vacuum- or ambient-based approaches. For vacuum-based TLC-MS, the TLC plate is cut into small pieces to fit in the ion source. However, ambient-based approaches allow the use of whole TLC plates for analysis.

10.4.1 Direct Sampling TLC-MS, Vacuum-based

The mass spectrometric ionization methods used for directly sampling TLC-MS techniques under vacuum conditions include FAB, LSIMS, LDI, MALDI, and surface-assisted laser desorption ionization (SALDI). In general, TLC plates are fixed on the sample probe using double-sided adhesive tape. The probe is then introduced into the vacuum chamber of the ionization source. Based on differences in DI mechanisms, direct sampling TLC-MS with vacuum-based ionization sources can be divided into two categories: (1) DI via surface bombardment with fast atoms or ions, and (2) DI via surface irradiation with a pulsed laser beam [15,16,35,36]. Organic and inorganic matrices must be preapplied to the zones of interest on the TLC plate to assist DI by FAB, LSIMS, MALDI, and

SALDI [15,16,37–39]. The techniques are particularly suitable for characterizing nonvolatile or thermally labile compounds. Volatile compounds are difficult to detect by this approach since they are easily pumped away under vacuum. In addition, several shortcomings are encountered for the operation of vacuum-based TLC-MS: (1) the introduction of the TLC plates from ambient to vacuum conditions is time consuming and labor intensive; (2) the size of the TLC plates is limited by the size of the ionization source in the vacuum chamber; (3) the desorbed TLC gel particles in the ionization source may damage the MS instrument, especially vacuum system that use turbo molecular pumps; and (4) high-throughput analysis is difficult to achieve.

10.4.1.1 Impact of TLC Plate Surfaces with Fast Atoms or Ions for DI

Previously, FAB and LSIMS have been used to characterize extracts recovered from TLC plates off-line. The same techniques are also suitable for directly characterizing analytes on TLC plates. A viscous matrix solution, such as glycerol, thioglycerol, threitol, *m*-nitrobenzyl alcohol, or triethanolamine is applied to sample spots on the TLC plate, after which the TLC plate is delivered into the ionization source for analysis. Unfortunately, the introduction of a viscous liquid matrix onto the sample spot will induce the lateral spreading of analyte zones throughout the layer, and thereby decreasing the chromatographic resolution and lowering the detection sensitivity. LSIMS was used to directly characterize organic halides, amines, and derivatized corticosterone separated on normal phase TLC plates [36–38]. The residual tetracycline antibiotics extracted from bovine tissues including those from the muscle, liver, and kidney were characterized via TLC-FAB, where antibiotic ions with 0.1 μg/spot were successfully detected [35] (Figure 10.2).

10.4.1.2 Using Pulsed Laser Irradiation for DI

LDI, MALDI and SALDI are laser-based mass spectrometric ionization techniques commonly used to desorb/ionize analytes on TLC plates. Both charged and neutral analytes are formed via pulsed laser irradiation. Instruments required to perform LDI, MALDI, and SALDI analysis include a pulsed laser beam and a mass spectrometer, in which a TOF mass analyzer is the most common. In LDI, the zone of interest on a developed TLC plate is irradiated with a pulsed laser beam. The analyte ions are formed in the desorption region through ion–molecule reactions (IMRs) with protons or charged matrix species. TLC-LDI/MS has been used to detect and identify small molecules such as berberine, plamatine, polycyclic aromatic hydrocarbons (PAHs), purines, sugars, and amines [24,39]. The chromatographic information of purine derivatives and PAHs was obtained after lane scanning, where the spatial resolution of LDI was

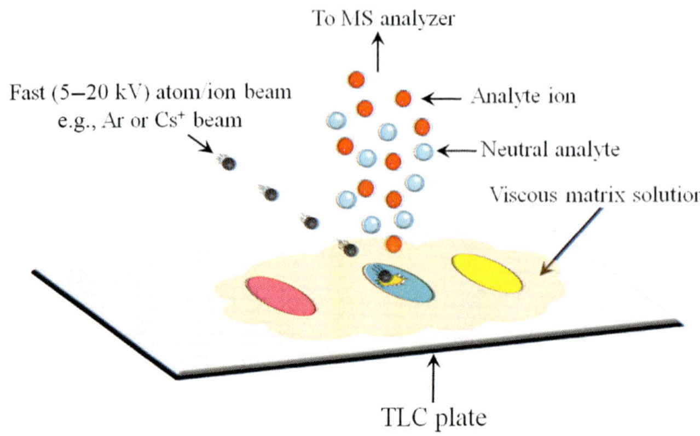

FIGURE 10.2 Analyses of chemical compounds on the surface of a TLC plate by FAB/MS or LSIMS. TLC, thin-layer chromatography; MS, mass spectrometry; FAB, fast atom bombardment; LSIMS, liquid secondary ion mass spectrometry.

limited by the laser spot diameter. In LDI, the analyte ions are formed in the absence of any matrix, this makes it only useful to desorb/ionize small molecules ($m/z < 2000$). However, DI of larger analytes in MALDI and SALDI requires the use of organic and inorganic matrices, respectively.

The interface to combine TLC with MALDI-TOF was first reported in 1995 for the characterization of small organic compounds and peptides [40,41]. A methanol solution is applied to TLC plate first to extract the analytes adsorbed on the gel particles. An organic matrix solution was then applied to the plate to form analyte-matrix crystals after drying. Different matrix application methods such as spotting, spraying, and evaporation in air or vacuum have been employed to apply the MALDI matrix to the plate; however, the application of the matrix solution on the TLC plate induces the spread of analytes from 0.5 mm up to 5 mm [40,41]. MALDI/MS has been used to directly characterize compounds such as dyes, peptides, proteins, alkaloids, gangliosides, lipids, and cyanobacterial toxins separated by TLC with detection limits in the ng–pg/spot range and linearity (R^2) of 0.994 for gangliosides [15,16,40–51]. Since laser irradiation provides innate characterization with high spatial resolution, the pulse laser can be used to scan the whole plate surface to obtain molecular images. Three-dimensional images showing the distribution of the guinea green B, rhodamine B, and angiotensin II spots on silica gel TLC plates were reported, although a relatively long acquisition time was needed [40,41].

Instead of using organic matrix in MALDI, an inorganic matrix is used in SALDI to assist DI of the analytes. Solutions containing inorganic materials, such as carbon, graphite, silicon, TiO_2, TiN, Co, or magnetic nanoparticles in nm to μm sizes are deposited on the TLC plate for SALDI analysis [37,52,53]. The matrix solution is prepared by suspending

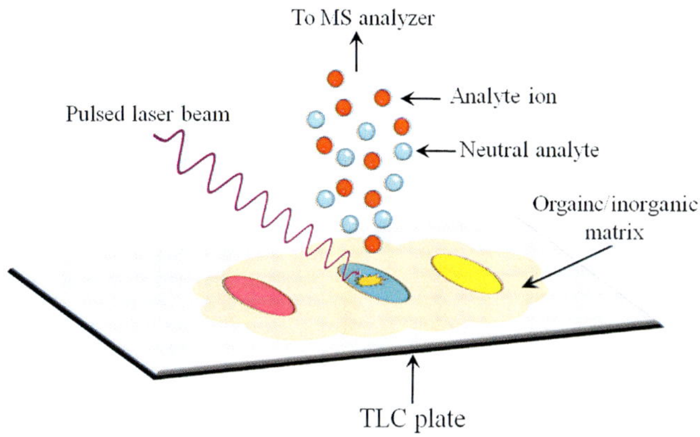

FIGURE 10.3 Analyses of chemical compounds on the surface of a TLC plate by MALDI/MS or SALDI/MS. TLC, thin-layer chromatography; MS, mass spectrometry; MALDI, matrix-assisted laser desorption/ionization; SALDI, surface-assisted laser desorption ionization.

the inorganic materials in a solution (e.g., sucrose/glycerol-methanol). Although inorganic particles are spread on the TLC plate surface instead of forming analyte–matrix cocrystals as in MALDI, the inorganic particles can still absorb the laser energy to induced DI of the analytes adsorbed on the sorbent matrix. Low interference from matrix ions and low ion suppression effects make SALDI suitable for the analysis of low mass compounds. TLC-SALDI/MS has been used to characterize peptides, herbicides (such as prometryn), diuretics, polymers, porphyrins, and tetracycline antibiotics [37,52–56]. The detection limits of bradykinin and porphyrins were 25 ng and 500 pg, respectively.

With the assistance of matrix for DI in MALDI and SALDI, detection limits at pico- to nanogram levels can be achieved, which are two to three orders lower than those of DI techniques that do not require a matrix. However, the deposition of matrix on TLC plates will induce the lateral spread of analytes, therefore decreasing chromatographic resolution. In addition, interference from matrix ions at lower mass ranges may be encountered in MALDI. Furthermore, the effects of "sweet spots" in MALDI severely reduce its reproducibility, rendering it impractical for quantitative analyses (Figure 10.3).

LD followed by CI in vacuum has been used to characterize analytes on TLC plates, where a stream of CI reagent gas, such as methane is employed to transport laser-desorbed analytes into a CI source [57]. Since analytes are desorbed by laser irradiation first and ionized later, the LD-CI approach is considered a two-step ionization technique. PAHs were directly characterized using LD-CI; the detection limit for phenanthrene was ca. 10 ng, and the reproducibility was within ±20% [57]. Another two-step ionization approach known as resonant two-photon ionization was used to characterize thermally labile biological molecules including indoleamines, catecholamines, peptides, and drugs on silica gel TLC plates [58]. With the use of internal standards, quantification of indole-3-acetic acid and imipramine was achieved, with dynamic ranges over four orders of magnitude and detection limits in the low nanogram range.

10.4.2 Direct Sampling TLC-MS under Ambient Conditions

Since TLC is performed under ambient conditions, it is therefore logical and convenient to characterize analytes in sample zones in their native environment. This can be achieved using atmospheric pressure ionization MS, where surface sampling and analyte ionization are both performed at room temperature and atmospheric pressure, the analyte ions are subsequently interrogated by the mass spectrometer for detection. Direct characterization of TLC plates under atmospheric pressure and room temperatures provides several benefits including: (1) the construction of an interface for TLC and MS under ambient conditions is much easier than for vacuum conditions; (2) the ion source will be compatible with TLC plates of different sizes; (3) sample switching is fast, allowing for high-throughput analysis; and (4) organic and inorganic components with different volatilities can be detected.

The surface sampling methods including elution and desorption approaches have been developed to remove the analytes from the TLC sorbent bed without tedious sample preparation. The analytes within the sorbent bed can be transferred into a solution using an elution system or overrun TLC development [59–62]. The analyte solution is then delivered to a solution-based atmospheric pressure ionization sources, such as ESI, APCI, or APPI for subsequent ionization. Desorption techniques including LD, thermal desorption, droplet impact, and laser-induced acoustic desorption have been used to sample the analytes on TLC plate surfaces [63–66].

One-step ionization means the analyte ions are formed simultaneously during desorption. Techniques such as atmospheric pressure (AP)-MALDI, direct analysis in real time (DART), desorption electrospray ionization (DESI), and easy ambient sonic spray ionization (EASI) are typical one-step ionization techniques. On the other hand, techniques such as electrospray laser desorption ionization (ELDI), laser ablation inductively coupled plasma (LA-ICP), laser desorption-atmospheric pressure chemical ionization (LD-APCI), laser-induced acoustic desorption electrospray ionization (LIAD-ESI), and plasma-assisted multiwavelength laser desorption ionization (PAMLDI) are two-step ionization techniques, where the desorption and ionization steps are separate events and occurring in tandem [63,64,66–72].

10.4.2.1 *Surface Sampling Devices Using Elution*

Solvent extraction has commonly been used to recover analytes sorbent removed from scrapped TLC plates. However, labor- and time-consuming sample preparation processes are the main obstructions for this approach for automated and high-throughput analysis. Approaches to conveniently extract analytes from TLC plates without sorbent removal, off-line extraction, solvent collection, filtration, and centrifugation have been developed to overcome the aforementioned drawbacks. The solvent extraction devices with coaxial or parallel tubes are mounted perpendicular to TLC plate for analyte extraction and recovery, followed by MS analysis using atmospheric pressure ionization sources such as ESI, APCI, and inductively coupled plasma (ICP) [59,60,73,74].

The solvent extraction device with a coaxial capillary tube, also known as a surface sampling probe (SSP), is used to characterize analytes on TLC plates [59]. The TLC plate is positioned at 90° relative to the SSP, where the distance between the TLC plate and the SSP tip is less than 0.1 mm. A syringe pump is used to deliver the extraction solvent through a space between the inner and outer tubes to establish a liquid microjunction on the TLC plate surface. The small gap between the SSP tip and analyte spot, the surface tension of the extracting solution, and the hydrophobicity of TLC plate surfaces (i.e., reversed-phase silica gel bed) confine the liquid junction and prevent the solution from flowing beyond the probe into the layer. Analytes adsorbed on the TLC layer are dissolved in the extraction solvent and the sample solution is then transported to an API source through the inner capillary tube of the SSP. After extraction, the chromatographic information of the analytes on the TLC plate is still retained. Compounds, such as dyes, caffeine, drugs, alkaloids, and peptides separated on reversed-phase or silica gel TLC plates have been detected and identified using SSP coupled with ESI/MS/MS or APCI/MS/MS [59,73,75–78]. Quantification of caffeine in drinks by selected reaction monitoring using a deuterium-labeled internal standard yielded a linear response from 1 to 50 ng, a coefficient of determination (r^2) of >0.9998, and a relative standard deviation (RSD) of less than 4.5%.

Another approach uses two parallel capillaries coupled to an elution head to transport the extract solution onto and away from the TLC plate surface [60,79]. The elution head consists of an inlet and outlet capillary, filter frit, and cutting ring. The elution head is pressed against the TLC plate so that the ring cuts into a defined area of the TLC layer, after which an extraction solvent pumped by an LC pump is delivered to the layer through the inlet capillary. Adsorbed analytes are dissolved by the extraction solvent flowing through the isolated portion of the layer. The sample solution is filtered through a frit and then transported to the ionization source through the outlet capillary. Since the elution head and the TLC plate are in physical contact, the liquid microjunction established in the aforementioned SSP device does not exist. The extraction device and filter frit are useful for recovering analytes and avoiding interference from gel particles. The resolution of the elution head-based solvent extraction device is influenced by the diameter of the ring-shaped cutting tool. The elution head-based solvent extraction device coupled with ESI/MS was used to characterize compounds, such as isopropyl thioxanthone, hormone, vitamins, caffeine, acrylamide, oligosaccharides, and antibiotics separated by TLC [79–86]. In addition, the parallel capillary device has been coupled with ICP-MS to characterize iodine-based X-ray reagents [74] (Figure 10.4).

10.4.2.2 Continuous Elution Devices for TLC-MS

Analytes can be eluted out of the TLC plate and detected by ESI/MS in a manner similar to LC-MS. This concept has been used by overrun TLC-MS. An aluminum-base TLC strip with a sharpened end is prepared for overrun TLC, where the gel bed at the sharpened end was scraped off [61]. The tip of the sharpened TLC plate is positioned 1–2 cm from the MS inlet, while the mobile phase is applied to the opposite end of the TLC plate. A high potential (4 kV) is applied to the mobile phase by inserting a platinum electrode into the mobile-phase reservoir. A makeup solution is delivered through a capillary by a syringe pump and introduced to the tip of the sharpened TLC strip to continuously induce ESI from the makeup solution containing the eluted analytes. Overrun TLC can also be performed in a small channel packed with silica-based C_{18} particles. One end of the channel is connected to a mobile-phase reservoir to migrate the analytes to the opposite end of the channel. Two bound optical fibers are buried in the layer so that they protruded from the channel to a distance of 5 mm. A high voltage is applied to a platinum wire inserted into the makeup solution reservoir to induce ESI of the solution flowing through

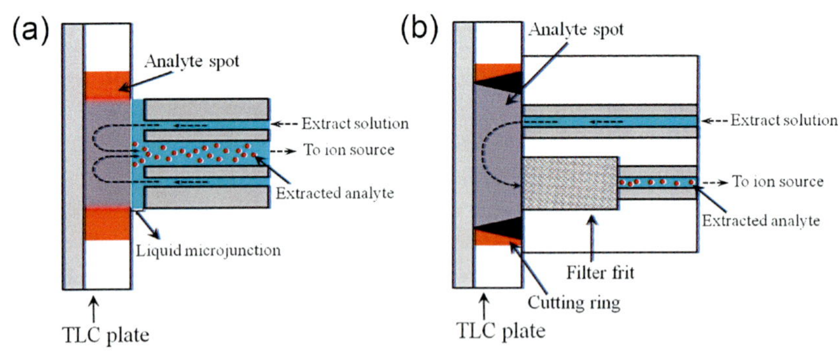

FIGURE 10.4 Analyses of chemical compounds in gel bed by ESI-based ambient TLC–MS using (a) a coaxial tube sampling probe and (b) a two-parallel-capillaries sampling devices. TLC, thin-layer chromatography; MS, mass spectrometry.

the bound optical fibers. The two overrun TLC-ESI/MS techniques have been used to characterize low mass compounds including drugs and ferrocene complexes [61].

There are still different manners to perform overrun TLC, such as overpressure TLC (OPLC) and rotation planar chromatography (RPC) [62,87]. In OPLC, a Teflon membrane is pressurized (e.g., 50 bar) on a TLC plate and the extraction solvent is pumped through the layer to elute the analytes to the end of the TLC plate. A short capillary tube continuously transports the solution from the OPLC layer to an ESI source. The OPLC-ESI-MS approach has been used to separate and characterize lipid mixtures with a sensitivity of 5 pmol for glycosphingolipids [87]. Instead of a pump, RPC utilizes a centrifugal force to move the eluent through the stationary phase. Samples are predeposited at the center of a circular plate, which is then spun at about 1000 rpm. The mobile phase is continuously applied at the center of the plate, resulting in a separation consisting of a series of concentric rings (circular development). The sample solution, eluted from the edge of the TLC plate is collected in a fraction collector for further MS analysis. A dye mixture containing Solvent Green 3, Solvent Blue 35, and Fat Red 7B was characterized by coupling a commercially available RPC system to APCI/MS [62]. Since the analytes are continuously eluted from the TLC plate, information used to identify substances such as the retardation factor (R_f) is lost, however, quantification of analytes with low detection limits is possible.

10.4.2.3 *Using a Pulsed Laser for Sampling and Ionization*

Direct sampling of analytes on TLC plates via a laser beam in a vacuum has been demonstrated for MALDI and SALDI. Laser-based sampling techniques can also be performed under ambient conditions; consequently, atmospheric pressure ionization methods, such as atmospheric pressure AP-MALDI, ICP, ESI, and APCI are employed to postionize analytes desorbed by laser irradiation [66,67,71,72].

The procedures for performing laser DI and the preparation of analyte-matrix cocrystals in AP-MALDI are similar to that for vacuum-MALDI except the analysis is performed under ambient conditions. In this case, fast sample switching and detection of volatile and semivolatile compounds are possible and the size of the TLC plate will not be limited by the dimensions of the vacuum chamber as is the case for a MALDI source. In addition, the displacement of layer particles will not cause damage to the turbo molecular pump in the MS system. In AP-MALDI, the TLC plates sprayed with a matrix solution are attached to a target plate set on a stage, which is placed near the MS inlet. Sample spots are irradiated with a pulsed laser beam to induce DI. Although the sensitivity of AP-MALDI is poorer than for vacuum-MALDI, TLC-AP-MALDI/MS has been used to separate and identify drug mixtures with detection limits in the pmol range and RSDs below 25% [88]. Nevertheless, AP-MALDI still has problems that are often encountered in vacuum-MALDI, such as the lateral spreading of analytes on the layer, sweet spot effect, and interferences from matrix ions in the low mass range.

ICP is an ionization method commonly used in inorganic MS to determine most elements in the periodic table. This ionization source has been combined with laser ablation to characterize elements on solid surfaces, such as geological samples, biological tissues, and wafers. During an LA-ICP-MS analysis of separations by TLC, the plates in an ablation cell are irradiated with a pulsed laser beam where the laser-ablated particles are carried to the ICP source by a flow of helium or argon [62,89,90]. Evaporation, atomization, and ionization of the particles then occur in a high-temperature plasma (6000–10,000 K). The use of a collision gas (i.e., He and H_2) or reaction gas (i.e., CH_4 or NH_3) reduces interferences from isobaric polyatoms (e.g., interference from $^{40}Ar^{35}Cl^+$ during analysis of $^{75}As^+$) via physical or chemical means. TLC-LA-ICP-MS has been used to characterize chromium and arsenic, with detection limits in the nanogram range and precision from 3 to 40%. In addition, the proportionality of V/Ni in an asphaltene fraction of crude oil has been determined by TLC coupled with femtosecond LA-ICP-MS [90].

ELDI is an ambient mass spectrometric technique that integrates LD with ESI [91]. The technique is useful for the characterization of organic and biological samples without tedious sample preparation. A pulsed laser beam and ESI source are combined to make an ELDI source. Analytes in solid or liquid state are desorbed by a pulsed laser beam and postionized in an ESI plume located several millimeters above the laser spot. The ionization mechanisms of ELDI are similar to those of fused-droplet ESI, where desorbed neutral compounds are ionized through IMRs or fused with the charged solvent species (e.g., protons, hydronium ions, and cluster solvent ions) in the ESI plume [92]. The hyphenation of TLC and ELDI was first reported in 2007, where the analyte zones on a silica-based C_{18} layer were analyzed without application of any organic or inorganic matrix [66]. The problems commonly encountered in TLC-MALDI/MS, such as the lateral diffusion of sample spots and interferences from matrix ions at the low mass range, were not observed in TLC-ELDI/MS. Samples such as dyes, amines, lipids and drug extracts were successfully analyzed by TLC-ELDI/MS, with detection limits in the ng range and a linear response ranging from several nanograms to micrograms ($R^2 = 0.9886$) [66]. In addition, the chromatographic information and molecular images showing the distribution of analytes on the TLC plates are retained with this technique (Figure 10.5).

Laser-desorbed analytes can also be postionized through APCI processes. PAMLDI hyphenates a pulsed laser beam and a DART source [71,93]. The setup and DI processes are similar to ELDI except that the

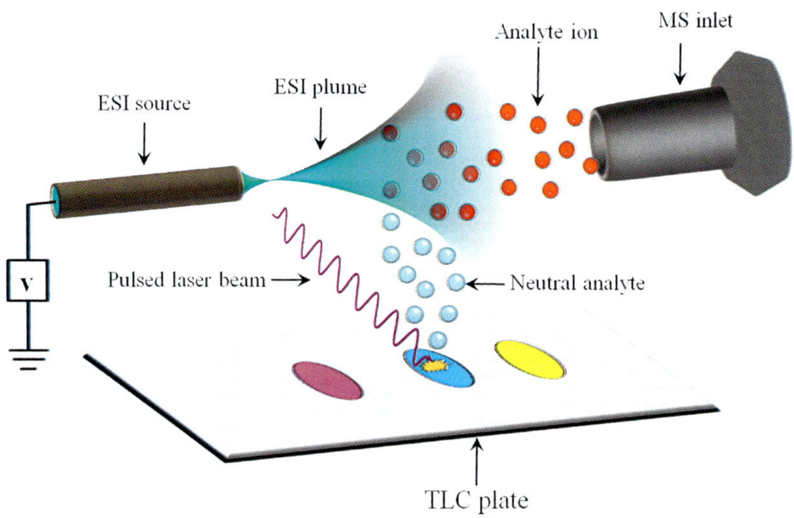

FIGURE 10.5 Analyses of chemical compounds on the surface of a TLC plate by ELDI/MS. TLC, thin-layer chromatography; MS, mass spectrometry; ELDI, electrospray laser desorption ionization.

ionization of the laser-desorbed analytes is by reactions with the reactive species generated by DART. A DART source is comprised of an electric needle, gas heater, and grid electrodes. A stream of inert gas (usually He) is flowed through an electrical discharge area to generate reactive species, such as ions, electrons, protons, and metastable atoms. The reactive species are heated and passed through grid electrodes to remove the ions in the plasma. It is suggested that the reaction of metastable He with water molecules (i.e., Penning ionization) generates hydronium (H_3O^+) and water cluster ions [$(H_2O)_nH_3O^+$], which further react with analytes to form ions via IMRs. Since analytes must be evaporated before IMRs, the temperature of the plasma gas is a critical factor for controlling the ionization efficiency of DART. PAMLDI/MS has been used to characterize analytes on silica gel and cellulose layers without sample pretreatment or matrix application. Samples including dye mixtures, drug standards, and tea extracts were separated and characterized with detection limits in the range of 5 ng/mm^2 [71] (Figure 10.6).

Laser-induced acoustic desorption/electrospray ionization mass spectrometry (LIAD-ESI/MS) characterizes analytes in a solid or liquid state without tedious sample preparation [63,94]. The instrument setup for LIAD-ESI is essentially similar to that of ELDI; however, the desorption mechanism of LIAD-ESI is different. In LIAD, a pulsed laser beam with a power density higher than 10^8 W/cm^2 is used to irradiate the rear of a thin metal foil (e.g., aluminum or titanium) to induce acoustic and shock waves. The acoustic and shock waves travel through the foil and

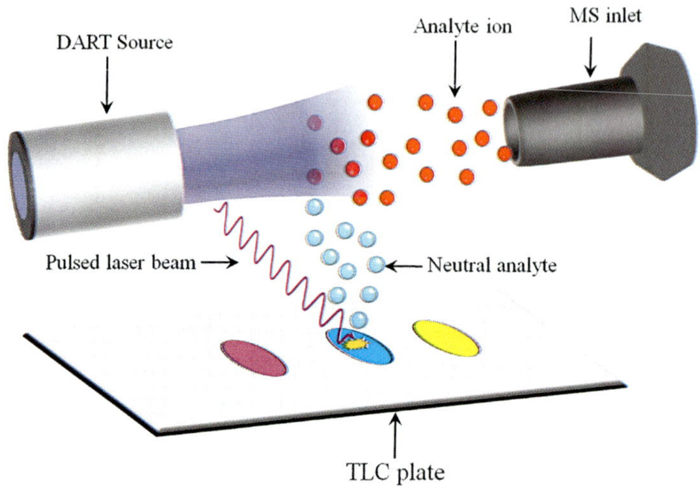

FIGURE 10.6 Analyses of chemical compounds on the surface of a TLC plate by PAMLDI/MS. TLC, thin-layer chromatography; PAMLDI, plasma-assisted multiwavelength laser desorption ionization; DART, direct analysis in real time.

glass slide to desorb analytes deposited on the other side of the foil. The desorbed analytes are subsequently ionized in an ESI plume. The sample zones on an aluminum TLC plate are analyzed by LIAD-ESI/MS using a pulsed IR laser beam. To efficiently transfer acoustic and shock waves to desorb the analytes, a glass slide is attached to the rear of the aluminum TLC plate. The space between the glass slide and aluminum TLC plate is filled with a viscous solution such as glycerol or polyethylene glycol. The ablated area at the front side of the silica-based C_{18} layer has a diameter approximately four times larger than the laser spot at the rear side of the aluminum plate (from 0.35 to 1.3 mm) [63]. Since the ablated area is larger than the laser spot, LIAD provides lower spatial resolution than ELDI. TLC-LIAD-ESI/MS was used to characterize drugs, dyes, and essential oils with detection limits in the nanogram range [63] (Figure 10.7). Similar to PAMLDI, a combination of LIAD and APCI/MS has been used to characterize analytes on TLC plates. A dielectric barrier discharge ionization source was used to generate the plasma by applying a high AC voltage to a nitrogen stream. The reactive species in the plasma then react with the analytes desorbed from the TLC plate via LIAD. TLC-LIAD-APCI/MS was used to characterize PAHs and essential oil.

10.4.2.4 Generation of Reactive Species for Desorption and Ionization under Ambient Conditions

DI of analytes by impacting the sample spot on TLC plate with fast atoms or ions (i.e., FAB and SIMS) in vacuum has been reported.

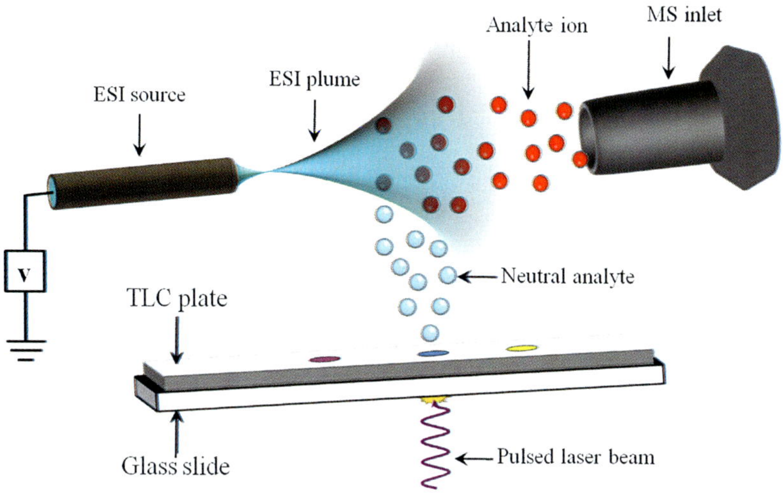

FIGURE 10.7 Analyses of chemical compounds on the surface of a TLC plate by LIAD-ESI/MS. TLC, thin-layer chromatography; MS, mass spectrometry; LIAD-ESI, laser-induced acoustic desorption electrospray ionization.

However, the shortcomings of using FAB or SIMS to characterize the analytes on TLC plate include: (1) the analytes are limited to polar and nonvolatile compounds; (2) applying a viscous matrix to the TLC plate will induce lateral spreading of the analyte zones; and (3) due to the restriction of the size of the ion source, the TLC plate must be cut to into small pieces to fit into the source. To overcome these problems, DI of the analytes by impacting the sample zones on the TLC plate with a stream of reactive species generated by several ambient ionization techniques have been developed. The ambient ionization sources include DESI, EASI, and DART.

DESI uses a stream of accelerated charged droplets generated from an ESI emitter to impact the analytes adsorbed on surfaces [95]. The analytes are desorbed and carried by the droplets, after which the ESI proceeds from the droplets to generate analyte ions. The use of a DESI source to characterize analytes on TLC plates was first reported in 2005, where reversed-phase TLC plates were fixed on a sample stage using double-sided tape. The plate surface was set at ~50° relative to the ESI emitter and ~10° from the axis of the sampling orifice of the MS [64]. Samples including peptides, dyes, drugs, lipids, and alkaloids were successfully separated and characterized by TLC-DESI/MS [64,96–102]. Calibration curves for berberine, palmatine, and hydrastinine standards indicated detection limits at low nanogram levels and coefficients of determination (R^2) larger than 0.94 [96]. Furthermore, an XY stage was combined with a DESI source to acquire chemical images of the TLC plates after 2-D separation. The speed of movement of the TLC plates and the line-to-line scanning distance were the two most important factors that influenced the resolution of the image [102] (Figure 10.8).

EASI, also known as desorption sonic spray ionization, is another spray-related sampling/ionizing technique that has a similar experimental setup and operation to DESI [103]. However, the generation of pneumatic droplets in EASI does not require the use of a high electric field. It has been suggested that an imbalance in the charge distribution among the pneumatic droplets induces the formation of charged droplets. The droplets generated by a sonic spray are subsequently directed onto the sample surface for desorption and ionization. EASI/MS has also been combined with TLC for characterization of zones of interest, where an acidic water/methanol solution is sprayed with the assistance of a nebulizing gas. The sonic spray emitter is fixed at an angle of 30° relative to the TLC plate, and the plate-to-emitter distance is c. 2 mm. Drug standards and biodiesel samples were separated on silica gel TLC plates and characterized by TLC-EASI/MS [70,104].

DART is a plasma-based APCI source which uses a stream of metastable, inert, and gaseous atoms and molecules to desorb and ionize volatile and semivolatile compounds on solid surfaces [105]. A TLC plate is cut so that the zones of interest are located between the DART source and the MS inlet. The zone of interest is impacted by the stream of excited gases from a DART source to allow DI of the analytes. TLC-DART has been used to characterize compounds, such as explosives, caffeine, and

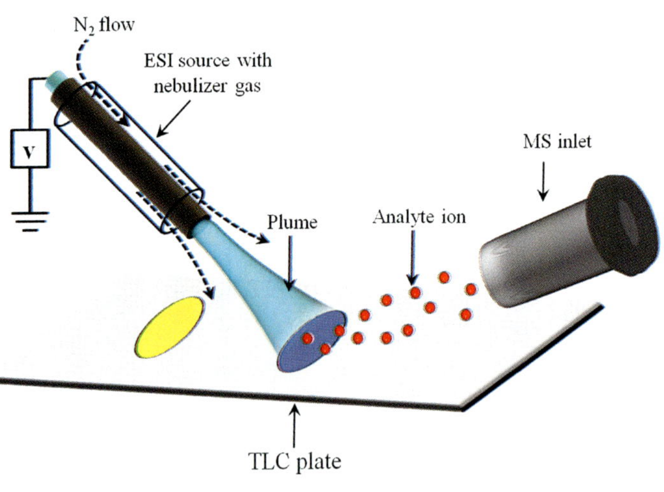

FIGURE 10.8 Analyses of chemical compounds on the surface of a TLC plate by DESI/MS. TLC, thin-layer chromatography; MS, mass spectrometry; DESI, desorption electrospray ionization.

drugs [69,82,106–109]. With the help of isotope-labeled internal standards, the coefficient of determination (R^2) and RSD of caffeine is 0.9892 and 5.4%, respectively (Figure 10.9).

10.4.2.5 Thermal Desorption TLC-MS

Thermal desorption can be applied in TLC-MS evaporate to analytes from TLC plates by heating [65,110]. In the proximal probe thermal desorption/secondary ionization MS, a TLC plate is mounted on a stage so that the edge of the plate is close to the sampling inlet [65]. A proximal probe set at 100–350 °C is positioned near the zones of interest to thermally evaporate the analytes. The analytes are then postionized via reactions with charged species generated from an ESI or an APCI sources. Since analytes are produced by heating, they are limited to thermally stable and volatile compounds. Samples separated by TLC plates, such as dyes, pharmaceuticals, explosives, and pesticides were successfully characterized [65]. Quantification of 2,4,6-trinitrotoluene has been performed, resulting in a determination coefficient (R^2) of 0.95 and a detection limit of 24 ng (Figure 10.10).

Irradiation of the sample surface with a continuous wave (CW) laser beam provides a feature of thermally heating at the focal point. The technique so-called LD-APCI combines a laser beam with a corona discharge needle to evaporate and ionize analytes on TLC plates [68,111].

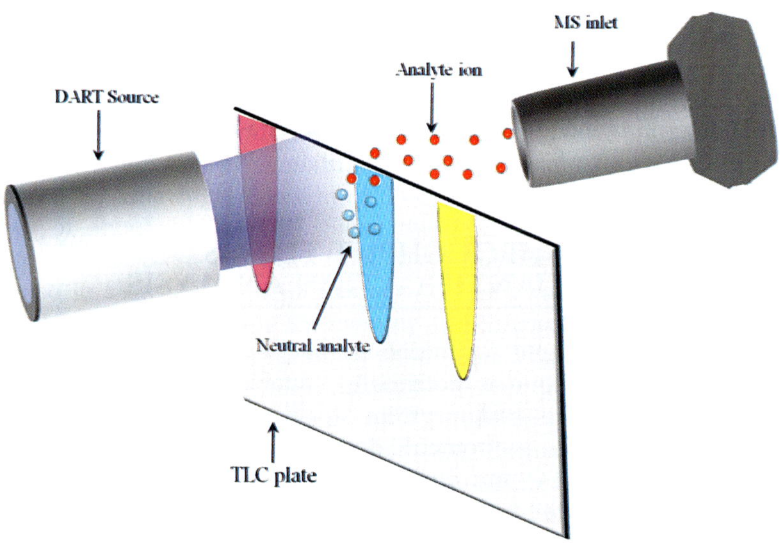

FIGURE 10.9 Analyses of chemical compounds on the surface of a TLC plate by DART/MS. TLC, thin-layer chromatography; MS, mass spectrometry; DART, direct analysis in real time.

approaches, such as external standard, internal standard, or standard addition. For direct sampling TLC-MS, since only a portion of the analytes in each zone are sampled and delivered into the MS for analysis, the technique is generally considered a poor technique for quantification when compared with GC-MS and LC-MS where all analytes are transported into the MS for detection. However, many strategies have been proposed to quantify analytes by TLC-MS. One of the more useful approaches uses isotope-labeled internal standards to improve the accuracy of quantification. In addition, techniques, such as overrun TLC and elution-based surface sampling device with a head that can cover the sample spot are employed to recover all of the analyte on the layer for subsequent quantification. The aforementioned efforts greatly improve the precision of TLC-MS quantification, and a RSD less than 5% can be achieved in some cases.

10.6 CONCLUSION

TLC is able to rapidly separate compounds using simple procedures with economical use of materials. The combination of TLC and MS provides a method to characterize compounds after their chromatographic separation. The evolution of TLC-MS techniques reveals several trends: (1) analytes are increasingly being detected in a direct manner without tedious sample pretreatment; (2) ambient ionization MS is being applied increasingly to characterize analytes on separated TLC plates rather than vacuum-based techniques; (3) instead of characterizing individual spots, TLC plates are increasingly being scanned to obtain chromatographic information and molecular images of compounds distributed on their surfaces.

Although many mass spectrometric ionization techniques have been developed for analysis of solid samples, the intensity of analyte ions and the detectable mass ranges are different when different mass spectrometric ionization methods are used. TLC-MS techniques using different sampling and ionization method can provide complementary information for analytes in the same spots. However, such approaches will inevitably increase the time for analysis, since the sample is analyzed repeatedly using different mass spectrometric systems. A TLC-MS system with multiple ion sources is presumed to obtain more information than that with single ion source. We believe that the TLC-MS system with multiple ion sources will be a developing trend in TLC-MS.

Although ambient MS has been widely and effectively used to characterize analytes on TLC plates, several problems remain unsolved. First, an interface or device must be developed to prevent TLC gel particles from entering the mass spectrometer during DI processes. Several DI techniques including ELDI, LIAD-ESI, DESI, and EASI, result in detached particles during surface sampling; which are then transferred by the charged droplet stream to the mass spectrometer. Second, the detection sensitivity of direct TLC-MS methods must be improved. Since analytes on the TLC plates are directly characterized by MS without extraction or concentration, the detection sensitivity by ambient MS is usually lower than that for indirect TLC-MS methods. Third, the separation efficiency of TLC must be improved. Developing an interface to connect TLC with a second separation/detection technique, such as LC-MS, GC-MS, or capillary electrophoresis-MS, will provide higher efficiencies for separation and detection.

References

[1] Gross JH. Mass spectrometry: a textbook. Springer; 2011.
[2] Hoffmann E, Stroobant V. Mass spectrometry: principles and applications. Wiley; 2007.
[3] Wilson ID. The state of the art in thin-layer chromatography-mass spectrometry: a critical appraisal. J Chromatogr A 1999;856:429–42.
[4] Poole CF. Thin-layer chromatography: challenges and opportunities. J Chromatogr A 2003;1000:963–84.
[5] Morlock G, Schwack W. Coupling of planar chromatography to mass spectrometry. Trends Anal Chem 2010;29:1157–71.
[6] Cheng SC, Huang MZ, Shiea J. Thin layer chromatography/mass spectrometry. J Chromatogr A 2011;1218:2700–11.
[7] Ghosh P, Reddy MMK, Ramteke VB, Rao BS. Analysis and quantitation of diazepam in cream biscuits by high-performance thin-layer chromatography and its confirmation by mass spectrometry. Anal Chim Acta 2004;508:31–5.
[8] Berkov S, Bastida J, Nikolova M, Viladomat F, Codina C. Rapid TLC/GC-MS identification of acetylcholinesterase inhibitors in alkaloid extracts. Phytochem Anal 2008; 19:411–9.
[9] Badea DN. Identification and quantification of active principles of products and preparations for tea and coffee. Counterfeiting detection of beverages by coupling with extracts xanthine TLC/GC/MS. Rev Chim Bucharest 2012;63:10–3.
[10] Kawai Y, Miyoshi M, Moon J-H, Terao J. Detection of cholesteryl ester hydroperoxide isomers using gas chromatography–mass spectrometry combined with thin-layer chromatography blotting. Anal Biochem 2007;360:130–7.
[11] Liu Q-T, Kinderlerer JL. Preparative thin-layer chromatographic separation and subsequent gas chromatographic-mass spectrometric analysis of monoacylglycerols derived from butter oil by fungal degradation. J Chromatogr A 1999;855:617–24.
[12] Minami Y, Yokoi S, Setoyama M, Bando N, Takeda S, Kawai Y, et al. Combination of TLC blotting and gas chromatography-mass spectrometry for analysis of peroxidized cholesterol. Lipids 2007;42:1055–63.
[13] Conchillo A, Cercaci L, Ansorena D, Rodriguez-Estrada MT, Lercker G, Astiasaran I. Levels of phytosterol oxides in enriched and nonenriched spreads: application of a thin-layer chromatography-gas chromatography methodology. J Agric Food Chem 2005;53:7844–50.
[14] Gorchein A. Characterization of porphyrins of rat Harderian gland by thin layer chromatography and mass spectrometry: no evidence for a tricarboxylic acid porphyrin. Biomed Chromatogr 2003;17:526–9.

[15] Fuchs B. Analysis of phospolipids and glycolipids by thin-layer chromatography-matrix-assisted laser desorption and ionization mass spectrometry. J Chromatogr A 2012;1259:62–73.

[16] Müthing J, Distler U. Advances on the compositional analysis of glycosphingolipids combining thin-layer chromatography with mass spectrometry. Mass Spectrom Rev 2010;29:425–79.

[17] Watanabe T, Kawasaki H, Kimoto T, Arakawa R. Characterization of polyether mixtures using thin-layer chromatography and matrix-assisted laser desorption/ionization mass spectrometry. Rapid Commun Mass Spectrom 2007;21:787–91.

[18] Therisod H, Labas V, Caroff M. Direct microextraction and analysis of rough-type lipopolysaccharides by combined thin-layer chromatography and MALDI mass spectrometry. Anal Chem 2001;73:3804–7.

[19] Schiller J, Süß R, Fuchs B, Müller M, Zschörnig O, Arnold K. Combined application of TLC and matrix-assisted laser desorption and ionisation time-of-flight mass spectrometry (MALDI-TOF MS) to phospholipid analysis of brain. Chromatographia 2003;57: S297–302.

[20] Sommerer D, Süß R, Hammerschmidt S, Wirtz H, Arnold K, Schiller J. Analysis of the phospholipid composition of bronchoalveolar lavage (BAL) fluid from man and minipig by MALDI-TOF mass spectrometry in combination with TLC. J Pharm Biomed Anal 2004;35:199–206.

[21] Ruh H, Sandhoff R, Meyer B, Gretz N, Hopf C. Quantitative characterization of tissue globotetraosylceramides in a rat model of polycystic kidney disease by PrimaDrop sample preparation and indirect high-performance thin layer chromatography-matrix-assisted laser desorption/ionization-time-of-flight-mass spectrometry with automated data acquisition. Anal Chem 2013;85:6233–40.

[22] Ji H, Sato N, Nonidez WK, Mays JW. Characterization of hydroxyl-end-capped polybutadiene and polystyrene produced by anionic polymerization technique via TLC/MALDI TOF mass spectrometry. Polymer 2002;43:7119–23.

[23] Lee CS, Kim YG, Joo HS, Kim BG. Structural analysis of lipid A from *Escherichia coli* O157:H7:K– using thin-layer chromatography and ion-trap mass spectrometry. J Mass Spectrom 2004;39:514–25.

[24] Shariatgorji M, Spacil Z, Maddalo G, Cardenas LB, Ilag LL. Matrix-free thin-layer chromatography/laser desorption ionization mass spectrometry for facile separation and identification of medicinal. Rapid Commun Mass Spectrom 2009;23:3655–60.

[25] Meisen I, Friedrich AW, Karch H, Witting U, Peter-katalinić J, Müthing J. Application of combined high-performance thin-layer chromatography immunostaining and nanoelectrospray ionization quadrupole time-of-flight tandem mass spectrometry to the structural characterization of high- and low-affinity binding ligands of Shiga toxin 1. Rapid Commun Mass Spectrom 2005;19:3659–65.

[26] Hoffmann P, Hulsewig M, Duvar S, Ziehr H, Mormann M, Peter-Katalinic J, et al. On the structural diversity of Shiga toxin glycosphingolipid receptors in lymphoid and myeloid cells determined by nanoelectrospray ionization tandem mass spectrometry. Rapid Commun Mass Spectrom 2010;24:2295–304.

[27] De Wasch K, De Brabander HF, van Ginkel LA, Spaan A, Sterk SS, Meiring HD. Confirmation of residues of thyreostatic drugs in thyroid glands by multiple mass spectrometry after thin-layer chromatographic screening. J Chromatogr A 1998;819:99–111.

[28] Meisen I, Peter-Katalinić J, Müthing J. Direct analysis of silica gel extracts from immunostained glycosphingolipids by nanoelectrospray ionization quadrupole time-of-flight mass spectrometry. Anal Chem 2004;76:2248–55.

[29] Oriňák A, Arlinghaus HF, Vering G, Oriňáková R. Modified chromatographic thin layer surface as an interface to couple thin layer chromatography with ToF-SIMS. Surf Interface Anal 2004;36:1122–5.

[30] Oriňák A, Arlinghaus HF, Vering G, Oriňáková R, Hellweg S. Introduction to time-of-flight secondary ion mass spectrometry application in chromatographic analysis. J Chromatogr A 2005;1084:113–8.
[31] Tames F, Watson ID, Morden W, Wilson ID. Detection and identification of morphine in urine extracts using thin-layer chromatography and tandem mass spectrometry. J Chromatogr B 1999;729:341–6.
[32] Chang TT, Lay JO, Francel RJ. Direct analysis of thin-layer chromatography spots by fast atom bombardment mass spectrometry. Anal Chem 1984;56:109–11.
[33] Taki T. An approach to glycobiology from glycolipidomics: ganglioside molecular scanning in the brains of patients with Alzheimer's disease by TLC-Blot/Matrix assisted laser desorption/ionization-time of flight MS. Biol Pharm Bull 2012;35:1642–7.
[34] Zaima N, Goto-Inoue N, Adachi K,Setou M. Selective analysis of lipids by thin-layer chromatography blot matrix-assisted laser desorption/ionization imaging mass spectrometry. J Oleo Sci 2011;60:87–92.
[35] Oka H, Ikai Y, Hayakawa J, Masuda K, Harada K-i, Suzuki M, et al. Improvement of chemical analysis of antibiotics. 18. Identification of residual tetracyclines in bovine tissues by TLC/FABMS with a sample condensation technique. J Agric Food Chem 1993;41:410–5.
[36] Parent AA, Anderson TM, Michaelis DJ, Jinang G, Savage PB, Linford MR. Direct ToF-SIMS analysis of organic halides and amines on TLC plates. Appl Surf Sci 2006; 252:6746–9.
[37] Chen Y-C, Shiea J, Sunner J. Thin-layer chromatography-mass spectrometry using activated carbon, surface-assisted laser desorption/ionization. J Chromatogr A 1998; 826:77–86.
[38] DiDonato GC, Busch KL. Phase-transition matrix for chromatography/secondary ion mass spectrometry. Anal Chem 1986;58:3231–2.
[39] Kubis AJ, Somayajula KV, Sharkey AG, Hercules DM. Laser mass spectrometric analysis of compounds separated by thin-layer chromatography. Anal Chem 1989;61: 2516–23.
[40] Gusev AI, Proctor A, Rabinovich YI, Hercules DM. Thin-layer chromatography combined with matrix-assisted laser desorption/ionization mass spectrometry. Anal Chem 1995;67:1805–14.
[41] Gusev AI,Vasseur OJ, Proctor A, Sharkey AG, Hercules DM. Imaging of thin-layer chromatograms using matrix-assisted laser desorption/ionization mass spectrometry. Anal Chem 1995;67:4565–70.
[42] Mehl JT, Gusev AI, Hercules DM. Coupling protocol for thin layer chromatography/matrix-assisted laser desorption ionization. Chromatographia 1997;46:358–64.
[43] Meisen I, Mormann M, Müthing J. Thin-layer chromatography, overlay technique and mass spectrometry: a versatile triad advancing glycosphingolipidomics. Biochim Biophys Acta 1811;2011:875–96.
[44] Fuchs B, Süß R, Nimptsch A, Schiller J. MALDI-TOF-SIMS combined with TLC: a review of the current state. Chromatographia 2009;69:S95–105.
[45] Santos LS, Haddad R,Hoehr NF, Pilli RA, Eberlin MN. Fast screening of low molecular weight compounds by thin-layer chromatography and "on-spot" MALDI-TOF mass spectrometry. Anal Chem 2004;76:2144–7.
[46] Dreisewerd K, Müthing J, Rohlfing A, Meisen I, Vukelić Z, Peter-Katalinić J, et al. Analysis of gangliosides directly from thin-layer chromatography plates by infrared matrix-assisted laser desorption/ionization orthogonal time-of-flight mass spectrometry with a glycerol matrix. Anal Chem 2005;77:4098–107.
[47] Ivleva VB, Sapp LM, O'Connor PB, Costello CE. Ganglioside analysis by thin-layer chromatography matrix-assisted laser desorption/ionization orthogonal time-of-flight mass spectrometry. J Am Soc Mass Spectrom 2005;16:1552–60.

[48] Panchagnula V, Mikulskis A, Song L, Wang Y, Wang M, Knubovets T, et al. Phosphopeptide analysis by directly coupling two-dimensional planar electrochromatography/thin-layer chromatography with matrix-assisted laser desorption/ionization time-of-flight mass spectrometry. J Chromatogr A 2007;1155:112–23.

[49] Distler U, Huelsewig M, Souady J, Dreisewerd K, Haier J, Senninger N, et al. Matching IR-MALDI-o-TOF mass spectrometry with the TLC overlay binding assay and its clinical application for tracing tumor-associated glycosphingolipids in hepatocellular and pancreatic cancer. Anal Chem 2008;80:1835–46.

[50] Souady J, Soltwisch J, Dreisewerd K, Haier J, Peter-Katalinić J, Müthing J. Structural profiling of individual glycosphingolipids in a single thin-layer chromatogram by multiple sequential immunodetection matched with direct IR-MALDI-o-TOF mass spectrometry. Anal Chem 2009;81:9481–92.

[51] Meisen I, Distler U, Müthing J, Berkenkamp S, Dreisewerd K, Mathys W, et al. Direct coupling of high-performance thin-layer chromatography with UV spectroscopy and IR-MALDI orthogonal TOF MS for the analysis of cyanobacterial toxins. Anal Chem 2009;81:3858–66.

[52] Crecelius A, Clench MR, Richards DS, Parr V. Thin-layer chromatography–matrix-assisted laser desorption ionisation–time-of-flight mass spectrometry using particle suspension matrices. J Chromatogr A 2002;958:249–60.

[53] Chen C-C, Yang Y-L, Ou C-L, Chou C-H, Liaw CC, Lin P-C. Direct monitoring of chemical transformations by combining thin layer chromatography with nanoparticle-assisted laser desorption/ionization mass spectrometry. Analyst 2013; 138:1379–85.

[54] Chen YC, Wu JY. Analysis of small organics on planar silica surfaces using surface-assisted laser desorption/ionization mass spectrometry. Rapid Commun Mass Spectrom 2001;15:1899–903.

[55] Wu JY, Chen YC. A novel approach of combining thin-layer chromatography with surface-assisted laser desorption/ionization (SALDI) time-of-flight mass spectrometry. J Mass Spectrom 2002;37:85–90.

[56] Chen Y-C. In situ determination of organic reaction products by combining thin layer chromatography with surface-assisted laser desorption/ionization time-of-flight mass spectrometry. Rapid Commun Mass Spectrom 1999;13:821–5.

[57] Ramaley L, Nearing ME, Vaughan MA, Ackman RG, Jamieson WD. Thin-layer chromatographic plate scanner interfaced with a mass spectrometer. Anal Chem 1983;55: 2285–9.

[58] Li L, Lubman DM. Resonant two-photon ionization spectroscopic analysis of thin-layer chromatography using pulsed laser desorption/volatilization into supersonic jet expansions. Anal Chem 1989;61:1911–5.

[59] Van Berkel GJ, Sanchez AD, Quirke JM. Thin-layer chromatography and electrospray mass spectrometry coupled using a surface sampling probe. Anal Chem 2002;74: 6216–23.

[60] Luftmann H. A simple device for the extraction of TLC spots: direct coupling with an electrospray mass spectrometer. Anal Bioanal Chem 2004;378:964–8.

[61] Hsu FL, Chen CH, Yuan CH, Shiea J. Interfaces to connect thin-layer chromatography with electrospray ionization mass spectrometry. Anal Chem 2003;75:2493–8.

[62] Van Berkel GJ, Llave JJ, De Apadoca MF, Ford MJ. Rotation planar chromatography coupled on-line with atmospheric pressure chemical ionization mass spectrometry. Anal Chem 2004;76:479–82.

[63] Cheng SC, Huang MZ, Shiea J. Thin-layer chromatography/laser-induced acoustic desorption/electrospray ionization mass spectrometry. Anal Chem 2009;81:9274–81.

[64] Van Berkel GJ, Ford MJ, Deibel MA. Thin-layer chromatography and mass spectrometry coupled using desorption electrospray ionization. Anal Chem 2005;77:1207–15.

[65] Ovchinnikova OS, Van Berkel GJ. Thin-layer chromatography and mass spectrometry coupled using proximal probe thermal desorption with electrospray or atmospheric pressure chemical ionization. Rapid Commun Mass Spectrom 2010;24:1721–9.
[66] Lin SY, Huang MZ, Chang HC, Shiea J. Using electrospray-assisted laser desorption/ionization mass spectrometry to characterize organic compounds separated on thin-layer chromatography plates. Anal Chem 2007;79:8789–95.
[67] Salo PK, Salomies H, Harju K, Ketola RA, Kotiaho T, Yli-Kauhaluoma J, et al. Analysis of small molecules by ultra thin-layer chromatography-atmospheric pressure matrix-assisted laser desorption/ionization mass spectrometry. J Am Soc Mass Spectrom 2005;16:906–15.
[68] Peng S, Ahlmann N, Kunze K, Nigge W, Edler M, Hoffmann T, et al. Thin-layer chromatography combined with diode laser desorption/atmospheric pressure chemical ionization mass spectrometry. Rapid Commun Mass Spectrom 2004;18:1803–8.
[69] Morlock G, Ueda Y. New coupling of planar chromatography with direct analysis in real time mass spectrometry. J Chromatogr A 2007;1143:243–51.
[70] Haddad R, Milagre HM, Catharino RR, Eberlin MN. Easy ambient sonic-spray ionization mass spectrometry combined with thin-layer chromatography. Anal Chem 2008;80:2744–50.
[71] Zhang J, Zhou Z, Yang J, Zhang W, Bai Y, Liu H. Thin layer chromatography/plasma assisted multiwavelength laser desorption ionization mass spectrometry for facile separation and selective identification of low molecular weight compounds. Anal Chem 2012;84:1496–503.
[72] Resano M, Ruiz EG, Mihucz VG, Móricz AM, Záray G, Vanhaecke F. Rapid screening method for arsenic speciation by combining thin layer chromatography and laser ablation-inductively coupled plasma-dynamic reaction cell-mass spectrometry. J Anal At Spectrom 2007;22:1158–62.
[73] Asano KG, Ford MJ, Tomkins BA, Van Berkel GJ. Self-aspirating atmospheric pressure chemical ionization source for direct sampling of analytes on surfaces and in liquid solutions. Rapid Commun Mass Spectrom 2005;19:2305–12.
[74] Meermann B, Möller I, Nowak S, Luftmann H, Karst U. Iodine speciation using thin-layer chromatography coupled to inductively coupled plasma-mass spectrometry by means of an extraction device. J Anal At Spectrom 2010;25:1654–8.
[75] Walworth MJ, Stankovich JJ, Van Berkel GJ, Schulz M, Minarik S, Nichols J, et al. Hydrophobic treatment enabling analysis of wettable surfaces using a liquid microjunction surface sampling probe/electrospray ionization-mass spectrometry system. Anal Chem 2011;83:591–7.
[76] Ford MJ, Van Berkel GJ. An improved thin-layer chromatography/mass spectrometry coupling using a surface sampling probe electrospray ion trap system. Rapid Commun Mass Spectrom 2004;18:1303–9.
[77] Ford MJ, Kertesz V, Van Berkel GJ. Thin-layer chromatography/electrospray ionization triple-quadrupole linear ion trap mass spectrometry system: analysis of rhodamine dyes separated on reversed-phase C_8 plates. J Mass Spectrom 2005;40:866–75.
[78] Ford MJ, Deibel MA, Tomkins BA, Van Berkel GJ. Quantitative thin-layer chromatography/mass spectrometry analysis of caffeine using a surface sampling probe electrospray ionization tandem mass spectrometry system. Anal Chem 2005;77:4385–9.
[79] Jautz U, Morlock G. Efficacy of planar chromatography coupled to (tandem) mass spectrometry for employment in trace analysis. J Chromatogr A 2006;1128:244–50.
[80] Aranda M, Morlock G. New method for caffeine quantification by planar chromatography coupled with electropray ionization mass spectrometry using stable isotope dilution analysis. Rapid Commun Mass Spectrom 2007;21:1297–303.

[81] Alpmann A, Morlock G. Improved online coupling of planar chromatography with electrospray mass spectrometry: extraction of zones from glass plates. Anal Bioanal Chem 2006;386:1543–51.
[82] Morlock G, Schwack W. Determination of isopropylthioxanthone (ITX) in milk, yoghurt and fat by HPTLC-FLD, HPTLC-ESI/MS and HPTLC-DART/MS. Anal Bioanal Chem 2006;385:586–95.
[83] Chen Y, Schwack W. Planar chromatography mediated screening of tetracycline and fluoroquinolone antibiotics in milk by fluorescence and mass selective detection. J Chromatogr A 2013;1312:143–51.
[84] Luftmann H, Aranda M, Morlock GE. Automated interface for hyphenation of planar chromatography with mass spectrometry. Rapid Commun Mass Spectrom 2007;21: 3772–6.
[85] Aranda M, Morlock G. Simultaneous determination of riboflavin, pyridoxine, nicotinamide, caffeine and taurine in energy drinks by planar chromatography-multiple detection with confirmation by electrospray ionization mass spectrometry. J Chromatogr A 2006;1131:253–60.
[86] Devenport NA, Reynolds JC, Weston DJ, Wilson ID, Creaser CS. Direct extraction of urinary analytes from undeveloped reversed-phase thin layer chromatography plates using a solvent gradient combined with on-line electrospray ionisation ion mobility-mass spectrometry. Analyst 2012;137:3510–3.
[87] Chai W, Leteux C, Lawson AM, Stoll MS. On-line overpressure thin-layer chromatographic separation and electrospray mass spectrometric detection of glycolipids. Anal Chem 2003;75:118–25.
[88] Salo PK, Vilmunen S, Salomies H, Ketola RA, Kostiainen R. Two-dimensional ultra-thin-layer chromatography and atmospheric pressure matrix-assisted laser desorption/ionization mass spectrometry in bioanalysis. Anal Chem 2007;79:2101–8.
[89] Lafleur JP, Salin ED. Speciation of chromium by high-performance thin-layer chromatography with direct determination by laser ablation inductively coupled plasma mass spectrometry. Anal Chem 2008;80:6821–3.
[90] Chirinos J, Oropeza D, González J, Ranaudo M, Russo RE. Determination of vanadium/nickel proportionality in the asphaltene fraction of crude oil using thin-layer chromatography with femtosecond laser ablation–inductively coupled plasma–mass spectrometry. Energy Fuels 2013;27:2431–6.
[91] Shiea J, Huang MZ, Hsu HJ, Lee CY, Yuan CH, Beech I, et al. Electrospray-assisted laser desorption/ionization mass spectrometry for direct ambient analysis of solids. Rapid Commun Mass Spectrom 2005;19:3701–4.
[92] Shiea J, Chang DY, Lin CH, Jiang SJ. Generating multiply charged protein ions by ultrasonic nebulization/multiple channel-electrospray ionization mass spectrometry. Anal Chem 2001;73:4983–7.
[93] Zhang J, Li Z, Zhang C, Feng B, Zhou Z, Bai Y, et al. Graphite-coated paper as substrate for high sensitivity analysis in ambient surface-assisted laser desorption/ionization mass spectrometry. Anal Chem 2012;84:3296–301.
[94] Cheng SC, Cheng TL, Chang HC, Shiea J. Using laser-induced acoustic desorption/electrospray ionization mass spectrometry to characterize small organic and large biological compounds in the solid state and in solution under ambient conditions. Anal Chem 2009;81:868–74.
[95] Takáts Z, Wiseman JM, Gologan B, Cooks RG. Mass spectrometry sampling under ambient conditions with desorption electrospray ionization. Science 2004;306: 471–3.
[96] Van Berkel GJ, Tomkins BA, Kertesz V. Thin-layer chromatography/desorption electrospray ionization mass spectrometry: investigation of goldenseal alkaloids. Anal Chem 2007;79:2778–89.

[97] Van Berkel GJ, Kertesz V. Automated sampling and imaging of analytes separated on thin-layer chromatography plates using desorption electrospray ionization mass spectrometry. Anal Chem 2006;78:4938–44.

[98] Pasilis SP, Kertesz V, Van Berkel GJ, Schulz M, Schorcht S. HPTLC/DESI-MS imaging of tryptic protein digests separated in two dimensions. J Mass Spectrom 2008;43: 1627–35.

[99] Harry EL, Reynolds JC, Bristow AWT, Wilson ID, Creaser CS. Direct analysis of pharmaceutical formulations from non-bonded reversed-phase thin-layer chromatography plates by desorption electrospray ionisation ion mobility mass spectrometry. Rapid Commun Mass Spectrom 2009;23:2597–604.

[100] Pasilis SP, Kertesz V, Van Berkel GJ, Schulz M, Schorcht S. Using HPTLC/DESI-MS for peptide identification in 1D separations of trypticprotein digests. Anal Bioanal Chem 2008;391:317–24.

[101] Wiseman JM, Li JB. Elution, partial separation, and identification of lipids directly from tissue slices on planar chromatography media by desorption electrospray ionization mass spectrometry. Anal Chem 2010;82:8866–74.

[102] Kennedy JH, Wiseman JM. Direct analysis of *Salvia divinorum* leaves for salvinorin A by thin layer chromatography and desorption electrospray ionization multi-stage tandem mass spectrometry. Rapid Commun Mass Spectrom 2010;24:1305–11.

[103] Haddad R, Sparrapan R, Kotiaho T, Eberlin MN. Easy ambient sonic-spray ionization-membrane interface mass spectrometry for direct analysis of solution constituents. Anal Chem 2008;80:898–903.

[104] Eberlin LS, Abdelnur PV, Passero A, de Sa GF, Daroda RJ, de Souza V, et al. Analysis of biodiesel and biodiesel-petrodiesel blends by high performance thin layer chromatography combined with easy ambient sonic-spray ionization mass spectrometry. Analyst 2009;134:1652–7.

[105] Cody RB, Laramée JA, Durst HD. Versatile new ion source for the analysis of materials in open air under ambient conditions. Anal Chem 2005;77:2297–302.

[106] Dytkiewitz E, Morlock GE. Analytical strategy for rapid identification and quantification of lubricant additives in mineral oil by high-performance thin-layer chromatography with UV absorption and fluorescence detection combined with mass spectrometry and infrared spectroscopy. J AOAC Int 2008;91:1237–43.

[107] Alpmann A, Morlock G. Rapid and sensitive determination of acrylamide in drinking water by planar chromatography and fluorescence detection after derivatization with dansulfinic acid. J Sep Sci 2008;31:71–7.

[108] Smith NJ, Domin MA, Scott LT. HRMS directly from TLC slides.a powerful tool for rapid analysis of organic mixtures. Org Lett 2008;10:3493–6.

[109] Howlett SE, Steiner RR. Validation of thin layer chromatography with AccuTOF-DART™ detection for forensic drug analysis. J Forensic Sci 2011;56:1261–7.

[110] Ovchinnikova OS, Kertesz V, Van Berkel GJ. Molecular surface sampling and chemical imaging using proximal probe thermal desorption/secondary ionization mass spectrometry. Anal Chem 2011;83:598–603.

[111] Peng S, Edler M, Ahlmann N, Hoffmann T, Franzke J. A new interface to couple thin-layer chromatography with laser desorption/atmospheric pressure chemical ionization mass spectrometry for plate scanning. Rapid Commun Mass Spectrom 2005;19: 2789–93.

[112] Cheng SC, Huang MZ, Wu LC, Chou CC, Cheng CN, Jhang SS, et al. Building blocks for the development of an interface for high-throughput thin layer chromatography/ambient mass spectrometric analysis: a green methodology. Anal Chem 2012;84: 5864–8.

CHAPTER

11

Effects-Directed Biological Detection: Bioautography

Irena Choma, Wioleta Jesionek

Chromatographic Method Department, M. Curie-Skłodowska University, Lublin, Poland

11.1 INTRODUCTION

Instrumental analytical chemistry is a powerful tool for the analysis of natural and man-made substances encountered in everyday life (food and water components, additives or pollutants, pharmaceuticals, cosmetics, etc.). Millions of substances have been identified, some of which are constantly monitored. Still, at present, the information on their biological role and impact on our health is insufficient. Moreover, we should be aware that although millions of identified compounds exist, problems are still possible from unidentified compounds (derivatives of known compounds or new compounds), which can be potentially dangerous to humans. There are also some natural compounds, for example, antibiotics, alkaloids, and mycotoxins, which possess high toxicity to living organisms. Effect-directed analysis (EDA) provides information for the biologically relevant substances in a mixture facilitating their detection in even complex matrices. The method is a bridge connecting bio- or toxicity-assays with instrumental analysis [1]. The targeted substances can be visualized, isolated, identified, classified, and quantified. In most cases, separation/fractionation is needed to obtain a proper substance/fraction response for the biological effect. Sometimes subsequent separations are necessary to eliminate fractions with no biological response. Bio- or toxicity-assays can be performed after each separation step using different test materials, e.g., enzymes, microorganisms, algae, daphnia, fishes, etc. Instrumental analysis should be the last step of EDA, enabling identification of the substances of

Instrumental Thin-Layer Chromatography
http://dx.doi.org/10.1016/B978-0-12-417223-4.00011-X

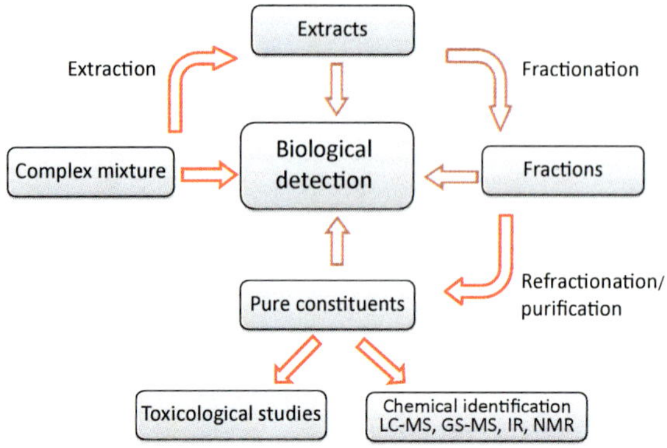

FIGURE 11.1 Scheme of effect-directed analysis of complex mixtures. LC-MS, liquid chromatography–mass spectrometry; GC-MS, gas chromatography–mass spectrometry; IR, infrared; NMR, nuclear magnetic resonance.

interest, mostly using mass spectrometry and/or nuclear magnetic resonance spectroscopy [2,3] combined with computer-aided tools [4] (Figure 11.1).

The term "EDA" is usually used in environmental analysis, while "biochemical detection," "bioassay-guided/directed fractionation," "(bio) activity-guided/directed fractionations," or "bioactivity/bioautographic screening" are mostly used in drug discovery [2]. However, there is a large flexibility in this phraseology. In this chapter, EDA will be used for environmental, food, and drug analysis, as well as for screening for known or potential toxicants.

EDA was introduced in the early 1980s [5] and, as shown by Hecker [6], the number of publications on EDA is constantly increasing. Partial information can be found in several reviews [1–8]. Also, two subsequent issues of Trends in Analytical Chemistry were devoted to this subject [9,10].

11.1.1 Bioautography

Bioautography belongs to the EDA methods in which the separation step is performed using planar chromatography techniques, such as thin-layer chromatography (TLC), high-performance thin-layer chromatography (HPTLC), overpressured-layer chromatography (OPLC) and planar electrochromatography. Bioautography can be based on any biological effect, e.g., antioxidant, enzymatic, antiestrogenic, antimutagenic, antibacterial, antifungal, etc. [11–17]. However, the term bioautography is used predominantly when antimicrobial or antifungal properties of the analytes are measured on (HP)TLC plates. Bioautography can be

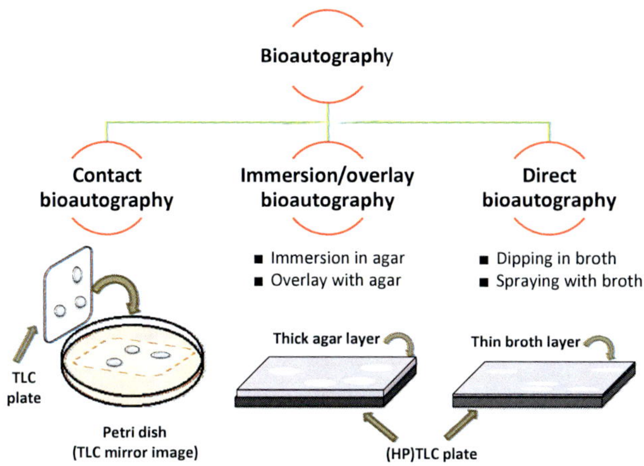

FIGURE 11.2 Scheme of bioautographic methods. (HP)TLC, (high performance) thin-layer chromatography.

subdivided into contact, immersion/agar overlay, or direct bioautography (DB) (Figure 11.2).

In contact bioautography, the developed (HP)TLC plate is placed on an inoculated agar surface for some time to allow diffusion [12,14,16–19]. Then, the plate is removed and the agar layer is incubated. The zones of inhibition growth appear in the places, where the antimicrobial compounds were in contact with the agar layer.

In immersion (agar overlay) bioautography, the developed plate is immersed in (or covered with) agar medium seeded with the test microorganism. The agar layer remains on the plate surface during incubation and visualization [12,14,16,17,20–23].

In DB [11–17,19,24] a developed TLC plate is dipped in a suspension of microorganisms grown in a specific broth and then incubated in a humid atmosphere. The microorganisms grow directly on the surface of the TLC layer excluding places where antimicrobial agents were located. In all types of bioautography, visualization of inhibition zones is mostly carried out using tetrazolium salts, such as MTT [3-(4,5-dimethylthiazol-2-yl)-2,5-diphenyltetrazolium bromide]. The dehydrogenases of living microorganisms convert tetrazolium salts into an intensely colored formazan. Cream-colored zones appearing on a purple-colored background are due to the presence of antimicrobial agents (Figure 11.3).

Bioautography, was introduced for paper chromatography by Goodall and Levi in 1946 [25]. Since then the method has undergone continuous development. The first review on bioautography concerning both paper and TLC was written by Betina, followed by a critical paper

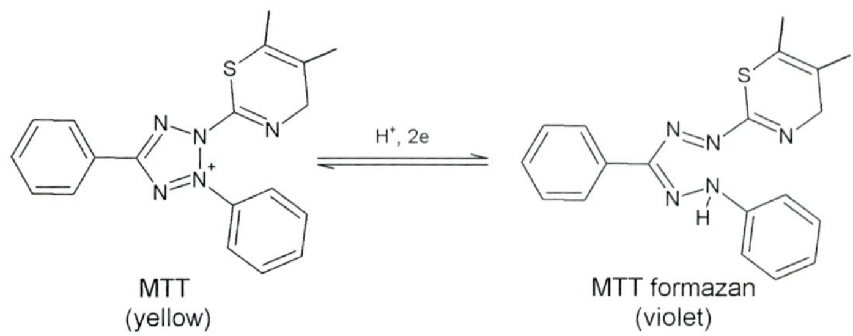

FIGURE 11.3 Reduction of tetrazolium salt by microorganism dehydrogenase. MTT, [3-(4,5-dimethylthiazol-2-yl)-2,5-diphenyltetrazolium bromide].

by Rios [19,26]. Rios discussed many parameters affecting bioautography and divided the method into contact, immersion, and direct versions. An early comprehensive review was written by Botz, Nagy, and Kocsis [14]. The authors describe the history, scope, and various applications of bioautography, focusing on the direct version. The full procedure for DB is illustrated in Figure 11.4. Moreover, optimization of thin-layer chromatography–direct bioautography (TLC-DB) is discussed, mostly the influence of extraction solvents and chromatographic systems (mobile and stationary phases, additives, etc.). Optimum conditions for test bacteria are also given. The problem of identifying the best bacteria and fungi growth conditions was discussed in papers by Nagy et al. [27–29]. One paragraph of the chapter [14] was devoted to a bioautographic test kit produced by Merck. The ready-to-use Chrom Biodip Antibiotic kit consisted of a culture medium, *Bacillus subtilis* spore suspension, nutrient medium, and MTT reagent [30]. As indicated by others [14,24], the incubation time for conversion of *B. subtilis* spores to their vegetative form was insufficient, resulting in poor colorization of the TLC plates with MTT. The test worked perfectly when the incubation time in a circulating air incubator was increased from 2 h at 35 °C (proposed in the Chrom Biodip instruction) to 4 h at 35–37 °C [14,24,31–33]. This inadequacy was probably the reason for the mild interest in purchasing the test. As a consequence Merck ceased production of the kit. Nowadays, there is no commercial bioautographic test, except for the bioluminiscent bacteria produced by Chromadex. Each research group uses its own bioautographic recipe or bioautographic system such as BioArena [34,35]. In our laboratory two bioautographic tests were developed and validated: one based on *Escherichia coli* and the other on *B. subtilis* [36,37]. Despite the lack of commercially available tests, bioautography has become increasingly popular over the last few decades. Literature on the subject is very rich, including several recent review papers [11–17,35].

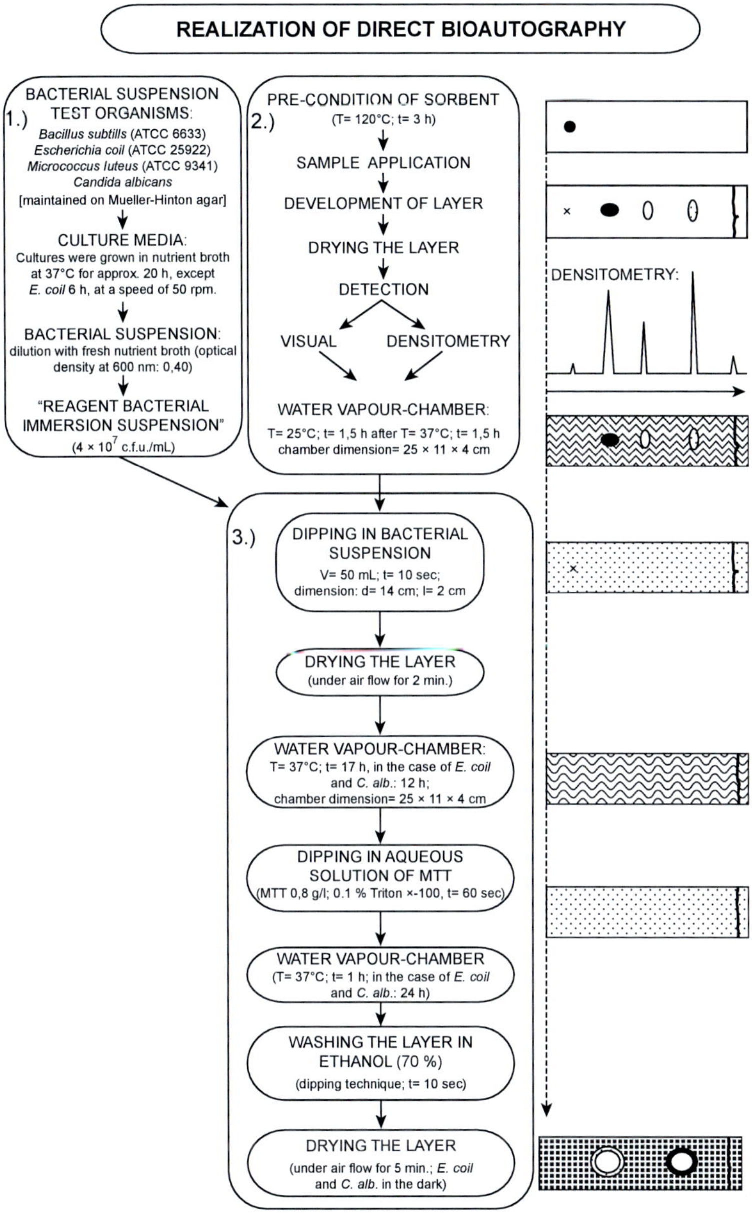

FIGURE 11.4 Realization of direct bioautography. *Reprinted from [14] with permission.*

11.2 APPLICATIONS OF BIOAUTOGRAPHIC TESTS

11.2.1 Antimicrobial Detection

Most thin-layer chromatography–bioautography (TLC-B) applications involve microbials, that is bacteria and fungi, as testing organisms. Because of the increasing microbial resistance observed in medicine and veterinary use of current antibiotics, the discovery of new antibiotics is an especially important objective. Bioautography can be used for detection of antimicrobials in environmental and food samples [24,31–33,38] as well as for searching for new antibacterials in plants [39–45].

Table 11.1 presents representative applications of TLC-B, mostly direct and immersion/overlay, for determination of antimicrobials in different sample matrices. Only selected papers from the twenty-first century are included to illustrate contemporary trends in this method. Over the last 14-years two main trends can be observed. The first is the estimation of food quality, which is nowadays particularly important. There are numerous papers describing the TLC-bioautographic determination of contaminants, such as antibiotics and pesticides in food products, such as milk, fruits, or juices, as well as water [31–33,46–50] (Figure 11.5). The second trend is related to the plant analysis. Owing to the increasing use of herbal medicines for the treatment of many diseases, bioautography of plant samples is increasingly popular. The richness of bioactive compounds in plant materials serves to stimulate the continuous development of bioassays for their identification. Numerous bacterial strains, including human and plant pathogenic microorganisms, are used to check the activity of isolated compounds. In case of complex matrices, biological screening performed on developed TLC plates is the best way to obtain reliable information for the activity of single compounds. This is not possible using typical diffusion and dilution microbiological methods, where antagonistic and synergic effects may be observed. Some bioscreening experiments are focused on optimization problems, such as finding the best solvent for extraction of active compounds from the matrix, or the selection of a mobile phase providing a suitable separation of the investigated samples [21,62–64,70,71]. Bioassays performed on the TLC plate allow comparison of the bioactivity of samples of various origin (e.g., different plant species or extracts from different parts of a plant) in parallel [21,63,64,70,71]. In the search for new interesting compounds bioautographic tests are often combined with analytical methods like LC-MS, NMR, or IR to determine the structure of active constituents [57,58,61,67,68,74,76]. The commercial available equipment (Camag TLC-MS Interface, DART - Direct Analysis in Real Time System) makes such hyphenation easier and gives new possibilities for detecting natural bioactive compounds directly from the TLC plate [48–50,80].

TABLE 11.1 Selected Examples of Antimicrobial TLC-B (Direct, Contact, Overlay) in Searching for Target Compounds in Different Matrices

Type of Sample	Investigated Compounds	Type of Microorganism	Bioassay	Ref.
Cow's milk	Cefacetril	*Bacillus subtilis*	TLC–direct bioautography (dipping)	[46]
Cow's milk	Chloramphenicol, benzylpenicilin, dicloxacilin, erythromycin, ampicillin	*B. subtilis*	HPTLC–contact bioautography	[47]
Sewage effluent, surface water, drinking water	Sucralose	*Vibrio fischerii*	HPTLC–direct bioautography (dipping)	[48]
Grape pomace, animal feed, wine, fruit juices	Pelargonidin-3-glucoside, peonidin-3-glucoside, malvidin-3-glucoside, cyanidin-3-glucoside, delphinidin-3-glucoside, malvidin-3,5-diglucoside	*V. fischerii*	HPTLC–direct bioautography (dipping)	[49]
Bilberry, blueberry	Malvidin	*Aliivibrio fischerii*	HPTLC–direct bioautography (dipping)	[50]
12 marine Sponges	Avarol, avarone aeroplysinin, dienon	*V. fischerii*	HPTLC–direct bioautography (dipping)	[51]
Marine sponge *Hymeniacidon perleve*	Norharman (beta-carboline alkaloid) from *Pseudoalteromonas piscicida* NJ6-3-1	*Staphylococcus aureus*	TLC–overlay bioautography	[22]
Artemisia annua *Artemisia dracunculus* *Eucalyptus globulus* *Mentha longifolia* var. *chorodictya*	Piperitone + nitrofurantoin	*Enterobacter cloacae*	TLC–overlay bioautography	[52]
Sclerocarya birrea	Acetone and aqueous extracts	*Helicobacter pylori*	TLC–overlay bioautography	[53]

Continued

TABLE 11.1 Selected Examples of Antimicrobial TLC-B (Direct, Contact, Overlay) in Searching for Target Compounds in Different Matrices—cont'd

Type of Sample	Investigated Compounds	Type of Microorganism	Bioassay	Ref.
Lentinula edodes	Oleic acid, linoleic acid, linolenic acid, stearic acid	*Saccharomyces cerevisiae, Micrococcus lysodeikticus*	TLC–direct bioautography	[54]
Cordia gilleti	n-hexane, ethyl acetate and methanolic extracts	*S. aureus,* Methicillin-resistant *S. aureus, V. fischerii*	TLC–direct bioautography (dipping)	[55]
Melaleuca alternifolia, Eucalyptus globulus, Syzygium aromaticum, Cinnamomum zeylanium, Thymus vulgaris	Essential oils; eucalyptol, eugenol, *trans*-cinnamic aldehyde, thymol	*Pseudomonas syringae* pv. *maculicola, V. fischerii*	TLC–direct bioautography (dipping)	[39]
Lavandula angustifolia, E. globulus, Mentha spicata, C. zeylanium, T. vulgaris	Essential oils; eugenol, *trans*-cinnamic aldehyde, linalool, thymol, carvacrol	*Xanthomonas campestis* pv. *vesicatoria, P. syringae* pv. *phaseolicola, Staphylococcus epidermidis, Staphylococcus saprophyticus, S. aureus,* Methicillin-resistant *S. aureus, V. fischerii*	TLC–direct bioautography (dipping)	[40]
T. vulgaris	Essential oil; Thymol, carvacrol, geraniol	*X. campestis* pv. *vesicatoria, P. syringae* pv. *phaseolicola*	TLC–direct bioautography (dipping)	[41]
T. vulgaris, Mentha x *piperita, Salvia officinalis*	Thymol, menthol, α,β-thujon	*B. subtilis, Escherichia coli*	TLC–direct bioautography (dipping)	[42]

Garcinia magnostana	α-magnostin	Methicillin-resistant *S. aureus*	TLC–overlay bioautography	[56]
Commercial water-soluble annatto extracts	9′-*cis*-Norbixin, *all-trans*-Norbixin	*S. aureus*	TLC–overlay bioautography	[57]
Croton lechleri	Essential oil	*Pseudomonas aeruginosa, Klebsiella oxytoca, E. coli, Enteroccocus foecalis, Micrococcus luteus, S. aureus*	HPTLC–overlay bioautography	[58]
Helichrysum italicum, Napeta cataria, Phytolacca dodecandra	Diethyl ether extracts	*S. aureus*	TLC–overlay bioautography	[59]
Actaea racemosa	Methanolic extract	*V. fischerii*	HPTLC–direct bioautography (dipping)	[60]
Eremophila duttonii	Serrulat-14-en-7,8,20-triol Serrulat-14-en-3,7,8,20-tetraol 4-hydroxyngaione	*S. aureus*	TLC–overlay bioautography	[61]
Matricaria recutita	Hexane, acetone, methanolic, 50% aqueous ethanolic extract	*B. subtilis, P. syringae* pv. *maculicola*	TLC–direct bioautography (dipping)	[43]
M. recutita	Cis-spiroether, *trans*-spiroether, α-bisabolol, bisabolol oxides, metricarin, herniarin, umbelliferone	*B. subtilis, P. syringae* pv. *maculiccla*	TLC–direct bioautography (dipping)	[44]
M. recutita	Cis-spiroether, *trans*-spiroether, α-bisabolol, herniarin	*V. fischerii*	TLC–direct bioautography (dipping)	[45]
Dodonaea viscosa	n-butanol, ethyl acetate, dichloromethane, n-hexane extracts	*S. aureus, M. luteus, B. subtilis, Bacillus*	TLC–contact bioautography	[62]

Continued

TABLE 11.1 Selected Examples of Antimicrobial TLC-B (Direct, Contact, Overlay) in Searching for Target Compounds in Different Matrices—cont'd

Type of Sample	Investigated Compounds	Type of Microorganism	Bioassay	Ref.
		cereus, E. coli, P. aeruginosa, Salmonella typhi		
Combretum vendae, Commiphora harveyi, Khaya anthotheca, Kirkia wilmsii, Loxostylis alata, Ochna natalitia, Protorhus longifolia	Hexane, acetone, dichloromethane, methanolic extracts	*E. foecalis, E. coli, S. aureus, Aspergillus fumigatus, Candida albicans, Sporothrix schenckii*	TLC–direct bioautography (spraying)	[63]
Curtisia denata	Acetone extract	*B. subtilis, B. cereus, E. coli, S. aureus, C. albicans*	TLC–direct bioautography (spraying)	[64]
5 Rhododendron sp., 2 Scrophularia sp., 3 Phlomis sp., Leonurus cardiaca, Morina persica, Asperula nitida, Putoria calabrica, Wendlandia ligustroides, Urtica dioica	Hexane, chloroform, dichloromethane, methanolic extracts	*M. luteus, B. subtilis, B. cereus, S. aureus, S. epidermidis, E. coli, P. aeruginosa, C. albicans*	TLC–overlay bioautography	[21]
Scaligeria tripartita	Essential oil	*Colletotrichum acutatum, Colletotrichum fragariae, Colletotrichum gloeosporioides*	TLC/oplc–direct bioautography (spraying)	[65]
Arnica longifolia, Aster hesperius, Chrysothamnus nauseosus var. *nauseosus*	Essential oils, α-bisabolol, carvacrol	*C. acutatum, C. fragariae, C. gloeosporioides*	TLC–direct bioautography (spraying)	[66]

Piper arboretum, Piper tuberculatum	13 amides	*Cladosporium sphaerospermum, Cladosporium cladosporioides*	TLC–direct bioautography (spraying)	[67]
Piper hispidum, P. tuberculatum	10 amides	*C. sphaerospermum*	TLC–direct bioautography (spraying)	[68]
Anethum graveolens, Tachyspermum copticum, Cymbopogan martini	Thymol, carvone, geraniol	*Colletotrichum lindemuthianum, Alternaria alternata*	TLC–direct bioautography (spraying)	[69]
Leaves extracts of 24 *Combretum* species	Hexane, dichloromethane, acetone, methanol extracts	*C. albicans, Cryptococcus neoformans, A. fumigatus, Microsporum canis, Sporothix schenckii*	TLC–direct bioautography (spraying)	[70]
Breonadia salicina, Olinia ventosa	Acetone, hexane, dichloromethane, methanol extracts	*Aspergillus niger, Aspergillus parasiticus, Colletotrichum gloeosporioides, Penicillium janthinellum, Penicillium expansum, Trichoderma harzianum, Fusarium oxysporus*	TLC–direct bioautography (spraying)	[71]
Aspergillus terreus	Ethyl acetate extract	*C. albicans*	TLC–direct bioautography (spraying)	[72]
Schinus molle, Schinus terebinthifolius, Anacardium occidentale	Aqueous extracts, supernatant	*C. albicans, C. neoformans*	TLC–overlay bioautography	[23]

Continued

TABLE 11.1 Selected Examples of Antimicrobial TLC-B (Direct, Contact, Overlay) in Searching for Target Compounds in Different Matrices—cont'd

Type of Sample	Investigated Compounds	Type of Microorganism	Bioassay	Ref.
Myrtus communis	Ethyl acetate, dichloromethane petroleum ether, menthanol extracts	*Microsphorum canis, Microsphorum gypseum, Tricophyton mentagrophytes*	TLC–overlay bioautography	[73]
Stemona pierrei	Dihydrophenanthrenes, dihydrostilbenes, stilbene	*Cladosporium herbarum*	TLC–direct bioautography (spraying)	[74]
Bazzania trilobata	Six sesquiterpenes: 5- and 7-Hydroxycalamenene, drimenol, drimenal, viridiflorol, gymnomitrol; three bisbibenzyls: 6′,8′-dichloroisoplagiochin C, isoplagiochin D, 6′-chloroisoplagiochin D	*Botrytis cinerea, Cladosporium cucumerinum, Phythophthora infestans, Pyricularia oryzae, Septoria tritici*	TLC–direct bioautography (spraying)	[75]
Erythrina vogelii	Isowighteone, 5,7,4′-trihydroxy-2′-methoxy-5′-(3-methylbut-2-enyl)-isoflavanone, isolupalbigenin, 6,8-diphenylgenestein, isoderone	*C. cucumerinum,*	TLC–direct bioautography (spraying)	[76]
Balantiopsis cancellata	*Trans*-β-methylthioacrylic acid	*C. herbarum*	TLC–direct bioautography (spraying)	[77]
Garcinia artoviridis	2-(butoxycarbonylmethyl)-3-butoxycarbonyl-2-hydroxy-3-propanolide, 1′,1″-dibutyl methyl hydroxycitrate	*C. herbarum*	TLC–direct bioautography (spraying)	[78]
Glycosmis pentaphylla	Dichloromethane extract	*Valsa ceratosperma*	TLC–direct bioautography (spraying)	[79]

TLC-B, thin-layer chromatography–bioautography; HPTLC, high-performance thin-layer chromatography; OPLC, overpressured-layer chromatography.

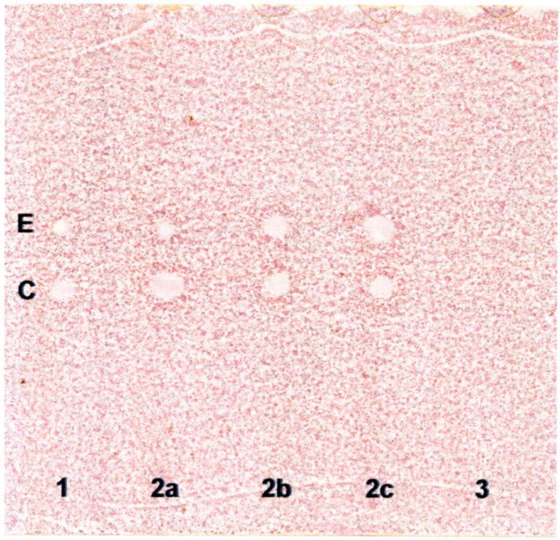

FIGURE 11.5 MSPD-TLC-DB of the eluates referring to the milk spiked at 0.05 ppm level (Maximum Residue Level) with ciprofloxacin (C) and enrofloxacin (E). The level of C and E in the Matrix Solid-Phase Dispersion eluates was 0.5 ppm (10 times higher then in the spiked milk, due to Matrix Solid-Phase Dispersion concentration). Tracks from left to right: 1—mixture of standards at 0.5 ppm, 2a–2c the eluates from MSPD cartridges (three repetitions), 3—blank. Applied volume—50 μL. TLC-DB, thin-layer chromatography–direct bioautography.

A special type of bioassay using luminescent bacteria as bioautographic indicators offers a new, promising type of DB. Most frequently the commercially available *Aliivibrio fischeri* (former name *Vibrio fischeri*) test (Bioluminex™) is used [81]. The mechanism of action is strongly related to a multistage bacterial luminescence reaction. The appropriate cell density enabling observation of the emission of blue-green light at 490 nm plays a key role in the process [51]. In the presence of bioactive substances, the changes in light emission (darker or brighter zones), related to disturbance in vital cell processes, are observed. The luminescence is decreased for cells exposed to, for example, toxic substances like pesticides or heavy metals. The hyphenation of TLC with bioluminescence detection offers several advantages over standard toxicity screening assays, e.g., cuvette test Microtox™, which is used in daily routine analysis to check toxicity of water samples [82]. What is most important, is that the bioautographic test shows a real correlation between separated single compounds and their activity. The integration of developed plates using image conversion and spreadsheet programs (ChromQuest, Gimp, VideoScan, Microsoft Excel, ImageMagick) enables very precise quantitative biodetection [83]. TLC–bioluminescence assays using *A. fischeri* are often used in the search for organic pollutants, degradation products, adulterants in various water samples, foodstuffs,

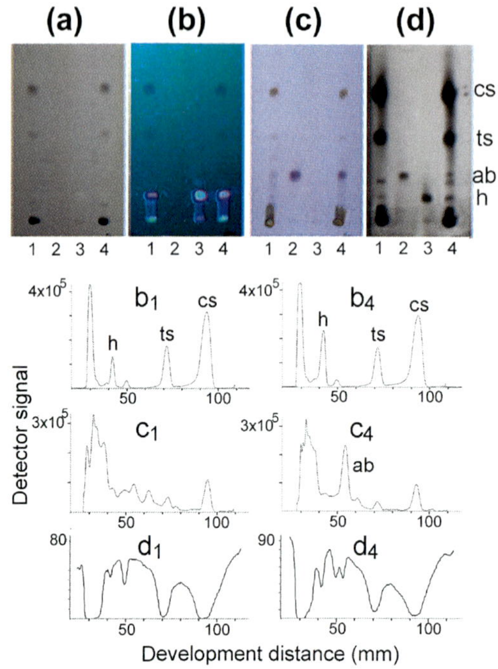

FIGURE 11.6 Detection of ingredients of 50% aqueous ethanol chamomile extract separated by infusion OPLC; 1—6 μL 50% aqueous ethanol chamomile (Herbaria) extract; 2—3 μg (−)-alpha-bisabolol; 3—3 μg herniarin; 4—6 μL 50% aqueous ethanol chamomile extract + 1.5 μg of each (−)-alpha-bisabolol and herniarin test substances. (a) and (b) The developed layers under UV light $\lambda = 254/365$ nm. (c) Visualization with vanillin-sulfuric acid reagent. (d) Bioautogram using *Vibrio fischeri* (dark spot = inhibition zone). b_1 and b_4 densitometric evaluation of tracks 1 and 4 in Figure 5b at 350 nm c_1 and c_4 densitometric evaluation of tracks 1 and 4 in Figure 5c at 600 nm d_1 and d_4 the measurement of the intensity of the luminescent light in tracks 1 and 4 in Figure 5d. cs = *cis*-spiroether; ts = trans-spiroether; ab = (−)-alpha-bisabolol; h = herniarin; OPLC, overpressured-layer chromatography. *Reprinted from [45] with permission.*

cosmetics, and botanical samples [8,39,45,60,84,85] (Figure 11.6). Another luminescent test, using the plant pathogenic bacterium *Pseudomonas syringae* pv. *maculicola*, was used to search for bioactive components in plant material. The detected active compounds may be used as natural agents against bacterial plant infections [39,43,44].

Tyihák et al. introduced a complex separation and detection system, called BioArena, which combines the advantages of layer chromatography (OPLC or TLC) and DB [34,35,86—92]. BioArena can be used for investigating biochemical interactions between microorganisms and biologically active compounds (e.g., antibiotics, mycotoxins, *trans*-resveratrol, etc.) in the adsorbent layer after chromatographic separation, as well as for studying cell proliferation promoting and retarding processes. Especially, model reactions of endogenous formaldehyde, hydrogen peroxide, and ozone, key molecules in biological systems, are reported

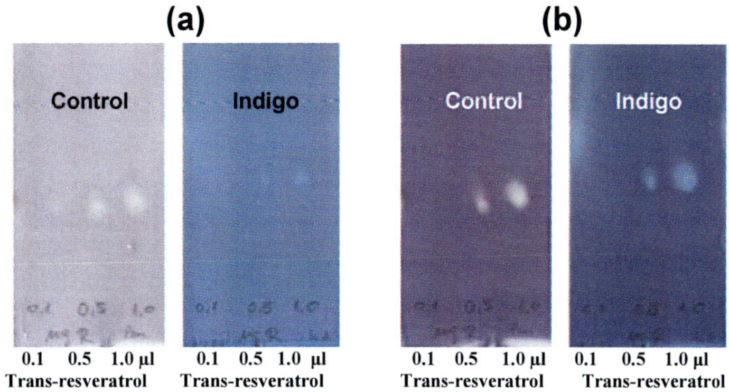

FIGURE 11.7 Effect of indigo carmine on the antibacterial activity of *trans*-resveratrol. 20 mg indigo carmine was added to 100 mL Psm. (*Pseudomonas savastanoi* pv. *Phaseolicola)* (a) 3 h between inoculation and staining (with MTT). (b) 18 h between inoculation and staining (with MTT). MTT, [3-(4,5-dimethylthiazol-2-yl)-2,5-diphenyltetrazolium bromide]. *Reprinted from [86] with permission.*

there. BioArena visualizes effects of HCHO-capturing species (e.g., L-arginine and glutathione) and its promoters (e.g., Cu^{2+} ions) (Figure 11.7). The inhibition zones can be examined quantitatively by use of in situ densitometry to obtain a calibration curve [35]. BioArena can be hyphenated with spectroscopic methods such as MS, (Fourier Transform) Infrared Spectroscopy, Fourier Transform-Raman or NMR [35,89].

As mentioned earlier, the growing interest in drugs of plant origin has been observed for almost 40 years. The isolation of active plant constituents from hundreds of primary and secondary metabolites requires a multidisciplinary approach of botanists, pharmacognosists, chemists, pharmacologists, and toxicologists [93]. Besides extraction and purification, usually two or more steps of separation from the matrix are necessary. The best approach is bioactivity-directed isolation by various chromatographic techniques combined with spectroscopic methods, such as UV, IR, MS, and NMR [93,94]. The most investigated plants are those with antifungal properties associated with the increasing incidence of mycoses due to antibiotic therapy, oncology chemotherapy, or AIDS. Direct bioautographic procedures were established by Homans and Fush for spore producing fungi such as *Aspergillus*, *Penicillum*, and *Cladosporium* [95]. The protocol concerning *Cladosporium* was later included in a strategy for the biological and chemical evaluation of plant extracts by the so-called Swiss school of Hostettman, Hamburger, Marston, Wolfender, and others [96,97]. Numerous antifungal compounds have been characterized using the *Cladosprium cucumerinum* TLC-DB assay. The search was usually accomplished with spectroscopic data (LC-UV, LC-MS, NMR) as well as with microfractionation [76,98–100]. An earlier developed agar overlay assay with *Candida albicans* [20] was often combined with the TLC-DB test based on *C.*

cucumerinum. The structures of active compounds were established by chemical and spectroscopic methods [101–103]. In some papers, besides antifungal activity against *C. cucumerinum* and *C. albicans*, activities against bacteria *B. subtilis* and *E. coli* were also measured [20,104] using the TLC-DB method established earlier [105]. Identification of bioactive compounds from *Blumea gariepina* was achieved by HPLC/UV/MS/NMR combined with TLC-DB assays against *C. cucumerinum* and acetylcholinesterase assay performed for fractions obtained by HPLC microfractionation [106]. A recent review paper by Favre-Godal et al. presents applications and advances in the screening for antifungal compounds from natural sources. Various methods were presented with a special emphasis on TLC, HPTLC, and bioautography, both agar overlay and direct. Figure 11.8 illustrates the hyphenation of TLC-DB with LC-microfractionation as well as with LC-UV and LC-NMR analysis. Additionally, the activities of about 100 antifungal compounds of

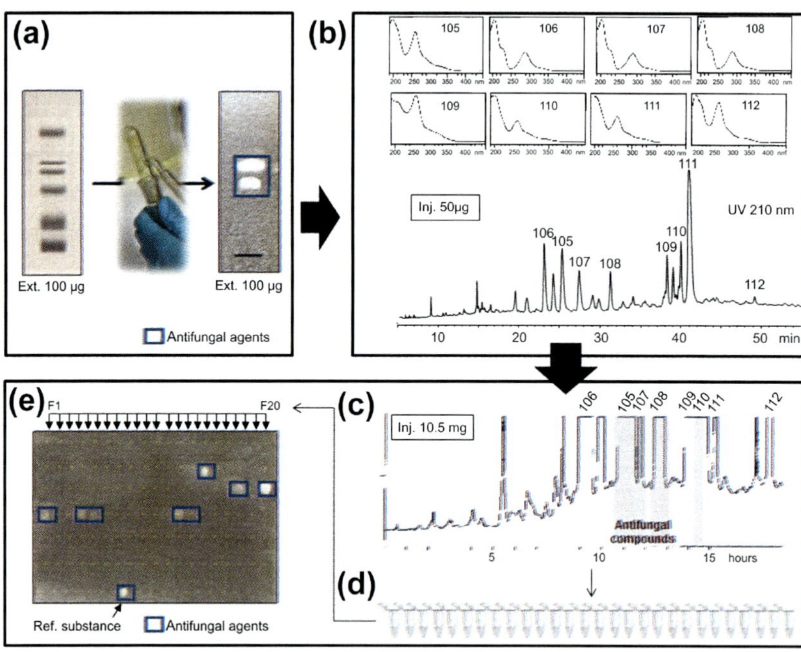

FIGURE 11.8 (a) TLC-direct bioautography with *Cladosporium cucumerinum*, showing the antifungal activity of the CH_2Cl_2 extract of *Erythrina vogelii*. (b) LC-photodiode array detector analysis of the crude CH_2Cl_2 extract under analytical conditions (50 μg on column). (c) LC-UV trace recorded during the LC-NMR analysis (10 mg on column). (d) LC microfractions collected in ependorff tubes after the LC-NMR analysis. (e) TLC-bioautography with *C. cucumerinum* with the LC microfractions. TLC, thin-layer chromatography; LC-UV, liquid chromatography–ultraviolet; LC-NMR, liquid chromatography–nuclear magnetic resonance. *Reprinted from [107] with permission.*

natural origin were presented with their minimum inhibitory quantity, determined using the microdilution method and/or bioautographic assays. Perspectives concerning future improvements in sensitivity and reproducibility of bioautography were also discussed [107].

The separate "school" of bioautography is represented by Wedge and coworkers. The principle objective of this group is the identification of natural products with antifungal properties to control agricultural fungal pathogens. TLC-DB against three agronomically important *Colletrichum* spp., i.e., *C. frageriae, C. gloeosporioides*, and *C. acutatum*, was used in the majority of reports [65,66,108–116]. The antifungal activity of several classes of quinones from natural products against three *Colletrichum* spp. was reported [108]. Also, extracts from 30 plants from the Greek flora were evaluated by TLC-DB against *Colletrichum* spp. Bioassay-guided fractionation led to the isolation of pure compounds evaluated using bioautography and a 96 well broth microdilution assay against a wide spectrum of fungi [109]. They were identifying by NMR spectroscopy. A similar approach was used for the bioassay-directed isolation of antifungal compounds from *Ruta graveolens* [110] as well as from *Gomphus floccosus* (additionally GC-MS and LC-MS analysis were performed) [111], and from *Conyza canadensis* (besides GC-MS and LC-MS analysis, bioassays against *Lactuca sativa, Agrostis stolonifera*, and *Lemna paucicostata* were performed) [112]. In addition, TLC-DB against the *Colletrichum* spp. was performed for *Hedychium, Angelica sinensis* and *archangelica*, as well as for *Haplopappus greenei* volatile constituents. Components were characterized by gas chromatography with flame ionization detection and GC-MS. In addition, their insecticidal properties were evaluated [113–115]. Volatile compounds from *Origanum onites* were analyzed by GC-FID, GC-MS, HPTLC, TLC, and OPLC. The mechanism of antibacterial activity was studied by Bio-Arena bioautography. The antifungal properties were characterized both by TLC-DB against three *Colletrichum* spp and 96-well microdilution broth assay [116]. Both one-dimensional (1D) and two-dimensional (2D) DB against *Colletrichum* and *Phomopsis* spp were described in a paper by Wedge and Nagle. Plant pathogenic fungi were evaluated for their sensitivity to commercial agricultural fungicides by 1D TLC-DB, while extracts of aquatic plants, algae and marine cyanobacteria were evaluated by 2D-TLC-DB. The 2D-TLC fingerprints and antifungal zones facilitated the identification of chemically unique chemotypes from microbial isolates, plants, and algae [117].

11.2.2 Enzymes Inhibitor Detection

11.2.2.1 Esterases Inhibitors

A number of enzyme inhibition bioautographic methods have been developed, mainly based on esterases (predominantly acetylcholinesterase,

AchE and butyrylcholinesterase, BchE), glucosidases, and xanthine oxidase (XO) inhibition [17]. The largest group of enzymatic assays is focused on identifying compounds with AchE inhibitory properties as potential drugs in the therapy of Alzheimer's disease. One of these drugs is galanthamine, an alkaloid isolated from various species of the Amaryllidaceae family [17,118–122]. Also plants from other families were tested for their acetyl(butyryl)cholinesterase activities, e.g., *Chelidonium majus* (Papaveraceae) [118], 40 herbal drugs used in traditional Iranian medicine [123], *Salvia militorrhiza* (Pamiaceae) [124], *Ferula gummosa* (oleo-gum-resin galbanum) (Apiaceae) [125] as well as seeds of plants from the genus *Peganum* (Nitrariaceae) used in traditional Chinese medicine [126] (Figure 11.9) and various Gabonese medicinal plants [127].

TLC-cholinesterase tests were also used to detect organophosphate and carbamate insecticides [128–131] as well as chemical warfare agents [132]. The most popular visualization technique used for acetyl(butyryl) cholinesterase inhibition is based on the diazotization reaction. The developed and dried TLC plate is sprayed with enzyme (AchE or BchE), incubated, and sprayed with a mixture of 1-naphthyl acetate and Fast Blue B Salt [106,118,119,123,125,126,128–131] (Figure 11.10). In a variant of this method, after spraying with enzyme the plate is first incubated with 1-naphthyl acetate and then sprayed with Fast Blue B Salt [124]. Active

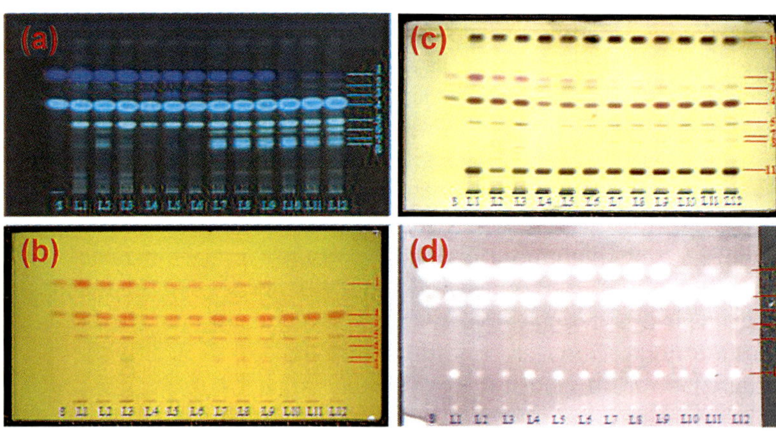

FIGURE 11.9 Comparison of the TLC chromatograms obtained from *Peganum harmala* Linn. (L1–L3: Sample Nos 1–3), *P. multisectum* (Maxim.) Bobr (L4–L6: Sample Nos 4–6), *Peganum* variety (L7–L9: Sample Nos 7–9), and *P. nigellastrum* Bunge (L10–L12: Sample Nos 10–12) viewed under different visualization conditions: (a) inspected under UV 366 nm directly, (b) visualized by sprayed Dragendorff's reagent, (c) sprayed with vanillin–sulfuric acid reagent, (d) visualized by the bioautographic technique (concentration of standard substances and samples in (d) are 1/10th of that in (a–c)). S: mixed standard substance solution with harmine (1) and harmaline (2) of 1 mg/mL in (a–c) and 0.1 mg/mL in (d). TLC, thin-layer chromatography. *Reprinted from [126] with permission.*

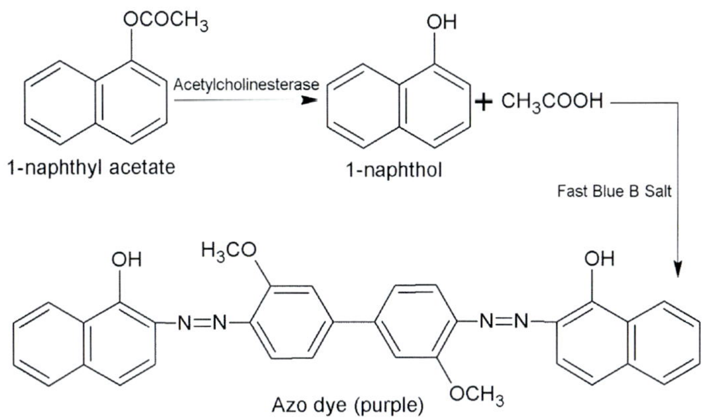

FIGURE 11.10 Reaction of acetylcholinesterase with 1-naphtyl acetate to give 1-naphthol and subsequent formation of a purple azo dye with a diazonium salt.

components are visible as white spots on a purple background. In a modification of the method 1-naphthyl acetate was replaced with the more expensive but giving a deeper purple background 2-naphthyl acetate applied through the mobile phase. After development and drying, the plate was sprayed with AchE, incubated, and then sprayed with Fast Blue B Salt [121,122]. The method was validated [121] and compared with bioautographic Ellman detection [122]. The Ellman method, popular in colorimetric determination of AchE activity, has been adopted for TLC-DB of AchE inhibitors. The developed and dried TLC plate is sprayed with a solution of acetylthiocholine and 5,5′-dithio-bis-(2-nitrobenzoic acid) followed by the enzyme. AChE inhibitors are observed as white spots against a pale yellow background [17,120,122]. The false-positive effects of some compounds in TLC-DB based on the Ellman reaction have been discussed [133].

11.2.2.2 Glucosidase Inhibitors

Glucosidase inhibitors are potential drugs for the treatment of diabetes, obesity, cancer, viral infections (including HIV), and lysosomal storage diseases. They can be detected by TLC bioautographic methods based on the glucosidase inhibitory activity [17,127,134,135]. The developed and dried TLC plate is sprayed with α- or β-D-glucosidase and incubated. α-D-Glucosidase needs longer incubation (60 min) than β-D-glucosidase (20 min). Then the mixture of substrate solution (2-naphthyl-α-D-glucopiranozide or 2-naphthyl-β-D-glucopiranozide) and Fast Blue B Salt is sprayed onto the plate. The 2-naphthol coming from the cleavage of 2-naphthyl-α/β-D-glucopiranozide reacts with the Fast Blue B Salt giving a purple coloration in a few minutes [17,135]. An alternative method is based on the hydrolysis of glucoside esculin into esculetin (aglicone) followed by reaction with ferric chloride to produce a brown complex. β-D-Glucosidase inhibitors prevent cleavage of esculin and are seen as white spots on a brown background [17,134].

11.2.2.3 Xanthine Oxidase Inhibitors

XO catalyzes the oxidation of hypoxanthine and xanthine to uric acid producing reactive oxygen species, superoxide radicals, and hydrogen peroxide. These products contribute to oxidative stress and are involved in many pathological processes. Many plant extracts show inhibitory effects for XO activity. The TLC bioautographic method is ideal for identifying these active constituents. The developed TLC plate is covered with agar solution containing XO and nitroblue tetrazolium. After solidification of the gel, the plate is immersed in xanthine solution. Superoxide radicals from oxidation of xanthine reduce the pale yellow tetrazolium salt to a purple formazan. XO inhibitors are detected as white spots against a purple background. Radical scavengers which are not inhibitors of XO also give positive results. To distinguish them from pure XO inhibitors the additional TLC assay should be performed in which superoxide radicals are generated chemically [17,136].

11.2.3 Determination of Estrogenic Activity

A wide range of natural and synthetic compounds have been found to be estrogenic. However, usually they show little structural resemblance to natural estrogens. These include phytoestrogenes, mycoestrogenes, some pesticides, herbicides, polychlorinated biphenyls, combustion pollutants, flame retardants, plasticizers, breakdown products of surfactants, preservatives, and UV filters. Many of these compounds are released into the environment and can be found in rivers, wastewater, or sewage treatment plants [137–141]. The yeast estrogen screen (YES) is one of the in vitro tests developed for the measurement of estrogenic activity. The assay is based on genetically modified yeast cells with the human estrogen receptor gene incorporated into the main genome. Activation of the receptor by estrogen-like compounds leads to expression of the reporter gene *Lac-Z*, which produces the enzyme β-galactosidase. The enzyme metabolizes the yellow chromogenic substrate, chlorophenol red-β-D-galactopyranoside (CPRG), to a red product. The original YES assay was carried out in microtiter plates using a yeast cell suspension [142]. In the TLC-DB modification of the method, the dried TLC plate is dipped into the test yeast culture and incubated in a moisture box. Two different substrates can be used: CPRG, which is either added to the cell suspension or sprayed onto the plate after incubation or 4-methylumbelliferyl-β-D-galactopyranoside (MUG), which is only sprayed onto the plate after incubation. Then, regardless of the substrate used, the plates are incubated a second time. Estrogenic compounds are observed as red spots on the yellow background (for CPRG) or fluorescent zones of 4-Methylumbelliferone from the enzymatic cleavage of MUG (to increase the fluorescence signal the plate after the second incubation can be exposed to ammonia vapor). In TLC-DB of estrogenic compounds usually MUG substrate is used as it provides higher sensitivity [139–141,143] (Figure 11.11). Similar to the microtiter method [142], dose response curves can be constructed from the peak areas obtained on the basis of the densitograms of the fluorescent zones and the amount of estrogenic substance. For a narrow range of masses (one order of magnitude) a linear dependence is obtained [139] while for wider

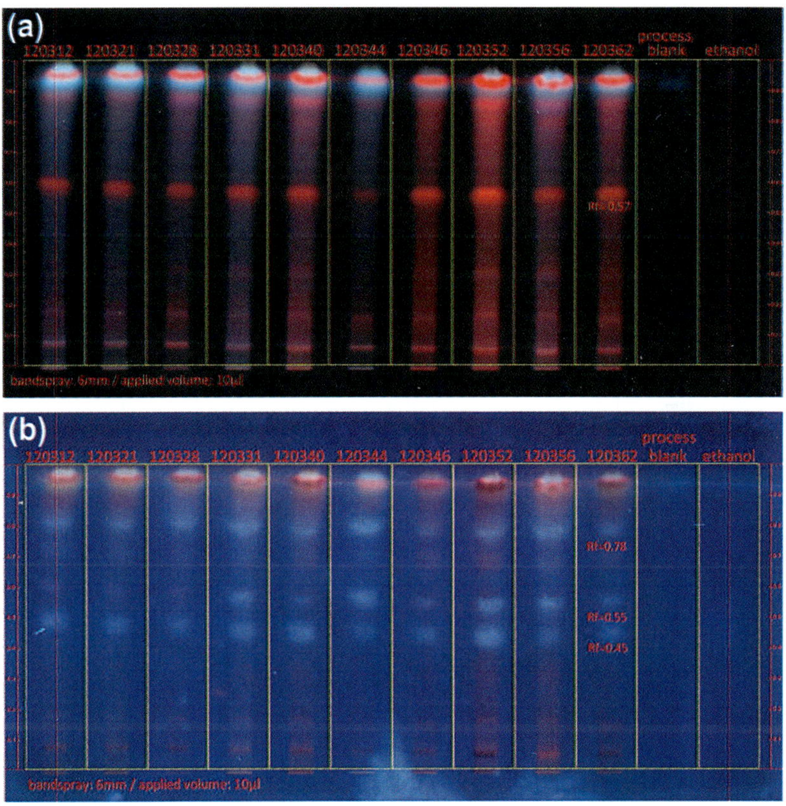

FIGURE 11.11 Analysis of several sediment extracts from the river Elbe with the p-YES. 10 μL of organic sediment extracts from the river Elbe were applied to a TLC plate. A blank sample which was directed to the complete extraction process and the solvent ethanol were added to the plate in lanes 11 and 12. (a) TLC plate after chromatographic development at 366 nm. (b) TLC plate after the p-YES with visible fluorescence signals of 4-MU at 366 nm. YES, yeast estrogen screen; TLC, thin-layer chromatography. *Reprinted from [141] with permission.*

ranges a sigmoidal dependency is observed [141,143]. It is worth mentioning here, that in antimicrobial TLC-DB assays a linear dependency of the area of the inhibition zone vs logarithm of amount/concentration applied is obtained for a narrow range of concentrations while an exponential dependency is observed for wider ranges (more than two orders of magnitude) [12,16,24,31–33,46].

11.2.4 Detection of Antioxidants

TLC-B for the identification of antioxidants and radical scavengers in natural samples plays an important role in routine analysis. This hyphenation affords several advantages, such as fast results, a reduction in the time-consuming sample preparation steps, and what is the most important, direct access to separated target compounds on the (HP)TLC plate [144]. Additionally, contrary to spectrophotometric assays, the negative influence of external factors that disturb the measurements, such as light, oxygen, and pH of the reaction mixture, can be neglected [145]. Two groups of TLC-DB tests are applied to determine either free radical scavenging or antioxidant activity [17].

Radical scavenging activity can be assessed using stable 2,2-diphenyl-1-picrylhydrazyl radical (DPPH) (Figure 11.12) or 2,2′-azino-bis(3-ethylbenzothiazoline-6-sulphonic acid) (ABTS), whose radical cation needs to be generated by reaction with potassium persulfate. The first TLC-DPPH$^{\bullet}$ analysis was carried out in 1967 to determine tocopherols in natural products [146]. Since then the method has been mainly applied to screen extracts from complex natural products for the presence of free radical scavengers as well as to look for new compounds with these properties. Generally, the developed (HP)TLC plate is sprayed or dipped in a DPPH$^{\bullet}$ solution and the color change is observed related to the reduction of DPPH. In a paper by Ciesla, various parameters influencing the DPPH• reaction, i.e., dissolution of DPPH•, type of TLC stationary phase, and documentation time were discussed [147]. Nonspecific sorbents like Reversed Phase-18 and nonpolar solvents,

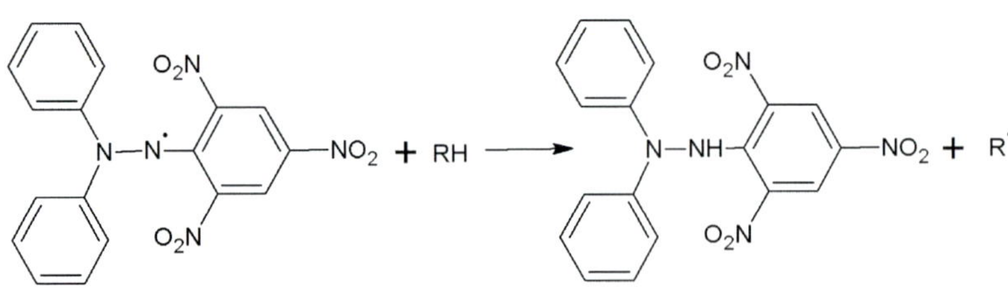

FIGURE 11.12 Reaction of 2,2-diphenyl-1-picrylhydrazyl radical with radical scavengers.

e.g., n-hexane were recommended as having no influence on the radical–antioxidant reaction. The use of silica gel plates and methanolic solutions of DPPH• (the most popular) gives the most intensive results. The reaction of the positively charged adsorbent (like silica gel) with the target compounds, e.g., polyphenolics, causes a strengthening of the radical–antioxidant reaction increasing the reaction rate. An investigation of compounds spotted directly on the TLC plate without development was also reported and named a "dot-blot" test. It allows the radical scavenging capacity of pure compounds, mostly, to be determined [147,148]. The TLC-DPPH• assay indicates the presence of radical scavengers as yellow spots/bands on a purple background. The plates should be scanned at 517 nm (DPPH• absorption maximum). Nowadays, TLC-DPPH• is the most frequently used test to detect antioxidants in natural products, mainly polyphenolics in plant extracts [147,149–152] (Figure 11.13). Experiments using this assay to check radical scavenging of other compound types, e.g., coumarins, curcuminoids as well as to detect activity of unidentified fractions, were also reported [21,152–154]. The TLC-ABTS$^{•+}$ assay is less popular then the TLC-DPPH• assay but is used for the same purposes, e.g., comparing radical scavenging/antioxidant capacity of natural products or guiding the isolation of bioactive compounds. Active compounds appear as colorless or pink zones against a green background. The plates need to be scanned at 734 nm (ABTS$^{•+}$ absorption maximum). The "dot-blot" TLC-ABTS$^{•+}$ test was used to compare the antioxidant capacity of different teas as well as hydrophilic and lipophilic standards [148,155]. In order to improve the sensitivity and stability of the method a new type of ABTS$^{•+}$ solution, performed in agar, was evaluated. A modified method was used to detect antioxidants in separated plant extracts [155]. However, a comparative analysis indicated higher stability for DPPH• than ABTS$^{•+}$ on the TLC plate [148,155].

Antioxidant activity can be performed with two β-carotene bleaching tests. According to the first report for the carotene method, the TLC plate was dipped in a chloroform solution of β-carotene and exposed to white light. Compounds showing antioxidant activity were observed as orange spots against a pale yellow/white background [17,21,156]. A modification of this basic β-carotene bleaching test was reported by Pratt et al. and used to detect antioxidants in peanuts (*Arachia hypogoea*) and cottonseeds (*Gossypium hirsutum*) [157,158]. The β-carotene solution was enriched with linoleic acid in ethanol, which enabled screening of the compound's ability to stabilize lipid oxidation. After exposure to daylight until a yellow background was obtained, the antioxidants appear as orange spots with an intensity related to the antioxidant activity. The β-carotene bleaching tests have been applied to evaluate the antioxidant activity of many plant extracts (with or without separation) [21,149].

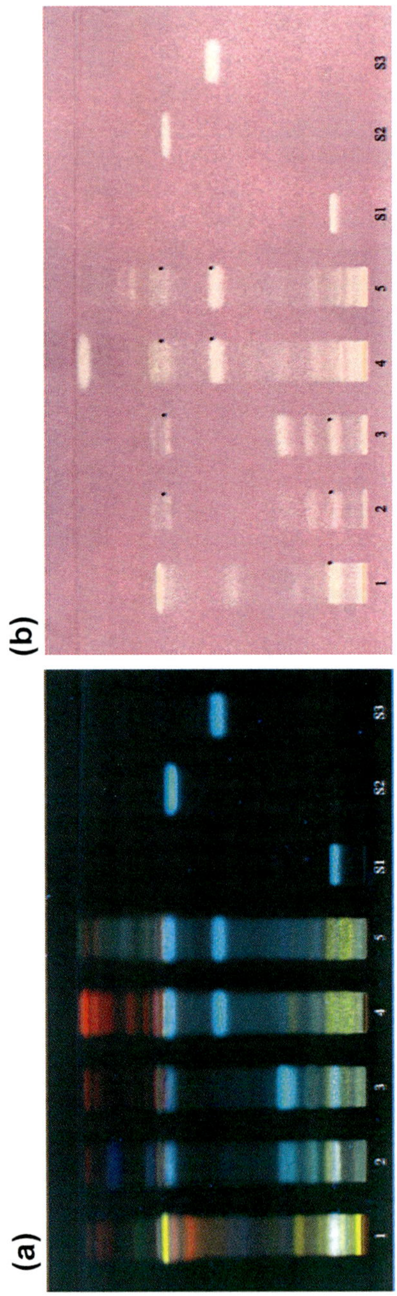

FIGURE 11.13 (a) TLC chromatogram of plant extracts after spraying with Natural Product/Polyethylene Glycol reagent. (b) TLC-DB using DPPH reagent. Tracks: 1—*Hypericum perforatum*, 2—*Metricaria chamomilla*, 3—*Achillea milefolium*, 4—*Thymus vulgaris*, 5—*Salvia officinalis*, S1—Chlorogenic acid, S2—Rosmarinic acid, S3—Caffeic acid. Standard solutions at 1 mg/mL. Applied volume: 5 μL. Silica gel plate, mobile phase: chloroform:ethyl acetate:formic acid:acetone (40:30:10:20). TLC-DB, thin-layer chromatogaphy–direct bioautography; DPPH, 2-diphenyl-1-picrylhydrazyl radical.

11.3 PERSPECTIVES

Despite the long history of TLC-B, which has lasted more than half-century, the method is still vital and expending. This success can be attributed to the growing interest in an EDA. The increasing number of environmental pollutants and increasing demand for new drugs has stimulated the need to identify both harmful and healing substances in complex matrices. Despite the most popular antimicrobial and antifungal tests, other assays are gaining in importance, in particular, bioluminescence, enzymatic, and estrogenic bioassays [159]. (HP)TLC, due to its separation power and open layer provides ideal hyphenation possibilities: analytical techniques with biological and/or spectroscopic and/or chemical detection [15,160,161]. The use of several TLC-B tests in parallel linked to chemical/spectroscopic methods provides complete information for either bioactivity or the structure of the target compounds. In many cases semiquantitative results can be achieved [12,14,16,24,31–33,35,46,129,139,141,143].

References

[1] Brack W, Ulrich N, Bataineh M. Separation techniques in effect-directed analysis. In: Brack W, editor. Effect-directed analysis of complex environmental contamination. Berlin Heidelberg: Springer-Verlag; 2011.

[2] Weller MGA. Unifying review of bioassay-guided fractionation, effect-directed analysis and related techniques. Sensors 2012;12:9181–209.

[3] Ballesteros-Gómez A, Rubio S. Recent advantages in environmental analysis. Anal Chem 2011;83:4579–613.

[4] Schymanski EL, Bataineh M, Goss K-U, Brack W. Integrated analytical and computer tools for structure elucidation in effect-directed analysis. Trends Anal Chem 2009; 28(5):550–61.

[5] Brack W. Effect-directed analysis: a promising tool for the identification of organic toxicants in complex mixtures? Anal Bioanal Chem 2003;377:397–407.

[6] Hecker M, Hollert H. Effect-directed analysis (EDA) in aquatic ecotoxicology: state of the art and future challenges. Environ Sci Pollut Res 2009;16:607–13.

[7] Brack W. Effect-directed analysis in ecotoxicology. In: Férard JF, Ch Blaise, editors. Encyclopedia of aquatic ecotoxicology. Dordrecht (Netherlands): Springer; 2013.

[8] Reemtsma T. Prospects of toxicity-directed wastewater analysis. Anal Chim Acta 2001; 426:279–87.

[9] Applying combinations of chemical analysis and biological effects to environmental and food samples – I. In: Barceló D, editor. Trends in analytical chemistry, vol. 28(5); 2009. pp. 515–626.

[10] Applying combinations of chemical analysis and biological effects to environmental and food samples – II. In: Barceló D, editor. Trends in analytical chemistry, vol. 28(6); 2009. pp. 627–832.

[11] Grzelak EM, Jesionek W, Majer-Dziedzic B, Choma IM. Applications of novel direct bioautography tests for analysis of antimicrobials: a review. J AOAC Int 2013;96:1167–74.

[12] Choma IM. The use of thin-layer chromatography with direct bioautography for antimicrobial analysis. LC–GC 2005;18:482–9.

[13] Weins Ch. Overview of bioactivity-based analysis by HPTLC. Bridging the gap between cause and effect – HPTLC detection of bioactive compounds in the environment and in food. J Planar Chromatogr 2008;6:405–10.

[14] Botz L, Nagy S, Kocsis B. Detection of microbiologically active compounds. In: Sz Nyiredy, editor. Planar chromatography a retrospective view for the third millennium. Budapest: Springer; 2001.

[15] Morlock G, Schwack W. Hyphenations in planar chromatography. J Chromatogr A 2010;1217:6600–9.

[16] Choma IM, Grzelak EM. Bioautography detection in thin-layer chromatography. J Chromatogr A 2011;1218:2684–91.

[17] Marston A. Thin-layer chromatography with biological detection in phytochemistry. J Chromatogr A 2011;1218:2676–83.

[18] Meyers E, Smith D. Bioautography of antibiotic spread-layer chromatograms. J Chromatogr 1964;14:129–32.

[19] Betina V. Bioautography in paper and thin-layer chromatography and its scope in the antibiotic field. J Chromatogr 1973;78:41–51.

[20] Rahalison L, Hamburger M, Hostettmann K, Monod M, Frenk E. A bioautographic agar overlay method for the detection of antifungal compounds from higher plants. Phytochem Anal 1991;2:199–203.

[21] Tasdemir D, Dönmez AA, Calis I, Ruedi P. Evaluation of biological activity of Turkish plants. Rapid screening for the antimicrobial, antioxidant and acetylcholinesterase inhibitory potential by TLC bioautographic methods. Pharm Biol 2004;42:374–83.

[22] Zheng L, Chen H, Han X, Lin W, Yan X. Antimicrobial screening and active compound isolation from marine bacterium NJ6-3-1 associated with the sponge *Hymeniacidon perleve*. World J Microb Biotechnol 2005;21:201–6.

[23] Schmourlo G, Mendonça-Filho RR, Alviano CS, Costa SS. Screening of antifungal agents using ethanol precipitation and bioautography of medicinal and food plants. J Ethnopharmacol 2005;96:563–8.

[24] Choma IM, Choma A, Komaniecka I, Pilorz K, Staszczuk K. Semiquantitative estimation of enrofloxacin and ciprofloxacin by thin-layer chromatography – direct bioautography. J Liq Chromatogr Relat Technol 2004;27(13):2071–85.

[25] Goodall RR, Levi AA. A microchromatographic method for the detection and approximate determination of the different penicillins in a mixture. Nature 1946; 158:675–6.

[26] Rios JL, Recio MC, Villar A. Screening methods for natural products with antimicrobial activity. J Ethnopharmacol 1988;23:127–49.

[27] Nagy S, Kocsis B, Koszegi T, Botz L. Optimization of growth condition for test bacteria cultures used in direct bioautography TLC detection. 1. Gram-positive test bacterium: *Bacillus subtilis*. J Planar Chromatogr 2002;15:132–7.

[28] Nagy S, Koszegi T, Botz L, Kocsis B. Optimization of growth condition for test bacteria cultures used in direct bioautography TLC detection. 2. Gram-negative test bacterium: *Escherichia coli*. J Planar Chromatogr 2003;16:121–6.

[29] Nagy S, Kocsis B, Koszegi T, Botz L. Optimization of growth condition for test fungus cultures used in direct bioautography TLC detection. 3. Test fungus: *Candida albicans*. J Planar Chromatogr 2007;20:385–9.

[30] Eymann R, Hauck HE. Chrom Biodip® antibiotics – a high-performance bioautographic test kit. In: Proceedings of the international symposium on planar separations, planar chromatography. Lillafüred; 2000.

[31] Choma I, Komaniecka I. Matrix solid-phase dispersion combined with thin-layer chromatography – direct bioautography for determination of enrofloxacin and ciprofloxacin residues in milk. J Liq Chromatogr Relat Technol 2005;28(16): 2467–78.

[32] Choma I. Screening of enrofloxacin and ciprofloxacin residues in milk by HPLC and by TLC with direct bioautography. J Planar Chromatogr 2006;19:104–8.

[33] Choma I. Thin-layer chromatography-direct bioautography of flumequine residues in milk. J Liq Chromatogr Relat Technol 2006;29(14):2083–93.

[34] Tyihák E, Botz L, Nagy S, Kocsis B, Mincovics E. BioArena™ as a complex bioautographic system. In: Proceedings of the international symposium on planar separations, planar chromatography. Lillafüred; 2001.

[35] Móricz ÁM, Adányi N, Horváth E, Ott PG, Tyihák E. Applicability of the BioArena system to investigation of the mechanisms of biological effects. J Planar Chromatogr 2008;21:417–22.

[36] Grzelak EM, Majer-Dziedzic B, Choma IM. Development of a novel direct bioautography-thin-layer chromatography test: optimization of growth condition for Gram-negative bacteria, *Escherichia coli*. J AOAC Int 2011;94:1569–72.

[37] Grzelak EM, Majer-Dziedzic B, Choma IM. Development of a novel direct bioautography-thin-layer chromatography test: optimization of growth conditions for gram-positive bacteria, *Bacillus subtilis*. J AOAC Int 2013;96:386–91.

[38] Choma IM, Grzelak EM, Majer-Dziedzic B. Comparison of deproteinization methods used before TLC-DB and HPLC analysis of flumequine residues in milk. Med Chem 2012;8:95–101.

[39] Horváth Gy, Kocsis B, Lemberkovics É, Böszörményi A, Ott PG, Móricz ÁM. Detection of antibacterial activity of essential oil components by TLC-bioautography using luminescent bacteria. J Planar Chromatogr 2013;26:114–8.

[40] Gy Horváth, Jámbor N, Végh A, Böszörményi A, Lemberkovics É, Héthelyi É, et al. Antimicrobial activity of essential oils: the possibilities of TLC-bioautography. Flavour Fragr J 2010;25:178–82.

[41] Gy Horváth, Szabó LGy, Lemberkovics É, Botz L, Kocsis B. Characterization and TLC-bioautographic detection of essential oils from some Thymus taxa. Determination of the activity of the oils and their components against plant pathogenic bacteria. J Planar Chromatogr 2004;17:300–4.

[42] Jesionek W, Grzelak EM, Majer-Dziedzic B, Choma IM. Thin-layer chromatography – direct bioautography for the screening of antimicrobial properties of plant extracts. J Planar Chromatogr 2013;26:109–13.

[43] Móricz AM, Tyihák E, Ott PG. Usefulness of transgenic luminescent bacteria in direct bioautographic investigation of Chamomile extracts. J Planar Chromatogr 2010;23:180–3.

[44] Móricz AM, Szarka S, Ott PG, Héthelyi ÉB, Szóke É, Tyihák E. Separation and identification of antibacterial chamomile components using OPLC, bioautography and GC-MS. Med Chem 2012;8:85–94.

[45] Móricz AM, Szarka S, Ott PG, Kertesy D, Héthelyi ÉB, Szóke É, et al. Application of direct bioautography and SPME-GC-MS for the study of antibacterial Chamomile ingredients. J Planar Chromatogr 2012;25:220–4.

[46] Choma IM, Kowalski C, Lodkowski R, Burmańczuk A, Komaniecka I. TLC-DB as an alternative to the HPLC method in the determination of cefacetril residues in cow's milk. J Liq Chromatogr R T 2008;31:1903–12.

[47] Ramirez A, Gutiérrez R, Diaz G, González C, Pérez N, Vega S, et al. High-performance thin-layer chromatography – bioautography for multiple antibiotic residues in cow's milk. J Chromatogr B 2003;784:315–22.

[48] Morlock GE, Schuele L, Grashorn S. Development of a quantitative high-performance thin-layer chromatographic method for sucralose in sewage effluent, surface water, and drinking water. J Chromatogr A 2011;1218:2745–53.

[49] Krüger S, Urmann O, Morlock GE. Development of a planar chromatographic method for quantitation of anthocyanes in pomace, feed, juice and wine. J Chromatogr A 2013;1289:105–18.
[50] Cretu GC, Morlock GE. Analysis of anthocyanins in powdered berry extracts by planar chromatography linked with bioassay and mass spectrometry. Food Chem 2014;146:104–12.
[51] Klöppel A, Grasse W, Brümmer F, Morlock GE. HPTLC coupled with bioluminescence and mass spectrometry for bioactivity-based analysis of secondary metabolites in marine sponges. J Planar Chromatogr 2008;21(6):431–6.
[52] Shahverdi AR, Abdolpour F, Monsef-Esfahani HR, Farsam H. A TLC bioautographic assay for the detection of nitrofurantoin resistance reversal compound. J Chromatogr B 2007;850:528–30.
[53] Njume C, Afolayan AJ, Ndip RN. Preliminary phytochemical screening and in vivo anti-Helicobacter pylori activity of acetone and aqueous extracts of the stem bark of *Sclerocarya birrea* (*Anacardiaceae*). Arch Med Res 2011;42:252–7.
[54] Molnár V, Billes F, Tyihák E, Ott PG. Bioautographic detection of antimicrobial compounds in the edible shiitake mushroom. J Planar Chromatogr 2008;21:423–6.
[55] Okusa NP, Stévigny C, Devleeschouwer M, Duez P. Optimization of the culture medium used for direct TLC-bioautography. Application to the detection of antimicrobial compounds from Cordia gilletii De wild (Boraginaceae). J Planar Chromatogr 2010;23:245–9.
[56] Chamnawang MT, Surassmo S, Wongsariya K, Bunyapraphatsara N. Antibacterial activity of Thai medicinal plant against methicillin-resistant *Staphylococcus aureus*. Fitoterapia 2009;80:102–4.
[57] Galindo-Cuspinera V, Rankin SA. Bioautography and chemical characterization of antimicrobial compound(s) in commercial water-soluble annatto extracts. J Agric Food Chem 2005;53:2524–9.
[58] Rossi D, Guerrini A, Maietti S, Bruni R, Paganetto G, Poli F, et al. Chemical fingerprinting and bioactivity of Amazonian Ecuador *Croton lechleri* Müll. Arg. (Euphorbiaceae) stem bark essential oil: a new functional food ingredient? Food Chem 2011;126:837–48.
[59] Nostro A, Germanò MP, D'Angelo V, Marino A, Cannatelli MA. Extraction methods and bioautography for evaluation of medicinal plant antimicrobial activity. Lett Appl Microbiol 2000;30:379–84.
[60] Verbitski SM, Gourdin GT, Ikenouye LM, McChesney JD. Detection of *Actaea racemosa* adulteration by thin-layer chromatography and combined thin-layer chromatography-bioluminescence. J AOAC Int 2008;91:268–75.
[61] Smith JE, Tucker D, Watson K, Jones GL. Identification of antibacterial constituents from the indigenous Australian medicinal plant *Eremophila duttonii* F. Muell. (Myoporaceae). J Ethnopharmacol 2007;112:386–93.
[62] Khurram M, Khan MA, Hameed A, Abbas N, Qayum A, Inayat H. Antibacterial activities of *Dodonaea viscosa* using contact bioautography technique. Molecules 2009;14(3):1332–41.
[63] Suleiman MM, McGaw LJ, Naidoo V, Eloff JN. Detection of antimicrobial compounds by bioautography of different extracts of leaves of selected South African tree species. Afr J Trad CAM 2010;7(1):64–71.
[64] Shai LJ, McGaw LJ, Eloff JN. Extracts of the leaves and twigs of the threatened tree *Curtisia dentata* (Cornaceae) are more active against *Candida albicans* and other microorganisms than the stem bark extract. S Afr J Bot 2009;75:363–6.
[65] Tabanca N, Demirci B, Baser KHC, Mincsovics E, Khan SI, Jacob MR, et al. Characterization of volatile constituents of *Scaligeria tripartite* and studies on the antifungal activity against phytopathogenic fungi. J Chromatogr B 2007;850:221–9.

[66] Tabanca N, Demirci B, Crockett SL, Baser KHC, Wedge DE. Chemical composition and antifungal activity of *Arnica longifolia, Aster hesperius* and *Chrysothamnus nauseosus* essential oils. J Agric Food Chem 2007;55:8430–5.

[67] Vasques da Silva R, Navickiene HMD, Kato MJ, Bolzani VS, Méda ChI. Young MCM. Antifungal amides from *Piper arboretum* and *Piper tuberculatum*. Phytochemistry 2002; 59:521–7.

[68] Navickiene HMD, Alécio AC, Kato MJ, Cavalheiro AJ, Furlan M. Antifungal amides from *Piper hispidum* and *Piper tuberculatum*. Phytochemistry 2000;55:621–6.

[69] Sridhar SR, Rajagopal RV, Rajavel R, Masilamani S, Narasimhan S. Antifungal activity of some essential oils. J Agric Food Chem 2003;51:7596–9.

[70] Masoko P, Eloff JN. Bioautography indicates the multiplicity of antifungal compounds from twenty-four southern African *Combretum* species (Combretaceae). Afr J Biotechnol 2006;5(18):1625–47.

[71] Mahlo SM, McGaw LJ, Eloff JN. Antifungal activity of leaf extracts from South African trees against plant pathogens. Crop Prot 2010;29:1529–33.

[72] Nagaraja BK, Sripathy R, Ramchand CN, Sundararaman M. Bioautography guided identification of anticandidal compounds from *A. terreus* st. 1. Am J Infect Dis 2011; 7(4):91–7.

[73] Mehrabani M, Abdolhassan K, Mousavi SAA, Rezaifar M, Alikhah H, Nosky A. Evaluation of antifungal activity of *Myrtus communis L.* by bioautography method. Jundishapur Microbiol 2013;6:1–7.

[74] Kostecki K, Engelmeier D, Pacher T, Hofer O, Vajrodaya S, Greger H. Dihydrophenanthrenes and other antifungal stilbenoids from Stemona cf. pierrei. Phytochemistry 2004;65:99–106.

[75] Scher JM, Speakman J-B, Zapp J, Becker H. Bioactivity guided isolation of antifungal compounds from the liverwort *Bazzania trilobata* (L.) S.F. Gray. Phytochemistry 2004; 65:2583–8.

[76] Queiroz EF, Wolfender J-L, Atindehou KK, Traore D, Hostettmann K. On-line identification of the antifungal constituents of Erythrina vogelii by liquid chromatography with tandem mass spectrometry, ultraviolet absorbance detection and nuclear magnetic resonance spectrometry combined with liquid chromatographic microfractionation. J Chromatogr A 2002;974:123–34.

[77] Labbé C, Faini F, Villagrán C, Cool P, Rycroft DS. Antifungal and insect antifeedant 2-phenylethanol esters from the liverwort *Balantiopsis cancellata* from Chile. J Agric Food Chem 2005;53:247–9.

[78] Mackeen MM, Ali AM, Lajis HJN, Kazawu K, Kikuzaki H, Nakatani N. Antifungal garcinia acid esters from the fruits of *Garcinia artovidiris*. Z Naturforsch C 2000;57:291–5.

[79] Islam N, Parveen SA, Nakazawa N, Marston A, Hostettmann K. Bioautography with the fungus *Valsa ceratosperma* in the search for antimycotic agents. Pharm Biol 2003; 41(8):637–40.

[80] Morlock GE, Ristivojevic P, Chernetsova ES. Combined multivariate data analysis of high-performance thin-layer chromatography fingerprints and direct analysis in real time mass spectra for profiling of natural products like propolis. J Chromatogr A 2014; 1328:104–12.

[81] Verbitsky S, McChesney JD, Gerald T, Gourdin GT, Larissa M, Ikenouye LM. Patent US 2007/0184514.

[82] International Organization for Standardization (ISO). ISO 13348; 2007.

[83] Baumgartner V, Schwack W. Enhanced quantitative evaluation of the HPTLC-bioluminescence detection. J Liq Chromatogr Relat Technol 2010;33:980–95.

[84] Verbitski SM, Hickey S, Gourdin GT, Ikenouye LM. Bioluminex™: an effective yet simple tool for screening mixtures. CBS CAMAG Bibliogr Serv 2007;99:11–3.

[85] Weber W, Seitz W, Aichinger A, Albert R. Luminographic detection of toxicity with *Vibrio fischeri* (luminescent bacteria). CBS CAMAG Bibliogr Serv 2005;94:2–4.

[86] Tyihák E, Móricz ÁM, Ott PG. Use of the BioArena system for indirect detection of endogenous ozone in spots after TLC or OPLC separation. J Planar Chromatogr 2008;21:77–82.

[87] Tyihák E, Móricz ÁM, Ott PG, Kátay G, Király-Véghely Z. The potential of BioArena in the study of the formaldehydome. J Planar Chromatogr 2005;18:67–72.

[88] Tyihák E, Mincsovics E, Kátay G, Király-Véghely Z, Móricz ÁM, Ott PG. BioArena: an unlimited possibility of biochemical interaction in the adsorbent layer after chromatographic separation. J Planar Chromatogr 2008;21:15–20.

[89] Tyihák E, Botz L, Ott PG, Nagy S, Kocsis B, Király-Véghely Z, et al. The combination of the overpressured layer chromatography and bioautography and its application to the analysis of molecules influencing cell proliferation. Chem Anal 2003;48:543–53.

[90] Tyihák E, Móricz ÁM, Ott PG, Hajnos ML, Głowniak K. New approach to mechanism of action of paclitaxel by means of BioArena studies. J Planar Chromatogr 2008;21: 331–6.

[91] Tyihák E, Móricz ÁM, Ott PG, Kátay G, Király-Véghely Z. Antibiosis, antibiotics, and the formaldehyde cycle: the unique importance of planar chromatographic techniques to progress in these fields. J Planar Chromatogr 2004;17:84–8.

[92] Móricz AM, Otta KH, Ott P, Tyihák E. Separation and detection of aflatoxins using overpressured-layer chromatography and bioautography. J Planar Chromatogr 2003; 16:417–20.

[93] Hostettmann K, Hamburger M, Hostettmann M, Marston A. In: Fischer NH, et al., editors. New developments in the separation of natural products in modern phytochemical methods. New York: Plenum Press; 1991.

[94] Hostettmann K, Marston A. Twenty years of research into medicinal plants: results and perspectives. Phytochem Rev 2002;1:275–85.

[95] Homans AL, Fusch A. Direct bioautography on thin-layer chromatograms as a method for detecting fungitoxic substances. J Chromatogr 1970;51:327–9.

[96] Hostettmann K, Marston A. Biological and chemical evaluation of plant extracts and subsequent isolation strategy. In: Bohlin L, Bruhn JG, editors. Bioassay methods in natural product research and drug development. Dordrecht: Kluwer Academic Publishers; 1999.

[97] Hostettmann K, Terreaux Ch, Marston A, Potterat O. The role of planar chromatography in the rapid screening and isolation of bioactive compounds from medicinal plants. J Planar Chromatogr 1997;10:251–7.

[98] Hostettmann K, Marston A. Bioactive constituents of plants used in African traditional medicine. In: Atta-ur-Rahman, editor. Studies in natural products chemistry. Amsterdam: Elsevier; 1990.

[99] Zhao W, Wolfender J-L, Hostettmann K, Xu R, Qin G. Antifungal alkaloids and limonoids derivatives from *Dictamnus dasycarpus*. Phytochemistry 1998;47:7–11.

[100] Hostettmann K, Marston A. Search for new antifungal compounds from higher plants. Pure Appl Chem 1994;66:2231–4.

[101] Tan RX, Wolfender J-L, Ma WG, Zhang LX, Hostettmann K. Secoiridoids and antifungal aromatic acids from *Gentiana algida*. Phytochemistry 1996;41:111–6.

[102] Tan RX, Wolfender J-L, Zhang LX, Ma WG, Fuzzati N, Marston A, et al. Acyl secoiridoids and antifungal constituents from *Gentiana macrophylla*. Phytochemistry 1996; 42:1305–13.

[103] Garcia Cortez DA, Young MCM, Marston A, Wolfender J-L, Hostettmann K. Xanthones, triterpenes and a biphenyl from *Kielmeyera coriacea*. Phytochemistry 1998;47: 1367–74.

[104] Gafner S, Wolfender J-L, Nianga M, Stoeckli-Evans H, Hostettmann K. Antifungal and antibacterial naphthoquinones from *Newbouldia laevis* roots. Phytochemistry 1996;47: 1315–20.

[105] Hamburger M, Cordell GA. A direct bioautographic TLC assay for compounds possessing antibacterial activity. J Nat Prod 1987;50:19–22.

[106] Queiroz EF, Ioset J-R, Ndjoko K, Guntern A, Foggin CM, Hostettmann K. On-line identification of the bioactive compounds from *Blumea gariepina* by HPLC-UV-MS and HPLC-UV-NMR, combined with HPLC-micro-fractionation. Phytochem Anal 2005; 16:166–74.

[107] Favre-Godal Q, Queiroz EF, Wolfender J-L. Latest developments in assessing antifungal activity using TLC-bioautography: a review. J AOAC Int 2013;96:1175–88.

[108] Meazza G, Dayan FE, Wedge DE. Activity o quinones on Colletotrichum species. J Agric Food Chem 2003;51:3824–8.

[109] Fokialakis N, Cantrell ChL, Duke SO, Skaltsounis AL, Wedge DE. Antifungal activity of thiophenes from *Echinops ritro*. J Agric Food Chem 2006;54:1651–5.

[110] Oliva A, Meepagala KM, Wedge DE, Harries D, Hale AL, Aliotta G, et al. Natural fungicides from *Ruta graveolens* L. leaves, including a new quinolone alkaloids. J Agric Food Chem 2003;51:890–6.

[111] Cantrell ChL, Case BP, Mena EE, Kniffin TM, Duke SO, Wedge DE. Isolation and identification of antifungal fatty acids from the *Basidiomycete Gomphus floccosus*. J Agric Food Chem 2008;56:5062–8.

[112] Queiroz SCN, Cantrell ChL, Duke SO, Wedge DE, Nandula VK, Moraes RM, et al. Bioassay-directed isolation and identification of phytotoxic and fungitoxic acetylenes from *Conyza canadensis*. J Agric Food Chem 2012;60:5893–8.

[113] Sakhanokho HF, Sampson BJ, Tabanca N, Wedge DE, Demirci B, Baser KHC, et al. Chemical composition, antifungal and insecticidal activities of *Hedychium* essential oils. Molecules 2013;18:4308–27.

[114] Wedge DE, Klun JA, Tabanca N, Demirci B, Ozek T, Baser KHC, et al. Bioactivity-guided fractionation and GC/MS fingerprinting of *Angelica sinensis* and *Angelica archangelica* root components for antifungal and mosquito deterrent activity. J Agric Food Chem 2009;57:464–70.

[115] Demirci B, Baser KHC, Tabanca N, Wedge DE. Characterization of volatile constituents of *Haplopappus greenei* and studies on the antifungal activity against phytopathogens. J Agric Food Chem 2006;54:3146–50.

[116] Altintas A, Tabanca N, Tyihák E, Ott PG, Móricz AM, Mincsovics E, et al. Characterization of volatile constituents from *Origanum onites* and their antifungal and antibacterial activity. J AOAC Int 2013;96:1200–8.

[117] Wedge DE, Nagle DG. A new 2D-TLC bioautography method for the discovery of novel antifungal agents to control plant pathogens. J Nat Prod 2000;63:1050–4.

[118] Marston A, Kissling J, Hostettmann K. A rapid TLC bioautographic method for the detection of acetylcholinesterase and butyrylcholinesterase inhibitors in plants. Phytochem Anal 2002;13:51–4.

[119] Kissling J, Ioset JR, Marston A, Hostettmann K. Bio-guided isolation of cholinesterase inhibitors from the bulbs of *Crinum* x *powellii*. Phytother Res 2005;19:984–7.

[120] Rhee IK, van de Meent M, Ingkaninan K, Verpoorte R. Screening for acetylcholinesterase inhibitors from Amaryllidaceae using silica gel thin-layer chromatography in combination with bioactivity staining. J Chromatogr A 2001;915:217–23.

[121] Mroczek T. Highly efficient, selective and sensitive molecular screening of acethylocholinesterase inhibitors of natural origin by solid-phase extraction-liquid chromatography/electrospray ionization-octopole-orthogonal acceleration time-of-flight-mass spectrometry and novel thin-layer chromatography-based bioautography. J Chromatogr A 2009;1216:2519–28.

[122] Mroczek T, Mazurek J. Pressurized liquid extraction and anticholinesterase activity-based thin-layer chromatography with bioautography of Amaryllidaceae alkaloids. Anal Chim Acta 2009;633:188–96.

[123] Adhami HR, Farsam H, Krenn L. Screening of medicinal plants from Iranian traditional medicine for acetylcholinesterase inhibition. Phytother Res 2011;25:1148–52.

[124] Yang Z, Zhang X, Duan D, Song Z, Yang M, Li S. Modified TLC bioauthographic method for screening acethylcholinesterase inhibitors from plant extracts. J Sep Sci 2009;32:3257–9.

[125] Adhami HR, Scherer U, Kaehling H, Hettich T, Schlitterbeck G, Reich E, et al. Combination of bioautography with HPTLC-MS/NMR: a fast identification of acethylocholinesterase inhibitors from Galbanum. Phytochem Anal 2013;24:395–400.

[126] Zheng X, Zhang L, Cheng X, Zhang Z, Wang Ch, Wang Z. Identification of acetylcholinesterase inhibitors from seeds of plants of genus Peganum by thin-layer chromatography-bioautography. J Planar Chromatogr 2011;24:470–4.

[127] Agnaniet H, Ankli A. Bioautographic HPTLC assays for screening of Gabonese medicinal plants used against Diabetes mellitus. CBS CAMAG Bibliogr Serv 2013; 110:5–7.

[128] Mendoza CE, Wales PJ, McLeod HA, McKinley WP. Enzymatic detection of ten organophosphorus pesticides and carbaryl on thin-layer chromatograms: an evaluation of indoxyl, substituted indoxyl and 1-naphthyl acetates as substrates of esterases. Analyst 1968;93:34–8.

[129] Weins Ch, Jork H. Toxicological evaluation of harmful substances by in situ enzymatic and biological detection in high-performance thin-layer chromatography. J Chromatogr A 1996;750:403–7.

[130] Akkad R, Schwack W. Effect on bromine oxidation on high-performance thin-layer chromatography multi-enzyme inhibition assay detection of organophosphates and carbamate insecticides. J Chromatogr A 2011;1218:2775–84.

[131] Akkad R, Schwack W. Multi-enzyme inhibition assay for detection of insecticidal organophosphates and carbamates by high-performance thin-layer chromatography. 1. Basics of method development. J Planar Chromatogr 2008;21:411–5.

[132] Witkiewicz Z, Mazurek M, Szulc J. Chromatographic analysis of chemical warfare agents. J Chromatogr A 1990;503:293–357.

[133] Rhee IK, van Rijn RM, Verpoorte R. Qualitative determination of false-positive effects in the acetylcholinesterase assay using thin layer chromatography. Phytochem Anal 2003;14:127–31.

[134] Salazar MO, Furlan RL. A rapid TLC autographic method for the detection of glucosidase inhibitors. Phytochem Anal 2007;18:209–12.

[135] Simões-Pires CA, Hmicha B, Marston A, Hostettmann K. A TLC bioautographic method for the detection of alpha- and beta-glucosidase inhibitors in plant extracts. Phytochem Anal 2009;20:511–5.

[136] Ramallo IA, Zacchino SA, Furlan RLE. A rapid TLC autographic method for the detection of xanthine oxidase inhibitors and superoxide scavengers. Phytochem Anal 2006; 17:15–9.

[137] Schmitt S, Reifferscheid G, Claus E, Schlüsener M, Buchinger S. Effect directed analysis and mixture effects of estrogenic compounds in a sediment of the river Elbe. Environ Sci Pollut Res 2012;19:3350–61.

[138] Grung M, Lichtenthaler R, Ahel M, Tollefsen KE, Langford K, Thomas KV. Effects-directed analysis of organic toxicants in wastewaters effluent from Zagreb, Croatia. Chemosphere 2007;67:108–20.

[139] Schönborn A, Grimmer GA. Coupling sample preparation with effect-directed analysis of estrogenic activity – proposal for a new rapid screening concept for water samples. J Planar Chromatogr 2013;26:402–8.

[140] Buchinger S, Spira D, Bröder K, Schlüsener M, Ternes T, Reifferscheid G. Direct coupling of thin-layer chromatography with a bioassay for the detection of estrogenic compounds: applications for effect-directed analysis. Anal Chem 2013;85:7248–56.

[141] Spira D, Reifferscheid G, Buchinger S. Combination on high-performance thin-layer chromatography with a specific bioassay – a tool for effect-directed analysis. J Planar Chromatogr 2013;26(5):395–401.

[142] Routledge EJ, Sumpter JP. Estrogenic activity of surfactants and some of their degradation products assessed using a recombinant yeast screen. Environ Toxicol Chem 1996;15:241–8.

[143] Müller MB, Dausend C, Weins C, Frimmel FH. A new bioautographic screening method for the detection of estrogenic compounds. Chromatographia 2004;60:207–11.

[144] Poole CF. Thin-layer chromatography: challenges and opportunities. J Chromatogr A 2003;1000:963–84.

[145] Noipa T, Srijaranai S, Tuntulani T, Ngeontae W. New approach for evaluation of the antioxidant capacity based on scavenging DPPH free radical in micelle systems. Food Res Int 2011;44:798–806.

[146] Glavind J, Holmer G. Thin-layer chromatographic determination of antioxidants by the stable free radical α,α′-diphenyl-β-picrylhydrazyl. J Am Oil Chem Soc 1967;44:539–42.

[147] Cieśla Ł, Kryszeń J, Stochmal A, Oleszek W, Waksmundzka-Hajnos M. Approach to develop a standardized TLC-DPPH• test for assessing free radical scavenging properties of selected phenolic compounds. J Pharm Biomed 2012;70:126–35.

[148] Soler-Rivas C, Espin JC, Wichers HJ. An easy and fast test to compare total free radical scavenger capacity of foodstuffs. Phytochem Anal 2000;11:330–8.

[149] Cieśla Ł. Thin-layer chromatography with biodetection in the search for new potential drugs to treat neurodegenerative diseases – state of the art and future perspectives. Med Chem 2012;8:102–11.

[150] Gu L, Wu T, Wang Z. TLC bioautography-guided isolation of antioxidants from fruit of *Perilla frutescens* var. *acuta*. Food Sci Technol-LWT 2009;42:131–6.

[151] Simonovska B, Vovk I, Andrenšek S, Valentová K, Ulrichová J. Investigation of phenolic acids in yacon (*Smallanthus sonchifolius*) leaves and tubers. J Chromatogr A 2009;1016:89–98.

[152] Xiao G, Li G, Chen L, Zhang Z, Yin J-J, Wu T, et al. Isolation of antioxidants from *Psoralea corylifolia* fruits using high-speed counter-current chromatography guided by thin layer chromatography-antioxidant autographic assay. J Chromatogr A 2010;1217:5470–6.

[153] Pozharitskaya ON, Ivanova SA, Shikov AN, Makarov VG. Separation and free radical-scavenging activity of major curcuminoids of *Curcuma longa* using HPTLC-DPPH method. Phytochem Anal 2008;19:236–43.

[154] Zhang X-J, Liu L-J, Song T-T, Wang Y-Q, Yang X. An approach based on antioxidant-efficacy relationship and TLC bioautography assay to quality evaluation of *Rubia cordifolia* from various sources. J Nat Med 2014;68(2):448–54.

[155] Zampini IC, Ordonez RM, Isla MI. Autographic assay for the rapid detection of antioxidant capacity of liquid and semi-solid pharmaceutical formulations using ABTS•+ immobilized by gel entrapment. AAPS Pharm Sci Technol 2010;11:1159–63.

[156] Sanchez-Medina A, Garcia-Sosa K, May-Pat F, Pena-Rodriguez LM. Evaluation of biological activity of crude extracts from plants used in Yucatecan traditional medicine – part I. Antioxidant, antimicrobial and beta-glucosidase inhibition activities. Phytomedicine 2001;8:144–51.

[157] Pratt DE, Miller EE. A flavonoid antioxidant in Spanish peanuts (*Arachia hypogoea*). J Am Oil Chem Soc 1984;61:1064–7.

[158] Whittern CC, Miller EE, Pratt DE. Cottonseed flavonoids as lipid antioxidants. J Am Oil Chem Soc 1984;61:1075–8.

[159] Bioeffective-linked analysis in modern HPTLC. In: Spangenberg B, Poole CF, Ch Weins, editors. Quantitative thin-layer chromatography. A practical survey. Berlin Heidelberg: Springer-Verlag; 2010.

[160] Cieśla Ł, Kowalska I. Detection and identification in TLC drug analysis. In: Komsta Ł, Waksmundzka-Hajnos M, Sherma J, editors. Thin-layer chromatography in drug analysis. Boca Raton: Taylor & Francis Group; 2014.

[161] Thin-layer chromatography hyphenated with bioassays. In: Choma IM, editor. J AOAC Int. Special Section, vol. 96; 2013. pp. 1165–232.

CHAPTER

12

Solvent Selection and Method Development

Colin F. Poole

Department of Chemistry, Wayne State University, Detroit, Michigan, USA

12.1 INTRODUCTION

A broad view of method development would include a discussion of all steps at which a selection among parameters and their optimization was a necessary part of the process. Compared to other chromatographic techniques, thin-layer chromatography has more of these decision steps and relies on the knowledge and experience of the analyst to make the correct choice or recognize a misdirection. Method development is the intellectual part of designing a separation process but is not easily taught. Guidelines based on experience can be summarized but there are few reliable mathematical relationships to describe the segments of the separation process that are suitable for predicting results without the need for experimental verification. It is the latter that assists and simplifies method development in column chromatography. The advantage that thin-layer chromatography has is that experiments can be performed quickly and in parallel, resulting in a large amount of information that if correctly processed allows rapid progress. My purpose in writing this contribution is to summarize a practitioner's guide to method development focused on the separation process without replacing the need to consult other chapters in this book for more detailed information. For an alternative and sometimes contrary view of the method development process, the following sources are recommended [1–9]. The general steps for developing an effective separation are outlined in Figure 12.1. Before experimental work commences, it is important to establish the goals of the separation. With this information, the needs of the method should now be clear and a strategy can be devised to go about meeting the goals. At this stage, a separation mode is selected and preliminary experiments

Instrumental Thin-Layer Chromatography
http://dx.doi.org/10.1016/B978-0-12-417223-4.00012-1

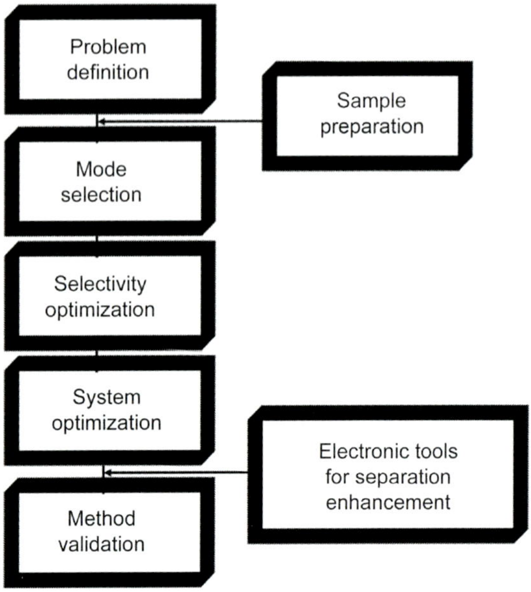

FIGURE 12.1 Outline of the sequence of operations in a general strategy for method development.

performed to establish whether the conditions are compatible with the goals. If the approach looks promising, the method is further optimized until the goals are reached. Finally, the method is validated to confirm that it is fit for purpose. It is convenient to treat these individual steps separately as part of a sequence; they thus become building blocks of the method. Validation is treated in a separate chapter.

12.2 PROBLEM DEFINITION

The amount of information available for a sample can vary with its origin. It is rare, however, for there to be no relevant sample information known by someone associated with the sample. Some poignant questions to ask at the outset of method development are summarized in Table 12.1. The most generally useful sample information is its composition. If the number of relevant detectable compounds is known then the complexity of the analysis can be established. The only way to know when method development is complete is to know how many detectable components should be visible in the sample chromatogram. Based on the general theory of statistical overlap the number of singlet components in a chromatogram is relatively a small fraction of the zone capacity [3]. For thin-layer chromatography, it will be difficult to obtain baseline

TABLE 12.1 Key Questions to Establish Goals at the Start of Method Development

How many detectable compounds are present in the sample?
Are all compounds equally relevant?
Are standards available for the relevant compounds?
What is the concentration range of relevant compounds?
What is the composition of the sample matrix?
How many samples are likely to require analysis by the method?
Will unknown compounds require identification?
Is the method to be primarily used for analytical of preparative separations?

separation of more than about 8–10 components in a single development with capillary-controlled flow [10]. For forced-flow and automated multiple development, the number of likely baseline separated single components rises to about 20. A larger zone capacity is possible for two-dimensional separations, around 100 compounds might be separated, but favorable conditions require that the separation mechanisms in the orthogonal directions are complementary, and this is rarely the case. It is more difficult to record two-dimensional chromatograms and to obtain quantitative information, and consequently two-dimensional separations are not in common use. The number of relevant sample components that require separation provides a useful indication of whether to continue with developing a thin-layer chromatographic method or to try another technique with a higher peak capacity. It also provides insight into the likely effort required for success and indicates the type of equipment required for the development step.

If not all sample components are relevant to the goals of the separation then a complete separation is no longer required. The problem is simplified because those components not of interest need only be separated from those of interest and not from each other. Those compounds not of interest will occupy a portion of the separation space, which is usually less than the separation space required for their baseline separation. The success in minimizing the separation space populated by compounds that need not be separated may be important in achieving method goals. Such considerations are not unusual for real samples, which are likely to be composed of target compounds of interest and other compounds considered to be part of the sample matrix.

Method development is facilitated if standards are available for the relevant compounds. Standard substances simplify peak tracking during optimization and are generally required for calibration in quantitative

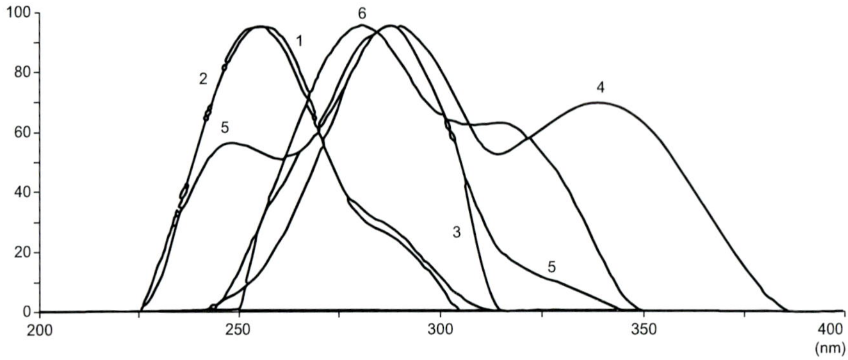

FIGURE 12.2 In situ UV−visible spectra of flavor compounds characteristic of cinnamon extracts. Identification: 1 = cinnamyl alcohol; 2 = cinnamyl acetate; 3 = cinnamaldehyde; 4 = 2-methoxycinnamaldehyde; 5 = eugenol; and 6 = coumarin.

methods. Standard substances also facilitate establishing detection properties and the construction of spectral libraries for identification. For routine methods UV−visible and fluorescence spectroscopy are widely used. Figure 12.2 provides an example of the UV−visible spectra for flavor compounds found in cinnamon extracts [11]. Cinnamyl alcohol and cinnamyl acetate have almost identical UV−visible spectra and so it is important that these two compounds are separated from each other in the chromatographic step. On the other hand, for 2-methoxycinnamalaldehyde there is a region of the spectra free from overlap and this compound could be determined without separation from other compounds. Since fewer compounds are naturally fluorescent, fluorescence detection can offer higher selectivity than absorption measurements for the selective detection of target compounds in mixtures. Higher specificity is offered by mass detection but cost considerations may limit access to this option (thin-layer chromatography-mass spectrometry (TLC-MS)) for some laboratories.

The concentration range of relevant compounds influences the range of detection options, could infer the need for a concentration or derivatization step, and may require greater resolution between major and minor components for adequate separation. Sample preparation steps reduce sample throughput so are used as a necessity based on the expected concentration of relevant compounds and their established detection characteristics or the likelihood of the matrix interfering in the analysis of the compounds of interest. Fortunately, a large number of derivatization reactions have been developed for thin-layer chromatography to enhance selectivity and detectability [4,5,7,12]. These reactions can be performed prior to or after the separation step. The latter takes advantage of the accessibility of the immobilized sample on the layer after separation. Some considerations for the selection of derivatizing techniques are

TABLE 12.2 Considerations for Optional Strategies for Methods Employing Derivatization

(1) PRECHROMATOGRAPHIC METHODS
Modifies sample properties (e.g., enhance stability, improve extraction efficiency)
Improved chromatographic resolution if derivatives are easier to separate or move relevant compounds away from matrix interferences
Simplify optimization of reaction conditions free of time, solvent, and reagent concentration constraints
Excess reagent and byproducts, catalysts, etc., may interfere with the detection of relevant compounds
Sample component differences are minimized and the difficulty of separation is increased
(2) POSTCHROMATOGRAPHIC METHODS
Byproducts of the reaction do not interfere in the separation
All samples and standards are derivatized simultaneously
The separation is unaffected by the reaction
Reagent and derivatives must have different detection characteristics
Reagent must be evenly distributed over the layer

summarized in Table 12.2. Prechromatographic derivatization almost always requires that the target compounds contain a reactive functional group. Postchromatographic derivatization employs a wider range of class specific as well as functional groups specific reactions.

Samples are applied to the layer in solution using a micropipette or spray-on devices. The sample must be soluble in a volatile solvent of low solvent strength to avoid prechromatographic sample fractionation that can distort separated zones. The sample solution must be of low viscosity to facilitate filling, emptying, and cleaning of the sample application device. The sample solution must be homogeneous and free of particle matter to facilitate quantitative sample transfer to the layer. Inhomogeneous sample solutions can be handled using layers with separate concentrating zones of low retention power [13]. These layers often provide a solution to the problem of analyzing biological fluids containing inorganic salts and proteins of low solubility in organic solvents. Quantification may be difficult, however, unless the target compounds are evenly distributed throughout the sample zone applied to the concentrating region. Aqueous solutions can be difficult to handle because of high solvent strength on inorganic oxide layers and low volatility limiting solvent evaporation. Addition of ethanol, or a similar solvent that forms a lower boiling azeotrope with water, can facilitate direct application to the layer as well as aliquot application with interspersed drying steps. Samples containing solvents like ethylene glycol, are difficult in all regards, and lack an obvious solution. They should be avoided or subject to solvent exchange before sample application. Liquid–liquid extraction on a small scale or solid-phase extraction can be used for solvent exchange.

Finally, the number of samples to be analyzed is an important consideration. For a one sample problem, it is reasonable to take shortcuts to reach a satisfactory result quickly. For methods to be used over an extended time statistically based robustness and validation procedures can be justified to identify weak points in the method for correction if possible. If identification of unknown components is required then TLC-MS is the preferred option. UV–visible spectral libraries are useful for confirming identification but rarely adequate on their own for an initial identification.

12.2.1 Sample Properties

If the relevant compounds are known substances then useful information about how they will behave in a chromatographic system can be obtained or inferred. From their structure, chromatographic and detection properties can be estimated by analogy to similar compounds. Since many derivatizing reagents are functional group selective, the selection of suitable derivatizing reagents, if to be employed, will be facilitated if the structures for the relevant compounds are known. For known compounds, solubility information in common solvents may be available, and in other cases, can be inferred from the characteristic properties of similar compounds. Reasonable solubility in a volatile solvent is required for sample application. The pK_a or conductivity of a compound indicates whether it is ionic or easily ionizable. The dissociation of weak acids and bases might be controlled by predicting suitable buffers for inclusion in the mobile phase. For strong acids and bases, there are few ion-exchange layers for thin-layer chromatography and it may be preferable to derivatize easily ionized functional groups. Alternatively, ion-pair separations might be explored as a solution. From structural information, the spectral properties of relevant compounds can be estimated through software or inferred by analogy to similar compounds. This is useful for the selection of detection conditions. Most software approaches predict the properties of solution spectra, which may be different to solid-phase spectra [7,14]. This is particularly true for fluorescence spectra, for which the solution and solid-phase spectra may have little in common, and could be misleading if detection conditions are selected based on a solution spectra.

With an indication of the structures of relevant compounds no attempt at method development should begin without using literature resources to establish whether the relevant compounds, or similar compounds, have been separated before. As a minimum this information may narrow the scope of the method development process or, if fortunate, suggest a suitable system, or a system that would be acceptable with some minor improvements. The scientific literature is easily interrogated electronically today and the experience of others is a valuable resource for method development. Handbooks can be useful sources of physicochemical properties for target compounds or similar compounds. In the practical world, the availability of instrumentation and materials may need to be considered in concluding the initial phase of method selection. At this point, the question of what needs to be done and how success will be established should be unambiguous. The identification of the best chromatographic system from all available systems for the relevant compounds may have been narrowed down but without experimental verification, it is only a starting point to the next phase of method development.

12.3 MODE SELECTION

The various options for selection of a separation mode in thin-layer chromatography are summarized in Figure 12.3 [8]. Normal-phase separations using silica gel, impregnated silica gel, or chemically bonded silica gel layers dominate the practice of thin-layer chromatography, and in the absence of conflicting information, would be the start point for initial sample screening [15]. For compounds of low-molecular-mass, soluble in organic solvents, silica gel is generally the first choice. Other inorganic oxides (alumina, kieselguhr, Florisil) are rarely used and reserved for specific problems. Alumina can be useful when separations on silica gel produce irregular shaped zones and chemically bonded layers lack the required selectivity.

Silica gel precoated layers are prepared with sorbents in two pore size ranges: 60 nm for general purpose applications and 5000 nm for compounds that bind strongly to silica gel or for high-molecular-mass compounds. Silica gel 60 layers are optimized for the separation of molecules with a molecular mass less than 1000 due to size exclusion effects, which limit access to the internal pore volume, and have optimum performance for much smaller molecules as indicated by the contrasting separations of poly(ethylene glycol) oligomers in Figure 12.4 [16]. The options for the separation of polymers and compounds with a molecular mass above 1000 are limited largely to precipitation chromatography for compounds soluble in organic solvents [17] and cellulose or silica gel 5000 layers for water-soluble compounds [18]. Separations by precipitation chromatography involve adsorption by the stationary phase at a critical condition for solubility in the mobile phase. Solvent selection is very important for this

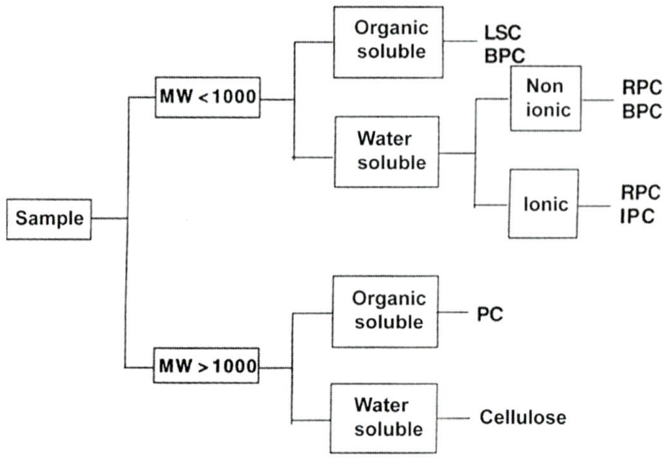

FIGURE 12.3 Separation mode selection guide for thin-layer chromatography. LSC = liquid–solid chromatography on inorganic oxide adsorbents (normal-phase chromatography); BPC = liquid–solid chromatography on a polar chemically bonded layer (normal-phase chromatography); RPC = reversed-phase chromatography on a chemically bonded layer with an aqueous mobile phase containing organic solvent and possibly further additives to control selectivity; MW = molecular weight; IPC = ion-pair chromatography; and PC = partition chromatography.

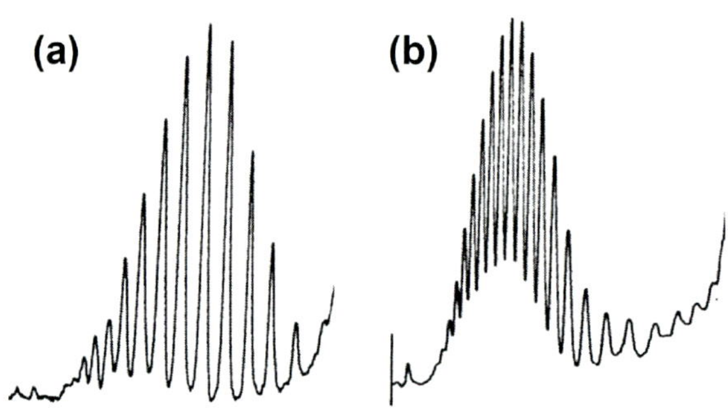

FIGURE 12.4 Separation of poly(ethylene glycol) oligomers as their 3,5-dinitrobenzyl esters on silica gel 60 with an average molecular mass of (a) 400 and (b) 1000. *Reproduced with permission from Ref. [16].*

technique. The general retention mechanism on cellulose layers is thought to be partition, and on silica gel 5000 partition with a contribution from adsorption. There are few known applications for synthetic polymers on these layers. More generally, cellulose is used for the separation of polar compounds with water-based mobile phases and is not generally useful for the separation of low-polarity compounds due to low retention.

Compounds of low polarity are difficult to separate on silica gel because of weak retention (mobile phase selection is limited because most solvents are too strong for these separations). In addition, highly polar compounds and compounds with ionic functional groups are difficult to separate on silica gel because of strong retention (mobile phase selectivity is limited because most solvents are too weak for these separations). Chemically bonded phases with nonaqueous mobile phases (normal-phase chromatography) or aqueous mobile phases (reversed-phase chromatography) are preferable in these cases and are the main alternative when silica gel has inadequate selectivity for a separation. The selectivity of silica gel can also be adjusted by impregnating the layer with substances that provide additional interactions. Some common examples are silver ions for the separation of compounds that differ in the number, configuration, or position of double bonds through formation of weak reversible charge transfer complexes with silver [19]; boric acid for the separation of compounds with vicinal hydrogen bonding functional groups [20]; caffeine for the separation of polycyclic aromatic compounds by ring number [21]; and ethylenediaminetetraacetic acid to mask undesirable sample interactions with metal impurities in the layer. The retention properties of silica gel can also be modified through the gas phase by preconditioning the layer with ammonia or acid vapors to improve zone shape and adjust selectivity [1].

There are two types of alkylsiloxane-bonded silica layers indicated as water-wettable for use in reversed-phase chromatography and layers with a higher bonding density and smaller particle size used with nonaqueous solvents for normal-phase chromatography. The latter are restricted mainly to the separation of compounds of low polarity and exhibit weak retention of polar compounds. The reversed-phase layers have a wider range of applications for compounds of different types with retention controlled by the amount and type of organic modifier in the mobile phase. Other suitable mobile phase additives include buffers to suppress ionization of weak acids and bases; ion-pair reagents for the separation of strong acids and bases; surfactants for separations by micellar liquid chromatography; and various complexing agents, such as cyclodextrins, which adjust selectivity by host–guest complex formation. These methods are not routine but for the right sample type may result in exceptional separations of intransigent problems.

The polar chemically bonded phases are generally considered more versatile for applications in thin-layer chromatography than is the practice in column chromatography. They are used in both the normal- and reversed-phase modes. 3-Cyanopropylsiloxane-bonded silica layers exhibit properties similar to a low capacity silica gel with strong dipole-type interactions. In the reversed-phase mode retention is weak and similar to short-chain alkylsiloxane-bonded silica layers. Dipole-type interactions are not important for retention with aqueous mobile phases due to selective solvation of the cyano group by the mobile phase [22].

3-Aminopropylsiloxane-bonded silica layers are mainly used in normal-phase chromatography as retention is weak in the reversed-phase mode. Normal-phase retention is dominated by hydrogen-bonding interactions but the selectivity is different to silica gel. 3-Aminopropylsiloxane-bonded silica layers function as weak anion exchangers in acidic mobile phases [23]. Ion-exchange is not a widely employed separation mechanism in thin-layer chromatography and the 3-aminopropylsiloxane-bonded layers are the most important high performance layers for anion separations. Spacer bonded propanediol silica layers are used predominantly in the normal-phase mode. Retention is weak in the reversed-phase mode but more useful than for 3-aminopropylsiloxane-bonded silica layers. In the normal-phase mode, it is more hydrogen-bond acidic and less hydrogen-bond basic than the 3-aminopropylsiloxane-bonded silica layers. The choice of which polar chemically bonded layer to pursue initially is based on an assessment of complementary interactions with the polar functional groups of the target compounds for normal-phase separations. The 3-cyanopropylsiloxane-bonded silica layers are the usual initial choice for reversed-phase chromatography if alkylsiloxane-bonded silica layers provide inadequate selectivity.

The separation of enantiomers employs a more specific mechanism than other separation techniques and is achieved by including a chiral selector in either the mobile or stationary phase. Compared with column chromatography there are comparatively few chiral sorbents for thin-layer chromatography. The most common are cellulose derivatives and reversed-phase layers impregnated with a copper–amino acid complex. In general, indirect methods employing derivatization to form diastereomers, which can be separated on achiral phases, is the more frequent approach [24,25]. Cellulose and its derivatives (triacetate, tribenzoate, triphenylcarbamate) with aqueous-organic mobile phases have been used to separate enantiomers of largely aromatic compounds with hydroxyl or ketone groups close to the stereogenic center. The ability of the enantiomers to fit into the chiral cavity of the crystalline polymer is often an important factor together with hydrogen-bonding and dipole-type interactions between the polar functional groups of the enantiomer and the carbonyl group of the cellulose ester.

Separations based on ligand exchange chromatography employ precoated reversed-phase layers impregnated with a copper (II) complex of (2S,4R, 2′RS)-4-hydroxyl-1-(2-hydroxydodecyl)proline [24]. A prime requirement for the separation of enantiomers is the capability of forming complexes with copper. Suitable compounds include amino acids and their derivatives, peptides, α-hydroxycarboxylic acids, β-blockers (drugs), and some lactones.

In summary, the main target applications for thin-layer chromatography are the separation of low-molecular-mass compounds (<750 molecular weight) by normal- or reversed-phase chromatography employing silica gel or chemically bonded silica gel layers. Thin-layer chromatography is not used extensively for the separation of ions by ion-exchange chromatography or polymers by size-exclusion chromatography. There are a few specialized approaches for the separation of ions or polymers but these are quite time consuming to develop and rather limited in scope. Consequently, in the next section the general discussion will be limited to normal- and reversed-phase chromatography covering the bulk of current applications.

12.4 MOBILE PHASE SELECTION

The selection of an optimum mobile phase in thin-layer chromatography would be an overwhelming task in the absence of a suitable classification scheme for solvents. Compared with column liquid chromatography the number of solvents employed in method development is larger. Volatile UV-absorbing solvents like acetone are widely used in thin-layer chromatography because detection occurs in the stationary phase after evaporation of the mobile phase. Solvents of low volatility are generally unsuitable and usually avoided. High purity solvents should be employed and attention paid to solvents that may contain stabilizers, since these may have different separation properties to the named solvent or interfere in the quantification of separated compounds [14].

The purpose of the mobile phase is to provide transport in a chromatographic system and to modify selectivity through competition with the stationary phase for intermolecular interactions. Solvents are usually classified for chromatography according to their strength and selectivity. Strength is a single parameter used to rank a solvent's capability to cause migration in a chromatographic system. Since it depends on the properties of the stationary phase, it is really a system property rather than a solvent property. The same solvent can have different strengths in different chromatographic systems. Water, for example, is considered a strong solvent on silica gel (causes most compounds to move with the mobile phase) and a weak solvent in reversed-phase chromatography (few compounds move with the mobile phase). It cannot be defined without reference to a stationary phase. Solvent selectivity is viewed as the factor that distinguishes individual solvents that have suitable solvent strength for a separation by virtue of their capability for different intermolecular interactions. Solvent selectivity is then a characteristic property of a solvent but cannot be defined by a single parameter scale since several and separate different types of interactions are involved in solute–solvent interactions. Solvents can have similar solvent strength but different selectivity resulting in significant differences in band spacing and possibly migration order. It is this property that allows variation of solvent type and composition to be used as a convenient and powerful tool for optimization following mode selection.

General strategies for solvent classification used in chromatography include the solubility parameter model, solvatochromic parameters, Snyder's solvent selectivity triangle, and the solvation parameter model [2,3,7,8,26]. The original solubility parameter concept was developed as an extension of regular solution theory in which the principal intermolecular interactions were dominated by dispersion forces. This approach was extended to polar solvents by decomposing the total solubility parameter into a series of polar partial solubility parameters treated as additive quantities. The larger the value for the polar partial solubility parameter the greater potential of the solvent to participate in that interaction with maximum solvency achieved for solutes and solvents with matched (similar values for the dispersion, induction and orientation parameters, and complementary values for the hydrogen-bonding) partial solubility parameters. The polar partial solubility parameters are only approximate, however, and no general agreement has emerged for how they should be calculated (reported values for the same polar partial solubility parameters differ significantly). The only contemporary use of the solubility parameter model in thin-layer chromatography is for the prediction of mobile phase compositions in precipitation chromatography.

The solvatochromic parameters are derived from UV–visible absorption measurements for probe compounds selected to determine, usually, only a single intermolecular interaction. They are generally average values and assumed to be independent of probe identity. The most comprehensive solvatochromic treatment of solvent selectivity is the π^* (solvent dipolarity/polarizability), α (solvent hydrogen-bond acidity), and β (solvent hydrogen-bond basicity) scales of Kamlet and Taft [3,8,27,28]. A significant problem with solvent classification schemes based on the solvatochromic parameters is that it considers only the polar interactions of the solvents and not their cohesive energy. The cohesive energy of solvents is far from constant and influences solvation through the energy required for cavity formation, which has to be compensated for by solute–solvent interactions established by transfer of the solute to the solvent. The three solvatochromic scales for polar interactions are normalized but have no common origin and so the relative contribution of each intermolecular interaction to selectivity is only approximate. The $E_T(30)$ scale has been used as a single parameter scale for solvent polarity and was used in reversed-phase chromatography as a measure of solvent strength [3,28]. The $E_T(30)$ scale is based on the position of the intermolecular charge transfer band of Reichardt's betaine dye and can be estimated reasonably well by a linear combination of the π^* and α solvatochromic parameters. The $E_T(30)$ scale attempts to measure a bulk solvent property described as "polarity" but fails to sample all polar interactions. In addition, since solvent strength is a system property, its value for use in thin-layer chromatography is doubtful.

Snyder's solvent selectivity triangle is the most enduring method for solvent classification for liquid chromatography [8,27–29]. Snyder classified solvents based on their interactions with three prototypical compounds determined by their gas–liquid partition coefficients corrected for differences in solvent size, polarizability, and dispersion interactions. Each value was then corrected empirically to give a value of zero for the polar partition coefficient in saturated hydrocarbon solvents. Snyder chose the compounds nitromethane, ethanol, and dioxane as prototypical compounds to assign a solvent's capability for dipole-type, hydrogen-bond base, and hydrogen-bond acid interactions, respectively. The sum of the three polar partition coefficients was used to provide a measure of the solvent strength (P') and the ratio of the individual polar partition coefficients to their sum as a measure of selectivity (x_n, x_e, and x_d). Representing each solvent by the three solvent selectivity coordinates and plotting the results on the surface of a triangle ($x_n + x_e + x_d = 1$) resulted in the classification of solvents into eight selectivity groups with solvents in the same group having similar properties, Figure 12.5 [3,7]. Representative solvents from different groups were recommended for the initial stage of method development, Table 12.3. A main strength of the solvent

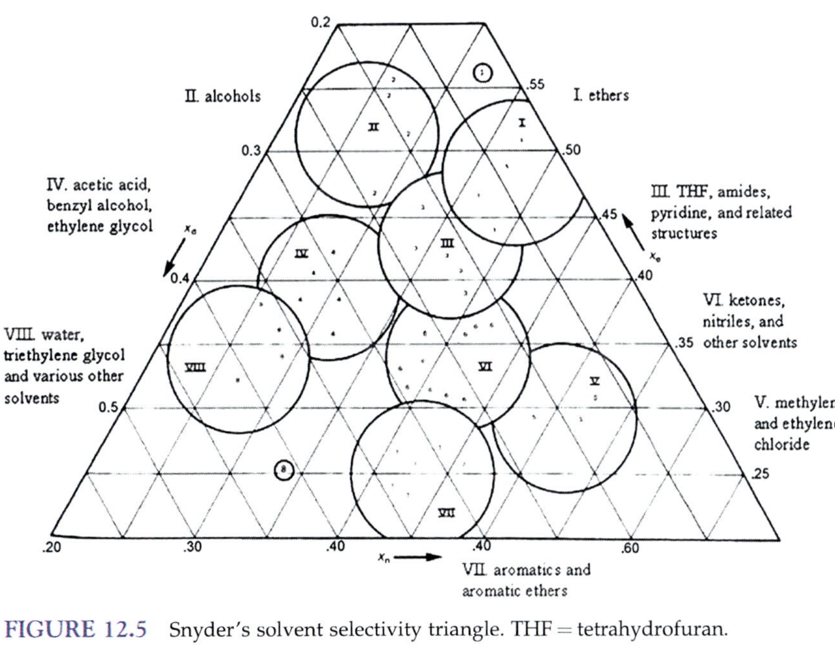

FIGURE 12.5 Snyder's solvent selectivity triangle. THF = tetrahydrofuran.

TABLE 12.3 Solvent Strength and Selectivity Parameters Based on Snyder's Selectivity Triangle

		Solvent Strength		Solvent Selectivity		
Solvent	**Selectivity Group**	P'	S_i	x_e	x_d	x_n
n-Butyl ether	I	2.1		0.44	0.18	0.38
Diisopropyl ether		2.4		0.48	0.14	0.38
Methyl *t*-butyl ether		2.7		0.49	0.14	0.37
Diethyl ether		2.8		0.53	0.13	0.34
n-Butanol	II	3.9		0.59	0.19	0.25
2-Propanol		3.9		0.55	0.19	0.27
1-Propanol		4.0		0.54	0.19	0.27
Ethanol		4.3	3.6	0.52	0.19	0.29
Methanol		5.1	3.0	0.48	0.22	0.31
Tetrahydrofuran	III	4.0	4.4	0.38	0.20	0.42
Pyridine		5.3		0.41	0.22	0.36
Methoxyethanol		5.5		0.38	0.24	0.38
N,N-Dimethylformamide		6.4		0.39	0.21	0.40
Dimethylsulfoxide		7.2		0.39	0.23	0.39

Acetic acid	IV	6.0		0.39	0.31	0.30
Formamide		9.6		0.38	0.33	0.30
Dichloromethane	V	4.3		0.27	0.33	0.40
1,2-Dichloroethane		3.5		0.30	0.21	0.49
Ethyl acetate	VI	4.4		0.34	0.23	0.43
Methyl ethyl ketone		4.7		0.35	0.22	0.43
Dioxane		4.8	3.5	0.36	0.24	0.40
Acetone		5.1	3.4	0.35	0.23	0.42
Acetonitrile		5.8	3.1	0.31	0.27	0.42
Toluene	VII	2.4		0.25	0.28	0.47
Benzene		2.7		0.23	0.32	0.45
Nitrobenzene		4.4		0.26	0.30	0.44
Cyclohexanone		4.7		0.36	0.22	0.42
Chloroform	VIII	4.3		0.31	0.35	0.34
Dodecaheptafluoroheptanol		8.8		0.33	0.40	0.27
Water		10.2	0	0.37	0.37	0.25

S_i is an empirical strength parameter for reversed-phase chromatography.
Reproduced with permission from Ref. [8].

selectivity triangle approach is its simple visual interpretation, although it was obvious that some solvents where incorrectly classified with respect to their neighbors [29].

The most significant limitation of the solvent selectivity triangle approach is the association of an individual intermolecular interaction with a single prototypical compound and the incomplete correction for differences in cohesive energy of the solvents [8,27,29]. The solubility of the prototypical compounds is the result of multiple intermolecular interactions and not due to a single dominant interaction. Ethanol, for example, is dipolar and hydrogen-bond acidic and basic, and could have a significant coordinate on the x_e scale without participating in solvent hydrogen-bond base interactions at all. Since there are no compounds that are strong hydrogen-bond acids or bases that are not simultaneously dipolar, it is impossible to characterize intermolecular interactions based on the properties of single (prototypical) compounds. The P′ scale of solvent strength is a bulk solvent property and cannot be expected to correctly describe solvent strength in chromatography, a system property, for which a single value scale is inappropriate for different stationary phases.

Linear solvation energy relationships, which do not depend on the properties of any single compound, but use a number of varied compounds to assign the contribution of individual intermolecular interactions and cohesive energy to the solvation process, have proven more successful for solvent classification and for modeling retention in separation processes. Of these approaches, the solvation parameter model has been used most often [3,8,22,26,30–32].

The solvation parameter model is based on a cavity model of solvation and uses a series of descriptors to define solute properties and a complementary series of system constants to represent solvent properties. These are defined in Table 12.4 [30–32]. The transfer of 1 mol of solute from an ideal gas phase to a solvent occurs in three steps: a cavity of the same size as the solute is created in the solvent; the solvent molecules reorganize themselves around the cavity to establish favorable interactions with the solute; and finally, the solute enters the cavity and establishes the appropriate intermolecular interactions with the solvent. Cavity formation requires disruption of solvent–solvent interactions and the free energy of cavity formation varies with the cohesive energy of the solvent. Reorganization of the solvent molecules around the cavity to new positions more favorable for solute interactions occurs with little change in the free energy of the system and can generally be ignored in estimating the change in free energy of solvation. The set up of solute–solvent interactions when the solute is placed in the cavity are delineated as dispersion, interactions of a dipole-type, and hydrogen-bonding. For

TABLE 12.4 Definition of Solute Descriptors and Their Complementary System Constants Used in the Solvation Parameter Model for Solvent Characterization

Solute Property	Solvent Property	Description
E	e	Solute–solvent electron lone pair interactions. The E descriptor is the excess molar refraction and can be calculated for liquids from a refractive index value and the characteristic volume of the solute. For solids it can be estimated or determined experimentally.
S	s	Solute–solvent, dipole-type interactions (both orientation and induction). The S descriptor is an experimental property.
A	a	Solute hydrogen-bond acid–solvent hydrogen-bond base interactions. The A descriptor is an experimental property.
B	b	Solute hydrogen-bond base–solvent hydrogen-bond acid interactions. The B descriptor is an experimental property.
L	l	Combination of the free energy required to create a cavity of the same size as the solute in the solvent and solute–solvent dispersion interactions. Cavity formation depends on the cohesive properties of the solvent and involves breaking solvent–solvent interactions. The L descriptor is the solute gas–liquid partition coefficient at 298 K on hexadecane.
V	v	For transfer between condensed phases it represents the difference in free energy required for cavity formation in the two phases together with residual dispersion interactions that are not self-cancelling. The V descriptor is McGowan's characteristic volume and can be calculated from structure.

quantitative calculations, it is necessary to parameterize the model, resulting in Eqn (12.1) for transfer from the gas phase to a solvent

$$\log K = c + eE + sS + aA + bB + lL \qquad (12.1)$$

where K is the gas–liquid partition coefficient. Calculation of the system constants for the solvent requires the determination of the experimental property ($\log K$) for a number of solutes with known and varied descriptor values using multiple linear regression analysis. There are certain chemical and statistical requirements to be met for use of Eqn (12.1), which are explained elsewhere [30–32].

The solvation parameter model characterizes solvent properties in terms of five system constants. As such it is not a simple matter to visualize the similarity among solvents to recognize redundancy and solvents with singular properties. The data cannot be represented in a three-dimensional space without reducing its dimensionality. Principal component analysis is ineffective for this purpose and hierarchical cluster analysis proved to be a better alternative for classification in which the Euclidian distance between solvents in five-dimensional space is used to compute a similarity matrix. Solvents that are near neighbors in hyperspace are grouped together and solvents further removed from each other, or from a group of solvents, are placed into a different class until all the solvents have been included in the analysis. The results are visualized as a dendrogram shown in Figure 12.6 for the 36 solvents in Table 12.5 [26]. The solvents are classified into seven

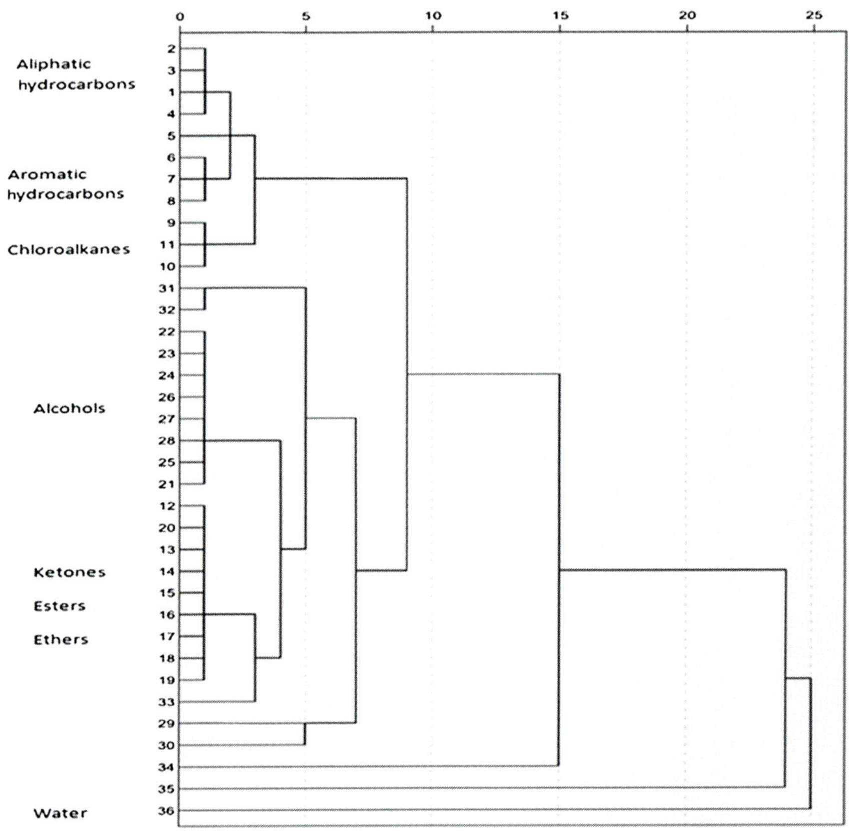

FIGURE 12.6 Nearest neighbor agglomeration cluster dendrogram for the 36 solvents indicated in Table 12.4. Identification: 1 = cyclohexane; 2 = *n*-hexane; 3 = *n*-heptane; 4 = 2,2,2--trimethylpentane; 5 = carbon tetrachloride; 6 = benzene; 7 = toluene; 8 = chlorobenzene; 9 = dichloromethane; 10 = chloroform; 11 = 1,2-dichloroethane; 12 = acetone; 13 = butan-2-one; 14 = cyclohexanone; 15 = ethyl acetate; 16 = butyl acetate; 17 = diethyl ether; 18 = methyl *t*-butyl ether; 19 = tetrahydrofuran; 20 = dioxane; 21 = methanol; 22 = ethanol; 23 = propan-1-ol; 24 = propan-2-ol; 24 = propan-2-ol; 25 = butan-1-ol; 26 = butan-2-ol; 21 = hexan-1-ol; 28 = octan-1-ol; 29 = ethylene glycol; 30 = formamide; 31 = acetonitrile; 32 = propylene carbonate; 33 = *N,N*-dimethylformamide; 34 = dimethyl sulfoxide; 35 = 2,2,2-trifluoroethanol; and 36 = water. *Reproduced with permission from Ref. [26].*

TABLE 12.5 System Constants for Gas to Solvent Transfer for Solvents Commonly Used in Thin-Layer Chromatography

Solvent	System Constants					
	e	*s*	*a*	*b*	*l*	*c*
Cyclohexane	−0.110	0	0	0	1.013	0.163
n-Hexane	−0.169	0	0	0	0.979	0.292
n-Heptane	−0.162	0	0	0	0.983	0.275
2,2,4-Trimethylpentane	−0.230	0	0	0	0.975	0.264
Carbon tetrachloride	−0.435	0.544	0	0	1.069	0.217
Benzene	−0.313	1.053	0.457	0.169	1.020	0.107
Toluene	−0.222	0.938	0.467	0.099	1.012	0.121
Chlorobenzene	−0.399	1.156	0.313	0.171	1.032	0.064
Dichloromethane	−0.572	1.492	0.460	0.847	0.965	0.192
Chloroform	−0.560	1.259	0.374	1.333	0.976	0.157
1,2-Dichloroethane	−0.337	1.600	0.774	0.637	0.921	0.017
Acetone	−0.387	1.733	3.060	0	0.866	0.127
Butan-2-one	−0.474	1.671	2.878	0	0.916	0.112
Cyclohexanone	−0.441	1.725	2.786	0	0.957	−0.086
Ethyl acetate	−0.352	1.316	2.891	0	0.916	0.182
Butyl acetate	−0.414	1.212	2.623	0	0.954	0.147
Diethyl ether	−0.379	0.904	2.937	0	0.963	0.288
Methyl *t*-butyl ether	−0.536	0.890	2.632	0	0.999	0.231
Tetrahydrofuran	−0.347	1.238	3.289	0	0.982	0.189
Dioxane	−0.354	1.674	3.021	0	0.919	−0.034
Methanol	−0.338	1.317	3.826	1.396	0.773	−0.039
Ethanol	−0.232	0.867	3.894	1.192	0.846	0.017
Propan-1-ol	−0.246	0.749	3.888	1.076	0.874	−0.042
Propan-2-ol	−0.324	0.713	4.036	1.055	0.884	−0.048
Butan-1-ol	−0.285	0.765	3.705	0.879	0.890	−0.004
Butan-2-ol	−0.387	0.719	3.736	1.088	0.905	−0.034

(Continued)

TABLE 12.5 System Constants for Gas to Solvent Transfer for Solvents Commonly Used in Thin-Layer Chromatography—cont'd

	System Constants					
Solvent	*e*	*s*	*a*	*b*	*l*	*c*
Hexan-1-ol	−0.205	0.583	3.621	0.891	0.913	−0.014
Octan-1-ol	−0.203	0.560	3.576	0.702	0.939	−0.120
Acetonitrile	−0.595	2.461	2.085	0.418	0.738	−0.007
Propylene carbonate	−0.413	2.587	2.207	0.455	0.719	−0.356
Ethylene glycol	0.132	1.657	4.457	2.355	0.565	−0.887
Formamide	0.310	2.292	4.130	1.933	0.442	−0.800
N,N-dimethylformamide	−0.869	2.107	3.774	0	1.011	−0.391
Dimethyl sulfoxide	0.131	2.811	5.474	0	0.734	−0.619
2,2,2-Trifluoroethanol	−0.547	1.339	2.213	3.807	0.645	−0.092
Water	0.822	2.743	3.904	4.814	−0.213	−1.271

clusters with four solvents behaving independently. The solvents in Table 12.5 are entered according to the selectivity groups indicated in Figure 12.6. Cluster 1 contains the apolar *n*-alkane and cycloalkane solvents of low cohesion and very weak or nonexistent polar interactions. These are separated from Cluster 2, which contains the aromatic hydrocarbons and chlorobenzene. These solvents have low cohesion and weak polar interactions. Cluster 3 contains the haloalkanes, which have low cohesion, weak polar interactions but are significantly more hydrogen-bond acidic than the solvents in Cluster 2. Cluster 4 contains acetonitrile and propylene carbonate, which are strongly dipolar, weakly hydrogen-bond acidic and basic, and of moderate cohesion. They are distinguished from the other strongly dipolar solvents by their characteristic s/a ratio. Cluster 5 contains the alcohols, which are amphiprotic solvents (moderately dipolar, strong hydrogen-bond bases, intermediate hydrogen-bond acids, and moderately cohesive). Cluster 6 contains the ketones, esters, and ethers, which are weakly cohesive solvents with moderate dipolarity and strong hydrogen-bond basicity but no hydrogen-bond acidity. Cluster 7 contains the amphiprotic solvents ethylene glycol and formamide, which are significantly more cohesive, dipolar, and hydrogen-bond acidic than the alcohols forming Cluster 5. Of the four solvents behaving independently only water is likely to be truly independent. It is easily the most cohesive solvent in Table 12.5 and the strongest hydrogen-bond acid. 2,2,2-Trifluoroethanol is significantly more hydrogen-bond acidic and less hydrogen-bond basic than a typical alcohol (Cluster 5) due to the effect of the electronegative fluorine atoms on the hydroxyl group. *N,N*-dimethylformamide is more dipolar and hydrogen-bond basic than the other aprotic dipolar solvents in Cluster 6.

The classification of solvents in Table 12.5 is more logical than the selectivity triangle approach in Table 12.3. Some of the subtlety in the classification of solvents in Figure 12.6 is masked at full scale. Figure 12.7 illustrates the variation of solvation properties for the eight alcohols of Cluster 5. Although these solvents have characteristics in common, it is obvious that there are small and logical differences that could be exploited for selectivity optimization. Methanol is more dipolar, hydrogen-bond acidic and cohesive than the other alcohols. The alcohols with two to four carbon atoms are slightly different to *n*-hexanol and *n*-octanol. For solvent screening purpose, it would be reasonable to take two solvents from this group, of which one would be methanol, and for the other, the choice is less critical, and as a compromise propan-2-ol would be a reasonable

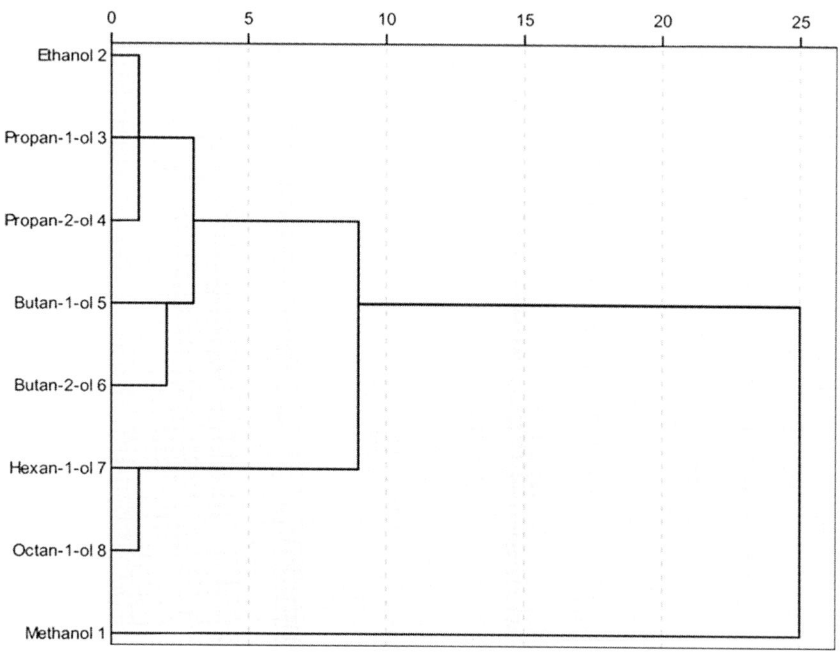

FIGURE 12.7 Average linkage agglomeration cluster dendrogram for the *n*-alcohols of Cluster 5 of Figure 12.6. *Reproduced with permission from Ref. [26].*

choice. The effect of the n-alkyl chain length on the system constants for a wider range of n-alcohols indicates a smooth change of the system constants with chain length, and in particular, a decrease in capability to participate in hydrogen-bond and dipole-type interactions while cavity formation is easier in alcohols with longer alkyl chains compared to those with shorter chains. For this reason, the characteristic properties of n-alkyl alcohols cannot be considered independent of the alkyl chain length and it would be logical to select different alcohols to fine-tune selectivity in separations having first identified Cluster 5 as the desired solvent selectivity group for mobile phase optimization.

Cluster 6 contains nine ketones, esters, and ethers. The dendrogram for this group is shown in Figure 12.8. The ketones form one reasonably homogeneous group including the ether dioxane, which is virtually selectivity equivalent to acetone. The two alkyl esters while separated from the ketones are only slightly less dipolar/polarizable and for screening purposes could be combined with the ketones. Of the three remaining ethers, tetrahydrofuran is more dipolar/polarizable and hydrogen-bond basic than diethyl ether and methyl t-butyl ether, but in terms of differentiation within Cluster 6, methyl t-butyl ether or diethyl ether would be a good second choice because of their larger a/s ratio.

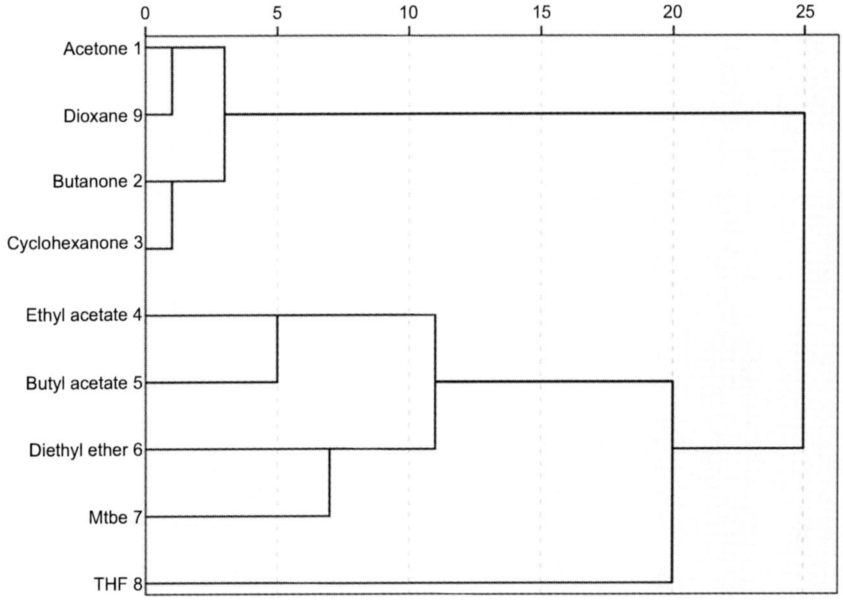

FIGURE 12.8 Average linkage agglomeration cluster dendrogram for the ketones, ethers, and esters of Cluster 6 of Figure 12.6. (Abbreviations: Mtbe = methyl t-butyl ether and THF = tetrahydrofuran.) *Reproduced with permission from Ref. [26].*

12.4.1 Solvent Strength Scales

Solvent strength scales are system properties and cannot be derived from solvent properties alone. For inorganic oxide sorbents, such as silica gel, the simplified form of the competition model leads to a useful empirical parameter for solvent strength, the solvent strength parameter ε°, defined as the free energy of adsorption of the solvent per unit surface area with pentane as the zero reference [1,3,7,33]. The organization of solvents in ascending order of ε° is known as an eluotropic series,

TABLE 12.6 Eluotropic Series for Common Solvents Used in Thin-Layer Chromatography for Inorganic Oxide Sorbents

		Solvent Strength Parameter	
Solvent	**Type**	$\varepsilon^\circ_{(\text{silica})}$	$\varepsilon^\circ_{(\text{alumina})}$
Hexane, heptane, isooctane	Nonlocalizing	0	0
Toluene	Nonlocalizing	0.22	0.30
Benzene	Nonlocalizing	0.25	0.32
Chloroform	Nonlocalizing	0.26	0.36
1-Chloropropane	Nonlocalizing	0.28	0.31
Dichloromethane	Nonlocalizing	0.30	0.40
Diisopropyl ether	Minor localizing	0.32	0.28
Methyl *t*-butyl ether	Localizing and basic	0.32	0.31
1,2-Dichloroethane	Nonlocalizing	0.34	0.44
Diethyl ether	Localizing and basic	0.38	0.38
Ethyl acetate	Localizing	0.48	0.60
Tetrahydrofuran	Localizing and basic	0.48	0.51
Dioxane	Localizing and basic	0.51	0.61
Acetonitrile	Localizing	0.52	0.65
Acetone	Localizing	0.53	0.58
Formamide		0.55	1.45
1- or 2-Propanol	Localizing and basic	0.60	0.80
2,2,2-Trifluoroethanol		0.61	0.99
Ethanol	Localizing and basic	0.65	0.88
Methanol	Localizing and basic	0.70	0.95
N,N-dimethylformamide		0.76	0.91
Water		0.82	1.29

Table 12.6. The ε° values are experimental but can be predicted with a similar uncertainty (±0.04) by the solvation parameter model [34]

$$\varepsilon^\circ_{(\text{silica})} = 0.272 + 0.199S + 0.384A + 0.535B - 0.264V \quad (12.2)$$

$$\varepsilon^\circ_{(\text{alumina})} = 0.230 + 0.359S + 0.938A + 0.475B - 0.226V \quad (12.3)$$

Retention of neutral compounds on silica gel results from hydrogen-bonding interactions mainly and to a lesser extent dipole-type interactions. Alumina is significantly more hydrogen-bond basic and dipolar/polarizable than silica gel while less hydrogen-bond acidic. Increasing solute size reduces retention on both adsorbents. For chemically bonded phases, the ε° values are small and compressed into a narrow range for different solvents. In this case, secondary solvent effects become important, and the solvent strength order is dependent on the identity of the solute. The concept of an eluotropic series is no longer useful and there is no solvent strength scale for these sorbents. For reversed-phase conditions, the solvent strength S_i scale is applied, Table 12.3. This is an experimentally derived scale and not completely independent of solute identity [3]. It is reasonably effective for first-order estimates of solvent strength for low-molecular-mass compounds and binary solvent mixtures containing water. It is used more commonly in column chromatography where solvent strength estimates are often better because the chromatographic system is at equilibrium and competing adsorption interactions are minimized.

Single solvents provide only limited options for method development and binary and higher-order solvent mixtures provide a more flexible approach for the simultaneous optimization of solvent strength and selectivity. The competition model allows calculation of the solvent strength for solvent mixtures, but requires empirical adjustments for polar solvents. For weak solvents, there is little preference for adsorption on any given part of the adsorbent surface and little tendency for weakly retained molecules to localize. On the other hand, polar compounds and solvents share a tendency to localize on specific adsorption cites and have a higher energy of adsorption than for the nonlocalized state [1,3,7,34]. Once localized strong solvent molecules cover most of the surface, the remaining spaces on the surface are unable to accommodate strong solvent molecules in the configuration required for localization, a process referred to as restricted-access delocalization. Restricted-access delocalization occurs for rigid inorganic oxide sorbents like silica gel with a high concentration of surface adsorption sites but is less important for polar chemically bonded sorbents due to the relatively low ligand density and the flexibility of the bonded chains. A practical consequence of restricted-access delocalization of solvent mixtures is that the solvent strength of the mobile phase will vary continuously with the mole fraction of the strong solvent.

The retention of polar solutes is also affected by site-competition delocalization. A moderately polar, nonlocalizing solvent molecule can interact laterally with sites upon which a solute molecule is localized. This added competition for the site by both solute and solvent molecules weakens the net interaction of the solute with the surface. In addition, especially for hydrogen-bonding compounds, solute–solvent interactions in the mobile and solvated sorbent are unlikely to be identical, as assumed in the competition model, requiring an empirical correction for solvent strength calculations.

It should come as no a surprise in the light of the above discussion that the relationship between solvent strength and solvent composition is nonlinear. The initial addition of a strong solvent to a strength adjusting solvent results in a large change in the solvent strength of the mixture while at high volume fractions of strong solvent increases in solvent strength are small. A number of empirical schemes to identify binary solvent mixtures of different composition and approximately equal solvent strength have been developed. These methods provide an alternative to complex calculations [1,3,7,35]. An example is shown in Figure 12.9. A vertical line drawn at a particular solvent strength identifies all combinations of binary solvent mixtures with the same solvent strength for selectivity optimization in thin-layer chromatography.

When multicomponent mobile phase are used in thin-layer chromatography, composition gradients in the direction of development can form due to demixing. Solvent demixing results from the selective uptake of the polar solvent by the dry sorbent with the region behind the advancing solvent front increasingly denuded in the polar solvent. The net result is that separations occur in regions of the layer with different mobile phase compositions. These effects are less apparent when saturated developing chambers are used and when the strength adjusting solvent and polar solvent are not too different in properties. The strength adjusting solvent for inorganic oxides need not be a weak solvent like *n*-heptane in every case, but can be a polar solvent, if that solvent when used as a developing solvent fails to cause significant migration of sample components from the sample origin. Solvent demixing and the presence of a vapor phase are two of the main reasons why selectivity optimization strategies based solely on the composition of the solvent added to the developing chamber are not as successful or as popular for thin-layer chromatography as for column chromatography.

12.4.2 Solvent Selectivity Scales

The general scheme for selectivity optimization involves the dilution of select solvents with a strength adjusting solvent to obtain a number of trial mobile phases of similar solvent strength with different separation characteristics. The select solvents are identified from solvent classification

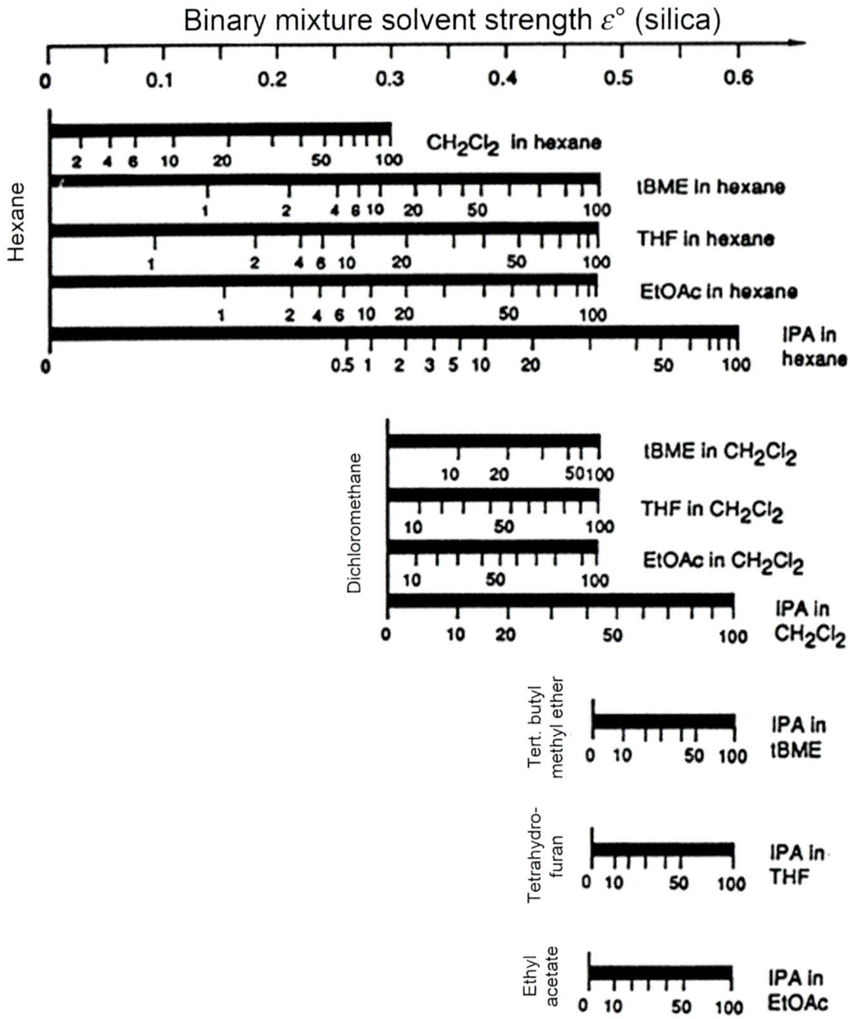

FIGURE 12.9 Method for the identification of binary solvent mixtures with similar solvent strength on silica gel. Solvent identity: tBME = methyl *t*-butyl ether; EtOAC = ethyl acetate; IPA = 2-propanol; and THF = tetrahydrofuran. *Reproduced with permission from Ref. [35].*

schemes. For normal-phase separations in thin-layer chromatography these are volatile solvents from the individual selectivity groups of Snyder's solvent selectivity triangle, Table 12.7, or from the clusters obtain by classification of the system constants of the solvation parameter model, Table 12.8. At the start of the selectivity screening process solvents should be selected to cover the full selectivity space, which is equivalent to selecting at least one solvent from each selectivity group (or cluster). In the case of the cluster

TABLE 12.7 Select Solvents for Method Development in Normal-Phase Thin-Layer Chromatography Based on Snyder's Selectivity Triangle Classification

Select Solvent	Selectivity Group	Solvent Type
Diethyl ether	I	Aliphatic ethers
2-Propanol Ethanol	II	Aliphatic alcohols
Tetrahydrofuran	III	Pyridine, tetrahydrofuran, glycol ethers, amides
Acetic acid Formamide	IV	Formamide, acetic acid, glycols
Dichloromethane	V	Dichloromethane, 1,2-dichloroethane
Ethyl acetate Dioxane	VI	Aliphatic ketones and esters, dioxane, acetonitrile
Toluene	VII	Aromatic hydrocarbons and ethers, aromatic halo- and nitro-compounds
Chloroform	VIII	Chloroform, water, nitromethane, phenols

TABLE 12.8 Select Solvents Based on the Cluster Analysis of the System Constants of the Solvation Parameter Model

Select Solvent	Cluster	Classification
n-Heptane	1	Apolar
Toluene	2	Apolar aromatic
Dichloromethane Chloroform	3	Haloalkane
Acetonitrile	4	Dipolar and weakly aprotic
Methanol 2-propanol	5	Amphiprotic
Acetone Methyl *t*-butyl ether	6	Polar and nonhydrogen-bond acidic
Formamide	7	Polar and cohesive
2,2,2-Trifluoroethanol *N*,*N*-dimethylformamide Dimethyl sulfoxide Water		Polar and independent

classification, the number of solvents exceeds the number of clusters for two reasons. The solvents in individual clusters are not selectivity equivalent but cover a narrow range of properties. Consequently, the properties of individual clusters are better represented by more than one solvent. The classification based on cluster analysis also identifies several solvents as behaving independently with properties different to those defined by the clusters and should be utilized for initial screening. The cluster classification did not include experimental data for acetic acid and triethylamine. These solvents are useful for the separation of weakly acidic and basic compounds and are included in the screening phase for these compounds.

The selection of select solvents for reversed-phase chromatography is carried out in the same way except that in this case the choice of solvents is restricted to those miscible with water. Suitable select solvents for the screening stage in reversed-phase chromatography are summarized in Table 12.9 [8,22,36].

12.4.3 Mobile Phase Optimization

Mobile phase optimization for mixtures of moderate complexity is not necessarily particularly difficult and can be achieved quite quickly using guided trial-and-error approaches [4–9]. A mobile phase with the correct solvent strength for a single development will migrate sample components into the R_F range 0.2–0.8, and mobile phases with the correct selectivity, will distribute the sample components within this range in accordance with some criteria established for their satisfactory separation. Solvent systems can be screened in parallel using several development chambers or by circular chromatography using an automated sample applicator. An example of the former method is the PRISMA method, the most widely used guided trial-and-error method for method development in thin-layer chromatography [3,8,37–39]. Alternatively, sample

TABLE 12.9 Select Solvents for Reversed-Phase Chromatography (Water as Strength Adjusting Solvent)

Select Solvent	Cluster	Classification
Acetonitrile	4	Dipolar and weakly aprotic
Methanol 2-Propanol	5	Amphiprotic
Acetone Tetrahydrofuran	6	Polar and nonhydrogen-bond acidic
2,2,2-Trifluoroethanol *N,N*-dimethylformamide Pyridine		Polar and independent

spots can be applied at suitable positions on a single layer and automatically developed in sequence with a small volume of solvent or solvent mixture using an automated sample applicator [38]. The individual circular chromatograms enable rapid identification of solvents with suitable strength and selectivity for the separation.

Initial solvent selection in the PRISMA model uses an empirical screening step based on the use of select solvents identified in Section 12.4.2. These experiments are carried out in parallel on thin-layer plates in unsaturated development chambers with single solvents. After these initial experiments, the solvent strength is adjusted so that the sample components are distributed within the optimum R_F range. If the sample components migrate into the upper third of the layer the solvent strength is reduced by dilution with a strength adjusting solvent. If the sample components remain in the lower third of the layer with the single solvents the solvent strength has to be increased by addition of a strong solvent. From these trial experiments, the mobile phases providing the best separation (maximum number of separated zones) are selected for further optimization in the second part of the PRISMA model. The screening step is illustrated by the example in Figure 12.10 and is achieved quite quickly because of the use of parallel separations.

In the second part of the PRISMA model between two and five solvents can be selected for mobile phase optimization. Modifiers such as acids, ion pair reagents, complexing agents, etc., can be added to improve the separation or to control zone shapes. Modifiers are generally used in low and constant concentration so that the effect on solvent strength can be neglected. The PRISMA model, which gives the optimization process its name, is shown in Figure 12.11. It consists of a three-dimensional geometrical

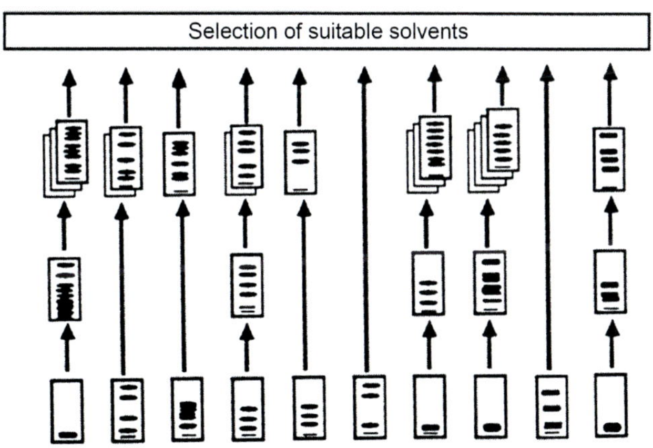

FIGURE 12.10 Schematic illustration of the solvent screening step used in the PRISMA model.

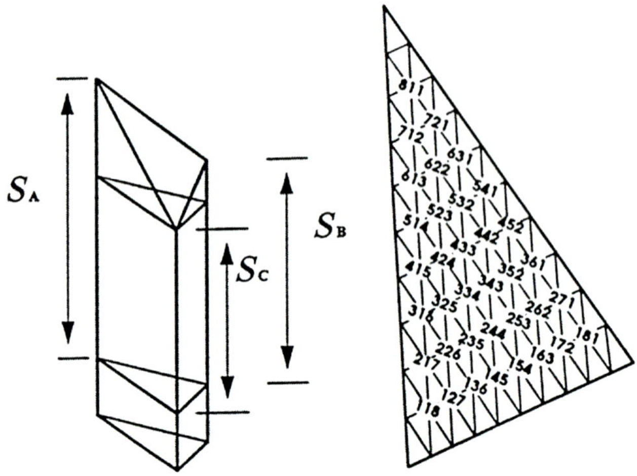

FIGURE 12.11 The PRISMA mobile phase optimization model showing the construction of the prism and the selection of selectivity points. *Reproduced with permission from Ref. [8].*

design, which correlates the solvent strength with the selectivity of the mobile phase. The model consists of three parts: the base or platform representing the modifier; the regular part of the prism with congruent base and top surfaces; and the irregular truncated top prism (frustum). The lengths of the edges of the prism (S_A, S_B, S_C) correspond to the solvent strengths of the single solvents (A, B, and C) selected from the screening step. Since the selected solvents usually have different solvent strengths, the edges of the prism may be of unequal length and the top plane of the prism will not be parallel and congruous with its base. Cutting the prism parallel to its base at the height of the lowest edge (the solvent strength corresponding to the weakest solvent) gives a regular prism, where the top and any planes representing weaker solvents diluted with a strength adjusting solvent are parallel equilateral triangles. The upper frustum is used for mobile phase optimization of polar compounds in normal-phase chromatography, while the regular part is used for the separation of nonpolar and moderately polar compounds for normal-phase chromatography and all applications using reversed-phase chromatography.

For polar compounds in normal-phase chromatography optimization is always started on the top irregular triangle of the model, either within the triangle, when three solvents are selected, or along one side, for binary mobile phases. Any solvent composition on the face of the triangle can be represented by three coordinate selectivity points, each coordinate corresponding to the volume fraction of the solvent at that position on the triangle. Optimization is commenced by selecting solvent combinations corresponding to the center point (333) and three other points close to the apexes of the triangle (811, 181, 118), see Figure 12.11. If the separation obtained is inadequate other selectivity points are tested around the solvent combination that gave the best separation. On changing the selectivity points on the top of the triangle the solvent strength changes as well and should be adjusted with the strength adjusting solvent to maintain the separation in the optimum R_F range.

The regular center portion of the prism is used to optimize the mobile phase composition for the separation of compounds of low to moderate polarity in the normal-phase mode and all compounds in the reversed-phase mode. The initial solvent composition corresponds to the center of the triangular top face of the regular prism (333); this composition is then diluted to bring all sample components into the optimum R_F range. At this solvent strength three more separations are performed corresponding to the selectivity points close to the apexes of the triangle. These initial separations are then used to choose selectivity points for further separations until the optimum solvent composition is located.

Empirical relationships can be built between the R_F values and the solvent strength, or at a constant solvent strength, between the R_F values and the selectivity points. These relationships can form the basis of a computer-aided optimization strategy but require extensive experimental input and are little used compared with the guided trial-and-error approach described above [40,41].

Optimization of the solvent selectivity by varying the selectivity points is carried out until the required separation is obtained. If no adequate separation is obtained then a different solvent or solvents identified as promising from the screening stage should be tried. Failing that, a different layer type should be tried and the new system optimized starting at the screening stage. If a suitable separation would result from a modest increase in efficiency then forced-flow development should be tried, if suitable equipment is available. For samples of a narrow polarity range that tends to migrate as a compact group (e.g., isomers, diastereomers, and analogs with minor structural variations) isocratic multiple development with incremental change in the solvent entry position is the preferred approach [42]. The most selective solvent for the separation can be identified from the screening stage of the PRISMA model. If the range of R_F values exceeds the optimum range for all single solvents at the screening stage then method development should continue as described for gradient methods (Section 12.4.4).

The solvation parameter model has been used as the basis of a structure-driven retention model for method development under reversed-phase conditions, and is presumably suitable for normal-phase separations with polar chemically bonded phases [8,22,36]. This model produces poor predictions for normal-phase chromatography on inorganic oxide sorbents as it fails to properly account for solute-size differences and site-specific interactions [43]. Method development for binary mobile phases using the solvation parameter model is based on the use of system maps. An example is shown in Figure 12.12. A system map is a continuous plot of the system constants of the solvation parameter model against the mobile phase composition. It provides a permanent record of system properties that is independent of solute identity. System maps for common binary solvent systems, Table 12.9, are available for octadecylsiloxane-bonded silica [36], 3-cyanopropylsiloxane-bonded silica [22], and spacer bonded propanediol [8] layers. Retention maps are created to judge separation performance for any set of compounds with known descriptor values. For other compounds descriptor values, Table 12.5, can be determined by experiment or estimated by fragment models [30–32]. A typical retention map for the separation of a mixture of analgesics on an octadecylsiloxane-bonded layer is shown in Figure 12.13 for 2,2,2-trifluoroethanol and acetonitrile binary mobile phase mixtures [36]. Solvent compositions resulting in suitable zone separation are easily identified by moving a vertical line along the composition axis and noting those regions of mobile phase composition where adequate zone separation is obtained on the y-axis. In this case, it is clear that several mobile phase compositions containing 2,2,2-trifluoroethanol result in an adequate separation while for acetonitrile no suitable separation is possible. This approach has been extended to ternary mobile phase compositions but requires additional experimental effort to construct retention surfaces [44]. Because the screening step is performed by

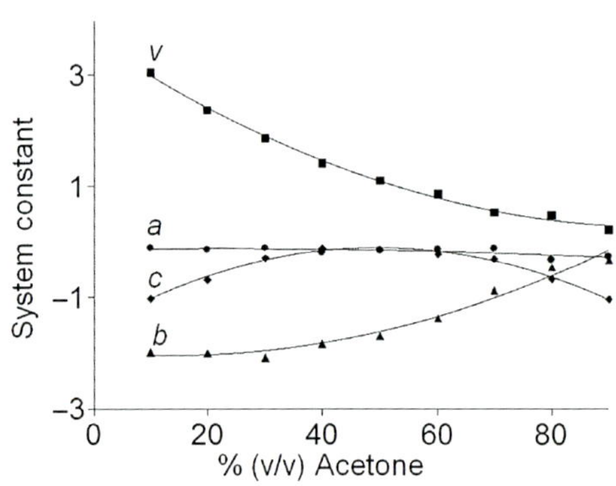

FIGURE 12.12 System map for acetone–water mobile phase compositions on an octadecylsiloxane-bonded silica layer. In this case, the system map is simple because $e = 0$ and $s = 0$ (except for 10% acetone where $s = -0.18$). This means that electron lone pair and dipole-type interactions are about the same in the solvated stationary phase and mobile phase and do not contribute to the distribution of compounds between the two phases. For identification of the other system constant see Table 12.4 and c is the model constant (dominated by the phase ratio).

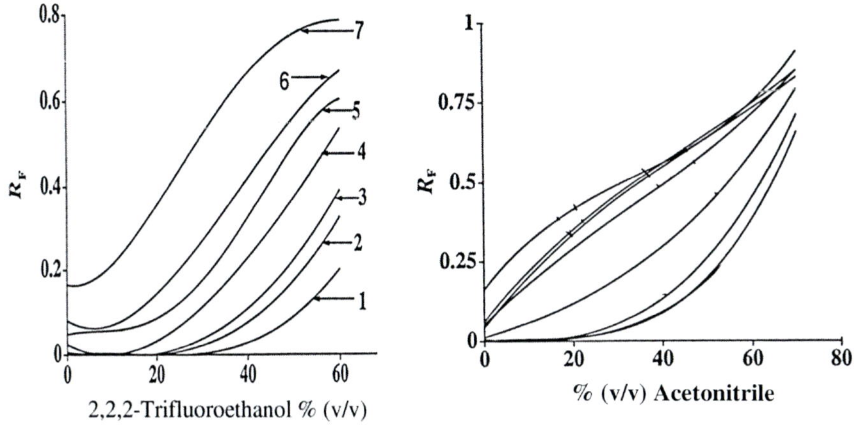

FIGURE 12.13 Retention maps for the separation of analgesics by reversed-phase thin-layer chromatography on an octadecylsiloxane-bonded with 2,2,2-trifluoroethanol and acetonitrile mobile phases. Identification: 1 = chlorophenIramine; 2 = ibuprofen; 3 = naproxen; 4 = phenacetin; 5 = aspirin; 6 = caffeine; and 7 = acetaminophen.

calculation and graphic procedures it is very fast and suitable systems quickly identified for experimental verification.

12.4.4 Mobile Phase Optimization in Gradient Methods

For mixtures containing components spanning a wide retention range, some form of gradient development is required to obtain a separation of all components in a single chromatogram, as in automated multiple development, or in separate chromatograms for successive developments, as in incremental multiple development. Automated multiple development employs incremental multiple development with a decreasing solvent strength gradient whereas conventional multiple development is performed with an increasing solvent strength gradient. In the latter approach, the sample is fractionated into a series of successive separations in which just a few compounds are separated at each or some of the successive development steps [45]. Compounds of interest are usually quantified by densitometry at an intermediate step where adequate separation is obtained and are likely to be remixed or pushed into the solvent front in later steps. The screening stage of the PRISMA model can be used to identify suitable solvents for the separation in each segment. In this case, a separation is developed as a series of developments of variable length and variable mobile phase composition combined into a sequence of successive developments with intermediate drying and scanning steps interspersed into the sequence. The process is difficult to automate and for complex mixtures can be tedious.

To simplify automation, the alternative approach of incremental multiple development with a decreasing solvent strength gradients has been preferred. In this case, the sample is developed for the shortest distance with the strongest mobile phase, and in each subsequent and longer development with mobile phases of decreasing solvent strength. This approach is the operating basis of the automated multiple development chamber, which allows the conditions for each development step to be preprogrammed as timed events, including solvent selection, development length, drying time, and layer conditioning for each development. Run times are long with a typical program of 20 steps requiring 2.5 h for execution but during this time the user can work on other problems.

Optimized gradients for automated multiple development are usually arrived at by pragmatic means and method development time can be lengthy, but once established, sample throughput is high because multiple samples are separated in parallel during one development sequence [8,45–49]. Two general approaches have been adopted for guided trial-and-error optimization of gradient conditions. The first is based on the use of a universal gradient which commences with methanol, ends with *n*-hexane, and uses dichloromethane or methyl *t*-butyl ether as the intermediate or base solvent (typically employing 25 steps). By scaling and superimposing the chromatogram of the separation above the theoretical gradient profile those regions affecting the chromatogram are easily identified. If necessary, new initial and final mobile phase compositions can be selected if the chromatogram contains empty space at the start or end of the gradient. In addition, the gradient shape can be modified to enhance the separation in those regions of the chromatogram where the separation is inadequate or to make better use of the zone capacity by minimizing regions devoid of sample zones. In those cases where the separation remains inadequate after making the above adjustments it is necessary to identify a different solvent combination for the separation. The solvent screening stage of the PRISMA model can be employed to identify more selective solvents for building the gradient as a replacement for the initial, terminal, or base solvent. It should be noted that the universal gradient (and most gradients) programmed in the automated multiple development chamber are not linear solvent strength gradients. The abrupt change in solvent strength occurring during the gradient can cause grouping of sample components that may be minimized by employing a shallower gradient or introducing isocratic steps in the gradient program [49]. Alternatively, if standards are available for the components of the mixture isocratic plots of R_M against mobile phase composition for binary solvent mixtures can be used to infer suitable gradient separation conditions [48]. Because method development is often lengthy, for mixtures with more than 15–20 components of interest, it is often faster to develop separate methods in which the least polar and most polar components are separated separately using different programs.

12.4.5 Computer-Aided Method Development

Computer-aided strategies for method development in thin-layer chromatography are not as commonly employed as is the case for column chromatography. The probable reasons for this are the greater number of variables that require control or optimization for separations by thin-layer chromatography and the lack of simple mathematical relationships to describe the variation of retention with system properties. An exception is the use of the solvation parameter model to simulate separations in reversed-phase thin-layer chromatography discussed above (Section 12.4.3). Computer-aided strategies for mobile phase optimization using window diagrams, overlapping resolution maps, simplex methods, and iterative procedures have found occasional use [8,37,40,50,51]. In these procedures, some form of statistical design is used to select a group of solvents for evaluation, or alternatively, the results obtained from the arbitrary selection of solvents are compared to indicate the best separation. It is perhaps surprising how little work has been done based on standard experimental design approaches in support of method development in thin-layer chromatography [44,52,53]. Experimental design approaches have the potential to provide a structured approach to method development combined with empirical modeling for the prediction of optimum separation conditions.

To rank separation quality for computer interpretation a single value numerical index is required. Suitable index scales are difficult to develop for thin-layer chromatography. Resolution based functions favored in column chromatography are not suitable for thin-layer chromatography because zone widths are difficult to predict and depend on the migration distance and the number of successive developments. Generally, separation quality has been defined as an index based on zone center separations, sometimes combined with a function to characterize the distribution of zones throughout the separation [51].

There have been few publications in the last decade on computer-aided approaches to method development suggesting that the guided trial-and-error approach is the standard approach adopted in practice and is reasonably efficient for simple and moderately complex mixtures. Approaches based on experimental design that use empirical equations to describe retention surfaces are probably the best approach for future developments. These approaches, however, require a significant number of preliminary experiments before predictions are made and would be most effective for difficult separations.

12.5 CONCLUSIONS

Method development is the intellectual part of thin-layer chromatography and tends to separate analysts by ability just as much as mixtures by chromatography. Almost all separations require an individual approach and proceed faster when the analyst has a sound knowledge of solvent and layer properties. The guided trial-and-error approaches are efficient but not rigid, and require multiple decisions as the process proceeds. To be efficient at method development, the analyst needs to be engaged in the process, informed about the goals of the analysis, knowledgeable about the properties of the compounds, and flexible in the selection of system properties. Once a satisfactory separation has been obtained, anyone can perform the separation so long as the method is adequately documented and the original method was demonstrated to be robust.

References

[1] Geiss F. Fundamentals of thin-layer chromatography. Heidelberg: Huethig; 1987.
[2] Nyiredy Sz, editor. Planar chromatography. A retrospective view for the third millennium. Budapest: Springer; 2001.
[3] Poole CF. The essence of chromatography. Amsterdam: Elsevier; 2003.
[4] Wall PE. Thin-layer chromatography. A modern approach. Cambridge: Royal Society of Chemistry; 2005.
[5] Hahn-Deinstrop E. Applied thin-layer chromatography. Weisbaden: Wiley-VCH; 2006.
[6] Reich E, Schibli A. High-performance thin-layer chromatography for analysis of medicinal plants. New York: Thieme Medical Publishers; 2007.
[7] Spangenberg B, Poole CF, Weins Ch. Quantitative thin-layer chromatography. A practical survey. Heidelberg: Springer; 2011.
[8] Poole CF, Dias NC. Practitioner's guide to method development in thin-layer chromatography. J Chromatogr A 2000;892:123–42.
[9] Shewiyo, Kaale E, Risha PG, Dejaegher B, Smeyers-Verbeke J, Vander Heyden Y. HPTLC methods to assay active ingredients in pharmaceutical formulations: a review of the method development and validation steps. J Pharm Biomed Anal 2012;66:11–23.
[10] Poole CF. Planar chromatography at the turn of the century. J Chromatogr A 1999;856: 399–427.
[11] Poole SK, Kiridena W, Miller KG, Poole CF. Planar chromatographic methods for determination of the quality of spices and flavors as exemplified by cinnamon and vanilla. J Planar Chromatogr 1995;8:257–68.
[12] Jork H, Funk W, Fischer W, Wimmer H. Thin-layer chromatography. Reagents and detection methods. Weinheim: VCH; 1990.
[13] Halpappp H, Krebs K-F. Thin-layer chromatographic and high-performance thin-layer chromatographic ready-for-use preparations with concentrating zones. J Chromatogr 1977;142:823–53.
[14] Poole CF, Poole SK, Dean TA, Chirco NM. Sample requirements for quantitation in thin-layer chromatography. J Planar Chromatogr 1989;2:180–9.
[15] Poole SK, Poole CF. High performance stationary phases for planar chromatography. J Chromatogr A 2011;1218:2648–60.

[16] Belay MT, Poole CF. Determination of molecular weight and oligomer distribution of poly(ethylene glycols) by thin-layer chromatography and automated multiple development. J Planar Chromatogr 1991;4:424–31.
[17] Litvinova LS. Thin-layer chromatography of poly(methyl methacrylate) using binary mobile phases with methylethylketone as displacer. J Planar Chromatogr 2012;25: 112–6.
[18] Hauck HE, Halpaap H. HPTLC on inert pre-coated layers of cellulose and silica 50000. Chromatographia 1980;13:538–47.
[19] Momchilova S, Nikolova-Damyanova B. Stationary phases for silver-ion chromatography of lipids: preparation and properties. J Sep Sci 2003;26:261–70.
[20] Panalu HA, Homayon A, Rahimi A, Mori E, Izadi A, Parvin MM. HPTLC separation and quantitative analysis of aspirin, salicylic acid and sulfosalicyclic acid. J Planar Chromatogr 2010;23:137–40.
[21] Jarne C, Cebolla VL, Membradio, Le Maplham K, Giusti P. High-performance thin-layer chromatography using automated multiple development for the separation of heavy petroleum products according to their number of aromatic rings. Energy & Fuels 2011;25:4586–94.
[22] Kiridena W, Poole CF. Structure-driven retention model for solvent selection and optimization in reversed-phase thin-layer chromatography. J Chromatogr A 1998;802: 335–7.
[23] Jost W, Hauck HE. High-performance thin-layer chromatographic pre-coated plates with amino modification and some applications. J Chromatogr 1983;261:235–44.
[24] Kowalska T, Sherma J, editors. Thin layer chromatography in chiral separations and analysis. Boca Raton: CRC Press; 2007.
[25] Del Bubba M, Checchini A, Lepri L. Thin-layer chromatography enantioseparations on chiral stationary phases: a review. Anal Bioanal Chem 2013;405:533–54.
[26] Poole CF, Karunasekara T. Solvent classification for chromatography and extraction. J Planar Chromatogr 2012;25:190–9.
[27] de Juan A, Fonrodona G, Casassas E. Solvent classification based on solvatochromic parameters: a comparison with the Snyder approach. Trends Anal Chem 1997;16:52–62.
[28] Barwick VJ. Strategies for solvent selection – a literature review. Trends Anal Chem 1997;16:293–309.
[29] Johnson AR, Vitha MF. Chromatographic selectivity triangles. J Chromatogr A 2011; 1218:556–86.
[30] Abraham MH, Ibrahim A, Zissimos AM. Determination of solute descriptors from chromatographic measurements. J Chromatogr A 2004;1037:29–47.
[31] Poole CF, Atapattu SN, Poole SK, Bell AK. Determination of solute descriptors by chromatographic methods. Anal Chim Acta 2009;652:32–53.
[32] Poole CF, Ariyasena TC, Lenca N. Estimation of the environmental properties of compounds from chromatographic measurements and the solvation parameter model. J Chromatogr A 2013;1317:85–104.
[33] Snyder LR. Solvent selectivity in normal-phase TLC. J Planar Chromatogr 2008;21: 315–23.
[34] Poole SK, Poole CF. A method for estimating the solvent strength parameter for liquid-solid chromatography. Chromatographia 2001;53. S-162–6.
[35] Meyer VR, Palamareva MD. New graph of binary mixture solvent strength in adsorption liquid chromatography. J Chromatogr 1993;641:391–9.
[36] Kiridena W, Poole CF. Structure-driven retention model for method development in reversed-phase thin-layer chromatography on octadecylsiloxane-bonded layers. J Planar Chromatogr 1999;12:13–25.
[37] Siouffi A-M. Some aspects of optimization in planar chromatography. J Chromatogr 1991;556:81–94.

[38] Reich E, George T. Method development in HPTLC. J Planar Chromatogr 1997;10:273–80.

[39] Nyiredy Sz. Planar chromatographic method development using the PRISMA optimization system and flow charts. J Chromatogr Sci 2002;40:1–11.

[40] Rozylo JK, Siembide R. Comparison of different methods of optimization of the TLC process. J Planar Chromatogr 1997;10:97–107.

[41] Nyiredy Sz, Dallenbach-Toelke, . Stricher O. Correlation and prediction of the k values for mobile phase optimization in HPLC. J Liq Chromatogr 1989;12:95–116.

[42] Poole CF, Poole SK, Belay MT. Insights into the mechanism and applications of unidimensional multiple development in thin-layer chromatography. J Planar Chromatogr 1993;6:438–45.

[43] Dias NC, Poole CF. Compliance of retention data on inorganic oxide adsorbents with the solvation parameter model. J Planar Chromatogr 2001;14:160–74.

[44] Dias NC, Poole CF. Optimization of ternary mobile phases in reversed-phase thin-layer chromatography by use of a mixture-design approach with the solvation parameter model. J Planar Chromatogr 2000;13:337–47.

[45] Poole CF, Poole SK, Fernando WPN, Dean TA, Ahmed HD, Berndt JA. Multidimensional and multimodal thin-layer chromatography: pathway to the future. J Planar Chromatogr 1989;2:336–45.

[46] Poole CF, Belay MT. Progress in automated multiple development. J Planar Chromatogr 1991;4:345–59.

[47] Summanen J, Hiltunen R, Vuorela H. The choice of parameters for the optimization of automated multiple development. J Planar Chromatogr 1998;11:16–24.

[48] Lodi G, Betti V, Brandolini E, Menziani E, Tosi B. Some aspects and examples of automated multiple development (AMD) gradient optimization. J Planar Chromatogr 1991;4:106–10.

[49] Colthup PV, Bell JA, Gadson DL. Use of linear solvent gradients in manual and automated multiple development for the separation and characterization of drugs and metabolites by HPTLC. J Planar Chromatogr 1993;6:386–93.

[50] Markowski W. Computer-aided optimization of gradient multiple development thin-layer chromatography. Part II. Multi-stage development. J Chromatogr 1993;635:283–9.

[51] Nurok D. Strategies for optimizing mobile phases in planar chromatography. Chem Rev 1989;89:363–75.

[52] Bayne CK, Ma CY. Optimization of solvent composition for high performance thin-layer chromatography. J Liq Chromatogr 1987;10:3529–46.

[53] Zhang D-S, Wen L, Li Y-P, Hu C-Q. Establishment and optimization of an HPTLC method for the analysis of gatifloxacin and related substances by design of experiment. J Planar Chromatogr 2013;26:219–25.

CHAPTER

13

Validation of Thin Layer Chromatographic Methods

D.H. Shewiyo [1], *B. Dejaegher* [2,3], *Y. Vander Heyden* [2]

[1] Tanzania Food and Drugs Authority, Dar es Salaam, Tanzania; [2] Department of Analytical Chemistry and Pharmaceutical Technology (FABI), Center for Pharmaceutical Research (CePhaR), Vrije Universiteit Brussel (VUB), Brussels, Belgium; [3] Laboratory of Instrumental Analysis and Bioelectrochemistry, Institute of Pharmacy, Université Libre de Bruxelles (ULB), Brussels, Belgium

13.1 INTRODUCTION

As it is the case with all analytical methods that are assaying compounds, thin-layer chromatographic and high-performance thin-layer chromatographic (TLC/HPTLC) methods must be validated to ensure that they are fit for their intended purpose and, where applicable, meet the strict regulatory requirements for controlled products, such as foods and pharmaceuticals. Validation procedures carried out for other analytical techniques can also be applied to TLC/HPTLC methods to evaluate their performance parameters. In this chapter, the discussion on validation procedures of TLC/HPTLC analytical methods will be based on the validation of methods for pharmaceutical analysis since validation in this area is considered to be quite strict.

Method validation, as observed in the literature, may be categorized into classical [1–5] and alternative [6–8] approaches. The classical approach, which determines a number of method performance parameters, is widely described in the literature and is the most commonly used, whereas the alternative approach was more recently introduced by the *Société Française des Sciences et Techniques Pharmaceutiques* (SFSTP) [6], and

Instrumental Thin-Layer Chromatography
http://dx.doi.org/10.1016/B978-0-12-417223-4.00013-3

is more frequently applied nowadays. It is based on the use of the so-called accuracy profiles.

13.2 METHOD VALIDATION USING THE CLASSIC APPROACH

The sources of information in the literature describing the classical validation procedures for analytical methods include among many others, the International Conference on Harmonization of Technical Requirements for the Registration of Pharmaceuticals for Human Use (ICH) guidelines [1], the Food and Drug Administration Guidance for Industry [2], the United States Pharmacopeia [3], the Association of Official Analytical Chemists guidelines for method validation for a single laboratory [4], and the Quality Assurance guideline in Analytical Chemistry [5].

According to the above references, method performance parameters to be evaluated are the linearity of the calibration line, precision, trueness, range, limits of detection and quantification, specificity, and robustness. The extent to which a TLC/HPTLC method is validated, i.e., which parameters are evaluated, depends on the intended goals to be achieved and the intended use. For example, a qualitative TLC/HPTLC procedure will be less extensively validated and its validation will be limited to evaluating the specificity and detection limit, while for a quantitative procedure, more parameters, such as linearity of the calibration line, precision, trueness, limits of detection and quantification, and specificity, will be evaluated. When the method is to be used in many different laboratories, it most probably also will be subjected to a robustness test. In a classical validation approach, generally the performance parameters are evaluated individually and the result for each one of them is compared with an acceptance criterion.

13.2.1 Linearity of the Calibration Line

Analysts generally prefer to work with linear equations allowing an easy estimation of the amount of sample ingredient(s). In these equations, a response related to the spot (e.g., peak area after scanning) on the plate is related to its concentration. With TLC/HPTLC methods, the concentration in the standard (sample) is often expressed as the amount (mass) spotted. For these methods, straight lines may be obtained when dilute solutions are used within limited concentration ranges. However, in some cases, due to the saturation effect taking place at the application point on the plate, quadratic relationships are reported to offer acceptable results. For instance, Habte et al. [9] developed and validated a method for the simultaneous separation and determination of lamivudine (LVD) and zidovudine in pharmaceutical formulations. The calibration data fitted well to the quadratic model with $r^2 = 0.9997$ and 0.9996 for LVD and zidovudine, respectively, as compared to a straight-line model, where the data fitted less well ($r^2 = 0.9776$ and 0.9616, respectively). The mean assay results obtained using the quadratic models were $99.74 \pm 0.84\%$ and $99.76 \pm 0.88\%$, for LVD and zidovudine, respectively.

Kaale et al. [10] participated in and described the results of an interlaboratory investigation on the use of HPTLC methods to perform assays of LVD–zidovudine, metronidazole, nevirapine (NVP), and quinine composite samples. A linearity study for each compound was performed at concentration levels covering 50–150% of the target concentration. The calibration data for all compounds fitted well to quadratic models with r^2 values above 0.9980. The assay results obtained using the quadratic models for the participating laboratories for all compounds ranged from 97.0 to 102.7% with the percent relative standard deviations (%RSD) $\leq$3.21%.

When linear calibration lines are desired, usually a coefficient of determination (r^2) of at least 0.99 is considered by many authors as an indicator of the linearity of the calibration lines. However, this is not a good criterion [11–14] since it has been demonstrated that values close to 1.0 can also be obtained for a nonlinear relationship [14,15]. Figure 13.1 shows calibration data that gave an r^2 close to one but the relationship is not a straight line. Therefore, visual observation of the plot of the response (y) against the concentration of the analyte or the amount spotted (x) is advised. It may be followed by plotting the y-residual values against the

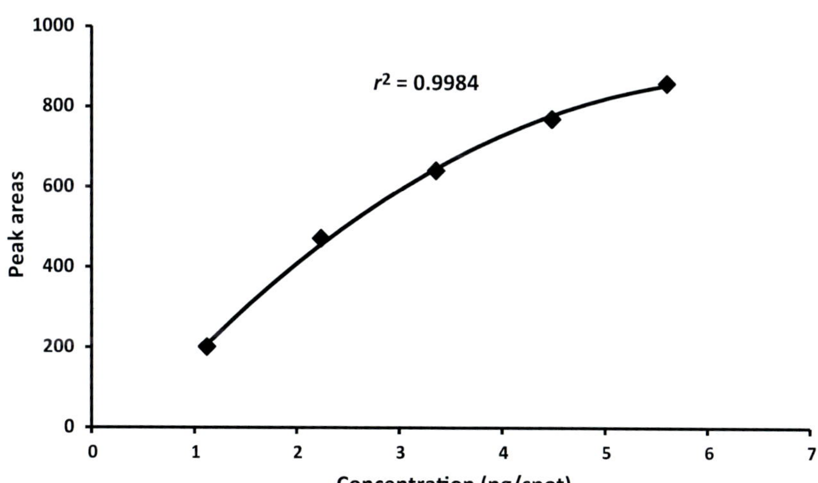

FIGURE 13.1 A nonlinear calibration curve showing a coefficient of determination (r^2) close to 1 (above 0.99).

concentration when there are enough measurements, in order to evaluate whether a trend can be seen [11,13,16]. For a straight-line situation the residuals do not show a trend, but they do when a curved behavior of the response is modeled by a straight line (i.e., by an improper model). However, statistical tests, such as a lack-of-fit (LOF) test (F-test) [16–18] or a test for the significance of the quadratic coefficient b_2 in the model $y = b_0 + b_1x + b_2x^2$ [16] are most appropriate for evaluating the linearity of the calibration line.

13.2.2 Precision (Repeatability and Time-Different Intermediate Precision)

The precision of an analytical procedure "expresses the degree of closeness of agreement (degree of scatter) between a series of measurements obtained from multiple sampling of the same homogeneous sample under the prescribed conditions" [1]. It is important to note that a precision study should always include all steps of the method from the sample preparation to the final measurements. Regulatory guidelines stipulate repeatability, intermediate precision, and reproducibility as the types of precisions possibly to be evaluated [1,2]. Precision studies for formulation assays should preferably be performed at three concentration levels with replicates at each level. Such concentration levels are 80, 100, and 120% of the expected sample concentration. In general, the number of concentration levels and their range will depend on the purpose of the method.

Repeatability expresses the precision under the same operating conditions (one instrument, one analyst) over a short interval of time, while the intermediate precision expresses within-laboratory variations possibly due to different days, different analysts, different equipment, different calibrations, or combinations of these factors. Reproducibility evaluates precision between laboratories, thus acquiring an interlaboratory study. Reproducibility is not covered in this chapter. Since analytical methods within a laboratory are used on different days and possibly by several analysts, sometimes also using different equipment, it is important that the repeatability and intermediate precision are evaluated. Literature search within published, in-house developed, and validated TLC/HPTLC methods for pharmaceutical analysis to assay active drugs in oral solid and liquid dosage forms, indicated appropriate repeatability and time-different intermediate precision values expressed as %RSD to be $\leq$3% and $\leq$5%, respectively [19].

In practice, the intermediate precision usually is studied by measuring on several days, and occasionally varying also the analysts and/or equipment [1]. Some authors recommend a simultaneous evaluation of the repeatability and the time-different intermediate precision using an experimental design with replicated, usually duplicated, measurements under repeatability conditions, at different days [1,17,20]. When such experimental design is employed, an analysis of variance (ANOVA) is used to estimate the individual variance components, i.e., the repeatability variance (s_r^2) and the between-days variance ($s^2_{\text{between-days}}$). Then the time-different intermediate precision ($s^2_{I(t)}$) is estimated as

$$s^2_{I(t)} = s_r^2 + s^2_{\text{between-days}} \tag{13.1}$$

It has been shown that the time-different intermediate precision is underestimated when the above experimental design is not used [20]. It is described that, without the experimental design, the time-different intermediate precision was estimated as %RSD = 5.2, while with the design and from the ANOVA it was estimated as %RSD = 6.0 for the same method.

13.2.3 Trueness

Trueness of an analytical method "expresses the closeness of agreement between the mean value obtained from a series of measurements and the value which is accepted either as a conventional true value or an accepted reference value" [21,22]. It is measured as recovery or bias. For a method assaying pharmaceutical formulations it is evaluated by spiking with reference standard(s) of the analytes of interest to the prepared blank matrix or to preanalyzed samples, i.e., by using a so-called standard-addition approach. The ICH guidelines [1] indicate that recovery studies for the above-mentioned assay methods are to be performed at a minimum of three concentration levels and three replicates at each level. Preferably, recoveries are determined at 80, 100, and 120% of the expected sample concentration.

13.2.4 Range

According to the ICH guidelines [1], the range of the method is "the interval between the upper and lower concentrations of the analyte in samples (including these concentrations) where methods demonstrate a suitable level of precision, trueness and linearity". The ICH guidelines propose concentration levels covering 80–120% of the target concentration to be evaluated as the range of a particular method to assay a drug substance or a finished (drug) product. In many cases, authors report the ranges as the lower and upper concentrations levels of the calibration lines or most extremely as the upper and lower quantitation limits, which usually exceed the 80–120% of the target sample concentration. The range of the method can be easily determined when a

method can be used in situations where some validation standard data are missing, therefore making the repeated data within concentration levels to be unbalanced. In such conditions, the variance estimation using, for example, ANOVA, would be biased [8].

However, because the validation procedure is normally well planned, the number of validation standards at each concentration level ($n = 1{:}w$) is usually balanced. In such a situation, the different variance components can be estimated using ANOVA [16]. For each concentration level of the validation standards ($m = 1{:}v$), the repeatability or intraseries variance, $s^2_{\mathrm{r},m}$, is estimated. The time-different intermediate-precision variance, $s^2_{\mathrm{I}(t),m}$, is then determined as follows (Eqn (13.10)):

$$s^2_{\mathrm{I}(t),m} = s^2_{\mathrm{r},m} + s^2_{\mathrm{between-days},m} \tag{13.10}$$

where $s^2_{\mathrm{between-days},m}$ represents the between-series variance. These precision estimates are then expressed as %RSD.

13.3.1.4 Estimation of Trueness

The trueness of an analytical method expresses the closeness of agreement between the average of a series of measurements $\overline{x}_m$ and the accepted reference value $\widehat{\mu}_m$ [7,8,21,22], and estimates the systematic error of an analytical method. At each concentration level of the validation standards ($m = 1{:}v$), trueness can be expressed as bias_m (Eqn (13.11)), absolute bias_m (Eqn (13.12)), $\%\mathrm{bias}_m$ (Eqn (13.13)), or as %recovery (Eqn (13.14)) [8,16]:

$$\mathrm{bias}_m = \overline{x}_m - \widehat{\mu}_m \tag{13.11}$$

$$|\mathrm{bias}_m| = |\overline{x}_m - \widehat{\mu}_m| \tag{13.12}$$

$$\%\mathrm{bias}_m = \frac{\overline{x}_m - \widehat{\mu}_m}{\widehat{\mu}_m} \times 100 \tag{13.13}$$

$$\%\mathrm{recovery}_m = \frac{\overline{x}_m}{\widehat{\mu}_m} \times 100 \tag{13.14}$$

where $\overline{x}_m$ is the average of the back-calculated concentrations, and $\widehat{\mu}_m$ is the average of the introduced amounts of the validation standards at each concentration level ($m = 1{:}v$).

13.3.1.5 Estimation of Accuracy

The accuracy of an analytical method expresses the closeness of agreement between the trial result x_{imn} and the accepted reference value μ_{imn} for each measured validation standard ($i = 1{:}s$, $m = 1{:}v$, $n = 1{:}w$) [8,16].

It includes both the trueness and precision of the method. Accuracy (Eqn (13.15)) and %accuracy (Eqn (13.16)) are estimated as follows:

$$\text{accuracy}_{imn} = x_{imn} - \mu_{imn} \tag{13.15}$$

$$\%\text{accuracy}_{imn} = \frac{x_{imn} - \mu_{imn}}{\mu_{imn}} \times 100 \tag{13.16}$$

13.3.1.6 Total Error

Each individual measurement, x_{imn} ($i = 1{:}s$, $m = 1{:}v$, $n = 1{:}w$), is the sum of the true value, μ_{imn}, the absolute bias, $|\text{bias}_m|$, and the intermediate precision standard deviation $s_{\text{I}(t),m}$:

$$\begin{aligned} x_{imn} &= \mu_{imn} + |\text{bias}_m| + s_{\text{I}(t),m} \\ \Leftrightarrow x_{imn} - \mu_{imn} &= |\text{bias}_m| + s_{\text{I}(t),m} = \text{absolute total error} \end{aligned} \tag{13.17}$$

The absolute total error of an analytical method thus is the sum of the trueness ($|\text{bias}_m|$) and precision ($s_{\text{I}(t),m}$) contributions, and indicates the ability of the analytical method to produce accurate results [8]. It enables the assessment of the quantitative performance of a method. The total error estimation is an essential step in the assessment of the validity of an analytical method, as it is used to calculate the 95% tolerance interval, which provides guarantee of what results will be produced by the method in the future, during routine use of the method [8].

13.3.1.7 Tolerance Interval and Accuracy Profile

The parameters $\%\text{bias}_m$, $s^2_{\text{r},m}$, $s^2_{\text{between-days},m}$, and $\%RSD_{\text{I}(t),m}$ are then used to calculate the expected proportion of observations that falls within the predefined bias acceptance limits $[-\lambda,+\lambda]$ (Figure 13.4). This expected proportion is calculated using the two-sided 95% β-expectation tolerance interval at each concentration level of the validation standards ($m = 1{:}v$). The lower L_m and the upper U_m tolerance interval limits are determined as follows:

$$L_m = \%\text{bias}_m - Q_{t\left(v,\frac{1+\beta}{2}\right)} \sqrt{1 + \frac{1}{swB_m^2}}\,\%\text{RSD}_{\text{I}(t),m} \tag{13.18}$$

$$U_m = \%\text{bias}_m + Q_{t\left(v,\frac{1+\beta}{2}\right)} \sqrt{1 + \frac{1}{swB_m^2}}\,\%\text{RSD}_{\text{I}(t),m} \tag{13.19}$$

where $Q_{t\left(v,\frac{1+\beta}{2}\right)} = \beta$ quantile of the Student t distribution with ν degrees of freedom, with $v = \frac{(R_m+1)^2}{\frac{(R_m+1/w)^2}{(s-1)} + \frac{(1-1/w)}{sw}}$ where s is the number of series, w the

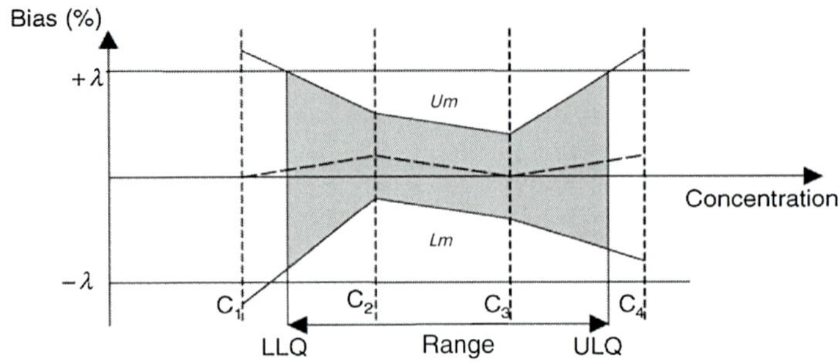

FIGURE 13.4 Accuracy profile showing the acceptance limits (+λ, −λ), the upper and the lower tolerance limits (continuous lines), and the intersection points that define the upper and lower quantification limits (ULQ and LLQ). C1−C4: concentration levels. *Reprinted with permission from Ref. [30].*

number of replicates for the validation standards, and $B_m^2 = \frac{R_m + 1}{wR_m + 1}$ with

$$R_m = \frac{s^2_{\text{between-days},m}}{s^2_{\text{r},m}}.$$

After calculating the two-sided 95% β-expectation tolerance limits at each concentration level of the validation standards ($m = 1{:}v$), the resulting L_m, U_m, $-\lambda$, and $+\lambda$ at the different concentration levels are then connected to construct the accuracy profile [7,8,20] (Figure 13.4). An accuracy profile allows visually observing the ability of an analytical method to fulfill its objectives and to control the risks associated with its routine application [6,7]. When the 95% β-expectation tolerance interval limits for the considered concentration levels of the validation standards fall within the bias acceptance limits, it is guaranteed that future results of the method will fall within the acceptable bias limits.

The accuracy profile reflects the adequacy of the response function used to calculate concentrations of the validation standards. In addition, the upper and the lower concentrations of the validation standards, considered during the validation study, define the upper and the lower quantification limits. In case the tolerance limits intersect with the bias acceptance limits, the intersection points will define the upper and lower quantification limits (Figure 13.4).

13.3.1.8 Measurement and Expanded Measurement Uncertainties

The measurement uncertainty is a parameter that characterizes the dispersion of the values that could reasonably be attributed to the measured compound [28]. It is used to indicate whether measurement result complies with specifications, such as regulatory or manufacturing limits [29]. The measurement uncertainty, u(Y), is calculated at each concentration level of each compound, according to Eqns (13.20), (13.21), and (13.22) [28]:

$$u(Y)^2 = s_{I(t)}^2 + u\left(\widehat{\delta}\right)^2 \tag{13.20}$$

with $s_{I(t)}$ the time-different intermediate precision, and $u(\widehat{\delta})$ the uncertainty associated with the bias δ of the method. The latter is derived as follows:

$$u\left(\widehat{\delta}\right)^2 = \frac{s_{I(t)}^2\left(1 - \gamma + \frac{\gamma}{w}\right)}{s} \tag{13.21}$$

with $\gamma = \frac{s_r^2}{s_{I(t)}^2}$, w the number of replicates, and s the number of series of the validation standards.

The measurement uncertainty, u(Y), is then used to derive the expanded measurement uncertainty U(Y) as follows:

$$U(Y) = 2u(Y) \tag{13.22}$$

where 2 is the coverage factor based on the 95% confidence level.

The expanded uncertainty provides an interval within which the value of the measurand is believed to lie with the above level of confidence.

13.3.2 Example of an HPTLC Method Validated Using Accuracy Profiles

A normal-phase HPTLC method for the analysis of sulfamethoxazole (SMX) and trimethoprim (TMP) in co-trimoxazole tablets was validated using accuracy profiles [30]. Validation protocol V2 was found to be the most appropriate. Calibration standard solutions were prepared without tablet matrix, while the validation standards were prepared independently in the tablet matrix. The number of series in V2 is three, thus this protocol requires at least 45 experiments. The low, intermediate, and high concentration levels for both calibration and validation standards were 80, 100, and 120% of the target concentration of each compound in the tablet formulation. For both calibration and validation standards the concentrations were 120, 150, and 180 ng/spot band for SMX and 150, 180, and 216 ng/spot band for TMP. Duplicate measurements were made for the calibration standards and triplicate for the validation standards at each concentration level, daily for 6 days. The calibration and validation standards were spotted on the HPTLC plates as bands. The calibration-standard amounts at each concentration level (x_{mn}) were also estimated using (Eqn (13.9)) and expressed as nanograms per band, i.e., 119.6, 149.4, and 181.1 for SMX and 150.2, 179.2 and 215.8 for TMP.

The straight-line models were evaluated by an F-test for LOF at 5% significance level and were found to be adequate (see Section 13.2.7.1). Therefore, a straight-line response function was built for both SMX and TMP. Then, the amount of each compound in the validation standards was estimated using Eqn (13.9).

Since the validation data were balanced, ANOVA instead of the maximum likelihood method was used to estimate the repeatability and time-different intermediate precision variances (Eqn (13.10)). For both SMX and TMP the %RSD values for repeatability and time-different intermediate precision did not exceed 1.63 and 1.95%, respectively. The repeatability, the between-days variances, and the %RSD of the time-different intermediate precision for each compound at each concentration level are indicated in Table 13.1, and were used to estimate the 95% tolerance intervals.

Trueness was expressed as %bias or %recovery. The %bias values calculated using Eqn (13.13) at each of the three concentration levels were −0.35, −0.42, and 0.62 for SMX; and 0.31, −0.46, and −0.08 for TMP. These values are much lower than the limits set for active ingredients in pharmaceutical formulations, i.e., ±5%, and indicate a very low bias and the absence of matrix effects. The %bias values are also used in the computation of the 95% tolerance interval. Trueness was also presented as %recovery (Eqn (13.14)) and the obtained values ranged from 99.0 to 100.6% for all compounds.

The accuracy and %accuracy for SMX and TMP were estimated using Eqns (13.15) and (13.16). The highest values of %accuracy for SMX and

TABLE 13.1 Precision Results: Repeatability, Between-Days Variances, and Time-Different Intermediate Precision for SMX and TMP at Each Concentration Level ($w = 3$, $v = 3$, $s = 6$)

		Repeatability			Time-Different Intermediate Precision	
Compound	Concentration Level (%)	Standard Deviation (ng/spot)	% RSD	Between-Days Variance	Standard Deviation (ng/spot)	% RSD
SMX	80	0.864	0.72	2.074	1.680	1.40
	100	1.274	0.85	0.418	1.429	0.95
	120	0.766	0.42	0.165	0.867	0.48
TMP	80	2.456	1.63	2.613	2.940	1.95
	100	1.739	0.97	4.895	2.814	1.57
	120	1.663	0.77	8.232	3.316	1.54

TMP were 2.18 and 3.56, respectively, and both were well within the bias limits of [−5%, +5%].

The absolute total error values were calculated using (Eqn (13.17)) at each concentration level. The values (in nanogram per band) obtained for SMX were 2.1, 2.1, and 2.0, whereas for TMP they were 3.4, 3.6, and 3.5. The values of the total error are also used in calculating the 95% tolerance intervals of the methods.

For each compound, the bias acceptance limits for the total errors were selected to be ±5%, which are often used for active ingredients in pharmaceutical formulations. The lower L_m and the upper U_m tolerance interval limits for each concentration level were calculated using Eqn (13.18) and (Eqn (13.19), respectively. Table 13.2 shows the two-sided 95% β-tolerance limits for each compound at every concentration level studied. The accuracy profiles of each compound are presented in Figure 13.5.

These accuracy profiles show that the tolerance limits for all compounds fall well within the bias acceptance limits set at ±5%. This means that the HPTLC method is capable of obtaining accurate results at the studied concentration ranges (120.0−180.0 ng/band for SMX and 150.0−216.0 ng/band for TMP). The risk of having future measurements falling outside the ±5% acceptance limits at the studied concentration levels is strictly controlled; hence, the method is valid to be applied in the routine assay of the compounds in the co-trimoxazole fixed-dose combination tablets.

Furthermore, this accuracy profile approach also confirms the selection of the straight-line response functions for the quantification of SMX and TMP in their tablets. The lowest and the highest concentrations studied for each compound (as indicated in the range above) can be considered as the lower and upper QLs, respectively. Most probably still lower and higher limits can be defined, but such values were not validated for the actual methods. Within the actual quantification limits, the measurement results have acceptable accuracy because their 95% tolerance limits do not

TABLE 13.2 The Two-Sided 95% Tolerance Limits for Each Compound at Each Concentration Level

Compound	Concentration (ng/spot)	95% Tolerance Limits
SMX	119.6	[−3.7; 3.0]
	149.4	[−2.5; 1.6]
	181.1	[−0.4; 1.6]
TMP	150.5	[−4.4; 5.0]
	179.2	[−4.0; 3.1]
	215.8	[−3.7; 3.5]

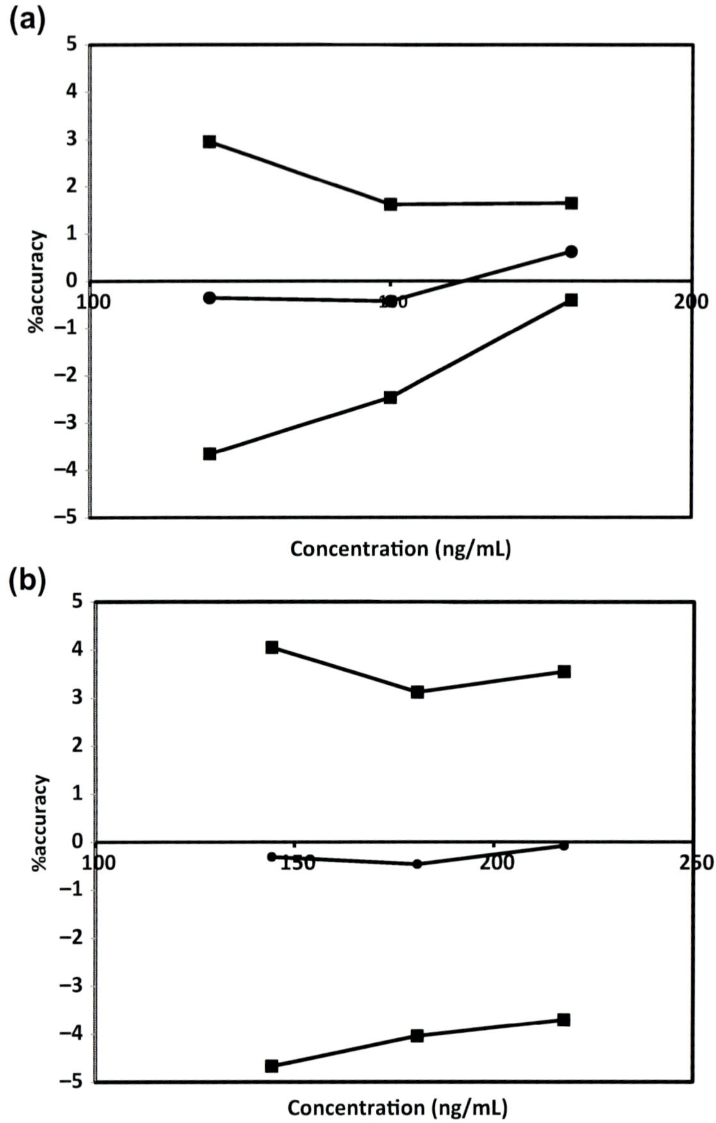

FIGURE 13.5 The accuracy profiles of (a) SMX and (b) TMP. Legend: ● %bias, ■ tolerance limit.

exceed the ±5% acceptable limits. Therefore, for each compound the interval from the lowest to the highest tested concentration can be regarded as the range of the corresponding method.

The measurement uncertainties and expanded measurement uncertainties (nanograms per band) for each compound at each concentration

TABLE 13.3 Measurement Uncertainty, u(Y) (ng/spot), and Expanded Uncertainty, U(Y) (ng/spot), for Each Compound at Each Concentration Level

Compound	Concentration (ng/spot)	Measurement Uncertainty u(*Y*) (ng/spot)	Expanded Uncertainty U(*Y*) (ng/spot)
SMX	119.6	1.8	3.6
	149.4	1.5	3.0
	181.1	0.9	1.8
TMP	150.5	3.1	6.1
	179.2	3.0	6.0
	215.8	3.5	7.1

level of SMX and TMP are presented in Table 13.3. The highest and the lowest values indicate the range of uncertainty over the concentration levels of each compound, and are used to determine compliance of results to specifications during routine use of the validated method. In determining the compliance of a sample characteristic to the set specifications, the measured result with the measurement uncertainty value are compared to the upper and lower value of the specification. For instance, suppose 40 ng/band is the measured result, 2 ng/band the measurement uncertainty of the method, and 35–45 ng/band the set specification. The result is spread as 40 ± 2 ng/band (i.e., 38–42 ng/band), therefore complying with the specifications. For further description of the compliance assessment using measurement uncertainty, the reader is referred to reference [29].

13.4 CONCLUSION

Analytical method validation is as important for quantitative TLC/HPTLC methods as for any other analytical method. Both the classic and alternative method validation approaches, the latter using accuracy profiles, can be used to validate these methods.

When properly optimized and validated, quantitative TLC/HPTLC methods result in adequate assays.

References

[1] Guidelines Prepared within the International Conference on Harmonization of Technical Requirements for the Registration of Pharmaceuticals for Human use (ICH). Validation of analytical procedures: text and methodology, Q2 (R1). p. 1–13, http://www.ich.org/; 2005 [accessed 04.11.2013].

[2] Food and Drug Administration. Guidance for Industry, analytical procedures and method validation (Draft guidance). Available at: http://www.fda.gov/downloads/Drugs/GuidanceComplianceRegulatoryInformation/Guidances/ucm122858.pdf; 2000 [accessed 15.10.2013].

[3] The United States Pharmacopeia 29, National Formulary 24, Unites States Pharmacopoeia Convention, Rockville MD USA; 2006.

[4] AOAC Guidelines for single laboratory validation of chemical methods for dietary supplements and botanicals. Available at: http://www.aoac.org/Official_Methods/slv_guidelines.pdf; [accessed 16.09.2013].

[5] Funk W, Dammann V, Donnevert G. Quality assurance in analytical chemistry. 2nd ed. Weinhein: Wiley-VCH; 2007.

[6] Hubert Ph, Nguyen-Huu J-J, Boulanger B, Chapuzet E, Chiap P, Cohen N, et al. Harmonization of strategies for the validation of quantitative analytical procedures: a SFSTP proposal – Part I. J Pharm Biomed Anal 2004;36:579–86.

[7] Hubert Ph, Nguyen-Huu J-J, Boulanger B, Chapuzet E, Chiap P, Cohen N, et al. Harmonization of strategies for the validation of quantitative analytical procedures: a SFSTP proposal – Part II. J Pharm Biomed Anal 2007;45:70–81.

[8] Hubert Ph, Nguyen-Huu J-J, Boulanger B, Chapuzet E, Cohen N, Compagnon P-A, et al. Harmonization of strategies for the validation of quantitative analytical procedures: a SFSTP proposal – Part III. J Pharm Biomed Anal 2007;45:82–96.

[9] Habte G, Hymete A, Mohammed AMI. Simultaneous separation and determination of lamivudine and zidovudine in pharmaceutical formulations using the HPTLC method. Anal Lett 2009;42:1552–70.

[10] Kaale E, Risha P, Reich E, Layloff T. An interlaboratory investigation on the use of high-performance thin-layer chromatography to perform assays lamivudine-zidovudine, metronidazole, nevirapine, and quinine composite samples. J AOAC Int 2010;93: 1836–43.

[11] Ferenczi-Fodor K, Renger B, Vegh Z. The frustrated reviewer–recurrent failures in manuscripts describing validation of quantitative TLC/HPTLC procedures for analysis of pharmaceuticals. J Planar Chromatogr 2010;23:173–9.

[12] Van Loco J, Elskens M, Croux C, Beernaert H. Linearity of calibration curves: use and misuse of correlation coefficient. Accred Qual Assur 2002;72:281–5.

[13] EURACHEM Guide. The fitness for purpose of analytical methods. Available at: http://www.eurachem.org/guides/pdf/valid.pdf; 1998 [accessed 25.11.2013].

[14] Kulikov AU, Verushkin AG, Loginova LP. Comparison of micellar and reversed-phase liquid chromatography for determination of sulfamethoxazole and trimethoprim. Chromatographia 2005;61:455–63.

[15] Yrjonen T. Extraction and planar chromatographic separation technique in the analysis of natural products [Academic Dissertation]. Faculty of Pharmacy, University of Helsinki; November, 2004.

[16] Massart DL, Vandeginste BGM, Buydens LMC, De Jong S, Lewi PJ, Smeyers-Verbeke J. Handbook of chemometrics and qualimetrics: part A. Amsterdam: Elsevier; 1997.

[17] Vovk I, Prosek M. Reproducibility of densitometric and image analyzing quantitative evaluation of thin-layer chromatograms. J Chromatogr A 1997;770:329–36.

[18] Morlock G, HPTLC – MS and other coupling techniques. Available at: http://nte-serveur.univ-lyon1.fr/afsep/fichier/271_14_11_2005.pdf; [accessed 15.11.2013].

[19] Shewiyo DH, Kaale E, Risha PG, Dejaegher B, Smeyers-Verbeke J, Vander Heyden Y. HPTLC methods to assay active ingredients in pharmaceutical formulations: a review of the method developments and validation steps. J Pharm Biomed Anal 2012;66:11–23.

[20] Rozet E, Marini RD, Ziemons E, Boulanger B, Hubert Ph. Advance in validation, risk and uncertainty assessment of bioanalytical methods. J. Pharm. Biomed. Anal. 2011;55: 848–58.

[21] Eurachem. The Fitness for purpose of analytical methods: a laboratory guide to method validation and related topics; 1998. p. 1–75.
[22] International Organization for Standardization (ISO). Statistical methods for quality control, vol. 2, 4th ed., Accuracy (Trueness and precision) of measurement methods and results – part 1-6, ISO 1994(E), 5725 – 1 till 6, www.iso.ch/; [accessed 23.12.2013].
[23] Hubert Ph, Chiap P, Crommen J, Boulanger B, Chapuzet E, Mercier N, et al. The SFSTP guide on the validation of chromatographic methods for drug bioanalysis: from the Washington conference to the laboratory. Anal Chim Acta 1999;391:135–48.
[24] Vander Heyden Y, Nijhuis A, Smeyers-Verbeke J, Vandeginste BGM, Massart DL. Guidance for robustness/ruggedness tests in method validation. J Pharm Biomed Anal 2001; 24:723–53.
[25] Hund E, Vander Heyden Y, Haustein M, Massart DL, Smeyers-Verbeke J. Robustness testing of a reversed-phase high-performance liquid chromatographic assay: comparison of fractional and asymmetrical factorial design. J Chromatogr A 2000;874:167–85.
[26] Vander Heyden Y, Jimidar M, Hund E, Niemeijer N, Peeters R, Smeyers-Verbeke J, et al. Determination of system suitability limits with a robustness test. J Chromatogr A 1999; 845:145–54.
[27] Shewiyo DH, Kaale E, Ugullum C, Sigonda MN, Risha PG, Dejaegher B, et al. Development and validation of a normal-phase HPTLC method for the simultaneous analysis of lamivudine, stavudine and nevirapine in fixed-dose combination tablets. J Pharm Biomed Anal 2011;54:445–50.
[28] EURACHEM/CITAC. Guide: quantifying uncertainty in analytical measurement. 2nd ed. Budapest: EURACHEM/CITAC. http://www.measurementuncertainty.org/pdf/QUAM2000-1.pdf; 2000 [accessed 21.11.2013].
[29] EURACHEM/CITAC. Guide: use of uncertainty information in compliance assessment. 1st ed. EURACHEM/CITAC. http://www.measurementuncertainty.org/pdf/Interpretation_with_expanded%20uncertainty_2007_v1.pdf; 2007 [accessed on 25.11.2012].
[30] Shewiyo DH, Kaale E, Risha PG, Dejaegher B, De Beer J, Smeyers-Verbeke J, et al. Accuracy profiles assessing the validity for routine use of high-performance thin-layer chromatographic assays for drug formulations. J Chromatogr A 2013;1293:159–69.
[31] Shrivastava A, Gupta VB. Methods for the determination of limit of detection and limit of quantitation of the analytical methods. Chron Young Sci 2011;2:21–5.

CHAPTER

14

Separation of (Phospho)Lipids by Thin-Layer Chromatography

Beate Fuchs, Yulia Popkova, Rosmarie Süß, Jürgen Schiller

Medical Department, University of Leipzig, Institute of Medical Physics and Biophysics, Härtelstr, Leipzig, Germany

14.1 INTRODUCTION

During the last decades, it was assumed that liquid chromatographic methods, such as high-performance liquid chromatography (HPLC), would completely replace thin-layer chromatography (TLC) for the separation of lipids. However, a revival of TLC seems nowadays very likely—particularly regarding lipid analysis, since TLC is a frequently used separation technique for lipids [1–5]. Although the terms "TLC" and "high-performance thin-layer chromatography (HPTLC) will be used nearly synonymously in this review, the reader should note that HPTLC is a more sophisticated version of TLC. Differences are primarily related to (1) the different particle size of the layers (5–6 μm in comparison to 10–12 μm), (2) the attention paid to sample application (manual vs automatic), and (3) the methods employed for data processing [6].

Although there are some potential concerns (e.g., the reduced resolution of TLC compared with HPLC and the potential, unwanted oxidation of lipids caused by air exposition on the layer surface), TLC is perhaps the most efficient and versatile technique for the separation of complex lipid mixtures. The following advantages are obvious [7]:

1. TLC is convenient and simple.
2. The (absolutely) necessary equipment is rather inexpensive.
3. The commercial availability of high quality TLC plates has strongly increased the reproducibility of the separation—the former weak point of TLC.

Instrumental Thin-Layer Chromatography
http://dx.doi.org/10.1016/B978-0-12-417223-4.00014-5

4. TLC is rapid since dozens of samples can normally be separated in less than an hour.
5. The identification of separated lipids can be performed easily by various staining reactions.
6. Individual lipid fractions can be re-extracted easily and studied by other analytical methods (such as mass spectrometry (MS) or nuclear magnetic resonance (NMR) spectroscopy) and/or isolated on a preparative scale. There are even "preparative" TLC plates with thick silica layers that can be used to fractionate sample amounts of up to 1 g!
7. TLC is not influenced by any "memory" effects since TLC plates are used for a single application only. This is a significant advantage compared with HPLC and one reason why (quantitative) TLC is certified for use in many different industrial and pharmaceutical processes.
8. TLC requires much smaller solvent volumes than HPLC.

TLC is used classically for routine separations, for the identification of individual lipid classes (normally according to their different headgroups) and for their quantitative determination, normally by densitometry. With the advent of the "automated multiple development" (AMD) technique, all steps from the mixing of the required solvents, development, and drying could be automated. In particularly, AMD supports applications that require reproducible solvent gradient development. The use of gradients (i.e., mixtures of different solvents and/or different salt concentrations varied in a time-dependent manner) is common in HPLC but was not widely used in TLC until the introduction of the AMD apparatus. One selected lipid-related (skin) example of this technique is available in Ref. [8] and more details (not only dedicated to lipids) can be found in the excellent book by Hahn-Deinstrop [9]. A survey of commercially available HPTLC equipment is available at the CAMAG's (a leading supplier of TLC equipment) homepage (http://www.camag.com).

14.1.1 The Structure of Lipid Types Relevant to this Review

"Lipids" are most probably the biomolecules with the highest structural variability. Thus, a detailed discussion of all potential lipids is clearly beyond the scope of this chapter. However, the reader interested in more "exotic" lipids is referred to the recently published book by Claude Leray, which provides an in-depth survey of virtually all known types [10]. In this chapter we will focus on animal-derived lipids with physiological and/or medical significance. A coarse overview of these lipids is presented in Figure 14.1. In particular we will highlight recent advances in the separation of glycero(phospho)lipids and sphingolipids (SL). The reader with little experience in the lipid field should note that lipids are

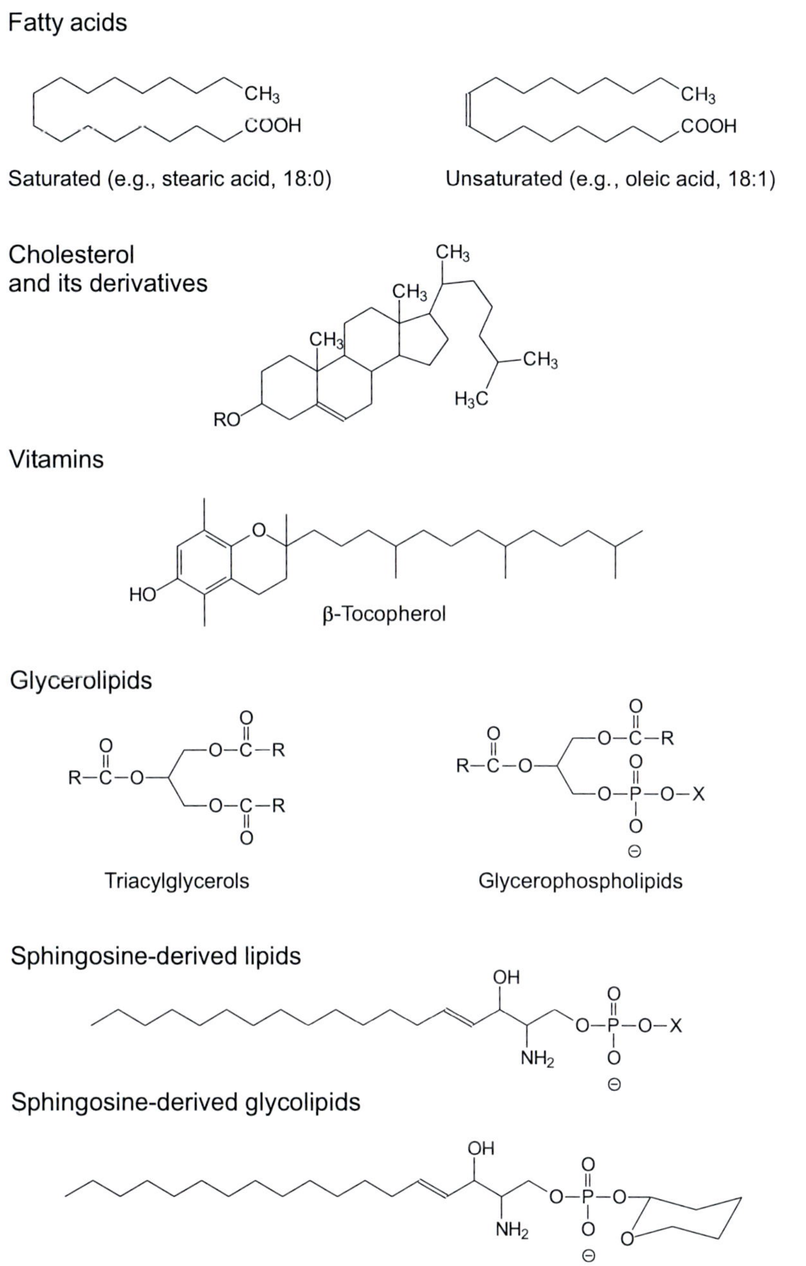

FIGURE 14.1 Short survey of the different classes of "lipids" relevant to this chapter, whereby only some selected examples of the most important lipid subclasses are provided. "R" indicates the presence of a fatty acyl residue, while "x" represents the (polar) headgroup.

not only of relevance for the storage of energy (fat tissue) and nutrition, but are also massively involved in signal transduction processes, for instance, cell differentiation, apoptosis, and phagocytosis [11]. Thus, the determination of selected lipid species is of considerable scientific interest for the characterization of metabolic processes. Therefore, lipid analysis has increasing diagnostic relevance, mainly but not exclusively, in the context of the lipoproteins (LP).

14.1.2 Extraction of Lipids from Biological Samples

Lipids are traditionally (and in the simplest way) defined as apolar compounds that are insoluble in water, and consequently, can be enriched by extraction with organic solvents, such as chloroform or hexane. The extraction of lipids from a given body fluid, cell culture, or tissue sample, therefore, is normally the first step of lipid analysis [12]. Despite its central role, lipid extraction does not engender major interest! This is surprising, since the "fine-tuning" of the extraction conditions helps to improve the extraction yield and influences the accuracy in the analysis of individual lipids. There is also the risk of losing specific lipids: while bulk lipids, such as the zwitterionic phosphatidylcholines (PCs), are extracted nearly quantitatively, major losses can be predicted if lipids with higher (e.g., lysolipids or phosphorylated phosphoinositides) or lower polarities (such as cholesteryl esters or triacylglycerols (TAGs)) are of interest. Some established methods of lipid extraction are summarized in Table 14.1, although there are more successful techniques [13–18].

Independent of the method used, the reader should keep in mind that lipid extraction is a very important step: in all subsequent steps only the lipids extracted from the biological material are observed. In addition, solid-phase extraction can afford a useful alternative method to solvent extraction for the isolation of lipids [22].

14.2 SEPARATION OF LIPIDS BY TLC

14.2.1 Stationary Phases

Silica gel, alumina, and kieselguhr are common stationary phases used in the separation of lipid mixtures by normal-phase chromatography, with silica gel being the most widely used. Silica gel modified by chemically bonded ligands, such as octadecylsiloxane-bonded groups, is suitable for reversed-phase separations.

In normal-phase TLC (most widely used for the separation of lipids), the stationary phase (often unmodified silica gel) is polar and the mobile phase is apolar (i.e., the used solvent system contains significant amounts

TABLE 14.1 Survey of Commonly Used Methods of Lipid Extraction (Note that All Methods have the Disadvantage that Some Lipids May Be Lost during the Extraction)

Solvent System	Particularly Suitable For	Comments/References
$CHCl_3/CH_3OH$ (1:1 v/v) "Bligh & Dyer method"	Useful for water-rich systems, particularly body fluids. Most widely used method.	Partial lipid losses may occur [13]. It is a disadvantage that $CHCl_3$ has a higher density than water because this may lead to impurities from the aqueous layer that has to be penetrated to obtain the lipids. For comparison with the "Folch" method see Ref. [14].
$CHCl_3/CH_3OH$ (2:1 v/v) "Folch method"	Lipids from animal, plant and bacterial tissues. Normally with lower water content in comparison to Bligh & Dyer.	The tissue water is the ternary component and its amount is very important in order to avoid lipid losses [15]. This paper is among the most cited ("top 10") papers worldwide and has received already more than 36,000 citations.
Butanol saturated with water	Plant lipids, i.e., lipids entrapped in starch and rather polar lipids. Products of phospholipase digestion.	Provides particularly good recovery of lysolipids, i.e., of polar lipids [16].
Hexane/2-propanol (3:2 v/v)	Low extraction yields of polar compounds (proteins, pigments, small molecules) as impurities since the solvent mixture is apolar.	In contrast to $CHCl_3$ (suspected carcinogen), hexane and 2-propanol are solvents of low toxicity [17]. Plastic material can be used without the release of plasticizers. However, the method seems less effective than chloroform extraction [18].
Chloroform/2-propanol (7:11 v/v)	Particularly suitable for erythrocytes with a high lipid content.	Indicated to provide higher lipid yields than other extraction methods [19].
Chloroform/methanol/ 12 N HCl (2:4:0.1 v/v/v)	Acidic phospholipids, such as phosphatidylserine and phosphoinositides (that are otherwise very difficult to extract quantitatively).	Addition of HCl leads to charge screening and improves extraction yields for acidic lipids [20]. This method leads to complete hydrolysis of plasmalogens (alkenyl-acyl lipids), however, which are very sensitive to acidic conditions [21].

of apolar solvents, such as hexane or chloroform). Layers prepared from smaller particles (HPTLC layers) result in higher separation quality and lower detection limits, i.e., smaller lipid amounts can be applied. Normal-phase chromatography is the standard method of (phospho)lipid separations based on polarity differences due to the different headgroups.

Silver nitrate and boric acid are common silica gel modifiers: silver nitrate is, in particular, useful for the separation of glycerolipids (or free fatty acids (FFA)) with different degrees of unsaturation [23]. The Ag^+ ion is able to form complexes with the π electrons of the lipid double bonds leading to the selective retardation of unsaturated compounds. Even the determination of the position of double bonds within an alkyl chain (that otherwise requires oxidation (for instance by ozone) and careful analysis of the oxidation products [24]) is possible by this approach [25].

In contrast, the impregnation of silica with boric acid is the method of choice to differentiate isomeric lipids, such as 1,2- and 1,3-monoacylglycerols [26] or lipids with carbohydrate moieties. Boric acid forms complexes with lipids containing vicinal hydroxyl groups resulting in the slower migration of these compounds. In addition, boric acid does also bind to acidic phospholipids (such as phosphatidylserine (PS)) and modifies their migration properties as well. Ethylene-diamine-tetraacetic acid (EDTA) acts in a similar manner and improves the detection of PS by forming more compact spots [27]. Ammonium sulfate can be used to improve the separation of phosphatidylinositol (PI) and PS. Using silica gel plates impregnated with 0.4% $(NH_4)_2SO_4$ and chloroform−methanol−acetic acid acetone−water (40:25:7:4:2 v/v/v/v/v) as mobile phase, five different phospholipids (PS, phosphatidylethanolamine (PE), PI, PC, and sphingomyelin (SM)) and three lysophospholipids (LPS (lysophosphatidylserine), LPE (lysophosphatidylethanolamine), and LPC (lysophosphatidylcholine)) can be successfully separated [28]. Some selected R_f values obtained by using boric acid-impregnated silica gel as the stationary phase and chloroform−ethanol−water−triethylamine (30:35:6:35 v/v/v/v) as the mobile phase are summarized in Table 14.2 [29].

14.2.2 Detection Systems

The majority of TLC-separated lipids can be easily identified by characteristic color reactions. A large number of spray agents are commercially available for this purpose [30]. These are normally sorted according to their specificity and if they are destructive or nondestructive reagents.

14.2.2.1 Nondestructive and Nonspecific Reagents

This is the preferred staining method because the lipid structure is not permanently modified and prior knowledge of the identity of the lipid is not required. One of the most frequently used (and oldest) methods is the

TABLE 14.2 Survey of Selected R_f Values for Important Lipids Using Boric Acid-Impregnated Silica Gel Layers and Chloroform–Ethanol–Water–Triethylamine (30:35:6:35 v/v/v/v) as Mobile Phase for the Separation. The Difference between Cerebroside Type I and II is from the Presence of Either a Glucose or a Galactose Residue, Respectively

R_f	Compound	R_f	Compound
0.08	Lysophosphatidylcholine (LPS)	0.47	Phosphatidylglycerol (PG)
0.11	Sphingomyelin (SM)	0.51	Phosphatidylethanolamine (PE)
0.21	Phosphatidylcholine (PC)	0.58	Phosphatidic acid (PA)
0.22	Cerebroside type I	0.68	Cardiolipin (CL)
0.25	Cerebroside type II	0.70	Ceramide (CER)
0.26	Phosphatidylinositol (PI)	0.74	Free fatty acids
0.32	Sulfatides	0.81	*N*-acylphosphatidylethanolamine
0.32	Lysophosphatidylethanolamine (LPE)	0.96	Cholesterol
0.38	Phosphatidylserine (PS)	0.98	Mono-, di-, and triacylglycerols

Reproduced with permission from Ref. [29].

exposure of the developed TLC plate to iodine vapors [31]. This leads to a (reversible) brown complex between iodine and the double bonds of the lipids; afterward the iodine can be removed by exposing the TLC plate to vacuum. There are differences in dependence on the double bond content, however, and fully saturated lipids (not very abundant in biological systems) are difficult to visualize. In addition, iodine can be difficult to remove from highly unsaturated lipids. "Mild" staining can be achieved also by using 2,7-dichlorofluorescein or rhodamine 6G [32] that results (under UV light) in yellow or pink spots, respectively. Both dyes can be easily removed if the polarity of the solvent is changed or the detected lipid is passed through a short column. Similar results can be obtained with primuline [33], which provides detection limits comparable to rhodamine (low nanomole range). Interestingly, primuline can be removed by exposing the TLC plate to high vacuum and is suitable for subsequent identification of separated lipids by MS [34].

It is known that polyunsaturated lipids give intense darkening under conditions employed in silver ion TLC. This is explained by the reduction of Ag^+ to colloidal silver [23]. However, this effect depends on the solvents used for the separation and is most marked in the presence of aromatic hydrocarbons, such as toluene.

14.2.2.2 Destructive and Nonspecific Reagents

Treating the entire TLC plate with a corrosive reagent (e.g., 50% H_2SO_4) and subsequent charring (about one hour at 120 °C) is an established, classical method of lipid detection [35]. However, although not yet investigated in detail, saturated and unsaturated lipids require different times to react. Measuring the intensities of the resulting black spots is also a widely used quantitative measure of individual lipids (detection limits about 25–50 ng per lipid class) [36].

14.2.2.3 Destructive and Specific Reagents

Many reagents (such as ninhydrin that reacts with the amino residue of PE) that react selectively with a particular functional group of phospholipids to generate a colored product are known. A detailed survey of these reagents is beyond the scope of this chapter but is available in Refs [2,37] and from the excellent Internet site www.cyberlipid.org. It should be noted also that complete oxidation of phospholipids (eluted from the silica gel) and subsequent phosphate determination (normally according to Bartlett [38]) is another established quantitative method. This method allows an unequivocal identification of phospholipids and their differentiation from other lipids. In a recent study selected dyes were compared in terms of their achievable detection levels [39]. The most sensitive stain was 0.2% amido black 10B in 1 M NaCl. About 15 ng for diacylglycerols (DAGs), TAGs, and PS, about 100 ng for FFA, and about 500 ng of phorbol esters were obtained. This is a clear indication that standards are an absolute necessity if quantitative data are to be obtained.

14.3 APPLICATIONS

Lipids represent a very complex class of biomolecules. In addition to the headgroup that determines the (phospho)lipid class, the fatty acid chain composition and its linkage type (acyl, alkyl, or alkenyl) add to the complexity. Thus, it is obvious that a single separation step is normally not sufficient to resolve all individual lipid species. However, this information is readily available if normal-phase TLC (i.e., separation of the different lipid classes) is combined with MS detection of the individual lipid species. Therefore, TLC–MS coupling is nowadays a hot topic of research [40].

The aim of this section is to provide a survey of the most important lipid classes and how they can be identified in complex mixtures. Lipids are sorted according to increasing complexity, i.e., the discussion will start with FFA and end with phospho- and glycolipids.

14.3.1 Fatty Acids

The concentration of FFA in biological tissues and body fluids is modest except for organs responsible for fat metabolism, such as the liver [41]. The majority of fatty acids are esterified with alcohols, particularly glycerol. It is common practice to hydrolyze the related lipids and to determine the released FFA. Although this is regularly the domain of gas chromatography–mass spectrometry (GC–MS) [42], FFA analysis may be performed also by TLC—even if the achievable accuracy is higher for GC–MS.

14.3.1.1 Determination of Differences in Chain Length and Number of Double Bonds

Our focus will be on normal-phase TLC because it is more widely used than reversed-phase TLC for this application [43]. However, silver nitrate-impregnated TLC plates are mandatory to differentiate FFA that differ in their double bond content and to distinguish *cis* and *trans* configurations [44]. Using silver nitrate-impregnated silica gel 60 TLC plates, it was possible to isolate polyunsaturated FFA fractions as their methyl esters to facilitate subsequent determination by GC–MS [45]. Toluene–acetonitrile (97:3 v/v) was used to develop the plates and resulted in an excellent separation between dienes, trienes, and tetraenes. It is particularly remarkable that even saturated FFA can be differentiated by TLC [46]. This method requires, however, initial conversion of the FFA to the monodansyl cadavaride derivatives [46]. Using methanol–acetonitrile–tetrahydrofuran (18:2:1 v/v/v), the following R_f values were obtained: 20:0 (0.28), 17:0 (0.45), and 15:0 (0.58). Using the *x*:*y* nomenclature, where "*x*" represents the number of carbon atoms, while "*y*" denotes the number of double bonds—without specifying the positions of the double bonds.

Unfortunately, one-dimensional TLC is not suitable for the separation of complex FFA mixtures that have to be analyzed by two-dimensional TLC. A suitable solvent for the first dimension development is hexane–diethyl ether (9:1 v/v) and hexane–diethyl ether (2:3 v/v) for the second dimension development.

14.3.1.2 Oxidation Products of FFA

Highly unsaturated FFA experience much higher interest than saturated or moderately unsaturated FFA. This is particularly true for arachidonic acid (AA, 20:4) which is released by the enzyme phospholipase A_2 from phospholipids. AA is the educt for many physiologically important products, such as leukotrienes, thromboxanes, etc. In addition, AA is a major target for reactive oxygen species (ROS) generated by inflammatory conditions [47]. An in-depth discussion of these aspects is beyond of the scope of this review. However, the reader should be aware that this is a challenging topic due to (1) the large variety of ROS that are generated under in vivo conditions and, thus, the complex product patterns and (2) the limited stability of some primary products [47]. Rao et al. [48] described a method to separate selected metabolites of AA on silica gel G plates. The mobile phase for the isolation of thromboxane B2 was diethyl ether–methanol–acetic acid (135:3:3 v/v/v). In addition, hydroxyl acids could be separated by using petrol ether–diethyl ether–acetic acid (60:39:1 v/v/v).

Threo- and erythro-isomers of vicinal dihydroxy esters could be separated on silica gel impregnated with boric acid using hexane–diethyl ether (60:40 v/v) as the mobile phase, in which, the threo isomer migrates more rapidly than the erythro isomer.

The hydrogenation of unsaturated fat (fat hardening) plays a significant industrial role but is accompanied by the partial isomerization of the *cis* configuration into the *trans* form. Since many enzymes recognize exclusively compounds with a *cis* configuration, *trans* compounds are assumed to have harmful effects on human health. Therefore, the differentiation of *cis* and *trans* isomers by silver ion TLC has many important applications [49].

14.3.2 Cholesterol and Cholesteryl Esters

Cholesterol is (in addition to phospholipids, such as PC) an important component of biomembranes, while cholesteryl esters are important for the transport of water insoluble FFA in the form of LP. Oxidized LP have been implicated as important in atherosclerosis research [4]. A fluorescence-based method to detect cholesterol in amounts as low as 5 ng was described in 1996 [50]. Organic extracts of human LP were separated on silica layers with hexane–diethyl ether–acetic acid (80/15/1 v/v/v) and afterward incubated in a filipin (a strong polyene fluorophore) suspension. The measured fluorescence intensity was linear between 5 and 3000 ng cholesterol. This approach is also suitable to determine oxidized forms of cholesterol. Combining TLC and GC–MS is a powerful tool for analysis of cholesteryl ester hydroperoxides [51]. Less than 1 nmol was detectable on silica gel TLC plates developed with *n*-hexane–diethyl ether–acetic acid (70/30/1 v/v/v) [51].

Cholesterol is an abundant constituent of food and assumed to be affected by processes such as cooking and freezing. Accordingly, there is significant interest in the determination of cholesterol oxidation products. TLC of the nonsaponifiable constituents of meat extracts was performed on silica gel with hexane–diethyl ether to separate oxysterols from the native sterols. After elution of the oxysterols, a second development with hexane–diethyl ether–ethyl acetate (1:1:1 v/v/v) was performed to separate the sterols. Subsequently, the sample was extracted from the layer and derivatized for detailed composition analysis by GC [52]. Cholesterol oxides can be identified also by TLC [53]. Finally, TLC is also suitable for the analysis of bile acids. These are major metabolites of cholesterol and facilitate its elimination in the feces by the formation of micelles that solubilize the cholesterol in the bile [54].

14.3.3 Glycerides

TAGs are extremely important for nutrition and the optimum storage of excessive energy in the organism since fat tissue is nearly free of water. Additionally, TAGs (normally from vegetable oils, such as palm or olive oil) are also important chemicals in the cosmetic and pharmaceutical industries. In contrast, DAGs are important second messenger molecules. Therefore, the condensation products between glycerol and FFA are of significant interest.

14.3.3.1 *Acylglycerols*

The enzyme lipoprotein lipase (LPase) hydrolyzes the TAG moiety of LP and its activity is of importance to establish risk factors associated with atherosclerosis [55]. Physiologically relevant LPase, however, generates a mixture of different isomers: first 1,2- and 2,3-DAGs that are further converted into 2-monoacylglycerols (MAGs). Finally, 2-MAGs undergo isomerization to 3-MAGs that are subsequently converted into FFA and glycerol. Fortunately, all relevant products can be separated by normal-phase TLC. The mobile phase hexane–diethyl ether–acetic acid (70:30:1 v/v/v) enables the separation of TAGs, FFA, 1,2- and 1,3-DAGs, and MAGs to be achieved. The approximate R_f values for these compounds are 0.7, 0.45, 0.26, 0.23, and 0.05, respectively [56]. Using diethyl ether–hexane–methanol (65:35:3, v/v/v) and Na_2CO_3-impregnated silica gel layers, FFA, MAGs, DAGs, and TAGs result in R_f values of 0.0, 0.18, 0.79–0.85, and 0.98, respectively [57]. It is important to note that the migration of the acyl groups from the *sn*-2 position to the *sn*-1 and *sn*-3 positions may occur if protic solvents (such as water or alcohols) are used [58]. Acyl migration, however, can be suppressed using boric acid-impregnated silica gel plates. Keep in mind that acyl migration can also occur in detergents [59] that are sometimes used to extract lipids or to characterize phospholipids by ^{31}P NMR spectroscopy [60].

14.3.3.2 *Separation by Degree of Unsaturation*

As already discussed, silver nitrate-impregnated silica gel layers are suitable for the separation of fatty acids that differ in the number of double bonds [25]. Since oleic acid (18:1), linoleic acid (18:2), and linolenic acid (18:3) predominate in TAGs from vegetable oils [61], up to nine double bonds may occur within a TAG. Denoting S = saturated, M = monoenoic, D = dienoic, and T = trienoic acids, the following order of R_f values (illustrated in Figure 14.2) is typically obtained:

$$\text{SSS} > \text{SSM} > \text{SMM} > \text{SSD} > \text{MMM} > \text{SMD} > \text{MMD} > \text{SDD} > \text{SST} > \text{MDD} > \text{SMT} > \text{MMT} > \text{DDD} > \text{SDT} > \text{MDT} > \text{DDT} > \text{STT} > \text{MTT} > \text{DTT} > \text{TTT}$$

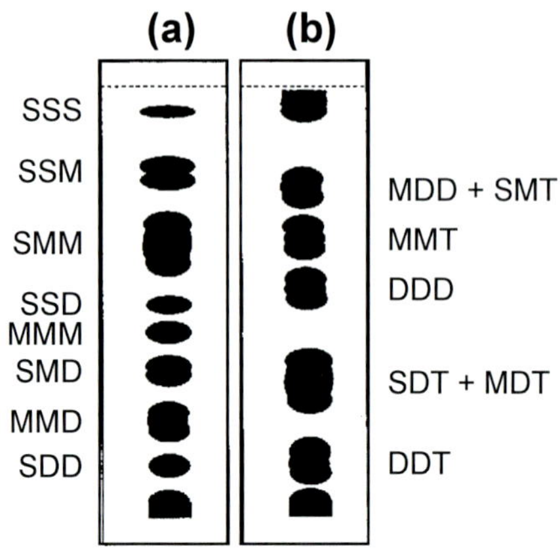

FIGURE 14.2 Schematic separation of soybean TAGs on silica gel G impregnated with 10% (w/w) silver nitrate. Plate (a) was developed with chloroform–methanol (99:1 v/v) while plate (b) was developed with chloroform–methanol (96:4 v/v). Abbreviations: S = saturated, M = monoenoic, D = dienoic and T = trienoic acid glycerol esters. *Reprinted with modification and permission from Christie and Han, Ref. [1].*

However, this separation quality can be hardly achieved by a single TLC development. It is common practice, therefore, to separate the less polar fractions with hexane–diethyl ether (80:20 v/v) or chloroform–methanol (197:3 v/v) first and subsequently the remaining fractions with more polar solvents such as diethyl ether alone or chloroform–methanol (96:4 v/v) [1].

In many papers a description such as "TAG 54:3" can be found. Using this nomenclature, all three fatty acyl residues are combined into a single hypothetical residue. This is done because the determination of the positions of the individual fatty acid substituents is even today a challenging task. Enzymatic or chemical degradation of the TAG of interest is normally needed. Unfortunately, most lipases are not regiospecific [62]. This also applies for pancreatic lipase that is normally used on account of its availability and relatively low cost.

The analysis of the TAG composition of vegetable oils to establish authenticity is an increasing issue in the European Union owing to the adulteration of edible vegetable oils (such as virgin olive oil). TLC is a useful method for this purpose. For instance, silver ion TLC of eight samples of sunflower oil (with different linoleoyl contents) with petrol ether–acetone (25:1 v/v) and petrol ether–acetone–ethyl acetate (100:5:2 or 50:3:2 v/v/v) exhibited the expected differences for the fatty acid side-chain compositions [63]. Combining TLC and GC–MS enabled Myher et al. [64] to quantify more than 100 TAG species in a butter sample. Changes in oil compositions induced by frying can be monitored also by

TLC [65]: however, silver ion TLC and reversed-phase TLC should be combined to obtain detailed information about the double bond contents and the different chain lengths, respectively.

Many common mobile phases contain solvents with strongly different polarities. This may result in "demixing" leading to poor reproducibility of individual TLC separations. The stable solvent mixture (dichloromethane–ethyl acetate–methanol–acetic acid (27:22:38:13 v/v/v/v)) [66] is particularly useful for the determination of the TAG composition of vegetable oils.

14.3.4 SL and Glycolipids

SL are "markers" of various diseases (see Ref. [67] for a review). SL are also common components of human skin [68]. Sphingomyelin, the most abundant, will be discussed later since it is normally detected together with common phospholipids. Chloroform–methanol–water mixtures (ratios between 70:30:4 and 50:40:10 v/v/v) are typically used for neutral SL, while gangliosides containing sugar or sialyl residues require the addition of salts for their separation. Mixtures such as 2-propanol–6 M aqueous ammonia–methyl acetate (15:5:1 v/v/v) are typically used to separate glycated SL (up to four carbohydrate residues). Figure 14.3 provides an illustration of a typical separation. Even isomeric glycopyranose residues, such as glucose and galactose, can be differentiated using this method [69]. It is also remarkable that normal fatty acid and 2-hydroxyfatty acid residues lead to a splitting of the separated zones and can, thus, be differentiated.

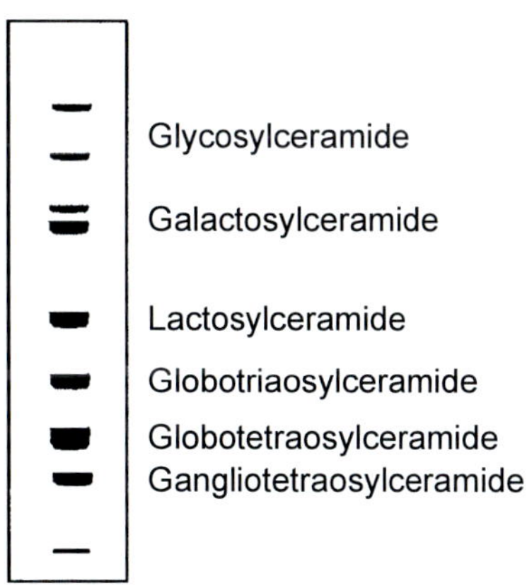

FIGURE 14.3 Schematic separation of neutral sphingolipids by HPTLC on silica gel using 2-propanol–6 M aqueous ammonia–methyl acetate (15:5:1 v/v/v) as mobile phase. *Reprinted with modification and permission from Christie and Han, Ref. [1].*

Chloroform–acetone–methanol–acetic acid–water (46:17:15:14:8 v/v/v/v/v) and chloroform–methanol–acetic acid (65:25:10 v/v/v) are established solvent systems for the identification of sulfatides, which are characteristic components of selected tissues (brain) or cells (spermatozoa). Up to nine different sulfatides were identified in kidney cell extracts [70] using normal-phase HPTLC.

Complex lipid extracts (for instance from HL-60 cells [71]) require the use of two-dimensional TLC. The cells are initially extracted with chloroform–methanol (2:1 v/v) and then separated in the first dimension with chloroform–methanol–water (65:25:4 v/v/v) and subsequently with tetrahydrofuran–dimethoxymethane–methanol–water (10:6:4:1, v/v/v/v) in the second dimension. Although excellent separation quality can be achieved, glycolipids are very challenging compounds. One particular problem is that newly discovered glycolipids often contain very long oligosaccharide chains that can significantly change the chromatographic properties. Fortunately, monoclonal antibodies directed against specific glycosphingolipids are now available. This simplifies their detection [72].

14.3.5 Phospholipids

Phospholipids constitute an important class of biomolecules, of which glycerophospholipids (GPLs) are of particular significance. All GPLs consist of a glycerol backbone, esterified with two fatty acids and one molecule of phosphoric acid. The resulting phosphatidic acid (PA) may formally react with different alcohols to give products, such as PC and PE, or negatively charged phospholipids, such as PS, phosphatidylglycerol, PI, and polyphosphoinositides (PPI) (e.g., PIP_2 (phosphatidylinositol-bisphosphate)), Figure 14.4. In addition, even more complex phospholipids, such as cardiolipin (CL) that may be considered as two PA units linked by one glycerol molecule, are possible [73]. One aspect is important regarding nomenclature: "PC", for instance, implies the presence of two ester residues [74]. As there are also alkyl and alkenyl lipids in biological systems, the more general term "glycerophosphocholine" should be used, even if "PC" is often (incorrectly) used to indicate all species.

Since phospholipids are ubiquitous compounds, there is an increasing interest in using either them or compounds derived from them, such as lysolipids that lack one fatty acid residue, as markers for disease [41]. Normal-phase TLC is commonly used to separate phospholipid mixtures and numerous excellent solvent systems have been developed [75]. Detection limits of about 20 ng per phospholipid are easily achieved [76]. There are one- and two-dimensional TLC methods to separate complex phospholipid mixtures and are discussed below.

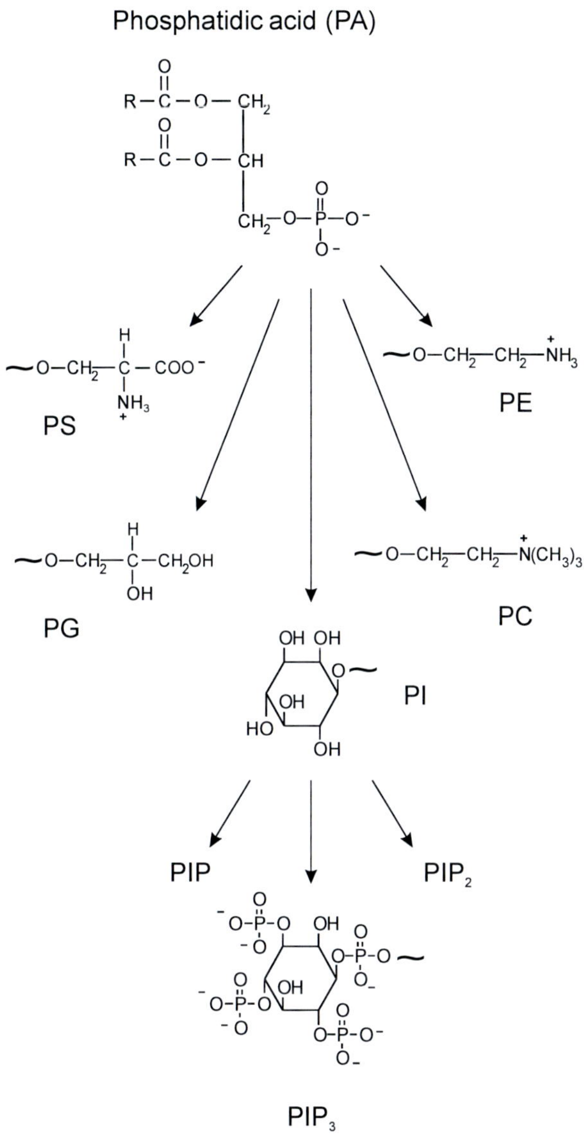

FIGURE 14.4 Selected structures of glycerophospholipids relevant to this review. All compounds are basically derived from glycerol by esterification at two positions with fatty acids ("R" represents varying fatty acyl residues) and in the third position with phosphoric acid. The resulting phosphatidic acid (PA) can formally react again with a variety of small organic molecules. Accordingly, the following compounds are formed: phosphatidylethanolamine (PE), phosphatidylcholine (PC), phosphatidylglycerol (PG), phosphatidylserine (PS), and phosphatidylinositol (PI). PI can be further phosphorylated on the inositol ring to give phosphatidylinositolmonophosphate (PIP), phosphatidylinositol-bisphosphate (PIP2) and phosphatidylinositol-trisphosphate (PIP3).

14.3.5.1 One-Dimensional Separations

One-dimensional TLC is regularly used if neutral or zwitterionic phospholipids are of particular interest. The separation quality achievable by one-dimensional TLC was illustrated more than 40 years ago for plasma and erythrocyte extracts [77]. Silica gel impregnated with 7.5% (w/w) magnesium acetate in combination with chloroform–methanol–ammonia (65:25:4 v/v/v) was used for the separation. If only small quantities of acidic phospholipids are expected chloroform–methanol–water (25:10:1 v/v/v) is widely used [78]. However, PS and PE, as well as PI and PC, may interfere under these conditions. A solvent system commonly used for complex mixtures of negatively charged phospholipids is methyl acetate–2-propanol–chloroform–methanol containing 0.25% aqueous KCl (25:25:25:10:9 v/v/v/v/v) [79]. This system is suitable for many complex lipid mixtures although PA and PE are not well resolved. A comparison of eight different solvent systems for the separation of phospholipids by one-dimensional TLC has been performed [80]. Chloroform–methanol–water (65:25:4 v/v/v) was recommended as providing the best overall separation quality for phospholipids. The effect of different solvent systems on the separation quality is illustrated in Figure 14.5. It should be noted, however, that excellent results are only obtained if great care is

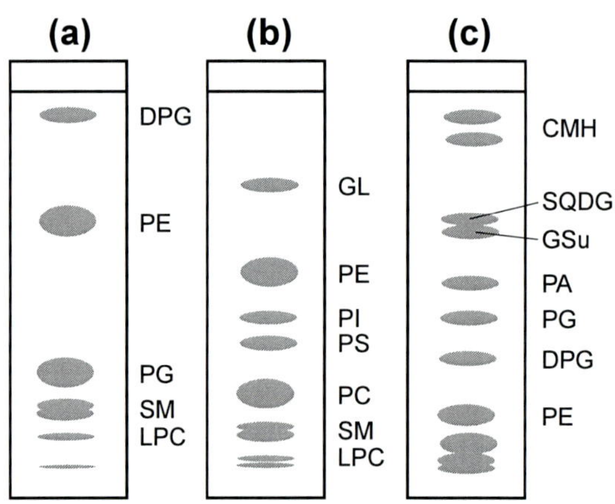

FIGURE 14.5 Schematic HPTLC separations of complex lipid mixtures from animal tissues. (a) Chloroform–methanol–water (25:10:1 v/v/v); (b) methyl acetate–2-propanol–chloroform–methanol–0.025% KCl (25:25:25:10:9 v/v/v/v/v); (c) first development with pyridine–hexane (3:1 v/v) and second development in the same direction with chloroform–methanol–pyridine–2 M ammonia (35:12:65:1 v/v/v/v). Abbreviations: DPG, diphosphatidylglycerol; GL, glycolipid; GSu, glycolipid sulfate; CMH, ceramide monohexoside; SQDG, sulfoquinovosyldiacylglycerol; LPC, lysophosphatidylcholine; PE, phosphatidylethanolamine; PG, phosphatidylglycerol; PI, phosphatidylinositol; PS, phosphatidylserine; PC, phosphatidylcholine; PA, phosphatidic acid; SM, sphingomyelin. *Reprinted with modification and permission from Christie and Han, [1].*

taken regarding the water content of the solvent mixture and the activity of the TLC plates. This important topic has been reviewed recently [3]. Another important point is that the thorough drying of the plate (between different developments) is crucial for consistent separation quality [81]. If glycolipid identification is also of interest, a two-step TLC is advisable. The phospholipids are first separated from other lipids and the glycolipids are separated in a second development. Reasonable resolution is obtained using diisobutyl ketone–acetic acid–water (40:25:3.7 v/v/v). Chloroform–ethanol–water–triethylamine (30:35:6:35 v/v/v/v) is another useful solvent system for this separation [29].

Although the resolution achievable by one-dimensional TLC is limited, it is remarkable that all major lipid classes from human plasma and crude liver extracts can be resolved in a single TLC separation following pre-development over a short distance with chloroform–methanol–water (65:30:5 v/v/v) to remove protein-bound lipids. Afterward, the TLC plate was fully developed with hexane–diethyl ether–formic acid (80:20:1.5 v/v/v) [82]. Although the lipid concentration of urine is comparatively low, TLC is sufficiently sensitive to characterize the lipid composition of human urine. A very sensitive detection technique is to spray the plate with copper sulfate reagent with subsequent charring [83].

14.3.5.1.1 DETERMINATION OF ENZYMATIC ACTIVITY

Phospholipases A_2, C, and D generate lysophospholipids, diacylglycerols, and phosphatidic acids, respectively (Figure 14.6). One important (clinical) task is the determination of the related enzyme

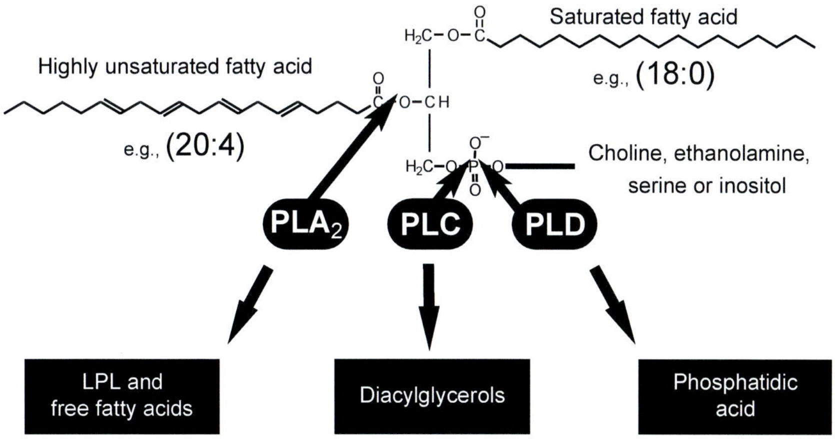

FIGURE 14.6 Schematic diagram of in vivo generation of lipid-derived second messengers from a selected phospholipid molecule, along with the enzymes involved: phospholipase A_2 (PLA_2), phospholipase C (PLC), and phospholipase D (PLD). LPL, lysophospholipid.

activity by determining the consumption of the substrate and/or the generation of related products, which can be accomplished using TLC. An elegant phospholipase C-related study was performed by Goldfine et al. [84]: these authors observed different rates of digestion for phospholipids using phospholipase C. Organic solvent extracts were separated using chloroform—methanol—acetic acid (65:25:8 v/v/v) and gave an excellent separation between the phospholipids and related diacylglycols [84].

Phospholipases A_2 activities were determined similarly by Wang and Gustafson [28]. Different phospholipids (PS, PE, PI, PC, and SM) and three lysophospholipids (LPS, LPE, and LPC) were separated on silica gel impregnated with 0.4% (w/w) ammonium sulfate using chloroform—methanol—acetic acid—acetone—water (40:25:7:4:4:2 v/v/v/v) as the mobile phase and subsequently detected by iodine staining. Finally, it was shown recently that a simple TLC assay allows the determination of ceramide concentration and phospholipases D activity [85]. Methods for chromatographic enzyme activity determination were recently reviewed [86] as well as the use of labeled compounds [87].

14.3.5.1.2 PHOSPHOLIPID OXIDATION

This is a very important topic since all inflammatory diseases (ending with "itis" such as "arthritis") are accompanied by the generation of ROS. This topic has been comprehensively reviewed recently [88]. Since many ROS react in a diffusion-controlled manner, the focus is normally on the "bulk" lipids, for example, PC hydroperoxides and products derived from them. These can be isolated by TLC from oxidized lysophospholipids using chloroform—methanol—water (10/5/1 v/v/v) [89]. The aldehyde group present in PC subsequent to oxidation [47] was visualized by spraying with Schiff's reagent while the hydroperoxides were detected by spraying with potassium iodide and starch.

The detection of lipid oxidation products in turkey meat was accomplished by TLC of the phospholipid hydroperoxides and their parent phospholipids (SM, PC, and PE) [90]. Hydroperoxide detection was performed by dipping the developed TLC plate into a freshly prepared solution of *N,N*-dimethyl-p-phenylenediamine and detecting the product by scanning densitometry at 654 nm.

TLC is also useful for the separation of headgroup-modified phospholipids, such as PE. Prior to reaction with HOCl (leading to the generation of chloramines), PE was eluted with chloroform—methanol—acetic acid (80:12:8 v/v/v) and FFA with diethyl ether—petrol ether—acetic acid (70:30:1 v/v/v) [91]. Air-dried plates were sprayed with ninhydrin and heated at 100 °C to visualize the amine group of PE, or were charred at 180 °C to visualize all lipids. HOCl reacts with the fatty acid residues with formation of chlorohydrins while the amino group is converted into a chloramine [91]. Therefore, the differentiation between the individual products is important.

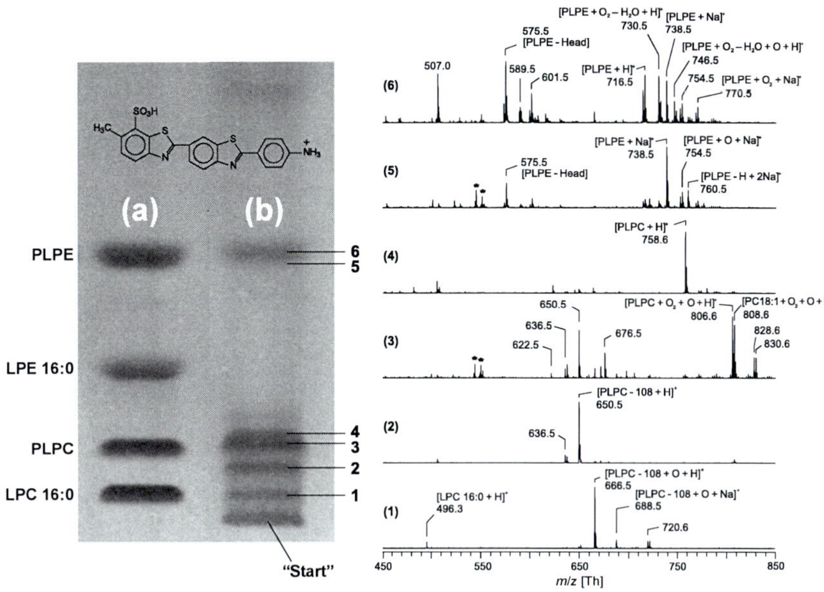

FIGURE 14.7 Positive ion MALDI–TOF mass spectra of the individual fractions of an air-oxidized mixture of PC 16:0/18:2 (1-Palmitoyl-2-linoleoyl-sn-phosphatidylcholine, PLPC) and PC 16:0/18:2 (1-Palmitoyl-2-linoleoyl-sn-phosphatidylethanolamine, PLPE). The developed TLC plate from which the spectra were recorded is shown on the left. Lane (a) represents a PLPC/PLPE/LPC (lysophosphatidylcholine) 16:0/LPE (lysophosphatidylethanolamine) 16:0 mixture as control while lane (b) represents a sample of air-oxidized PLPC/PLPE. Positions of the acquired mass spectra directly from the TLC plate are labeled by numbers. Ions are labeled according to their *m/z* ratios and the most prominent peaks are structurally assigned. The structure of the most abundant ion ($m/z = 454.1$) from the primuline dye is also indicated (left top). Due to the presence of the sulfonic acid residue, primuline is detected with low sensitivity as a positive ion. The reasons of the dark background in lane (b) are so far unknown. PC, phosphatidylcholine; LPE, lysophosphatidylethanolamine; TLC, thin-layer chromatography. *Reprinted from Ref. [92].*

Finally, it was shown that TLC combined with matrix-assisted laser desorption and ionization (MALDI) MS detection gives good results for the oxidation products of polyunsaturated PE or PC species illustrated in Figure 14.7 [92].

14.3.5.2 Two-Dimensional Separations

Two-dimensional TLC is an excellent tool for the separation of complex lipid mixtures. A typical example (lipids from brain mitochondria) is shown in Figure 14.8 [93]. However, two-dimensional TLC has serious drawbacks also: first, only a single sample can be investigated by two-dimensional TLC. Second, the simultaneous application of the sample and a lipid standards mixture is impossible. Due to these disadvantages, multiple developments in a single dimension are often used as an

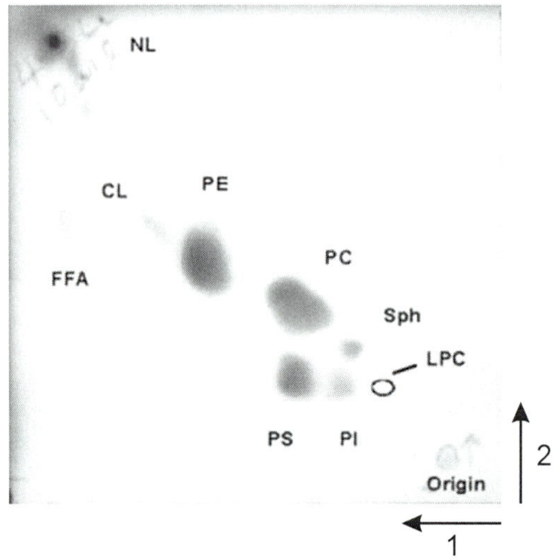

FIGURE 14.8 Typical two-dimensional HPTLC separation of total lipids extracted from the cortical P2 fraction from brain mitochondria. The silica plate was first developed with a solvent system consisting of chloroform–methanol–28% ammonia (65:25:5 v/v/v). After drying the plate was developed in the orthogonal direction with a solvent system consisting of chloroform–acetone–methanol–glacial acetic acid–water (50:20:10:10:5 v/v/v/v/v). Phospholipids were visualized by exposure to iodine vapor. "NL" means neutral lipids such as triacylglycerols. CL, cardiolipin; PE, phosphatidylethanolamine; PC, phosphatidylcholine; PI, phosphatidylinositol; PS, phosphatidylserine; FFA, free fatty acids Sph, sphingomyelin. *Reprinted with modification and permission from Ref. [93].*

alternative to two-dimensional TLC. In this case, a solvent mixture of high elution power is used first, followed by further developments of increasing length with solvents of decreasing elution power. This approach confers the advantage of first concentrating the sample components while the gradient development helps to overcome resolution problems. This approach is often used for glycolipid analysis [94].

Two-dimensional TLC is the method of choice to screen for PPI in complex mixtures [95]: After extraction with chloroform–methanol–HCl, aliquots are subjected to normal-phase TLC on silica gel impregnated with 1% (w/w) potassium oxalate. The first development with chloroform–methanol–4.3 M ammonia (90:65:20 v/v/v) is used to separate the PPI. After drying, a second development with chloroform–methanol–concentrated ammonia (130:50:10 v/v/v) was used to separate lysophosphatidylethanol from PC. Next, the plate was rotated and developed in the orthogonal direction with chloroform–methanol–acetic acid–water (100:30:35:3 v/v/v/v) to resolve the remaining phospholipids. The individual lipids were visualized by charring [95].

Two-dimensional TLC is also useful for the separation of lipid oxidation products in complex lipid mixtures [96]. Two-dimensional TLC of PC,

PE, PI, PS, SM, CL, LPC, and LPE was performed on silica gel impregnated with 7.5% (w/w) magnesium acetate using chloroform–methanol–ammonia (5:25:5 v/v/v) for the first development and chloroform–acetone–methanol–acetic acid–water (6:8:2:2:1 v/v/v/v/v) for the second development in the orthogonal direction. Afterward thiobarbituric acid reactive substances were identified in individual fractions to determine the extent of oxidation. This approach allowed the determination of whether all phospholipids are equally sensitive to oxidation.

14.3.6 Phosphoinositides

PPI are involved in signal transduction and are of significant physiological interest. Unfortunately, PPI are difficult to analyze because they occur in only very small amounts and are difficult to extract with organic solvents on account of their high polarity. Accordingly, radioactive labeling with ^{32}P (on the phosphate residues) or ^{3}H (on the inositol ring) is often used to detect low abundant PPI with sufficient sensitivity. The separation of PPI from residual phospholipids is not difficult (due to their increased polarity) and can be performed in one- or two-dimensional TLC [97]. One-dimensional TLC, for example, was able to resolve the majority of major and minor phospholipid species in extracts from human erythrocytes and platelets on silica gel with chloroform–methanol–acetic acid (55:25:5 v/v/v). Normal-phase TLC is also capable of resolving different PI isomers phosphorylated in the 3-, 4-, or 5 position [98]. The most common method uses TLC plates impregnated with boric acid and 1-propyl acetate–2-propanol–ethanol–6% aqueous ammonia (3:9:3:9 v/v/v/v) as the mobile phase as shown in Figure 14.9.

14.4 MALDI FOR MS DETECTION

MALDI is fast, simple and has the advantage of producing nearly exclusively singly charged ions. In addition, MALDI is tolerant of relatively high sample contamination from salts and other materials. For methodological details see Ref. [99] or the second edition of the excellent book by Franz Hillenkamp and Jasna Peter-Katalinić [100].

14.4.1 Glyco- and Sphingolipids

Considerable progress has been made recently in the TLC–MALDI MS of Glyco- and sphingolipids [101]. In place of a detailed survey we will mention some landmarks since this topic was recently reviewed [102]. It is remarkable that only very little fragmentation occurs when the analytes are directly desorbed from a conventional TLC plate—even if complex

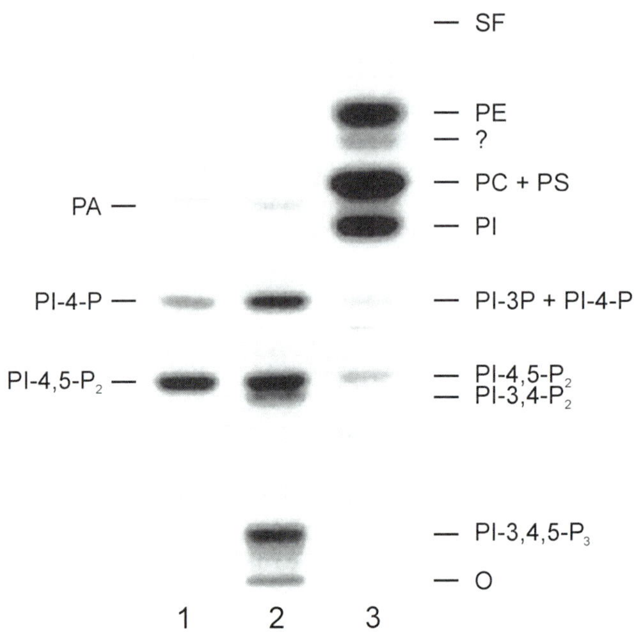

FIGURE 14.9 Phosphor screen autoradiography of lipid compounds separated on boric acid-impregnated HPTLC silica gel 60 layers developed with 1-propyl acetate–2-propanol–absolute ethanol–6% aqueous ammonia (3:9:3:9 v/v/v/v). ^{32}P-labeled erythrocyte (lane 1), ^{32}P-labeled erythrocyte phospholipids incubated additionally with PI 3-kinase γ and Mg–[γ-^{32}P] ATP (adenosine triphosphate) (lane 2), and ^{32}P-labeled A431 cell phospholipids (lane 3). Abbreviations: O, origin; PI 3,4,5-P_3, phosphatidylinositol 3,4,5-trisphosphate; PI 3,4-P_2, phosphatidylinositol 3,4-bisphosphate; PI 4,5-P_2, phosphatidylinositol 4,5-bisphosphate; PI 3-P, phosphatidylinositol 3-phosphate; PI 4-P, phosphatidylinositol 4-phosphate; PI, phosphatidylinositol; PA, phosphatidic acid; PC, phosphatidylcholine; PS, phosphatidylserine; PE, phosphatidylethanolamine; SF, solvent front. "?" indicates the presence of an unidentified product. *Reprinted from Ref. [98] with permission.*

molecules such as gangliosides are analyzed [103]. However, one important aspect is the application of the matrix (normally a small organic molecule that has the task of absorbing the laser energy) [104]. The homogeneity of the matrix/analyte cocrystals determines the reproducibility of the measurements and the matrix layer must be applied as homogeneous as possible without compromising the chromatographic resolution. This last point is particularly important because applying an organic solution of the matrix may lead to the blurring of compounds with similar R_f values. Fortunately, special spray devices that meet these requirements are now available from several companies.

The majority of studies of glycolipids have used UV lasers with a smaller number employing infrared lasers primarily with noncommercial spectrometers self-built instruments. The use of infrared lasers has advantages and disadvantages: on the one hand, glycerol (the most common infrared matrix) is a liquid and, thus, there are no problems regarding inhomogeneous matrix/analyte cocrystallization. On the other hand, infrared MALDI is characterized by more complex adduct ions in contrast to results for UV lasers, which might complicate data analysis. Using an infrared MALDI source, Dreisewerd and coworkers were able to show that even minor gangliosides could be unequivocally identified on a TLC plate, as there was virtually no fragmentation of molecular ions that could be determined with high accuracy [105]. Similar results were obtained using a UV laser. In a recent study, 2,5-dihydroxybenzoic acid (DHB) in acetonitrile–water (1:1 v/v) was used as matrix for the analysis of glycosphingolipids. Detection limits of about 50 pmol were obtained [106]. It is a significant advantage that antibodies, i.e., oligosaccharide-specific proteins, are often available to assist in the identification of specific glycosphingolipids. Using this approach in combination with TLC–MALDI MS, detection limits less than 1 ng were obtained [107]. The application of antibodies allows the direct use of crude lipid extracts without prior separation of glycosphingolipids in some cases.

14.4.2 Glycerophospholipids

As for glycolipids, two different approaches are in use for the identification of glycerophospholipids by TLC–MALDI MS. Using an infrared laser and glycerol as a matrix has the advantage that quantitative results for selected model phospholipids (with defined acyl groups) can be obtained even if abundant glycerol adducts (and to a minor extent even NaCl adducts) of the lipids were also detectable [108]. In another approach the authors used basically the same setup with a nitrogen laser and 2,5-DHB as matrix [34]. An extract from hen egg yolk was used because it is readily available and contains an abundance of phospholipids. A selected single track of a TLC-separated hen egg yolk extract and some selected positive ion MALDI mass spectra (directly recorded from the TLC plate) are shown in Figure 14.10 [109]. Two aspects of this analysis should be emphasized: first, even low abundant lipids (e.g., PI) that make up less than 1% (w/w) of the phospholipids from egg yolk can be easily detected [34]. Thus, the detection limit is about 400 pmol [34]. Second, different mass spectra are obtained that depend on the position where the laser beam hits the spot of a lipid fraction. This is particularly evident for the PE fraction where shorter and longer chain fatty acid groups can be distinguished. This clearly indicates that changes in the fatty acid chain composition slightly affect the migration properties of the phospholipids. This difference would never have been resolved by the visual inspection of the TLC plate.

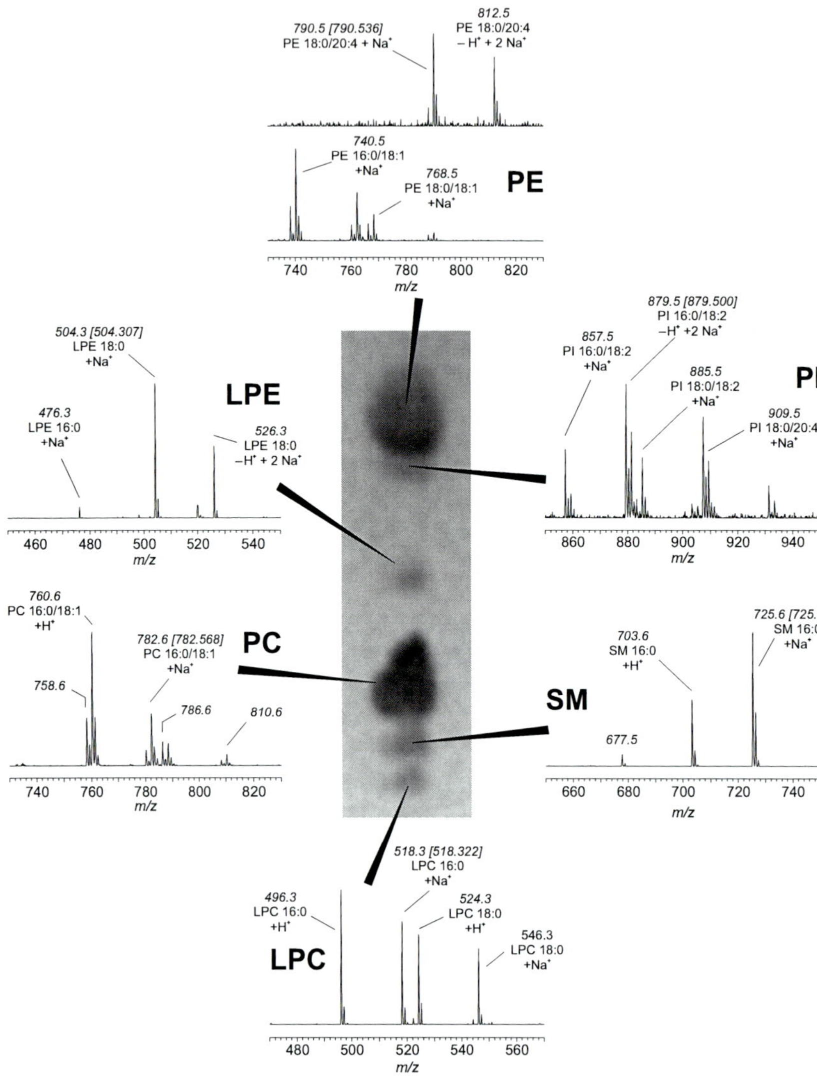

FIGURE 14.10 Expanded region of a TLC-separated egg yolk extract and the corresponding positive ion MALDI–TOF mass spectra recorded directly from the indicated positions on the plate. Only the relevant mass regions of each phospholipid class are shown and assignments are provided directly in the individual traces. Data given in parentheses correspond to theoretical masses and were introduced to enable comparisons with the experimental data in selected cases. Note that the PE fraction provides different mass spectra, depending on the position where the laser beam hits the PE spot. The only marked fragmentation is the loss of the headgroup of SM (leading to $m/z = 677.5$). PE, phosphatidylethanolamine; PI, phosphatidylinositol; LPE, lysophosphatidylethanolamine; PC, phosphatidylcholine; LPC, lysophosphatidylcholine; SM, sphingomyelin. *Reprinted with permission from Ref. [109].*

14.5 SUMMARY AND OUTLOOK

There is an unequivocally considerable and even increasing interest in (phospho)lipid analysis. It may even be expected that lipids will experience additional interest in the future because an increasing number of diseases (such as atherosclerosis or rheumatic diseases) are recognized to be accompanied by alterations of the lipid compositions of the affected tissues and/or body fluids [110]. Another advantage is surely the ubiquity of lipids: since the same lipids occur in human and animals, the same biomarkers can be used, and there is no need to produce different antibodies.

Hopefully, we were able to provide sufficient evidence that (HP)TLC is a powerful tool for lipid analysis and can be applied to all relevant lipid classes of physiological and diagnostic interest. Although many other analytical methods [111] can be used for the same purpose, (HP)TLC is accepted as a time-saving and economical method that may be used with a minimum of trouble shootings. Finally, (HP)TLC may be applied to "suspicious" samples (for instance, from food or cosmetics) that may easily plug or even damage an HPLC column.

The scope of the hyphenation of HPTLC to other analytical techniques—particularly MS—appears to hold considerable promise for the analysts who previously had reservations concerning the use of planar chromatography. Although a lot of different methods are already commercially available, further significant progress can be expected in this field. So far, there are basically methods based on the extraction of analytes of interest prior to MS and different desorption methods that allow the characterization of analytes directly on the TLC plate. Of course, the selection of the most appropriate method depends on the analytical problem and access to particular instrument platforms. To date, methods based on extraction seem to provide more reliable quantitative data, while desorption methods provide higher resolving power. For instance, lipids with different acyl groups can be identified within a single spot on a TLC plate. This obviously opens a new dimension that makes HPTLC highly competitive with more common LC–MS methods.

Acknowledgments

The authors wish to thank all colleagues and friends who helped them in writing this review. Particularly the kind and helpful advice of Dr Suckau and Dr Schürenberg (Bruker Daltonics, Bremen) as well as Dr Griesinger and Dr Mattheis (Merck KGaA, Darmstadt) is gratefully acknowledged.

This work was supported by the German Research Council (DFG Schi 476/12-1 and FU 771/1-2 as well as TR 67 project A2 & A8) and the Federal Ministry of Education and Research of the Federal Republic of Germany ("The Virtual Liver", 0315735).

[83] Massa DR, Vasta JD, Fried B, Sherma J. High-performance thin-layer chromatographic analysis of the polar lipid content of human urine and urine from BALB/c mice experimentally infected with *Echinostoma caproni*. JPC: J Planar Chromatogr 2008;21: 337–41.
[84] Goldfine H, Johnston NC, Knob C. Nonspecific phospholipase C of Listeria monocytogenes: activity on phospholipids in Triton X-100-mixed micelles and in biological membranes. J Bacteriol 1993;175:4298–306.
[85] Liu G, Kleine L, Hébert RL. A direct method for the simultaneous measurement of ceramide and phospholipase D activity. Prostaglandins Leukot Essent Fatty Acids 2000; 63:187–94.
[86] Hendrickson HS. Phospholipase A_2 and phosphatidylinositol-specific phospholipase C assays by HPLC and TLC with fluorescent substrate. Methods Mol Biol 1999;109: 1–6.
[87] Deranieh RM, Joshi AS, Greenberg ML. Thin-layer chromatography of phospholipids. Methods Mol Biol 2013;1033:1–27.
[88] Kuksis A, Suomela JP, Tarvainen M, Kallio H. Lipidomic analysis of glycerolipid and cholesteryl ester autooxidation products. Mol Biotechnol 2009;42:224–68.
[89] Itabe H, Yamamoto H, Suzuki M, Kawai Y, Nakagawa Y, Suzuki A, et al. Oxidized phosphatidylcholines that modify proteins. Analysis by monoclonal antibody against oxidized low density lipoprotein. J Biol Chem 1996;271:33208–17.
[90] Bruun-Jensen L, Colarow L, Skibsted LH. Detection and quantification of phospholipid hydroperoxides in turkey meat extracts by planar chromatography. JPC: J Planar Chromatogr 1995;8:475–9.
[91] Carr AC, van den Berg JJ, Winterbourn CC. Differential reactivities of hypochlorous and hypobromous acids with purified *Escherichia coli* phospholipid: formation of haloamines and halohydrins. Biochim Biophys Acta 1998;1392:254–64.
[92] Bischoff A, Eibisch M, Fuchs B, Süß R, Schürenberg M, Suckau D, et al. A simple TLC/MALDI method to monitor oxidation products of phosphatidylcholines and ethanolamines. Acta Chromatogr 2011;23:365–75.
[93] Bayir H, Tyurin VA, Tyurina YY, Viner R, Ritov V, Amoscato AA, et al. Selective early cardiolipin peroxidation after traumatic brain injury: an oxidative lipidomics analysis. Ann Neurol 2007;62:154–69.
[94] Müthing J. Improved thin-layer chromatographic separation of gangliosides by automated multiple development. J Chromatogr B 1994;657:75–81.
[95] Mallinger AG, Yao JK, Brown AS, Dippold CS. Analysis of complex mixtures of phospholipid classes from cell membranes using two-dimensional thin-layer chromatography and scanning laser densitometry. J Chromatogr 1993;614:67–75.
[96] Pikul J, . Kummerow Fred A. Thiobarbituric acid reactive substance formation as affected by distribution of polyenoic fatty acids in individual phospholipids. J Agric Food Chem 1991;39:451–7.
[97] Singh AK, Jiang Y. Quantitative chromatographic analysis of inositol phospholipids and related compounds. J Chromatogr B 1995;671:255–80.
[98] Hegewald H. One-dimensional thin-layer chromatography of all known D-3 and D-4 isomers of phosphoinositides. Anal Biochem 1996;242:152–5.
[99] Fuchs B, Süss R, Schiller J. An update of MALDI-TOF mass spectrometry in lipid research. Prog Lipid Res 2010;49:450–75.
[100] Hillenkamp F, Peter-Katalinić J. MALDI MS – a practical guide to instrumentation, methods and application. 2nd ed. Weinheim: Wiley-VCH; 2013.
[101] Müthing J, Distler U. Advances on the compositional analysis of glycosphingolipids combining thin-layer chromatography with mass spectrometry. Mass Spectrom Rev 2010;29:425–79.

[102] Fuchs B. Analysis of phospholipids and glycolipids by thin-layer chromatography-matrix-assisted laser desorption and ionization mass spectrometry. J Chromatogr A 2012;1259:62–73.

[103] Ivleva VB, Sapp LM, O'Connor PB, Costello CE. Ganglioside analysis by thin-layer chromatography matrix-assisted laser desorption/ionization orthogonal time-of-flight mass spectrometry. J Am Soc Mass Spectrom 2005;16:1552–60.

[104] Fuchs B, Schiller J. Recent developments of useful MALDI matrices for the mass spectrometric characterization of apolar compounds. Curr Org Chem 2009;13:1664–81.

[105] Dreisewerd K, Müthing J, Rohlfing A, Meisen I, Vukelić Z, Peter-Katalinić J, et al. Analysis of gangliosides directly from thin-layer chromatography plates by infrared matrix-assisted laser desorption/ionization orthogonal time-of-flight mass spectrometry with a glycerol matrix. Anal Chem 2005;77:4098–107.

[106] Nakamura K, Suzuki Y, Goto-Inoue N, Yoshida-Noro C, Suzuki A. Structural characterization of neutral glycosphingolipids by thin-layer chromatography coupled to matrix-assisted laser desorption/ionization quadrupole ion trap time-of-flight MS/MS. Anal Chem 2006;78:5736–43.

[107] Souady J, Soltwisch J, Dreisewerd K, Haier J, Peter-Katalinić J, Müthing J. Structural profiling of individual glycosphingolipids in a single thin-layer chromatogram by multiple sequential immunodetection matched with direct IR-MALDI-o-TOF mass spectrometry. Anal Chem 2009;81:9481–92.

[108] Rohlfing A, Müthing J, Pohlentz G, Distler U, Peter-Katalinić J, Berkenkamp S, et al. IR-MALDI-MS analysis of HPTLC-separated phospholipid mixtures directly from the TLC plate. Anal Chem 2007;79:5793–808.

[109] Fuchs B, Schiller J, Süß R, Nimptsch A, Schürenberg M, Suckau D. Capabilities and disadvantages of combined matrix-assisted laser desorption/ionization time-of-flight mass spectrometry (MALDI-TOF MS) and high-performance thin-layer chromatography (HPTLC): analysis of egg yolk lipids. JPC: J Planar Chromatogr 2009;22:35–42.

[110] McIntyre TM. Bioactive oxidatively truncated phospholipids in inflammation and apoptosis: formation, targets, and inactivation. Biochim Biophys Acta 2012;1818:2456–64.

[111] Hou W, Zhou H, Elisma F, Bennett SA, Figeys D. Technological developments in lipidomics. Brief Funct Genomic Proteomic 2008;7:395–409.

CHAPTER

15

Applications in Food Analysis

S. Krüger, G. Morlock

Institute of Nutritional Science, Justus Liebig University, Giessen, Germany

15.1 INTRODUCTION

Foodstuffs for human consumption originate, with a few exceptions (e.g., minerals, water), from living organisms like animals or plants [1]. First and foremost, food is essential to maintain the body's energy balance, but it can also serve as a beverage or stimulant. Food is usually consumed for its taste (e.g., sweets, soft drinks, salty snacks) or because it may have certain effects on the human body (e.g., alcohol, caffeine, tobacco) [2].

As an energy supplier, the food's nutrient components (e.g., proteins, fat, polysaccharides) have to be made up from building blocks that are universal to all biological systems (e.g., amino acids, fatty acids, sugars). Furthermore, endogenous enzymes have to be able to break down these nutrients, releasing enough energy to sustain the consuming organism. In terms of physics, biological systems are open systems (no thermodynamic equilibrium), which means a constant energy input is necessary "to keep the systems running". Also, because most endogenous compounds have a more or less long half-life, the building blocks consumed are used by the organism to create new endogenous, species-specific structures by relinking the individual components in a—for the organism—characteristic way. And as food is obtained from biological systems, these individual structures can be used as an analytical tool for identification of the food's source [1,2].

Biological organisms are complex, and since food is derived from them, they are equally complex mixtures [1]. Most products on the market these days are processed foods, which in themselves are complex mixtures of various base products. Moreover, to extend the shelf life of these food products, various technological methods (e.g., heating, fermenting, drying) are used that can change their natural chemical makeup compared with the source products. In addition, additives, nonendogenous to the

http://dx.doi.org/10.1016/B978-0-12-417223-4.00015-7

base products, are used to improve the products' properties such as color, taste, texture, or shelf life [3]. This complexity of food products together with rapid changes in consumer preferences is what makes the analysis of food challenging [4].

The food industry is one of the fastest growing industries worldwide [5,6]. More people are consuming preprepared and convenience foods, making this one of the main driving forces in the food industry [2,6]. Never was the food supply as diverse as today. In addition to the industrial production of usual and well-known foods, there is also a growing sector for the production of vegetarian, vegan, halal, koscher and weight loss foods. Over the past few years, the demand for value-added foods of superior quality, for example, functional food, dietary supplements, and advanced medical nutrition, as well as non-genetically modified organisms (GMO) foods has steadily increased [2,4]. In 2013 alone the functional food sector is expected to increase by over 6%, making up close to 10% of the overall world nutritional product market with a value of over 235 billion dollars. [7].

It is this large variety of food products on the market coupled with their ephemerality that shapes and challenges the current methods of food analysis [4]. Furthermore, the focus these days is no longer on food safety alone and the food's main ingredients and basic aspects of food chemistry are just as important. Due to a broadened global scope, food analysis includes new aspects like the targeted investigation of specific contaminants, food fraud, and problems related with food chain safety. The importance of effect-directed analysis (EDA) increases, as potential falsifications, frauds, and contaminants of the global commodity flow are beyond the scope of any multimethod of target analysis. Also, important aspects are the provision of a healthy diet and aspects concerning food production as well as GMO food, issues that have increasingly attracted the interest of the public in recent years. Moreover, food analysis provides assistance in matters concerning convenience and novel food technology, as well as molecular gastronomy. Due to increased capabilities of detection, thanks to modern equipment coupled with lower detection limits, not only a more detailed and deeper analysis of food but also trace and ultratrace analysis has become possible, for example, the migration of components from food packaging into the food and their potential reaction products with food compounds [4].

Overall, most modern methods of food analysis are chromatographic methods, especially gas chromatography (GC) and liquid chromatography. Although only a small percentage is done by thin-layer chromatography (TLC), the method is well suited for food analysis due to the possibility of side-by-side sample analysis and multiple detections, as well as high flexibility, important especially for the challenges of the rapidly changing issues mentioned above. Still, for many years, TLC was mostly used as a mere screening tool. The reason behind this neglect was the poor reputation of TLC with regard to reproducibility and the lack of information and training in high-performance thin-layer chromatography (HPTLC). Due to recent advances in automation and standardization, HPTLC is nowadays a reliable, reproducible, quantitative, and highly flexible analytical technique that allows high sample throughput, simple sample preparation, and multiple detections [4].

As mentioned above, food is a complex and complicated mixture of a wide variety of substance classes. Extraction of an analyte from such complicated matrices often necessitates long, complicated, and in the end, expensive sample preparation procedures. Compared with other chromatographic techniques, TLC/HPTLC is a relatively matrix-robust method and mostly does not need time-consuming sample preparation steps. For example, interfering matrix effects can be reduced by employing two-step development methods.

15.2 HPTLC IN FOOD ANALYSIS

In [4] it was noted that of the about 600 publications on TLC in food and feed analysis published between 1987 and 2007 only about 10% used HPTLC techniques and most of the remaining TLC applications rarely proceeded beyond the screening stage. When looking at TLC this is not really surprising, but HPTLC has a greater potential. In an example from 2011, it was shown that the results obtained by HPTLC differed substantially from those obtained by TLC in the separation of sugars [8]. HPTLC offers clearly better separation, sharper spots, and thus better resolution. HPTLC is therefore better suited to quantitative analysis. Validated methods, such as the one described above, reach reproducibility values comparable to those associated with high-performance liquid chromatography (HPLC) or GC methods, and furthermore, they also provide satisfactory calibration curve correlations, as well as good detection and quantification limits.

As food contains many substances, the analysis has to be diverse. A little under one-third of these analyses is concerned with basic ingredients. About another third deals with the analysis of additives and the remainder with residues and contaminants. Analyses of dyestuffs and of lipids, phospholipids, and fatty acids are named as two main areas of HPTLC analysis, followed by the analysis of mycotoxins [4]. Furthermore, HPTLC is used for the analysis of plant extracts used as food additions or in food supplements. Due to the traditional use in many cultures of medicinal herbs, there is a huge amount of work to be done to categorize these botanicals.

Validated HPTLC methods exist for water-soluble food dyes in various foodstuffs and pharmaceutical preparations on silica gel layers [9,10]. One

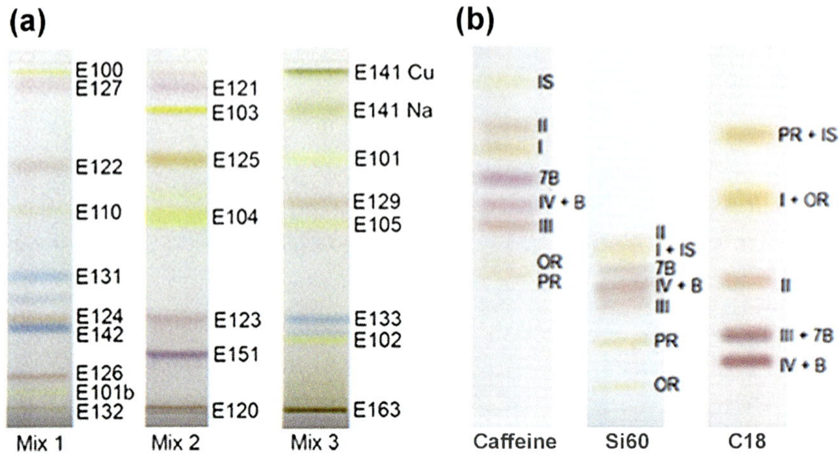

FIGURE 15.1 HPTLC chromatograms for authorized and nonauthorized dyes: separation of 25 water-soluble food dyes divided into three mixtures (a; purity of some dyes ≤85%) [9,17], and separation of Sudan dyes, used for food falsification (consumer fraud), on three different stationary phases (b) [12].

method allowed the separation of 25 food dyes (in three mixes) in one run, resulting in about 1.5 min overall time per run including application and evaluation, as well as a calculated solvent consumption of about 200 μL. The method is thus very time- and cost-efficient (Figure 15.1) [9]. Moreover, methods were also described that use reversed phases (RP) for the separation of food dyes [11], e.g., in jelly dessert powders. Over the past few years, azo dyes such as Sudan I–IV, Sudan Red B and 7B, Sudan Orange G, or Para Red have come into public focus. Beginning with the discovery of Sudan I in hot chili and hot chili products originating from India in 2003, several TLC/HPTLC methods for their determination have been developed (Figure 15.1) [12–16]; interestingly most of this work was accomplished by research groups from Asia (China and Japan) [14–16].

Contrary to dyes, there are many substances that cannot be detected in the visual range or under ultraviolet (UV) light because they lack a chromophore. Analysis of these compounds, e.g., fatty acids, lipids, and phospholipids [18,19], require derivatization with suitable reagents such as primuline (Figure 15.2) [20]. For the separation of lipids the choice of

FIGURE 15.2 HPTLC chromatogram of phospholipids from chocolate lecithin samples originating from soy (1) and sunflower (2) derivatized with primuline (recorded at 366/>400 nm) (a) and their respective mass spectra obtained by the elution head-based TLC-MS Interface: samples 1 and 2 (b, c), as well as standard substances phosphatidylethanolamine (PE) (d) and phosphatidyl choline (PC) (e); two different molecular species for each phospholipid could be identified (PC/PE 36:4 and PC/PE 34:2), modified according to [20].

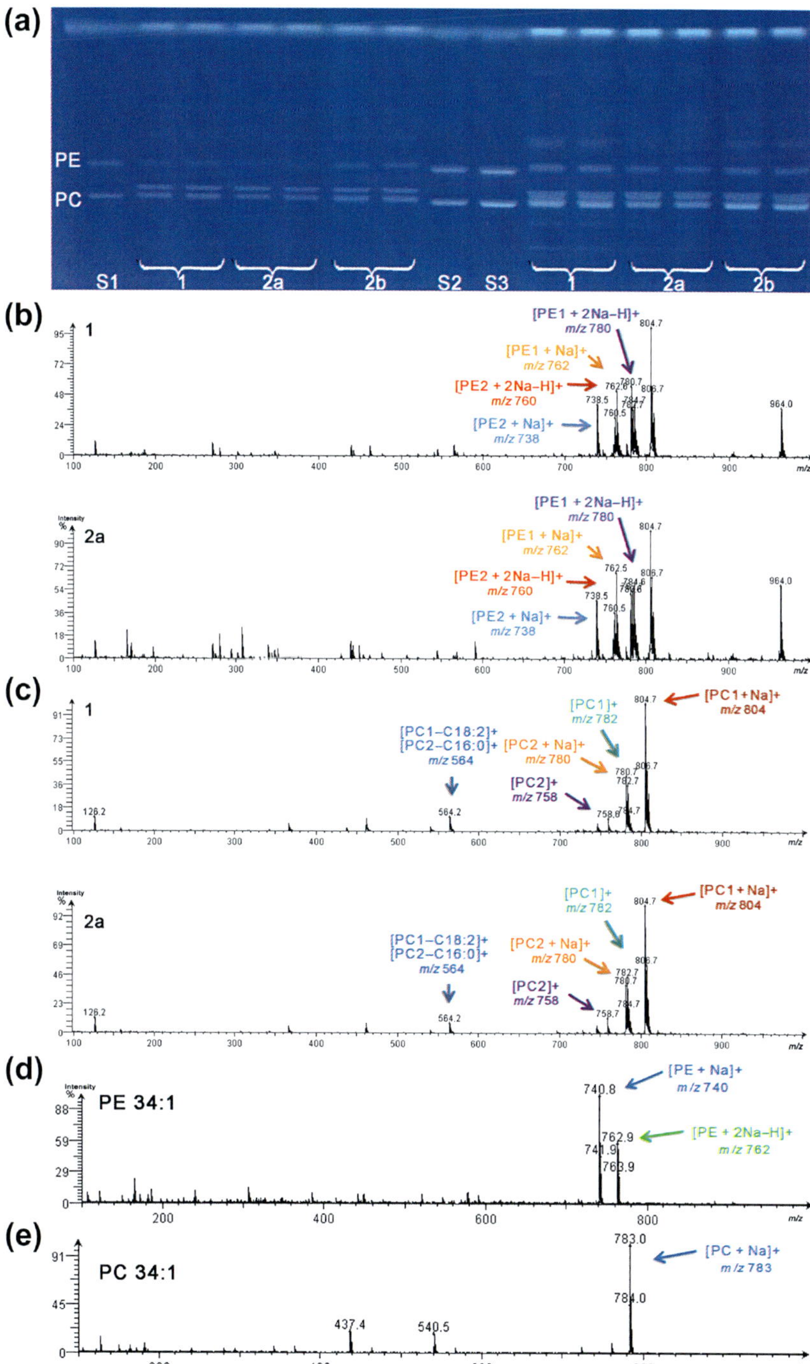
(a)
PE
PC
S1
1
2a
2b
S2
S3
1
2a
2b
(b)
1
[PE1 + 2Na–H]+
m/z 780
[PE1 + Na]+
m/z 762
[PE2 + 2Na–H]+
m/z 760
[PE2 + Na]+
m/z 738
804.7
964.0
2a
[PE1 + 2Na–H]+
m/z 780
[PE1 + Na]+
m/z 762
[PE2 + 2Na–H]+
m/z 760
[PE2 + Na]+
m/z 738
(c)
1
[PC1–C18:2]+
[PC2–C16:0]+
m/z 564
[PC1]+
m/z 782
[PC2 + Na]+
m/z 780
[PC2]+
m/z 758
[PC1+Na]+
m/z 804
2a
[PC1–C18:2]+
[PC2–C16:0]+
m/z 564
[PC1]+
m/z 782
[PC2 + Na]+
m/z 780
[PC2]+
m/z 758
[PC1+Na]+
m/z 804
(d)
PE 34:1
740.8
762.9
[PE + Na]+
m/z 740
[PE + 2Na–H]+
m/z 762
(e)
PC 34:1
783.0
784.0
437.4
540.5
[PC + Na]+
m/z 783
m/z

the stationary phase is of great importance. For example, phospholipids can be separated either according to class, i.e., functional groups (normal-phase chromatography), or according to lipophilicity (reversed-phase chromatography) [21]. Especially the so-called *argentation* HPTLC, where the layers are impregnated with silver nitrate, is a helpful tool. It enables the separation of sterols or fatty acids not only according to the number of double bonds but also by their location within the molecule (*cis*- or *trans*-fatty acids) [4,22,23].

The group of sugars, oligomeric carbohydrates, carbohydrate derivatives, and glycosides can be detected, for example, by the 2-naphthol reagent, the *p*-aminobenzoic acid reagent, or the diphenylamine/aniline/*o*-phosphoric acid reagent [24–26]. The use of the layer for both separation and derivatization is convenient, as demonstrated by amino phases used for the separation and detection of sugars or sugar derivatives, such as the sweetener sucralose [27]. Even the analysis and quantitation of hydrolyzed or methanolyzed polysaccharides was possible [28,29]. The group of amino acids and hydrolyzed peptides are also favorable candidates for detection by derivatization. The selective reaction with ninhydrin is suitable for their rapid quantification [30].

Mycotoxins are naturally occurring, toxic, secondary metabolites produced by fungi. Of the several hundred structurally diverse mycotoxins discovered so far, the most frequently occurring toxins are aflatoxins, ochratoxins, patulin, zearalenone, trichothecenes, fumonisins, and ergot alkaloids. These potent toxins can act as cyto-, nephro- and neurotoxins and can affect the health of mammals as carcinogenic, mutagenic, immunosuppressive, and estrogenic compounds [31,32]. There are certain food commodities—all of vegetable origin—that are prone to mycotoxin contamination and are the main sources of contamination of the food chain. For many years TLC was the method of choice for these analytes, because it allows a high sample throughput combined with acceptable detectability, less expensive sample preparation procedures (e.g., isolation and separation on the plate), and variation of the application volume [33]. There are still a reasonable number or reports every year that use (HP) TLC for the determination of mycotoxins, although HPLC is becoming an increasingly important technique for these compounds. While only wishing to give a quick survey of recent publications, it appears that most work is done investigating aflatoxins in foodstuffs, such as boiling water used to cook rice [34], peanut products [35], spices [36], milk [37,38], eggs [38], soybean oil [35], and cereals [39]. It is also striking that corn/maize seems to be infected by a great variety of fungi producing mycotoxins like fumonisins [40], zearalenone [41], trichothecene [42], and aflatoxins [43].

Rapidly developing applications of TLC appear in health food-related topics [44–49], with research done mostly by the Asian science community. The commercial and financial benefit linked to adulteration of functional food products makes this a growing area of concern. Therefore, screening for illegally added drugs and pharmaceutical preparations like glibenclamide [45] or other sulfonylurea-type antidiabetics/hypoglycemic agents [44,46–48] and authenticity control, for example, for aloe products [49,50], are growing topics of interest in food analysis. Questions about authenticity do not only arise concerning health foods. For example, absinthe—a green-colored spirit drink prohibited for many years—has seen a revival in recent years coupled with a rise of inferior products on the market. Traditional absinthe owes its bitter taste to substances like sesquiterpenlactone absinthin found in the wormwood plant (*Artemisia absinthium* L.). Absinthin can be used as a marker substance to confirm the authenticity of absinthe [51–53].

In 2008, the steviol glycoside rebaudioside A was approved as a sweetener in the United States. In 2011, the European Union followed and allowed *Stevia* extracts containing at least 95% steviol glycosides as food additives (E960) for certain foodstuffs [54]. Since then the need for suitable methods for its analysis has grown. There are several HPTLC methods for the analysis of steviol glycosides in *Stevia* leaves and powder (Figure 15.3) [55,56] and in various foodstuffs like tea, candies and yogurt [57], and health drinks [58]. Another recently EU-approved sweetener is sucralose (E955), an artificial product with sucrose as the source substance. By now there are several HPTLC methods for its determination in food products such as soft drinks [59], milk-based products [27], dietetic products [25], and drinking water [26].

Further issues are food supplementation with vitamins [60] and their derivatives [61], with γ-aminobutyric acid [60] or with caffeine, e.g., in energy drinks and pharmaceutical preparations [63–65], as well as monitoring of the deterioration of process contaminants such as acrylamide, e.g., in coffee or water [62,66], 5-hydroxymethylfurfural (HMF) in products like honey [67,68], heterocyclic aromatic amines (HAA), e.g., in meat [69], isopropylthioxanthone (ITX), e.g., in milk and milk products [70], and phthalates, e.g., in food packaging [71].

15.3 SAMPLE PREPARATION, HYPHENATION, AND NEW POSSIBILITIES

As previously stated, an advantage of HPTLC is the relative short analysis time per sample due to the possibility of simultaneous separation of samples on up to over 40 tracks under identical conditions (Figure 15.4) [4,27]. Another advantage is that in most cases simple sample preparation steps, like dissolving, centrifuging, and filtering are sufficient, since HPTLC is extremely matrix robust. Moreover, since plates are used only once, it does not matter if the matrix is either held at the application origin

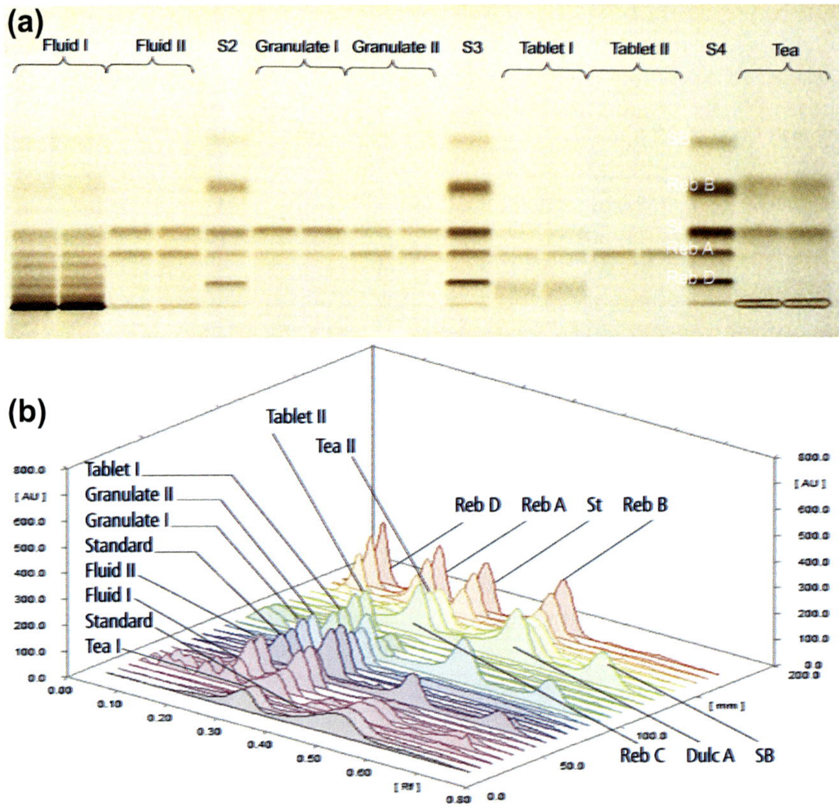

FIGURE 15.3 HPTLC chromatogram of steviol glycosides analyzed in various *Stevia* products on the market (fluids, granulates, tea, and tablets) after derivatization with β-naphthol reagent (a) and the respective three-dimensional densitogram (b) [5].

or eluted with the solvent front [4]. Also, once the separated analytes are evaluated, the unseparated matrix compounds are still accessible for further investigation. In case the matrix compounds endogenous to the food sample interfere, e.g., by coeluting with the analyte, HPTLC's flexibility comes in handy. For example, in a so-called prerun, the analyte and matrix can be separated from one another, e.g., when investigating the HMF content in honey a two-step chromatography was used. The mobile phase for the first step (pre-run) was selected to elute the HMF with the solvent front thus separating the analyte from the matrix that stays at the application zone [67]. A new approach to TLC was first used by Oellig in 2011 [24,72]. In the so-called high-throughput planar solid-phase extraction, TLC plates are utilized to separate pesticides from the matrix and to focus them into sharp zones before elution into a mass spectrometer with

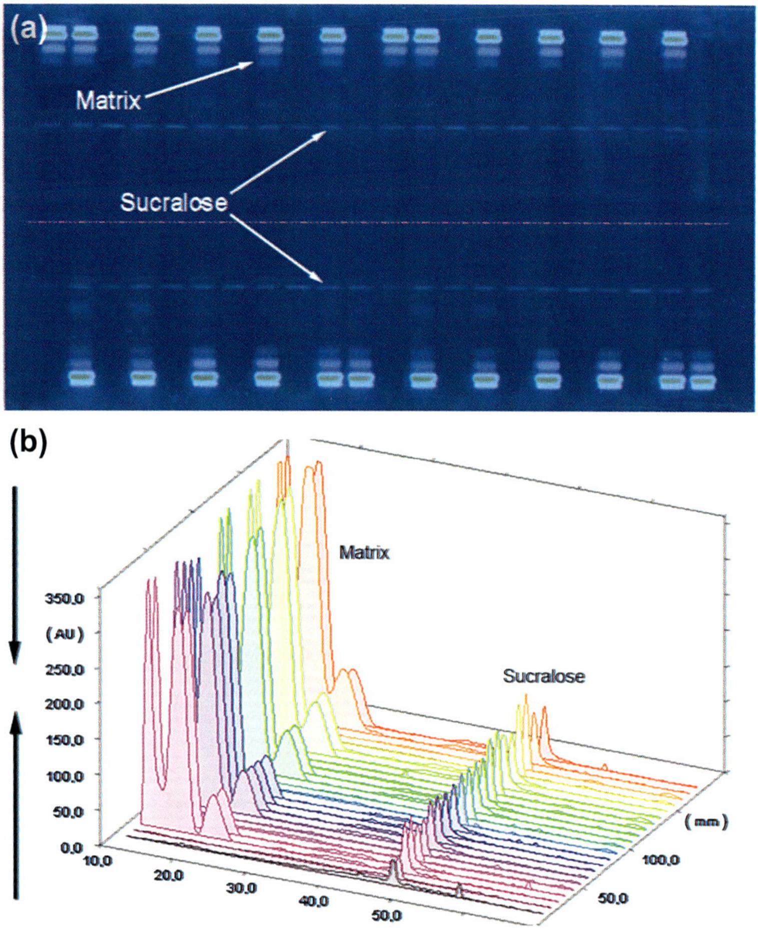

FIGURE 15.4 Antiparallel chromatography (HPTLC chromatogram with 44 runs developed from both sides (a) took 15 min), minimized sample preparation (matrix was left at the starting zone), and simultaneous derivatization of 44 runs (merely heating the plate) contribute to the high-throughput capacity of quantitative HPTLC (densitograms of one side (b)) [4,27].

the TLC-mass spectrometry (MS) Interface. Thus, TLC is used as an effective step for simultaneous sample preparation for many samples in parallel and has attracted the attention of residue analysts. In the analysis of sucralose in dietetic products like cakes and biscuits [25] and milk-based confectionery [27] simplified sample preparation was used, separating the remaining matrix and the analyte directly on the plate. This reliable method allowed a cost-effective quantification in a difficult matrix (Figure 15.4) [4].

15.3.1 Derivatization and Densitometry

It is possible in the separation of complex mixtures that certain compounds coelute. The solution is either modification of the mobile phase or—if possible—selective derivatization. A good example is the analysis of bioactive compounds in energy drinks. HPTLC allows the detection of caffeine, taurine, nicotinamide, riboflavin, and pyridoxine in a single track by employing different detection principles (absorption and fluorescence) for evaluation. The densitometric evaluation is done with a multiple wavelength scan followed by the selective derivatization of taurine with ninhydrin [4,64]. A multiwavelength scan in absorption and fluorescence mode was also used in the analysis of HAA in meat [4,69]. Further, there are analytes without chromophores that require derivatization with suitable reagents for visualization before an evaluation is possible. Such compounds are, for example, lipids like phospholipids (Figure 15.2). Using primuline as a visualization reagent allows detection via fluorescence measurement at UV wavelength 366/>400 nm. Also, when using primuline for visualization, a follow-up confirmation by MS is possible, since the (phospho)lipid primuline binding is noncovalent [20]. Sucralose, an artificial sweetener, is a further example [4]. Depending on the stationary phase used—silica gel or amino phases—the derivatization process is different. When using amino phases, merely heating the plate is sufficient to make the zone visible [27]. When using silica gel layers, derivatization with reagents like 2-naphthol/sulfuric acid or aniline/diphenylamine/orthophosphoric acid is necessary [25]. It is also possible to analyze several sweeteners like saccharin, cyclamate, aspartame, sucralose, steviol gylcosides, and acesulfame-K by multidetection via a reagent sequence, meaning by the successive application of reagents to the same samples on the same plate. In summary, using suitable reagents, HPTLC offers easy, selective, and cost-effective ways to differentiate and visualize compounds.

In the past few years, alternate documentation and evaluation options have emerged. Thanks to high-performance twelve-bit industrial charge-coupled device cameras whose capture settings are standardized and whose chips are highly linear, digital evaluation is advancing fast. These cameras enable images to be acquired at various illumination settings (UV/Vis in transmission and reflectance mode, or both combined) with precise and accurate color reproduction of the sample (Figure 15.5). Further, in quantitative methods, the conversion of image pixels to analog curves is highly effective and the background noise can be reduced by various options [4].

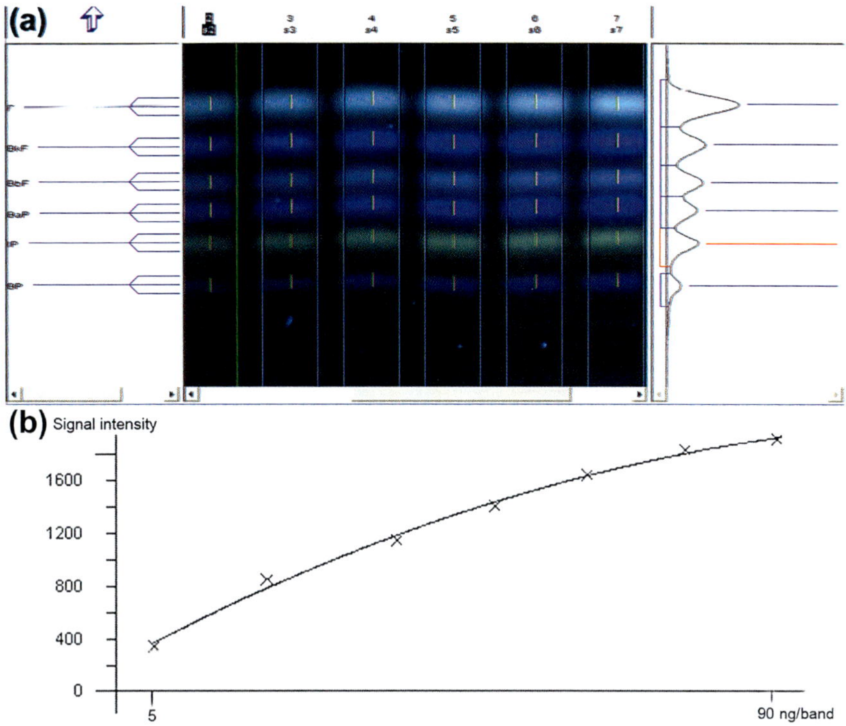

FIGURE 15.5 Digital evaluation of a chromatogram with six polyaromatic hydrocarbons ranged from 5 to 90 ng/band (a) and calibration curve of benzo[a]pyrene ($y = -0.136x^2 + 31.3x + 217$) with a relative standard deviation of 3.7% (b) [4].

15.3.2 Bioactivity-Based Detection

The identification of bioactive compounds, and also of potential toxins, in foods is of great interest. Moreover, the investigation of degradation products and metabolites is important, since they are often more toxic than the initially targeted substance [4]. (HP)TLC separation into individual substances allows direct assignment of bioactivity to a substance. An easy way to assess if a substance has toxic potential is to dip a TLC plate post separation in a suspension of the luminescent bacteria *Aliivibrio fischeri*. In Europe, *A. fischeri* is used to test drinking water for toxic substances (DIN EN ISO 11348-1 (2009), Part 1, Section 5). After HPTLC separation substances that are *A. fischeri* toxic, i.e., have inhibitory effects on the bacteria, are indicated as dark spots on a blue-green luminescent background, while substances with promoting effects show higher luminescence (Figure 15.6) [72,73]. Various substances in food and feed products, functional foods, plants, medicinal herbs, and consumer goods have been investigated (Figure 15.7) [26,74–80].

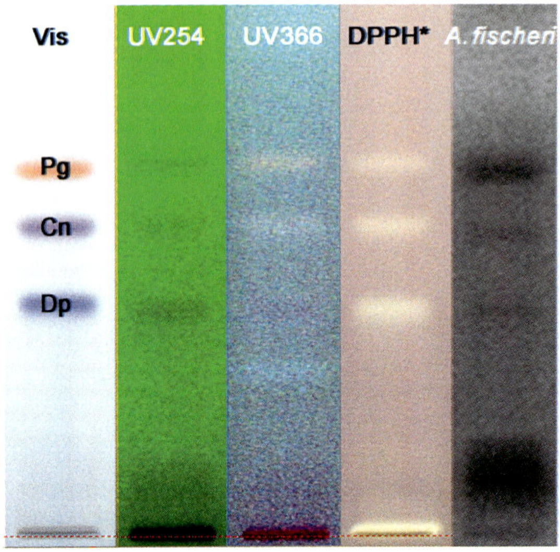

FIGURE 15.6 Example of multidetection in HPTLC: the three anthocyanidins delphinidin (Dp), cyanidin (Cn), and pelargonidin (Pg) were recorded with the TLC Visualizer at UV/Vis (natively and after derivatization with DPPH* reagent) and with bioluminizer for bioactivity detection with *Alliivibrio fischeri* bacteria.

To assess a substance's antioxidant potential HPTLC plates can be dipped into a solution of 1,1-diphenyl-2-picrylhydrazyl (DPPH*) reagent. Compounds with radical scavenging activity are indicated as yellow-colored zones on a purple background (Figure 15.6) [74]. Many herbs and plants have been examined for compounds with antioxidant potential, like anthocyanes [74], hesperidin from citrus peel [81], curcuminoids from *Curcuma longa* [82] and phenolics in *i.a.* propolis [83].

A further example of EDA is the so-called planar yeast estrogen screen assay which was substantially optimized to obtain sharp-bounded zones [84]. This method utilizes yeast cells sensitive to estrogenic substances. When such substances are present, the production of the enzyme β-galactopyranoside is induced, which—after reaction with a substrate—produces a fluorescent product that can be determined by densitometry [85]. Samples investigated so far include pore water, wastewater, and sunscreens [86].

There are also the so-called direct bioautographic TLC assays used to detect substances with antibiotic effects using *Bacillus subtilis* [87,88] or substances with inhibitory effects by glucosidase [89,90] or esterase [91].

15.3.3 Hyphenation with Mass Spectrometry

Since the development of the TLC-MS Interface device it has been possible to elute zones of interest directly from the plate to the

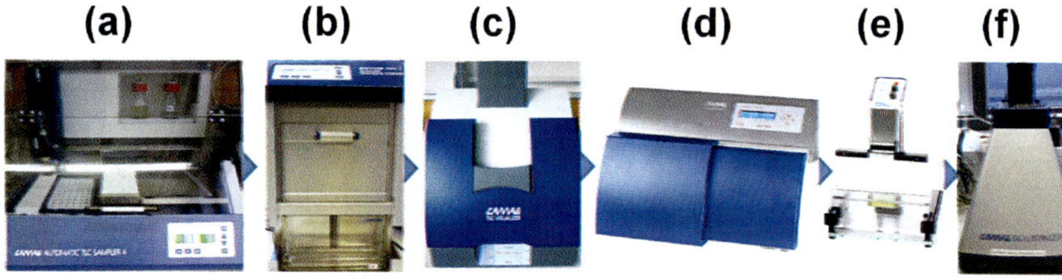

FIGURE 15.7 Example of the EDA process of many samples in parallel: application (a), development (b), documentation (c), densitometry (d), dipping in bioassay (e), and evaluation of bioluminescence (f) (all equipment from CAMAG).

ionization source of a mass spectrometer. This hyphenation of planar chromatography with MS is a helpful tool that facilitates comparison of not-clearly-assignable sample zones with standard substances, as a confirmation method. Moreover, zones of unknown origin, e.g., in research of bioactive compounds, can be examined via MS, providing the first hints to a substance's identity [92]. Examples of the application of HPTLC-MS coupling techniques include HPTLC/electrospray ionization (ESI)-MS [9,20,26,62–66,69,70,93], HPTLC/direct analysis in real time (DART)-MS [62,70,76,94], and HPTLC/matrix-assisted laser desorption ionization (MALDI)-MS [19].

For the identification of the fatty acid composition of phospholipids in lecithin from sunflower and soy HPTLC/ESI-MS (Figure 15.2) [20] as well as HPTLC/MALDI-time-of-flight (TOF) MS [19] methods were employed. For the analysis of acrylamide in water [62,65] and dialkyl phosphates as breakdown products of organophosphorus insecticides in fruit juices [93], HPTLC/ESI-MS provides the possibility for identity confirmation. Examples of the application of HPLC/DART-MS are confirmation of ITX in milk-based products [70] and the investigation of acrylamide in water [62].

An advantage of HPTLC, namely, the possibility of combining detection methods, can be seen in the determination of secondary metabolites in marine sponges. The sample after separation was subjected to EDA with *A. fischeri* (Figure 15.7) and only zones of interest, i.e., zones that show bioactivity, are specifically targeted for further investigation by HPTLC/DART-MS (Figure 15.8) [76]. The possibility of precisely targeting the so-called *hot zones* (positive findings) enables time-effective, cost-reduced sample analysis [4].

Further hyphenation possibilities are the coupling of HPTLC with nuclear magnetic resonance [62] or ATR/FTIR. After identification of zones of interest, these can be eluted from the plate with the TLC-MS interface into a vial. The solutions can then be used directly for measurement or after additional preparation steps like solvent evaporation and redissolving in a different solvent.

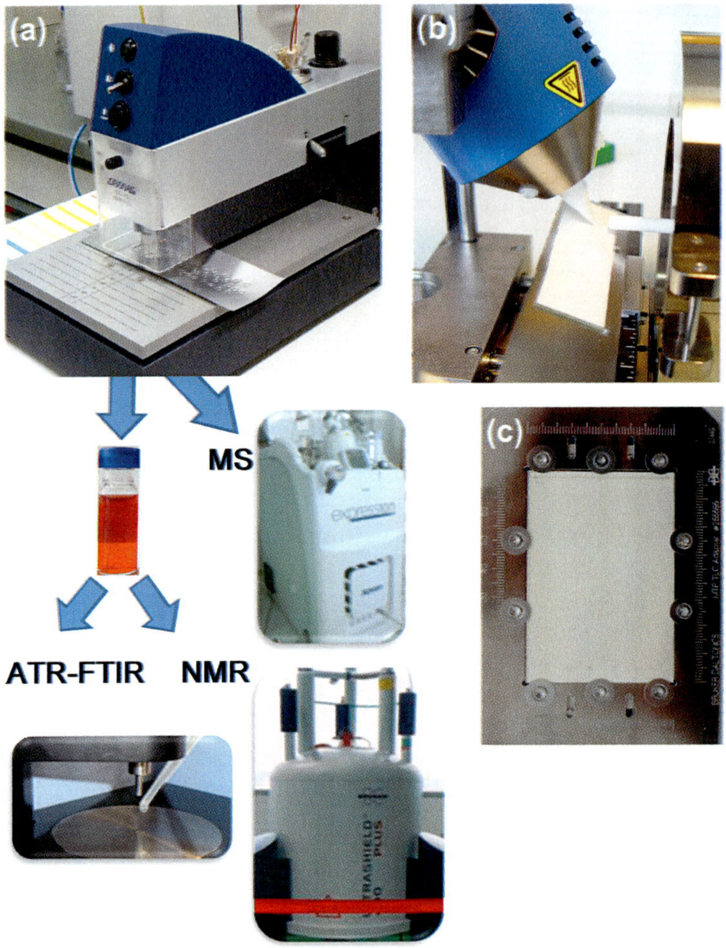

FIGURE 15.8 Example for various hyphenation possibilities in HPTLC: analysis of the zones of interest via ATR-FTIR, NMR, or ESI-MS after elution with the TLC-MS interface (a), analysis of a track with DART-MS (b), or MALDI-MS (aluminum foil 5 × 7.5 mm) (c).

15.3.4 Office Chromatography

The idea behind office chromatography is the use of office peripherals like common printers and scanners instead of the usual (HP)TLC equipment. Analyte solutions are applied by printing rather than spraying onto HPTLC plates or foils [95] or onto ultrathin-layer chromatography (UTLC) [95–98] layers. The efficacy of modified printers for applying reagents onto HPTLC plates was investigated for postchromatographic derivatization of taurine with ninhydrin in 2007 [95]. The comparison of a piezoelectric-driven and a bubble jet printer indicated that chip control of certain mechanisms of some printers can cause problems. The spatial resolution and repeatability were satisfactory. The use of miniaturized ULTC plates led to even shorter separation times and very low sample and reagent volume consumption [96]. Mixtures of food dyes were applied on two different UTLC plate types, monolithic UTLC plates (M-UTLC) and nanostructured plates (NS-UTLC), with an inkjet printer and two commercially available systems, a piezoelectric ejector and an aerosol applicator (ATS4). It was concluded that while there is still work to be done, the UTLC layers are a strong new asset and that the easily modified inkjet printer performed as well or slightly better than conventional sample application systems. In 2013, the so-called glancing angle deposition ultrathin layers were successfully used for the analysis of sugars in chocolate [97] and for food dyes [98], both with ESI-MS detection.

15.4 SUITABILITY AND CAPABILITIES

Most (HP)TLC methods developed in the past stopped at the screening stage, posing the question whether (HP)TLC is really a suitable quantitative tool. In the following sections some examples are given to illustrate the capabilities of HPTLC in food analysis.

15.4.1 Performance

TLC methods acquired a reputation as providing poor repeatability, which became a reason for using TLC for qualitative analysis only. With the following data it is shown that these prejudices are no longer true. For example, the mean repeatabilities for the determination of the anthocyanin malvidine-3-glucoside in grape seed/marc meal and supplemented animal feed samples were $\leq$1.8% ($n = 2$) [74]; for the analysis of the steviol glycoside stevioside in yogurt, $\leq$5.4% ($n = 3$) [58]; and for sucralose in milk-based confectionery, cakes, and biscuits, $\leq$4.4% ($n = 3$) [25,27]. Even for complicated matrices like honey, with optimized conditions, repeatabilities (as intraday precision, $n = 3$ per plate, $n = 2$ per day) of 3.2% can be reached for the analysis of HMF [67]. In addition, for caffeine in energy drinks a repeatability <3.8% [65]; for the trace analysis of ITX in milk, yogurt, and fat, $\leq$6.4% [94]; and for ultratrace levels of acrylamide in water, <4.8% ($n = 3$) [62] and in coffee, <3.0% ($n = 6$) [66] was obtained. Certainly, these values prove that HPTLC is a valuable and appropriate analytical tool. Further, the correlation coefficients for the calibration plots of validated methods are usually $\geq$0.999 [27,57,63–67,70,74]. The levels reached for limits of detection (LODs) are also satisfactory. For example, the LOD for stevioside in yogurt is as low as 40 mg/kg and fully meets the regulatory maximum amount for E960 of 100 mg/kg in fermented milk products and most other products [54,57]. LOD levels of 1 mg/kg can be reached for sucralose in milk-based confectionery, also meeting regulatory limits set by the European Union (10–3000 mg/kg in various foodstuffs) [3,4,27]. For the HMF content in honey the LOD values were <4 mg/kg and meet the lowest restriction level of 15 mg/kg [67,99]. As of December 2014, caffeinated beverages with a content exceeding 150 mg/L have to be labeled as "High caffeine content. Not recommended for children or pregnant or breast-feeding women" (Regulation (EU) No. 1169/2011). A LOD of 75 μg/L reached by HPTLC is appropriate for the EU stipulated limit [63]. Also, for acrylamide in drinking water the LOD by HPTLC is as low as 0.025 μg/L, which is more than adequate for checking the EU- or World Health Organization- and US Environmental Protection Agency-set limits of 0.1 μg/L or 0.5 μg/L, respectively [4,62].

15.4.2 Method Comparisons

HPTLC and HPLC are liquid chromatographic methods with overlapping sample application domains. Method comparisons have shown that results obtained by HPTLC are comparable with those obtained by HPLC. Further interlaboratory trials have demonstrated good method accuracy, leading to ISO/DIN standards for the analysis of, e.g., polycyclic aromatic hydrocarbons [100] or pesticides [101] in water. The commonly used herbicide glyphosate and its principal degradation product aminomethylphosphonic acid were analyzed in water samples by HPTLC with fluorescence detection. Compared with findings from an HPLC-MS interlaboratory trial the method showed good correlation ($r^2 = 0.986$) with the true values. Furthermore, the Z_u-scores for three concentrations levels in the ultratrace range was satisfactory ($Z_u = 0.05–0.55$) [4,102]. Comparison of HPTLC and HPLC-evaporative light scattering detector (ELSD) in the phospholipid analysis of lecithin samples showed that the values obtained by both methods (accuracy and precision) were comparable, giving neither method an advantage over the other [103]. For HAA, a process contaminant in meat production, the results from HPLC/UV-fluorescence detection (FLD) method were similar to those obtained by a newly developed HPTLC/UV-FLD method [4,104]. Similar findings were also obtained for the analysis of ultratrace levels of acrylamide in groundwater by HPTLC with fluorescence detection and HPLC-MS/MS. Even at these very low levels both methods showed satisfactory agreement [4,66]. Moreover, the paper describes another advantage of HPTLC, the selective derivatization of a compound for visualization. With a simple step HPTLC allows solving of a challenging task, something that cannot be done so easily by other chromatographic techniques. Moreover, due to the fact that HPTLC allows multiple detection, online mass spectra were recorded after zone elution from the plate, allowing confirmation of the zone's identity [4]. Results with satisfactory correlation were also obtained for analysis of aflatoxin M_1 in milk by TLC-UV and HPLC-FLD, giving comparable results for recoveries and quantification limits (both 0.01 μg/L), as well as good values for precision (relative standard deviations ≤7.7%) [37]. Lin et al. [105] compared various methods for mycotoxins like enzyme-linked immunosorbent assay (ELISA), HPLC, and GC with TLC. It is striking that TLC produces results that are mostly comparable and in some cases better than those obtained by other methods. For example, comparing HPTLC with ELISA yielded more robust and reliable results for aflatoxins in corn-based foodstuffs by HPTLC [43].

Another question when comparing different chromatographic methods is the relatively low separating power of HPTLC compared with HPLC and GC. This results from the inability of capillary forces to provide a constant and optimum mobile phase velocity during the separation. In many cases, however, the selectivity of HPTLC is perfectly adequate for a given task and can overcome its lower kinetic performance. This is demonstrated for the quantification of lactose in lactose-free dairy products (Figure 15.9 [106]). As shown above, the results obtained were comparable with those obtained by other chromatographic methods. Nowadays, zone purity, and therefore sufficient selectivity, can be confirmed by recording of mass spectra [20].

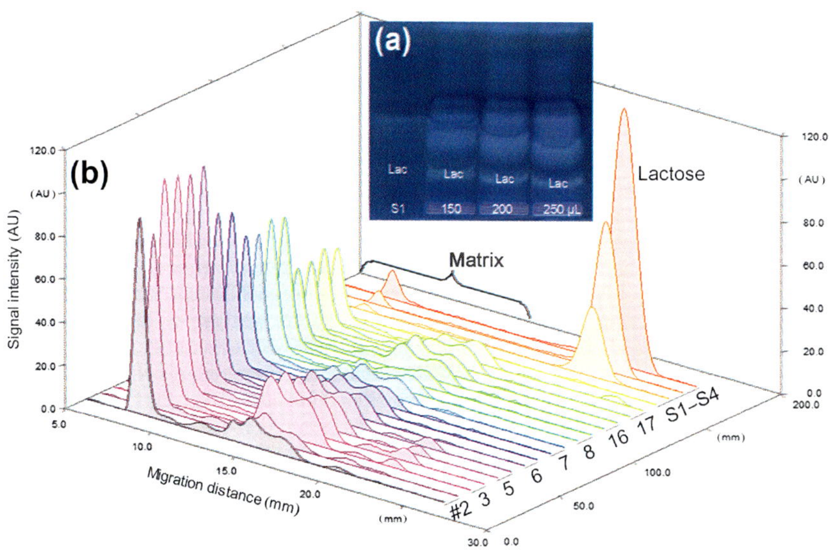

FIGURE 15.9 Streamlined analysis of lactose-free dairy products by HPTLC: application of up to 250 μL allows LODs down to 0.04 mg/L (a); screening and quantitation of lactose in eight dairy products (#2–17) applied twice (b) [106].

[75] Baumgartner V, Hauri U, Hohl C. Bio-activity based analysis of irradiated sunscreens using HPTLC and in situ detection with *Vibrio fischeri*. CBS 2008;100:2–5.

[76] Klöppel A, Grasse W, Brümmer F, Morlock G. HPTLC coupled with bioluminescence and mass spectrometry for bioactivity-based analysis of secondary metabolites in marine sponges. J Planar Chromatogr 2008;21:431–6.

[77] Morlock G, Kopacz S. Fast and precise SBSE-HPTLC/FLD method for quantification of six polycyclic aromatic hydrocarbons frequently found in water. J Liq Chromatogr Rel Technol 2008;31:1925–42.

[78] Misra K, Tulsawani R, Shyam R, Meena DM, Morlock G. Hyphenated high-performance thin-layer chromatography for profiling of some Indian natural efficiency enhancers. L Liq Chromatogr Rel Technol 2012;35:1364–87.

[79] Baumgartner V, Hohl C, Schwack W. Screening for antimicrobials in mouthwashes using HPTLC-Bioluminescence detection. Chromaographia 2013;76:1315–25.

[80] Seigel A, Schrock A, Hauser R, Spangenberg B. Sensitive quantification of diclofenac and ibuprofen using thin layer chromatography coupled with Vibrio fischeri bioluminescence assay. J Liq Chromatogr Rel Technol 2011;34:817–28.

[81] Al-Ashaal HA, El-Sheltawy ST. Antioxidant capacity of hesperidin from citrus peel using electron spin resonance and cytotoxic activity against human carcinoma cell lines. Pharm Biol 2011;49:276–82.

[82] Pozharitskaya ON, Ivanova SA, Shikov AN, Makarov VG. Separation and free radical-scavenging activity of major curcuminoids of *Curcuma longa* using HPTLC-DPPH method. Phytochem Anal 2008;19:236–43.

[83] Yrjöenena T, Peiwub L, Summanena J, Hopiab A, Vuorelaa H. Free radical-scavenging activity of phenolics by reversed-phase TLC. JAOCS 2003;80:9–14.

[84] Morlock G, Klingelhöfer I, Liquid chromatography-bioassay-mass spectrometry for profiling of pharmacologically active food, Anal Chem, 2014, in print.

[85] Müller MB, Dausend C, Weins C, Frimmel FH. A new bioautographic screening method for the detection of estrogenic compounds. Chromatographia 2004;60:207–11.

[86] Buchinger S, Spira D, Bröder K, Schlüsener M, Ternes TA, Reifferscheid G. Direct coupling of thin layer chromatography with a bioassay for the detection of estrogenic compounds—applications for effect directed analysis. Anal Chem 2013; 85:7248–56.

[87] Hamburger MO, Ordell GC. A direct bioautographic TLC assay for compounds possessing antibacterial activity. J Nat Prod 1987;50:19–22.

[88] Bories GF, Baradat MM, Corpet DE. Bioautographic method for evaluation of glycopeptide actaplanin in milk. J Assoc Off Anal Chem 1986;69:938–40.

[89] Salazar MO, Furlan RLE. A rapid TLC autographic method for the detection of glucosidase inhibitors. Phytochem Anal 2007;18:209–12.

[90] Simões-Pires CA, Hmicha B, Marston A, Hostettmann K. A TLC bioautographic method for the detection of α- and β-glucosidase inhibitors in plant extracts. Phytochem Anal 2009;20:511–5.

[91] Akkad R, Schwack W. Multi-enzyme inhibition assay for the detection of insecticidal organophosphates and carbamates by high-performance thin-layer chromatography applied to determine enzyme inhibition factors and residues in juice and water samples. J Chromatogr B 2010;878:1337–45.

[92] Morlock G. Kopplung der Planar-Chromatography mit der Massenspektrometrie. CLB 2006;57:343–8.

[93] Schwack W, Zeisler T, Stiefel C. Determination of dialkyl phosphates as breakdown products of organophosphorus insecticides in fruit juices by HPTLC with fluorescence detection. J AOAC Int 2009;92:691–8.

[94] Morlock G, Ueda Y. New coupling of planar chromatography with direct analysis in real time mass spectrometry. J Chromatogr A 2007;143:243–51.

[95] Morlock G, Stiefel C, Schwack W. Efficacy of a modified printer for application of reagents in planar chromatography. J Liq Chromatogr Rel Technol 2007;30:2171–84.
[96] Morlock GE, Oellig C, Bezuidenhout LW, Brett MJ, Schwack W. Miniaturized planar chromatography using office peripherals. Anal Chem 2010;82:2940–6.
[97] Kirchert S, Wang Z, Taschuk MT, Jim ST, Brett MJ, Morlock GE. Inkjet application, chromatography, and mass spectrometry of sugars on nanostructures thin films. Anal Bioanal Chem 2013;405:7195–203.
[98] Wannenmacher J, Jim SR, Taschuk MT, Brett MJ, Morlock GE. Ultrathin-layer chromatography on SiO_2, Al_2O_3, TiO_2, and ZrO_2 nanostructured thin films. J Chromatogr A 2013;1318:234–43.
[99] Codex Alimentarius, Alinorm 01/25. Draft revised standard for honey at step 8 of the Codex procedure and EU Directive/1/110/2001 of 02/12/2001 (L 10/47); 2000.
[100] Deutsches Institut für Normierung, ISO 7981-1. Berlin: Beuth Verlag; 2005.
[101] Deutsches Institut für Normierung, DIN 38047-11. Berlin: Beuth Verlag; 1995.
[102] Weber W, Seitz W, Aichinger A. Ultra trace analysis of glyphosate and AMPA in water with HPTLC. CBS 2005;95:5–7.
[103] Schärer C. Comparison of HPLC-ELSD and modern planar chromatography in today's quality control of phospholipids. CBS 2002;89:12–5.
[104] Jautz U, Gibis M, Morlock GE. Quantification of Heterocyclic Aromatic Amines in fried meat by HPTLC/UV-FLD and HPLC/UV-FLD: a comparison of two methods. J Agri Food Chem 2008;56:4311–9.
[105] Lin L, Wang P, Wang Y, Chen J, Zhang J. Thin-layer chromatography of mycotoxins and comparison with other chromatographic methods. J Chromatogr A 1998;815: 3–20.
[106] Morlock G, Morlock L, Lemo C. Streamlined analysis of lactose-free dairy products. J Chromatogr A 2014;1324:215–23.

CHAPTER

16

Environmental Applications

Wolfgang Schwack

Institute of Food Chemistry, University of Hohenheim, Stuttgart, Germany

16.1 INTRODUCTION

Environmental chemistry aims to study the sources, mobility, reactions, and fate of chemicals in the atmosphere, in waters, and in soils, including the effects on organisms. Analytical methods afford important tools to provide data for contaminants and pollutants in environmental samples, almost always constituted as complex matrices. Environmental chemicals, on the one hand, refer to anthropogenic chemical species and their degradation products, which intentionally or unintentionally are introduced into the environment, like pesticides and nutrients from agricultural applications, metals from industrial processes and rust protection, or polychlorinated biphenyls from leakages. Wastes, industrial effluents, and sewage are other sources of environmental contaminants, including pharmaceuticals and their metabolites, estrogens and xenoestrogens, heavy metals, flame retardants, plasticizers, sunscreens, sweeteners, and surfactants, etc. On the other hand, environmental contaminants derive from processes like combustions releasing polychlorinated dibenzodioxins and furanes or polycyclic aromatic hydrocarbons; disinfection products releasing trihalomethanes and halogenated phenols; or volcanic events releasing heavy metals. Finally, metabolic processes afford organometallic compounds from inorganic species like arsenic, selenium, or mercury.

Environmental contaminants are almost always complex multi-compound substance classes, for example, dioxins or polychlorinated biphenyls. Therefore, high-resolution chromatography coupled to selective mass spectrometric detection is generally required for monitoring purposes [1]. In other cases, the composition of contaminant classes may be simpler depending on regional circumstances. Not all of the hundred

Instrumental Thin-Layer Chromatography
http://dx.doi.org/10.1016/B978-0-12-417223-4.00016-9

or so known pesticides are applied in a single region or country. Thus, runoff from field applications leading to contaminated surface or groundwaters is possibly restricted to only a dozen or so active ingredients which require monitoring. As compared to municipal effluents, for industrial or hospital effluents, it is also almost known, which chemicals are present. Contaminants in sewage are strongly reduced by wastewater treatment plants (WWTP), resulting in lower concentrations and reduced complexity of effluents entering surface waters. However, microbial transformation inside the WWTP may result in contaminants quite different from those present in the influent wastewater. In the case of heavy metals, only a few are of high priority, like lead, cadmium, arsenic, and mercury.

Instrumental TLC is suitable for monitoring environmental contaminants, if at least sample history is slightly known. Additionally, selective staining (see Chapter 8) and effect-directed detection (see Chapter 11) provides some compensation for low resolution and interference from comigrating (matrix) compounds not of interest, as the mass selective detector also does. The direct coupling of TLC with mass spectrometry (see Chapter 10) provides an attractive and optional tool to confirm positive findings.

The general planar chromatography literature was reviewed by Sherma on a biannual basis for many years [2], while he also regularly reviewed the literature on advances in TLC of pesticides [3]. The following examples of environmental applications of TLC were selected from the literature to show the options and limits in the field of environmental analysis.

16.2 HPTLC IN ENVIRONMENTAL ANALYSIS

16.2.1 Pesticides

The determination of pesticide residues in water encountered a severe challenge when the EU guidelines for drinking water became effective in 1986, setting limits of 0.1 μg/L for individual pesticides and 0.5 μg/L for the sum of all pesticides. Their analysis by HPTLC became possible in the 1990s after the introduction of the automated multiple development techniques (AMD, see Chapter 4) [4]. Gradients starting with basic conditions and ending with an acidic solvent mixture increased the selectivity of the separation when used in combination with multiwavelength scanning (Figure 16.1) [5,6]. Positive findings were confirmed by comparison of UV spectra and separation on a second plate with an AMD gradient of a different selectivity [7,8]. HPTLC data were collected for nearly 300 pesticides isolated by solid-phase extraction from water,

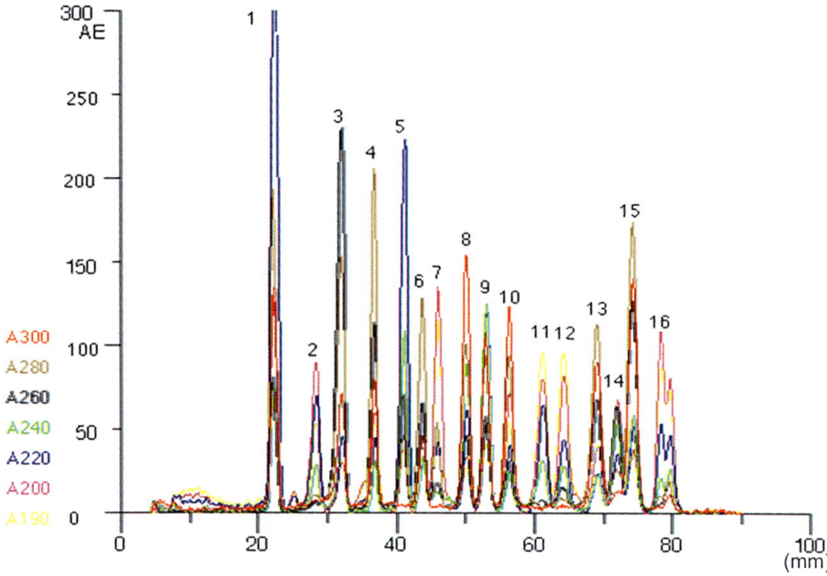

FIGURE 16.1 HPTLC–AMD separation and multiwavelength scanning for different pesticides (100 ng each) at seven wavelengths from 190 to 300 nm (1 naphthalene-l-sulfonic acid, 2 procloraz, 3 triazoxid, 4 ethidimuron, 5 simazine, 6 bromazil, 7 carbofuran, 8 metribuzin, 9 azinphos methyl, 10 coumaphos, 11 prosulfocarb, 12 dichlofluanid, 13 parathion, 14 fenthion, 15 dinoseb, 16 prothiofos). *(With permission reprinted from Ref. [6].)*

allowing the determination of all active compounds in use at that time, including all pesticides not amenable to gas chromatography (GC). The method finally led to a German standard [9]. This HPTLC–AMD method was later adapted to soils extracted by supercritical fluid extraction providing clean extracts, which could be applied to an HPTLC plate without further cleanup [10].

During the same time, HPTLC methods (without AMD) were developed for multipesticide residue analysis of fruits and vegetables, and used for conformation of the results obtained by GC electron capture detection and nitrogen phosphorus detection (GC-ECD-NPD) [11–13]. However, a significant effort was required for the clean up of extracts by gel permeation chromatography and fractionation on a silica gel column to minimize interferences by coextracted matrix components. Additionally, separation was performed on both silica gel and C18 plates using different staining reagents for detection. The confirmatory procedure was laborious and time consuming in the absence of mass selective detectors. However, in 2005 Ambrus et al. published extensive studies on the screening of pesticide residues in fruits, vegetables, and cereals [14,15]. They also performed a multistep cleanup procedure with separations on normal and reversed phase layers, and applied eight reagent systems for detection, including four biotests. Their aim was to provide a cost-effective alternative for laboratories that are unable to invest in modern GC–MS and LC–MS technologies. However, they concluded that HPTLC screening for pesticide residues in samples of unknown origin is not to recommend. Nevertheless, the above cited sources provide a treasure of HPTLC data and selective detection techniques. Furthermore, many publications dealing with the TLC analysis of pesticides, but only a limited number of pesticides, have been reported, but often without application to real samples.

While nowadays highly sensitive and selective GC–MS and LC–MS systems are the work horses for multipesticide screening in environmental samples, HPTLC–AMD is described as an important tool for the widest possible multidimensional screening [16–19], including bioactivity-based detection with luminescent bacteria (*Aliivibrio fischeri*) [20]. Also, instrumental planar chromatography should be considered when the sample's history is known for its high sample-throughput capability, for example, during supervised field trials to follow the degradation of a pesticide during washing and cooking of harvested vegetables [21]. In a vineyard it is almost always known which pesticides have been applied, encouraging the use of HPTLC to study their fate from vine to wine. The same holds true for a tea garden, allowing the determination of pesticide residues in tea by HPTLC [22]. The study of the fate of triazine herbicides in soils is a further example [23], in which detection limits (LOD) of 10 pg/zone were obtained after derivatization with dansyl chloride [24]. In any case, it should be kept in mind that coextracted matrix components from soils or plants can give rise to severe interferences. However, with the focus on only a few pesticides, matrix-tolerant layers can be used for simultaneous cleanup as shown for the acetonitrile extracts of different fruits [25]. For the first development, a mobile phase was selected to migrate the pesticides with the solvent front, while most matrix compounds were left behind (Figure 16.2).

In addition, the great advantage of postchromatographic selective derivatizations [12,14] highly compensates for the low separation numbers of HPTLC layers as compared with HPLC columns. For quaternary ammonium pesticides, the so-called "quats", a new staining reagent consisting of sodium tetraphenylborate was recently published [26], resulting in strong fluorescent zones with a low LOD (ng/zone). The reagent also worked well with some neonicotionoids [27]. Phenols released by alkaline hydrolysis of pesticides on the plate can be detected with diazotized aniline derivatives [28] or 4-aminoantipyrine in the presence of potassium hexacyanoferrate(III) [29], resulting in deeply colored zones. One of the most selective and sensitive detection modes for organophosphate and carbamate pesticides is choline esterase inhibition. In terms of an effect-directed analysis, enzyme inhibition is observed after separation on an HPTLC plate. Choline esterases from different sources,

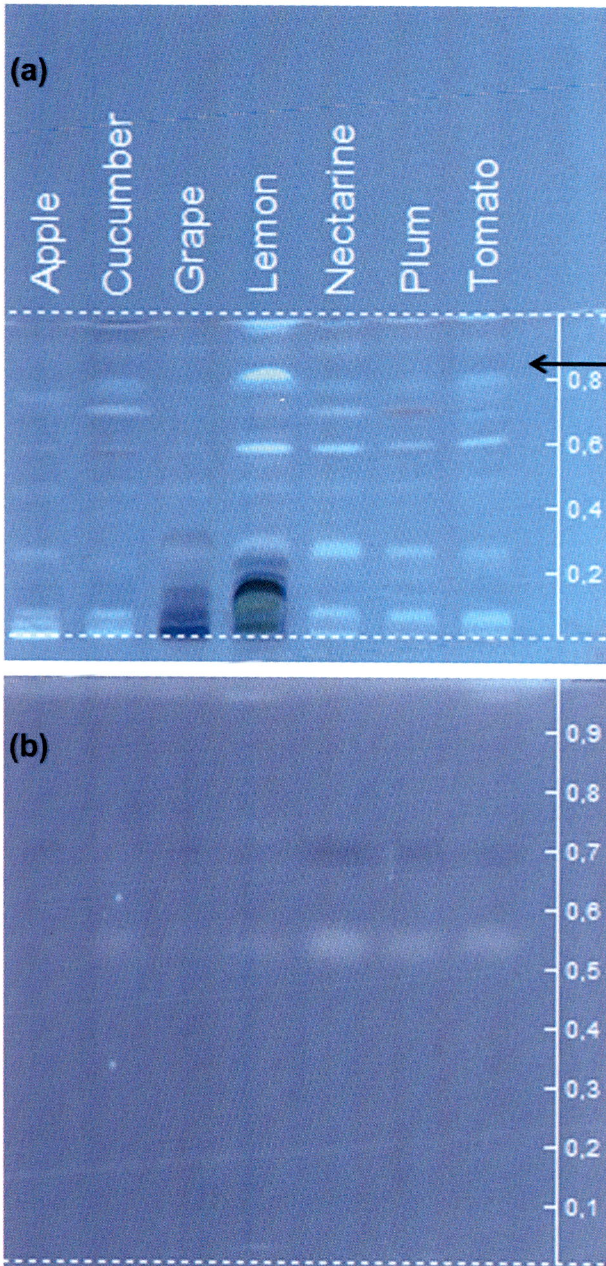

FIGURE 16.2 HPTLC clean up of QuEChERS blank extracts of fruit and vegetable samples. (a) Extracts (10 μL) were developed with methanol–dichloromethane (10 + 90, v/v) to a distance of 30 mm. (b) After cutting the plate at 25 mm from the lower edge (arrow) and focusing the zones with acetone, the plate was redeveloped with n-hexane–ethyl acetate–dichloromethane (65 + 20 + 15, v/v/v). Separations were visualized by dipping into primuline reagent. *(Modified Figure 1 from Ref. [25], permission granted by AOAC International.)*

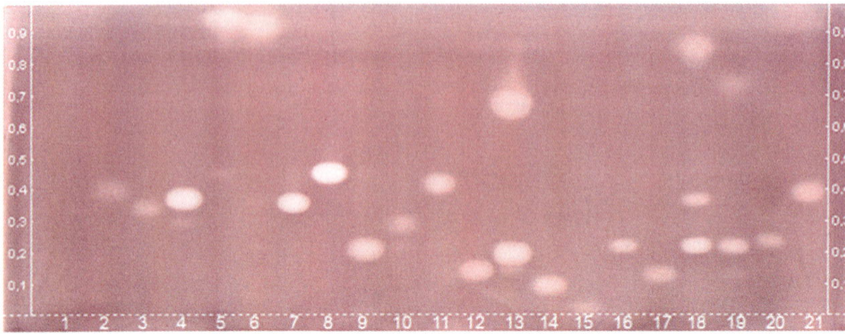

FIGURE 16.3 Organophosphate and carbamate pesticides, separated on a silica gel HPTLC plate (n-hexane/ethyl acetate/dichlormethane, 65 + 20 + 15) and detected by an enzyme inhibition assay with rabbit liver esterase as the enzyme source (1 acephate 1 μg, 2 carbaryl 10 ng, 3 carbofuran 100 ng, 4 chlorfenvinfos 100 pg, 5 chlorpyrifos 200 ng, 6 chlorpyrifos-methyl 1 μg, 7 chlorpyrifos-methyl oxon 1 ng, 8 chlorpyrifos oxon 100 pg, 9 demeton-S-methyl 50 ng, 10 dichlorvos 10 pg, 11 ethiofencarb 50 ng, 12 malaoxon 10 ng, 13 malathion 2 μg, 14 methomyl 50 ng, 15 monocrotofos 50 ng, 16 paraoxon 10 pg, 17 paraoxon-methyl 100 pg, 18 parathion 20 ng, 19 parathion-methyl 50 ng, 20 pirimicarb 50 ng, 21 propoxur 70 ng), also showing the esterase-active impurities of commercial standards. *(With permission reprinted from Ref. [31].)*

as well as other esterases, have been used (Figure 16.3) [30], providing for multienzyme assays [31,32].

Dialkyl phosphates and dialkyl thiophosphates (DAPs) are common degradation products of organophosphorus pesticides, resulting from a simple hydrolysis. They are widely used as urinary biomarkers to monitor human exposure to different organophosphorus pesticides [33]. The presence of DAPs in urine samples, however, cannot answer the question, as to whether they are metabolites of ingested pesticides or if they are already present in drinking water or food [34]. After extraction from water, juice or urine samples and separation on silica gel TLC plates, staining with 4-(p-nitrobenzyl) pyridine resulted in LODs of 0.05 and 10 μg/zone for diethyl and dimethyl phosphate, respectively [35]. Sherma et al. separated dialkyl thiophosphates (containing P=S groups) on RP18 plates, detected them with N-2,6-dichlorobenzoquinoneimine as chromogenic reagent and obtained LODs of 200 ng/zone [36]. LODs and LOQs of approximately 0.5 and 1 ng/zone, respectively, were obtained after prechromatographic derivatization with 1-(bromoacetyl)pyrene (BAP) for both dialkyl phosphates and dialkyl thiophosphates [37]. While dialkyl phosphates afforded light blue fluorescent zones, the dialkyl thiophosphates gave yellow–brown zones which turned light blue after a short irradiation with UV light (Figure 16.4). In combination with the different reactivities of DAPs toward BAP [37], these effects provide useful selectivity for the identification of dialkyl phosphates and thiophosphates.

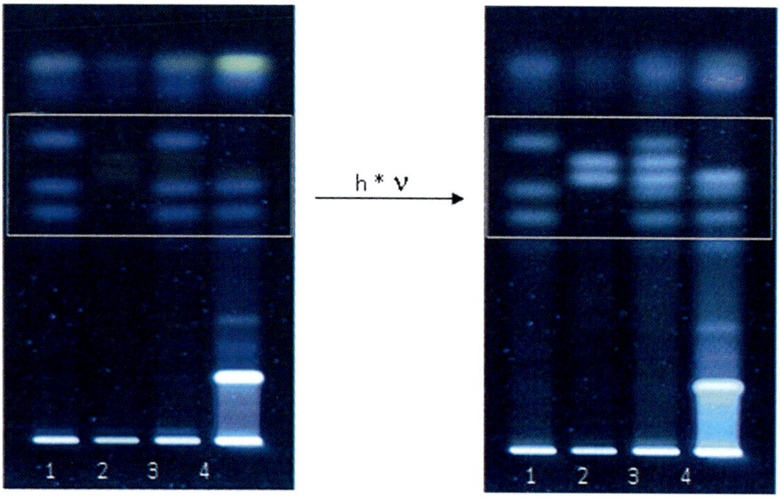

FIGURE 16.4 Plate images of the separation of dialkyl phosphates and dialkyl thiophosphates as 1-(bromoacetyl)pyrene derivatives on HPTLC amino plates (1: dibutyl phosphate = internal standard, diethyl phosphate, and dimethyl phosphate with descending R_f; 2: diethyl thiophosphate and dimethyl thiophosphate with descending R_f; 3: standard mix; 4: apple juice sample spiked with three DAPs) before and after UV irradiation. *(Reprinted Figure 4 from Ref. [37], permission granted by AOAC International.)*

16.2.2 Polycyclic Aromatic Hydrocarbons

Polycyclic aromatic hydrocarbons (PAHs) are products of incomplete combustions of organic matter and enter the environment from different sources. They are found as contaminants in the atmosphere, soil, surface water and sediments. They are also found as process contaminants in heated, grilled or smoked food. Several PAHs are known or suspect carcinogens and require monitoring, especially in water and food samples. As they generally exhibit strong fluorescence, they can be detected on HPTLC plates with LODs $<<$ 1 ng/zone, while the separation of PAHs mixtures may be a challenge. Additionally, PAHs are not only present in samples as hydrocarbons, they are also found as oxygenated derivatives, such as phenols and quinones, and also as nitrogen- or sulfur-containing derivatives. On caffeine-impregnated silica gel layers developed with dichloromethane, only 6 of the 16 Environmental Protection Agency (EPA) PAHs are detectable by fluorescence after excitation at 366 nm (Figure 16.5). The separation can be optimized by performing the development in a freezer at -20 °C [38,39].

This method first became a German Standard followed by an international standard for the determination of PAHs in drinking water [40]. Later, the mobile phase dichloromethane was substituted by the chlorine-free solvent isopropyl acetate [41] or isopropyl acetate/acetonitrile (7/3)

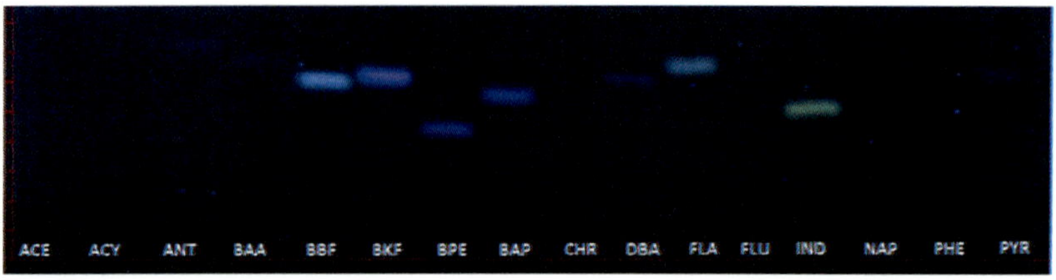

FIGURE 16.5 Separation of 16 EPA polycyclic aromatic hydrocarbons (10 ng/zone) on an HPTLC silica gel caffeine-impregnated layer, developed with isopropyl acetate/ acetonitrile (7/3). Plate image obtained by illumination at 366 nm (anthracene ANT, benzo[b] fluoranthene BBF, benzo[k]fluoranthene BKF, pyrene PYR, acenaphthene ACE, benzo[a] anthracene BAA, benzo[a]pyrene BAP, benzo[ghi]perylene BPE, chrysene CHR, dibenzo[a,h] anthracene DBA, indeno [1,2,3-c,d]pyrene IND, fluorine FLU, fluoranthene FLA, phenanthrene PHE).

[42]. The liquid–liquid extraction of water samples initially used was subsequently replaced by an elegant stir bar sorptive extraction (SBSE), for which a 10-mL water sample was sufficient to determine PAHs at 8–52 ng/L [42]. On a reversed-phase plate (RP18), all PAHs are generally detectable by fluorescence, except for naphthalene and acenaphthylene which are too volatile for stable detection. However, all 16 EPA PAHs cannot be fully separated by either multistep [43] or AMD [44] developments (Figure 16.6), but the resolution of comigrating pairs can be obtained by multiwavelengths detection selecting optimal wavelengths of excitation [43–46]. The selective quenching of alternant PAHs by dipping the plate into nitromethane can help to overcome the comigration of alternant and nonalternant PAHs [46].

Recently, the separation of PAHs on (homemade) silver thiolate silica gel plates was shown to depend mainly on the number of rings [47]. The silver thiolate plates provided a principle of separation quite comparable to silver ion silica materials, which are stable to light during storage. The transfer of this approach to HPTLC plates with the use of instrumental TLC could provide an attractive alternative for the analysis of PAHs.

Compared to drinking water, extracts of soils, airborne particulate matter, food or commodities like toys require a cleanup, preferably on a silica gel cartridge, to remove lipids and polar fluorescent interferents [43,46,48]. For the determination of polar, functionalized PAHs like nitro-, amino- or azarenes, generally present in lower amounts in environmental samples, enrichment and clean up on silica gel columns, followed by separation on RP18 plates with different solvent systems, are required as demonstrated for the analysis or airborne particulate matter [49], sewage sludge [50,51], and grilled meat [52].

Last, but not least, it should be pointed out that PAHs are strongly affected by light and oxygen. Protection from light should be considered

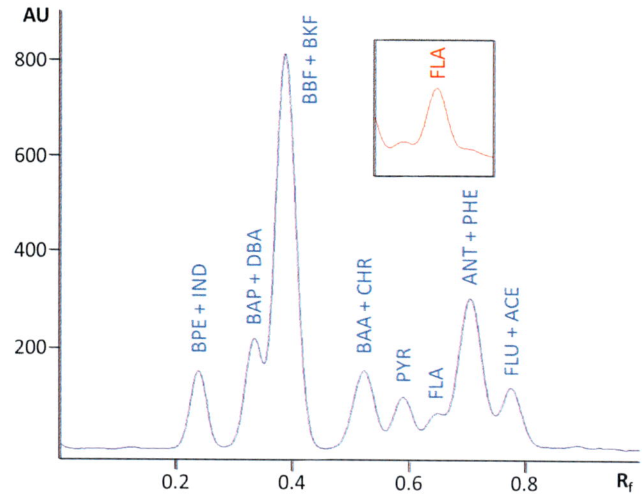

FIGURE 16.6 Separation of 14 EPA polycyclic aromatic hydrocarbons (10 ng/zone) on an HPTLC RP18 plate after a three-fold development (migration distances 45, 55, and 65 mm) with acetonitrile/water (90 + 10) [46]. Fluorescence scan recorded at 250 nm/ >320 nm; the insert shows the selective determination of fluoranthene at 270 nm/>400 nm. Abbreviations are explained in Figure 16.5.

during all steps of the analysis, including separation by HPTLC, especially if a multistep development is undertaken. Here the closed and nitrogen flushed AMD system is a useful instrument.

16.2.3 Inorganic Contaminants

The analysis of heavy metals in environmental, feed, and food samples is the domain of atomic absorption spectrometry and inductively coupled plasma (ICP) optical emission spectrometry or ICP—MS today. Therefore, it may come as a surprise that TLC can be used to determine metal ions as well. There are numerous publications dealing with more or less simple TLC experiments to separate two or three metals on different layers with different solvent systems, reporting only R_f values and possibly LODs, but not presenting any application to real samples. However, these publications can be used as a source of information concerning layer/solvent systems and detection reagents. The old literature (before 1995) was reviewed by Mohammad [53], but up until now instrumental HPTLC applications have rarely been published. Heavy metal ions can be directly separated on silica gel plates, if the mobile phase contains a suitable coordinating agent, such as picolinic acid [54] or an amino acid [55,56]. In addition, micellar mobile phases can help to enhance the separation efficiency [57,58]. Selective and sensitive detection is possible with dithizone (Zn^{2+}, Cd^{2+}, Hg^{2+}, Pb^{2+}), dimethylglyoxime (Ni^{2+}, Co^{2+}), or alkaline hydrogen peroxide (Mn^{2+}) [57–59]. Another approach consists of the extraction of metal ions from aqueous samples into an organic solvent with the aid of a complexing reagent like dithizone [60,61], sodium diethyldithiocarbamate [62], or ammonium pyrrolidine dithiocarbamate [63], which simultaneously includes an enrichment step. Using dithizone, for example, allowed mercury to be detected in aqueous samples down to 20 μg/L [60]. Recently, porphyrins were proposed as chelating agents to determine heavy metals by instrumental HPTLC and applied to water samples [64]. The metal chelates were enriched on C18 solid-phase extraction cartridges, resulting in LODs below 1μg/L for mercury, zinc, copper, and cobalt.

In environmental and biological samples, some metals occur in different oxidation states and also as organometallic derivatives. To identify the distinct species (called speciation), a chromatographic separation is generally required. By HPTLC, the mercury species (Hg^{2+}, CH_3Hg^+, $C_2H_5Hg^+$, $C_6H_5Hg^+$) extracted from water by dithizone were separated on silica gel plates [65]. Preconditioning the plate with ammonia vapor prevented interference by other coextracted metals, allowing a selective and sensitive determination of mercury species at levels of approximately 1 μg/L. Diantipyrylmethane (DAM) has been widely used as a complexing agent for the spectrophotometric determination of many metals [66]. It has also been used in TLC by overspotting samples prior to development [67]. DAM salts of inorganic acids extracted into chloroform were used for TLC separation of different anions [68]. Coupling ICP–MS with HPLC separation is widely used, but was also coupled with TLC separations. For example, speciation of arsenic and chromium was demonstrated using a combination of HPTLC and laser ablation ICP–MS [69,70]. Recently, a plasma jet desorption atomization source was presented as a novel interface to couple an HPTLC plate with an atomic fluorescence spectrometer [71]. For this device, the determination of mercury species extracted from water and urine by dithizone and separated by TLC, LODs of a few pg/zone were obtained.

16.2.4 Estrogens

With the EU directive 2013/39 concerning priority substances in the field of water policy, the European Commission introduced a watch list of substances, for which monitoring data are to be gathered. Besides diclofenac, the first list contains natural 17-β-estradiol (E2), and the synthetic contraceptive 17-α-ethinylestradiol (EE2), which contaminate surface waters at very low ng/L levels through wastewater treatment plant effluents, together with other (xeno)estrogenic compounds. The determination of such low concentrations is a challenge for modern LC–MS systems, and the more so for HPTLC. Applying a 2D-development on a cyanopropyl plate in the normal and reversed phase mode, EE2 could be separated from other steroids and after treatment with dilute sulfuric acid gave fluorescent spots with an LOD of 1 ng/spot [72]. To reach the low concentrations in surface waters, 5 L samples were enriched on solid-phase extraction cartridges.

For the detection of estrogenic compounds in environmental samples, the yeast estrogen screen (YES) has been used for effect-directed analysis for many years. YES is a reporter gene assay based on genetically modified yeast cells providing the human estrogen receptor. Activation of the receptor induces the reporter gene lacZ encoding for β-galactosidase, whose activity is determined with a suitable chromogenic substrate and is a measure of the total estrogenic activity of a sample. With the advantages of planar chromatography in terms of cleanup and separation, YES was successfully combined with HPTLC, called "planar-YES (p-YES)" (Figure 16.7), and applied to extracts from sediments [73] and sediment pore water and wastewater samples [74].

As the substrate for β-galactosidase, 4-methylumbellifery-β-D-galactoside (MUG) [75] results in blue fluorescence enabling LODs of 1.3 and 0.5 pg/zone for E2 and EE2, respectively, to be obtained. Comparing influents and effluents from a wastewater treatment plant by p-YES, the transformation of influent estrogenic substances to more polar derivatives was indicated, which cannot be achieved with the micro-titer plate YES assay. For environmental extracts containing strongly fluorescent interferents, chlorophenol red-β-D-galactoside can be used as substrate affording a red colored product in the presence of estrogenic compounds (Figure 16.8). Thus, p-YES is a highly promising tool still under development for monitoring the efficiency of wastewater treatment plants and the contamination of surface waters.

16.2.5 Cleanup by TLC

A significant advantage of TLC methods is that they allow all compounds of interests and coextracted matrix components to be easily located. A good example of a combined cleanup and quantitation of a target analyte is the determination of the synthetic sweetener sucralose, rather resistant to biodegradation, in sewage effluents and river water [76]. Using an optimized separation system combined with a selective post-chromatographic derivatization reaction, sucralose was separated from coextracted matrix compounds and could be determined down to 100 ng/L. In other applications, the TLC plate was used in place of a solid-phase extraction cartridge or column for cleanup. On the one hand, plate cleanup can be performed as for preparative TLC [77] using the full width of the plate to separate a single extract into multiple zones which are

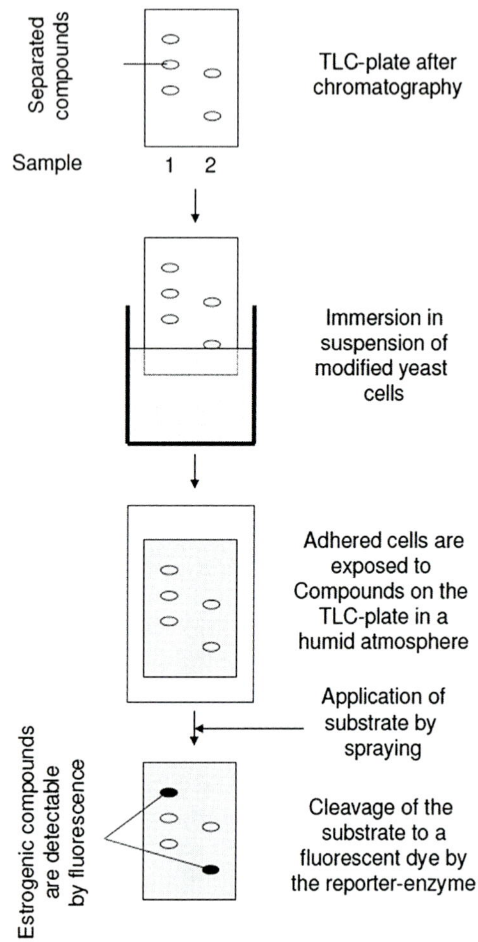

FIGURE 16.7 Flow chart for planar-yeast estrogen screen. After development, the HPTLC plate is dipped into a suspension of yeast cells (*S. cerevisiae*) that are genetically modified to express the reporter gene *lacZ* (encoding the enzyme β-galactosidase) depending on an estrogenic stimulus. During the incubation in a humid atmosphere, estrogenic compounds enter the cells at the respective zones and induce the heterologously expressed estrogen receptor, activating *lacZ*. The substrate applied (4-methylumbellifery-β-D-galactoside, MUG) is hydrolyzed by β-galactosidase to a fluorescent product to be detected. *(With permission reprinted from Ref. [74].)*

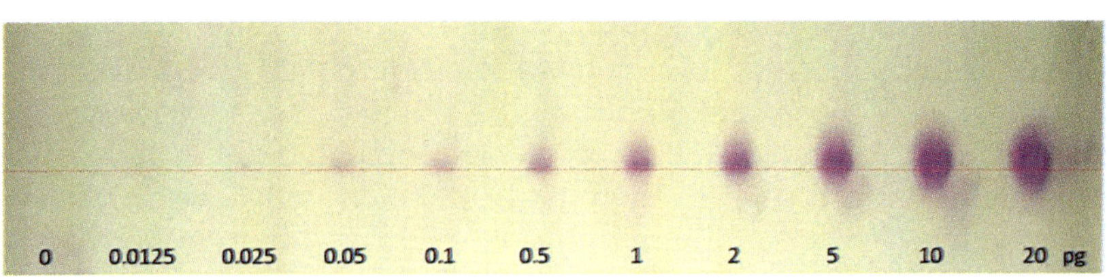

FIGURE 16.8 Plate image after planar-yeast estrogen screen of different amounts of 17-α-ethinylestradiol (EE2) per zone on an HPTLC silica gel plate, following the procedure of Figure 16.7 without development, but instead of 4-methylumbellifery-β-D-galactoside, chlorophenol red-β-D-galactoside was applied as substrate producing a red colored product.

subsequently scraped-off the layer, extracted, and analyzed by different methods. Examples include pesticide residue analysis by HPTLC [78,79] and GC–ECD [80] and the determination of PAHs in sewage sludge [81] or marine samples by GC–MS [82]. On the other hand, the application of small bands or spots contrarily permits a multisample TLC cleanup on one plate, which was used for the determination of PAHs in aerosol samples by GC-isotope ratio mass spectrometry to identify the source of contamination [83]. The $\delta^{13}C$ values of individual PAHs differed slightly depending on the location of sample collection (e.g., a bus station or a kitchen chimney). The extraction and clean up of PAHs from vegetable oils is a challenge. TLC was described as an efficient tool to separate PAHs from coextracted compounds before GC–MS analysis [84]. Instead of solvent extraction of the scraped-off zones, the material was loaded in a thermal desorption unit for analysis by GC–MS. For naphthalenes, PAHs and some pesticides, identification at levels of 20 ng/spot was possible [85].

The efficient clean up of pesticide extracts from fruits and vegetables was recently shown using planar solid-phase extraction (pSPE) [86]. After a two-fold development, pesticides were collected in a sharp zone separated from coextracted matrix compounds (Figure 16.9). Applying the TLC–MS interface coupled to an LC–MS system equipped with a monolithic RP18 column to prevent high back-pressure, extraction of the target zones resulted in the highest sensitivity. Alternatively, the TLC–MS extracts were collected in auto sampler vials followed by LC–MS separation. The obtained chromatograms were nearly free of matrix interferences, thus matrix effects were completely eliminated (Figure 16.10). Therefore, calibration could simply be performed with a solvent standard

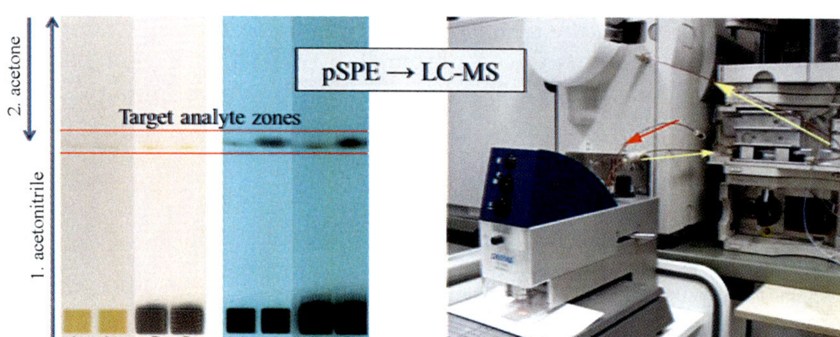

FIGURE 16.9 Principle of planar solid-phase extraction directly coupled to LC–MS. QuEChERS acetonitrile extracts (50 μL) of apples (A) and grapes (G), raw and spiked (+), where applied to an amino TLC layer and developed two times. Pesticides are focused within a sharp zone visible by the addition of Sudan II to the extracts. Target zone extraction was performed by the TLC–MS interface connected to an HPLC column for separation and detection by a mass selective detector.

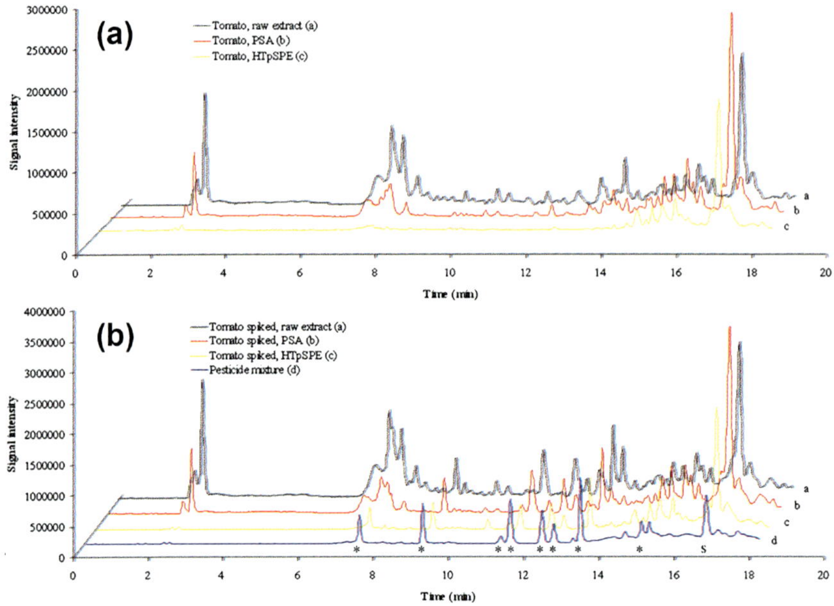

FIGURE 16.10 Total ion current LC–MS chromatograms for a tomato blank (A) and spiked (B) QuEChERS acetonitrile extracts after different sample cleanup techniques (a) raw, (b) dispersive SPE with a primary secondary amine (PSA), and (c) planar SPE (HTpSPE) on an amino layer. Track (d) refers to a pesticide standard including internal standards. The high degree of clean up by pSPE is clearly visible. SPE, planar solid-phase extraction. *(With permission reprinted from Ref. [86].)*

instead of matrix-matched standards generally strongly recommended for routine analysis. The pSPE application for fruits and vegetables was later modified for residue analysis in black and green teas, where coextracted caffeine proved to be a problem during GC–MS and LC–MS analysis [87]. However, the pSPE methodology is generally more attractive when an automated TLC–MS interface is available.

16.3 CONCLUSION

Although an HPTLC plate cannot compete with the high resolving power of capillary GC or ultra high-performance liquid chromatography (UHPLC) columns, it is a suitable alternative for the analysis of many environmental samples. The plate is rather matrix-tolerant (only one use), strongly reducing efforts of clean up, which is essential to protect expensive LC columns and to lengthen maintenance intervals for MS detectors. The extensive possibilities of postchromatographic staining reactions including effect-directed analysis, together with multiwavelength scanning, compensate to some extent for the less efficient separation capability of HPTLC layers. HPTLC is generally an inexpensive, efficient and rapid multisample screening tool to separate the "good" from the "bad" samples, which then have to be re-analyzed by GC–MS or LC–MS to confirm the findings and to check if legal requirements are fulfilled. These great advantages of HPTLC deserve greater recognition so as to reserve high-price MS systems for more important jobs than to analyze many samples free of contaminants or residues.

References

[1] Lebedev AT. Environmental mass spectrometry. Annu Rev Anal Chem 2013;6:163–89.

[2] Sherma J. Biennial review of planar chromatography: 2009–2011. J AOAC Int 2012;95: 992–1009.

[3] Sherma J. Review of advances in the thin layer chromatography of pesticides: 2010–2012. J Environ Sci Health, Part B 2013;48:417–30.

[4] Burger K, Koehler J, Jork H. Application of AMD to the determination of crop protection agents in drinking water. Part 1: fundamentals and method. J Planar Chromatogr–Mod TLC 1990;3:504–10.

[5] de la Vigne U, Jaenchen DE, Weber WH. Application of high-performance thin-layer chromatography and automated multiple development for the identification and determination of pesticides in water. J Chromatogr 1991;553:489–96.

[6] Morlock GE. Analysis of pesticide residues in drinking water by planar chromatography. J Chromatogr A 1996;754:423–30.

[7] Burger K, Jork H, Koehler J. Application of AMD to the determination of crop protection agents in drinking water. Part IV. Fundamentals of a confirmatory test. Acta Hydrochim Hydrobiol 1996;24:6–15.

[8] Butz S, Stan HJ. Screening of 265 pesticides in water by thin-layer chromatography with automated multiple development. Anal Chem 1995;67:620–30.

[9] Deutsches Institut für Normung (DIN). German standard methods for the examination of water, waste water and sludge—Jointly determinable substances (group F) - Part 11: Determination of selected organic plant protecting agents by Automated Multiple Development (AMD)-technique (DIN 38407). Berlin: Beuth; 1995.

[10] Koeber R, Niessner R. Screening of pesticide-contaminated soil by supercritical fluid extraction (SFE) and high-performance thin-layer chromatography with automated multiple development (HPTLC/AMD). Fresenius J Anal Chem 1996;354:464–9.

[11] Gardyan C, Thier HP. Application of modern thin-layer chromatographic techniques for confirmation of results in pesticide multi-residue analysis. Fresenius J Anal Chem 1991;339:338–9.

[12] Gardyan C, Thier HP. Identification and quantitation of currently important pesticides by HPTLC. Z Leb–Unters Forsch 1991;192:40–5.

[13] Gardyan C, Thier HP. HPTLC for the confirmation of results in pesticide multi-residue analysis. Z Leb Unters Forsch 1992;194:344–50.

[14] Ambrus A, Fuezesi I, Susan M, Dobi D, Lantos J, Zakar F, et al. A cost-effective screening method for pesticide residue analysis in fruits, vegetables, and cereal grains. J Environ Sci Health Part B 2005;40:297–339.

[15] Ambrus A, Fuezesi I, Lantos J, Korsos I, Szathmary M, Hatfaludi T. Application of TLC for confirmation and screening of pesticide residues in fruits, vegetables, and cereal grains: Part 2. Repeatability and reproducibility of Rf and MDQ values. J Environ Sci Health, Part B 2005;40:485–511.

[16] Weber WH, Mueller A, Weiss SC, Seitz W, Schulz W. Benzotriazoles in the aquatic environment. Multidimensional screening using HPTLC/AMD and nano-HPLC/QTOF-MS 265. GIT Spez September 2010;30:26–8.
[17] Weber W, Seitz W, Aichinger A. Ultra trace analysis of glyphosate and AMPA in water with HPTLC. CAMAG Bibliogr Serv (CBS) 2005;95:5–7.
[18] Weber WH, Schulz W, Mueller A, Weiss SC. Drinking water treatment - identification of reaction by-products of 4- and 5-methyl-1H-benzotriazole formed during ozonation. CAMAG Bibliogr Serv (CBS) 2012;109:13–5.
[19] Weber WH, Mueller A, Weiss SC, Seitz W, Schulz W. 1H-benzotriazole and tolyltriazoles in the aquatic environment. Occurrence in ground, surface and wastewater 293. Vom Wasser 2009;107:16–24.
[20] Weber W, Seitz W, Aichinger A, Albert R. Luminographic detection of toxicity with Vibrio fischeri (luminescent bacteria). CAMAG Bibliogr Serv (CBS) 2005;94:2–4.
[21] Kar A, Mandal K, Singh B. Decontamination of chlorantraniliprole residues on cabbage and cauliflower through household processing methods. Bull Environ Contam Toxicol 2012;88:501–6.
[22] Yue Y, Zhang R, Fan W, Tang F. High-performance thin-layer chromatographic analysis of selected organophosphorous pesticide residues in tea. J AOAC Int 2008;91:1210–7.
[23] Cobzac SC, Olah NK, Gocan S. TLC determination of triazinic pesticides from soils - a comparative study of some extraction methods. J Planar Chromatogr–Mod TLC 2012;25:97–100.
[24] Hamada M, Wintersteiger R. Rapid screening of triazines and quantitative determination in drinking water. J Biochem Biophys Methods 2002;53:229–39.
[25] Akkad R, Schwack W. Determination of organophosphorus and carbamate insecticides in fresh fruits and vegetables by high-performance thin-layer chromatography-multienzyme inhibition assay. J AOAC Int 2012;95:1371–7.
[26] Goecer M, Hoferer K, Zipfel J, Spangenberg B. A new TLC method for quantification of paraquat, diquat, difenzoquat, mepiquat and chloromequat in water. J Planar Chromatogr–Mod TLC 2009;22:59–63.
[27] Milz B, Idros IB, Spangenberg B. Limits of quantification of some neonicotinoid insecticides measured by thin-layer chromatography. J Liq Chromatogr Relat Technol 2012;35:1404–14.
[28] McGinnis SC, Sherma J. Determination of carbamate insecticides in water by C-18 solid phase extraction and quantitative HPTLC. J Liq Chromatogr 1994;17:151–6.
[29] Bhatia J, Sharma JD. Thin-layer chromatographic detection of carbosulfan by 4-aminoantipyrine reagent. J Planar Chromatogr–Mod TLC 2011;24:545–6.
[30] Weins C, Jork H. Toxicological evaluation of harmful substances by in situ enzymic and biological detection in high-performance thin-layer chromatography. J Chromatogr A 1996;750:403–7.
[31] Akkad R, Schwack W. Multi-enzyme inhibition assay for the detection of insecticidal organophosphates and carbamates by high-performance thin-layer chromatography applied to determine enzyme inhibition factors and residues in juice and water samples. J Chromatogr B 2010;878:1337–45.
[32] Akkad R, Schwack W. Multi-enzyme inhibition assay for detection of insecticidal organophosphates and carbamates by high-performance thin-layer chromatography. 1. Basics of method development. J Planar Chromatogr–Mod TLC 2008;21:411–5.
[33] Lu C, Barr DB, Pearson MA, Waller LA. Dietary intake and its contribution to longitudinal organophosphorus pesticide exposure in urban/suburban children. Environ Health Perspect 2008;116:537–42.
[34] Zhang X, Driver JH, Li Y, Ross JH, Krieger RI. Dialkyl phosphates (DAPs) in fruits and vegetables may confound biomonitoring in organophosphorus insecticide exposure and risk assessment. J Agric Food Chem 2008;56:10638–45.

[35] Katagi M, Tsuchihashi H, Hanada S, Jinmori H, Otsuki K. Determination of dichlorvos, trichlorfon, and their metabolites and degradation products in environmental water, drinks, and human urine. Jpn J Toxicol Environ Health 1993;39:459–68.
[36] Sherma J, Harnett HD, Jain AV. Thin layer chromatography analysis of dialkyl phosphate degradation products of organophosphate pesticides on C-18 chemically bonded plates. J Liq Chromatogr Relat Technol 1999;22:137–41.
[37] Schwack W, Zeisler T, Stiefel C. Determination of dialkyl phosphates as breakdown products of organophosphorus insecticides in fruit juices by HPTLCc with fluorescence detection. J AOAC Int 2009;92:691–8.
[38] Funk W, Glueck V, Schuch B, Donnevert G. Polynuclear aromatic hydrocarbons (PAHs): charge transfer chromatography and fluorometric determination. J Planar Chromatogr–Mod TLC 1989;2:28–32.
[39] Funk W, Donnevert G, Schuch B, Glueck V, Becker J. Quantitative HPTLC determination of six polynuclear aromatic hydrocarbons (PAH) in water. J Planar Chromatogr–Mod TLC 1989;2:317–20.
[40] International Organisation for Standardization. Water quality—determination of polycyclic aromatic hydrocarbons (PAH)—Part 1: Determination of six PAH by high-performance thin-layer chromatography with fluorescence detection after liquid-liquid extraction (ISO 7981–1:2005). Geneva: Switzerland; 2005.
[41] Hegewald H. Chlorine-free mobile phase for determination of PAH in water extracts. CAMAG Bibliogr Serv (CBS) 2007;98:9–11.
[42] Morlock G, Kopacz S. Fast and precise SBSE-HPTLC/FLD method for quantification of six polycyclic aromatic hydrocarbons frequently found in water. J Liq Chromatogr Relat Technol 2008;31:1925–42.
[43] Reimers C, Zielonka B, Stegmann R, Steinhart H. Determination of PAH in soil samples by high-performance thin-layer chromatography (HPTLC). J Soils Sediments 2001;1: 159–63.
[44] Jänchen D. Analysis of polycyclic aromatic hydrocarbons. CAMAG Bibliogr Serv (CBS) 1999;83:6–7.
[45] Spangenberg B, Klein KF. New evaluation algorithm in diode-array thin-layer chromatography. J Planar Chromatogr–Mod TLC 2001;14:260–5.
[46] Behringer S, Schwack W. Determination of PAHs in toys by HPTLC. CAMAG Bibliogr Serv (CBS) 2012;108:15.
[47] Dillon JT, Aponte JC, Tsai YJ, Huang Y. Thin layer chromatography in the separation of unsaturated organic compounds using silver-thiolate chromatographic material. J Chromatogr A 2012;1251:240–3.
[48] Poole SK, Dean TA, Poole CF. Preparation of environmental samples for the determination of polycyclic aromatic hydrocarbons by thin-layer chromatography. J Chromatogr 1987;400:323–41.
[49] Janoszka B, Bodzek D, Szotek A, Warzecha L. HPLC and TLC identification of nitrogen-containing polynuclear aromatic compounds in airborne particulate matter. Acta Chromatogr 1997;7:149–58.
[50] Janoszka B, Tyrpien K, Bodzek D. Application of TLC and GC-MS to the identification of amino-PAHs in sewage sludges. J Planar Chromatogr–Mod TLC 1996;9:450–5.
[51] Bodzek D, Dobosz C, Janoszka B, Konecki J. Quantification of nitrogen-containing polycyclic aromatic compounds in sewage sludges by TLC - densitometry. J Planar Chromatogr–Mod TLC 1999;12:265–8.
[52] Janoszka B. Densitometric TLC analysis of azaarenes in grilled meat. J Planar Chromatogr–Mod TLC 2007;20:221–6.
[53] Mohammad A, Ajmal M, Anwar S, Iraqi E. Twenty-two years report on the thin-layer chromatography of inorganic mixtures: observations and future prospects. J Planar Chromatogr–Mod TLC 1996;9:318–60.

[54] Wanjari SM, Deshmukh L, Paliwal LJ. Chromatographic separation of heavy metal cations on a silica gel-G. Pharma Chem 2011;3:425–31.

[55] Wanjari SM, Deshmukh L, Paliwal LJ. Rapid separation of some toxic heavy metal ions by thin layer chromatography. J Chem Pharm Res 2012;4:140–5.

[56] Wanjari SM, Deshmukh L, Paliwal LJ. Separation of toxic heavy metal cations on a silica gel-G with amino acid by TLC. Chem Sin 2011;2:222–7.

[57] Mohammad A, Hena S, Moheman A. Micellar TLC of inorganic ions: simultaneous separation of lead(II), zinc(II), and cobalt(II) and spectrophotometric estimation of zinc(II). J Planar Chromatogr–Mod TLC 2010;23:162–5.

[58] Bhawani SA, Hena S, Ibrahim MNM, Sulaiman O, Hashim R, San KL. Identification and separation of lead (II), nickel (II), and cobalt (II) on silica gel 60 F254 high-performance thin-layer chromatographic plates with mixed aqueous sodium dodecyl sulfate-oxalic acid solvent system. J Planar Chromatogr–Mod TLC 2012;25:355–7.

[59] Mohammad A, Agrawal V. Use of cationic micellar mobile phases in normal-phase TLC for enhanced selectivity in the separation of transition metal ions: simultaneous separation of mixtures of zinc, nickel, mercury, and cadmium or manganese cations. Acta Chromatogr 2002;12:177–88.

[60] Agarwal R, Behari JR. Screening for mercury in aqueous environmental samples and urine samples using thin layer chromatography. Water Environ Res 2007;79: 2457–63.

[61] Gurkan R, Savasci S, Ozkan MH, Akcay M. Investigation of separation and identification possibilities of some metal-dithizonate complexes by sequential TLC-IR system. J Anal Chem 2005;60:800–4.

[62] Bruno P, Caselli M, Traini A. HPTLC separation and determination of cobalt, copper and nickel as diethyldithiocarbamates. J High Resolut Chromatogr Chromatogr Commun 1985;8:366–7.

[63] Savasci S, Akcay M, Erguel S. Thin-layer chromatographic specification and separation of Cu^{1+}, Cu^{2+}, Ni^{2+}, and Co^{2+} cations. J Chromatogr Sci 2010;48:473–7.

[64] Bozkurt SS, Cavdar IK, Kurtbay HM, Merdivan M. Determination of trace metal ions using porphyrins as chelating agents by high performance thin layer chromatography-densitometry. J Liq Chromatogr Relat Technol 2010;33:748–60.

[65] Bruno P, Caselli M, Traini A. Inorganic and organic mercury determination by HPTLC and in situ densitometric detection. Application to real samples. J High Resolut Chromatogr Chromatogr Commun 1985;8:135–9.

[66] Li Z, Hua B, Pan J. Advances of derivatives of diantipyrinylmethane as reagents in analytical chemistry. Rev Anal Chem 2006;25:83–116.

[67] Timerbaev A, Shadrin O, Zhivopistsev V. Diantipyrylmethane as complex-forming reagent in the thin-layer chromatographic determination of metals. Chromatographia 1990;30:436–41.

[68] Shadrin O, Zhivopistsev V, Timerbaev A. Thin-layer chromatographic determination of inorganic anions as counter-ions of metal-diantipyrylmethane cationic complexes and diantipyrylmethane cations. Chromatographia 1993;35:667–70.

[69] Lafleur JP, Salin ED. Speciation of chromium by high-performance thin-layer chromatography with direct determination by laser ablation inductively coupled plasma mass spectrometry. Anal Chem 2008;80:6821–3.

[70] Resano M, Garcia Ruiz E, Mihucz VG, Moricz AM, Zaray G, Vanhaecke F. Rapid screening method for arsenic speciation by combining thin layer chromatography and laser ablation-inductively coupled plasma-dynamic reaction cell-mass spectrometry. J Anal Spectrom 2007;22:1158–62.

[71] Liu Z, Zhu Z, Zheng H, Hu S. Plasma jet desorption atomization-atomic fluorescence spectrometry and its application to mercury speciation by coupling with thin layer chromatography. Anal Chem 2012;84:10170–4.

[72] Milz B, Spangenberg B. 2D-thin layer chromatography (2D-TLC) flash test of 17α-ethinylestradiol and related steroids detected by fluorescence densitometry. J Liq Chromatogr Relat Technol 2013;36:2378–86.
[73] Spira D, Reifferscheid G, Buchinger S. Combination of high-performance thin-layer chromatography with a specific bioassay - a tool for effect-directed analysis. J Planar Chromatogr–Mod TLC 2013;26:395–401.
[74] Buchinger S, Spira D, Broeder K, Schluesener M, Ternes T, Reifferscheid G. Direct coupling of thin-layer chromatography with a bioassay for the detection of estrogenic compounds: applications for effect-directed analysis. Anal Chem 2013;85:7248–56.
[75] Mueller MB, Dausend C, Weins C, Frimmel FH. A new bioautographic screening method for the detection of estrogenic compounds. Chromatographia 2004;60:207–11.
[76] Morlock GE, Schuele L, Grashorn S. Development of a quantitative high-performance thin-layer chromatographic method for sucralose in sewage effluent, surface water, and drinking water. J Chromatogr A 2011;1218:2745–53.
[77] Kowalska T, Sherma J, editors. Preparative layer chromatography. Chromatogr Sci Ser, Vol. 95. Boca Raton, London, New York: CRC Press, Taylor & Francis; 2006.
[78] Tuzimski T, Soczewinski E. Use of database of plots of pesticide retention (RF) against mobile-phase compositions for fractionation of a mixture of pesticides by micropreparative thin-layer chromatography. Chromatographia 2004;59:121–8.
[79] Tuzimski T. Two-stage fractionation of a mixture of 10 pesticides by TLC and HPLC 482. J Liq Chromatogr Relat Technol 2005;28:463–76.
[80] Sukul P. Extraction and clean up procedures for the analysis of permethrin, cypermethrin, deltamethrin and fenvalerate in crops by GLC-ECD. Toxicol Environ Chem 1994; 44:217–23.
[81] Tyrpien K, Bodzek D, Janoszka B. Application of thin layer chromatography for isolation of polycyclic aromatic hydrocarbons from sewage sludges. Acta Chromatogr 1995; 4:102–8.
[82] Filipkowska A, Lubecki L, Kowalewska G. Polycyclic aromatic hydrocarbon analysis in different matrices of the marine environment. Anal Chim Acta 2005;547:243–54.
[83] Liu X, Bi X, Mai B, Sheng G, Fu J. Separation of PAHs in aerosol by thin layer chromatography for compound-specific stable carbon isotope analysis. Talanta 2005;66: 487–94.
[84] Diletti G, Scortichini G, Scarpone R, Gatti G, Torreti L, Migliorati G. Isotope dilution determination of polycyclic aromatic hydrocarbons in olive pomace oil by gas chromatography-mass spectrometry. J Chromatogr A 2005;1062:247–54.
[85] Chen X, Smart RB. Direct analysis of thin-layer chromatography spots by thermal extraction-gas chromatography-mass spectroscopy. J Chromatogr Sci 1992;30:192–6.
[86] Oellig C, Schwack W. Planar solid phase extraction-A new clean-up concept in multiresidue analysis of pesticides by liquid chromatography-mass spectrometry. J Chromatogr A 2011;1218:6540–7.
[87] Oellig C, Schwack W. Planar solid phase extraction clean-up for pesticide residue analysis in tea by liquid chromatography-mass spectrometry. J Chromatogr A 2012;1260: 42–53.

CHAPTER

17

Pharmaceutical Applications of High Performance Thin Layer Chromatography

Sunil R. Dhaneshwar [1,2]

[1] Department of Pharmaceutical Chemistry, RAK College of Pharmaceutical Sciences, Ras Al Khaimah Medical and Health Sciences University, Ras Al Khaimah, UAE; [2] Poona College of Pharmacy, Bharati Vidyapeeth University, Erandwane, Pune, India

17.1 INTRODUCTION

The introduction of high performance thin layer chromatography (HPTLC) in pharmaceutical analysis represents an important milestone having superseded conventional TLC where it has several distinct advantages. Before we discuss the pharmaceutical applications of HPTLC it would be appropriate to highlight these advantages, as it will help us understand why HPTLC is becoming more popular amongst pharmaceutical analysts.

1. Being an open method, sitting substances (spots which do not move from the point of application) can be identified easily and corrective action taken (e.g., indicates the need for a different mobile phase or stationary phase).
2. Since development and detection are separate steps, ultraviolet (UV) absorbing mobile phases do not cause any problem.
3. Samples can be detected by either universally applied stain reagents, such as iodine vapors, or specifically applied reagent for a particular sample, such as ninhydrin for amino acids or 2,4,6-trinitrophenylhydrazine for ketones.

Instrumental Thin-Layer Chromatography
http://dx.doi.org/10.1016/B978-0-12-417223-4.00017-0

4. A large number of samples (e.g., 30–40) can be applied to a single TLC plate and separated simultaneously.
5. Since HPTLC uses disposable stationary phases, time and solvent consuming column cleanup procedures are avoided.
6. Simultaneous analysis of samples and standards under identical experimental conditions is possible. This is a unique feature of HPTLC which is not possible with any other form of chromatography.
7. Choice of solvent for sample preparation is not very critical as the sample solvent and mobile phase do not come in contact with each other.
8. Mobile phase can be prepared with analytical grade solvents and not chromatographic grade solvents. Moreover no prior treatment such as filtration or degassing is required.
9. Different volumes of standard or sample solutions can be applied on the same TLC plate. Thus it is possible that during a single run, construction of a calibration curve and sample quantification can be achieved simultaneously.
10. All the commonly used stationary phases of High Performance Liquid Chromatography (HPLC) are also available for HPTLC.
11. Use of different development methods, such as ascending, descending, two dimensional, circular etc., is possible in HPTLC which is not possible in other forms of chromatography. This flexibility is important in method development for compounds of similar polarity.
12. Since the number of samples applied on a plate is large and the mobile phase used is very small, the per sample cost is low, thereby providing an economical method of analysis.
13. The development of mass spectrometer interfaces for High Performance Thin Layer Chromatography - Mass Spectrometry (HPTLC–MS) is a milestone in the evolution of HPTLC, which now has similar capabilities to High Performance Liquid Chromatography - Mass Spectrometry (HPLC-MS), in terms of separation and identification of samples. The use of modern spectroscopic methods such as UV, Fourier Transform Infra Red Spectroscpy (FT-IR), or Raman spectroscopy has increased the specificity and detection power of HPTLC.
14. Recent developments in ultraperformance TLC (UPTLC) using layers 10–12 μm thick and shorter development lengths of up to 5–6 cm has further reduced separation times and mobile phase consumption, and increased sample throughput by increasing the number of samples that can be simultaneously separated.

17.2 QUANTITATIVE ANALYSIS

Quantitative analysis by HPTLC is normally performed by one of two methods.

17.2.1 Area Comparison Method

In this method, replicate spots ($n = 3$ or 6) of known concentration of pure drug are spotted on a TLC plate along with replicate spots of the sample solution. The areas of the standard and sample spots, after development, are calculated and averaged. The area of the standard is compared with the area of the sample and since concentration of the standard is known, the concentration of sample can be calculated. This is a simple and quick method.

17.2.2 Calibration Curve Method

In this method, five varied concentrations of the standard are spotted in replicate ($n = 3$ or 6) along with the sample in replicate ($n = 3$ or 6). After development of the chromatogram, the average area of each concentration is plotted against the concentration and a calibration curve is constructed. The concentration of sample can be found by calculating the average area for the sample and with the aid of the calibration curve transposing the area to a concentration. This method is time consuming but more reliable.

17.3 PHARMACEUTICAL APPLICATIONS

17.3.1 Single Drug Formulation Analysis

An HPTLC method can be developed for formulations containing a single active component by identifying a mobile phase that separates the active ingredient from any matrix components. Ideally an R_f value between 0.3 and 0.7 is selected. The peak area for the active ingredient should be reproducible and the peak shape sharp and without tailing. The method is then validated as per International Conference on Harmonization (ICH) guidelines. The results can be compared with the official method or a reported method to further confirm the accuracy of the developed method.

The following two examples illustrate this kind of application:

1. Analysis of metformin tablets [1] (Figure 17.1; Tables 17.1 and 17.2):

HPTLC Conditions	
Mobile phase	Ammonium sulfate (0.5%):2-propanol:methanol (8.0:1.6:1.6 v/v/v)
Detection wavelength	238 nm
R_f	0.50 ± 0.03
Position of solvent front	8 cm
Saturation time	30 min
TLC plates	Merck silica gel 60 F_{254} on an aluminum support (20 × 10 cm)

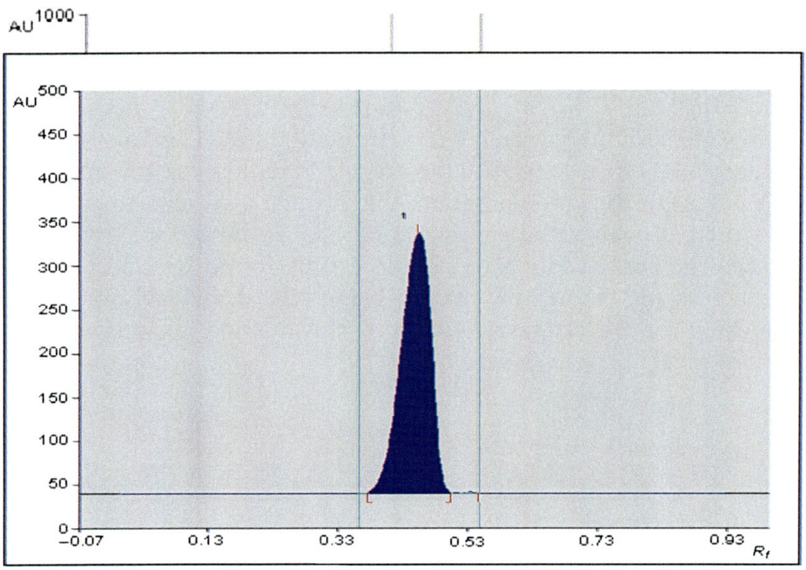

FIGURE 17.1 Densitogram of metformin hydrochloride.

TABLE 17.1 Assay Results for Metformin

Formulation	Label Claim (mg)	Drug Content (%)[a]	% RSD
Bigomet	250	100.24	0.36
Glycomet	500	99.82	0.98
Biciphage	850	101.00	1.30

[a]n = 6.

TABLE 17.2 Recovery Study for Metformin

Formulation	Amount of Drug Added (%)	Theoretical Content (ng)	Amount of MET Recovered (ng)[a]	% Recovery
Glcomet	80	18,000	18,018	100.10
	100	20,000	19,846	99.93
	120	22,000	21995.6	99.98
Bigomet	80	9000	9003.6	100.04
	100	10,000	9997	99.97
	120	11,000	11031.9	100.29
Biciphage	80	30,600	30544.92	99.82
	100	34,000	33874.2	99.63
	120	37,400	37171.86	99.39

[a]n = 6.

2. Analysis of levocetirizine hydrochloride tablets [2] (Figure 17.2; Tables 17.3 and 17.4):

HPTLC Conditions	
Mobile phase	Ethyl acetate:methanol:ammonia (9:2.5:1.5 v/v/v)
Detection wavelength	230 nm
R_f	0.50 ± 0.02
Position of solvent front	8 cm
Saturation time	20 min
TLC plates	Merck silica gel 60 F_{254} on an aluminum support (20 × 10 cm)

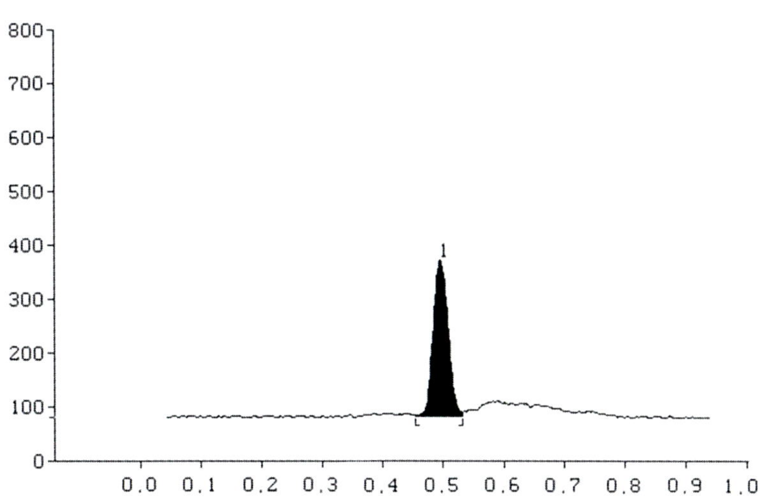

FIGURE 17.2 Densitogram of levocetirizine hydrochloride.

TABLE 17.3 Assay Results for Levocetirizine Hydrochloride ($n = 6$)

	Levocetirizine Found (mg per Tablet)	
Okacet-L Levocetirizine (5 mg)	**Mean ± SD (n = 6)**	**Recovery (%)**
First lot	4.92 ± 0.32	98.41
Second lot	4.93 ± 0.09	98.50

TABLE 17.4 Recovery Study for Levocetirizine Hydrochloride

Label Claim (mg per Tablet)	Amount Added (mg)	Total Amount (mg)	Amount Recovered (mg) ± % RSD	% Recovery[a]
5	4 (80%)	9	9.00 ± 1.00	100.05
5	5 (100%)	10	9.83 ± 0.76	98.38
5	6 (120%)	11	11.11 ± 0.99	101.09

[a]n = 6.

17.3.2 Two-Component Formulation Analyses

Method development for a two-component formulation requires more skill and expertise as two drugs of likely different polarity require separation with adequate resolution with a single mobile phase. The peak purity of the resolved drugs should be checked by overlaying the spectra of the pure drugs obtained in the same development. Further the ICH guidelines need to be applied to the developed method for validation.

The following are two examples of this kind of application:

1. Analysis of tablets of simvastatin and ezetimibe [3] (Figure 17.3; Tables 17.5 and 17.6):

Mobile phase	Toluene:2-propanol (8:2 v/v)
Detection wavelength	240 nm
R_f	0.48 ± 0.01 (simvastatin)
	0.53 ± 0.01 (ezetimibe)
Position of solvent front	15 cm
Saturation time	20 min
TLC plates	Merck silica gel 60 F_{254} on an aluminum support (20 × 10 cm)

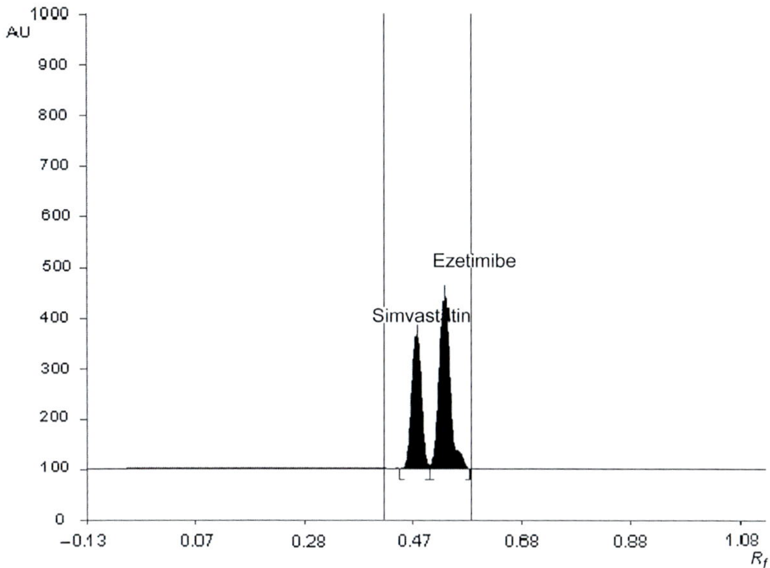

FIGURE 17.3 Densitogram of simvastatin and ezetimibe.

TABLE 17.5 Assay Results for Simvastatin and Ezetimibe

Drug	Label Claim (mg per Tablet)	Amount Found (mg)	Drug Content (%)[a]	RSD (%)
Simvastatin	10	9.69	96.93 ± 0.28	0.28
Ezetimibe	10	9.74	97.73 ± 0.25	0.25

[a]n = 3.

TABLE 17.6 Recovery Study for Simvastatin and Ezetimibe

Drug	Label Claim[a] (mg per Tablet)	Amount Added		Amount Recovered (mg)	Recovery (%)[b]
		(%)	(mg)		
Simvastatin	10	80	08	07.86	98.14 ± 0.26
		100	10	09.86	98.36 ± 0.17
		120	12	11.89	99.00 ± 0.26
Ezetimibe	10	80	08	07.93	99.16 ± 0.44
		100	10	09.90	98.97 ± 0.28
		120	12	11.87	98.85 ± 0.40

[a]*Formulation: Simvas-EZ, Micro Labs, Pondicherry, India.*
[b]*Mean ± standard deviation (n = 6).*

2. Analysis of tablets of paracetamol and aceclofenac [4] (Figure 17.4; Table 17.7):

Mobile phase	Toluene:2-propanol:25% ammonia(8:7:1 v/v/v)
Detection wavelength	254 nm
R_f	0.48 ± 0.01 (aceclofenac)
	0.67 ± 0.01 (paracetamol)
Position of solvent front	15 cm
Saturation time	20 min
TLC plates	Merck silica gel 60 F_{254} on an aluminum support (10 × 10 cm)

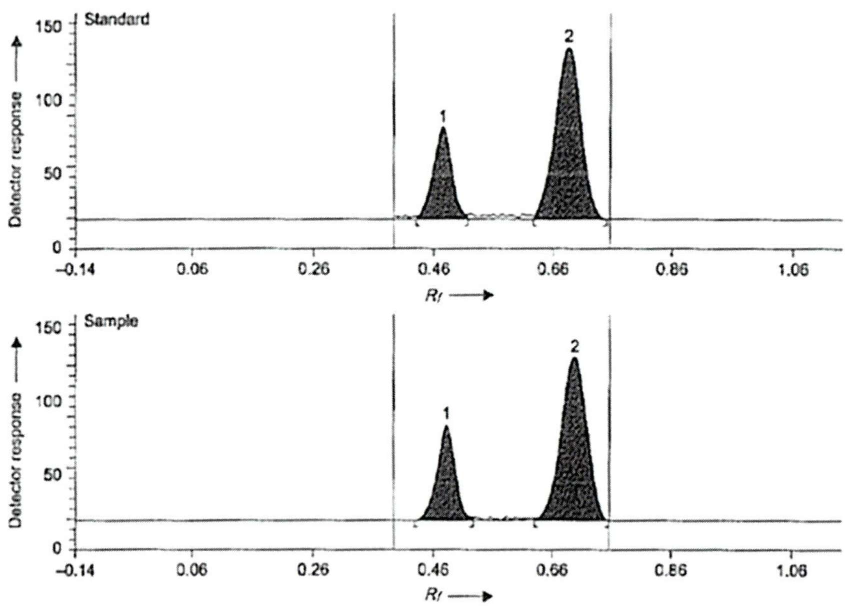

FIGURE 17.4 Densitogram of aceclofenac (1) and paracetamol (2). This is based on the work carried out at Anchrom Test Lab Pvt. Ltd. Mumbai, India.

TABLE 17.7 Assay Results for Paracetamol and Aceclofenac

Drug	Drug Content (%)[a]	RSD (%)
Paracetamol	98.75 ± 0.28 to 102.25 ± 0.25	0.28
Aceclofenac	98.75 ± 0.28 to 102.25 ± 0.25	0.25

[a]n = 6.

17.3.3 Three-Component Formulation Analysis

Method development for a three-component formulation is more challenging as the mobile phase should be able to adequately resolve the three drugs, which should each have appropriate R_f values. The method should be validated according to ICH guidelines. Although method development may be lengthy, the per sample analysis time for formulations is short increasing the productivity of quality control laboratories.

1. Simultaneous analysis of atenolol, hydrochlorothiazide, and amlodipine [5] (Figure 17.5; Tables 17.8 and 17.9):

Mobile phase	Chloroform:methanol:acetic acid (8:2:0.2 v/v/v)
Detection wavelength	232 nm
R_f	0.22 ± 0.02 (atenolol)
	0.36 ± 0.02 (hydrochlorothiazide)
	0.55 ± 0.02 (amlodipine besylate)
Position of solvent front	8 cm
Saturation time	30 min
TLC plates	Merck silica gel 60 F_{254} on an aluminum support (20 × 10 cm)

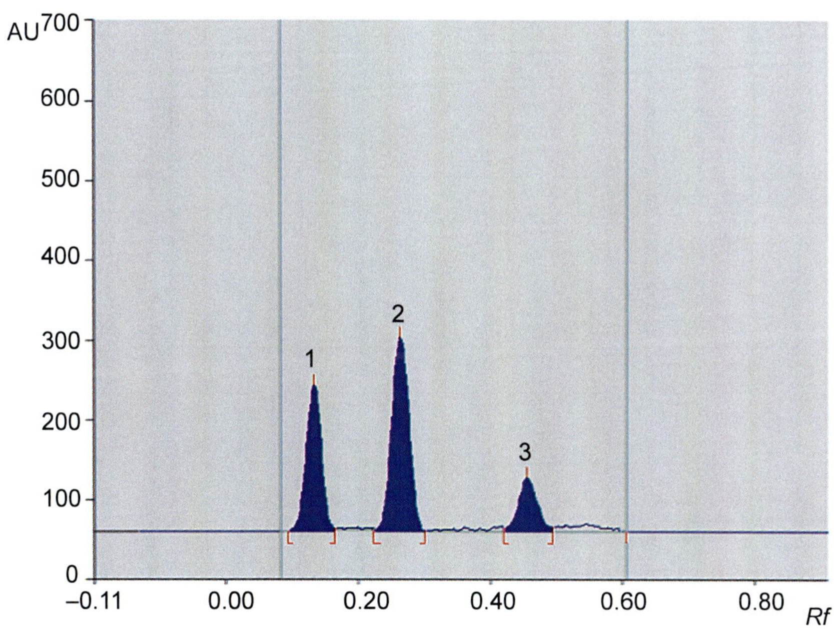

FIGURE 17.5 Densitogram of atenolol (1) R_f (0.22), hydrochlorothiazide (2) R_f (0.36), and amlodipine besylate (3) R_f (0.55).

TABLE 17.8 Assay Results for Atenolol, Hydrochlorothiazide and Amlodipine ($n = 6$)

	Atenolol Found (mg per Tablet)	
Atenolol (50 mg)	**Mean ± SD (n = 6)**	**Recovery (%)**
First lot	49.93 ± 6.73	99.86
Second lot	50.04 ± 6.60	100.08
	Hydrochlorothiazide Found (mg per Tablet)	
Hydrochlorothiazide (25 mg)	**Mean ± SD (n = 6)**	**Recovery (%)**
First lot	24.89 ± 3.67	99.56
Second lot	24.95 ± 3.55	99.80
	Amlodipine Besylate Found (mg per Tablet)	
Amlodipine Besylate (5 mg)	**Mean ± SD (n = 6)**	**Recovery (%)**
First lot	5.01 ± 1.21	100.20
Second lot	4.99 ± 1.18	99.80

TABLE 17.9 Recovery Study for Atenolol, Hydrochlorothiazide and Amlodipine Besylate ($n = 6$)

Label Claim (mg per Tablet)	Amount Added (mg)	Total Amount (mg)	Amount Recovered (mg) ± % RSD	% Recovery
ATENOLOL				
50	40 (80%)	90	89.92 ± 0.62	99.91
50	50 (100%)	100	100.21 ± 1.01	100.21
50	60 (120%)	110	109.66 ± 0.73	99.69
HYDROCHLOROTHIAZIDE				
25	20 (80%)	45	44.78 ± 0.34	99.51
25	25 (100%)	50	50.23 ± 0.67	100.46
25	30 (120%)	55	54.89 ± 0.54	99.80
AMLODIPINE BESYLATE				
5	4 (80%)	9	8.98 ± 0.63	99.77
5	5 (100%)	10	9.95 ± 0.87	99.58
5	6 (120%)	11	11.03 ± 0.98	100.27

2. Simultaneous analysis of paracetamol, diclofenac potassium, and famotidine [6] (Figure 17.6; Tables 17.10 and 17.11):

Mobile phase	Toluene:acetone:methanol:formic acid (5:2:2:0.01, v/v/v/v).
Detection wavelength	274 nm
R_f	0.17 ± 0.03 (famotidine)
	0.62 ± 0.03 (paracetamol)
	0.75 ± 0.02 (diclofenac potassium)
Position of solvent front	8 cm
Saturation time	30 min
TLC plates	Merck silica gel 60 F_{254} on an aluminum support (10 × 10 cm)

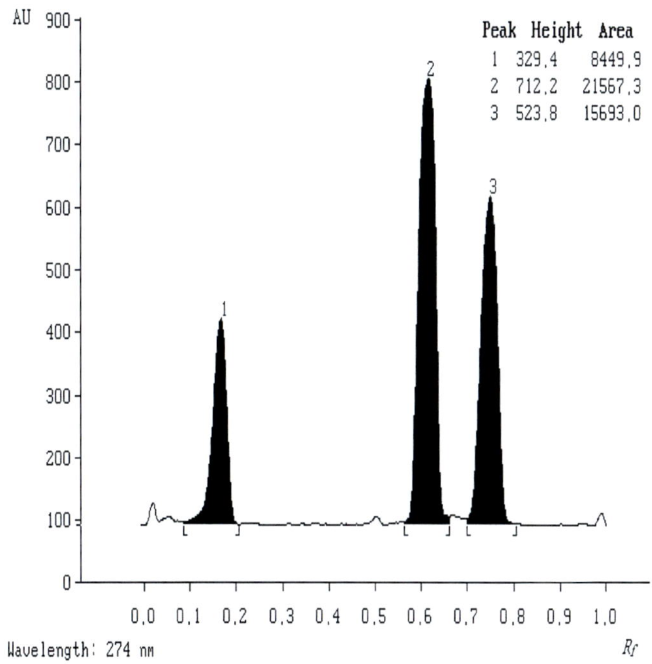

FIGURE 17.6 Densitogram of famotidine (1), paracetamol (2), diclofenac potassium (3). *Reprinted from The Journal of AOAC INTERNATIONAL, J. AOAC Int. 2010, 93(3), 765–771. Copyright by AOAC INTERNATIONAL.*

TABLE 17.10 Assay Results for Paracetamol, Diclofenac Potassium, and Famotidine

Drug	Label Claim	Amount Found, %[a]
Paracetamol	325	97.5
Diclofenac potassium	50	96.3
Famotidine	20	97.1

[a]n = 6. *Two different batches of the same brand were used for analysis by the developed HPTLC method.*

TABLE 17.11 Recovery Study for Paracetamol, Diclofenac Potassium, and Famotidine

Drug	Label Claim (mg per Tablet)	Amount Added (%)	Total Amount (mg)	Amount Recovered (mg) ± % RSD	% Recovery[a]
Paracetamol	325	80	585	564 ± 1.27	96.41
		100	650	638 ± 0.61	98.15
		120	715	690 ± 0.45	96.50
Diclofenac potassium	50	80	90	87.6 ± 0.52	97.33
		100	100	96.3 ± 1.14	96.30
		120	110	105. ±0.74	95.45
Famotidine	20	80	36	34.5 ± 1.02	96.48
		100	40	38.3 ± 1.32	95.75
		120	44	42.1 ± 0.85	95.68

[a]n = 6.

3. Simultaneous analysis of paracetamol, caffeine citrate, and propyphenazone [7] (Figure 17.7; Table 17.12):

Mobile phase	Dioxane:chloroform (2.5:7.5 v/v).
Detection wavelength	273 nm
R_f	0.25 ± 0.03 (paracetamol)
	0.50 ± 0.03 (caffeine)
	0.85 ± 0.02 (propyphenazone)
Position of solvent front	5 cm
Saturation time	10 min
TLC plates	Merck silica gel 60 F_{254} on an aluminum support (10 × 10 cm)

17.3.4 Stability Indicating Methods

The ICH guidelines recommend that every drug should be exposed to stress conditions (acid, alkali, dry heat, wet heat, oxidation and photostability, both sunlight and artificial light of known intensity) to assess its inherent stability and the possible mechanisms of its break down. A method which is capable of separating the drug in the presence of its degraded products is referred to as a stability indicating analytical method. The

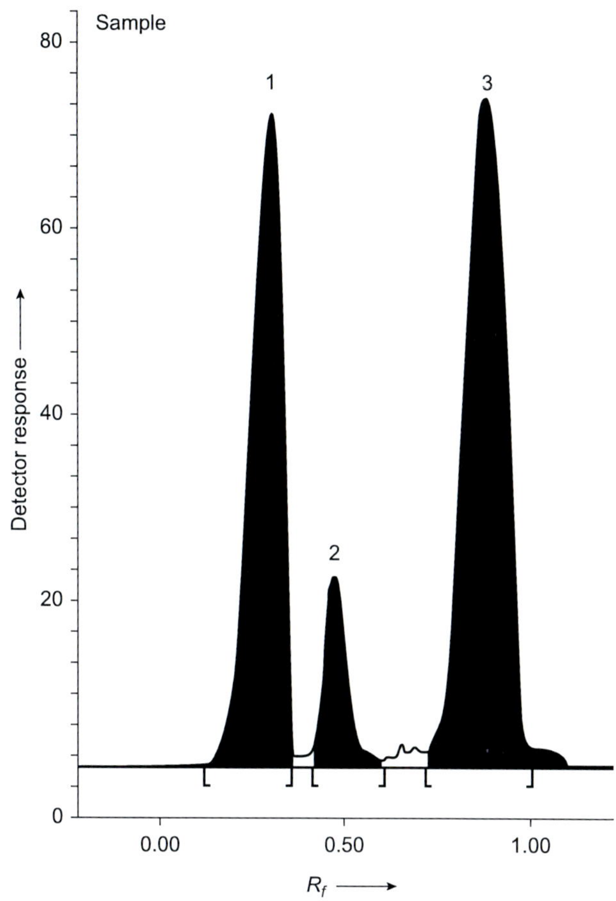

FIGURE 17.7 Densitogram of paracetamol (1), caffeine citrate (2) and propyphenazone (3). This is based on the work carried out at Anchrom Test Lab Pvt. Ltd. Mumbai, India.

TABLE 17.12 Assay Results for Paracetamol, Caffeine Citrate and Propyphenazone

Drug	Label Claim	Amount Found, %[a]
Paracetamol	325	97.5
Caffeine citrate	50	96.3
Propyphenazone	20	97.1

[a] *Average of 6 readings* n=6

development of stability indicating method is a challenging task since the degradants obtained usually have similar polarities to the drug.

1. Stability indicating method for tenatoprazole [8] (Figure 17.8(a and b); Tables 17.13 and 17.14):

Mobile phase	Toluene:ethyl acetate:methanol (6:4:1 v/v/v)
Detection wavelength	306 nm
R_f	0.34 ± 0.02
Position of solvent front	8 cm
Saturation time	30 min
TLC plates	Merck silica gel 60 F_{254} on an aluminum support (10 × 10 cm)

2. Stability indicating method for desvenlafaxine [9] (Figure 17.9(a and b); Tables 17.15 and 17.16):

HPTLC Conditions	
Mobile phase	Ethyl acetate:toluene:methanol:ammonia (7:2:0.5:0.5, v/v/v/v)
Detection wavelength	228 nm
R_f	0.48
Run distance	8 cm
Saturation time	30 min
TLC plates	Merck silica gel 60 F_{254} on an aluminum support (10 × 10 cm)

3. Stability indicating method for tenofovir disoproxil fumarate [10] (Figure 17.10(a and b); Tables 17.17 and 17.18):

HPTLC Conditions	
Mobile phase	Chloroform:methanol (9:1 v/v)
Detection wavelength	260 nm
R_f	0.49 ± 0.03
Run distance	8 cm
Saturation time	30 min
TLC plates	Merck silica gel 60 F_{254} on an aluminum support (10 × 10 cm)

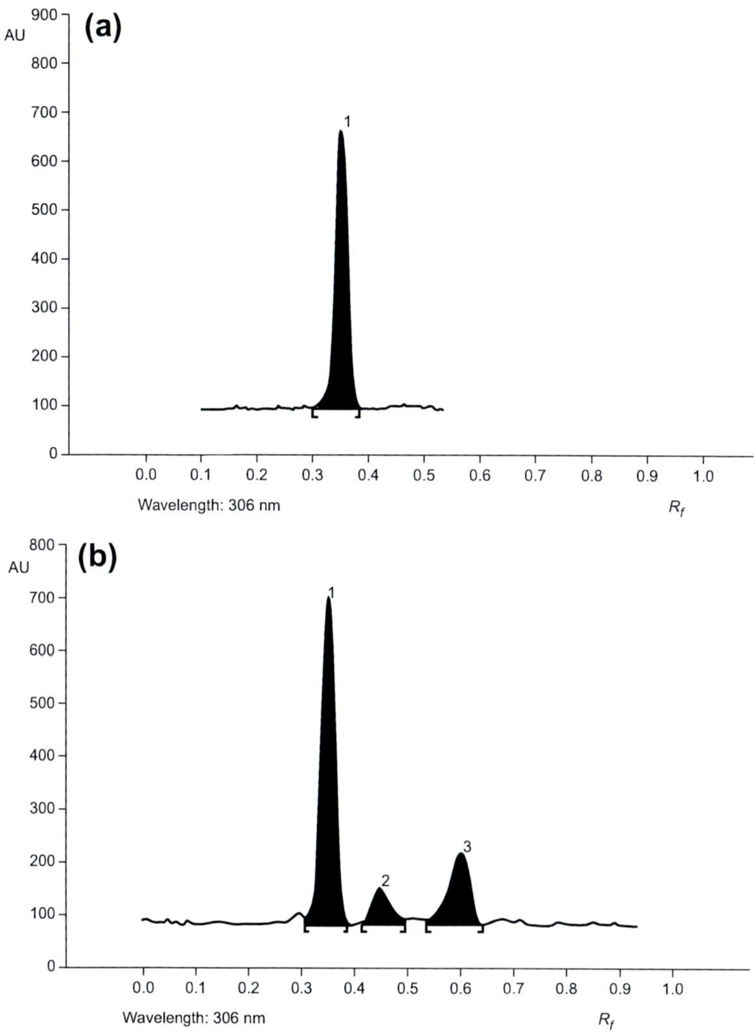

FIGURE 17.8 (a) Densitogram of tenatoprazole. (b) Densitogram of base-treated tenatoprazole 1000 ng/spot; condition: 1 M NaOH at 80 °C for 4 h; peak 1 (tenatoprazole, R_f: 0.35), peak 2 (degraded, R_f: 0.43), peak 3 (degraded, R_f: 0.59). *Reprinted from The Journal of AOAC INTERNATIONAL, J. AOAC Int. 2009, 92(2), 387–93. Copyright by AOAC INTERNATIONAL.*

TABLE 17.13 Assay Results for Tenatoprazole

Standard Added to 20 mg Sample, mg	Found, mg[a]	Recovery, %	RSD, %
16	35.7	99.16	0.95
20	39.3	98.25	1.67
24	43.6	99.09	1.72

[a]n = 6.

TABLE 17.14 Recovery Study for Tenatoprazole

	Tenatoprazole Found[b]	
PCP Formulation[a]	Mean ± SD	% Label Claim
Lot (PCP-TNZ 0307)	19.22 ± 1.2	96.1
Lot (PCP-TNZ 1007)	19.12 ± 1.4	95.6

[a]*label claim: 20 mg tenatoprazole*
[b]*average of six readings* n=6

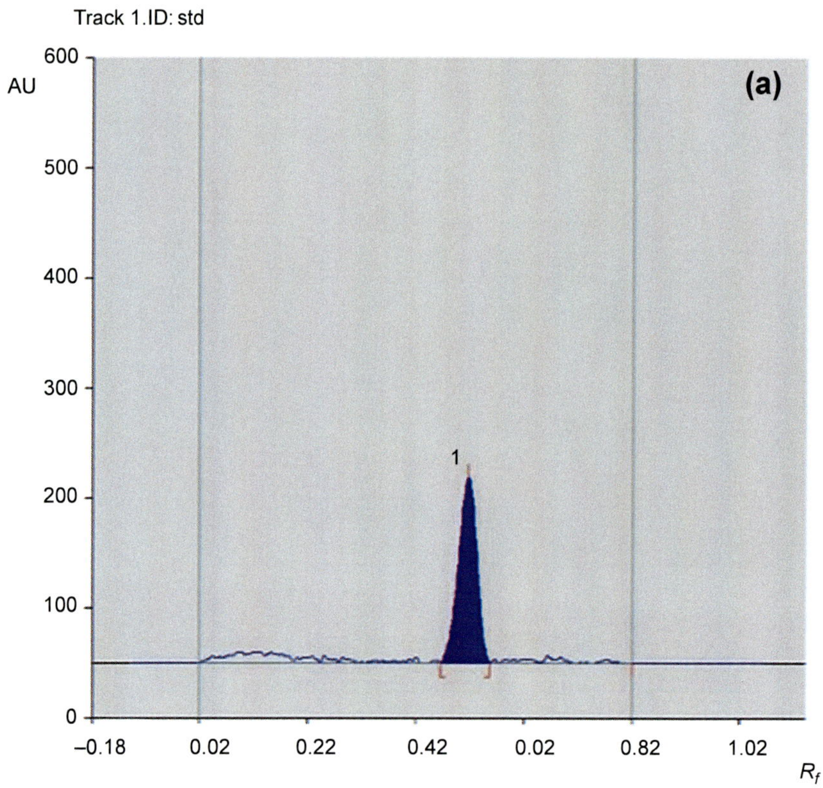

FIGURE 17.9 (a) Densitogram of desvenlafaxine. (b) Densitogram of acid degradation product of desvenlafaxine, condition: 1 M HCl at 80 °C for 8 h.

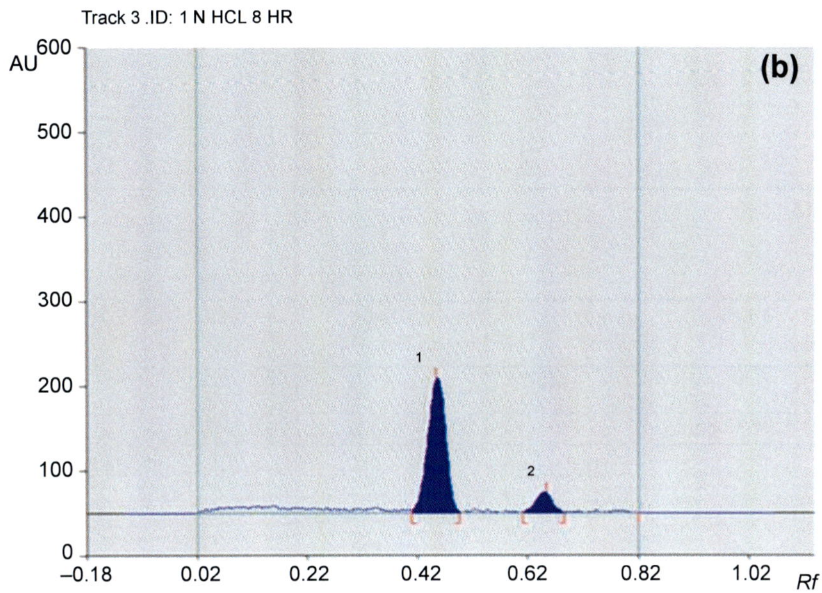

FIGURE 17.9 *(Continued)*

TABLE 17.15 Assay Results for Desvenlafaxine ($n = 6$)

Commercial Formulation D-VENIZ (50 mg)	Desvenlafaxine Found (mg per Tablet) Mean ± SD, RSD (%) (n = 3)	Recovery (%)
First Lot	49.87 ± 0.38, 0.76	99.46
Second Lot	50.08 ± 0.60, 0.89	100.18

TABLE 17.16 Recovery Study for Desvenlafaxine ($n = 3$)

Label Claim (mg per Tablet)	Amount Added (mg)	Total Amount (mg)	Amount Recovered (mg) ± SD, RSD (%)	Recovery (%)
50	40 (80%)	90	89.18 ± 0.74, 0.83	99.09
	50 (100%)	100	99.66 ± 1.89, 1.90	99.66
	60 (120%)	110	109.04 ± 1.19, 1.09	99.12

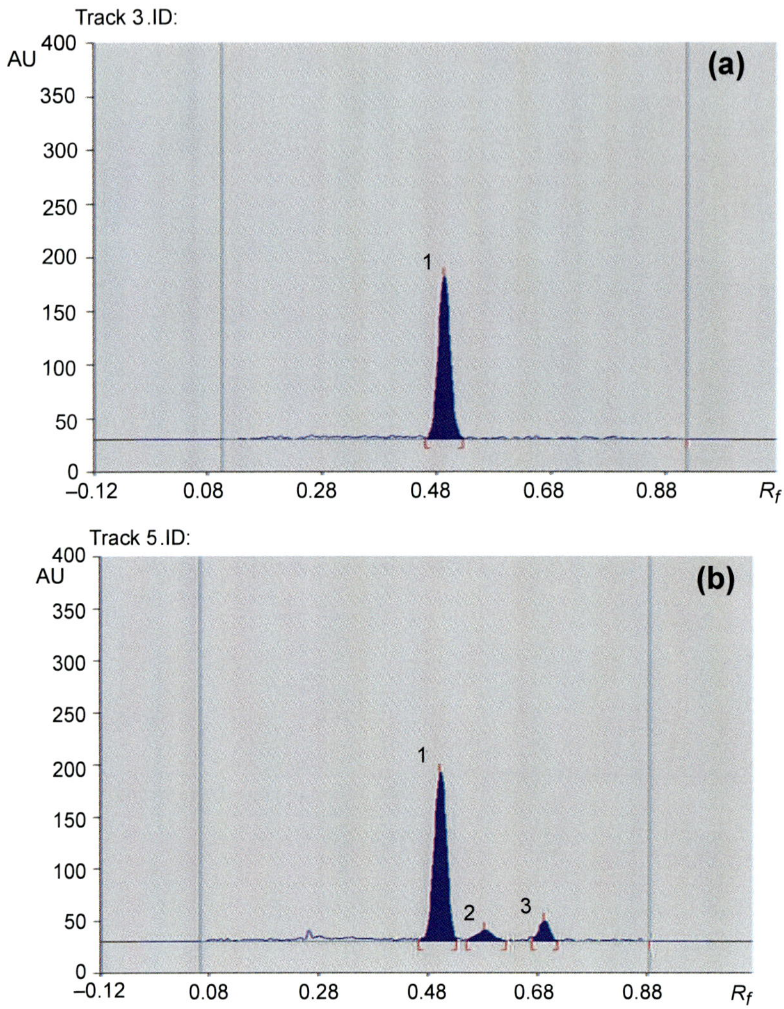

FIGURE 17.10 (a) Densitogram of tenofovir disoproxil fumarate (30 μg/mL). Mobile phase: chloroform: methanol (9.0: 1.0 v/v); detection wavelength: 260 nm; R_f: 0.49 ± 0.03; position of solvent front: 8 cm; saturation time: 30 min; TLC plates: Merck silica gel 60 F_{254} aluminum (20 × 10 cm). (b) Densitogram of photo degraded tenoforvir disoproxil fumarate (150 ng/spot) peak 1: tenoforvir disoproxil fumarate, peak 2 and 3 degradants.

TABLE 17.17 Assay Results for Tenofovir Disoproxil Fumarate and Comparison with HPLC

		HPLC		HPTLC	
Label Claim (mg)	**Sample**	**Drug Content (%)**	**% RSD**	**Drug Content (%)**	**% RSD**
300	Tentide	99.69	0.002	99.47	0.05
300	Tavin	99.80	0.006	99.91	0.009

TABLE 17.18 Recovery Study for Tenofovir Disoproxil Fumarate

Label Claim (mg)	Amount of Drug Added (%)	Total Amount of Drug Present (mg)	Amount Found (mg)	% Recovery[a]
TENTIDE				
300	80	540	536.65	99.38
	100	600	597.06	99.51
	120	660	655.05	99.25
TAVIN				
300	80	540	536.27	99.31
	100	600	599.04	99.84
	120	660	654.12	99.11

[a]n = 3.

17.3.5 Determination of Content Uniformity

All national pharmacopoeia require that each dosage form is tested for content uniformity. For tablets, as an example, 10–20 tablets are taken and assayed individually. The batch passes or fails the test depending upon how many tablets fall outside the specification. The number of tablets which must pass the test is given by the pharmacopoeia depending upon the amount of active ingredient or average weight of the tablets.

Thus all other methods of analysis would require the repetition of the same experiment 10 or 20 times, which is very tedious and time consuming. Using HPTLC, once the individual tablet solutions are ready, all solutions can be applied to a single plate and separated simultaneously, saving time and reducing the release time of the batch.

The prerequisite is the development of a validated HPTLC method for the drug. Following two examples will illustrate this application:

1. Determination of content uniformity of alprazolam tablets [11] (Figures 17.11 and 17.12; Table 17.19):

HPTLC Conditions	
Mobile phase	Chloroform:ethanol (9:1 v/v)
Detection wavelength	225 nm
R_f	0.40
Run distance	7 cm
Saturation time	20 min
TLC plates	Merck silica gel 60 F_{254} on an aluminum support (10 × 10 cm)

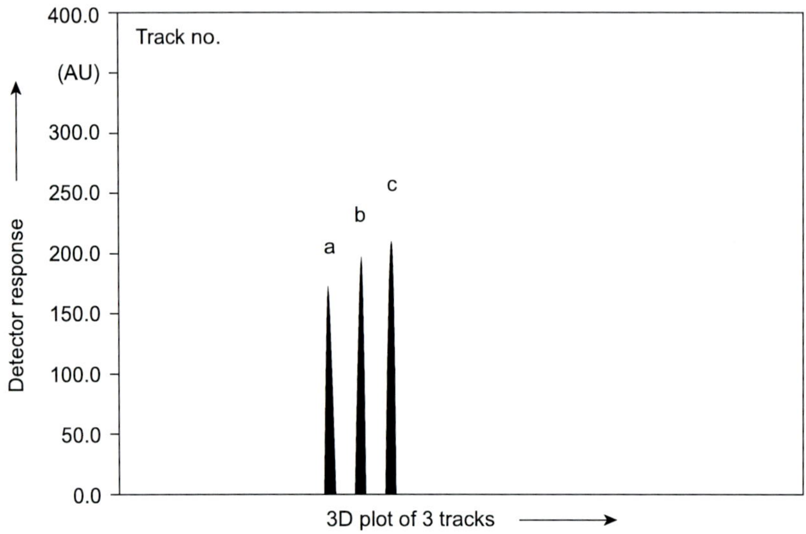

Densitogram of alprazolam standard & sample,
standard (90%): track-a; pooled sample: track-b; standard (110%): track-c

FIGURE 17.11 Densitogram of alprazolam standard and sample. This is based on the work carried out at Anchrom Test Lab Pvt. Ltd. Mumbai, India.

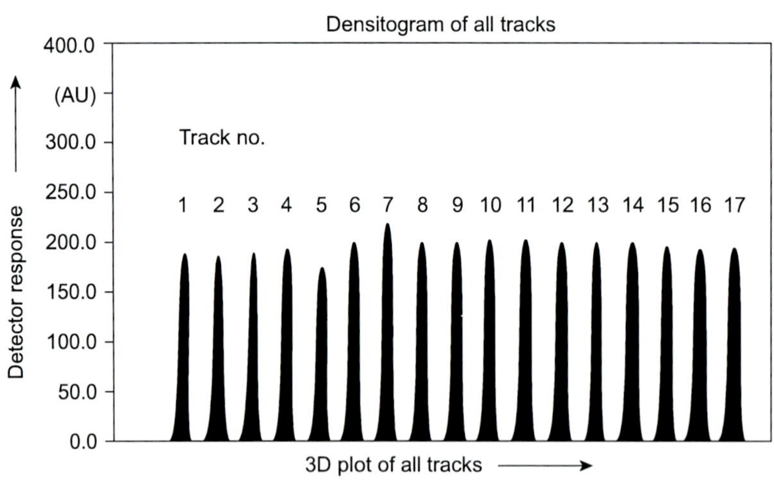

Densitogram of alprazolam standards(light) and the samples (dark) For detailed experimental and chromatographic conditions, refer to protocol no: 1 Tracks identification: Standard (100%) :3,8,10,13,15, Standard (85%): 5 and Standard (115%): 7, samples:1,2,4,6,9,11,12,14,16,17

FIGURE 17.12 Densitogram of alprazolam tablets. This is based on the work carried out at Anchrom Test Lab Pvt. Ltd. Mumbai, India.

TABLE 17.19 Assay Results for Alprazolam Tablets

	Assay %[a]	Minimum Found	Maximum Found
Commercial formulation alprazolam (1 mg)	Mean = 99.34 RSD (%) = 1.28	96.76	101.76

[a]n = 5.

2. Determination of content uniformity of hydrochlorothiazide and lisinopril tablets [12] (Figures 17.13 and 17.14; Table 17.20):

HPTLC Conditions	
Mobile phase	n-Butanol:methanol:water (3:1:1 v/v/v)
Detection wavelength	200 nm
R_f	Lisinopril 0.12
	Hydrochlorothiazide 0.73
Run distance	7 cm
Saturation time	20 min
TLC plates	Merck silica gel 60 F_{254} on an aluminum support (10 × 10 cm)

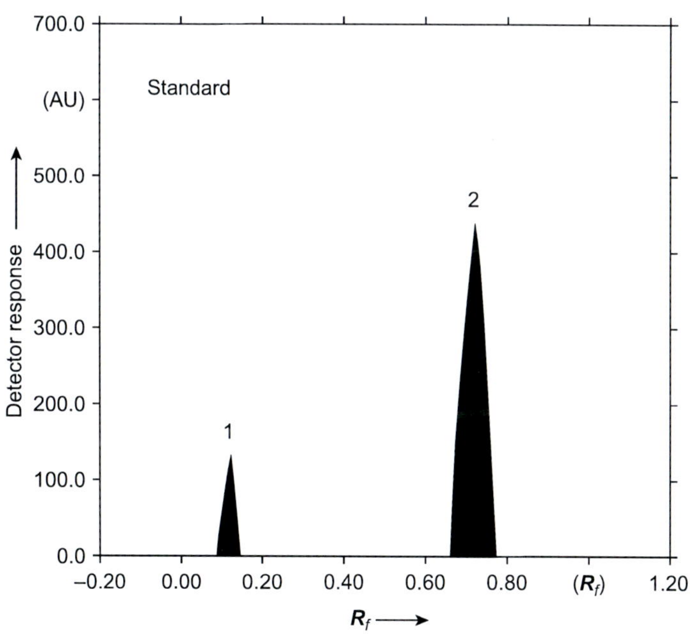

FIGURE 17.13 Densitogram of lisinopril (1) and hydrochlorothiazide (2) standard. This is based on the work carried out at Anchrom Test Lab Pvt. Ltd. Mumbai, India.

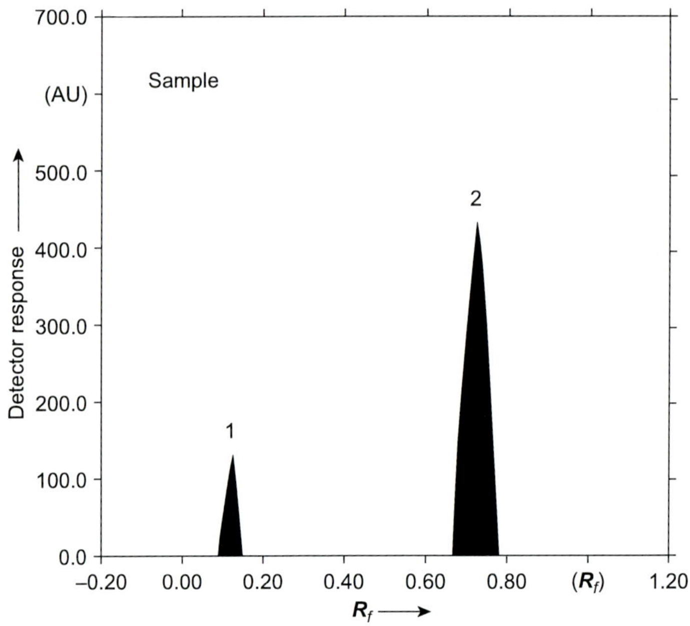

FIGURE 17.14 Densitogram of lisinopril (1) and hydrochlorothiazide (2) tablets. This is based on the work carried out at Anchrom Test Lab Pvt. Ltd. Mumbai, India.

TABLE 17.20 Assay Results for Hydrochlorothiazide and Lisinopril Tablets

Name of the Drug	Assay (%)[a]	Minimum Found	Maximum Found
Lisinopril (label claim 5 mg)	Mean = 98.98 RSD (%) = 1.81	96.36	101.30
Hydrochlorothiazide (label claim 12.5 mg)	Mean = 97.53 RSD (%) = 4.44	86.64	101.07

[a]n = 5.

17.3.6 Bulk Drug Purity Profiles

Bulk drugs contain impurities even if they pass meet the criteria set out in national pharmacopeias. Consequently, impurity profiling is a major interest in pharmaceutical analysis. HPTLC has been used successfully for this purpose. A high concentration of sample is applied to the layer, which is developed in a mobile phase capable of providing adequate resolution between the pure drug and the impurities present. The impurities can be isolated by preparative TLC and the structure of the impurities determined by conventional spectroscopic methods, such as Nuclear Magnetic Resonance(NMR), Mass Spectrometry(MS), Infra Red Spectroscopy (IR) and Ultraviolet Spectroscopy(UV). TLC–MS is increasingly being used for this application.

The following example of famotidine illustrates this application [13] (Figure 17.15 (a) and (b)):

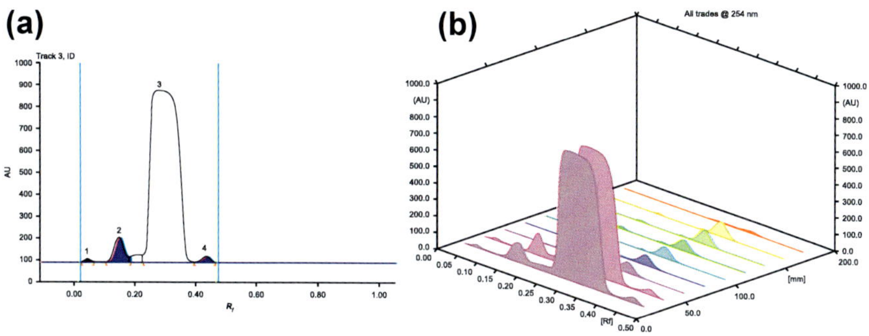

FIGURE 17.15 (a) Densitogram of impurity profiling of Famotidine: 1,2 and 4 impurities;3 famotidine (b) 3-D plot of densitometric scans of above samples at UV 254 nm. This work has been carried out at Anchrom Test Lab Pvt. Ltd. Mumbai, India

17.3.7 Standardization of a Herbal Drug Formulation

TLC has been used for decades for identification of different constituents of herbal drugs. However, the ability of HPTLC to perform quantitative analysis is particularly helpful for both identification (e.g., authentication) and quantitation to standardize herbal formulations. Due to their complexity herbal formulations are difficult to standardize by other approaches.

Typically, a prominent biomarker/chemical marker is identified in the authentic sample of the crude drug and an HPTLC method is developed for its determination. Once a validated method is available, then either a crude drug sample can be authenticated or a herbal formulation can be standardized by determining the presence of the maker qualitatively and quantitatively.

1. Example: Rapid densitometric method for estimation of alizarin and betulinic acid in a polyherbal formulation [14] (Figures 17.16 and 17.17(a and b); Table 17.21):

HPTLC Conditions	
Mobile phase	Toluene:ethyl acetate:formic acid (9:1.5:0.5 v/v/v)
Detection wavelength	287 nm
R_f	Alizarin −0.58 ± 0.02
	Betulinic acid −0.52 ± 0.02
Run distance	8 cm
Saturation time	30 min
TLC plates	Merck silica gel 60 F_{254} on an aluminum support (10 × 10 cm)

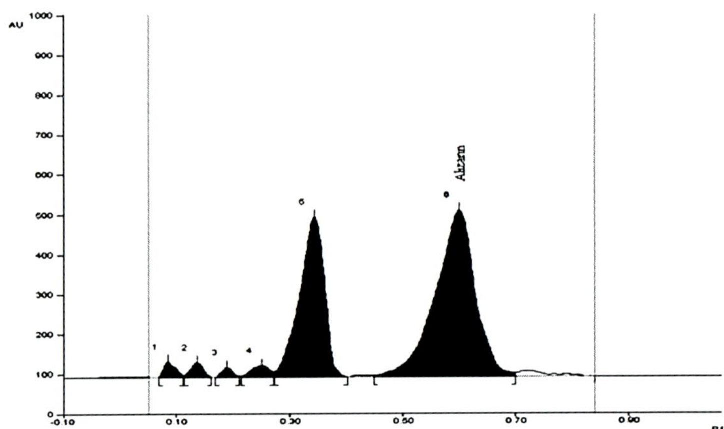

FIGURE 17.16 Densitogram of crude extact of *Rubia Cordifolia* Linn. (287nm)

2. Development and validation of an HPTLC method for the simultaneous estimation of corilagin, gallic acid, and ellagic acid in a polyherbal formulation [15] (Figures 17.18 and 17.19; Table 17.22):

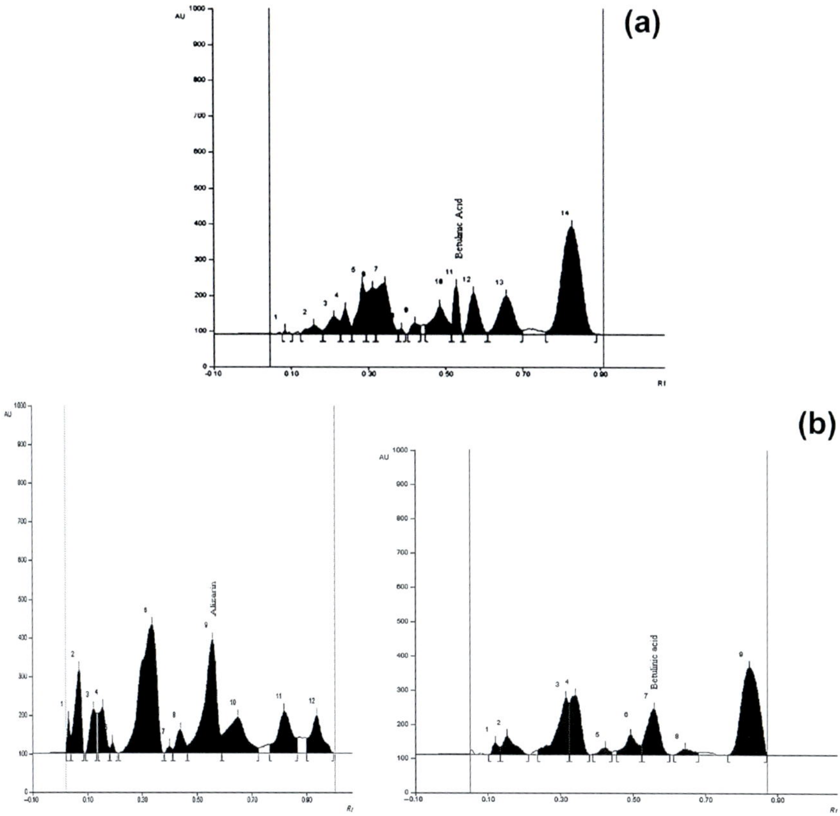

FIGURE 17.17 (a) Densitogram of crude extract of *Syzygium jambolanum* DC (595nm).

TABLE 17.21 Assay Results for Formulations Containing Alizarin and Betulinic Acid ($n = 3$)

	Drug Content (%w/w ± SD)	
Samples	**Alizarin**[a]	**Betulinic Acid**[a]
Crude extract	1.77 ± 0.120	1.58 ± 0.166
Marketed formulation	0.40 ± 0.022	0.39 ± 0.012

[a] *Values for Alizarin and Betulinic acid are positive only the sign of + and − together is for indicating Standard Deviation*

HPTLC Conditions	
Mobile phase	n-Butanol:water methanol:formic acid (6:1:0.1:0.8 v/v/v/v)
Detection wavelength	283 nm
R_f	Corilagin −0.44
	Ellagic acid −0.63
	Gallic acid −0.73
Run distance	8 cm
Saturation time	30 min
TLC plates	Merck silica gel 60 F_{254} on an aluminum support (10 × 10 cm)

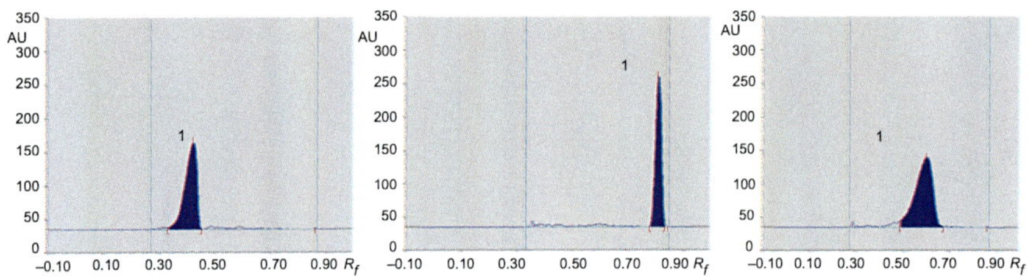

FIGURE 17.18 Densitogram of pure corilagin (R_f 0.44), gallic acid (R_f 0.79), and ellagic acid (R_f 0.63) at 600 ng spot (1).

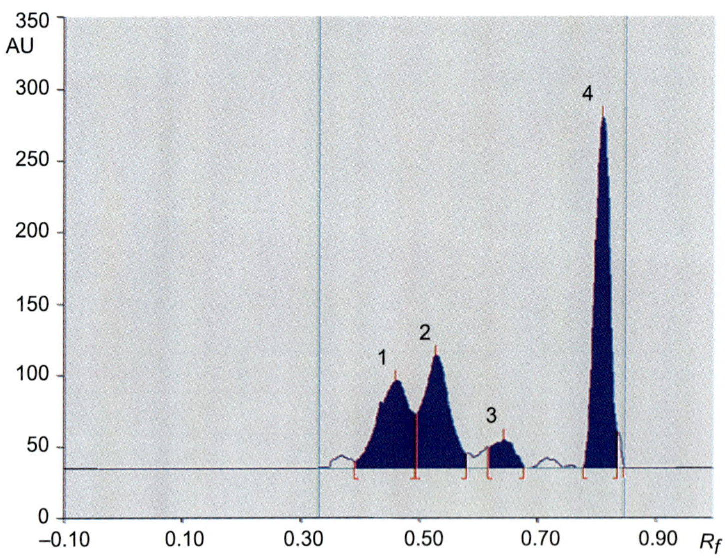

FIGURE 17.19 Densitogram of polyherbal formulation of *Phyllanthus amarus* containing corilagin (R_f 0.45), gallic acid (R_f 0.80), and ellagic acid (R_f 0.65).

TABLE 17.22 Assay Results for the Poly Herbal Formulation Containing Corilagin, Gallic Acid, and Ellagic Acid

Samples	Corilagin[a]	Gallic Acid[a]	Ellagic Acid[a]
Commercial crude powder Marketed formulation	0.030%	0.035%	0.072%
Formulation 1	0.355%	0.373%	0.121%
Formulation 2	0.013%	0.080%	0.104%

[a]*Mean ± standard deviation* (n = 3).

17.4 CONCLUSIONS

HPTLC is being increasingly used for applications in pharmaceutical analysis. Among its advantages are rapid analysis, simpler sample preparation, reduced cleanup requirements, and most important of all, simultaneous analysis of several samples and standard under identical conditions, making system suitability test unnecessary. The availability of different stationary phases has increased the applications of HPTLC to diverse samples incompatible with separation on bare silica gel. The application of HPTLC to content uniformity testing, bulk drug impurity profiling, and polyherbal formulations demonstrates how modern TLC has succeeded in expanding its scope to more complex sample types. The ability of HPTLC to separate a large number of samples simultaneously is unique and contributes to decreased per sample analysis costs. The future will definitely see more application of HPTLC in pharmaceutical analysis and increasing adoption of this method by national pharmacopoeias.

References

[1] Havele S, Dhaneshwar S. Estimation of metformin in bulk drug and formulation by HPTLC. J Nanomed Nanotechnol 2010;1:102. http://dx.doi.org/10.4172-7439, 1000102.

[2] Bhusari VK, Dhaneshwar SR. Application of stability indicating TLC method for the quantitative determination of Levocetirizine in pharmaceutical dosage forms. Intern J Adv Pharm Sci 2010;1:387–94. http://dx.doi.org/10.5138/ijaps,(2010), 0976.1055.01048.

[3] Dhaneshwar SS, Deshpande P, Patil M, Vadnerkar G, Dhaneshwar SR. Development and Validation of a method for Simultaneous Densitometric Analysis of Simvastatin and Ezetimibe as the bulk drug and in the tablet dosage form. Acta Chromatogr 2008;20:71–9. http://dx.doi.org/10.1556/A.Chrom.20.2008.1.6.

[4] Sethi PD. HPTLC high performance thin layer chromatography-quantitative analysis of pharmaceutical formulations, vol. 2. New Delhi: CBS Publishers and Distributors; 2013. p. 518–19.

[5] Bhusari VK, Dhaneshwar SR. Validated HPTLC method for simultaneous estimation of Atenolol, Hydrochlorothiazide and Amlodipine Besylate in bulk drug and formulation. Intern J Anal Bioanal Chem 2011;1(3):70–6.

[6] Khatal LD, Kamble AY, Mahadik MV, Dhaneshwar SR. Validated HPTLC method for simultaneous quantitation of Paracetamol, Diclofenac Potassium and Famotidine in tablet formulation. JAOAC Intern 2010;93(3):765–71.

[7] Sethi PD. HPTLC high performance thin layer chromatography quantitative analysis of pharmaceutical formulations, vol. 2. New Delhi: CBS Publishers and Distributors; 2013. p. 620–21.

[8] Dhaneshwar SR, Bhusari VK, Mahadik MV. Application of a stability indicating Thin Layer Chromatographic method to determination of Tenatoprazole in pharmaceutical dosage forms. JAOAC Intern 2009;92(2):387–93.

[9] Pawar SM, Dhaneshwar SR. Application of a stability indicating Thin Layer Chromatographic method for quantitation of Desvenlafaxine in pharmaceutical dosage forms. J Liq Chromatogr Rel. Technol 2012;35:499–510.

[10] Havele S, Dhaneshwar SR. Stress studies of Tenofovir Disoproxil Fumarate by HPTLC in bulk drug and formulation. Sci World J; 2012:1–6.

[11] Sethi PD. HPTLC high performance thin layer chromatography content uniformity of pharmaceutical formulations. New Delhi: Kongposh Publications Pvt. Ltd.; 2011. p. 1–2.

[12] Sethi PD. HPTLC high performance thin layer chromatography content uniformity of pharmaceutical formulations. New Delhi: Kongposh Publications Pvt. Ltd.; 2011. p. 98–101.

[13] Anchrom Test Lab Pvt. Ltd. Mumbai, India.

[14] Kulkarni PV, Sathiyanarayanan L, Nikam AR, Mahadik KR. (M.Pharm. thesis) submitted to Bharati Vidyapeeth Deemed University, Pune India; 2011.

[15] Kakade T, Sathiyanarayanan L, Mahadik KR. M.Pharm. thesis submitted to Bharati Vidyapeeth Deemed University, Pune India; 2011.

CHAPTER

18

Utility of Thin-Layer Chromatography in the Assessment of the Quality of Botanicals

Ahmed M. Galal[1], B. Avula[1], Ikhlas A. Khan[1,2,3]

[1] National Center for Natural Products Research, School of Pharmacy, University of Mississippi, University, MS, USA; [2] Department of Pharmacognosy, School of Pharmacy, University of Mississippi, University, MS, USA; [3] Cancer Institute, University of Mississippi Medical Center, Jackson, MS, USA

18.1 INTRODUCTION

At the present time instrumental thin-layer chromatography (ITLC), which is based on standardized methodologies, is regarded as superior to conventional TLC on account of its higher speed, simplicity, reproducibility, cost effectiveness, and of being environmentally friendly owing to the minimal consumption of organic solvents. In herbal analysis TLC provides several essential functions including: assisting in identification; authentication of medicinal herbs; quality control of raw materials and finished products; discrimination of plant species; in conjunction with other techniques serving as an adulteration detection method; studying stability of herbal products and elucidation of degradation products; standardization of herbal preparations; and identification of plant constituents. The analysis of herbal drugs by conventional TLC has long been an essential method of the pharmacopoeial monographs for monitoring quality and related parameters of herbs, extracts, and finished products. Furthermore this technique is routinely used in industry to monitor the

http://dx.doi.org/10.1016/B978-0-12-417223-4.00018-2

multiple stages in the manufacture of botanicals/herbal products. Sorbents used in high-performance thin-layer chromatography (HPTLC) include predominantly normal phase as well as reversed-phase silica-based materials. The combined use of normal- and reversed-phase layers allows the analysis of natural products with a broad range of polarity [1]. In addition, the advent of instrumental techniques, has tremendously facilitated documentation of separations by providing image-saving features [2–4]. Compared to high-performance liquid chromatography (HPLC) and gas chromatography (GC), HPTLC has lower resolving power, nevertheless, it is characterized by having several distinctive advantages over the aforementioned techniques, and these are:

1. Parallel separations for high sample throughput.
2. The samples to be compared, unlike HPLC and GC, are analyzed at the same time and under identical conditions, which allows proper and reliable comparison.
3. HPTLC is simple and cost effective relative to HPLC and GC.
4. In contrast to HPLC and GC each separation is performed with fresh sorbent, which further substantiates the existence of identical experimental conditions.
5. The planar nature of an HPTLC chromatogram confers the advantage of multiple detection methods including measurement by UV absorption or fluorescence with the further possibility of chemical derivatization to generate a suitable chromophore.
6. The function of generating saved chromatographic images provides features that facilitates documentation for future utility in comparing the chemical profiles of several herbs with samples from different geographical regions, different seasons, or different maturities for research purposes or for selecting raw materials for industrial production.
7. Analysis of substances with poor detection features.
8. The analysis time is relatively short compared with GC or HPLC.

The resolution of compounds by HPTLC depends on several factors including the type of stationary phase, mobile phase composition, layer thickness, type of binder, solvent purity, developing chamber size, chamber saturation, sample application volume, dimensions of the initial spot/band, solvent level in the chamber, relative humidity, temperature, separation distance, and mode of development.

The use of HPTLC in quantitative analysis necessitates that each method be validated in accordance with the appropriate regulatory guidelines. Initially, the validation has to show the suitability of the system and its capability to deal with the analytical questions to be addressed. HPTLC has been reported to provide sufficient resolution of substances for qualitative and quantitative analysis of a wide range of compounds, such as constituents of botanicals and ingredients in dietary supplements. Interfacing advanced supplementary devices to TLC has made HPTLC a competitive, or even superior, tool for identification, quantification, adulteration, and impurity detection in routine analysis [5]. These additional devices include, video scanners, densitometers, and advanced chromatographic chambers, as well as more effective elution techniques, high-resolution sorbents, hyphenated techniques, and the development of computer programs for method optimization. Method development and validation are the primary steps to attain reliable analysis.

The International Conference on Harmonization (ICH) guidelines achieved a great deal in harmonizing the principles and methods used for validation. However, ICH provides only a basis for a general discussion of the validation parameters, their calculation and interpretation. It is the responsibility of the analyst to identify validation parameters that are relevant to a given analytical procedure, as well as to design proper validation protocols and to perform an appropriate evaluation. Method development necessitates acquiring the physicochemical properties of compounds, including structure, polarity, solubility, volatility, and stability. To pursue HPTLC method development, essential steps are to be addressed, these are; selection of the stationary and mobile phases, method of sample application, development of the plates loaded with samples, visualization techniques, quantitative analysis, and documentation of work done. Instrument installation, operation qualification, and good laboratory practice/good manufacturing practice (GLP/GMP) requirements in regulated laboratories are met by many of the latest TLC instruments, such as sample applicators, developing chambers, and densitometers [6].

Sample preparation for HPTLC can be simplified compared with GC or HPLC as multiple sample cleanup steps can often be avoided. Although the sample can be a crude plant extract with several undesirable substances, most or many of these are retained at the baseline. This facilitates fast sample processing aided by the single use of layers for the separation. However, sample pretreatment may be necessary in some instances. This may comprise sampling, mechanical size reduction, an extraction step, filtration, and sometimes enrichment of components or removal of undesired impurities [2].

Selection of stationary phases for the separation of plant extracts is usually limited to silica-based sorbents for normal- or reversed-phase chromatography. Normal-phase separations on silica gel are used predominately, being suitable for compounds with a wide range of physicochemical properties. Commercially available precoated HPTLC plates with silica gel $60F_{254}$ on either glass or aluminum backing are used in the majority of published studies. In addition to silica gel, chemically bonded silica layers with C2, C8, and C18 modification are used occasionally. Other types of stationary phases are used for special applications, such as cellulose layers (normal-phase partition); polar chemically bonded layers (diol, cyanopropyl, and aminopropyl); ion-exchange layers; impregnated layers (e.g., argentation TLC for unsaturated organic compounds); thicker preparative layers (usually 0.5 ~ 1 mm) for isolation of target compounds in amounts suitable for further evaluation; and chiral layers for analysis of enantiomers.

Inspection of plates under visible and UV light is necessary to detect possible damage or the presence of impurities on the silica layer. Silica plates can be used without pretreatment if a blank plate is developed and does not exhibit unacceptable baseline disturbances that would interfere in the analysis. Nevertheless, it is a good practice to prewash plates to improve the reproducibility and robustness of the analysis. For prewashing, plates are developed in methanol, but mixtures of methanol and ethyl acetate or even the mobile phase of the method may be used. Prewashed plates should be dried at 120 °C for 30 min in a clean oven and equilibrated with laboratory atmosphere (temperature and relative humidity) in a container that provides protection from dust and fumes [2,7,8].

Sample application is an important variable that affects separation quality. Samples are typically applied to a layer using either contact application or spray-on techniques. The selection of a sample application technique depends on factors that include the nature of the analytical task, the type of sample matrix, the nature of the sorbent, and the sample volume. When using contact application techniques, the sample solvent migrates to the periphery of the applied zone creating a pattern similar to "circular chromatography." This may result in an irregular distribution of the sample components across the zone. In addition, the zone may undergo broadening forming a nonsymmetrical zone. The spray-on technique was designed to overcome these disadvantages and provides a homogenous distribution of the sample over the entire length of the zone [2]. Furthermore, the spray-on technique greatly improves the resolution and detection limit through formation of compact zones [2].

A specific advantage of HPTLC is that separated zones can be derivatized in situ using different approaches. A general method is to react the substance with a specific reagent to introduce a chromophore for UV–visible or fluorescence visualization. In many cases, substances or classes of substances can be identified by reacting with one of the many specific reagents designed for this purpose. Derivatization can be performed by either immersing the plates or by spraying the plates with a suitable reagent. For better reproducibility, immersion is the preferred technique [8].

18.2 METHOD VALIDATION

The purpose of method validation is to ascertain whether the performance characteristics of the method adhere to the requirements for the intended application. Validation should establish the performance characteristics while proving the suitability of a method for its intended purpose. The necessity for validation in analytical laboratories originates from the requirements of quality and accreditation standards and regulations, i.e., current GMP (cGMP), GLP, good clinical practices, ICH, Food and Drug Administration, Environmental Protection Agency, and United States Pharmacopeia (USP). Other regulatory requirements can be found in the International Standards Organization series and the European Norm [9]. Validation procedures are carried out following the recommendations of the regulatory agencies. Method validation is one part of the entire validation process that involves four main steps: (1) software validation, (2) hardware (instrumentation) validation/qualification, (3) method validation, and (4) system suitability. The validation procedure starts with validated software and a validated/qualified system. The second step is to develop a method and validate it using the qualified system. There are two guideline documents essential for conducting proper method validation: USP Chapter 1225: Validation of Compendial Methods; and the ICH guideline-Validation of Analytical Procedures-Text and Methodology Q2 (R1). Both the USP and ICH guidelines are general as they can be applied to any analytical procedure or technique used in a regulated laboratory [10].

Validation of these steps should be done through error analysis and a suitability test, which includes sample preparation, preparation of reference standards, application of samples, HPTLC separation, detection mode, quantification, and calculation of the final results [11].

A revalidation is necessary whenever a method is modified and the altered parameters are outside the operating range. Revalidation is also required if the sample matrix changes or if the analyst switches to a different instrument. Partial or full revalidation may also be considered if system suitability tests or the results of sample analysis are out of established acceptance criteria and the source of error cannot be traced back to instrument or other causative factors [9,12].

Validation parameters typically monitored for HPTLC are: (1) specificity; (2) linearity; (3) precision; (4) limit of detection (LOD) and limit of

of the HPTLC plates; (6) the chamber geometry; and (7) the degree of drying of the plate. Robustness is examined at an early stage of method development.

The ruggedness of a method is defined as the extent of the reproducibility of test results acquired by the analysis of the same samples under variable conditions, such as different laboratories, analysts, instruments; and different grades of solvents and reagents. Ruggedness can be determined for a proposed method by intentional variation of the experimental conditions.

18.2.6 Stability

Stability is evaluated to characterize the influence of environmental factors on the quality of herbs/herbal products over the course of time. Environmental factors include temperature, humidity, and oxygen. In addition, the determination of shelf life is also necessary. The stability of the analyte should be examined prior separation (in solution and on the plate) and while undergoing separation to detect partial or complete decomposition. The analyte solution is prepared in duplicate and applied four times on the plate. Freshly prepared analyte solution is also applied to the plate. The analyte solution and the plate are stored at ambient conditions for a defined time. After which a second analyte solution is prepared and immediately applied to the plate alongside the previously applied sample. This arrangement is used to examine the stability of the analyte on the plate and in solution by comparison of the aged samples to the freshly prepared analyte solution [8].

As a general practice, the effect of storage conditions on stability of substances should be assessed periodically to avoid possible degradation or deterioration of a single entity or a multiple-entity substance. Since heat and moisture are the most influential factors contributing to diminishing stability of substances, it is reasonable to investigate the effect of these factors during storage for a period that covers the duration from the production until the actual use of these substances.

18.3 HYPHENATED TECHNIQUES AND CHEMOMETRICS IN HERBAL ANALYSIS

The concept of hyphenation refers to the coupling of TLC to one or more spectroscopic methods in order to obtain additional information about the analyte [13]. Common spectroscopic methods in include optical spectroscopy (UV–visible) [14–16]; mass spectrometry [17–19]; FTIR and Raman [20,21]; and nuclear magnetic resonance (NMR) [22].

Chemometrics provides a complementary approach to assist in evaluating the quality, efficacy, and safety of medicinal herbs [23]. In addition, chemometric methods find use in chemical profile-based authentication of herbs to overcome the disparity of chemical profiles from similar samples obtained from different sources. These variations are attributed to exposure of the plants to different climatic, cultivation conditions, growing seasons, harvesting times, storage, and processing. In particular multivariate "empirical" modeling by principal component analysis (PCA) and partial least squares-discriminant analysis (PLS-DA), often shows unexpected patterns because the combined effect of all variables is taken into account, in contrast to traditional chemical and physical relationships, which usually consider just one or a few variables simultaneously [24]. Chemometrics can be classified into two main categories; pattern recognition methods (unsupervised and supervised) when a qualitative evaluation is considered and multivariate calibration for quantitative purposes [23]. HPTLC along with the statistical method PCA was applied to identify and discriminate botanicals for quality control [25].

18.3.1 Quality Assessment of the Traditional Chinese Herbal Medicine Chaihu (Bupleuri Radix)

Chaihu (Bupleuri Radix), roots of *Bupleurum chinense* and *Bupleurum scorzonerifolium*, is an authenticated plant of the Chinese traditional medicine (CTM), reported in the Chinese pharmacopoeia. Some other species, such as the roots of *Bupleurum falcatum*, *Bupleurum bicaule*, and *Bupleurum marginatum* var. stenophyllum similar to Chaihu are also occasionally found in local raw herb markets. The quality of 33 lots of authenticated Chaihu samples vs 31 lots of commercial samples were evaluated by both HPLC-evaporative light scattering detection and HPTLC analyses of its principal bioactive components (saikosaponins) [25,26]. The pretreated data acquired from both HPLC fingerprints and HPTLC fluorescent images were processed by chemometrics for similarity and pattern recognition, including artificial neural networks, *k*-nearest neighbor (*k*-NN), and an expert's panel. It was apparent that *k*-NN classifier exhibited good performance with sufficient flexibility for processing HPTLC fingerprint images, which otherwise were not easily dealt with by other algorithms due to the shift of R_f values and varying hue/saturation of the band colors between different TLC plates. These two chromatographic fingerprint methods can be considered complementary techniques for quality control. The roots of Chaihu from different species of the genus *Bupleurum* were easily distinguished from each other. Chaihu collected from several major herbal distribution centers was found to belong to *B. chinense* with great variation in the content of its major saikosaponins [25,26].

18.3.2 Metabolomics of Traditional Chinese Herbal Medicine Products

A recent metabolomics study involved the multivariate analysis of HPTLC fingerprints of nonpolar phytochemicals in the dried roots of four popular medicinal herbs, *Aster tataricus, Atractylodes lancea, Gentiana rigescens,* and *Gentiana macrophylla*. These herbal medicines are used in traditional Chinese medicine for treating several ailments. The extraction of the nonpolar phytochemicals was carried out using petroleum ether and analyzed by HPTLC. For data exploration PCA, PLS-DA, and orthogonal PLS-DA. The score plots indicated that all three methods provided a good spatial distributions with clear clusters/grouping of each herb [27].

18.4 APPLICATIONS OF TLC IN THE FIELD OF HERBAL PRODUCTS/BOTANICALS

Common uses of TLC in the analysis of herbal products and botanicals include: the identification and authentication of plant material, in conjunction with macro- and/or microscopic characterization; detection of adulterants; standardization of herbal products; assessment of stability [28]; discrimination of plant cultivars; selection of herbal raw material for industrial production; studying plant metabolomics; characterizing the influence of seasonal and geographical variations on the chemical profile of herbs; studying the shelf life of herbal products; monitoring manufacturing of herbal products/dietary supplements at different stages and all through the process; dereplication of natural products by HPTLC coupled with mass spectrometry (MS)/NMR [22]; and HPTLC combined with bioautography/bioassays for identification of certain bioactivities of natural products [22,29,30]. Some representative examples follow in the sections below.

18.4.1 Anthocyanins in Acai (*Euterpe oleracea* Mart.) Berries and Commercial Products

Two predominant anthocyanins, namely cyanidin-3-*O*-rutinoside and cyanidin-3-*O*-glucoside in acai berries (*Euterpe oleracea*), and in its commercial preparations were determined by HPTLC on silica gel 60 F_{254} plates using ethyl acetate—formic acid—acetic acid—water (10:1.1:1.1:2.6)

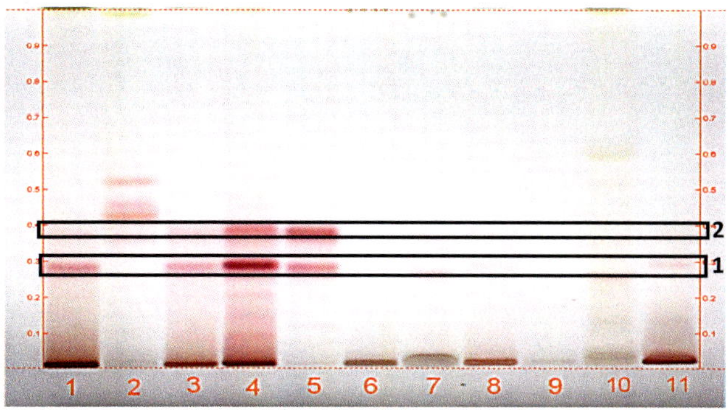

FIGURE 18.1 High-performance thin-layer chromatography image Track 1: authenticated berries of acai; Track 2–4: commercial samples of acai; Track 5: a mixture of two reference standards ((1) cyanidin-3-*O*-rutinoside and (2) cyanidin-3-*O*-glucoside); Track 6–11: commercial products under white light.

as the mobile phase (Figure 18.1). The anthocyanins were quantified by scanning densitometry at 520 nm. Compounds **1** and **2** were well separated from excipients with R_f of 0.29 ± 0.03 and 0.37 ± 0.04, respectively. It is noteworthy that acai fruit is an excellent source of antioxidants [31].

18.4.2 Analysis of *Passiflora*

Passiflora incarnata is used to treat stress, anxiety, depression, and insomnia. The aerial parts contain indole alkaloids and flavonoids, which are believed to be responsible for the medicinal effects of *Passiflora*. An HPTLC method together with a fingerprinting technique was used to identify the putative marker/biomarkers of this medicinal herb. Quantitative estimation of four flavonoids, namely orientin, isoorientin, vitexin, and isovitexin was the basis for the HPTLC analysis on silica gel with a mobile phase of ethyl acetate–formic acid–water (16.5:1.4: 1.2 v/v/v). Six species of *Passiflora*, namely *P. incarnata*, *Passiflora violaceae*, *Passiflora morifolia*, *Passiflora suberosa*, *Passiflora guadrangularis*, and *Passiflora edulis* were included in this study. In addition, a number of commercial products of *P. incarnata* were also examined for the presence and level of marker compounds. Visualization was achieved by derivatization with a methanolic solution of diphenylboric acid-β-ethylaminoester reagent, followed by inspection under UV light at 366 nm (Figure 18.2). Although the zone intensities of the chromatograms for different commercial preparations compared to the authenticated plant specimens were varied, the fingerprint profiles were similar [32].

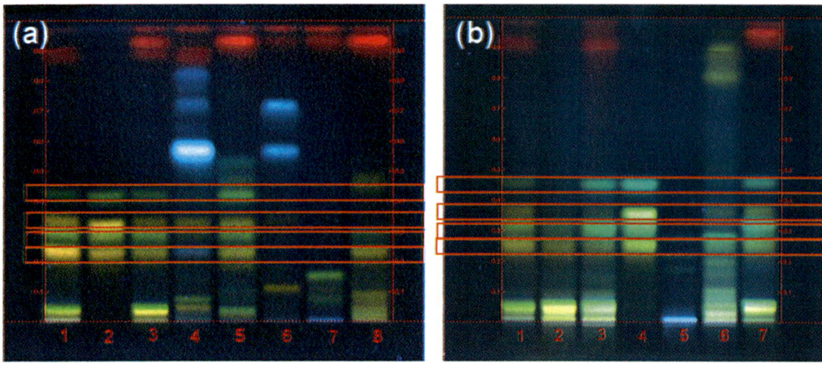

FIGURE 18.2 (a) HPTLC (normal phase) fingerprints of *Passiflora* species: Track assignment: (1) *Passiflora incarnata*; (2) a mixture of four reference standards, in order of decreasing polarity (isoorientin, isovitexin, orientin, and vitexin); (3) *P. incarnata*; (4) *Passiflora violaceae*; (5) *Passiflora morifolia*; (6) *Passiflora suberosa*; (7) *Passiflora guadrangularis*; (8) *Passiflora edulis*. (b) HPTLC (normal phase) fingerprints of various products of *Passiflora* and dietary supplements (DP1–DP4): Track assignment: (1) *P. incarnata*; (2) DP1; (3) DP2; (4) a mixture of four reference standards, in order of decreasing polarity (isoorientin, isovitexin, orientin, and vitexin); (5) DP3; (6) DP4; (7) *P. incarnata*.

18.4.3 Estimation of Parthenolide in Market Products

Representatives of feverfew market products were standardized by an HPTLC method on the basis of their parthenolide content. Feverfew is the common name for *Tanacetum parthenium*, and parthenolide is considered the putative biomarker of this medicinal herb. It is accountable, in part, for the antimigraine effect of feverfew via suppression of serotonin release from blood platelets. In this method parthenolide was determined by derivatization with *p*-anisaldehyde and detected at 543 nm [33].

18.4.4 Evaluation of *Echinacea* Species

Echinacea is one of the best selling herbal products in developed countries. The popularity of *Echinacea* is attributed to its immunostimulatory effects, which in turn might contribute to treatment and prevention of the common cold, influenza, and upper respiratory infections.

Echinacea includes three species, namely *Echinacea angustifolia*, *Echinacea pallida*, and *Echinacea purpurea*. Commercial *Echinacea* products may contain one or more of these species. The chemistry of *Echinacea* species is well known, and the constituent profile was shown to vary, qualitatively and quantitatively, from one species to another. The constituents implicated in the induction of immunostimulatory effect of *Echinacea*

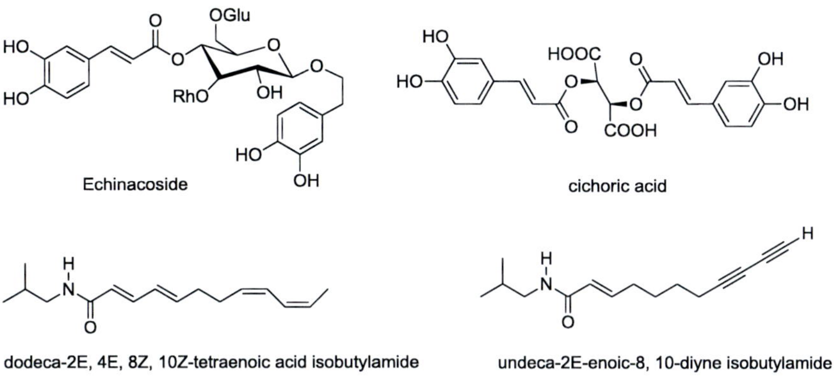

FIGURE 18.3 Putative bioactive constituents of *Echinacea* species.

include alkylamides, echinacoside, cichoric acid, and polysaccharides (Figure 18.3).

The identity of *Echinacea* species, namely *E. angustifolia*, *E. pallida*, and *E. purpurea*, and detection of their common adulterants was accomplished by adopting HPTLC methods that utilize phenylpropanoids (echinacoside, chlorogenic acid, caftaric acid, chicoric acid, caffeic acid, and cynarin), alkylamides, and fructofuranoids (inulin and saccharose) as markers. In one method, analysis of the phenylpropanoids was the basis for assessing the quality of *Echinacea*, [34]. The second method selected alkylamides, such as dodeca-2*E*,8*Z*,10*Z*-tetraenoic acid isobutylamide and undeca-2*E*-enoic-8, 10-diyne isobutylamide as markers to determine the identity and quality of *Echinacea* species. Since alkyamides, a class of fatty acid amides, have a poor UV response, they were derivatized with anisaldehyde reagent and subsequently visualized at 366 nm. A third method for assessment of *Echinacea* species was based on the HPTLC analysis of fructofuranoids, including inulin and saccharose. This work demonstrated that dodeca-2*E*,4*E*,8*Z*,10*E*/*Z*-tetraenoic acid isobutylamide were present in the root of all species with the exception of *E. pallida*. The discrimination of *Echinacea* species is possible since their HPTLC fingerprints are significantly different, which in turn facilitates detection of adulteration [35].

18.4.5 Chromatographic Fingerprinting of *Hoodia* Species

Dietary supplements claimed to contain *Hoodia gordonii*, an appetite suppressant, together with other *Hoodia* species, can be analyzed by HPTLC [36]. Market samples were compared with an authenticated sample and 11 pregnane glycosides as the putative markers. HPTLC was performed on precoated silica 60 F_{254} plates with dichloromethane–methanol–water (75:17:2.2 v/v/v) as the mobile phase. Out of

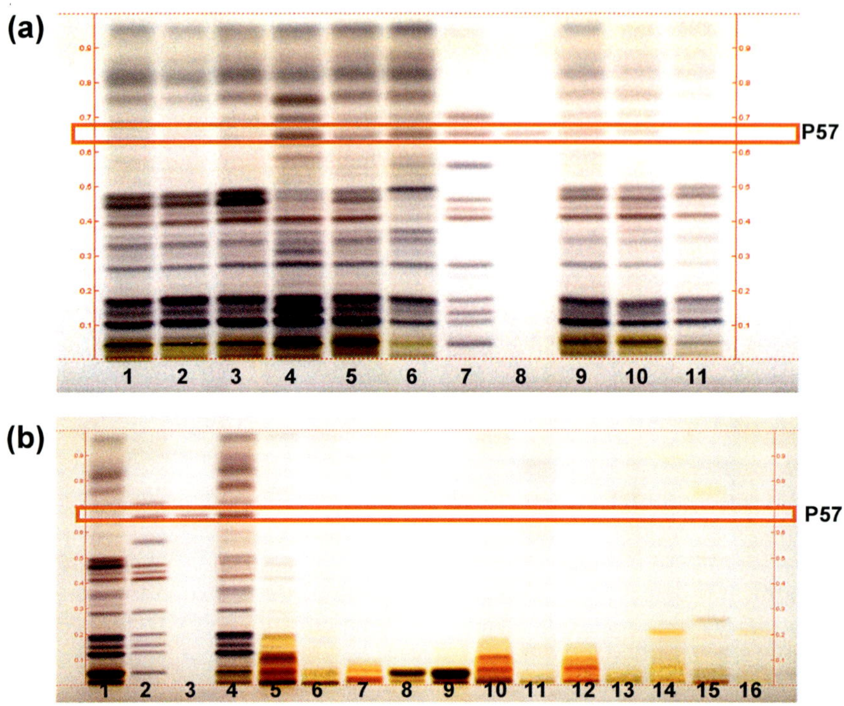

FIGURE 18.4 (a) HPTLC chromatogram of various *Hoodia gordonii* samples and its standard compounds. Evaluation by HPTLC on silica gel 60 F_{254} with dichloromethane/methanol/water 75:17:2.2 by volume as a mobile phase; derivatization with anisaldehyde reagent: Track 1–6: *H. gordonii*; Track 7: a mixture of 11 reference standards (hoodigoside M; hoodigoside L; hoodigoside P; hoodigoside U; hoodigoside O; hoodigoside E; hoodigoside F; hoodigoside J; hoodigoside N; P57; hoodigoside C); Track 8: P57; Track 9–11: *H. gordonii.* (b) Comparison of *H. gordonii* with dietary supplements (DS-1 to DS-13) of various dosage forms. Tracks: (1) *H. gordonii*; (2) Standard mix-11; (3) P57; (4), (6), (12–16) capsules of DS-1, DS-3, DS-9 to DS-13; (5), (7) tablets of DS-2 and DS-4; (8) juice of DS-5; (9) snack bar (DS-6); (10) tea bag of DS-7; (11) spray of DS-8.

13 dietary supplements, 11 did not contain any of the 11 pregnane glycosides. The method is also suitable for rapid, decisive authentication, and comparison of the subtle differences among samples from various plant sources (Figure 18.4(a) and (b)) [36].

18.4.6 Quantification of Oxyreservatrol in the Different Parts *Morus* species

Oxyresveratrol (Figure 18.5) is a natural stilbene occurring in grapes, peanuts, and mulberries. It has several beneficial medicinal properties such as antioxidant, antiinflammatory, and for the treatment of

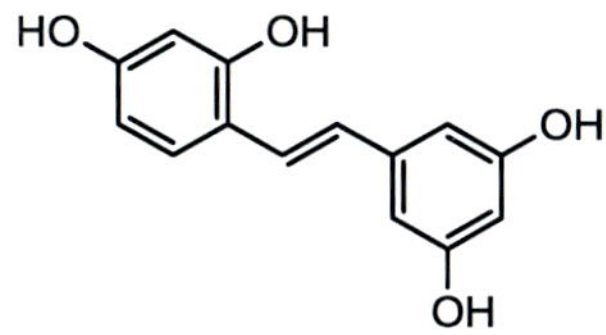

FIGURE 18.5 Structure of oxyresveratrol.

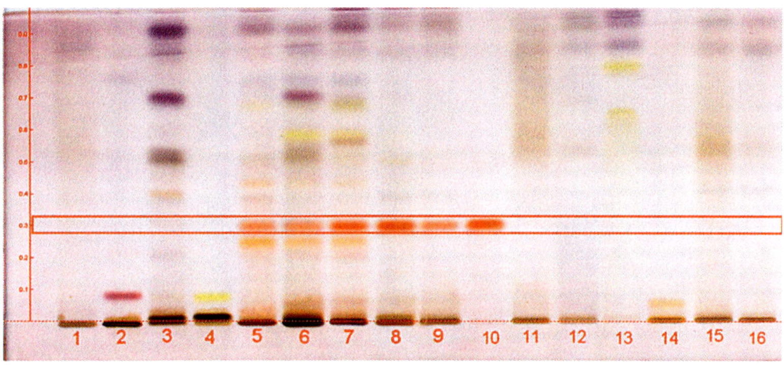

FIGURE 18.6 Comparison of *Morus species* with other related genera under Visible light. Tracks: (1) leaves of *Morus alba* L.; (2) wood of *M. alba* L.; (3) stems of *M. alba* L.; (4) fruits of *M. alba* L.; (5) roots of *M. alba* L.; (6) bark (BK) *M. alba* L.; (7) bark of *M. alba* L.; (8) stems of *Morus rubra* L.; (9) stems-leaves of *M. rubra* L.; (10) oxyresveratrol; (11) leaves of *M. rubra* L.; (12) leaves of *Maclura pomifera*; (13) fruits of *M. pomifera*; (14) roots of *M. pomifera*; (15) leaves-twigs of *Trymatococcus amazonicus*; and (16) stems of *Ficus maxima*.

hyperpigmentation disorders. Oxyresveratrol was originally isolated from the stem of *Morus rubra* L. HPTLC-based analysis, using hexanes–ethyl acetate–chloroform–methanol (3.0:2.0:3.4:1:6 v/v/v/v) as mobile phase on silica gel and detection at 327 nm, revealed the presence of oxyreveratrol in the root and bark of *Morus alba*. In *M. rubra*, it was detected in the stem and leaves. In addition, oxyresveratrol was identified in leaves, fruits, and root of *Maclura pomifera*, stem of *Ficus maxima*, and leaves/twigs of *Trymatococcus amazonicus*. The chromatographic fingerprints generated by HPTLC enabled the different species of *Morus* to be distinguished and the identification of the genus *Morus* from related ones [37] (Figure 18.6).

18.4.7 Quantification of Caffeic Acid and Quercetin in Cranberry Extracts

Cranberry contains several bioactive constituents. The substances that are apparently, or presumably, responsible or associated with health benefits of cranberry include caffeic acid, quercetin, catechin, and proanthocyanidins, PAC-A_2 and PAC-B_1. In a recent work cranberry was

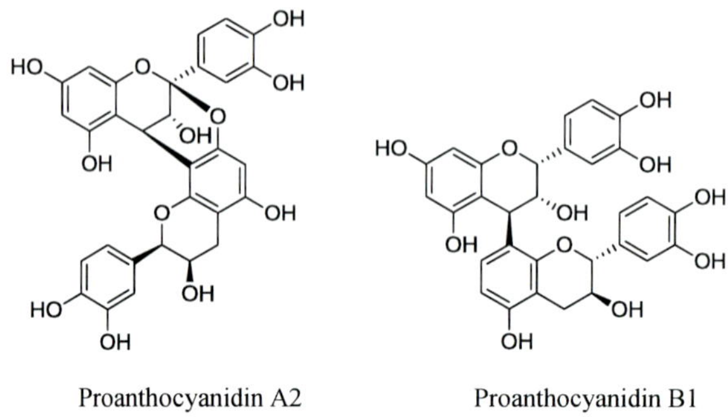

FIGURE 18.7 Structure of proanthocyanidins.

standardized on the basis of their contents of caffeic acid and quercetin [38]. Caffeic acid is reported to possess antioxidant and lipid oxidation suppressing properties. Quercetin is a flavonol, abundantly present in edible plants, such as onion, garlic, apples, and wine. Quercetin has antioxidant and antiinflammatory effects. It was also reported to exhibit anticarcinogenic activity. A rapid HPTLC was developed for the quantitative analysis of caffeic acid and quercetin in cranberry extract (*Vaccinium macrocarpon*) [38]. The compounds were extracted from cranberry with a mixture of methanol and water. The plant material (frozen berries of *V. macrocarpon*) was extracted with hot ethanol at 40–60 °C over 12 h. The extracts were separated on silica gel layers using *n*-hexanes–ethyl acetate–acetic acid (5:3:1, v/v/v) as mobile phase. Quantification was based on the fluorescence emission excited at 366 nm.

Cranberry extracts have also been standardized on the basis of its contents of catechin and proanthocyanidins, PAC-A_2 and PAC-B_1 (Figure 18.7). Cranberry (fresh fruits) were ground in a blender and macerated in a mixture of ethyl acetate and water (90:10 v/v) for different times, filtered, and the solvent removed by distillation. The residue was dissolved in petroleum ether, and the solution centrifuged to separate the insoluble part (proanthocyanidins), which was collected and dried to afford a crude proanthocyanidins extract. For analysis of the extract on silica gel 60 F_{254} the mobile phase was methylene chloride–ethyl acetate–formic acid (6:10:1 v/v/v). Visualization of the bands was facilitated by derivatization with vanillin/hydrochloric acid.

18.4.8 Quantitative Estimation of *Trans*-resveratrol in Herbal Products

Trans-resveratrol (Figure 18.8), is a stilbene compound shown to be a strong antioxidant. It possesses a protective effect against atherosclerosis, coronary heart diseases, and postmenopausal conditions and platelet

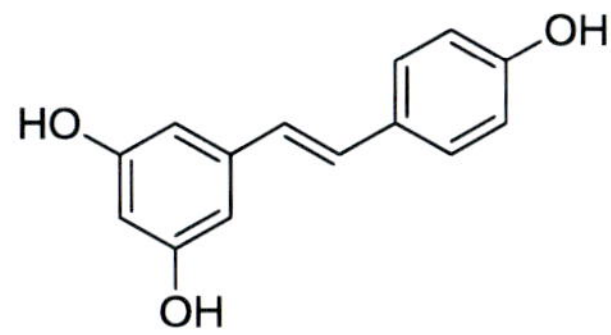

FIGURE 18.8 Structure of *Trans*-resveratrol.

aggregation. A simple and selective HPTLC method was developed to estimate *trans*-resveratrol in *Polyganum cuspidatum* root on silica gel 60F_{254} layers with chloroform–ethyl acetate–formic acid (2.5:1:0.1) as mobile phase. *Trans*-resveratrol was detected at 313 nm (LOD and LOQ 9 and 27 ng/spot, respectively).

18.4.9 Seasonal and Geographical Variations in Chemical Constituents in *Aerva lanata* Juss. Ex Schultes

Aerva lanata Juss. Ex Schultes is a common weed, widely spread in the dried parts of the tropics and subtropics of Africa and Asia. This plant possesses several curative uses in traditional medicine. An HPTLC method was developed to explore the influence of seasonal and geographical variation on the chemical profile of this herb. Extracts were separated on precoated silica 60F_{254} layers with ethyl acetate–toluene (7:3 v/v) as the mobile phase. *p*-Hydroxybenzoic acid was used as a reference standard. The chromatographic fingerprints showed the presence of *p*-hydroxybenzoic acid in samples from all seasons, but at varied concentrations. Samples from two Indian states were observed to contain *p*-hydroxybenzoic acid while samples from Kerala contained a different compound. This work displayed the qualitative and quantitative variation in chemical profile of *A. lanata* due to changes in the location and season of plant growth [39].

18.4.10 Assessment of Antioxidant Activity of Green Tea and Guava

HPTLC was used to identify the following substances in green tea extract: (-)-epicatechin, (-)-epicatechin-3-gallate (ECG), (-)-epigallocatechin, and (-)-epigallocatechin-3-gallate, caffeine, and rutin (Figure 18.9). In guava, the following markers were detected; (-)epicatechin, (-)ECG, gallic acid, ellagic acid, and quercetin. The guava leaf extract was indicated as a suitable alternative source to green tea extract for antioxidants. To visualize the compounds on the HPTLC plates derivatization with a mixture of 2-aminoethyldiphenyl borinate and polyethylene glycol 4000 reagent solution (1:1) was used. In a later study an HPTLC

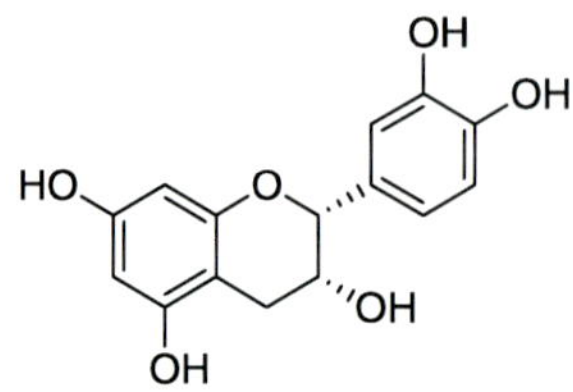

(-)-epicatechin

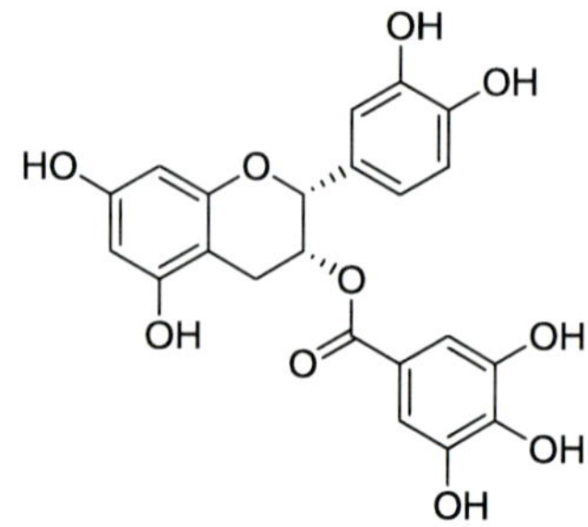

(-)-epicatechin-3-gallate

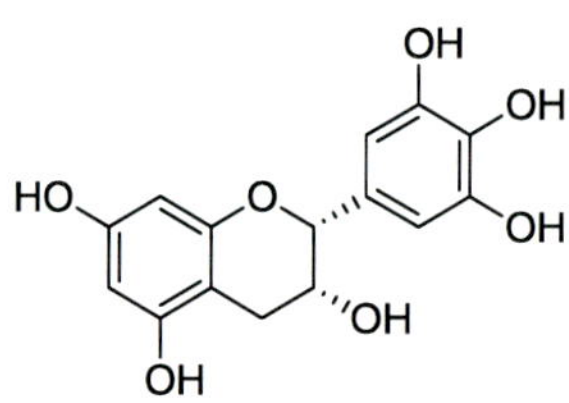

(-)-epigallocatechin

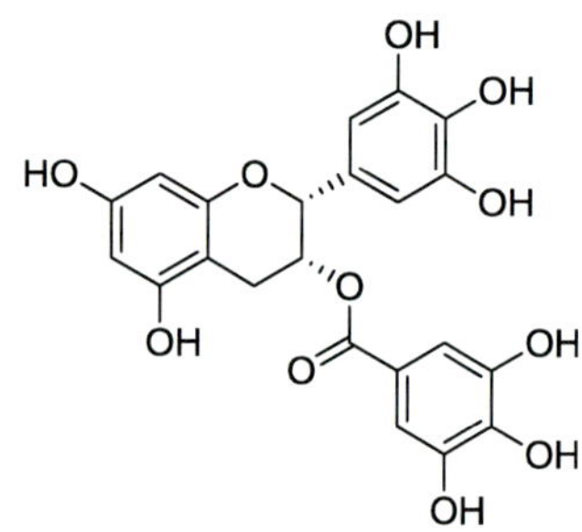

(-)-epigallocatechin-3-gallate

FIGURE 18.9 Structure of catechins of green tea.

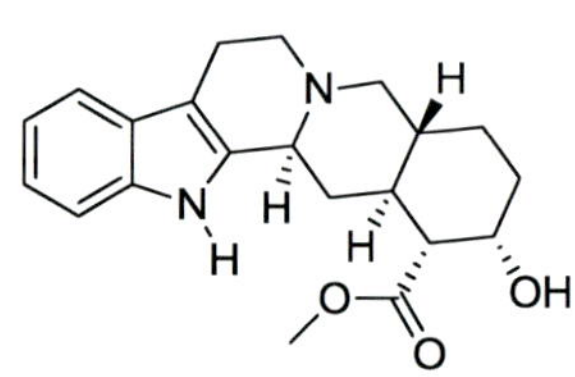

FIGURE 18.10 Structure of yohimbine.

method was employed for the quantitative analysis of 10 marker compounds (including catechins, flavonoids, and phenolic compounds) in green tea and guava extracts with detection at either 282 or 285 nm [40].

18.4.11 Quantification of Yohimbine in Plant Materials

Yohimbine (Figure 18.10) is an indole alkaloid isolated from the stem bark of *Pausinystalia yohimbe,* as the major constituent among several other alkaloids in the stem bark of the plant. It is used medicinally for the treatment of erectile dysfunction. This alkaloid occurs also in *Rauwolia serpentina*. Yohimbe is sold as a prescription drug in capsule or pill form combined with other nonactive ingredients. For quantification of yohimbine a normal-phase HPTLC method was developed using silica gel $60F_{254}$ as the stationary phase with toluene–ethyl acetate–diethylamine (7:2:1 v/v) as the mobile phase. Yohimbine was quantified at 285 nm. This method was able to discriminate genuine commercial raw materials of yohimbe from adulterated samples [41].

18.4.12 Discrimination of Cultivars of Jamaican Ginger (*Zingiber officinale* Roscoe)

HPTLC was utilized to categorize four cultivars of Jamaican ginger (*Zingiber officinale* Roscoe) [42]. Ginger is an underground stem or rhizome with a number of medicinal applications (e.g., the treatment of nausea and vomiting). It is also used as a spice in cooking. It is one of the 20 top-selling herbal supplements in the USA. HPTLC was used to detect the variations in the pungent principles of ginger, namely 6-gingerol, 8-gingerol, 10-gingerol, and 6-shogaol, and to assess differences attributed to different stages of maturation and storage conditions using the fingerprints of the major pungent principles of ginger in the four cultivars [42].

18.4.13 Selection of a Herbal Raw Materials for Further Processing

For the purpose of selecting *Vitex agnus-castus* fruits to manufacture a herbal product, a qualitative and semiquantitative HPTLC-based screening method was applied to 23 samples of raw material from different geographical locations [43]. The aim of this study was to identify a cultivar(s) with high content of flavonoid and diterpenes as the putative pharmacologically active constituents, suitable for cultivation. It was found that the differences in flavonoid content between various samples are considerable, based on the evaluation of the chromatographic fingerprints of the different cultivars. The separations were performed on silica gel with a mobile phase of tetrahydrofuran–toluene–formic acid-water (8:4:1:0.5 v/v/v/v). Samples were derivatized by dipping in natural products reagent, drying, and spraying with 5% (w/v) macrogol in methanol. The separated compounds were visualized at 366 nm.

18.4.14 TLC in the Detection of Adulteration and Determination of Adulterants

In recent years, botanicals/herbal products have gained widespread popularity all over the world. However, this trend was associated with the

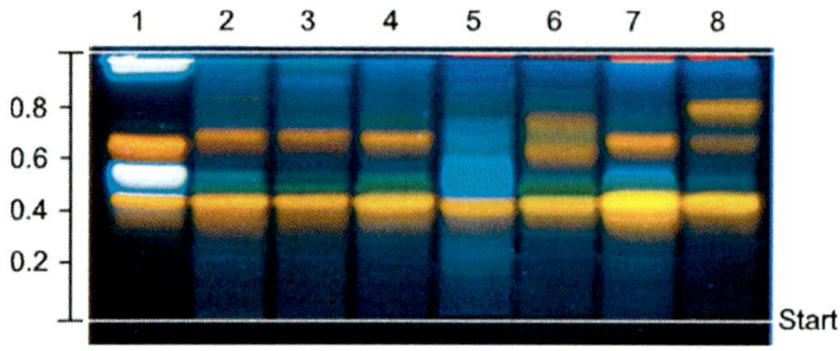

FIGURE 18.11 TLC fingerprint of different *Illicium* species: (1) reference solution (rutin, chlorogenic acid, hyperoside, caffeic acid); (2) *Illicium henryi* Diels; (3) *Illicium micranthum* Dunn; (4) *Illicium simonsii* Maxim; (5) *Illicium floridamun* Ellis; (6) *Illicium verum* Hook f.; (7) *Illicium lanceolatum* A. C. Sm.; (8) *Illicium anisatum* L.

emergence of adulteration as a problem challenging the botanicals/herbal products industry. The adulterants can potentially pose a health threat by inflicting adverse effects. This phenomenon has prompted researchers to investigate methods for detection and quantification of adulterants, as well as methods for identification and authentication of plant materials [44].

An HPTLC method was developed (Figure 18.11) as a component of the quality control of Chinese star anise [45]. This technique was used to categorize various *Illicium* species including Chinese and Japanese star anise. Black cohosh (*Cimicifuga racemosa)* is used to treat discomfort during menopause and as a substitute for synthetic drugs in hormone replacement therapy. The mostly wildcrafted plant is ranked among the top-selling herbs in the United States. There is a risk for adulteration with the similar-looking *Cimicifuga americana,* which grows in the same habitats of the Eastern United States. Other adulterants found in the global marketplace are the three Asian *Cimicifuga* species *Cimicifuga foetida, Cimicifuga heracleifolia,* and *Cimicifuga dahurica.* A practical, rapid, and reliable HPTLC method was developed for identification of *C. racemosa* and detection of its most common adulterants by chemical fingerprint profiles (Figure 18.12). With specific derivatization reagents, mixtures of *C. racemosa* with a minimum of 5% of one of the adulterants can be detected [46].

Curcumin, the principal curcuminoid of turmeric, is responsible for its yellow color and serves as a measure of turmeric quality. The Prevention of Food Adulteration Act of India allows only *Curcuma longa* L. for the production of turmeric powder and prohibits addition of any foreign matter/artificial color, but it does not specify a minimum curcumin content. The present surveillance was undertaken to study the quality of loose vs branded turmeric powders vis-a-vis curcumin content and the presence of unwarranted extraneous colors from city markets in India using a newly developed two-dimensional high-performance thin-layer

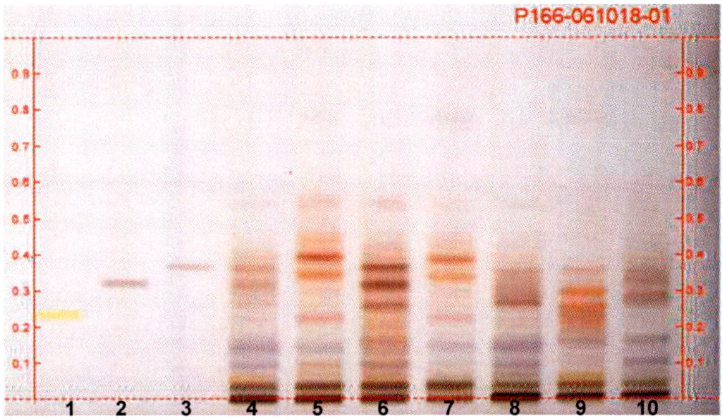

FIGURE 18.12 High-performance thin-layer chromatogram of black cohosh and common adulterants; white light; Tracks (1) cimifugin; (2) 23-epi-26-deoxyactein; (3) actein; (4) *Cimicifuga racemosa*; (5) *Cimicifuga heracleifolia*; (6) *Cimicifuga foetida*; (7) *Cimicifuga dahurica*; (8) *Cimicifuga rubra*; (9) *Cimicifuga americana*; (10) *Actaea pachypoda*.

chromatography method. The results showed that curcumin content in branded samples ranged from 2.2% to 3.7% and in nonbranded samples from 0.3% to 2.6%. Although none of the branded turmeric powders contained artificial colors, 17% of loose powders showed the presence of extraneous color metanil yellow, in the range 1.0–8.5 mg/g, which may pose a health threat. Low curcumin content in the analyzed samples may be due to mixing of other Curcuma species or their curcumin-depleted matrixes and foreign starches as low-cost alternatives. This is supported by the fact that major Indian turmeric trade types are known to contain curcumin [47].

Currently, slimming herbal medicinal products (HMPs) are in high demand. As a result, adulteration with synthetic weight reduction substances has dramatically increased to boost the efficacy of the herbal products. Sibutramine is an anorexic drug acting on the neurotransmitter serotonin but is banned in many countries due to its adverse effects on the heart. Sibutramine in slimming HMPs can be determined by HPTLC with UV detection at 223 nm [48]. The separation is performed on precoated silica gel G60 F_{254} layers with chloroform–acetone–methanol (8.5:1.0:0.55 v/v/v) as the mobile phase. The method was validated and demonstrated to be suitable for identifying herbal products adulterated with sibutramine.

Sildenafil is an orally administered synthetic drug for treatment of erectile dysfunction in men. A number of herbal preparations indicated for improving male erectile disorders were found to contain sildenafil or its derivatives (e.g., thiosildenafil, tadalafil, verdenafil, hydroxyhomosildenafil, and/or thiomethisosildenafil). These dietary supplements are claimed to be free of any synthetic additives. An HPTLC method has been developed for the detection and quantification of sildenafil citrate at 305 nm in pharmaceutical products and aphrodisiac herbal preparations. Separations were performed on silica gel layers with chloroform–methanol–diethylamine (90:10:1 v/v/v) as mobile phase [49–51]. Sildenafil was identified in one aphrodisiac herbal product at a level of 85.0 mg per capsule.

18.5 USE OF HPTLC BY THE AMERICAN HERBAL PHARMACOPOEIA

TLC is among the key tools for testing single- and multiple-entity substances in most pharmacopoeial monographs. Pharmacopoeial standards are used in industry as the basis for inspecting whether a given substance meets the QC requirements and cGMPs. TLC is also employed for developing raw material specifications, as well as for the finished products. TLC is useful for screening plant samples that are reviewed in the monographs of the pharmacopoeias and also for developing botanical reference materials. HPTLC has been included recently in the American Herbal Pharmacopoeia as an advanced method of analysis for multiple purposes [52].

18.6 RECENT ADVANCES IN HPTLC-BASED QUALITY CONTROL OF BOTANICALS AND DIETARY SUPPLEMENTS

Recent years have witnessed growing popularity in the consumption of herbal products/botanicals all over the world. Nevertheless, the outbreak of adulteration and contamination, poses a potential source for inflicting serious adverse effects on consumers. Added to this, the complex nature of herbal products/botanicals makes the assessment of their quality an ongoing challenge. Modern hyphenated techniques have added a new capability to assess the identity, quality, consistency, chemical profile, and stability of herbal products. Current research in instrumental planar chromatography has as it aims the development of new techniques and methods that can be exploited to serve progress in herbal medicine on several fronts. These include development of new products, the production of new and existing products, and postmarket monitoring. In this regard recent papers focused on coupling HPTLC to MS and/or NMR and the coupling HPTLC to bioautography are noteworthy. For example, HPTLC was coupled to MS and NMR to accomplish structure elucidation and identification of the chemical constituents accounting for the acetylcholine esterase inhibiting activity of a plant material in the search for potential therapeutic agents for Alzheimer's disease [22]. In a related article on TLC-coupled bioassays, TLC was combined with bioautography to identify the antibacterial, antifungal, antioxidant, anticancer, or other bioactivities in a complex mixture of compounds, such as a crude extract from a plant [30]. These recent advances in planar chromatography should contribute to improving the quality of botanicals.

18.7 CONCLUSIONS

HPTLC has been a widely accepted method for qualitative as well as quantitative assessment of herbs and herbal products. This is attributable to its simplicity, speed, low cost, and the provision of chemical fingerprints with additional information absent from other related techniques. A major advantage of instrumental TLC is that it can provide images compatible with visual assessment of product quality as well as for more detailed analysis by chemometric techniques. It has been demonstrated that HPTLC can provide efficient and rapid methods for authentication and quality control of botanicals/herbal products. Its multipurpose utility has extended to several applications, for example, identification of plant constituents, taxonomic discrimination of plants, identification of closely related species, detection of adulterants, inspection of purity, and stability assessment. Instrumental TLC has been implemented as a technique in GMPs for the analysis of botanical dietary supplements. On the other hand, the irreproducibility and complexity of the chromatographic fingerprints, as well as the quantification due to, primarily, variation in the factors affecting accumulation of secondary plant metabolites, including genotype, climatic conditions, agronomic techniques, such as irrigation and fertilization, and the developmental stage of the plant, in addition to seasonal variations, necessitates the introduction of chemometrics, as indispensable approach. Furthermore, chromatographic analysis in conjunction with chemometrics permits evaluation of the similarity of various chemical fingerprints. More studies on the detection and determination of the level of adulterants and contaminants in the raw materials as well as in the final herbal products need to be done.

References

[1] Gocan S. Stationary phases for thin-layer chromatography. J Chromatogr Sci 2002;40(10):538–49.

[2] Sherma J. Review of HPTLC in drug analysis: 1996–2009. J AOAC Int 2010;93(3):754–64.

[3] Reich E, Widmer V. Plant analysis 2008-planar chromatography. Planta Med 2009;75(7):711–8.

[4] Reich E, Sherma J. Thin layer chromatography in botanical analysis. J AOAC Int 2010;93(5):1347–8.

[5] Krzek J, Sherma J. New planar chromatography-densitometry methods for determination of drug substances and formulations. J AOAC Int 2010;93(3):751–3.
[6] Rashmin P, et al. HPTLC method development and validation: strategy to minimize methodological failures. J Food Drug Anal 2012;20(4):794–804.
[7] Waksmundzka-Hajnos M, et al. Thin layer chromatography in phytochemistry. Chromatogr Sci Ser 2008;99:871. CRC Press.
[8] Koll K, et al. Validation of standardized high-performance thin-layer chromatographic methods for quality control and stability testing of herbals. J AOAC Int 2003;86(5): 909–15.
[9] Sherma J. Chromatographic methods of analysis: thin layer chromatography. In: Swarbrick J, editor. Encyclopedia of pharmaceutical technology. 3rd ed. London (UK): Informa Healthcare; 2007. pp. 538–50.
[10] Swartz M, W. ID, Poole C. Chromatography: method validation. In: Handbook of methods and instrumentation in separation science. (CA, USA): Academic Press; 2009. pp. 614–20.
[11] Ferenczi-Fodor K, Renger B, Vegh Z. The frustrated reviewer—recurrent failures in manuscripts describing validation of quantitative TLC/HPTLC procedures for analysis of pharmaceuticals. J Planar Chromatogr—Mod TLC 2010;23(3):173–9.
[12] Huber L. Validation of analytical methods and processes. In: Nash RA, Wachter AH, editors. Pharmaceutical process validation third edition—revised and expanded. New York (USA): Marcel Dekker, Inc; 2003.
[13] Morlock G, Schwack W. Hyphenations in planar chromatography. J Chromatogr A 2010;1217(43):6600–9.
[14] Vasta JD, Sherma J. Analysis of lycopene in nutritional supplements by silica gel high-performance thin-layer chromatography with visible-mode densitometry. Acta Chromatogr 2008;20(4):673–83.
[15] Arthur B, Sherma J. Quantification of beta-carotene in a dietary supplement by reversed phase high performance thin layer chromatography using a ChromImage flatbed scanner in the visible mode. Acta Univ Cibiniensis, Ser F Chem 2007;10(1):3–9.
[16] Mathon C, et al. Screening and determination of sibutramine in adulterated herbal slimming supplements by HPTLC-UV densitometry. Food Addit Contam Part A 2014;31(1):15–20.
[17] Morlock GE, Ristivojevic P, Chernetsova ES. Combined multivariate data analysis of high-performance thin-layer chromatography fingerprints and direct analysis in real time mass spectra for profiling of natural products like propolis. J Chromatogr A 2014;1328:104–12.
[18] Morlock G, Ueda Y. New coupling of planar chromatography with direct analysis in real time mass spectrometry. J Chromatogr A 2007;1143(1–2):243–51.
[19] Haribabu K, et al. Quantification of bergenin from *Mallotus philippinensis* by HPTLC-MS and study on different extraction methods. J Planar Chromatogr—Mod TLC 2012;25(5): 445–9.
[20] Bauer GK, Kovar K-A. Direct HPTLC-FTIR online coupling: interaction of acids and bases with the binder in precoated HPTLC plates. J Planar Chromatogr—Mod TLC 1998;11(1):30–3.
[21] Koglin E. Combining TLC(HPTLC) and SERS. GIT Fachz Lab 1994;38(6). p. 627–8, 630–2.
[22] Adhami H-R, et al. Combination of bioautography with HPTLC-MS/NMR: a fast identification of acetylcholinesterase inhibitors from galbanum. Phytochem Anal 2013; 24(4):395–400.
[23] Du J, et al. Application of chemometrics in quality evaluation of medicinal plants. J Med Plants Res 2011;5(17):4001–8.

[24] Wold S. Chemometrics; what do we mean with it, and what do we want from it? Chemom Intell Lab Syst 1995;30(1):109–15.

[25] Tian R-t, Xie P-s, Liu H-p. Evaluation of traditional Chinese herbal medicine: Chaihu (Bupleuri Radix) by both high-performance liquid chromatographic and high-performance thin-layer chromatographic fingerprint and chemometric analysis. J Chromatogr A 2009;1216(11):2150–5.

[26] Gad HA, et al. Application of chemometrics in authentication of herbal medicines: a review. Phytochem Anal 2013;24(1):1–24.

[27] Ogegbo OL, et al. Metabolomics of four TCM herbal products: application of HPTLC analysis. Anal Methods 2012;4(8):2522–7.

[28] Singh M, et al. Development and validation of a stability-indicating HPTLC method for analysis of arjunolic acid in a herbal formulation. J Planar Chromatogr–Mod TLC 2011; 24(2):172–5.

[29] Cretu GC, Morlock GE. Analysis of anthocyanins in powdered berry extracts by planar chromatography linked with bioassay and mass spectrometry. Food Chem 2014;146: 104–12.

[30] Choma I. Thin-layer chromatography hyphenated with bioassays. J AOAC Int 2013; 96(6):1165–6.

[31] Rumalla CS, et al. Densitometric-HPTLC method development and analysis of anthocyanins from acai (*Euterpe oleracea* Mart.) berries and commercial products. J Planar Chromatogr–Mod TLC 2012;25(5):409–14.

[32] Avula B, et al. Simultaneous determination of alkaloids and flavonoids from aerial parts of *Passiflora* species and dietary supplements using UPLC-UV-MS and HPTLC. Nat Prod Commun 2012;7(9):1177–80.

[33] Abourashed EA. Validation and application of an HPTLC method for the determination of parthenolide in feverfew herbal products. J Planar Chromatogr–Mod TLC 2004;17(5):375–8.

[34] Barnes J, et al. Echinacea species (*Echinacea angustifolia* (DC.) hell., *Echinacea pallida* (Nutt.) Nutt., *Echinacea purpurea* (L.) Moench): a review of their chemistry, pharmacology and clinical properties. J Pharm Pharmacol 2005;57(8):929–54.

[35] Reich E, et al. An AOAC peer-verified method for identification of *Echinacea* species by HPTLC. J Planar Chromatogr–Mod TLC 2002;15(4):244–51.

[36] Rumalla CS, et al. Chemical fingerprint of *Hoodia* species, dietary supplements, and related genera by using HPTLC. J Sep Sci 2008;31(22):3959–64.

[37] Ayinampudi SR, et al. Quantitative analysis of oxyresveratrol in different plant parts of *Morus* species and related genera by HPTLC and HPLC. J Planar Chromatogr–Mod TLC 2011;24(2):125–9.

[38] Boudesocque L, et al. High performance thin layer chromatography-densitometry: a step further for quality control of cranberry extracts. Food Chem 2013;139(1–4):866–71.

[39] Mammen D, Daniel M, Sane RT. Rapid and parallel analysis using HPTLC to detect seasonal and geographical variation in *Aerva lanata* Juss. ex schultes. J Planar Chromatogr–Mod TLC 2011;24(5):388–93.

[40] Khan I, et al. Ten marker compounds-based comparative study of green tea and guava leaf by HPTLC densitometry methods: antioxidant activity profiling. J Sep Sci 2011; 34(7):749–60.

[41] Singh A, Alvi NI. High-performance thin-layer chromatographic quantification of yohimbine in the stem bark of *Pausinystalia yohimbe*. J Planar Chromatogr–Mod TLC 2011;24(3):253–6.

[42] Salmon CNA, et al. Characterisation of cultivars of Jamaican ginger (*Zingiber officinale* Roscoe) by HPTLC and HPLC. Food Chem 2012;131(4):1517–22.

[43] Meier B, Spriano D. Modern HPTLC–a perfect tool for quality control of herbals and their preparations. J AOAC Int 2010;93(5):1399–409.

[44] Kannan R, M.a. SK. A note on unintentional adulterations in ayurvedic herbs. Ethnobot Leafl 2007;11:11–5.

[45] Lederer I, et al. Combination of TLC and HPLC-MS/MS methods. Approach to a rational quality control of Chinese star anise. J Agric Food Chem 2006;54(6):1970–4.

[46] Ankli A, Reich E, Steiner M. Rapid high-performance thin-layer chromatographic method for detection of 5% adulteration of black cohosh with *Cimicifuga foetida, C. heracleifolia, C. dahurica*, or *C. americana*. J AOAC Int 2008;91(6):1257–64.

[47] Dixit S, et al. Surveillance of the quality of turmeric powders from city markets of India on the basis of curcumin content and the presence of extraneous colours. Food Addit Contam Part A 2009;26(9):1227–31.

[48] Bhatia L, Thumar K, Tanna R. Determination of adulteration of anti-obesity drug in Indian herbal medicinal product: development and validation of analytical method. Inven Impact: Pharm Anal Qual Assur 2012;3:177–82.

[49] Nicoletti M. Identification of thiosildenafil in a health supplement. Nat Prod Commun 2011;6(7):1003–4.

[50] Reddy TS, Reddy AS, Devi PS. Quantitative determination of sildenafil citrate in herbal medicinal formulations by high-performance thin-layer chromatography. J Planar Chromatogr–Mod TLC 2006;19(112):427–31.

[51] Abourashed EA, Abdel-Kader MS, Habib A-AM. HPTLC determination of sildenafil in pharmaceutical products and aphrodisiac herbal preparations. J Planar Chromatogr–Mod TLC 2005;18(105):372–6.

[52] Upton RT. Use of high-performance thin layer chromatography by the American herbal pharmacopoeia. J AOAC Int 2010;93(5):1349–54.

CHAPTER

19

Analysis of Plant Material

Monika Waksmundzka-Hajnos, Mirosław A. Hawryl, Łukasz Cieśla

Department of Inorganic Chemistry, Medical University of Lublin, Lublin, Poland

19.1 INTRODUCTION

The classic (horizontal or vertical) development of thin-layer plates (from one edge of the plate to the opposite edge) is generally applied for the separation of plant extracts. But given that plant extracts are often multicomponent mixtures their separation by classical single development strategies is a difficult task. Other special techniques of development are often more appropriate.

All techniques for the development of thin-layer plates can be categorized according to the following approaches [1]: one-dimensional development; forced-flow development; and two-dimensional development. One-dimensional development is performed as a single development in one direction, horizontally in opposite directions simultaneously, or in a circular or anticircular direction by locating the mobile phase application point at the center or as a peripheral circle on the layer. One-dimensional development is also performed as a series of (multiple) developments over the same distance or in a stepwise manner using increasing or decreasing distances with either the same mobile phase or with a solvent gradient. The last method can be carried out automatically in an automated multiple development (AMD) chamber.

Development with a forced flow of mobile phase is achieved using overpressured layer chromatography (OPLC), high-pressure planar liquid chromatography, or rotation planar chromatography (RPC). Thin-layer electrochromatography also belongs to the forced-flow techniques but requires a specialized apparatus not currently available in most laboratories.

Instrumental Thin-Layer Chromatography
http://dx.doi.org/10.1016/B978-0-12-417223-4.00019-4

Two-dimensional development can be performed in a single-solvent or two-solvent system utilizing the same plate. Alternatively, bilayer plates can be used with two different adsorbent layers. The method separation in 1st dimension, chemical reaction, separation in 2nd dimension (also known as SRS) is also used for the separation of complex mixtures.

19.2 SINGLE DEVELOPMENT

Most thin-layer chromatograms are realized in one dimension, and even today, it is difficult to obtain quantitative results from plates developed in more than one dimension.

In most cases, a single development is all that is necessary to obtain a suitable separation if the mixture is not too complex. Linear development can be performed in classical glass chambers ("N" chambers also known as "tanks"). In chambers without a solvent-saturated filter paper lining, it can take several hours to attain chamber saturation.

One-dimensional linear development is easily performed in conventional chambers or horizontal chambers of all types. The mobile phase in a reservoir contacts the adsorbent layer and migration of the solvent front occurs driven by capillary forces. Development of the chromatogram is stopped when the mobile phase front reaches a desired position. In the isocratic mode plates of different sizes are used to separate plant extracts (typically 5×5 cm, 10×10 cm, or 10×20 cm plates and migration distances of 4, 9, or 18 cm, respectively). The mobile phase can be supplied to the layer simultaneously from opposite edges (Horizontal Developing Chamber from Camag or Horizontal DS Chamber from Chromdes) doubling the number of plant extracts separated in comparison with classical edge-to-edge development [2].

19.2.1 Applications Using Linear Development

There are many examples of the use of one-dimensional development for the separation of secondary metabolites in plant extracts using instrumental thin-layer chromatography (TLC). Single development techniques on various layers were used for the separation of phenolic acids, a group of abundant compounds found in plant materials. They are important phytoalexins and possess antioxidant activity, as well as exhibiting a wide range of pharmacological activities. Reversed-phase RP-18 layers developed with aqueous mobile phases containing acetic acid as an ionization-suppressing agent were used for the separation of phenolic acids from *Lycopus europaeus* L. [3], while normal-phase solvent systems on NH_2-silica or diol layers developed with nonaqueous mobile phases are also suitable [4]. Silica layers and nonaqueous mobile phases were used for separation of extracts from *Sambucus nigra* flos [5].

Like phenolic acids, flavonoids are secondary plant metabolites with a polyphenolic structure. They can be classified into six groups: (1) flavones (luteonin, apigenin, tangeritin); (2) flavonols (quercetin, kaemferol, myricetin, isorhamnetin, pachypodol); (3) flavanones (hesteretin, naringenin, eriodictyol); (4) flavan-3-ols: catechins and epicatechins; (5) isoflavones (genistein, daidzein, glycitein); and (6) anthocyanidins (cyanidin, delphinidin, malvidin, pelargonidin, peonidin, petunidin). To date, about 6000 flavonoids have been isolated from plants, and many are common in higher plants. Some possess antioxidant and anticancer properties. For the separation of aglycones nonaqueous mobile phases containing formic or acetic acid are often used with silica gel layers. For the separation of glycosides more polar mobile phases (e.g., mixtures of alcohols, chloroform with addition of water) are frequently reported. For example, the flavonoid fraction of *Calendulae* flos was separated in such systems [6].

Coumarins with many different structures (simple coumarins, furanocoumarins and pyranocoumarins) are found in plant materials. They are an important class of pharmacological agents with a range of physiological activities including anticancer, antioxidant, anti-inflammation, anti-HIV, anticoagulant, antibacterial, analgesic, etc. Coumarins, also belonging to polyphenols, are often separated by one-dimensional TLC. Silica gel layers are the first choice for the separation of coumarins, for example, as demonstrated by the separation of furanocoumarins from *Peucedanum* species [7,8]. However, the use of other layers such as Florisil [9], polyamide, and alumina [10] was also reported. Solvents used in normal-phase separations of pyrano- and furanocoumarins consist of weak polar binary mobile phases containing a hydrocarbon or aromatic solvent with diethyl ether or ethyl acetate [11,12]. For simple coumarins a medium polar mobile phase with dichloromethane as the strength-adjusting solvent is frequently used. Alkyl-bonded silica layers [8,13] with aqueous organic mixtures have also been reported for the separation of coumarins.

Lignans, a major class of phytoestrogens, are dimeric phenylpropanoids with a wide range of pharmacological activities (antiviral, anticancer, anti-inflammatory, antimicrobial, antioxidant, immunosuppressive, hepatoprotective, etc.). They are present in flax seed, sesame seed, oils, cereals, soybeans, tea, and other plant materials. Silica gel layers, often fluorophore impregnated (F_{254}), are commonly used for the separation of lignans with various multicomponent mobile phases of different composition reflecting the wide range of polarity of these compounds [14]. Often an acidic additive like formic acid is required. As examples, lignans from *Phyllanthus amarus* were separated in the above system [15] and lignans from *Podophyllum* sp. in reversed-phase systems [16]. Complex mixtures of lignins usually require multidimensional (MD) techniques for adequate separation.

A "technical" problem for the separation of volatile terpenes is the possibility of their evaporation at room temperature. At low temperatures (−15 °C) separations of saturated mono- and sesquiterpenes with one to three double bonds are possible. Usually, silica gel layers and *n*-hexane are used for the separation of hydrocarbons. For oxygenated hydrocarbons mixtures of n-hexane with a medium polar solvent (ethyl acetate) are required. Essential oils from plants of the *Lamiaceae* family were separated using low-temperature TLC (see Figure 19.1) [17]. The separation of *cis-trans* and positional isomers has been achieved on silver-ion impregnated silica gel [18]. The separation occurs according to the degree of unsaturation of the compound. *Cis*-olefinic structures, complex more tightly with silver ions than do the *trans*-isomers and exhibit lower R_F values. Such systems were used for the separation of essential oil of *Cyperus rotundus* [19].

Diterpenes are a large and diversified group of compounds suitable for the systematic characterization of plant and marine products. For instance, cembranes and prenylsesquiterpenes are important chemosystematic distinctive features for different populations of *Dictyota dichotoma*—a cosmopolitan species of marine brown algae [20]. Diterpenes from the kaurene group are characteristic for the *Aspilia* genus of the *Asteraceae* family. Ginkgolides are a group of compounds occurring only in *Ginkgo biloba* (*Ginkgoaceae*) [21]. Taxoids are found only in plants of the *Taxus* genus [22]. As a consequence of the diversity of structures for diterpenes many different separation systems have been employed in

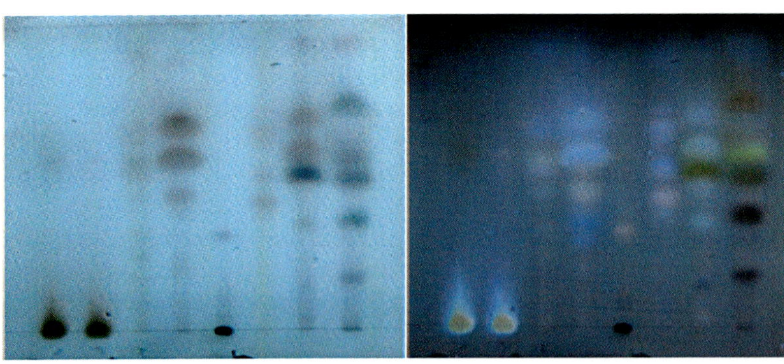

FIGURE 19.1 Low-temperature TLC of essential oils of Lamiaceae species (from left to right): Thymus vulgaris, Rosmarinus officinalis, Mentha pipperita, Lavandula officinalis, Salvia officinalis, Melissa officinalis, Majorana hortensis, Ocimum basilicum on silica gel with n-heptane–ethyl acetate (7:4) as mobile phase and UV detection at 254 nm (left) and 366 nm (right). TLC, thin-layer chromatography.

their qualitative and quantitative analysis. An important application is the quality control of known plant materials (e.g., triptolide analysis for plant material from *Tripterygium* species [23]) or fingerprint analysis (e.g., the identification of plants in chemotaxonomic investigations). Silica gel layers with multicomponent mobile phases of medium polar together with various derivatization methods are commonly used [24].

Most triterpenoids occur in plant material as multicomponent mixtures of glycosidic forms. Sugars can be substituted in one, two, or three chains. The number of sugars can be as high as 8–10 and the multiplicity of possible combinations results in high levels of complexity (one plant may contain 30–40 different triterpenoids). The number of sugar residues determines the polarity of these compounds, which can cover a wide range. Thus, their analysis is a challenge and a selection of separation system is crucial. Normal-phase systems with multicomponent mobile phases, often containing a small amount of water, were used for the separation of polar saponins from *Albizja* species [25], *Trifolium repens* [26] and *Arnica* sp. [27]. Reversed-phase layers with aqueous mobile phases were used for the separation of *Arnica* sp. [27] and *Panax* sp. [28].

A key structural feature of carotenoids, the long conjugated double-bond chromophore, is a major factor that determines a number of practical aspects of their separation, i.e., the light absorption properties and hence color that enables easy detection, the susceptibility to oxidative breakdown, and the strong affinity of some carotenoids for basic inorganic adsorbents. Carotenoids generally are less stable than most other isoprenoids, and are sensitive to oxygen, light, heat, acids, and, in some cases, alkali. Stringent precautions must be taken to minimize the possibility of losses of material or structural changes [29,30]. The risk of degradation is enhanced when the carotenoids are adsorbed on an active surface. However, successive separations on silica gel, alumina, and then magnesium oxide will separate most carotenoid mixtures. The most widely used mobile phases for the separation of carotenoids on silica gel are mixtures of hexane or petroleum ether with a solvent of higher polarity, usually diethyl ether or acetone. For example, extracts from green leaves after saponification (to destroy chlorophyll) can be fractionated on silica gel layers using a polar solvent to move the carotenoids close to the solvent front. Rechromatography on a magnesium oxide-Kiselguhr layer provides separation of the individual compounds [31].

Plants often contain phytosterols in various conjugated forms, which enhance their structural diversity even more. The most frequent are steryl fatty acid esters, phenylpropanoate steryl esters, steryl glycosides, and acylated steryl glycosides. Silica gel layers developed with *n*-hexane and diethyl ether containing various portions of methanol is a useful general mobile phase [32]. Also plant materials after fractionation into several groups were separated by TLC for isolation or determination of individual steroids. In this case, silica gel layers developed with more polar solvents containing methanol, acetone, and chloroform as diluents are used. Sometimes the addition of water to the mobile phase is reported. For example, cardenolides present in bulbs of *Scilla maderensis* were separated by silica TLC developed with ethyl acetate–methanol–water or chloroform–methanol–water [33]. Luyt et al. [34] used TLC on silica gel, developed with ethyl acetate–methanol–water for the qualitative comparison of bufadienolides in natural and tissue culture-derived plants of *Drimia robusta*.

Iridoids are a large group of secondary metabolites useful as chemotaxonomic markers for numerous botanical families. Silica gel is the most efficient adsorbent for their separation with octadecylsiloxane-bonded silica layers used occasionally. Typical mobile phases contain mixtures of chloroform or methylene chloride with methanol and/or water for the separation of iridoid glycosides. In the case of *Gentiana* iridoids normal- and reversed-phase systems are commonly used [35,36]. For determination of *Valeriana* species iridoids were separated on silica gel layers with nonaqueous mobile phases [37].

The alkaloids are a large group of compounds, categorized according to their structure and/or source (plant material). Because alkaloids contain basic nitrogen-ring atoms they are often challenging to separate. The basic nitrogen groups interact strongly with silica gel layers or the residual silanol groups on silica-based chemically bonded layers. Specific mobile phase additives are generally used for their separation with good peak shapes.

Indole alkaloids contain an indole ring system in their structure. They are found in genera of *Apocynaceae*, *Loganiaceae* and *Rubiaceae*. They are important from the pharmacological, ecological, and systematic viewpoints. Usually silica gel layers developed with polar mobile phases containing medium polar diluents (usually chloroform) and polar modifier are used for their separation. For example, harmine and harmaline from *Banistereopsis caapi* stem are separated on silica gel using a chloroform–methanol–water mobile phase [38]. Ergot alkaloids can be separated on silica gel with a mobile phase containing toluene–chloroform–ethanol [39]. Basic additives like diethylamine, other amines, or aqueous ammonia are often added to the mobile phase to suppress alkaloid dissociation. Mobile phases containing a mixture of ethanol and ammonia give good separations of *Psilocybe* alkaloids [39] and *Strychnos* alkaloids [39]. In the case of reversed-phase systems a buffered aqueous mobile phase incorporating a basic additive or ion-pair reagent is frequently used. Ion-pair additives were used in separation of *Strychnos* alkaloids [39].

Isoquinoline- or tetrahydroisoquinoline-containing alkaloids can be categorized into one of eight types. Isoquinoline alkaloids are found mainly in the following plants: *Ranunculales, Papaverales, Geraniales, Rutales, Plumboginales, Myrtiflorae,* and *Rosales*. Many TLC systems have been proposed for their separation based on silica gel or alumina layers [40] and polar chemically bonded stationary phases, such as aminopropyl-silica gel layers [41] and cyanonopropyl-silica gel layers [42], with nonaqueous mobile phases. Usually silica gel layers and polar mobile phases are used for these separations—polar modifiers (methanol, ethanol) with medium polar diluents (chloroform or dichloromethane) [43–45], or occasionally benzene or toluene [46], are commonly used. Addition of diethylamine or aqueous ammonia results in a marked improvement of spot shapes and symmetry. The addition of diethylamine to nonaqueous mobile phases usually results in increased system efficiency and improved peak symmetry [42,47]. Both effects are caused by blocking of surface silanol groups and suppression of the ionization of the alkaloids. Silica gel high-performance thin-layer chromatography (HPTLC) plates developed with multicomponent mobile phases containing aqueous ammonia were used for the separation of extracts from *Papaver somniferum* [48]. Similar systems were used for the separation of *Chelidonium majus* alkaloids [47,49]. The use of reversed-phase systems for the separation of alkaloids with buffered aqueous mobile phases containing amines or ion-pair reagent is less frequently reported [50].

Tropane alkaloids, containing the heterocyclic base tropane, occur mainly in *Solanaceae, Erythroxylaceae, Convolvulaceae* plant families. These compounds have been separated on aluminum oxide and silica gel layers, principally, or on reversed-phase layers. Mobile phase types are similar to those used for the separation of isoquinoline alkaloids, consisting of medium polar diluents and a polar modifier, often with a basic additive. As examples, the separation of tropane alkaloids from *Datura* with a mobile phase containing aqueous ammonia [51,52] and desitometric determination without derivatization should be mentioned [53]. Alkaloids from *Atropa belladonna* [51] and the *Solanaceae* family [54,55] can be separated using similar chromatographic conditions.

There are also many examples for separation of alkaloids from other groups, such as quinoline, pyrrolidine, pyrrolizidine, pyridine, and piperidine derivatives, and quinolizidine alkaloids containing xanthine or imidazole derivatives. Similar systems are used for their separation, sometimes containing an acid. Figure 19.2, illustrates the normal-phase separation of alkaloids from *Cinchona* with diethylamine added to the mobile phase as a basic additive [56].

More than 700 different polyacetylenes and related compounds have been isolated from higher plants. The majority have been isolated from

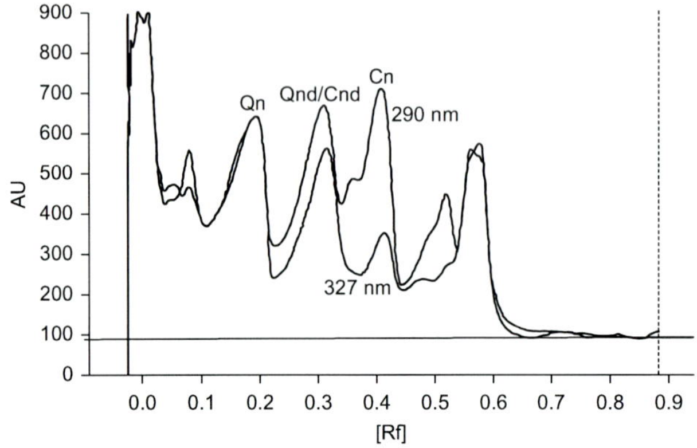

FIGURE 19.2 Separation of alkaloids in *Chinae* extract on silica gel with toluene–chloroform–diethyl ether–diethylamine (40:15:35:10 v/v) as mobile phase and a development distance of 9 cm. Densitograms recorded at $\lambda = 327$ and 290 nm. Qn, quinine; Qnd, quinidine; Cnd, cinchonidine; Cn, cinchonine.

the *Apiaceae*, *Araliaceae*, and *Asteraceae* plant families [57]. The limited distribution of polyacetylenes in the plant kingdom makes them important as chemotaxonomic markers. The polyacetylenes isolated from *Apiaceae* and *Araliaceae* are characterized by aliphatic acetylenes, whereas the structural diversity of polyacetylenes from the *Asteraceae* family is considerable. The neurotoxic effects of some polyacetylenes have long been known (for example, cicutoxin and related polyacetylenic derivatives, falcarinol). Some of them exhibit anti-inflammatory and antiplatelet-aggregatory or antibacterial, antimycobacterial, antifungal, and antiviral effects. The separation of polyacetylenes by TLC is usually performed on silica gel layers with a suitable mobile phase containing *n*-hexane or petroleum ether. Modifiers are of medium polarity such as dichloromethane, diisopropyl ether or ethyl acetate. Derivatizing reagents are usually employed for visualization.

TLC is widely used in the investigation of quinone derivatives. Their determination is simple due to their favorable chromophore groups that give characteristic colors for each compound (yellow to red). Quinone derivatives have a cholagogic action and their bitter taste stimulates gastric mucosa to hypersecretion improving digestion [58]. Benzoquinones are found in mushrooms, naphthoquinones in *Drosera* species, and anthraquinones in the *Polygonaceae* family (*Rheum* sp., *Frangulae* sp.). Normal-phase systems are usually used for their separation on silica gel layers with medium polar mobile phases as described in [59] for *Aloe* anthraquinones and in [60] for *Rheum* anthraquinones. Figure 19.3 illustrates the separation of anthraquinone derivatives from *Aloe* [61].

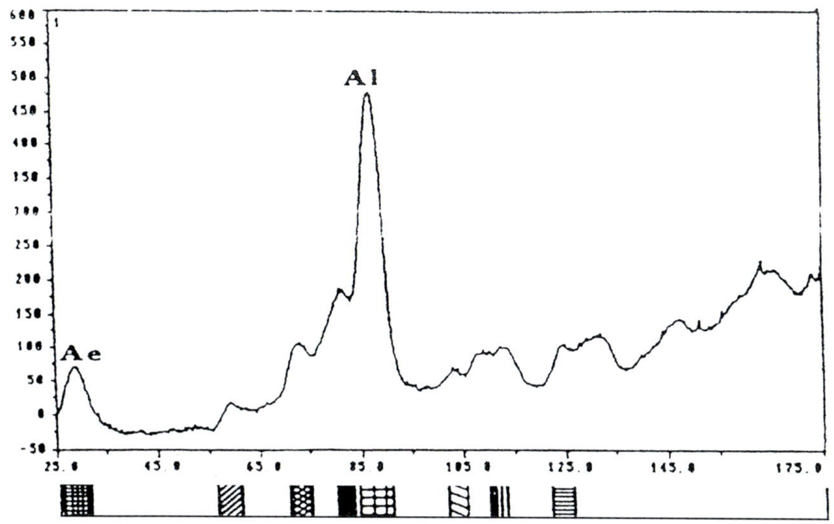

FIGURE 19.3 Micropreparative TLC of *Aloë*. Effective separation of aloine (Al) and aloeemodine (Ae) on a silica gel plate with preconcentration zone. Mobile phase: ethyl acetate–methanol–water (77:13:10 v/v).

19.2.2 Applications Using Circular Development

In circular development, the migration of the zones occurs radially from the plate center to the sides. The R_F values in circular development are higher than their linear counterparts, with the exception of $R_F = 0$ and $R_F = 1$, where they must, of course, be the same. The increase in R_F values is greater in the lower range [62]. This favors the separation of components with small R_F values without disadvantaging the separation of components with higher R_F values.

If a plate is developed from the outside toward the center, this is called centripetal (anticircular) development. In a centripetal development, the solvent is applied in a circle from the plate sides and flows toward the center. The sample is applied on an outer circle of large diameter near the plate sides. This allows the simultaneous separation of a larger number of samples than linear development. Forty-eight samples applied as spots on a 10 × 10 cm HPTLC plate can be separated simultaneously within about 4 min [63]. In linear and circular development, the mobile phase flow decreases quadratically with time. In centripetal development the solvent front velocity is constant since the unwetted area of the plate decreases quadratically with the time.

There are only a few examples of the use of circular chromatography in the separation of plant secondary metabolites. It was used for the separation of quaternary isoquinoline alkaloids from *Jatrorrhiza palmata* on silica gel layers developed with isopropanol-dioxane-1 M ammonium carbonate [64]. Circular TLC was also used for the separation of coumarins in extracts from *Peucedanum palustre* [65].

19.3 SPECIAL DEVELOPMENT TECHNIQUES

19.3.1 Applications Using AMD

AMD is based on the same principle as manual multiple development. An AMD chromatogram is the result of several (usually at least 10) automated separation steps. In most cases the number of developments varies between 10 and 40. The migration distance of the solvent front in each run is progressively longer by a constant increment. Typical increments are from 1 to 3 mm. The migration distance of the solvent front is programed via computer control of the instrument. After each solvent run is completed, the developing tank is drained, and the layer is dried by vacuum. Before the next run begins, solvent vapor is pumped into the chamber to precondition the layer. As the early solvent mixtures will be most polar (normal-phase development), the sample components of interest will be focused or sharpened into tight bands.

There are a number of examples using AMD for the separation of secondary plant metabolites: natural phenolic compounds [66], phenolic compounds, and oil components for vanilla, cinnamon, and other spices [67], vanilla extracts and vanilla-flavored foods [68], vanilla bean extract [69], terpenoids of *Boswellia serrata* [70], pollen carotenoids [71], extracts from *Artemisia* sp. [72], ginsenosides from *Panax quinquefolium* [73], cannabinoids from hemp [74], alkaloid standards and Opium extracts [75], glycoalkaloids—calystegines and tropanes [76] and anthraquinones from *Aloe* [77].

19.3.2 Manual Multiple Development

Unidimensional multiple development (UMD) procedure depends on the repeated development of the layer in the same direction with a mobile phase of constant composition [78]. By UMD, differences in R_F values are increased for compounds in the lower R_F range and reduced for those in the higher range. Consequently, mobile phases of low solvent strength should be used. Other aspects of UMD are described by Poole et al. [79]. As seen in Figure 19.4 [80]. each development step induces spot narrowing with an increase in resolution, because the lower edge of the spot is contacted earlier by the solvent front then the upper edge. The effect of UMD [81], [82] is presented in Figures 19.5 and 19.6. The increase of resolution of neighboring bands after consecutive

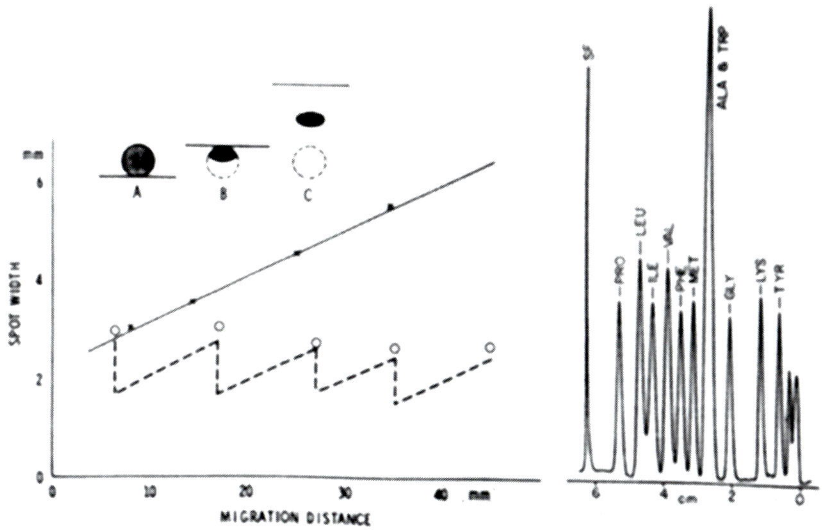

FIGURE 19.4 An illustration of the zone refocusing mechanism (left) and its application to the separation of a mixture of phenylthiohydantoin-amino acids (right). The broken line on the left-hand side of the figure represents the change in spot size due to the expansion and contraction stages in multiple development and the solid line depicts the expected zone width for a zone migrating the same distance in a single development. For the abbreviared names of the analyzed compounds please refer to Ref. [80].

development steps is clearly seen for alkaloids of *Chelidonium maius* [81] and of *Fumaria officinalis* [82].

Incremental multiple development (IMD) is a UMD technique in which the solvent front migration distance is increased incrementally in 10 or 20 mm steps with evaporation of the mobile phase from the layer after each step. The mobile phase composition is the same for each step. When the increment in development distance is constant, the R_F value after n developments can be predicted by the following equation:

$$R_{Fn}^{\text{lin}} = 1 - (1 - R_F)\frac{1 - (1 - R_F)^n}{n \cdot R_F} \tag{19.1}$$

Some furanocoumarins were separated by this method [83]. The separation of furanocoumarins in an extract from *Heracleum sphondylium* fruit after a single development and by IMD, with the use of the same mobile phase is shown in Figure 19.7. The resolution is markedly increased by multiple development.

Bivariant multiple development (BMD) involves the step-by-step changing of both the development distance and the mobile phase composition. The development distance is increased and the solvent strength of the mobile phase is reduced for consecutive steps. The improved version of

An advantage of planar separation methods compared with columns is that each TLC separation is performed in a previously unused stationary phase. Therefore, it is possible to fractionate or separate complex samples characterized by a heavy matrix loading. Miniaturization of TLC methods, micro-TLC, facilitates a reduction in solvent consumption using plates of a smaller size, typically less than 7 × 7 cm [137,138]. The separation time is shorter compared with conventional TLC methods using plates of 10–20 cm. There are a few examples in the literature using this approach for the separation of plant extracts [139–144].

19.4 COUPLING PLANAR CHROMATOGRAPHY WITH COLUMN CHROMATOGRAPHY

Another means of realizing a MD separation is by combining two complementary separation techniques using different separation modes. In such multimodal separations, planar chromatography is generally used as the second dimension. Some possible combinations are: different extraction methods (e.g., thermal and solid phase) with TLC; gas chromatography (GC) and TLC; supercritical fluid chromatography and TLC; HPLC and TLC; OPLC and TLC; capillary electrophoresis and TLC; and different countercurrent chromatographic methods and TLC.

Current interest is, however, mainly in the coupling of HPLC and TLC, to which considerable attention has been devoted for difficult separation problems. Since Boshoff et al. [145] first described the direct coupling of HPLC and TLC, several papers [146–149] have described alternative interfaces for this purpose. The off-line coupling of reversed-phase column chromatography and normal-phase TLC was described for the analysis of coumarins, Figure 20.12 [150] and flavonoids [151].

19.5 FORCED-FLOW DEVELOPMENT

19.5.1 Applications Using OPLC

OPLC, also known as overpressurized liquid chromatography, uses a constant volume pump to force the mobile phase through a sorbent layer sandwiched between a glass plate and a flexible polymer membrane forced into close contact with the layer by application of hydraulic pressure to the opposite face of the membrane [152]. With this arrangement the plate length and solvent migration velocity for a separation can be optimized independently [149,153]. The two basic parts of a forced-flow TLC instrument are the upper and lower support blocks, which form a sandwich with the thin-layer plate. The bottom support block may

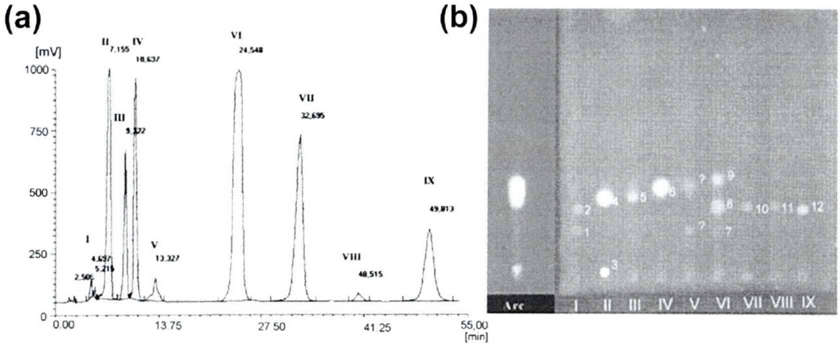

FIGURE 19.12 (a) Chromatogram obtained using methanol-water (6:4 v/v). Roman numerals indicate the fraction numbers collected during the HPLC separation. (b) The Camag Reprostar 3 photography of the TLC plate with applied fractions collected during the HPLC separation from *Archangelica officinalis*. Extract scanned in black and white mode. Number of substances: 1 - Xanthotoxol, 2 - Umbelliferone, 3 - Archangelicin, 4 - Xanthotoxin, 5 - Isopimpinellin, 6 - Berapten, 7 - Pimpinellin, 8 - Imperatorin, 9 - Phellopterin, 10 - Osthol, 11 - Umbelliprenin, 12 - Isoimperatorin, ?: unidentified substances. Arc: *Archangelica officinalis* extract. Roman numbers of fraction as in (a) [150]. TLC, Thin-layer chromatography; HPLC, high-performance liquid chromatography.

contain a thermostatting jacket. The upper support block contains the membrane, the mobile phase inlet, pressure inlet, and pressure gauge. The sorbent layer on a glass plate is positioned facing the membrane. Clamps hold the upper and bottom support blocks together.

For a single linear development in HPTLC the spot capacity is limited to about 20–25 at most. A spot capacity approaching 80 can be achieved using OPLC with commercial instruments. Further improvements in the spot capacity are only possible by using longer plates, two-dimensional techniques, or through implementing AMD in an overpressured system.

OPLC techniques are often used for phytochemical analysis, especially in for the separation of closely related compounds, for example, the separation of coumarin standards [154] and coumarins from the genus *Angelica* [105] and from *P. palustre* [65] using silica gel layers and nonaqueous mobile phases of medium polarity. Coumarins, psoralens and polymethoxylated flavones were also identified by this technique in citrus essential oils [155]. In addition, flavonoid glycosides were analyzed in *Betula folium* using silica gel with a multicomponent mobile phase [156] as well as anthocyanins in natural products [157]. Other examples include the separation of essential oils extracted from *Origanum onites* [158] and *Thymus vulgaris* [159], glycoside-bound components of fruits [160], iridoid glycosides from *Veronica officinalis* [161], and ginsenosides from the roots of *P. quinquefolium* [162] all on silica gel layers with nonaqueous mobile phases of suitable solvent strength. OPLC has been widely used to separate standard mixtures of alkaloid [163,164], quinolizidine alkaloids

[165], ergot alkaloids [166], quaternary alkaloids in various plant materials [167], and alkaloids from *Catharanthus roseus* [168]. The determination of anthraquinones (sennosides A and B) in *folium Sennae* was also reported [169].

19.5.2 Applications Using RPC

RPC, also known as centrifugal TLC, is a forced-flow technique mainly used for preparative separations [170]. The substances are separated on a rotating disk by a solvent mixture continuously fed to the layer. The homemade layers, typically silica gel, are spun at about 800 rpm and can handle flow rates up to about 10 mL/min. In most cases the separation is observed by illuminating the disk with short-wave UV light, so that the desired fractions can be collected. Extracts and crude samples from 0.5 to 2 g are commonly used for fractionation. RPC was used for the isolation of phenolic compounds, such as phenolic acids from *C. maius* [171], C-prenylated derivatives of 2,4-dihydroxybenzoic acid from *Louchocarpus nicon* roots [172], phaseolidin isoflavan [173], iridoids and flavones from *Gentiana depressa* [174] and iridoid from *G. depressa* [175], as well as flavonoids and xanthones form *Gentiana lactea* [176]. In these examples polar mobile phases containing mixtures of chloroform and methanol or ethanol were generally used as well as water or acetic acid as additives. Medium polar mobile phases were used for the isolation of sesquiterpenes from stem bark of *Cleistofolis glanca* [177], from *Eriostemon fitzgeraldii* [178], limonoids from *Fagaropsis glabra* [179], and neo-clerodane diterpenoids from *Teucrium yemense* [180]. Indole and heteroyohimbine alkaloids were isolated from *Mitragyna speciosa* [181], isoquinoline alkaloids from *Papaver* and *Fumaria* species [182], diterpenoid alkaloids from *Aconitum septentrionale* [183] and cyclopeptide alkaloids from *Zizyphus lotus* [184] by RPC.

19.6 CHEMICAL FINGERPRINTING OF PLANT MATERIALS

The principal difference between plant and synthetic drugs (usually a single substance or simple mixture) is the greater complexity of herbal medicines. Consequently, their quality control is becoming an increasingly important issue. For example, according to the European legislation (directive 2004/24/EC concerning "traditional herbal medicinal products") a quality and purity of these products should be strictly controlled [185]. Today, plant chemosystematics is an interdisciplinary field in which chemical composition is used to determine plant taxa [186].

Earlier, it was common practice to select one or more compounds as either active or "markers" for identification and quality assessment of plant materials [186]. Since many plants contain the same compounds, such an approach is neither able to confirm the identity of a specific plant, nor suitable as a quality criterion [187]. As a consequence, it is necessary to develop a quality assessment system that adequately accommodates the complex characteristics of plants. Separation techniques are good tools to build specific patterns for the recognition of multiple compounds in herbs that can be utilized as a chemical fingerprint. The full pattern of compounds can then be evaluated to determine not only the absence or presence of desired markers but the complete set of ratios of all detectable analytes.

Since 1991 WHO accepts chromatographic fingerprint as identification and quality evaluation technique for medicinal herbs [185]. A chromatographic fingerprint is a chromatogram that represents the chemical characteristics of a herb. Identity and quality can be derived from fingerprint chromatograms. These fingerprints can be defined as "a chromatographic pattern of a herbal extract showing some common pharmacologically active and/or chemical characteristic compounds".

The attractive features of HPTLC for chemical fingerprinting result from the following: HPTLC is the only chromatographic technique that facilitates the presentation of results as a picture-like image; HPTLC has the ability to allow many samples to be compared side by side; all samples on the plate are analyzed under the same conditions (impossible in case of sequential separation techniques using columns); the possibility to obtain rapid results based on simultaneous; sequential detection strategies with or without derivatization are possible since the sample is immobilized on the stationary phase when the separation is concluded; and HPTLC is more tolerant of matrix components minimizing sample preparation requirements.

Feng et al. [188] demonstrated that HPTLC, HPLC, and high-performance capillary electrophoresis (HPCE) were all successful for fingerprinting of *Pueraria* radix. The fingerprints obtained by all three techniques were effectively used to distinguish between authentic and conuterfeit plant materials based on similarity with authentic standard materials.

19.6.1 Application to Compounds of Wide Polarity

For compounds of wide polarity isocratic separation methods are unlikely to provide optimum results. One solution utilizes stepwise gradient elution, which in the case of TLC is achieved by introducing a stronger solvent to the solvent reservoir in discrete batches or continuously. Gradient elution markedly improves the separation of spots owing to stronger displacement effects under conditions of numerous adsorption–desorption processes. Because the lower edge of the spot is overtaken earlier than the upper edge by the solvent front together with an increase in solvent strength of the mobile phase, the spot becomes narrower than its starting profile. Figure 19.13 illustrates a separation of

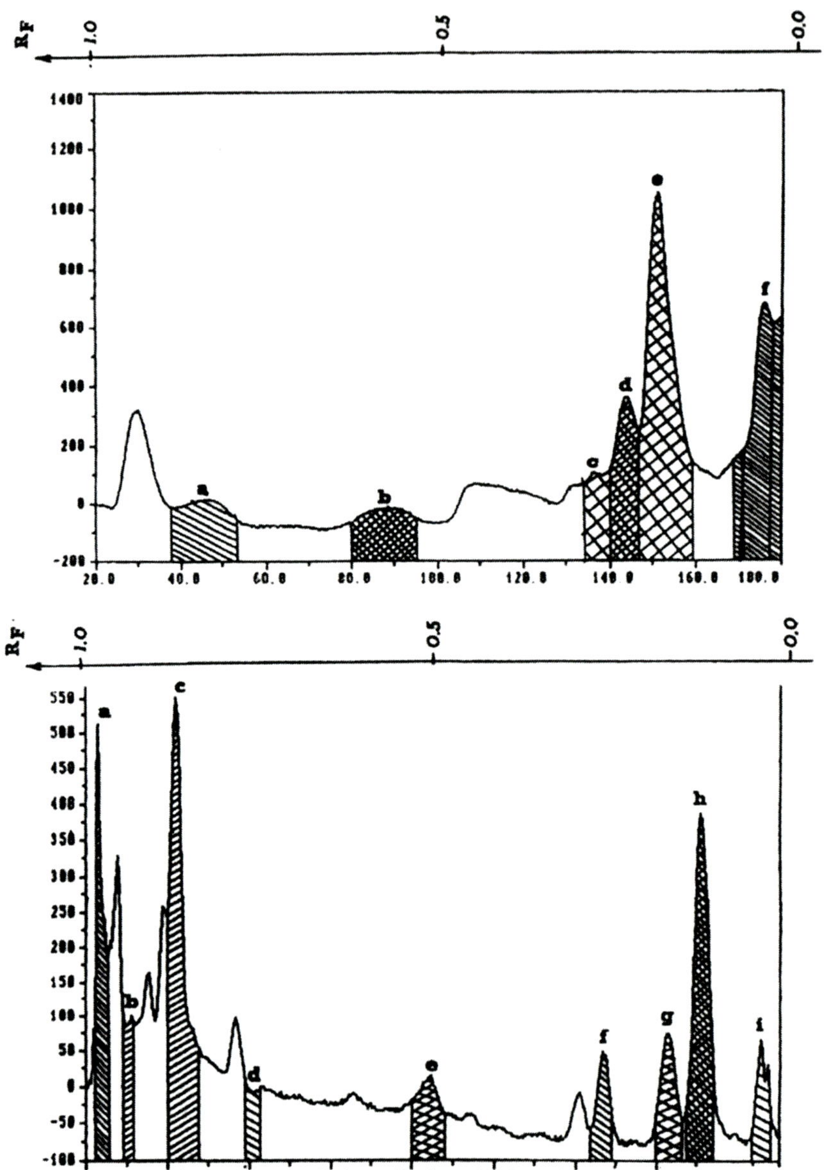

FIGURE 19.13 *Matricaria chamomilla* extracts separated on silica gel plates isocratically and with a five-step gradient using mixtures of chloroform-ethyl acetate.

Chamomilla recutita extract with a fixed solvent composition (A) and by a five-step gradient of ethyl acetate in chloroform (B). In the case of the stepwise development the zones are sharper, more numerous, and better spaced along the chromatogram [189]. Similar results were obtained for the separation of extract from *Erigeron canadensis* [190]. The use of stepwise gradient development enabled separation of eight cardenolide glycosides [191] and the separation of lanatosides A and C from *Digitalis lanata* [192].

Alternatively, mobile phase gradient separations in TLC can be performed by multiple gradient development (MGD). In this case, each stage of rechromatography is performed with a mobile phase of a different composition. The development distances for consecutive steps are usually identical. The stepwise separation of substances into groups of different polarity can be achieved using a mobile phase gradient of increasing solvent strength. This method was used for the separation of anthraquinone glycosides and aglycones in extracts from *Radix rhei* and from *Cortex Frangulae* on silica gel layers with MGD development using chloroform, ethyl acetate, and methanol in different concentrations [193]. The technique was also used for the separation of phenolic acids from *L. europaeus* on silica and diol layers [3], from *Aquilegia vulgaris* on silica gel layers [194], and from *Vaccinium* sp. on various polar chemically bonded layers [195]. The MGD technique was also used for the separation of flavonoid aglycones from *Vaccinium* species [196] and flavonoid derivatives from *Pseudotsuga menziessi* on diol layers [197].

The method of binary fingerprinting uses a mobile phase of moderate strength for the separation of compounds of intermediate polarity, for example, flavonoid aglycones while leaving polar compounds, such as flavonoid glycosides, immobilized at the position of sample application. Multiple wavelength scanning at this stage provides information about those compounds that are mobile in the selected mobile phase. After drying, the plate can be developed with a mobile phase of higher solvent strength. The weakly retained components move close to the solvent front and form a nonseparated band. Polar compounds (e.g., flavonoid glycosides) are separated in this step and can be detected by any suitable method. As a result, the separation of two fractions on the same plate has been achieved. This method was used to obtain "multiple chromatographic fingerprints" to distinguish between *Radix Puerariae lobatae* and *Radix Puerariae thomsonii* [187].

Binary fingerprinting was used to distinguish *Salvia* species. As the methanol extracts are complex, two different chromatographic systems were used, focusing on resolution of the low, medium, and high polarity compounds. To separate the less polar fraction, a mobile phase consisting of toluene–ethyl acetate–formic acid (60:40:1 v/v/v) was used. For the medium and highly polar substances, a mobile phase of

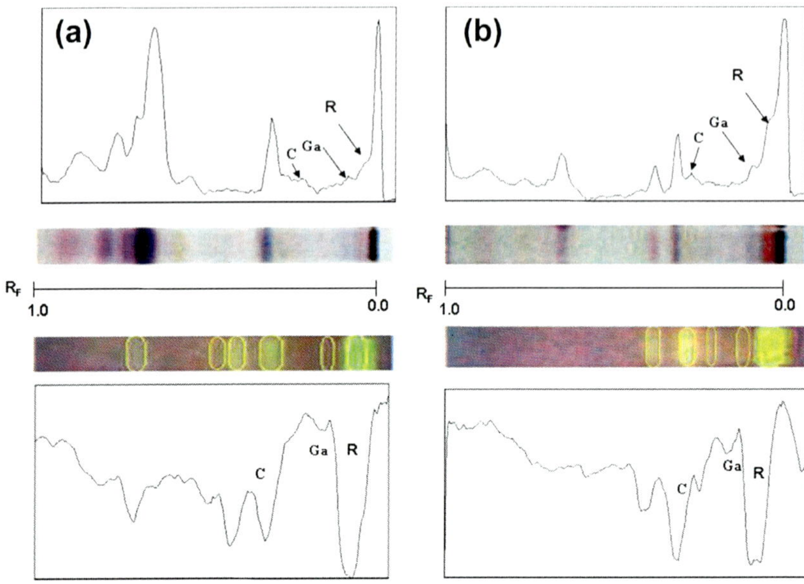

FIGURE 19.14 Comparison of chemical and free radical scavenging fingerprints obtained with solvent system I for extracts from two *Salvia* species: (a) *Salvia lavandulifolia* and (b) *Salvia atropatana*.

ethyl acetate–water–formic acid–acetic acid (100:26:11:11 v/v/v/v) was used. The results are summarized in Figure 19.14. A computer search of the features in the densitograms was implemented for species identification [198].

19.7 EFFECT-DIRECTED ANALYSIS OF PLANT MATERIALS

TLC has become an important tool for the detection and identification of secondary plant metabolites exhibiting biological activity. Such analysis may be considered as an additional dimension to chemical fingerprint analysis of plant material, and can be referred to as biological fingerprinting [199]. A biological fingerprint is used to identify active compounds present in a complex sample, such as plant extracts. Together with a chemical fingerprint, biological profiles deliver additional information about the samples, at the same time indicating which of the analyzed species or fractions contain compounds with potential pharmacological activity.

Biological fingerprinting is an example of effect-directed analysis (EDA) [200,201]. The focus of EDA is the detection of compounds with a specific activity (e.g., free radical scavengers, enzymes' inhibitors, antimicrobials, etc.) while other compounds are treated as excipients or part of the matrix. The abundance of plant species and the large number of possible plant-derived targets has necessitated the development of high-throughput analytical approach to facilitate the handling of many samples in reasonable time. Liquid chromatography (both HPLC and TLC) has become an attractive alternative to spectrophotometric and microplate assays, commonly used in phytochemistry laboratories [202].

Chromatographic techniques outperform spectrophotometric assays, as they are well suited for screening complex samples to indicate fractions or single compounds with biological activity. In case of spectrophotometric methods they can only measure the activity of the whole extract or fractions after the completion of the isolation procedure. In the traditional approach biological screening assays constitute the final step preceded by fractionation, isolation, and structure elucidation procedures. Such an approach is relatively time consuming and not always crowned with success—the identification of a new compound that may be considered as a drug candidate. During the aforementioned steps an active principle may be irretrievably lost. In addition, pharmacologically active metabolites may be present in minor quantities in plant extracts and such compounds may be overlooked by the traditional approach. In EDA the biological screening of the sample is the first step in a procedure aiming at identification of new drug leads (Figure 19.15).

In order to check the activity of fractions or individual compounds using the EDA approach, separation techniques are required. For phytochemical analyses, TLC was often considered a poor cousin of HPLC on account of its lower separation capacity, but in the specific case of biological detection it may be considered a princess of separation methods. The following are just some of the advantages of TLC, making it an optimal tool in screening complex samples for the presence of drug candidates: use of crude plant extracts, without the need for purification procedures; analysis of several samples on the same plate, under the same chromatographic conditions; direct access to separated compounds; numerous derivatizing agents; lack of interference between mobile phase solvents and enzymes used in bioassays, etc. [203]. As demonstrated in Figure 19.16, TLC is the only chromatographic method that can be applied to search for antimicrobial compounds, as there is no possibility to grow bacteria or fungi inside a chromatographic column. In addition, HPLC has severe limitations regarding the analysis of potential enzymes' inhibitors. Enzymes, like other proteins, are sensitive to high concentrations of organic solvents, which induce denaturation. The use of mobile phase additives is also

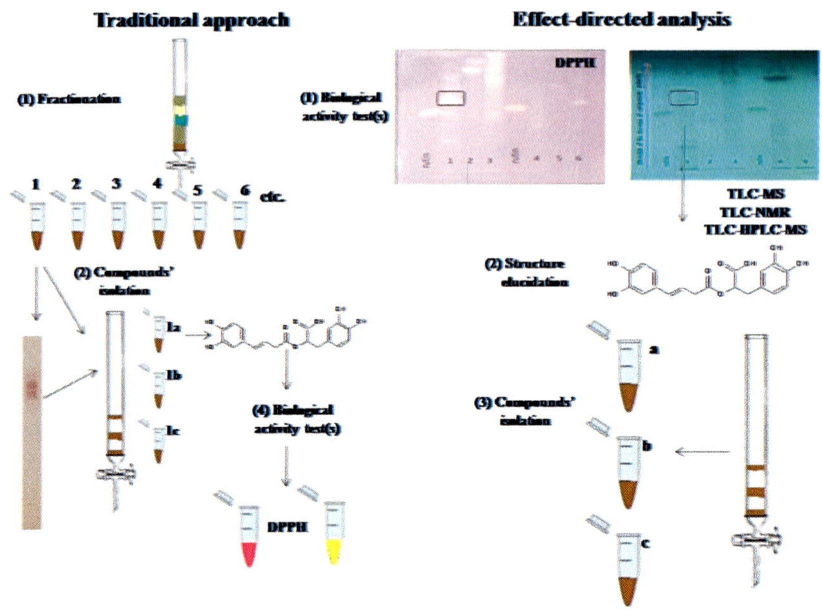

FIGURE 19.15 Traditional approach vs effect-directed analysis in the screening of plant extracts for the presence of biologically active compounds. DPPH, 1,1-diphenyl-2-picrylhydrazyl; TLC, Thin-layer chromatography; HPLC, high-performance liquid chromatography; MS, mass spectrometry; NMR, nuclear magnetic resonance.

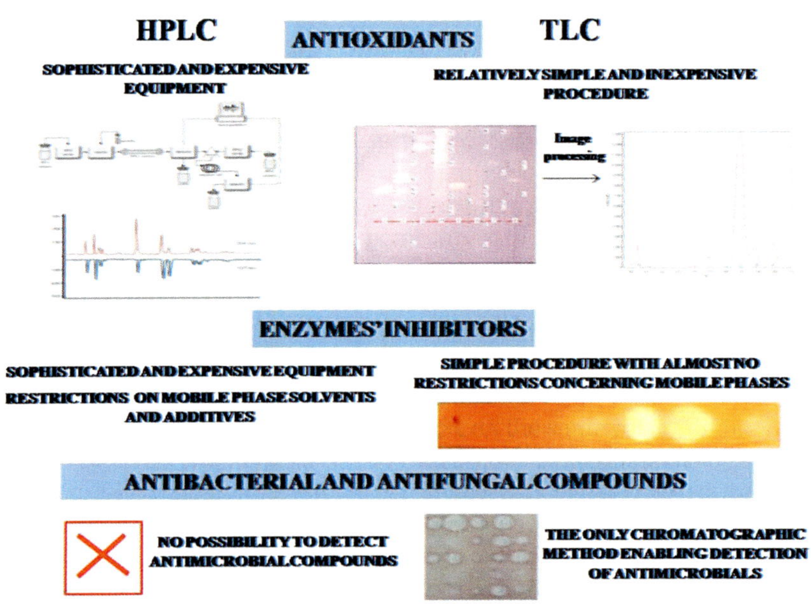

FIGURE 19.16 The use of thin-layer chromatography and high-performance liquid chromatography in effect-directed analysis. TLC, thin-layer chromatography; HPLC, high-performance liquid chromatography.

restricted as they may influence enzyme activity (e.g., high buffer concentration may cause desalting of proteins). The recent development of interfaces for thin-layer chromatography-mass spectrometry (TLC-MS), which enable direct recording of MS spectra under ambient conditions from any location on the layer, has the potential to revolutionize bioautography. For example, see Figure 19.16 [204].

There are some disadvantages associated with biological detection in TLC as well. The results may be affected by external factors such as temperature and humidity. The influence of the adsorbent's active sites on the observed biological activity seems to be overlooked in the majority of TLC-based bioassays [17,205]. Recent research indicates that the adsorbent type may strongly influence the results of TLC assay for natural free radical scavengers [17,205]. It was also observed that some simple amines, which may be used as mobile phase additives in the analysis of basic compounds (e.g., alkaloids), may be responsible for false-positive results in the TLC-acetylcholinesterase inhibitory assay [206].

There are several book chapters [207–209] and review articles [199,202,203,210,211] that summarize the use of TLC-bioautography assays. In order not to duplicate information from these sources the focus of this section will be on applications to plant materials as well as on the influence of modern TLC instrumentation on the development of the method. Currently TLC-based EDA includes techniques aimed at detection and identification of antioxidants, inhibitors of selected enzymes, antibacterial and antifungal compounds [209].

19.7.1 Practical Considerations

The first step in all types of TLC-bioautography assays is the optimization of the separation conditions for the plant metabolites present in the extract. For reliable and reproducible results the samples should be applied by means of an automatic applicator, preferably as bands. Applying samples as bands provides better resolution and a more-or-less uniform concentration distribution across the band compared with spots [210]. In the majority of cases silica gel is used as the stationary phase in TLC-bioautography assays [202]. A useful guideline for choosing the mobile phase for the separation natural product mixtures has been given in a book by Reich and Schibli [212]. To obtain reproducible separations the use of automatic developing chambers is advised. Such chambers provide the possibility to maintain comparable separation conditions, e.g., relative humidity, chamber saturation conditions, migration distance, etc., for multiple samples.

After the separation and drying steps, the plate is derivatized with a suitable reagent to visualize active zones. Depending on the bioautography assay different procedures have to be followed. Some considerations follow, but for detailed instructions see [203,204,209]. In order to evenly cover the layer with derivatizing agent, it is advised to use a dipping device, which is more likely to result in reproducible data. Some reactions require incubation or heating to reach completion. Commercially available plate heaters evenly heat the plate's surface increasing the reproducibility of TLC-based bioassays. The use of plate heaters is not always possible, as for example in the case of the detection of enzyme inhibitors or antimicrobial compounds. Detection reagents based on proteins and living microorganisms require proper humidity and are best performed in microbiological incubators or laboratory ovens.

The final step is the evaluation of the results and documentation. Direct access to the separated compounds is one of the great advantages of TLC-bioassays when compared to HPLC. A single plate can be sequentially evaluated by all possible means, e.g., under white and UV light, with the use of a densitometer, videoscanner or BioLuminizer®, etc. Using densitometry provides the possibility of recording a compounds spectra, which along with MS, is useful for compounds identification or confirmation. The purity of the separated zones can also be checked with the use of a TLC scanner. When analyzing crude extracts, the probability that a separated zone contains only a single compound is relatively low. However, screening crude extracts for biological activity by means of TLC can be useful in the identification of new compounds with pharmacological activity. Dereplication, which aims at excluding already known compounds from further analysis, can also be easily performed. Unfortunately, scanning densitometers have an important limitation for use in TLC-bioassays. Currently available scanners employ lane scanning so as to evaluate the layer track by track. The results of biological tests, on the surface of TLC plate, change with time. Therefore, in order to be able to compare results for compounds separated on the same plate, all the tracks should be evaluated at the same time. This is possible using videoscanning (digital photography) and image processing procedures. The on-line coupling of TLC-MS is becoming indispensible in EDA for compound identification [213]. For thin-layer chromatography-nuclear magnetic resonance (TLC-NMR) on-line connection is not possible, but the TLC-MS interface may be used to extract the compound into a vial for introduction into an NMR spectrometer after solvent exchange [214]. The commercially available TLC-MS interface has found a number of applications in phytochemical analysis, for example: the identification iridoids in *Gardenia jasminoides* fruits [215]; acetylcholinesterase inhibitors (coumarins) from *Galbanum* [214]; flavanols from *Juniperus communis* and *Punica granatum* [216]; anthocyanes [201], ursolic, oleanolic, and betulinic acids [217], phenolic and volatile compounds of *Salvia* spp. [218,219]; and saponins [220].

19.7.2 Antioxidant Assays

In recent years there has been an increase in the interest in natural antioxidants due to the fact that free radicals and oxidative stress have been implicated in age-related diseases. Plant-derived antioxidants have been used in the prevention and treatment of these illnesses. EDA based on TLC is an important tool for the identification of potent antioxidants. The most popular test utilizes a solution of a relatively stable nitrogen radical: 1,1-diphenyl-2-picrylhydrazyl (DPPH$^{\bullet}$). Its popularity is likely due to its simplicity. The TLC-DPPH$^{\bullet}$ assay, however, seems to be overused and sometimes exaggeratedly interpreted. The TLC-DPPH$^{\bullet}$ test is only a preliminary assay and in order to confirm antioxidant activity, other tests should be performed. TLC-MS may be used to identify compounds possessing direct antioxidant activity, in which case, another plate has to be developed, and the zone with the desired R_F value (read in the preceding test with DPPH$^{\bullet}$ solution) extracted and directed into mass spectrometer. It is often overlooked that analysis conditions may influence the results of the TLC-DPPH$^{\bullet}$ assay, as shown for selected polyphenolic compounds on the surface of different TLC adsorbents [205].

TLC-DPPH$^{\bullet}$ assay is usually performed for the detection and identification of polyphenolic free radical scavengers (phenolic acids, flavonoids, coumarins, anthocyanidins, etc.). Such compounds are considered potent free radical scavengers. Their activity is due to the presence of aryl hydroxyl groups, whose number and position influence their free radical scavenging properties. These hydroxyl groups interact with the adsorbents' active sites, mainly by formation of hydrogen bonds. Such interactions definitely influence the results of the free radical scavenging activity assay, and until more information is available, the use of nonspecific, e.g., RP-18 W plates, is recommended.

Polyphenols are weak acids and exist in both an ionized and unionized form, which is responsible for peak tailing, and is highly undesired in TLC-bioassays. The yellowish spots of free radical scavengers are diffused or they sometimes overlap. In order to overcome this problem, addition of organic (formic, acetic, trifluoroacetic) or mineral (*o*-phosphoric) acids is recommended to suppress the ionization of phenolic compounds. Attention should be paid to the influence of these mobile phase additives on the free radical scavenging activity of polyphenols.

Another group of secondary plant metabolites that possesses free radical scavenging activity are terpenoids (mono-, di-, sesquiterpenes, and carotenoids). In case of these compounds their activity is due to the presence of double bonds. Conjugated double bonds were shown to enhance direct antioxidant activity of terpenoids (unpublished results). Monoterpenes are volatile compounds and their analysis by TLC is problematic. The use of low-temperature TLC is better suited to the detection of free radical scavengers for these compounds [17]. Due to the presence of π bonds, terpenes may form π complexes with, for example, metal ions. Such complexes may be formed on the surface of some adsorbents, e.g., aluminum oxide. Therefore free radical scavenging activity of terpenes may be diminished on the surface of the aforementioned adsorbent (this hypothesis needs experimental verification).

Good examples of using TLC-DPPH$^{\bullet}$ test in EDA can be found in Kassim et al. [221], Rajauriaand Abu-Ghannam [222], Zhao et al. [223] and Gu et al. [224]. Apart from TLC-DPPH$^{\bullet}$ assays there are tests utilizing other reagents, namely 2,2′-azino-bis(3-ethylbenzothiazoline-6-sulfonic acid radical) or β-carotene/linoleic acid [209].

19.7.3 Enzyme Inhibition Assays

TLC has been applied for the detection and identification of inhibitors of the following enzymes: acetyl- (AChE) and butyrylcholinesterases (BuChE), α- and β-glucosidases, xanthine oxidase, lipase, and tyrosinase [209]. In all methods, after development and drying of the layer, the layer is covered with an enzyme solution. In order to stabilize the enzyme two approaches have been proposed: entrapping the enzyme in agar (gel-entrapment method) or the use of bovine serum albumin. In both cases, the enzyme solution is prepared in a buffered media [209]. To evenly cover the plate with enzyme, the dipping method is recommended, e.g., with the use of dipping device. Subsequently, the plate is incubated in a humid atmosphere at a controlled temperature. The enzyme activity is visualized by covering the layer with a solution of a substrate. The enzyme converts the substrate into a compound, which reacts with derivatizing agent forming a colored product. The inhibitors appear as pale spots/bands against a colored background. For more details regarding the development of different methods for screening complex samples refer to the book chapter by Cieśla and Kowalska [209].

TLC-bioautography tests for AChE and BuChE inhibitors are the most popular assays used to detect enzyme inhibitors in plant samples. There are several methods to detect AChE and BuChE inhibitors on the surface of TLC plates, as described elsewhere [209]. These tests were used mainly to detect alkaloids in selected *Amaryllidaceae* plants [225]; in *Galanthus nivalis, Leucojum vernum, Narcissus pseudonarcissus, C. majus, Ismene festalis, Nerine bowdenii, Crinum powellii, Sprekelia formosissima* [226]; 11 *Narcissus* species [227], etc. Some terpenes [17], coumarins [214], and ursolic acid [228] have been identified as the inhibitors of these enzymes as well. It was shown that some simple aldehydes and amines (e.g., diethylamine) produce false-positive results in TLC-bioassays focused on detection of AChE inhibitors [229].

Two methods are available for detection of glucosidase inhibitors developed by Salazar and Furlan [230] and Simões-Pires et al. [231]. Salazar and Furlan used esculin as an enzyme substrate, which cleaves to esculetin and forms brown complexes with iron (III) ions. Since esculin is a coumarin and naturally occurring secondary plant metabolite, this assay cannot be used to detect glucosidase inhibitors in extracts containing coumarins. The following plant extracts and natural products were found to be active in the TLC glucosidase-inhibitory tests: conduritol β-epoxide (known glucosidase inhibitor, used as a standard), *Brassica rapa, Solanum diflorum, Setaria parviflora* [230]; and castanospermine (standard) *Tussilago farfara, Urtica dioica* [231]. Ramallo et al. used the TLC-bioautographic method to detect β-glucosidase inhibitors in chemically engineered plant extracts [232]. Apigenin and ethyl *p*-cumarate were shown to inhibit the activity of β-glucosidase.

Ramallo et al. developed a method enabling identification of xanthine oxidase inhibitors and superoxide radical scavengers in plant extracts [233]. Taxifolin (a flavonoid) was used as a standard in this assay. Wangthong et al. elaborated a test for the detection of tyrosinase inhibitors [234]. Kojic acid, kojic acid dipalmitate, arbutin, glabridin were used as positive controls. The method was suitable to detect tyrosinase inhibitors in complex samples, as exemplified by *Glyzyrrhiza glabra* extracts. Hassan proposed a TLC-based bioassay for detection of lipase inhibitors [235]. *Camellia sinensis* and *Rosmarinus officinalis* extracts were shown to contain metabolites with lipase inhibitory properties.

19.7.4 Antimicrobial Assays

Detection of antimicrobial compounds was the first example of biological detection in TLC [236]. Recently several review articles and book chapters have been published with a focus on TLC antimicrobial assays [200,203,207–211]. TLC-bioassays were also used to detect compounds with estrogenic properties [237]. This assay employs yeast cells that produce a specific compound only in the presence of estrogenic substances.

There are several ways to perform TLC assays for antimicrobial activity. A suspension of bacterial or fungal cells can be poured directly onto the surface of the TLC plate (direct bioautography) or a suspension prepared in agar can be poured onto the TLC plate and left to solidify. Different microorganisms are used, for example, those forming color colonies, e.g., *Serratia marcescens,* as well as those with a bioluminescence gene, such as *Vibrio fischeri* [209].

TLC-bioautography tests have been widely used to screen plant species for biologically active compounds. It is nearly impossible to enumerate all of the species, which were investigated by this means, and only some recent examples are mentioned here: *Urtica urens*, *Lamium amplexicaule*, and *Sida rombifolia* [238]; *Phebalium squamulosum* [239]; *Artemisia adamsii* essential oil [240]; *O. onites* [241], *Geijera parviflora* [242]; volatile fractions from *Salvia officinalis*, *T. vulgaris*, and *Mentha* × *piperita* [243]; and many others. Some secondary plant metabolites, for example, *trans*-resveratrol or genistein may produce false-positive results when tetrazolium salts are used to detect growth inhibition zones [244].

19.8 IMAGE PROCESSING

The principles of image analysis in TLC are summarized in a book chapter by Vovk et al. [245]. In this section only the use of image processing programs for the analysis of plant extracts will be discussed. Image processing programs have been found useful for the analysis of chemical fingerprints [198,246–250], EDA [205,251–257], and to quantify the amount of secondary metabolites in plant extracts [258].

After completion of the separation and derivatization procedures, images of the layer can be acquired by means of a digital camera, flatbed scanner, or charge coupled device (CDD) camera, Figure 19.17. The first two methods are effectively restricted to the visible region. When using a

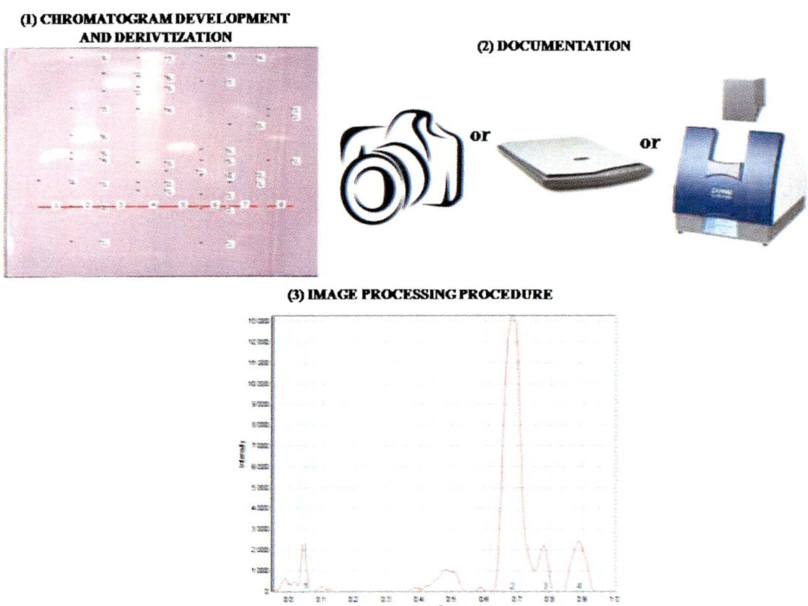

FIGURE 19.17 Documentation of separation and data acquisition by means of image processing procedures in a free radical scavenging test.

digital camera it is important to take note of: (1) it is only possible to compare results for samples analyzed on the same plate (this is not the case for flatbed scanners or in an illumination chamber, e.g., TLC Visualizer); (2) homogenous illumination is important as well as the plate should be photographed parallel to the photo boarders; (3) the images should be saved at the highest available resolution; and (4) it is crucial to apply the samples as bands and not spots [251].

Modern instrumental TLC provides the possibility to obtain high-quality images, with high reproducibility, by use of a CCD camera in conjunction with an illuminating chamber (e.g., TLC Visualizer). Visualization in the visible or UV (254 and 366 nm) region is possible. The software used with this equipment allows evaluation of images taken by the CCD camera.

Captured images can be further processed by available software, e.g., ImageJ (Wayne Rasband, National Institutesof Health, USA; http://rsbweb.nih.gov/ij/), Sorbfil TLC Videodensitometer (Sorbpolymer SCC, Krasnodar, Russian Federation, http://www.sorbfil.com), JustTLC (Sweday, Sweden, http://www.sweday.com). ImageJ and Sorbfil have been used most often for plant materials. In the ImageJ program the option designed for the analysis of one-dimensional electrophoresis gels should be used. Prior to obtaining semiquantitative data, noise removal and baseline drift correction are recommended. ImageJ contains built-in options to accomplish these steps [251]. Other image denoising approaches were characterized by Komsta [259,260]. The Sorbfil Videodensitometer offers even simpler solutions for obtaining quantitative data from videoscans [254].

In fingerprint analysis many samples have to be compared in order to fish out similarities and differences between samples. Sometimes it is difficult to observe these differences in the absence of suitable image processing software. It was proposed that for each analyzed species an "average fingerprint" be constructed and archived [249]. This facilitates the construction of fingerprint databases for sample identification and quality control. Image processing procedures are especially useful when the fingerprints are obtained from two-dimensional separations and/or when derivatizing agents are applied. After derivatization the results may change with time, and therefore all the samples from one plate should be evaluated at the same time. A good example of the use of modern TLC equipment and image processing for fingerprinting plant-derived extracts is described by Ristojević for the classification of samples of propolis [250]. Komsta et al. proposed a procedure for processing images of two-dimensional separations to characterize differences between closely related plant species [248]. Chemometric analysis of denoised and background-subtracted images is another possibility for characterizing plant species [248].

Image processing plays an important role in EDA. It was shown that for the free radical scavenging assays using DPPH˙ of pharmaceutical [261,262] and plant samples [198,248,249,251], the assay results change quickly with time and lane-scanning densitometers were unsuitable for quantification purposes. Image capture provides the certainty that all tracks are evaluated for the same reaction conditions. This is crucial for reliable and reproducible data.

References

[1] Hahn-Deinstrop E. Applied thin-layer chromatography. Weinheim: WILEY-VCH Verlag GmbH & Co. KGaA; 2007.

[2] Waksmundzka-Hajnos M, Sherma J, Kowalska T, editors. Thin layer chromatography in phytochemistry. Boca Raton London New York: CRC Press; 2008.

[3] Fecka I, Cisowski W. Multiple gradient development TLC analysis of complex phenolic acids from *Lycopus europaeus*. Chromatographia 1999;49:256–60.

[4] Wójciak-Kosior M, Skalska A. Thin-layer chromatography of phenolic acids on aminopropylsilica. J Planar Chromatogr 2006;19:200–3.

[5] Maleš Ž, Medić-Šarić M. Investigation of the flavonoids and phenolic acids of *Sambuci flos* by thin-layer chromatography. J Planar Chromatogr 1999;12:345–9.

[6] Matysik G, Wójciak-Kosior M, Paduch R. The influence of Calendulae officinalis flos extracts on cell cultures, and the chromatographic analysis of extracts. J Pharm Biomed Anal 2005;38:285–92.

[7] Głowniak K, Bartnik M, Mroczek T, Zabża A, Wierzejska A. Application of column chromatography and preparative TLC for isolation and purification of coumarins from *Peucedanum tauricum* Bieb. fruits. J Planar Chromatogr 2002;15:94–100.

[8] Kozyra M, Głowniak K, Zabża A, Zgórka G, Mroczek T, Cierpicki T, et al. Column chromatography and preparative TLC for isolation and purification of coumarins from *Peucedanum verticillare* L. Koch. J Planar Chromatogr 2005;18:224–7.

[9] Zgorka G. Retention behaviour of coumarin compounds on silica and Florisil TLC layers. J Liq Chromatogr 2001;24:1397–410.

[10] Bieganowska ML, Petruczynik A. Retention behaviour of coumarins and flavonoids on polyamide and alumina layers. Part I. J Planar Chromatogr 1999;12:135–9.

[11] Qin L, Zhang W, Zhang H, Viang R, Su Zh. Study of the interspecific difference and intraspecific variation of coumarins in *Cnidium* species by TLC. J Chin Trad Med 1996;21:525–6.

[12] Oliva A, Meepagala KM, Wedge DE, Harris D, Hale AL, Aliotta G, et al. Natural fungicides from *Ruta graveolens* L. leaves. J Agric Food Chem 2003;51:890–6.

[13] Bieganowska ML, Petruczynik A, Zobel MA. Retention parameters of coumarins and flavonoids in binary reversed-phase HPTLC systems. J Planar Chromatogr 1996;9:273–9.

[14] Opletal L, Sovová H, Bártlová M. Dibenzo[a,c]cyclooctadiene lignans of the genus Schisandra: importance, isolation and determination. J Chromatogr B 2004;812:357–71.

[15] Khatoon S, Rai V, Rawat AKS, Mehrotra S. Comparative pharmacognostic studies of three Phyllanthus species. J Ethnopharmacol 2006;104:79–86.

[16] Mishra N, Acharya N, Gupta AP, Singh B, Kaul VK, Ahuja PS. A simple microanalytical technique for determination of podophyllotoxin in *Podophyllum hexandrum* roots by quantitative RP-HPLC and RP-HPTLC. Curr Sci 2005;88:1372–3.

[17] Cieśla Ł, Kryszeń J, Stochmal A, Oleszek W, Waksmundzka-Hajnos M. Low-temperature thin-layer chromatography preliminary bioautographic tests for

detection of free radical scavengers and acetylcholinesterase inhibitors in volatile samples. J Planar Chromatogr 2012;25:225–31.

[18] Koch A, Basar S, Richter R. TLC of mono- and sesquiterpenes. In: Waksmundzka-Hajnos M, Sherma J, Kowalska T, editors. Thin layer chromatography in phytochemistry. Boca Raton London New York: CRC Press; 2008.

[19] Sonwa MM, König WA. Chemical study of the essential oil of *Cyperus rotundus*. Phytochemistry 2001;58:799–810.

[20] Teixeira VL, Almeida SA, Kelecom A. Chemosystematic and biogeographic studies of the diterpenes from the marine brown alga *Dictyota dichotoma*. Biochem Syst Ecol 1990; 18:87–92.

[21] Tallevi SG. Detection of ginkgolides by thin-layer chromatography. J Nat Prod 1991;54: 624–5.

[22] Parmar VS, Amitabh J, Bisht KS, Taneja P, Singh SK, Kumar A, Denmarkpp, Jain R, Olsen CE. Constituents of the yew trees. Phytochemistry 1999;50:1267–304. http://www.sciencedirect.com/science/article/pii/S003194229800702X.

[23] Zhou C, Peng S, Liao W. A method for the quality control of the penetrating liquid of Leigongteng, *Tripterygium wilfordii*. Chin J Herb Med (Zhongcaoyao) 1993;24: 191–2.

[24] Hajnos M. TLC of diterpenes. In: Waksmundzka-Hajnos M, Sherma J, Kowalska T, editors. Thin layer chromatography in phytochemistry. Boca Raton London New York: CRC Press; 2008.

[25] Haddad M, Miyamoto T, Laurens V, Lacaille-Dubois MA. Two biologically active triterpenoidal saponins acylated with salicylic acid from *Albizia adianthifolia*. J Nat Prod 2002;66:376–82.

[26] Oleszek W, Kapusta I, Stochmal A. TLC of triterpenes (including saponins). In: Waksmundzka-Hajnos M, Sherma J, Kowalska T, editors. Thin layer chromatography in phytochemistry. Boca Raton London New York: CRC Press; 2008.

[27] Schmidt TJ, von Raison J, Willuhn G. New triterpene esters from flowerheads of *Arnica lonchophylla*. Planta Med 2004;70:968–73.

[28] Yoshikawa M, Morikawa T, Kashima Y, Ninomiya K, Matsuda H. Structures of new dammarane-type triterpene saponins from the flower buds of *Panax notoginseng* and hepatotoprotective effects of principal ginseng saponins. J Nat Prod 2003;66: 926–31.

[29] Britton G. TLC of carotenoids. In: Waksmundzka-Hajnos M, Sherma J, Kowalska T, editors. Thin layer chromatography in phytochemistry. Boca Raton London New York: CRC Press; 2008.

[30] Schiedt K. Chromatography: Part III. Thin-layer chromatography. In: Britton G, Liaaen-Jensen S, Pfander H, editors. Carotenoids. Isolation and analysis, vol. 1A. Basel: Birkhäuser; 1995.

[31] Britton G. Example 1: higher plants. In: Britton G, Liaaen-Jensen S, Pfander H, editors. Carotenoids. Isolation and analysis, vol. 1A. Basel: Birkhäuser; 1995.

[32] Dinan L, Harmatha J, Lafont R. Chromatographic procedures for the isolation of plant steroids. J Chromatogr A 2001;935:105–23.

[33] Dias C, Borralho JA, Lurdes Gonçalves M. *Scilla maderensis*, TLC screening and positive inotropic effect of bulb extracts. J Ethnophormacol 2000;71:487–92.

[34] Luyt RP, Jäger AK, van Staden J. Bufadienolides in in vitro derived *Drimia robusta* plants. S. Afr J Bot 1999;65:443–5.

[35] Ma WG, Fuzzati N, Wolfender JL, Yang CR, Hostettmann K. Further acylated secoiridoid glucosides from *Gentiana rhodantha*. Phytochemistry 1996;43:805–10.

[36] Menković N, Šavikin-Fodulović K, Savin K. Chemical composition and seasonal variations in the amount of secondary compounds in *Gentiana* lutea leaves and flowers. Planta Med 2000;66:178–80.

[37] Bos R, Woerdenbag HJ, Pras N. Determination of valepotriates. J Chromatogr A 2002; 967:131–46.
[38] Schwarz MJ, Houghton PJ, Rose S, Jenner P, Lees AD. Activities of extract and constituents of *Banistereopsis caapi* relevant to parkinsonism. Pharmacol Biochem Behav 2003; 75:627–31.
[39] Houghton PJ. TLC of indole alkaloids. In: Waksmundzka-Hajnos M, Sherma J, Kowalska T, editors. Thin layer chromatography in phytochemistry. Boca Raton London New York: CRC Press; 2008.
[40] Küpeli E, Kosar M, Yesilada E, Basser KHC. A comparative study on the anti-inflammatory, antinociceptive and antipyretic effects of isoquinoline alkaloids from the roots of Turkish *Berberis* species. Life Sci 2002;72:645–57.
[41] Waksmundzka-Hajnos M, Petruczynik A. Retention behavior of some alkaloids in thin-layer chromatography with bonded stationary phases and binary mobile phase. J Planar Chromatogr 2001;14:364–70.
[42] Petruczynik A, Waksmundzka-Hajnos M, Michniowski T, Plech T, Tuzimski T, Hajnos M, Gadzikowska M, Jóźwiak G. Thin-layer chromatography of alkaloids on cyanopropyl bonded stationary phases. Part I. J Chromatogr Sci 2007;45:447–54.
[43] Pothier J, Galand N. Automated multiple development thin-layer chromatography for separation of opiate alkaloids and derivatives. J Chromatogr A 2005;1080:186–91.
[44] Sarkozi A, Janicsak G, Kursinszki L, Kery A. Alkaloid composition of *Chelidonium majus* L. studied by different chromatographic techniques. Chromatographia 2006;63:S81–6.
[45] Bringmann G, Wenzel M, Rübenacher M, Schäffer M, Rückert M, Laurent AA. Dioncophylline and 8-O-methyldioncophylline D, 7,8′-coupled naphthylisoquinoline alkaloids from *Triphyophyllum peltatum*. Phytochemistry 1998;49:1151–5.
[46] Agyangar MR, Biswas SS, Tambe AS. Separation of opium alkaloids by thin-layer chromatography combined with flame ionization detection using the peak pyrolysis method. J.Chromatogr 1991;547:538–43.
[47] Petruczynik A, Waksmundzka-Hajnos M, Hajnos MŁ. The effect of chromatographic conditions on the separation of selected alkaloids on silica layers. J Planar Chromatogr 2005;18:78–84.
[48] Szabo B, Lakatos A, Käszegi T, Botz L. HPTLC and HPLC determination of alkaloids in poppies subjected to stress. J Planar Chromatogr 2003;16:293–7.
[49] Szumiło H, Flieger J. Application of differently modified silica gel in the TLC analysis of alkaloids. J Planar Chromatogr 1999;12:466–70.
[50] Waksmundzka-Hajnos M, Petruczynik A. TLC of isoquinoline alkaloids. In: Waksmundzka-Hajnos M, Sherma J, Kowalska T, editors. Thin layer chromatography in phytochemistry. Boca Raton London New York: CRC Press; 2008.
[51] Gadzikowska M, Petruczynik A, Waksmundzka-Hajnos M, Hawryl M, Jóźwiak G. Two-dimensional planar chromatography of tropane alkaloids from *Datura inoxia* mill. J Planar Chromatogr-Mod TLC 2005;18:127–31.
[52] Duez P, Chamart S, Hanocq M, Molle L, Vanhaelen M, Vanhaelen-Fastre R. Comparison between thin-layer chromatography-densitometry and high-performance liquid chromatography for determination of hyoscyamine and hyoscyne in leaves, fruits and seeds of *Datura* (*Datura* spp.). J Chromatogr 1985;329:415–21.
[53] Mroczek T, Głowniak K, Kowalska J. Solid-liquid extraction and cation-exchange solid-phase extraction using a mixed-mode polymeric sorbent of *Datura* and related alkaloids. J Chromatogr A 2006;1107:9–18.
[54] Gupta AP, Gupta MM, Kumar S. Estimation of tropane alkaloids using high performance thin layer chromatography. J Indian Chem Soc 1999;76:277–89.
[55] Berkov S, Pavlov A. A rapid densitometric method for the analysis of hyoscyamine and scopolamine in *Solanaceous* plants and their transformed root cultures. Phytochem Anal 2004;15:141–5.

[56] Flieger J. TLC of alkaloids from the other biosynthetic groups. In: Waksmundzka-Hajnos M, Sherma J, Kowalska T, editors. Thin layer chromatography in phytochemistry. Boca Raton London New York: CRC Press; 2008.
[57] Christensen LP, Jacobsen HB. Polyacetylenes: distribution in higher plants, pharmacological effect and analysis. In: Waksmundzka-Hajnos M, Sherma J, Kowalska T, editors. Thin layer chromatography in phytochemistry. Boca Raton;London New York: CRC Press; 2008.
[58] Matysik G, Skalska-Kamińska A, Matysik-Woźniak A. Quinone derivatives in plant extracts. In: Waksmundzka-Hajnos M, Sherma J, Kowalska T, editors. Thin layer chromatography in phytochemistry. Boca Raton London New York: CRC Press; 2008.
[59] Metzger W, Reif K. Determination of 1,8-dihydroxyanthranoids in senna. J Chromatogr A 1996;740:133–8.
[60] Singh NP, Gupta AP, Sinha AK, Ahuja PS. High-performance thin layer chromatography method for quantitative determination of four major anthraquinone derivatives in *Rheum emodi*. J Chromatogr A 2005;1077:202–6.
[61] Wawrzynowicz T, Waksmundzka-Hajnos M, Mulak-Banaszek K. Isolation of aloine and aloeemodine from *Aloe* (Liliaceae) by micropreparative TLC. J Planar Chromatogr 1994;7:315–8.
[62] Geiss F. Fundamentals of thin-layer chromatography. Heidelberg: Hüthig; 1987.
[63] Kaiser RE. Anticircular high performance thin-layer chromatography. J High Resolut Chromatogr 1978;1:164–8.
[64] Baerheim Svendsen A, van Kepipen-Verleun R, Verpoorte R. Circular thin-layer chromatography of quaternary alkaloids. J Chromatogr 1984;291:389–91.
[65] Vuorela H, Dallenbach-Tölke K, Sticher O, Hiltunen R. Separation of the main coumarins of *Peucedanum palustre* with various planar chromatographic methods. J Planar Chromatogr 1988;1:123–7.
[66] Menziani E, Tosi B, Bonora A, Reschiglian P, Lodi G. Automated multiple development HP-TLC analysis of natural phenolic compounds. J Chromatogr 1990;511:396–401.
[67] Poole SK, Kiridena W, Miller KG, Poole CF. Planar chromatographic methods for determination of the quality of spices and flavors as exemplified by cinnamon and vanilla. J Planar Chromatogr 1995;8:257–68.
[68] Belay MT, Poole CF. Determination of vanillin and related flavor compounds in natural vanilla extracts and vanilla-flavored foods by TLC and AMD. Chromatographia 1993;37:365–73.
[69] Lavoine S, Amaudo JF, Coutiere D. HPTLC analysis of the major components in Vanilla bean extracts. CBS 1998;81:14–5.
[70] Pozharitskaya O, Ivanova S, Shikov A, Makariv V. Separation and quantification of terpenoids of *Boswellia serrata* extracts by planar chromatography techniques (TLC and AMD). J Sep Sci 2006;29:2245–50.
[71] Schulte F, Maeder J, Kroh W, Panne U, Kneipp J. Characterization of pollen carotenoids with in-situ and HPTLC resonant Raman spectroscopy. Anal Chem 2009;81:8426–33.
[72] Olah NK, Muresan L, Compan G, Gocan S. Normal-phase HPTLC and AMD of hydroalcohic extracts of *Artemisia* species. J Planar Chromatogr 1998;11:361–4.
[73] Markowski W, Ludwiczuk A, Wolski T. Analysis of ginsenosides from *Panax quinquefolium* by AMD. J Planar Chromatogr 2006;19:115–7.
[74] Viel C, Galand N, Dollet J, Pothier J. Analysis by planar chromatography (TLC and AMD) of cannabinoids. In: Proc. intern. symp. on planar separations, planar chromatography; 2001. pp. 309–12.
[75] Pothier J, Galand N, Viel C. Separation of opiates and derivatives – analgesic, antitussives and narcotic compounds by different techniques of planar chromatography ITLC, OPLC, AMD. In: Proc. intern. symp. on planar separations, planar chromatography; 2001. pp. 321–6.

[76] Scholl Y, Asano N, Drager B. Automated multiple development TLC for calycystegines and their biosynthetic precursors. J Chromatogr A 2001;928:217–24.
[77] Kiridena W, Poole SK, Meller KG, Poole CF. Determination of aloin and aloeemodin in *Aloe* consumable products by TLC. J Planar Chromatogr 1995;8:416–9.
[78] Poole CF, Belay MT. Progress in automated multiple development. J Planar Chromatogr 1991;4:345–59.
[79] Poole CF, Poole SK, Belay MT. Insights into the mechanism and applications of unidimensional multiple development in thin layer chromatography. J Planar Chromatogr 1993;6:438–45.
[80] Poole CF, Poole SK, Fernado WPN, Dean TA, Ahmed HD, Berndt JA. Multidimensional and multimodal thin-layer chromatography: pathway to the future. J Planar Chromatogr 1989;2:336–45.
[81] Waksmundzka-Hajnos M, Gadzikowska M, Hajnos M. Strategy for preparative separation of quaternary alkaloids from *Chelidonium majus* L. by thin-layer chromatography. J Planar Chromatogr 2002;15:289–93.
[82] Jóźwiak G, Wawrzynowicz T, Waksmundzka-Hajnos M. Optimization of the preparative chromatography of an extract of alkaloids from *Fumaria* officinalis. J Planar Chromatogr 2000;13:447–51.
[83] Wawrzynowicz T, Czapińska K, Markowski W. The use of special development techniques for micropreparative TLC separation of closely related compounds. J Planar Chromatogr 1998;11:388–93.
[84] Mulinacci N, Bardazzi C, Romani A, Pinelli P, Vincieri FF, Cotantini A. HPLC-DAD and TLC-densitometry for quantification of hypericin in *Hypericum perforatum* L. extracts. Chromatographia 1999;49:197–201.
[85] Sawicka U, Cisowski W, Matysik G, Kowalczyk A. HPTLC of phenolic acids in *Bistortae rhizoma, Polygoni avicularis* herba, *Rhei radix* with desitometric determination. J Planar Chromatogr 2002;15:442–8.
[86] Waksmundzka-Hajnos M, Gadzikowska M, Gołkiewicz W. Special modes of development of *Chelidonium maius* L. alkaloids in systems of the type polar adsorbent – multicomponent mobile phase by TLC. J Palanr Chromatogr 2000;13:205–9.
[87] Mroczek T, Ndjojko-Joset K, Głowniak K, Mietkiewicz-Capała A, Hostettmann K. Investigation of *Symphytum cordatum* alkaloids by liquid-liquid partitioning, TLC and liquid chromatography – ion trap mass spectrometry. Anal Chim Acta 2006;566:157–66.
[88] Andayi AW, Yenesew A, Derese S, Midiwo JO, Gitu PM, Jondiko JI, et al. Antiplasmoidal flavonoids from *Erythrina sacleuxii*. Planata Med 2006;72:187–9.
[89] Giddings JC. Two-dimensional separations: concept and promise. Anal Chem 1984;56: 1258A–70A.
[90] Giddings JC. Concepts and comparisons in multidimensional separation. J High Resolut Chromatogr 1987;10:319–23.
[91] Cortes HJ, editor. Multidimensional chromatography. Techniques and applications. New York: Marcel Dekker; 1990.
[92] Giddings JC. Use of multiple dimensions in analytical separations in multidimensional chromatography. In: Cortes HJ, editor. Multidimensional chromatography. Techniques and application. New York: Marcel Dekker; 1990.
[93] Nyiredy Sz. Multidimensional planar chromatography. In: Kaiser RE, Günther W, Gunz H, Wulff G, editors. Proceedings of the Dünnschicht-Chromatographi (in memoria Prof. Dr. Hellmut Jork). Düsseldorf: InCom Sonderband; 1996.
[94] Nyiredy Sz. In: Mondello, Lewis, Bartle, editors. Multidimensional chromatography. Baffins Lane; Chichester: JOHN WILEY & SONS; 2002.
[95] Nyiredy Sz. Multidimensional planar chromatography. LC–GC Eur 2003;16:52–9.
[96] Gocan S. Two-dimensional thin-layer chromatography. J Liq Chromatogr Rel Technol 2005;27:1105–13.

[97] Nyiredy Sz. Planar chromatography – a retrospective view for the third millennium. Budapest: Springer; 2001.

[98] Zakaria M, Gonnord MF, Guiochon G. Applications of two-dimensional thin-layer chromatography. J Chromatogr 1983;271:127–92.

[99] Gonnord MF, Levi F, Guiochon G. Computer assistance in the selection of the optimum combination of systems for two-dimensional chromatography. J Chromatogr 1983;264:1–9.

[100] Streinbrunner JE, Johnson EK, Habibi-Goudarzi S, Nurok D. In: Kaiser RE, editor. Planar chromatography, vol. 1. Heidelberg: Hüthig; 1986.

[101] Nurok D, Habibi-Goudarzi S, Kleyle R. Statistical approach to solvent selection as applied to two-dimensional thin-layer chromatography. Anal Chem 1987;59:2424–8.

[102] Tuzimski T, Soczewiński E. Correlation of retention parameters of pesticides in normal- and reversed-phase systems and their utilization for the separation of a mixture of 14 triazines and urea herbicides by means of two-dimensional thin-layer chromatography. J Chromatogr A 2002;961:277–83.

[103] Tuzimski T, Soczewiński E. Correlation of retention data of pesticides in Normaland reversed-phase systems and utilization of the data for separation of a mixture of ten urea herbicides by two-dimensional thin-layer chromatography on cyanopropyl-bonded polar stationary phase and on a two-adsorbent-layer multi-K SC5 plate. J Planar Chromatogr 2003;16:263–7.

[104] Tuzimski T. Separation of a mixture of eighteen pesticides by two-dimensional thin-layer chromatography on a cyanopropyl-bonded polar stationary phase. J Planar Chromatogr 2004;17:328–34.

[105] Harmala P, Botz L, Sticher O, Hiltunen R. Two-dimensional planar chromatographic separation of a complex mixture of closely related coumarins from the genus *Angelica*. J Planar Chromatogr 1990;3:515–20.

[106] Soczewiński E, Hawrył MA, Hawrył A. Retention behavior of some flavonoids in 2D-TLC system on cyano-bonded polar phases. Chromatographia 2002;54:789–94.

[107] Sawicki E, Stanley TW, Elbert WC. Analysis of the urban atmosphere and air pollution source effluents for Phenalen-1-one and 7 H-benz(de)anthracen-7-one. Microchim Acta 1965;53:11101123.

[108] Sawicki E, Stanley TW, McPherson S, Morgan M. Use of gas-liquid and thin-layer chromatography in characterising air pollutants by fluorometry. Talanta 1966;13:619–29.

[109] Kawatsaki JA, Frasch DL. J Assoc Off Anal Chem 1969;52:1108.

[110] Sherma J, Zweig G. Chromatographic separation and identification of chloroplast pigments in *Chlorella pyrenoidosa*. J Chromatogr 1967;31:589–91.

[111] MullickD B. Thin layer chromatography of anthocyanidins : I. Techniques and solvents for two-dimensional chromatography. J Chromatogr 1969;39:291–301.

[112] Saitoh K, Kobayashi M, Suzuki N. High-performance thin-layer chromatography of metal tetraphenylporphyrin chelates. Anal Chem 1981;53:2309–13.

[113] Lepri L, Desideri PG, Heimler D. High-performance thin-layer chromatography of 2,4-dinitrophenyl-amino acids on layers of RP-a8, RP-18 and ammonium tungstophosphate. J Chromatogr 1982;235:411–6.

[114] Macek K, Deyl Z, Smrz M. Two-dimensional thin-layer chromatography of Dns-amino acids on reversed-phase silica gel. J Chromatogr 1980;193:421–6.

[115] Waksmundzka-Hajnos M, Petruczynik A, Hawrył A. Comparison of chromatographic properties of cyanopropyl-, diol- and aminopropyl- polar-bonded stationary phases by the retention of model compounds in normal-phase liquid chromatography systems. J Chromatogr A 2001;919:39–50.

[116] Waksmundzka-Hajnos M, Petruczynik A, Hajnos MŁ, Tuzimski T, Hawrył A, Bogucka-Kocka A. Two-dimensional thin-layer chromatography of selected coumarins. J Chromatogr Sci 2006;44:510–7.

[117] Łuczkiewicz M, Migas P, Kokotkiewicz A, Walijewska M, Cisowski W. Two-dimensional TLC with adsorbent gradient for separation of quinolizidine alkaloids in the herb and in-vitro cultures of several *Genista* species. J Planar Chromatogr 2004;17:89–94.

[118] Hawrył MA, Hawrył A, Soczewiński E. Application of normal- and reversed chase 2D-TLC on cyanopropyl bonded polar stationary phase for separation of phenolic compounds from the flowers of *Sambucus nigra* L. J Planar Chromatogr 2002;15:4–10.

[119] Hahn-Deinstrop E, Koch A. *Ginko biloba* trans the brain. Bioforum 1998;7-8:428–34.

[120] Mikanagi X, Yokoi M, Ueda Y, Saito N. Flower flavonol and anthocyanin distribution in subgenus Rosa. Biochem. Syst Ecol 1995;23:183–200.

[121] Shi-Bao J, Yokoi M, Saito N, Mao LS. Distribution of anthocyanins in *Asteracea* leaves. Biochem. Syst Ecol 1992;20:771–81.

[122] Koupai-Abyazani MR, Bohm BA. The condensed tannins of *Leptrarrhena pyrolifolia*. Phytochemistry 1991;33:1485–7.

[123] Schmidt PC, Vogel K. Investigation on the stability of chamomile preparations. Dtsch Apoth Ztg 1992;132:462–8.

[124] Taylor AW, Barofsky E, Kennedy JA, Deinzer ML. Hop (*Humulus lupulus* L.) proanthocyanidins characterized by mass spectrometry, acid catalysis and gel permeation chromatography. J Agric Food Chem 2003;51:4101–10.

[125] Ma W-G, Wang DZ, Zeng Y-L, Yang CR. Three triterpenoid saponins from *Triplostegia grandiflora*. Phytochemistry 1991;30:3401–4.

[126] Matysik G, Skalska-Kamińska A, Stefańczyk B, Wójciak-Kosior M, Rapa D. Application of New technique in two-dimensional TLC separation of multicomponent mixtures. J Planar Chromatogr 2008;21:233–6.

[127] Tuzimski T, Bartosiewicz A. Correlation of retention parameters of pesticides in Normal and RP systems and their utilization for the separation of a mixture of ten urea herbicides and fungicides by two-dimensional TLC on cyanopropyl-bonded polar stationary phase and two-adsorbent-layer multi-K plate. Chromatographia 2003; 58:781–8.

[128] Sherma J, Fried B. Handbook of thin-layer chromatography. New York: Marcel Dekker; 1996.

[129] Levitt RE, Touchstone JC. Separation of conjugated bile acids by two-dimensional thin-layer chromatography. J High Resolut Chrom Chrom Comm 1979;2:587–8.

[130] Issaq HL. Two-phase thin-layer chromatography. J Liq Chromatogr 1980;3:841–4.

[131] Glensk M, Sawicka U, Mażol I, Cisowski W. 2D TLC – graft planar chromatography in the analysis of a mixture of phenolic acids. J Planar Chromatogr 2002;15:463–5.

[132] Glensk M, Czekalska M,Cisowski W. Resolution of saponins from a ginseng preparation by 2D TLC with an adsorbent gradient. J Planar Chromatogr 2001;14:454–6.

[133] Cieśla Ł, Petruczynik A, Hajnos M, Bogucka-Kocka A, Waksmundzka-Hajnos M. Two-dimensional thin-layer chromatography of structural analogs. Part I: graft TLC of selected coumarins. J Planar Chromatogr 2008;21:237–41.

[134] Cieśla Ł, Bogucka-Kocka A, Hajnos M, Petruczynik A, Waksmundzka-Hajnos M. Two-dimensional thin-layer chromatography with adsorbent gradient as a method of chromatographic fingerprinting of furanocoumarins for distinguishing selected varieties and forms of *Heracleum* spp. J Chromatogr A 2008;1207:160–8.

[135] Hawrył MA, Waksmundzka-Hajnos M. Separation of phenolic compounds by NP and RP two-dimensional thin layer chromatography on connected plates. J Planar Chromatogr 2006;19:92–7.

[136] Glensk M, Biały Z, Jurzysta M, Cisowski W. Two-dimensional TLC with a sorbent gradient for the analysis of *Alfalfa* root saponin. Chromatographia 2001;54:669–72.

[137] Poole CF, Poole SK. Modern thin layer chromatography. Anal Chem 1989;61:1257–69.

[138] Poole CF, Fernando WPN. Some musings on the measurement and interpretation of theoretical plate height in thin-layer chromatography. J Planar Chromatogr 1992;5: 323–33.
[139] Zarzycki PK, Ślaczka MM, Zarzycka MB, Bartoszuk MA, Włodarczyk E, Baran MJ. Temperature-controlled micro-TLC: a versatile green chemistry and fast analytical tool for separation and preliminary screening of steroids fraction from biological and environmental samples. J Steroid Biochem. Mol Biol 2011;127:418–27.
[140] Zarzycki PK, Ślaczka MM, Zarzycka MB, Włodarczyk ME, Baran MJ. Application of micro-thin-layer chromatography as a simple fractionation tool for fast screening of raw extracts derived from complex biological, pharmaceutical and environmental samples. Anal Chim Acta 2011;688:168–74.
[141] Zarzycki PK. Simple horizontal chamber for thermostated micro-thin-layer chromatography. J Chromatogr A 2008;1187:250–9.
[142] Zarzycki PK, Ohta H, Harasimiuk FB, Jinno K. Fast separation and quantification of C60 and C70 fullerenes using thermostated micro thin-layer chromatography. Anal Sci 2007;23:1391–6.
[143] Heinrich M, Barnes J, Gibbons S, Williamson EM. Fundamentals of pharmacognosy and phytotherapy. Livingstone: Churchill; 2004.
[144] Fraga CG, editor. Plant phenolics and human health. Biochemistry, nutrition, and pharmacology. J. Wiley & Sons, Inc.; Publication; 2010.
[145] Boshoff PR, Hopkins BJ, Pretorius V. Thin-layer chromatographic transport detector for high-performance liquid chromatography. J Chromatogr 1976;126:35–41.
[146] Hofstraat JW, Griffioen S, van de Nesse RJ, Brinkman UAT. Coupling of narrow-bore column liquid chromatography and thin-layer chromatography. Interface optimization and characteristics for normal-phase liquid chromatography. J Planar Chromatogr 1988;1:220–6.
[147] Mincsovics E, Garami M, Tyihák E. Direct coupling of OPLC with HPLC: cleanup and separation. J Planar Chromatogr 1991;4:299–303.
[148] Burger K. Online coupling HPLC–AMD (automated multiple development). Analysis 1990;18:1113–6.
[149] Queckenberg OR, Fraham AW. Chromatographic and spectroscopic coupling: a powerful tool for the screening of wild amaryllidaceae. J Planar Chromatogr 1993;6:55–61.
[150] Hawrył MA, Soczewiński E, Dzido TH. Separation of coumarins from *Archangelica officinalis* in high-performance liquid chromatography and thin-layer chromatography systems. J Chromatogr A 2000;886:75–81.
[151] Hawrył MA, Soczewiński E. Separation of some flavonoids using RP-HPLC-NP-TLC off-line coupled system. Chromatographia 2000;52:175–8.
[152] Tyihak E, Mincsovics E. New planar liquid chromatographic technique: overpressured thin-layer chromatography. J Chromatogr 1979;174:75–81.
[153] Wall PE. Thin-layer chromatography. A modern practical approach. Cambridge: RSC; 2005.
[154] Vuorela P, Rahko EL, Hiltunen R, Vuorela H. Overpressured layer chromatography with thin-layer and high-performance liquid chromatography for determination of coumarins with reference to composition of mobile phase. J Chromatogr 1994;670: 191–8.
[155] Dugo P, Mondello L, Lamonica G, Dugo G. OPLC analysis of heterocyclic oxygen compounds from ci trud fruit essentials oils. J Planar Chromatogr 1996;9:120–5.
[156] Dallenbach-Tolke K, Nyiredy S, Sticher O. The quality of birch leafs. Comparison of single and total determination methods of the flavonoid glycosides in *Betula folium*. Dtsch Apoth Ztg 1987;127:1167–71.
[157] Ligor M, Buszewski B. Application of TLC and OPLC in the determination of pigments from natural products. J Planar. Chromatogr 2001;14:334–7.

[158] Mincsovics E, Tabanca N, Moricz AM, Wedge DE, Tyihak E. Preliminary investigation of *Origanum onites* essential oil by OPLC and Bio-Arena. J Planar. Chromatogr 2010;23: 225–6.

[159] Moricz AM, Horvath G, Molnar P, Kocsis B, Boszormenyi A, Lemberkovics J, et al. Investigation of thyme (*Thymus vulgaris*) essential oil by use of bioArena system. J Planar Chromatogr 2010;23:406–10.

[160] Ch Salles, Jallogens JC, Crouzet J. Chromatographic separation and partial identifcation of glycosidally bound volatile components of fruits. J Chromatogr 1990;522: 255–65.

[161] Dallenbach-Tolke K, Nyiredy S, Sticher O. Application of various planar chromatographic techniques for the separation of iridoid glycosides from *Veronica officinalis*. J Chromatogr 1990;404:365–71.

[162] Ludwiczuk A, Nyiredy S, Wolski T. Separation of the ginsenosides obtained from the roots of *Panax quinquefolium*. J Planar Chromatogr 2005;18:104–7.

[163] Malinowska I, Gadzikowska M, Waksmundzka_Hajnos M. Mobile phase velocity – a tool for separation of alkaloids by OPLC. J Planar Chromatogr 2005;18:176–80.

[164] Pothier J, Galand N, Viel C. Separation of alkaloids in plant extracts by OPLC with ethyl acetate as mobile phase. J Planar Chromatogr 1991;4:392–6.

[165] Pothier J, Petit-Paly G, Nontagu M, Galand N, Chenieux JC, Rideau M, et al. Separation of quinolizidine and difuloro(2,3-b)quinolinium alkaloids by OPLC. J Planar Chromatogr 1990;3:356–8.

[166] Botz L, Nyiredy S, Sticher O. Separation of ergot alkaloids by HPTLC, OPLC and rotation planar chromatography (RPC) methods. J Planar Chromatogr 1990;3:193–5.

[167] Pothier J, Galand N, Tivollier P, Viel C. Separation of quaternary alkaloids in plant extracts by OPLC. J Planar Chromatogr 1993;6:220–2.

[168] Nagy-Turak A, Vegh Z. Extraction and in situ densitometric determination of alkaloids from *Catharanthus roseus* by means of OPLC on amino-bonded silica layers. I. Optimization and validation of the separation system. J Chromatogr 1994;668:501–7.

[169] David AZ, Mincsovics E, Papai K, Ludanyi K, Antal I, Klebovich I. OPLC comparison of methods for aqueous extraction of *Sennae folium* and *Tilia flos* plant samples. J Planar Chromatogr 2008;21:119–23.

[170] Moricz AM, Kalasz H. Centrifugal layer chromatography – rotation planar chromatography. J Planar Chromatogr 2010;23:415–9.

[171] Hahn R, Nahrstedt A. Hydroxycinnamic acid derivatives, caffeoylmalic and caffeoylaldonic esters from *Chelidonium maius*. Planta Med 1993;59:71–5.

[172] Kaonadji M. Two new C-prenynalted derivatives of 2,4-dihydroxy benzoic acid from *Louchocarpus nicon* roots. J Nat Prod 1988;51:376–8.

[173] Briggs DR, Fielder S. Synthesis and properties of phaseollidin isoflavan. J Nat Prod 1994;57:663–4.

[174] Chulia AJ, Vercauberen J, Mariotte AM. Iridoids and flavones from Gentiana depressa. Phytochemistry 1996;42:139–43.

[175] Chulia AJ, Vercauberen J, Waterman PG. Depresteroide, a mixed iridoid-secoiridoid structure from *Gentiana depressa*. Phytochemistry 1994;36:377–82.

[176] Schaufelbereger D, Hostettmann K. Chemistry and pharmacology of *Gentiana lactea*. Planta Med 1988;54:219–21.

[177] Etse JT, Gray AI, Waterman PG, Bates CD, Watson DG, Thomas DW. Chemistry in the annonaceae. Part XXV. Sesquiterpenes from stem bark of *Cleistopholis* glanca. J Nat Prod 1988;51:314–8.

[178] Sarker SD, Armstrong JA, Waterman PG. (Á)-1,12-oxaguai-10(15)-ene: a sesquiterpene from *Eriosternon fitzgeraldii*. Phytochemistry 1995;40:1159–62.

[179] Boustie J, Gleye J, Blaise A, Fouraste I. Limonoids of *Fagaropsis glabra*. Planta Med 1992;58:228–32.

[180] Sattar EA, Mossa JS, Muhammad I, El-Faraly FS. Neo-clerodane diterpenoids from *Teucrium yemenese*. Phytochemistry 1995;40:1737–41.

[181] Houghton PJ, Said IM. 3-dehydromitragynine, an alkaloid from *Mitragyna speciosa*. Phytochemistry 1986;25:2910–2.

[182] Ferrari M, Verotta L. Centrifugal TLC of alkaloids from vegetable sources using aluminium oxide layer. J Chromatogr 1988;437:328–32.

[183] Sayed HM, Desai HK, Ross SA, Pelletier SW. New diterpenoid alkaloids from the roots of *Aconitum septentrionale*: isolation by an ion-exchange method. J Nat Prod 1992;55: 1595–606.

[184] Ghedira K, Chemli R, Richard B, Nuzillard JM, Zeches M, Le Men-Olivier L. Two cyclopeptide alkaloids from *Zizyphus lotus*. Phytochemistry 1993;32:1591–4.

[185] Vander Heyden Y. Extracting information from chromatographic herbal fingerprints. LCGC Eur 2008;21(9):438–43.

[186] Zidorn Ch. Plant chemosystematics. In: Waksmundzka-Hajnos M, Sherma J, Kowalska T, editors. Thin layer chromatography in phytochemistry. Boca Raton;London New York: CRC Press; 2008.

[187] Chen S-B, Liu H-P, Tian -T, Yang D-J, Chen S-L, Xu H-X, et al. High performance thin-layer chromatographic fingerprints of isoflavonoids for distinguishing between *Radix Puerariae* lobate and *Radix Puerariae thomsonii*. J Chromatogr A 2006;1121: 114–9.

[188] Feng C, Wan X, Jiang Ch, Tan H, Hu Y, Can H. Comparison of HPTLC, HPLC and HPCE for fingerprinting of *Pueraria* radix. J Planar Chromatogr 2006;19:348–54.

[189] Matysik G, Soczewiński E, Polak B. Improvement of separation in zonal preparative thin-layer chromatography by gradient elution. Chromatographia 1994;39:497–504.

[190] Matysik G, Soczewiński E. Computer-aided optimization of stepwise gradient TLC of plant extracts. J Planar Chromatogr 1996;9:404–12.

[191] Matysik G, Markowski W, Soczewiński E, Polak B. Computer-aided optimization of stepwise gradient profiles in thin-layer chromatography. Chromatographia 1992;34: 303–7.

[192] Matysik G, Kowalski J, Strzelecka H, Soczewiński E. Investigation of the accumulation of lanatosides in digitalis-ianata by gradient elution TLC and densitometry. J Planar Chromatogr 1994;7:129–32.

[193] Matysik G. Modified programmed multiple gradient development (MGD) in the analysis of complex plant extracts. Chromatographia 1996;43:39–43.

[194] Szaufer-Hajdrych M, Bylka W, Matławska I, Wójciak-Kosior M, Matysik G, Jodynis-Liebert J. Densitometric HPTLC analysis of phenolic acids from *Aquillegia vulgaris*. Acta Chromatogr 2008;20:685–95.

[195] Migas P, Cisowski W. TLC-MGD analysis of phenolic acids from tissue cultures of *Vaccinium corymbosum* var. bluecrop. In: Nyiredy S, Kakuk A, editors. Planar chromatography; 2000. pp. 171–2. Hungary 200.

[196] Smolarz HD, Matysik G, Wójciak-Kosior M. HPTLC and densitometric determination of flavonoids in *Vaccinium myrtillus* and *Vaccinium vits-idaea*. J Planar Chromatogr 2000;13:101–5.

[197] Krauze-Baranowska M, Malinowska I, Skwierawska J. TLC of flavonol truxnic esters from *Pseudotsuga* menziessi. J Planar Chromatogr 2002;15:437–41.

[198] Cieśla Ł, Staszek D, Hajnos M, Kowalska T, Waksmundzka-Hajnos M. Development of chromatographic and free radical scavenging activity fingerprints by thin-layer chromatography for selected *Salvia* species. Phytochem Anal 2011;22:59–65.

[199] Cieśla Ł. Biological fingerprinting of herbal samples by means of liquid chromatography. Chromatogr Res Int 2012;2012:9. http://dx.doi.org/10.1155/2012/532418. Article ID 532418.

[200] Choma IM, Grzelak EM. Bioautography detection in thin-layer chromatography. J Chromatogr A 2011;1218:2684–91.

[201] Krüger S, Urmann O, Morlock GE. Development of planar chromatographic method for quantitation of anthocyanes in pomace, feed, juice and wine. J Chromatogr A 2013; 1289:105–18.

[202] Cieśla Ł. Thin-layer chromatography with biodetection in the search for new potential drugs to treat neurodegenerative diseases – state of the art and future perspectives. Med Chem 2012;8:102–11.

[203] Marston A. Thin-layer chromatography with biological detection in phytochemistry. J Chromatogr A 2011;1218:2676–83.

[204] Morlock GE, Schwack W. The contribution of planar chromatography to food analysis. J Planar Chromatogr 2007;20:399–406.

[205] Cieśla Ł, Kryszeń J, Stochmal A, Oleszek W, Waksmundzka-Hajnos M. Approach to develop a standardized TLC-DPPH test for assessing free radical scavenging properties of selected phenolic compounds. J Pharm Biomed Anal 2012;70:126–35.

[206] Di Giovanni S, Borloz A, Urbain A, Marston A, Hostettmann K, Carrupt P-A, et al. In vitro screening assays to identify natural or synthetic acetylcholinesterase inhibitors: thin layer chromatography versus microplate methods. Eur J Pharm Sci 2008;33: 109–19.

[207] Botz L, Nagy S, Kocsis B. Detection of microbiologically active compounds. In: Nyiredy Sz, editor. Planar chromatography. A retrospective view for the third millennium. Budapest: Springer Scientific Publisher; 2001. pp. 489–516.

[208] Tyihak E, Moricz A, Ott PG. Biodetection and determination of biological activity of natural compounds. In: Waksmundzka-Hajnos M, Sherma J, Kowalska T, editors. Thin layer chromatography in phytochemistry. Boca Raton: CRC Press; 2008. pp. 193–213.

[209] Cieśla Ł, Kowalska I. Detection and identification in TLC drug analysis. In: Komsta Ł, Waksmundzka-Hajnos M, Sherma J, editors. Thin layer chromatography in drug analysis. Boca Raton: CRC Press; 2013. pp. 193–224.

[210] Tyihak E, Mincsovics E, Moricz AM. Overpressured layer chromatography: from the pressurized ultramicro chamber to BioArena system. J Chromatogr A 2012;1232:3–18.

[211] Cheng Z, Wu T. TLC bioautography: high throughput technique for screening of bioactive natural products. Comb Chem High Throughput Screen 2013;16:531–49.

[212] Reich E, Schibli A. High-performance thin-layer chromatography for the analysis of medicinal plants. Stuttgart: Thieme; 2006.

[213] Gössi A, Scherer U, Schlotterbeck G. Thin-layer chromatography- nuclear magnetic resonance spectroscopy – a versatile tool for pharmaceutical and natural products analysis. Chimia 2012;66:347–9.

[214] Adhami HR, Scherer U, Kaehlig H, Hettich T, Schlotterbeck G, Reich E, et al. Combination of bioautography with HPTLC-MS/NMR: a fast identification of acetylcholinesterase inhibitors from *Galbanum*. Phytochem Anal 2013;24:395–400.

[215] Coran SA, Mulas S, Vasconi A. Profiling of components and validated determination of iridoids in *Gardenia jasminoides* Ellis fruit by a high-performance thin-layer chromatography/mass spectrometry approach. J Chromatogr A 2014;1325:221–6.

[216] Smrke S, Vovk I. Comprehensive thin-layer chromatography mass spectrometry of flavanols from *Juniperus communis* and *Punica granatum* L. J Chromatogr A 2013;1289: 119–26.

[217] Naumoska K, Simonovska B, Albreht A, Vovk I. TLC and TLC-MS screening of ursolic, oleanolic and betulinic acids in plant extracts. J Planar Chromatogr 2013;26:125–31.

[218] Sajewicz M, Staszek D, Natić M, Wojtal Ł, Waksmundzka-Hajnos M, Kowalska T. TLC-MS versus TLC-LC-MS fingerprints of herbal extracts. Part II. Phenolic acids and flavonoids. J Liq Chromatogr Relat Technol 2011;10-11:864–87.

[219] Sajewicz M, Wojtal Ł, Natić M, Staszek D, Waksmundzka-Hajnos M, Kowalska T. TLC-MS versus TLC-LC-MS fingerprints of herbal extracts. Part I. essential oils. J Liq Chromatogr Relat Technol 2011;10-11:848–63.

[220] Han QB, Song JZ, Qiao CF, Wong L, Xu H-X. Preparation isolation of cyclolanostane-type saponins from AstragalusmembranaceusBge. var. mongholicus (Bge.) Hsiao by TLC-MS/MS guided high-speed counter-current chromatography. J Sep Sci 2007;30: 135–40.

[221] Kassim NK, Rahmani M, Ismail A, Sukari MA, Ee GCL, Nasir NM, et al. Antioxidant activity-guided separation of coumarins and lignan from *Melicope glabra* (Rutaceae). Food Chem 2013;139:87–92.

[222] Rajauria G, Abu-Ghannam N. Isolation and partial characterization of bioactive fucoxanthin from himanthaliaelongata brown seaweed: a TLC-based approach. Int J Anal Chem; 2013. Article number 802573.

[223] Zhao J, Zhang JS, Yang B, Lv GP, Li SP. Free radical scavenging activity and characterization of sesquiterpenoids in four species of curcuma using a TLC bioautography assay and GC-MS analysis. Molecules 2010;15:7547–57.

[224] Gu L, Wu T, Wang Z. TLC bioautography-guided isolation of antioxidants from fruit of *Perillafrutescens* var. acuta. LWT – Food Sci Technol 2009;42:131–6.

[225] Rhee IK, Van De Meent M, Ingkaninan K, Verpoorte R. Screening for acetylcholinesterase inhibitors from *Amaryllidaceae* using silica gel thin-layer chromatography in combination with bioactivity staining. J Chromatogr A 2001;915:217–23.

[226] Marston A, Kissling J, Hostettmann K. A rapid TLC bioautographic method for the detection of acetylcholinesterase and butyrylcholinesterase inhibitors in plants. Phytochem Anal 2002;13:51–4.

[227] Cahlíková L, Ločárek M, Benešová N, Kučera R, Chlebek J, Novák Z, et al. Isolation and cholinesterase inhibitory activity of *Narcissus* extracts and *Amaryllidaceae* alkaloid. Nat Prod Com 2013;8:781–5.

[228] Oliveira AP, Cruz RAS, Botas GS, Kuster RM, Rocha L. Bio-assay guided isolation and structural identification of acetylcholinesterase inhibitors from *Sideroxylon obtusifolium* (Roem. &Schult.) T.D. Penn. Lat Am J Pharm 2013;32:765–8.

[229] Rhee IK, van Rijn RM, Verpoorte R. Qualitative determination of false-positive effects in the acetylcholinesterase assay using thin layer chromatography. Phytochem Anal 2003;14:127–31.

[230] Salazar MO, Furlan RLE. A rapid TLC autographic method for the detection of glucosidase inhibitors. Phytochem Anal 2007;18:209–12.

[231] Simões-Pires CA, Hmicha B, Marston A, Hostettmann KA. TLC bioautographic method for the detection of α- and β-glucosidase inhibitors in plant extracts. Phytochem Anal 2009;20:511–5.

[232] Ramallo IA, Sierra MG, Furlan RLE. Discovery of β-glucosidase inhibitors from a chemically engineered extract prepared through ethanolysis. Med Chem 2012;8: 112–7.

[233] Ramallo IA, Zacchino SA, Furlan RLE. A rapid TLC autographic method for the detection of xanthine oxidase inhibitors and superoxide scavengers. Phytochem Anal 2006; 17:15–9.

[234] Wangthong S, Tonsiripakdee I, Monhaphol T, Nonthabenjawan R, Wanichwecharungruang SP. Post TLC developing technique for tyrosinase inhibitor detection. Biomed Chromatogr 2007;21:94–100.

[235] Hassan AMS. TLC bioautographic method for detecting lipase inhibitors. Phytochem Anal 2012;23:405–7.

[236] Godall RR, Levi AA. A microchromatographic method for the detection and approximatedetermination of the different penicillins in a mixture. Nature 1946; 158:675–6.

[237] Műller MB, Dausend C, Frimmel FH. A new bioautographic screening method for the detection of estrogenic compounds. Chromatographia 2004;60:207–11.

[238] Salazar MO, Viarengo G, Sciara MI, Kieffer PM, Garcia Vescovi E, Furlan RLE. A thin-layer chromatography autographic method for the detection of inhibitors of the salmonella PhoP-PhoQ regulatory system. Phytochem Anal 2014;25:155–60.

[239] Sadgrove NJ, Telford IRH, Greatrex BW, Jones GL. Composition and antimicrobial activity of the essential oils from the *Phebalium squamulosum* species complex (Rutaceae) in New South Wales, Australia. Phytochem 2014;97:38–45.

[240] Horváth G, Ács K, Kocsis B. TLC-direct bioautography for determination of antibacterial activity of *Artemisia adamsii* essential oil. J AOAC Int 2013;96:1209–13.

[241] Altintas A, Tabanca N, Tyihák E, Ott PG, Móricz AM, Mincsovics E, et al. Characterization of volatile constituents from *Origanum onites* and their antifungal and antibacterial activity. J AOAC Int 2013;96:1200–8.

[242] Sadgrove NJ, Jones GL. Characterization and bioactivity of essential oils from *Geijer-aparviflora* (Rutaceae): a native bush medicine from Australia. Nat Prod Com 2013;8: 747–51.

[243] Jesionek W, Grzelak EM, Majer-Dziedzic B, Choma IM. Thin-layer chromatography - direct bioautography for the screening of antimicrobial properties of plant extracts. J Planar Chromatogr 2013;26:109–13.

[244] Moricz AM, Tyihak E, Ott PG. Usefulness of transgenic luminescent bacteria in direct bioautographic investigation of chamomile extracts. J Planar Chromatogr 2010;23: 180–3.

[245] Vovk I, Prošek M, Kaiser RE. Image analysis. In: Nyiredy Sz, editor. Planar chromatography. A retrospective view for the third millennium. Budapest: Springer Scientific Publisher; 2001. pp. 464–88.

[246] Cieśla Ł, Hajnos M, Staszek D, Wojtal L, Kowalska T, Waksmundzka-Hajnos M. Validated binary high-performance thin-layer chromatographic fingerprints of polyphenolics for distinguishing different *Salvia* species. J Chromatogr Sci 2010;48: 421–7.

[247] Cieśla Ł, Waksmundzka-Hajnos M. Application of thin-layer chromatography for the quality control and screening the free radical scavenging activity of selected pharmaceutical preparations containing *Salvia officinalis* L. extract. Acta Pol Pharm 2010;67: 481–5.

[248] Komsta Ł, Cieśla Ł, Bogucka-Kocka A, Józefczyk A, Kryszeń J, Waksmundzka-Hajnos M. The start-to-end chemometric image processing of 2D thin-layer videoscans. J Chromatogr A 2011;1218:2820–5.

[249] Cieśla Ł, Hajnos M, Waksmundzka-Hajnos M. Application of hydrophilic interaction TLC systems for separation of highly polar glycosidic compounds from the flowers of selected *Verbascum* species. J Planar. Chromatogr 2011;24:295–300.

[250] Ristivojević P, Andrić FL, Trifković JD, Vovk I, Stanisavljević LZ, Tešić ZL, et al. Pattern recognition methods and multivariate image analysis in HPTLC fingerprinting of propolis extracts. J Chemom; 2014. http://dx.doi.org/10.1002/cem.2592.

[251] Olech M, Komsta Ł, Nowak R, Cieśla Ł, Waksmundzka-Hajnos M. Investigation of antiradical activity of plant material by thin-layer chromatography with image processing. Food Chem 2012;132:549–53.

[252] Cieśla L, Kowalska I, Oleszek W, Stochmal A. Free radical scavenging activities of polyphenolic compounds isolated from *Medicago sativa* and *Medicago truncatula* assessed by means of thin-layer chromatography DPPH˙ rapid test. Phytochem Anal 2013;24:47–52.

[253] Kowalska I, Jedrejek D, Cieśla Ł, Pecio Ł, Masullo M, Piacente S, et al. Isolation, chemical and free radical scavenging characterization of phenolics from *Trifolium scabrum* L. aerial parts. J Agric Food Chem 2013;61:4417–23.

[254] Kowalska I, Cieśla Ł, Oniszczuk T, Waksmundzka-Hajnos M, Oleszek W, Stochmal A. Comparison of two TLC-DPPH•-image processing procedures for studying free radical scavenging activity of compounds from selected varieties of *Medicago sativa*. J Liq Chromatogr Relat Technol 2013;36:2387–94.

[255] Cieśla Ł, Staszek D, Kowalska T, Waksmundzka-Hajnos M. The use of TLC-DPPH• test with image processing to study direct antioxidant activity of phenolic acid fractions of selected *Lamiaceae* family species. J AOAC Int 2013;96:1228–32.

[256] Hawrył MA, Nowak R, Waksmundzka-Hajnos M, Świeboda R, Robak M. Two-dimensional thin layer chromatographic separation of phenolic compounds from *Eupatorium cannabinum* extracts and their antioxidant activity. Med Chem 2012;8: 118–31.

[257] Hawrył MA, Nowak R, Waksmundzka-Hajnos M. Two-dimensional thin-layer chromatographic determination of phenolic antioxidants from *Eupatorium cannabinum* extracts on cyano-bonded polar stationary phases. J Planar Chromatogr 2012;25: 394–402.

[258] Chewchinda S, Ruangwises N, Gritsanapan W. Comparative analysis of rhein content in *Cassia fistula* pod extract by thin-layer chromatographic-densitometric and TLC image methods. J Planar Chromatogr 2014;27:29–32.

[259] Komsta Ł. Suppressing the charged coupled device noise in univariate thin-layer videoscans: a comparison of several algorithms. J Chromatogr A 2009;1216:2548–53.

[260] Komsta Ł. Dealing with charged-coupled device noise in thin-layervideodensitometry. Optimization of several image-denoising techniques. Acta Chromatogr 2009;21:355–67.

[261] Sima IA, Casoni D, Sârbu C. Simultaneous determination of carbidopa and levodopa using a new TLC method and a free radical as detection reagent. J Liq Chromatogr Relat Technol 2013;36:2395–404.

[262] Sima IA, Casoni D, Sârbu C. High sensitive and selective HPTLC method assisted by digital image processing for simultaneous determination of catecholamines and related drugs. Talanta 2013;114:117–23.

CHAPTER

20

Analysis of Dyes and Inks

Virginia Coman, Florina Copaciu

Raluca Ripan Institute for Research in Chemistry, Babeş-Bolyai University, Cluj-Napoca, Romania

20.1 INTRODUCTION

Synthetic dyes and pigments are colored substances because they absorb more wavelengths of light than others. They are extensively used in a large number of industries (textile, food, plastics, cosmetics, pharmaceuticals, paper, writing and printings inks, etc.) and also in medicine, painting, agriculture, research, water science, and technology. It is very important in these applications to have information about purity (presence or absence of impurities or intermediates), identity, and uniformity (homogeneity) of colored compounds [1–3].

Conventional thin-layer chromatography (TLC) is a qualitative or semiquantitative method of analysis characterized by simplicity, rapidity and efficiency, low-cost analysis, not requiring sophisticated instrumentation. Also, TLC offers minimal sample cleanup, reduced number of sample preparation steps, simultaneous analysis of a considerable number of samples since all sample components are located on the same chromatoplate, many possibilities for detection even for the compounds with poor detection characteristics that need a postchromatographic treatment, and so on [4–7]. Based on the full capabilities of conventional TLC [8,9], high-performance TLC (HPTLC) is an instrumental and automation technique controlled by computer devices given precise quantitative results provided by accurate separations with higher efficiency and improved detection limits based on in situ measurements using a variety of fine particle layers and a recorded chromatogram as a fingerprint of the separation process [5,10–15]. Modern TLC is not a fully automated method, but various steps are automated by the use of instruments allowing optimization of each step of analysis of individual sample components [10,14,16]. TLC can be used to separate and identify the compounds from a mixture, to determine the substance purity, to monitor

Instrumental Thin-Layer Chromatography
http://dx.doi.org/10.1016/B978-0-12-417223-4.00020-0

the progress of a reaction, to find the proper conditions for column chromatography, and to analyze the fractions obtained from column chromatography [1,6–8,14,17].

In order to improve the lower separation capacity due to the isocratic TLC conditions and the efficacy of the chromatographic development, multidimensional TLC methods (gradient elution [5,6,16], unidimensional programmed or automated multiple development [6,12,17–20], two-dimensional development [20]) were welcomed. Other two main innovations of instrumental TLC are the interfaces to spectroscopic instruments and emerging techniques for video densitometry [6,10–14]. Instrumental TLC (Figure 20.1) usually involves four main stages: automated application of standards and samples to the layer as bands or spots, development of the layer with the mobile phase in an automatic developing chamber or automated multiple development (AMD), detection and identification of the separated analyte zones by image analysis using illumination unit and a digital camera, and quantification by TLC-densitometry using integrated software, which collects and evaluates all data for generating analysis reports [5,11–13,15]. Instrumental TLC offers automation, reproducibility, and accurate quantification for a wide variety of applications [5,13,20].

The off-line or on-line coupling of TLC with some instrumental detection methods [11,15], such as Ultraviolet-Visible (UV-Vis) [5,12], Fourier transform infrared spectroscopy (TLC-FTIR) [11–13], near-infrared FT-Raman spectroscopy [15] or mass spectrometry (TLC-MS) [5,11,13], generated the hyphenated techniques that received the attributes of both coupled techniques being helpful for structural elucidation. On-line TLC-MS coupling opened new possibilities of identification. These techniques allow the successful identification of unknown compounds from different matrices without using standards as in conventional TLC. By coupling two complementary separation techniques,

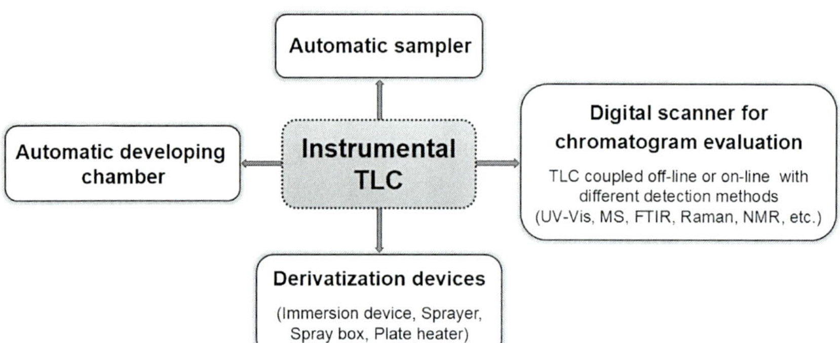

FIGURE 20.1 **Instrumental thin-layer chromatography.**

column liquid chromatography to TLC (LC-TLC, multimodal separation technique), an increase in the separation capacity is obtained [8,9,20].

TLC retains a historic link with the characterization of dyes [21,22] and inks [23,24].

20.2 DYES

20.2.1 About Dyes

The second part of the nineteenth century has been revolutionized by the numerous dyestuffs for textile industry starting with the accidental discovery of historical importance of the first synthetic dye (mauve, 1856, W.H. Perkin). A dye is usually a colored organic compound or a mixture that may be used to give color to a substrate (food, fabric, textile cloth, paper, plastic, leather, inks, others) having resistance to normal laundry or cleansing procedures and stability to light. All the dyes may not necessarily be colored substances. Therefore, optical brighteners or whiteners, which may be called as white dyes, may be included in the term "dye" [2,3]. The dye absorbs light strongly in the visible region and can firmly attach to the support by physical and chemical bonds between the functional groups of the dye and those on the support [25]. This absorption is caused by the energy alteration of delocalized electron systems of the aromatic structures of dyes by certain atomic configurations called chromophore groups. These configurations include alternate single and double bonds incorporating nitrogen, carbon, oxygen, or sulfur. Dye color can be enhanced by the addition of some functional groups (hydroxyl, carboxyl, amino, and sulfonyl) called auxochrome groups [23].

The number of dyes reaches thousands. Dyes can be classified [2,3,25] according to chemical structure, tinctorial principle, and dyeing support as given in Figure 20.2. Classification by application (dyeing support) is taken into consideration by the Colour Index, which covers all colorant classes and assigns for each dye a number by which it can be identified on market.

Due to their easy manufacture, broad spectrum of colors, and high dyeing power, azo dyes are the most widely used synthetic dyes. They are used for attraction and decoration purposes in various fields of everyday life including food production, textile industry, paper production, plastics, cosmetics, inks, paintings, biology, medicine, agriculture, and so on [3,22,25]. Most of them contain diazo and amino groups and/or metal complexes and produce skin diseases or have carcinogenic effects [26]. As a result, their identification, separation, and quantification have a particular importance for analytical chemistry [1,21,22].

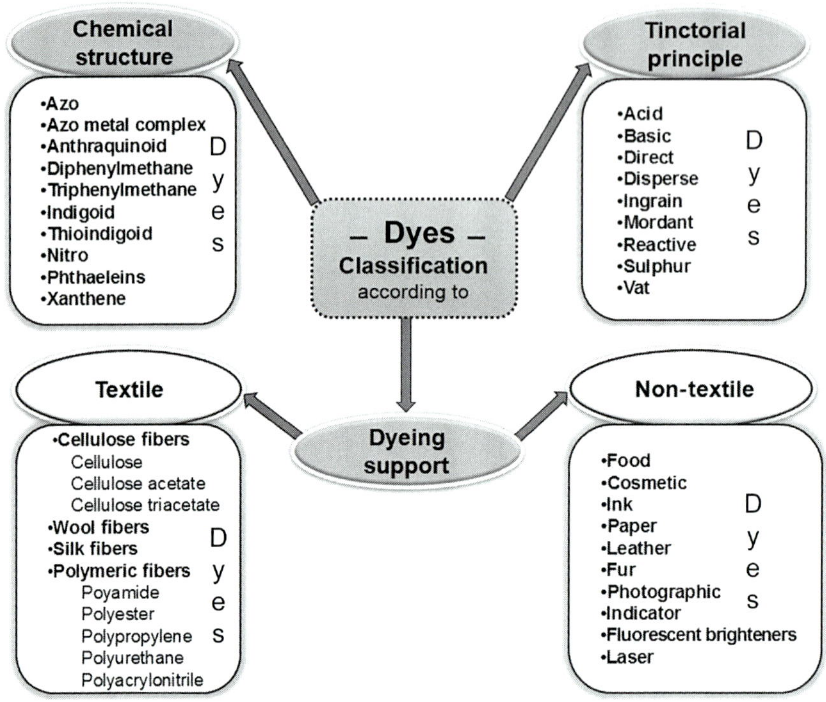

FIGURE 20.2 **Classification of dyes.**

20.2.2 Instrumental TLC for Dye Separation

Separation of dyes presents difficulties due to the close Rf values and spot tailing, the separation mechanism being not well known [21]. Conventional TLC is a valuable tool for the purity control and identification of new synthetic dyes and the analysis of dyes in complex matrices [1,21,22], but it is not usually applied to the analysis of samples containing traces of synthetic dyes due to its low sensitivity [27–29]. These analyses require performance TLC equipments mainly for the detection step. Instrumental TLC due to its various capabilities of detection is suitable to identify and quantify compounds at low levels. The combination of TLC with GC-MS, MS, FTIR, Raman or high-performance liquid chromatography (HPLC) gives new possibilities for the detection (identification and possible quantification) of traces of dyes found in different matrices [1,15,22]. In the TLC/HPTLC analysis of dyes, the use of video imaging techniques has great importance because it usually involves a large number of samples [1]. Usually, before TLC analysis, the extraction of dyes from the matrix is required. This can be done by solvent extraction, solid-phase extraction, solid-phase microextraction, or accelerated solvent extraction

[15,27,29]. Most of the reported TLC methods for dye analysis, both normal and reversed phase, use as mobile phase a mixture of organic solvents (methanol, ethanol, butanol, ethyl acetate, acetone, etc.) [1,21].

An overview of the research data from the literature focused on the instrumental TLC conditions of dyes from different food matrices (energy drinks, soft drinks, jelly, sugar candy, food, etc.) is presented in Table 20.1. Further, Table 20.2 shows applications of instrumental TLC for the textile, cosmetic, and other dyes. Though for dye analysis from food and textiles the TLC instrumentation is usually represented by samplers, densitometers, and scanners, more complex instruments are found when dyes are used to demonstrate the performances of the TLC methods under research. Overpressured layer chromatography (OPLC) technique is characterized by higher resolution, shorter analysis time, and improved separation efficiency than conventional TLC, combining the advantages of TLC/HPTLC and HPLC [5,12,15]. These applications for the dyes, called test dyes, are given in Table 20.3.

20.3 INKS

20.3.1 About Inks

Ink is a liquid or paste containing dyes and pigments, used for writing or drawing by a pen, brush, or quill or for marking a text, design, image, or colored surface. Inks include writing, typewriting, stamp, printing, printer, jet, and toner inks. Thicker inks, in the form of paste, are used in letterpress and lithographic printing. Composition of inks is complex. They contain dyes, pigments, solvents, lubricants, resins, surfactants, particulate matter, fluorescers, and other materials (Figure 20.3) [23,67]. The components of inks are responsible for many purposes such as dyeing, carrying, flow and thickness of ink, and its appearance when dry. All inks are made of dyeing materials found in a vehicle, which can be liquids (writing and jet inks), paste (printing inks), or solids (typewriting ribbon inks, toners). Nowadays, synthetic dyes are widely used in inks, their development being connected to the progress in writing, drawing, and printing [2]. Dyes are distinguished from pigments by their solubility in ink vehicles [23,24,67–69]. The selection of dyes and formulation of inks [23,24] depend on the ink application, the specific technology used (aqueous, solvent-based, or hot melt), and the printer type [2].

Usually, ink dyes are selected from food, acid, direct, sulfur, and reactive dyes for ink-jet applications and for writing, drawing, or marking materials. For the ink analysts, four classes of dyes present interest, namely: arylmethane (methyl violet, methyl blue), azo dyes (monoazo: acid orange 10, solvent black 47, solvent black 46, solvent yellow 162,

TABLE 20.1 Instrumental TLC Conditions and Sample Preparation for Analyzing Food Dyes from Different Matrices

Matrix	Dyes	Sample Preparation	Instrumental TLC	TLC Conditions (Stationary Phase; Mobile Phase (v/v); Detection; Quantification). Others	Observation	Reference
Mustard	Tartrazine	Simple liquid extraction followed by filtration	TLC-PD; BioDit TLC scanner in visible mode; Digitalization by image decipher-TLC software.	Silica gel 60 TLC plate; Isopropanol–ammonia (7:3).	Tartrazine from mustard determined by TLC-photodensitometry.	[30]
Drinks and drops	Tartrazine, Quinoline yellow, Azorubine, Ponceau 4R, Allura red AC, Patent blue V, Brilliant blue FCF	Bakerbond C18 SPE columns	HPTLC-DAD scanner.	HPTLC plates: RP-18W F_{254S} or CN F_{254}; Methanol–acetic or citric buffer with DEA/OSA-Na; Horizontal Teflon DS chamber; DAD scanning densitometry. Satisfactory LOD and LOQ values for all dyes (ng per zone).	HPTLC-DAD method for identification and quantification of dyes in food samples.	[31]
Energy drinks	Brilliant black BN, Tartrazine, Ponceau 6R, Resorcin yellow, Fast yellow AB, Orcein, Allura red, Green S, Amaranth, Quinoline yellow, Acid blue, Erythrosine, Sunset yellow FCF, Indigo carmine, Ponceau 4R, Azorubine, Brilliant blue FCF, Carmine, Scarlet GN, etc.	Dilution	Working station: application, development, and plate evaluation.	Silica gel $60F_{254}$ HPTLC plates; Ethyl acetate–methanol–water–acetic acid (65:23:11:1); Horizontal developing chamber; UV-Vis densitometry.	A rapid planar chromatographic method for identification and quantification of 25 water-soluble dyes used as food additives.	[28]
Beverages	Tartrazine, Sunset yellow FCF, Allura red AC, Ponceau 4R, Brilliant blue FCF, Indigotine,	Bakerbond C18 SPE columns	Scanner diode-array spectrophotometer.	HPTLC plates: silica gel $60F_{254}$, RP-18W F_{254S}, CN F_{254}; Methanol–acetate buffer pH 3.5	Quantitative determination of	[32]

	Brilliant black PN, Quinoline yellow, Patent blue V, Brilliant green BS, Azorubine, Brown HT			(4:1); Horizontal Teflon DS chamber; Visible and UV light at 254 and 366 nm.	tartrazine and sunset yellow in beverages.	
Soft drinks	Tartrazine, Azorubine, Sunset yellow		HPTLC with image processing of scanned chromatograms; TrueColor setting of the HP ScanJet Driver.	Nano-SIL NH_2/UV_{254} plates; Isopropanol–diethyl ether–ammonia (2:2:1); Quantitative determination by: Flatbed HP ScanJet; TLC scanner software for digital processing; LOD and LOQ calculation by SMAC; Statistical data treatment. Validation. Data processing on any computer of medium configuration.	Instrumentation is simple, easy to handle and acquire. Reliable quantitative evaluation in short time.	[33]
Food dye standard mixture	Azorubine, Amaranth, Ponceau 4R, Tartrazine, Sunset yellow FCF, Yellow 2G, Patent blue V, Brilliant blue FCF, Brilliant black BN, Indigo carmine		Flatbed scanner with optical resolution in a TrueColor mode.	Sorbfil PTSKh-P-V; Mobile phase OST (Branch Standard) 10298-2002; Ascending TLC; Computer-based handling of chromatogram images in various color channels.	Software processing of scanned chromatogram images.	[34]
Food dye standard	Tartrazine		Flatbed scanner as a detector in quantitative TLC analysis.	Ascending TLC. Computational procedure by flatbed scanner operating in the color mode. Chromatograms scanned and processed by Adobe Photoshop package.	TLC chromatogram treatment in the color mode can reduce errors.	[35]
Jelly, sugar candy	Ponceau 4R, Tartrazine, Sunset yellow	Dyes eluted with acetone–	TLC; Spectrophotometer	Silica gel 60 plate; Isopropanol–ammonia–water (10:1:1); Concentration of dyes found by	A simple first-derivative spectrophotometric	[36]

Continued

TABLE 20.1 Instrumental TLC Conditions and Sample Preparation for Analyzing Food Dyes from Different Matrices—cont'd

Matrix	Dyes	Sample Preparation	Instrumental TLC	TLC Conditions (Stationary Phase; Mobile Phase (v/v); Detection; Quantification). Others	Observation	Reference
		ammonia by a polyamide. Extraction with Soxhlet apparatus		measuring the absorbance of solutions obtained from the scraped adsorbents after development.	method for dye determination from sweets.	
48 Food samples	Coal tar dyes		HPTLC; Densitometry.	C18 layers; Acetonitrile or methyl ethyl ketone–methanol–5% aq. sodium sulfate (1:1:1); Measurement of visible absorption spectra with sample concentration techniques to improve detection limits 10- to 20-fold.	Identification of traces of coal tar dyes.	[37]
Food	Quinoline yellow, Sunset yellow, Cochineal red A, Indigo carmine, Tartrazine, Amaranth, Erythrozine		OPLC; Densitometry.	Solvent system: Ammonia–methanol–ethy1 acetate (1:3:6), ammonia–methyl ethyl ketone–1-butanol (2:3:5); Quantified by densitometry.	Separation of dyes by overpressured layer chromatography and their quantification by densitometry.	[37]
Food	12 Dyes	SPE XAD-2 column with acetone, methanol, and water	HPTLC; Densitometry.	Silica gel plates; 2-Propanol–1-propanol–1-butanol–ammonia–water (8:4:4:2:1); Densitometry at the maximum absorption wavelength of each dye; Detection limit 4–10 ng/zone.	Identification and quantification of dyes in food extracts.	[37]

Candy and powdered juice	27 Food dyes (12 dyes permitted for use in foods and 15 unlawful dyes in Japan)	Samples dissolved in water and acidified with acetic acid	TLC-FAB-MS method with a sample condensation technique. JMS-AX505W mass spectrometer with a TLC-FAB ion source allowed the measurement of all spectra.	C18-modified silica gel TLC plates; Solvent systems: (A) methanol–acetonitrile–5% sodium sulfate (3:3:10); (B) methanol–methyl ethyl ketone–5% sodium sulfate (1:1:1); Matrix, magic bullet; Condensation of spot with methanol.	TLC-FAB-MS method for identification of unlawful dyes in imported foods.	[38]
Food	Erythrosine B, Phloxine B, Rose bengal, Sulforhodamine B, Brilliant blue FCF, Fast green FCF, Indigo carmine, New Coccine, Sunset yellow FCF, Amaranth, Tartrazine	Solvent extraction	TLC/LSI-(+)MS and TLC-FAB techniques. Both LSI-MS and FAB-MS were performed using a double focusing mass spectrometer fitted with a high field magnetic and a data base system.	Reversed-phase silica gel aluminum-backed homemade TLC plate; Solvent systems: (A) methanol–methyl ethyl ketone–10% sodium sulfate (1:1:3.5) for the xanthenes, and (B) methanol–acetonitrile–10% sodium sulfate (3:3:10) for others; Condensation of spot by thioglycerol; TLC/LSI mass spectra of the separated dyes were measured by "magic bullet" as matrix.	TLC/LSI-MS method for separation and identification of 11 food dyes permitted in Japan and one unknown food dye (azorubine prohibited in Japan) from a candy bought in a foreign country.	[39]
Food	Xanthene dyes (Acid red, Erythrosine, Phloxine, Rose bengal) and triphenylmethane dyes (Brilliant blue FCF, Fast green FCF)		TLC/SIMS with the condensation technique of "magic bullet."	RP-TLC plate: Silica gel aluminum plate impregnated with liquid paraffin; Solvent systems: (A) methanol–acetonitrile–10% aqueous sodium sulfate solution (3:3:10);	TLC/SIMS with a condensation is suitable for practical analysis of food dyes.	[40]

Continued

TABLE 20.1 Instrumental TLC Conditions and Sample Preparation for Analyzing Food Dyes from Different Matrices—cont'd

Matrix	Dyes	Sample Preparation	Instrumental TLC	TLC Conditions (Stationary Phase; Mobile Phase (v/v); Detection; Quantification). Others	Observation	Reference
				(B) methanol–methyl ethyl ketone–10% aqueous sodium sulfate (1:1:3.5). To obtain SIMS spectra with good sensitivities (0.1–0.5 pg/spot), reconcentration of developed spot was performed with thioglycerol. SIMS spectra of the separated dyes were measured on the TLC plate by "magic bullet" as a matrix.		
Food	Group of eight dyes: Tartrazine, Amaranth, Indigo carmine, New Coccine, Sunset yellow FCF, Allura red AC, Fast green FCF, Brilliant blue FCF. Group of five dyes: Acid red, Eosin, Erythrosin, Rose bengal, Phloxine		RP-TLC and UV densitometry.	C18 TLC plate; Solvent systems: (A) methanol-acetonitrile–5.0% aqueous sodium sulfate (3:3:10) (determination of group 8 dyes and screening of group 5 dyes at 400 nm), (B) methanol–methyl ethyl ketone–5.0% aqueous sodium sulfate (1:1:1) (determination of group 5 dyes screened by solvent A at 500 nm). Double elution, two successive developments.	Efficient RP-TLC method for routine analysis of 13 food dyes by a combination of two solvent systems on C18 plate.	[41]

TLC—thin-layer chromatography; TLC-PD—TLC photodetection; HPTLC—high-performance thin-layer chromatography; HPTLC-DAD—HPTLC-diode array scanning densitometry; LOD—limit of detection; LOQ—limit of quantification; SPE—solid-phase extraction; OSA-Na—octane-1-sulfonic acid sodium salt; DEA—diethylamine; TLC/LSI-(+)MS—TLC-liquid secondary ion-(+)mass spectrometry; TLC-FAB-MS—TLC-fast atom bombardment mass spectrometry; TLC/SIMS—thin-layer chromatography-secondary ion mass spectrometry; RP-TLC—reversed-phase TLC.

TABLE 20.2 Instrumental TLC Conditions and Sample Preparation for Analyzing Textile, Cosmetic, and Other Dyes from Different Matrices

Matrix	Dyes	Sample Preparation	Instrumental TLC	TLC conditions (Stationary Phase; Mobile Phase (v/v); Detection; Quantification). Others	Observation	Reference
TEXTILE DYES						
Effluent wastewater from a textile factory	Lanasyn blue, Lanasyn dark brown, Lanasyn red, Nylosan dark brown, Nylosan red	SPE on WAX/ NH_2 Strata cartridges	Desaga TLC system: AS 30 applicator; CD-60 densitometer.	Precoated Alugram RP-18W/UV_{254} plates; *n*-Butanol–ethyl acetate–ammonia 5% (4:4:1); Quantification by UV scanning densitometry at 550 nm. Validated SPE–HPTLC procedure.	SPE–HPTLC procedure applied to monitor textile dyes at ppm level in effluent wastewater.	[27]
Wool, textile fibers	Artist dyestuffs: Alizarin, Purpurin, Carminic acid	Solvent extraction in methanol–formic acid (85:15)	Coupling of thin-layer chromatography and Surface-Enhanced Raman Spectroscopy (TLC-SERS).	Activated silica gel TLC plates; Toluene–acetic acid–methanol (9:2:2). After development and plate drying, silver colloid is directly deposited on the spots of interest, followed by a SERS recording. The dyes which co-elute to nearly the same spot (purpurin + alizarin) can be distinguished by SERS from each other.	Similar molecules with poor resolution and detection limits can be distinguished by coupling TLC with SERS.	[42]
Synthesized dyes; Polymer mixtures	Naphthalimide dyes (16), Benzanthrone dyes (17), Reactive triazine dyes (20), and Fluorescent brighteners (17)	Standard solutions: naphthalimide dyes and fluorescent brighteners in acetone; benzanthrone dyes and reactive triazine dyes in DMF	Camag TLC system: Linomat IV spotting device equipped with a Hamilton microsyringe, Scanner II, SP 4290 Integrator, UV detection.	Naphthalimide dyes: Silica gel $60F_{254}$ plates; *n*-Heptane–acetone (1:1, 3:1), Chloroform–methanol (1:1, 2:1); UV-Vis detection. Benzanthrone dyes: Silica gel $60F_{254}$ plates; *n*-Heptane–acetone (1:1), *n*-Heptane–benzene–chloroform (3:2:1), *n*-Heptane–chloroform–acetone (2:2:1); UV–Vis detection. Benzanthrone dyes: Aluminum oxide $60F_{254}$ plates; *n*-Heptane–ethyl acetate–acetone (1:1:1); UV–Vis detection.	Procedure suitable and reliable to monitor the synthesis and purity of dyes, and to perform dye analysis during application (dyeing and/or stability) on polymers.	[43]

Continued

TABLE 20.2 Instrumental TLC Conditions and Sample Preparation for Analyzing Textile, Cosmetic, and Other Dyes from Different Matrices—cont'd

Matrix	Dyes	Sample Preparation	Instrumental TLC	TLC conditions (Stationary Phase; Mobile Phase (v/v); Detection; Quantification). Others	Observation	Reference
				Triazine dyes: Silica gel $60F_{254}$ plates; *n*-Propanol–ammonia 25% (1:1, 2:1); *n*-Butanol–acetic acid–water (4:1:5); UV–Vis detection. Fluorescent brighteners: Silica gel $60F_{254}$ plates; *n*-Butanol–pyridine–ammonia 25% (1:1:1); Chloroform–methanol–ammonia 25% (11:5:1); UV detection. Identification by comparing the Rf values of dyes from samples with those of standards. Quantification by UV scanning densitometry at 254 nm.		
COSMETIC DYES						
53 Red lipsticks with very similar shades	Coloring agent	Solvent extraction	A combination of TLC and GC-MS can discriminate the indistinguishable lipstick samples by visual analysis.	Silica gel F_{254} aluminum plates; Solvent systems: (A) methylene chloride, (B) chloroform–methanol–water (50:15:2), (C) ethyl acetate–methanol–ammonia (5:1:1). Examination under white and UV lights.	TLC used for coloring agent analysis while GC-MS confirms the chemical structures of organic compounds.	[29]
Commercial cosmetics: nail enamel (a, b), hair stick, pomade	Oil-soluble coal tar dyes: Yellow 404, Yellow 405, Yellow 204, Violet 201, Orange 403, Red 215, Red 225, Blue 403, Green 202, Red 501, Red 505	Solvent extraction	TLC; Scanning densitometry in visible.	RP-$18F_{254S}$ plates; *n*-Hexane–2-butanone (5:1), Acetonitrile–methanol (5:1). Scanning range: 370–700 nm. Identified dyes: Red 501 in nail enamel a, Red 225 in nail enamel b, Violet 201 and Green 202 in hair stick, Yellow 204 and Green 202 in pomade.	Method can be applied to the identification of oil-soluble coal tar dyes in commercial cosmetics.	[44]

Aftershave lotion	Cogilor orange 213.11, Cogilor jaune 113.11, Cogilor rouge 311.00, Cogilor rouge 318.11, Cogilor vert 613.12, Hexacol dep red, Hexacol brilliant blue	Solvent extraction	TLC-FTIR off-line, Jasco 610 FTIR spectrometer.	Tuff-NH_2 (β-aminoethyl-γ-aminopropyl) plates; *n*-Butanol–methanol–water–glacial acetic acid (5:2:2:1). Tuff-C12 plates; Methanol–acetone–water (6:2:2). FTIR spectra using KBr pellet technique.	Cogilor jaune 113.11, Cogilor rouge 318.11, and Cogilor vert 613.12 were found in the cosmetic sample.	[45]
OTHER DYES						
Fuels	Diesel fuel dyes: Morton red 26, C.I. Solvent red 26, C.I. Solvent red 24 (NIST SRM 2037)	For NMR analysis of TLC fractions, the extracted material in ethanol was dissolved in deuterated chloroform	Flatbed scanner, un-scan-it gel software. Coupling of TLC and Nuclear Magnetic Resonance (TLC-NMR).	TLC: Silica gel GHLF plates; Toluene. HPTLC: Silica gel HLF plates with pre-adsorbent zone; Two sequential developments with toluene or a single development with toluene–ethyl acetate (5:5). Visualization by fluorescence quenching. Morton Red 26 is actually C.I. Solvent Red 24, not C.I. Solvent Red 26 as expected from the certification.	TLC-NMR is useful for determination of dye purity and verification of dye structure.	[46]
	Primuline (Direct yellow 59)—Fluorescence dye		Combination of TLC, MS, UV, and fluorescence spectroscopy. Video densitometer.	HPTLC silica gel 60 plates aluminum backed; Chloroform–methanol–acetic acid (80:12:8) or Chloroform–ethanol–water–triethylamine (30:35:7:35). Phosphoslipids visualized by spraying with primuline. After UV light irradiation (366 nm), individual phosphoslipids become detectable as violet spots. Scanning by video densitometry. Combination of TLC, UV and fluorescence spectroscopy shown that dichloramines of	Primuline is used in lipid research for TLC-MS evaluations. In the presence of dichloramines, it should be attentively used if the quantitative evaluations are based on its fluorescence intensity.	[47]

Continued

TABLE 20.2 Instrumental TLC Conditions and Sample Preparation for Analyzing Textile, Cosmetic, and Other Dyes from Different Matrices—cont'd

Matrix	Dyes	Sample Preparation	Instrumental TLC	TLC conditions (Stationary Phase; Mobile Phase (v/v); Detection; Quantification). Others	Observation	Reference
				phosphatidylethanolamine are responsible for the primuline fading in acidic conditions.		
Chemical synthesis of polymerizable dyes	Triazinylaminobenzanthrone dyes		Camag TLC system: Linomat IV applicator, Scanner II (densitometer) with an SP 4290 Integrator.	Aluminum oxide $60F_{254}$ plates aluminum-backed. Dye 1, *n*-Heptane–acetone (1:1); Dyes 2–5, *n*-Heptane–ethyl acetate–acetone (2:2:1). Scanning densitometry.	Synthesis of five triazinylaminobenzanthrone dyes was monitored by quantitative TLC.	[48]
Histological and cytological samples	Triphenylmethane dyes: Crystal violet, Ethyl violet, Victoria blue, Malachite green, Alkali blue, Aniline blue, etc.		Camag Nano-applicator; Spectrodensitometer Scanner II connected to an IBM-compatible PC loaded with CATS software.	Silica gel 60 HPTLC glass plates, Silica gel 60 TLC plastic sheet, Silica gel 60 RP18 HPTLC glass plates, Silica gel 60 RP8 HPTLC glass plates, Silica gel 60 RP18W HPTLC glass plates; Different solvent mixtures: 1-Butanol–acetic acid–water, Ethanol–water–formic acid–PSA, Ethanol–water–acetic acid–PSA, Ethyl acetate–methanol–ammonia–water, 1-Butanol–ethanol–ammonia–water.	Procedure applied to determine the real content of crystal violet, ethyl violet, and Victoria blue from commercial samples.	[49]
Commercial dyes	Thiazin and azure dyes		Camag Nano-applicator; Spectrodensitometer Scanner I.	Silica gel 60 TLC plastic sheet, Silica gel 60 HPTLC glass plates; Solvent systems: (A) 1-butanol–1-butanone–acetic acid–water (30:20:10:10), (B) 1-butanol–acetic acid–water (50:5:10).	HPTLC determination of thionine and azure concentration in commercial dyes.	[50]

TLC—thin-layer chromatography; SPE on WAX—Solid phase extraction with anion exchange; DMF—dimethylformanide; TLC-SERS—TLC Surface-Enhanced Raman Spectroscopy; TLC-NMR—TLC Nuclear Magnetic Resonance; PSA—pentanesulfonic acid sodium salt; FTIR—Fourier transform infrared spectroscopy; HPTLC—high-performance thin-layer chromatography.

TABLE 20.3 Conditions for Analyzing Dyes Used for Testing the Performances of Developed Instrumental TLC Methods

Test Dyes	Instrumental TLC	TLC Conditions (Stationary Phase; Mobile Phase (v/v); Detection; Quantification). Others	Observation	Reference
Rhodamine 5G, Fat green, Fat brown, 1-Aminoanthraquinone, and 4-(2-Pirydylazo)resorcinol, Patent blue, Green S, Azorubin, Allura red, Brilliant black	Horizontal developing chambers: DS-II-5 × 10 and DS-M (Chromdes, Poland); Digital camera Kodak EasyShare C913.	RP-18W HPTLC plates; Methanol–water in different ratios; Isocratic and gradient stepwise HPTLC elution.	New horizontal developing chamber (DS-M) with stepwise gradient elution showed practical advantages in RP-HPTLC.	[51]
Sudan II, Sudan III, *p*-Methoxyazobenzene, *p*-Aminoazobenzene (Group 1) Sudan II, Sudan III, Azobenzene, *p*-Hydroxyazobenzene, *p*-Aminoazobenzene (Group 2)	Planar electrochromatography (PEC) chamber; Sample applicator.	(Group 1) Plates: (1) Silica bonded with phenyl without any end-capping, (2) Silica bonded with octadecyl and end-capped with phenyl; Acetonitrile–water (95:5). (Group 2) Plates: RP-18 sintered silica gel plate; Acetonitrile–water with buffer salt (85:15).	RP-sintered plates bonded or end-capped for PEC.	[52,53]
Lipophilic test dye: Indophenol blue, Sudan red G, 4-Dimethylaminoazobenzene	Planar dielectrochromatography (PDEC) using vertical PDEC chamber; Source of high alternating electric current; Desaga AS-30 applicator; Desaga CD 60 densitometer.	Alox-25 UV_{254} glass plates; Toluene. Experiments with different types of armature. The enhancement of electric field intensity in the chromatographic layer, especially for the spherical and conical armatures, generates electric fields at the TLC plate surface with reduced electric losses.	The different electric field geometries which do not completely cross the glass support can be used in TLC.	[54,55]

Continued

TABLE 20.3 Conditions for Analyzing Dyes Used for Testing the Performances of Developed Instrumental TLC Methods—cont'd

Test Dyes	Instrumental TLC	TLC Conditions (Stationary Phase; Mobile Phase (v/v); Detection; Quantification). Others	Observation	Reference
Mixture 1: Crystal violet, Neutral blue, Xylene cyanol, Methanyl yellow, Acridine orange, Bromothymol dark blue, Indophenol, Ariabel red, Sudan blue II, Sudan IV, Dimethylaminoazobenzene Mixture 2: Dark violet, Bright orange, Yellow, Dark red, Violet	Three-dimensional TLC (3D TLC) on plates with open (OL) and sealed (SL) adsorption layers. Chromatogram processing by video densitometer and software.	Plates: (1) Sorbfil PTLC-P-V-UV, (2) Silica gel $60F_{254}$; Mobile phases of separation type: Mixture 1 (11 dyes), 1D ethanol–acetic acid (10:1) (three separated dyes), 2D acetone (six separated dyes), 3D toluene (11 separated dyes); Mixture 2 (five dyes), 1D tetrahydrofuran–benzene (9:1) (2 separated zones), 2D dichloromethane–benzene (3:1) (8 separated zones), 3D toluene (11 separated zones).	3D TLC is proper for complex mixtures. 3D TLC (OL) has a good efficiency and selectivity. 3D TLC (SL) is more complex, having higher speed and efficiency.	[56]
Rhodamine dyes: Rhodamine 6G, Rhodamine B, Rhodamine 123 FD&C dyes: Erythrosine B (FD&C Red no. 3), Brilliant blue FCF (FD&C Blue no. 1), Fast green FCF (FD&C Green no. 3), Sunset yellow FCF (FD&C Yellow no. 6)	Thin-layer chromatography and mass spectrometry coupled using desorption electrospray ionization (TLC/DESI-MS system). Spot scanning by moving the TLC plate under computer control.	Rhodamine dyes: RP C8 plates, RP C2 plates; Methanol–water with ammonium acetate (8:2); Detection in positive-ion mode using selected reaction monitoring; Fluorescence images of plates by a fluorescence imaging system operating in UV transillumination mode. FD&C dyes: Wettable RP C18 plates; Water–acetone with ammonium acetate (7:3); Detection in negative-ion full scan mode. Plate photos by a digital camera using white light illumination for both cases.	Use of DESI for coupling TLC and MS. DESI emitter position, TLC plate surface, and MS atmospheric sampling orifice play a very important role to obtain maximum analyte signal levels.	[57]

Fluorescein (1), Allyloxylfluorescein (1a), Eosin (2), Allyloxyeosin (2a), Rhodamine B (3), Allyloxyrhodamine B (3a), Erythrosine (4), Allyloxyerythrosine (4a)	Linomat IV device with Hamilton microsyringe, Camag Scanner II with integrator.	Silica gel 60 plates; Benzene–methanol (5:1) for dyes 1, 1a, 2, 2a; Acetonitrile–water (8:2) for dyes 3, 3a; *n*-Butanol–petroleum ether (5:1) for dyes 4, 4a. UV_{254} detection (sensitivity of 0.1 μg for dyes 1, 1a, 2, 2a and 0.2 μg for dyes 3, 3a, 4, 4a). Iodine vapor detection (sensitivity of 0.2 μg for dyes 1, 1a, 2, 2a and 0.5 μg for dyes 3, 3a, 4, 4a).	TLC method can be used to monitor the synthesis, progress, and dye purity.	[58]
Sudan IV, 4-(Diethylamino)-azobenzene, 1-(3-Pyridylazo)-2-naphthol, 1-(4-Chlorophenylazo)-2-naphthol, 1-(4-Hydroxyphenylazo)-2-naphthol, 4-Nitroaniline, 2-Naphthol, 1-Nitronaphthalene	Planar electrochromatography (PEC), Horizontal DS chamber adapted for PEC, Desaga CD-60 densitometer.	Prewetted RP-8 and RP-18F_{254S} HPTLC plates; Acetonitrile–buffer (8:2); Development in open and closed pressurized systems. UV detection at 420 or 254 nm.	PEC separation efficiency is higher than in conventional HPTLC.	[59]
1-Aminoanthraquinone, Fat brown, 4-Diethylaminoazo-benzene, 4-Hydroxybenzeneazo-naphthol-2, 4-(4-(N,N-Ethylethanol)benzeneazo)-N-methylphthalimide, 4-Nitroaniline	Planar electrochromatography (PEC), Horizontal DS chamber, Desaga CD-60 densitometer.	RP-18F_{254S} plate; Ethanol–pH 10 buffer (8:2), PEC at 2 kV; acetonitrile–pH 10 buffer (9:1), PEC at 2.5 kV. PEC development time is less than half that in HPTLC. Densitometric scanning at 440 nm.	Horizontal DS chamber in PEC separations leads to very good results if the chromatographic plate is equipped with a counterplate to eliminate the excessive eluent flow to adsorbent layer.	[60]

Continued

TABLE 20.3 Conditions for Analyzing Dyes Used for Testing the Performances of Developed Instrumental TLC Methods—cont'd

Test Dyes	Instrumental TLC	TLC Conditions (Stationary Phase; Mobile Phase (v/v); Detection; Quantification). Others	Observation	Reference
Acid light yellow G, Acid red B, Disperse deep blue H-GL, Disperse scarlet BWFL, Disperse orange 2BFL, Disperse balas 2GFL, Disperse blue BGL, Disperse red 3B, Disperse yellow 54, Disperse turquoise GL, Disperse fluorescent yellow 10GN, Disperse blue S-BGL, Disperse black 2B	Multidimensional relay development using the half-way development device transfers the unseparated spots to a new thin-layer plate on the basis of the first separation, then redevelops this new plate.	Silica gel G TLC plate (layer 0.5 mm); Elution steps: (1) cyclohexane–ethyl acetate (1:1), (2) diethyl ether–chloroform (2:1), (3) diethyl ether–chloroform (1:1) plate 1 and cyclohexane–ethyl acetate (1:1) plate 2, (4) cyclohexane–ethyl acetate (3:1) plate 3 and ethanol–chloroform (2:1) plate 4. Development and transfer can be repeated.	A mixture of 13 test dyes was used to evaluate a new TLC development mode based on half-way and relay development with satisfactory results. The method is effective and efficient for complex mixtures.	[61]
Test dye mixture III (Analtech)	Camag TLC system: Automatic sampler III; Horizontal developing chamber; Scanner II. Scanning using slit densitometry. Digital image capture and video densitometry.	Silica gel $60F_{254}$ (0.25 mm) plates; Ethyl acetate. Video densitometry was carried out on the digital data obtained and compared with slit densitometry measurements. Images of high quality, easy to process and archive were obtained.	A novel, low-cost imaging device for TLC use—a commercial flatbed scanner modified with a 366-nm UV source.	[62]

Anthocyanins and Chlorophyll from leaves	Overpressured layer chromatography (OPLC) technique. OPLC apparatus from Cobrabid.	Silica gel $60F_{254}$ plates aluminum foil-backed. 2-Propanol–acetone–water for anthocyanins more polar than (8:1:1) and for chlorophyll less polar than (4:1:5). Nitrogen pressure, 0.6 MPa, flow rate, 1 ml/min. Radial chromatograms developed for 5 min.	Better resolution and greater number of separated compounds in OPLC than TLC.	[63]
Violet 1, Sudan red G, Indophenol blue red, Butter yellow	OPLC technique Chrompres 25 OPLC apparatus (labor MIM). Linomat IV applicator. Desaga CD 60 densitometer.	Empore silica plates. Toluene. Pressured ultramicro chambers. Water pump pressure, 2 bar; flow rate, 0.6 ml/min; initial volume, 0.7 ml; eluent pump pressure, 6 bar.	Empore TLC sheets in forced-flow TLC gave similar performances to preimpregnated HPTLC plates.	[64]
Test dye mixture III (Camag): Ariabel red, Sudan blue, Yellow II, Oracet red, Butter yellow	Long-distance OPLC Linomat III TLC spotter, Chrompres 10 and 25 OPLC apparatus. Camag TLC scanner II with HP-9816 software.	Silica gel $60F_{254}$ flexible backed HPTLC plates. Toluene. Special arrangement of layers for long-distance OPLC. Overpressure: 11 and 25 bar, flow rate, 1.25 cm/min. Vis detection at 480 nm. Fully off-line separation of dyes.	Increased spot capacity. Long-distance OPLC successfully applied for biological complex extracts.	[65]
Test dye mixture I (Camag): Butter yellow, Sudan G, Indophenol	OPLC technique. In situ quantitative evaluation of spots by Zeiss PMQ II chromatogram spectrophotometer.	Silica gel $60F_{254}$ and silica gel 60 TLC glass plates and silica gel $60F_{254}$ HPTLC aluminum sheets; Methylene chloride. Normal, ultramicro, and pressured ultramicro chambers.	Spot diameters are significantly smaller for the separations in pressured ultramicro chamber due to decreased diffusion.	[66]

TLC—thin-layer chromatography; HPTLC—high-performance thin-layer chromatography; OPLC—overpressured layer chromatography.

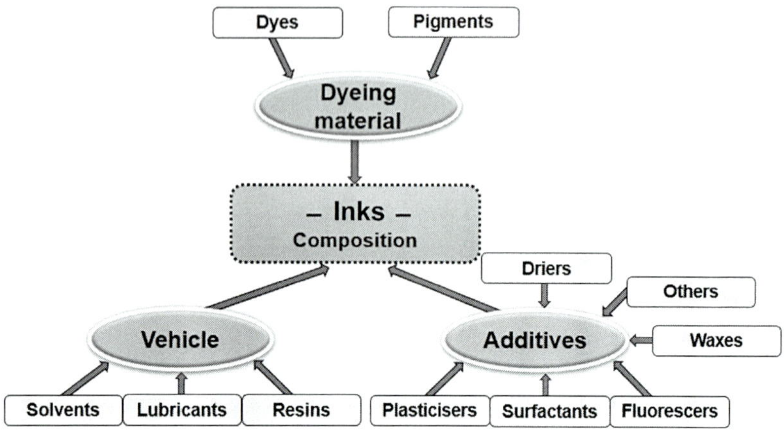

FIGURE 20.3 **Materials used in ink composition.**

solvent yellow 82, reactive black 5; diazo: solvent black 3, direct black 168, reactive black 31, amido black 10B, reactive yellow 37, reactive red 180), phthalocyanine dyes (copper phthalocyanine, solvent blue 38, solvent blue 64, solvent blue 70), and nigrosine dyes (solvent black 5, acid dyes of azine, or aniline chemical classes) [2,23].

20.3.2 Instrumental TLC for Dye Separation from Inks

In document analysis, the most frequent tasks consist of distinguishing inks on the same or different documents and identifying the source or date of a particular ink [23,24,68–71]. In this order, the chemical analyses including spectroscopic (UV-Vis, IR, MS) and chromatographic (TLC, HPLC, GC-MS, CE-capillary electrophoresis) methods are intensively used for the characterization, comparison, determination of source, and dating of ink [67,70–75]. For the dyeing materials that include dyes and pigments with the same solubility in the extraction solvent TLC is a proper technique to discriminate inks and to identify ink formulas, in comparison with a complete collection of standards (library specimens), consequently jet and writing inks and toners can be analyzed by this technique [23,24,68,69]. TLC can separate dyes from the noncolored organic components of the ink in the same plate. Most separated components of inks are easily detected on the developed plate due to their own color. The other separated spots as colorless vehicle components can be visualized by UV light or by spraying the plate with a suitable visualizing reagent. For the inks that cannot be distinguished by human eye or present small differences in their dye components, scanning TLC densitometry has a very important role [23]. The densitograms of the developed plates provide valuable information which is much more accurate than the visual examination of plate and can be considered as a fingerprint for various inks. The UV-Vis spectrum of each dye component of a mixture can give its positive authentication [76]. Also, by scanning TLC plate, the relative concentrations of dyes present in a specific ink can be found. TLC and HPTLC can be applied for the discrimination of different kinds of inks using normal- or reversed-phase mechanism, gradient elution by AMD, visualization by postchromatographic derivatization of colorless components and digital scanning densitometry for chromatogram evaluation. As a result, instrumental TLC can be successfully used for the determination of ink age on documents and for the characterization of aging ink [67,71,77–79]. The age or source of inks is determined by means of a collection of reference standards or by detecting tags or other unique components added by the manufacturer to know its production date, formulation, and origin. Estimation of how long inks stay on paper requires tests for comparison of the relative age of inks of the same formula and written on the same type of paper with the same storage conditions. An important rule in the TLC analysis to match inks with library standard inks is that all ink samples must be investigated under identical conditions using the same methods. A match is considered as an identity only if it is known to be unique and/or the library is complete [23,24,68–70]. Due to the complex composition of inks, these data can be obtained for the nonvolatile compounds by TLC densitometry and for the volatiles by GC-MS. For nonvolatile compounds, a comparison of the rates and extents of extraction of questioned and known dated inks in organic solvents and of the changes in dye concentrations by TLC densitometry should be done. Also, for dating inks, it is important to have knowledge about the first production date for each type of ink and/or certain ingredients. For the cases of comparison of unknown dated writings with questioned inks, accelerated aging can be used to estimate the age of ink using the mentioned techniques [23,68,69].

The inks whose dyes can be separated by TLC are soluble inks, namely ink-jet inks [73,80,81], writing inks [71,77,82–84], stamp pad inks [78], and laser printer toners [73,85]. Jet inks contain generally textile (direct, acid, basic) dyes and food dyes that are water-soluble and were adapted for ink-jet systems. Amines and carboxylic functional groups were added to improve the binding of dyes for the printed substrates. All writing inks consist of dyeing materials and a vehicle containing solvents, lubricants, and resins. Writing inks can be classified into two groups: ballpoint and nonballpoint inks. The latter are subdivided into water-based inks (fountain pens, felt-tip markers, fluorescent marker pens) and solvent-based inks (permanent markers). For ballpoint inks, the vehicle consists of a mixture of glycols, alcohols of low relative molecular mass, and resin binders which result in a pastelike consistency. Most writing inks contain complex mixtures of mostly soluble dyes and occasionally pigments resulting in unique formulations. Among the most used dyes and pigments are rhodamine, nigrosine, and methyl violet, and phthalocyanine blue, respectively. Dry toners are very fine powders based on dyeing materials and a polymer resin binder. The dyes and pigments are selected according to their chroma, hue, and purity color, the most used being nigrosine, respectively carbon black, copper phthalocyanines, quinacridone, and azo pigments. Sometimes dyes and pigments are added as additives to alter the color [23,69].

The TLC separation of soluble inks follows the steps of sampling, extraction, spotting, development, visualization/detection, comparison, and interpretation of the developed plate, when it is compared with the chromatograms within the ink library, followed by the establishment of a list of possible matches [69]. For the unknown mixtures of dyes, the best chromatographic system is the one which gives the largest number of spots [84]. A brief presentation of the main procedures for analyzing dyes from inks by TLC is given further as described by Pagano et al. [69]. The most used dyeing material for inks are dyes because pigments of less than 1-μm particle size are not frequently commercially available, even their properties (stability, light- and water-fastness, optical density) recommend them. The analysis starts with the solvent extraction of inks in dried state from documents. The most used method is the determination of ink solubility in pure solvents, or different solvent combinations in dimple plates. The often-used solvent for the extraction of jet inks is ethanol–water (1:1, v/v). For writing inks, systems like ethyl acetate–ethanol–water (70:35:30, v/v) (System I) or *n*-butanol–ethanol–water (41:35:32, v/v) (System II) are successfully used. For highly polar inks or containing nigrosine, systems such as cyclohexane–chlorobenzene–ethanol (10:2:1, v/v) (System III) or ethyl acetate–ethanol–chlorobenzene (2:2:10, v/v) (System IV) are used. For toners, chloroform is suitable. The sample concentration should be within the relative range of the library specimens. As stationary phase, silica gel has proven to be the most used. For the dyes with fluorescent characteristics, fluorescent indicators should be absent from the plate. Prior to spotting, it is very important to clean the plates by running the chosen solvent for the length and then drying to remove any contaminants. Due to the complexity of inks, for the separation of dyes, solvent mixtures must be used as mobile phase. These mixtures of solvents are chosen on the basis of their selectivity and strength according to Snyder group selectivity and they should be prepared just before analysis [12]. The conditions of TLC development for jet and writing inks and toners are similar: Whatman polyester silica gel plates, mobile phase System I, and saturated vertical chamber. The jet ink samples were spotted by means of a Camag Nanomat sampler. For comparison and interpretation, precleaned Whatman HPKF silica gel 60 plates were used for jet and writing inks and Merck silica gel 60 plates for toners. All ink samples, the possible library specimens and questioned ink, were spotted and then eluted in saturated horizontal chamber with System I. The dried plates were studied in UV–Vis light. To confirm any matches to the library, a second plate with questioned samples and matching library specimen was chromatographed: for jet ink, Whatman HPKF silica gel 60 plate eluted with water–acetic acid–*n*-butanol–butyl acetate (32:17:41:10, v/v) (System V); for writing ink, Whatman HPKF silica gel 60 plate eluted with System II; and for toners, Merck silica gel 60 plates eluted with System IV. The match is confirmed by analyzing the plates in UV–Vis light. TLC was also used to study the chromatographic separation of anionic cyan dye in ink-jet coating structures [86].

Until 2009, the search and identification of inks in criminalistics were practically performed manually [67,87]. Due to the limited level of information provided by ink evidence to the criminal and civil justice system, Newman and Margot [88–90] developed a complete research project in three stages regarding a generic methodology to provide ink identification and automatic comparison of ink samples with better objectivity, reliability, and efficiency. They report on the development of improved standardization procedures to ensure the best possible reproducibility between analyses. HPTLC method was chosen as being the most widely used method for ink analysis because it combines rapidity, efficiency, and low-cost analysis in the screening of dye composition of ink samples. The important role of the TLC instrumentation on the obtained results over the classical methods should be mentioned. To ensure the HPTLC quality, the acquirement and storage of ink samples were requested. The successful implementation of the new calibration method for the use of ink evidence in forensic science allowed the development of mathematical algorithms for the automatic comparison of ink samples analyzed by HPTLC and of a software package to complete the existing ink library founding the Digital Ink Library [87–90].

Of great importance in the last years is the use of instrumental TLC in forensic science in the field of ink analysis. An overview of the research data from the literature focused on the instrumental TLC conditions of ink dyes is presented in Table 20.4.

TABLE 20.4 Instrumental TLC Methods Used to Analyze Dyes from Different Types of Ink

Ink Matrix	Dyes, Inks	Sample Preparation	Instrumental TLC	TLC Conditions (Stationary Phase; Mobile Phase (v/v); Detection; Quantification). Others	Observation	Reference
13 Inks with different HPTLC dye profiles. 96 Samples per ink analyzed in different conditions with influence on their chromatograms	(a) Standard ladder of dyes for prototype: Acid red 18, Basic blue 26, Solvent orange 3 (b) Standard ladder of dyes for Digital Ink Library: Crystal violet, Rhodamine 6G, Metanil yellow, Acid red 52	For prototype: Mixture (a) in methanol (6:1.5:1:60, w/w/w/v) Ink extraction in ethanol–water 1:1 (v/v) For Digital Ink Library: Mixture (b) in tetrahydrofuran–water 4:1 (v/v) Ballpoint and nonballpoint inks extracted in tetrahydrofuran–water 4:1 (v/v)	Camag equipments: semiautomatic Linomat IV/automatic sampler, automatic developing chamber (ADC 2), TLC Scanner III with WinCats software for digital capture of spectral information.	Prototype development: Merck silica gel 60 HPTLC plates precleaned with methanol. 1-Butanol–ethanol–water–acetic acid (60:10:20:1.5). Horizontal developing chambers. Digital Ink Library development: Merck silica gel $60F_{254}$ HPTLC plates, precleaned with methanol. 1-Butanol–ethanol–water (50:10:15). Automatic developing chamber 2.	Method proposed for standardization and calibration of the HPTLC analysis of inks, based on standard dye ladders which contribute double to the quality assurance process of the ink forensic analysis (prototype). Digital Ink Library (DIL) developed and implemented.	[87–90]
33 Ballpoint pens (21 black and 12 blue)	Black and blue inks commonly found on the market	Inks extracted with absolute ethanol	Differentiation of ballpoint inks achieved by common analytical techniques: UV-Vis, TLC, and FTIR.	Merck HPTLC silica gel plates without fluorescence indicator. 1-Butanol–1-propanol–water–acetic acid (10:5:5:0.5). Horizontal developing chamber.	Discrimination potential (DP) is 100% for black and 98% for blue inks. FTIR technique is complementary to TLC and UV-Vis.	[91]

41 Blue ballpoint pen inks of different trade markets	Blue inks	Ink components dissolved in methanol	Processing color images by an IBM PC with a graphic card and an ordinary office scanner for digital images of TLC plates. Data acquisition by designing specific image analysis in Matlab software for evaluating TLC chromatograms.	Merck silica gel 60 TLC plastic sheets without fluorescent indicator. Ethyl acetate–ethanol–water (70:35:30). Horizontal developing chamber. Scanning TLC plates as digital images. New and effective software designed for specific image analysis (IA) based on the intensity profile of RGB (red, green, and blue) characteristic for the discrimination of pen inks after TLC analysis.	DP is 92.8% thus the TLC-IA method is able to forensic discrimination of a significant number of blue ballpoint pen-pair samples.	[92]
25 Black ballpoint pens	Standard dyes: Crystal violet and Methyl violet 2B Black ballpoint pen inks	Ink extracted from each pen and from ink lines on white photocopier paper in ethanol	TLC chromatograms examined under visible light, UV fluorescence, and IR luminescence were imaged by VSC4C video spectral comparator. Thermo-Electron Genesys 10 UV-Vis spectrometer.	Silica gel $60F_{254}$ TLC aluminum sheets. Ethyl acetate–ethanol–water (50:25:25). Rf value and intensity (scale 1–5) of components were noted. For ink classification, number and intensities of violet bands (methyl violet) and of other colored bands from the complementary dyes were considered for grouping pen inks according to TLC profiles.	For classification and individualization of black ballpoint pen inks, TLC results were compared with principal component analysis of UV–Vis data.	[93]
100 Pen inks from various sources	Writing inks: black (53), blue (31), red (14), green (2) Ballpoint and nonballpoint	Inks extracted in pyridine (ballpoint) and ethanol–water (1:1) (nonballpoint)	Optical examination by video spectral comparator.	Silica gel TLC plates. Ethyl acetate–ethanol–water (70:35:30). TLC followed by optical examination shows no significant differences in the IR reflectance and IR luminescence properties of the questioned and known inks.	85 Inks analyzed by TLC and optical examinations were categorized into 44 different ink formulations.	[77]
31 Blue ballpoint pens	Standard dye: Methyl violet Dyes: Basic violet	Dye extraction in methanol	Camag Linomat IV spot applicator. Camag TLC Scanner III with	Merck HPTLC silica gel plate. Horizontal developing chamber. Two sequential solvent systems: (1) 1-butanol–	Discrimination power 0.92 by HPTLC vs 0.99 by LDI-MS (laser desorption/	[94]

Continued

TABLE 20.4 Instrumental TLC Methods Used to Analyze Dyes from Different Types of Ink—cont'd

Ink Matrix	Dyes, Inks	Sample Preparation	Instrumental TLC	TLC Conditions (Stationary Phase; Mobile Phase (v/v); Detection; Quantification). Others	Observation	Reference
	3, Basic violet 4, Basic blue 2, Basic blue 26, Basic blue 7, and Copper phthalocyanine		WinCATS™® software with Savitsky-Golay 7 filter factor.	2-propanol—water—acetic acid (10:5:5:0.5) and (2) 1-butanol—ethanol—water—acetic acid (15:3:3.9:0.45). Densitometric scanning at 590 nm. 31 Blue ballpoint pen inks were classified by HPTLC in 12 classes on qualitative bases (spot number, Rf, color) and in 18 classes by calculating their relative peak intensities.	ionization mass spectrometry). Anionic dyes detected by HPTLC that is not evident in LDI mass spectra.	
49 Blue and 42 black ballpoint pens of different brands, models, and batches	Blue and black ballpoint pen inks Rhodamine 6G as reference	Ink extracted from paper support with pyridine	Polilight as forensic light source. Poliview imaging system.	Whatman Linear-K Preadsorbent 19 and 6 channel plates. Solvent systems: (1) ethyl acetate—ethanol—water (70:35:30), (2) *n*-butanol—absolute ethanol—water (50:10:15). Fluorescence: 350—650 nm. Rf values and band colors were charted on graph paper. Rf, band color, and band intensity were used to determine discernible inks.	TLC discriminating power: blue inks 0.98, black inks 0.99.	[95]
Ink from the script of electronic typewriter/ribbon	Ink of seven electronic typewriters	Ink extraction on pyridine	Camag HPTLC system: Linomat IV, Scanner II-V 3.15 coupled with computer PC AT 386 with Camag software.	Silica gel $60F_{254}$ aluminum sheet (Merck). *n*-Butanol—ethanol—water (5:2:1.5)/ *n*-Butanol—ethyl acetate—ethanol—water (5:2:2:1). Densitogram scanning at 540 nm. Integrated chromatograms scanned for the in situ visible spectra in the range of 400—800 nm.	HPTLC and in situ visible spectrum of each resolved band.	[96]

Aqueous solutions Inks: ballpoint, fiber tip, fountain pen, printing	Acid brilliant blue Z, Acid anthraquinone brilliant blue	Dye extraction in 50% aqueous methanol	Camag system: microapplicator I, TLC/HPTLC scanner connected to an SP4100 integrator. HP computer with own software.	Merck silica gel 60 TLC plates without fluorescent indicator. Linear ascending development. Ethyl acetate−isopropanol−water−glacial acetic acid (30:15:10:1). Densitograms (580 nm) recorded for each sample and the peak high values calculated by integrator and then computed by the own software.	Results show discrimination between samples with similar appearance and quantitative composition.	[79]
Ballpoint pen inks	Black and blue (10 trademarks), Blue and red (10 trademarks)	Extraction on paper with pyridine	Spectrophotometric scanner connected to a microcomputer to act as integrator.	Merck silica gel 60 TLC plates. Linear development. Scanning at 540 nm. Black-blue inks, ethyl acetate−ethanol −water (70:35:30); Blue-red inks, methanol−*n*-propanol−1-pentanol−water (2:10:26:4).	Good separation of dyes.	[83]
Fountain pen inks	Black, blue-black, royal blue, blue (38 trademarks) Red (six trademarks) Green (six trademarks)	Ink sample diluted 1:5	Nanoapplicator Spectrophotometric scanner connected to a microcomputer to act as integrator.	Merck silica gel 60 TLC plates. Linear development. Scanning at 540 nm. Blue-black inks, isobutanol−ethanol−acetic acid 99%−water (20:5:5:10); Red inks, iso-propanol−1-pentanol−water (12:22:6); Green inks, formic acid−butanol saturated with water (3:97).	Good separation for black-blue and red dyes. Average separation for green dyes.	[83]
Fiber-tip pen inks	Black and blue (14 trademarks) Black (six trademarks)	Direct application to the plate of the tip of writing instrument	Spectrophotometric scanner connected to a microcomputer to act as integrator.	Merck silica gel 60 TLC plates. Linear development. Scanning at 540 nm. Black-blue inks, isopropanol−1-pentanol−water (10:25:5); Black inks, isobutanol−isopropanol−water (23:10:7).	Good separation of dyes for black inks. Average separation of dyes for blue inks.	[83]
Board, felt markers and ballpoint pens	Black inks	Dye extraction in methanol	Vernon TRD 2 spectrophotodensito-meter with a tungsten lamp. No filter used.	Merck RP-TLC plates (RP-2, RP-8, RP-18). Phosphate buffer solution (pH 2.8)−ethanol (3.5:6.5). Best results on RP-18 plates.	RP-TLC separation of black ink dyes on alkyl-silica plates.	[76]

Continued

TABLE 20.4 Instrumental TLC Methods Used to Analyze Dyes from Different Types of Ink—cont'd

Ink Matrix	Dyes, Inks	Sample Preparation	Instrumental TLC	TLC Conditions (Stationary Phase; Mobile Phase (v/v); Detection; Quantification). Others	Observation	Reference
23 Color ink-jet cartridges of different brands	Yellow, cyan, and magenta inks	Inks extracted from paper with pyridine–water (4:3) and after drying reconstitution with ethanol–water (1:1)	Images of developed plates photographically recorded. Complementary HPLC investigations.	Alugram silica gel plate/UV_{254}. Mobile phases: (1) ethyl acetate–ethanol–water (70:35:30), (2) *n*-butanol–ethanol–water (10:2:3). Visualization under white and long-wavelength UV light. TLC indicated more components for magenta ink than for yellow and cyan counterparts. Rf values and color of spots generated a database of the ink profiles of different color ink-jet cartridges.	TLC method gave similar results from the ink profiles as those obtained by HPLC. These chromatographic methods can be used for differentiation of color inks of different brands of ink cartridges.	[80]
Magenta inks of two office printers	Magenta inks: Canon BJC 8500 and Hewlett–Packard DeskJet 690C		Camag apparatus: Linomat applicator, densitometer.	Silica gel 60 TLC glass plates. Ethyl acetate–ethanol–water (52:26:22). Dyes identified and quantified in both magenta inks: acid red 052 (0.15%) and acid red 249 (0.02%).	Determination of dyes in magenta inks.	[81]
28 Processed color toners and 10 raw color toners	Yellow, cyan, magenta, and black	Extraction of the four toner colors with chloroform after a thermal process	TLC followed by the complementary techniques UV and FTIR leads to the discrimination of color toners.	Merck silica gel 60 TLC plate without fluorescence. Mobile phases: (1) ethyl acetate–ethanol–water (70:35:30), (2) chloroform–methanol–*n*-hexane–acetic acid (85:25:5:1). Visualization by exposure to shortwave and longwave UV and visible light to record the color and fluorescence of spots.	TLC has more discrimination potential than UV and FTIR due to the dye presence in toners. All three methods are able to differentiate almost all of the analyzed toners.	[85]

FTIR—Fourier transform infrared spectroscopy; HPTLC—high-performance thin-layer chromatography; TLC—thin-layer chromatography; RP-TLC—reversed-phase TLC.

20.4 CONCLUSION

The instrumental TLC is a powerful tool with the capability to discriminate samples of a similar appearance which are indistinguishable from other applied techniques and to determine their quantitative composition. It is very useful and efficient for the analysis of dyes and inks given qualitative and quantitative information unavailable through the nondestructive spectroscopic techniques.

The advances obtained in the separation, identification, and quantification of dyes from various matrices like food, textiles, cosmetics, inks, and so on are due to the instrumental TLC, especially to the modern possibilities for development, video densitometry, digital scanning, and of the high quality and variety of TLC/HPTLC plates. Also, dyes are used as test compounds to demonstrate the performances of different instrumental TLC methods under research. The densitograms of the developed plates provide valuable information that are much more accurate than the visual examination of plate and can be considered as a fingerprint for the matrix. The UV–Vis spectrum of each dye component of a mixture can give its positive authentication. The coupling of TLC with instrumental detection methods such as TLC-MS or TLC-FTIR brings the attributes of both ones opening new possibilities for identification and structural elucidation due to the additional information as it can see in the cases of dyes and inks.

Ink analysis occupies a very important place in the examination of questioned documents allowing the determination of authenticity or validity of a document in forensics. TLC-UV–Vis analysis of ink dyes provides helpful information that can differentiate the dyes found in inks. Using complementary techniques like FTIR, additional information on dyes and/or other components can be obtained increasing the discrimination potential in the elucidation of ink formulation. Instrumental TLC is generally accepted as a scientific methodology applied to compare and characterize dyes from different ink formulations having the highest discrimination power for the individual techniques.

References

[1] Wall PE. Dyes/thin-layer (planar) chromatography. In: Willson ID, Cooke M, Poole CF, Adlard ER, editors. Encyclopedia of separation science. New York: Academic Press; 2000. p. 2619–31.

[2] Hunger K. Industrial dyes. Chemistry, properties, applications. Weinheim: Wiley-VCH Verlag GmbH & Co. KGaA; 2003.

[3] Chatwal GR. Synthetic dyes. Mumbai: Himalaya Publishing House; 2009.

[4] Kirchner JG. Thin-layer chromatography. 2nd ed. In: Perry ES, editor. New York: John Wiley & Sons Inc. 1978.

[5] Sherma J, Fried B. Handbook of thin-layer chromatography. 3rd ed. New York: Marcel Dekker Inc.; 2003.

[6] Poole CF. The essence of chromatography. Amsterdam: Elsevier; 2003. p. 499–567.

[7] Bele AA, Khale A. An overview on thin-layer chromatography. IJPSR 2011;2:256–67.

[8] Poole CF. Thin-layer chromatography: challenges and opportunities. J Chromatogr A 2003;1000:963–84.

[9] Poole CF. Planar chromatography at the turn of the century. J Chromatogr A 1999;856: 399–427.

[10] Poole CF, Poole SK. Modern thin-layer chromatography. Anal Chem 1989;61:1257A–69A.

[11] Poole CF, Poole SK. Instrumental thin-layer chromatography. Anal Chem 1994;66: 27A–37A.

[12] Nyiredy Sz. Planar chromatography. A retrospective view for the third millennium. Budapest: Springer Scientific Publisher; 2001.

[13] Srivastava MM. High-performance thin-layer chromatography. Heidelberg: Springer; 2011.

[14] Sherma J. Biennial review of planar chromatography 2011-2013. Cent Eur J Chem 2014; 12:427–52.

[15] Spangenberg B, Poole CF, Weins Ch. Quantitative thin-layer chromatography. A practical survey. Heidelberg: Springer; 2011.

[16] Grinberg N. Modern thin-layer chromatography. New York: Marcel Dekker Inc.; 1990.

[17] Hahn-Deinstrop E. Applied thin-layer chromatography. Best practice and avoidance of mistakes. 2nd, revised and enlarged ed. Weinheim: Wiley-VCH Verlag GmbH & Co. KGaA; 2007.

[18] Poole CF, Belay MT. Progress in automated multiple development. J Planar Chromatogr 1991;4:345–59.

[19] Poole CF, Poole SK, Belay MT. Insights into the mechanism and applications of unidimensional multiple development in thin-layer chromatography. J Planar Chromatogr 1993;6:438–45.

[20] Poole CF, Poole SK. Multidimensionality in planar chromatography. J Chromatogr A 1995;703:573–612.

[21] Gupta VK. Synthetic dyes. In: Sherma J, Fried B, editors. Handbook of thin-layer chromatography. 3rd ed. New York: Marcel Dekker Inc.; 2003. p. 935–68.

[22] Cserháti T. Liquid chromatography of natural pigments and synthetic dyes. 1st ed. Amsterdam: Elsevier; 2007. p. 374–402.

[23] Brunelle RL, Crawford KR. Advances in the forensic analysis and dating of writing ink. Springfield: Charles C. Thomas Publisher Ltd; 2003.

[24] Brunelle RL, Reed RW. Forensic examination of ink and paper. Springfield: Charles C. Thomas Publisher; 1984.

[25] Iqbal M. Textile dyes. Karachi: Rehbar Publishers; 2008.

[26] Alves de Lima RO, Bazo AP, Salvadori DMF, Rech CM, Oliveira DP, Umbuzeiro GA. Mutagenic and carcinogenic potential of a textile azo dye processing plant effluent that impacts a drinking water source. Mutat Res 2007;626:53–60.

[27] Copaciu F, Coman V, Vlassa M, Opriş O. Determination of some textile dyes in wastewater by solid phase extraction followed by high-performance thin-layer chromatography. J Planar Chromatogr 2012;25:509–15.

[28] Morlock GE, Oellig C. Rapid planar chromatographic analysis of 25 water-soluble dyes used as food additives. J AOAC Int 2009;92:745–56.

[29] Abdullah AFL, Marimuthu Y, Haw CK, Mohamad Said NF, Md Muslim NZ, Nik Hassan NF, et al. Forensic discrimination of lipsticks by thin layer chromatography and gas chromatography-mass spectrometry. Malaysian J Forensic Sci 2011;2:22–8.

[30] Cobzac SC, Casoni D, Pop D. Tartrazine determination from mustard sample by TLC-photodensitometry and TLC-digital processing of images. J Planar Chromatogr 2012; 25:542–7.
[31] Tuzimski T. Determination of sulfonated water-soluble azo dyes in foods by SPE coupled with HPTLC-DAD. J Planar Chromatogr 2011;24:281–9.
[32] Tuzimski T, Woźniak A. Application of solid-phase extraction and planar chromatography with diode-array detection to the qualitative and quantitative analysis of dyes in beverages. J Planar Chromatogr 2008;21:89–96.
[33] Saponar F, Moţ AC, Sârbu C. Quantitative determination of some food dyes using digital processing of images obtained by thin-layer chromatography. J Chromatogr A 2008; 1188:295–300.
[34] Gerasimov AV. Use of the software processing of scanned chromatogram images in quantitative planar chromatography. J Anal Chem 2004;59:348–53.
[35] Gerasimov AV. Use of computer-assisted color treatment of chromatograms in the TLC determination of tartrazine. J Anal Chem 2003;58:209–10.
[36] Sayar S, Ozdemir Y. First-derivative spectrophotometric determination of ponceau 4R, sunset yellow and tartrazine in confectionery products. Food Chem 1998;61: 367–72.
[37] Sherma J. Thin-layer chromatography in food and agricultural analysis. J Chromatogr A 2000;880:129–47.
[38] Oka H, Ikai Y, Ohno T, Kawamura N, Hayakawa J, Harada K-I, et al. Identification of unlawful food dyes by thin-layer chromatographic-fast atom bombardment mass spectrometry. J Chromatogr A 1994;674:301–7.
[39] Harada K-I, Masuda K, Suzuki M, Oka H. Separation and identification of food dyes by thin-layer chromatography/liquid secondary ion mass spectrometry. Biol Mass Spectrom 1991;20:522–8.
[40] Masuda K, Harada K-I, Suzuki M, Oka H, Kawamura N, Yamada M. Identification of food dyes by TLC/SIMS with a condensation technique. Org Mass Spectrom (OMS Letters) 1989;24:74–5.
[41] Oka H, Ikai Y, Kawamura N, Yamada M, Inoue H, Ohno T, et al. Simple method for the analysis of food dyes on reversed-phase thin-layer plates. J Chromatogr 1987;441: 437–44.
[42] Brosseau CL, Gambardella A, Casadio F, Grzywacz CM, Wouters J, Richard P, et al. Ad-hoc surface-enhanced Raman spectroscopy methodologies for the detection of artist dyestuffs: thin layer chromatography-surface enhanced Raman spectroscopy and in situ on the fiber analysis. Anal Chem 2009;81:3056–62.
[43] Konstantinova TN, Lazarova RA, Miladinova PP, Venkova AY. Thin-layer chromatographic study of some dyes and fluorescent brighteners for polymers. J Planar Chromatogr 2004;17:444–8.
[44] Ohno T, Mikami E, Matsumoto H. Identification of oil-soluble coal tar dyes in cosmetics using reversed-phase TLC/scanning densitometry. J Health Sci 2003;49:401–4.
[45] Măruţoiu C, Coman V, Grecu R, Filip C. Separation of some cosmetic dyes on bonded volcanic tuff by thin layer chromatography. Acta Universitatis Cibiniensis Seria F Chemia 2001;4:25–31.
[46] Harvey SD, Buchko GW, Lucke RB, Wright CW, Melville AM, Scott AJ, et al. The structure and purity of a reference dye standard used for quantification of C.I. Solvent Red 164 in fuels. Dyes Pigm 2009;82:307–15.
[47] Richter G, Schober C, Süß R, Fuchs B, Müller M, Schiller J. The reaction between phosphatidylethanolamines and HOCl investigated by TLC: Fading of the dye primuline is induced by dichloramines. J Chromatogr B 2008;867:233–7.

[48] Konstantinova TN, Lazarova RA. Synthesis of some polymerizable triazinylaminobenzotriazole stabilizers and benzanthrone dyes containing a stabilizer fragment. Dyes Pigm 2007;74:208–14.
[49] Wall PE. The value of planar chromatography for the analysis of triphenylmethane dyes. J Planar Chromatogr 1993;6:394–403.
[50] Wall PE. Thin layer chromatographic separation of thiazines: problems and solutions. J Planar Chromatogr 1991;4:365–9.
[51] Markowski W, Wróblewski K, Dzido TH. Stepwise gradient elution in RP HPTLC with a new horizontal developing chamber. J Planar Chromatogr 2012;25:200–7.
[52] Ding F, Wang D, Song S, Xu S. Preliminary investigation of sintered plates bonded or end-capped with phenyl for planar electrochromatography. J Planar Chromatogr 2011;24:10–5.
[53] Wang J, Wang D, Zhang H, Zhang Y, Zhou S. A new plate for planar electrochromatography. J Planar Chromatogr 2006;19:313–8.
[54] Coman V, Kreibik Şt, Vlassa M, Filip M. Study of electric field geometry using the vertical planar dielectrochromatographic chamber. J Planar Chromatogr 2010;23:434–9.
[55] Coman V, Kreibik Şt, Vlassa M. Planar dielectrochromatography in vertical chamber. J Planar Chromatogr 2008;21:373–8.
[56] Berezkin VG, Kulakova NY, Khrebtova SS. Three-dimensional thin-layer chromatography on plates with open and closed adsorption layers. J Planar Chromatogr 2009; 22:313–9.
[57] Van Berkel GJ, Ford MJ, Deibel MA. Thin-layer chromatography and mass spectrometry coupled using desorption electrospray ionization. Anal Chem 2005;77:1207–15.
[58] Konstantinova TN, Neicheva AS, Venkova AY. TLC and HPTLC study of new 9-phenylxanthene dyes. J Planar Chromatogr 2004;17:369–71.
[59] Dzido TH, Mróz J, Jóźwiak GW. Adaptation of a horizontal DS chamber to planar electrochromatography in a closed system. J Planar Chromatogr 2004;17:404–10.
[60] Dzido TH, Majewski R, Polak B, Golkiewicz W, Soczewiński E. Application of a horizontal DS chamber to planar electrochromatography. J Planar Chromatogr 2003; 16:176–82.
[61] Lan M, Wang D, Wei W, Lu W. Multidimensional relay development in TLC. J Planar Chromatogr 2003;16:461–4.
[62] Mustoe S, McCrossen S. TLC image capture and analysis by use of a prototype device for visualizing fluorescence. J Planar Chromatogr 2001;14:252–5.
[63] Ligor M, Buszewski B. Application of TLC and OPLC in the determination of pigments from natural products. J Planar Chromatogr 2001;14:334–7.
[64] Regnault C, Delvordre P, Postaire E. Direct use of Empore sheets in overpressured thin-layer chromatography. J Chromatogr 1991;547:403–9.
[65] Botz L, Nyiredy S, Sticher O. Applicability of long distance overpressured layer chromatography. J Planar Chromatogr 1991;4:115–22.
[66] Mincsovics E, Thihák E, Kalász H. Resolution and retention behaviour of some dyes in overpressured thin-layer chromatography. J Chromatogr 1980;191:293–300.
[67] Zlotnick JA, Smith FP. Chromatographic and electrophoretic approaches in ink analysis. J Chromatogr B 1999;733:265–72.
[68] Brunnelle RL. Ink analysis. In: Wilson ID, Cooke M, Poole CF, Adlard ER, editors. Encyclopedia of forensic sciences. New York: Academic Press; 2000. p. 591–7.
[69] Pagano LW, Surrency MJ, Cantu AA. Inks: forensic analysis by thin-layer (planar) chromatography. In: Wilson ID, editor. Encyclopedia of separation science. New York: Academic Press; 2000. p. 3101–9.
[70] Aginsky V. Analytical methods. In: Wilson ID, Cooke M, Poole CF, Adlard ER, editors. Encyclopedia of forensic sciences. New York: Academic Press; 2000. p. 566–70.

[71] Ezcurra M, Góngora JMG, Maguregui I, Alonso R. Analytical methods for dating modern writing instrument inks on paper. Forensic Sci Int 2010;197:1–20.
[72] Williams MR, Moody C, Arceneaux L-A, Rinke C, White K, Sigman ME. Analysis of black writing ink by electrospray ionization mass spectrometry. Forensic Sci Int 2009; 191:97–103.
[73] Sharma R, Baggi TR, Chattree A, Kesharwani L, Gupta AK. Detection and identification of printer inks – a review report on laser and inkjet printer ink analysis. Int J Res Rev 2013;5:46–51.
[74] Szafarska M, Wietecha-Posłuszny R, Woźniakiewicz M, Kościelniak P. Application of capillary electrophoresis to examination of color inkjet printing inks for forensic purposes. Forensic Sci Int 2011;212:78–85.
[75] Cantu AA, LaPorte GM, Wilson JD. Differentiation of black gel inks using optical and chemical techniques. J Forensic Sci 2004;49:364–70.
[76] Siouffi A, Guiochon G. Use of reversed-phase thin-layer chromatography for the identification of black inks from board felt markers and ball-point pens. J Chromatogr 1981; 209:441–5.
[77] LaPorte GM, Arredondo MD, McConnell TS, Stephens JC, Cantu AA, Shaffer DK. An evaluation of matching unknown writing inks with the United States international ink library. J Forensic Sci 2006;51:689–92.
[78] Jasuja OP, Singla AK, Seema BL. Thin layer chromatographic analysis on Indian stamp pad ink. J Forensic Sci 1989;42:255–62.
[79] Aginsky VN. A new application of instrumental planar chromatography in forensic analysis. J Planar Chromatogr 1991;4:167–9.
[80] Poon NL, Ho SSH, Li CK. Differentiation of coloured inks of inkjet printer cartridges by thin layer chromatography and high performance liquid chromatography. Science&Justice 2005;45:187–94.
[81] Muck T, Hladnik A. Evaluation of the quality ink-jet printed paper by planar chromatography. J Planar Chromatogr 2003;16:154–9.
[82] Verma RS, Prasad KN, Gopal J, Misra GJ. Thin layer chromatographic analysis of fibre-tip pen inks. Forensic Sci Int 1979;13:65–70.
[83] Tappolet JA. The high-performance thin layer chromatography (HPTLC). Its application to the examination of writing inks. Forensic Sci Int 1983;22:99–109.
[84] Matysyk G, Soczewsiński E. On-line extraction and preconcentration of solid samples in equilibrium sandwich chambers for thin-layer chromatography. Analysis of ink from ball-point pens. J Chromatogr 1986;355:363–6.
[85] Saini K, Saroa JS. Differentiation of color photocopy toners using TLC, UV, and FTIR techniques. JFI 2011;61:561–80.
[86] Lamminmäki TT, Kettle JP, Puukko PJT, Gane PAC. The chromatographic separation of anionic dye in inkjet coating structures. J Colloids Surf A: Physicochem. Eng. Aspects 2011;377:304–11.
[87] Neumann C, Ramotowski R, Genessay T. Forensic examination of ink by high-performance thin layer chromatography—The United States Secret Service Digital Ink Library. J Chromatogr A 2011;1218:2793–811.
[88] Neumann C, Margot P. New perspectives in the use of ink evidence in forensic science: Part I. Development of a quality assurance process for forensic ink analysis by HPTLC. Forensic Sci Int 2009;185:29–37.
[89] Neumann C, Margot P. New perspectives in the use of ink evidence in forensic science: Part II. Development and testing of mathematical algorithms for the automatic comparison of ink samples analysed by HPTLC. Forensic Sci Int 2009;185:38–50.
[90] Neumann C, Margot P. New perspectives in the use of ink evidence in forensic science: Part III. Operational applications and evaluation. Forensic Sci Int 2009;192: 29–42.

[91] Causin V, Casamassima R, Marega C, Maida P, Schiavone S, Marigo A, et al. The discrimination potential of ultraviolet-visible spectrophotometry, thin layer chromatography, and Fourier transform infrared spectroscopy for the forensic analysis of black and blue ballpoint inks. J Forensic Sci 2008;53:1468–73.

[92] Djozan D, Baheri T, Karimian G, Shahidi M. Forensic discrimination of blue ballpoint pen inks based on thin layer chromatography and image analysis. Forensic Sci Int 2008;179:199–205.

[93] Adam CD, Sherratt SL, Zholobenko VL. Classification and individualisation of black ballpoint pen inks using principal component analysis of UV–Vis absorption spectra. Forensic Sci Int 2008;174:16–25.

[94] Weyermann C, Marquis R, Mazzella W, Spengler B. Differentiation of blue ballpoint pen inks by laser desorption ionization mass spectrometry and high-performance thin-layer chromatography. J Forensic Sci 2007;52:216–20.

[95] Roux C, Novotny M, Evans I, Lennard C. A study to investigate the evidential value of blue and black ballpoint pen inks in Australia. Forensic Sci Int 1999;101:167–76.

[96] Varshney KM, Jettappa T, Mehrotra VK, Baggi TR. Ink analysis from typed script of electronic typewriters by high performance thin layer chromatography. Forensic Sci Int 1995;72:107–15.

CHAPTER

21

Analysis of Dietary Supplements

Irena Vovk, Vesna Glavnik

Laboratory for Food Chemistry, National Institute of Chemistry, Ljubljana, Slovenia

21.1 INTRODUCTION

As reported by the World Health Organization the majority of the world population chooses medicinal plants or their products as a first choice in an occurrence of a disease. According to the well-known principle "it is better to prevent than cure" and also to the Hippocrates' saying "let food be thy medicine and medicine be thy food", an interdisciplinary area of functional foods and dietary supplements, which is intertwined between food science and pharmacy, is rapidly evolving in the scientific and the entrepreneurial fields. Functional food might still primarily be considered as food, but on the other hand, a dietary supplement undeniably looks more like a pharmaceutical product, although both products may contain the same compound or mixture of compounds with beneficial effects on human health. Production and trade of functional foods and dietary supplements in the world is increasing each year. This is supported by the trend that the number of consumers who regularly consume dietary supplements originating from plant materials (also referred as botanicals) is increasing due to the consumer's increased awareness of health concerns and in some parts of the world, also the aging of the population.

The definition of "dietary supplement" varies worldwide. In the United States, "dietary supplement" according to The Dietary Supplement Health and Education Act of 1994 (DSHEA) [1] is defined as "a product (other than tobacco) intended to supplement the diet that bears or contains one or more of the following dietary ingredients: a vitamin, a mineral, an herb or other botanical, an amino acid, a dietary substance for use by man to supplement the diet by increasing the total dietary intake, or a concentrate, metabolite, constituent, extract, or combination of these ingredients. It is intended for ingestion in tablet, capsule, powder, softgel, gelcap, or liquid form and is not represented for use as a conventional

http://dx.doi.org/10.1016/B978-0-12-417223-4.00021-2

food or as a sole item of a meal or the diet and is labeled as a dietary supplement."

In the European Union, instead of the term "dietary supplement" "food supplement" is used. According to the EU Directive 2002/46/EC, June 10, 2002 [2]: "food supplements means foodstuffs the purpose of which is to supplement the normal diet and which are concentrated sources of nutrients (vitamins and minerals) or other substances with a nutritional or physiological effect, alone or in combination, marketed in dose form, namely forms such as capsules, pastilles, tablets, pills and other similar forms, sachets of powder, ampoules of liquids, drop dispensing bottles, and other similar forms of liquids and powders designed to be taken in measured small unit quantities." In the Directive 2002/46/EC, only vitamins (e.g., retinol, folic acid, L-ascorbic acid) and minerals (e.g., calcium chloride, magnesium carbonate) are specified for the use in manufacture of dietary supplements, while the use of other ingredients such as amino acids, essential fatty acids, pre- and probiotics, miscellaneous bioactive compounds as well as plants and plant extracts is the subject of the national legislations of each EU member state. Different authorities around the world, especially the Food and Drug Administration (FDA) in the United States and the European Food Safety Authority (EFSA) in Europe, are proposing regulation issues to the manufacturers. Data about the regulations concerning dietary supplements and functional food in different countries (United States, European countries, Russia, Japan, China, India, Malaysia, Australia, New Zealand, and African countries) are summarized in two books [3,4], which also provide the data concerning manufacturing compliance and analytical validation, importance of safety assessment, new dietary ingredients, health claims, and other related topics.

Everywhere around the globe dietary supplements are considered as food and therefore, in spite of the formulations or appearance being more pharmaceutical than food, the regulation is not as strict as for pharmaceuticals. In other words, the manufacturers of dietary supplements do not need to follow the guidelines similar to those prescribed for the manufacturers of pharmaceuticals, but they have to follow the Current Good Manufacturing Practices (cGMP) in manufacturing, packing, labeling, or holding operations [5]. Manufacturers are by law responsible for the safety of their dietary supplements, labeling of all the ingredients, validity of all the labeled statements and for providing the corresponding evidence. The International Alliance of Dietary/Food Supplement Associations (IADSA) was formed in 1998, as an alliance of dietary supplement associations from six continents, became the leading international expert association. IADSA's role is important as it represents an international network that can implement the development of the dietary supplement sector worldwide by suggesting regulatory changes.

Presently the authorities are more focused on the health claims written on the packages with far stricter control than in the past, and by this control at least protecting consumers from being misled by the advertising of health claims without appropriate scientific evidence. However, the consumers and the analysts must also pay attention to the tricky labeling of the bioactive ingredients per formulation (per single capsule, multiple capsules, and per daily dose). Safety and quality of dietary supplements are very much connected with the manufacturers' lack of knowledge, especially concerning plant-derived materials (possible unintentional adulterants, different chemotypes), as well as a lack of knowledge about chemical incompatibility and the pharmacological activity of the products' constituents. From the safety point of view, undefined relatively narrow therapeutic window of active ingredients is also critical. Another problem in quality control is that analytical methods that are essential for quality assurance are often not available because of the endless combinations of dietary supplement ingredients. In other words, dietary supplements contain active ingredients from plant materials, synthetic compounds (mostly vitamins), minerals or a mixture thereof, but due to the presence of plant materials, these products might also contain high content of contaminants like mycotoxins, pesticides, heavy metals, and so forth. Unfortunately unsafe adulterants (e.g., plant materials as well as drugs) can also be found in dietary supplements. These many obstacles regarding the quality and safety of these products for the average consumer are harder to control than pharmaceuticals. Therefore, they are still called to the attention of regulatory authorities and the scientific community.

Development and validation of new analytical methods, identification and characterization of biomarkers, and the generation of standard reference materials (SRMs) globally represent a big challenge for analytical chemistry. The combined use of thin-layer chromatography (TLC), high-performance liquid chromatography (HPLC), and gas chromatography (GC) and their hyphenation to mass spectrometry and spectrophotometry, as indispensable tools in the research and development in this field, also provides support to the manufacturers of these products as well as to regulatory authorities. The role of TLC from the conception through to the final dietary supplement product starts with the selection of the ingredients (plant materials, pure compounds, etc.), selection of plant varieties (giving the best yields), quality control of the raw materials (identification, chemical, and biological fingerprint), and final products (purity, uniformity, content, stability testing) (Figure 21.1). In other words, TLC is an important technique in quality by design of the dietary supplements product.

This chapter describes the state-of-the-art and potential of instrumental TLC for the analysis of dietary supplements. To be discussed are the analytical challenges related to the endless possible combinations of

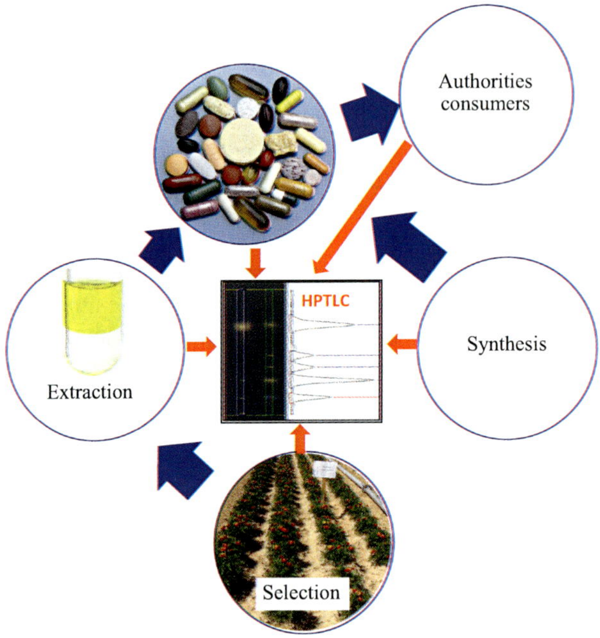

FIGURE 21.1 **The role of TLC in all the stages: from the conception to the design of the final dietary supplement product and its control by the regulatory authorities.**

bioactive ingredients and excipients, chemically undefined plant ingredients, adulterants, and consequently, the lack of the analytical methods, reference standards, and SRMs. Additionally, the chapter presents the instrumental possibilities, especially those used for the detection (densitometry, image analysis, mass spectrometry, Raman spectroscopy) as well as selected published and some new applications. In this chapter, the term "TLC method" is used for methods carried out on TLC and high-performance thin-layer chromatography (HPTLC) layers.

21.2 ANALYTICAL CHALLENGES

The analytics of dietary supplements is difficult due to the endless possible combinations of the bioactive ingredients (bioactive compounds, vitamins, minerals, plant extracts or even several plant extracts in one dietary supplement product, etc.) and excipients in different formulations such as hard/soft capsules, tablets, and liquids. Bioactive ingredients can also be chemically modified compounds (e.g., encapsulated). Additionally, the required level of the analytical accuracy is different depending on the analyte. What makes analytics even harder is the lack of available reference standards, SRMs, and marker compounds of particular plant materials, plant extracts, or plant extract fractions, as well as a variety of chemical structures, including isomeric compounds. Analytical methods are available and are validated for some of the dietary supplements' bioactive ingredients (carotenoids, flavanols, stilbenes, retinol, ascorbic acid, etc.) known also to be present in food. However, the application of these methods for the analysis of dietary supplements is not straightforward, because of different and in most cases complex dietary supplement matrices. Another problem in quantitative analysis of dietary supplements is that bioactive components from plant materials or plant derived materials are mostly unknown (the same is also valid for food). Therefore, analytical methods are not developed and no reference standards or SRMs are available.

Consequently, the lack of reference standards results in the lack of standardized analytical methods for the analysis of dietary supplements. Because of the complexity of the composition of dietary supplements regarding different formulations with different also modified bioactive ingredients from different sources (synthetic, plant derived, etc.), the sample preparation step is crucial in the analysis of these products. As a part of method development the stability of bioactive ingredients during analysis (from the sample preparation step till detection) has to be taken into account. Frequently the advertised main bioactive ingredient is present in a dietary supplement as a minor compound together with other bioactive compounds present in much higher concentrations. This should be considered during the extraction process (the optimization of the sample preparation). High accuracy is required for certain vitamins, minerals, and other nutrients, because of their importance for human health (e.g., folic acid, vitamin D, calcium), for estimation of total daily intake, while in case of the plant derived materials, the required accuracy depends on the bioactive compound. The optimization of extraction of the target analyte from different matrices, where it can be present at different concentration levels, is crucial for the recovery of the method.

It is clear that the complexity of the analytical tasks in the field of dietary supplements analysis requires a good collaboration of parties, including AOAC International, National Institute of Standards and Technology (NIST), USP, manufacturers, and scientists. AOAC and USP have published priority lists of components for which analytical methods should be developed and validated.

21.2.1 Reference Standards and SRMs

Due to the lack of the analyte-certified reference materials (CRMs provided by metrological laboratories), which are the best reference materials for determination of the recovery, standard materials certified by commercial suppliers are used in the analysis in most cases. Additionally, the lack of the reference standards can be overcome by the isolation of target compounds from the plant material using preparative TLC, preparative rotation planar chromatography (RPC), preparative HPLC, countercurrent chromatography, or other techniques followed by structural elucidation by mass spectrometry (MS) and nuclear magnetic resonance spectroscopy (NMR).

Some dietary supplement reference standards are available from the USP. Currently, more than 200 are listed in their catalog [6] as amino acids, botanicals, and miscellaneous dietary supplements (carotenoids, vitamins, fish oil, and krill oil components). The list of botanicals includes compounds from about 50 plants: aloe, American ginseng, Asian ginseng, bilberry, *Boswellia serrata*, cat's claw, chamomile, cranberry, *Echinacea*, garlic, ginkgo, grape (seed), green tea, hawthorn, holy basil, horse chestnut, kava, licorice, maritime pine, milk thistle, red clover, rosemary, saw palmetto, soy, St John's wort, tomato, valerian, etc.

The situation concerning the SRMs with certified values for selected organic ingredients and trace elements is even more complex. NIST currently provides 23 dietary supplements (including nutraceuticals and herbs) SRMs (Table 21.1) [7] with certificates of analysis available online. It is evident from these certificates that analytical methods based on other analytical techniques (not TLC) were used for the analyses. However, TLC methods (application notes F16C for ginkgolic acid [8] and F16B for flavonoids [9]) developed by Camag were applied for further characterization of *Ginkgo biloba* SRMs (leaves, SRM 3246; extract, SRM 3247; and ginkgo-containing tablets, SRM 3248). The SRMs can support method development, assessment of consistency and quality in final products, and as control materials, they can also support the characterization of raw materials for authenticity, contamination or adulteration.

21.2.2 Problems of Labeling Dietary Supplements

Although the label of a dietary supplement product is required to be truthful and not misleading, there are two major reasons why this is not always the case. The first reason is chemical. For example, a statement "guaranine is a type of caffeine" is wrong. According to the PubChem database, guaranine is one of 25 synonyms for caffeine [10]. Inappropriate translation to different languages has led to misleading advertising of the active ingredient. Such misleading advertising can easily present too high intake for sensitive groups.

The second reason is botanical. For example, the name "Marigold" is the English language common name for *Calendula officinalis* L. and *Tagetes*

TABLE 21.1 List of NIST standard reference materials (SRM)

Description	NIST SRM Number
Ginkgo biloba (leaves)	3246
Ginkgo biloba extract	3247
Ginkgo-containing tablets	3248
Serenoa repens (fruit)	3250
Serenoa repens extract	3251
Camellia sinensis (green tea) leaves	3254
Camellia sinensis (green tea) extract	3255
Green tea-containing solid, oral dosage form	3256
Catechin calibration solution	3257
Bitter orange (fruit)	3258
Bitter orange (extract)	3259
Bitter orange-containing solid, oral dosage form	3260
Hypericin calibration solution	3266
Botanical oils containing omega-3 and omega-6 fatty acids	3274
Omega-3 and omega-6 fatty acids in fish oil	3275
Multivitamin/multielement tablets	3280
Cranberry (fruit)	3281
Low-calorie cranberry juice cocktail	3282
Cranberry extract	3283
Cranberry-containing solid, oral dosage form	3284
Mixed berry-containing solid, oral dosage form	3285
Organic acids calibration standard	3286
Bilberry extract	3291

Adopted from Ref. [7].

erecta L. This can lead to a wrong translation if the Latin name is not included on the label. In one case, a dietary supplement was advertised with the picture of *Calendula officinalis* L., the advertising indicating a high amount of carotenoid lutein, whereas the actual amount is only 10% of total carotenoids. In fact, the main bioactive compound lutein was from *Tagetes erecta* L. The picture should have correctly advertised *Tagetes erecta* L. as the main source of lutein presenting 90% of total carotenoids.

21.3 INSTRUMENTAL TLC

TLC is a very versatile technique that can be applied in a simple version without any instrumental equipment or as almost fully instrumental. Therefore, it has spread all over the world. TLC has a specific advantage over the complementary HPLC – i.e., the possibility of separating up to 74 samples simultaneously on one plate and detecting them individually before or after the postchromatographic derivatization, applying multiple detection techniques described in other chapters of this book and in Refs [11,12]. The available instrumentation – starting with semiautomatic applicators or autosamplers, automatic developing chambers (also in the version of automatic multiple development chamber), slit-scanning densitometers, diode-array densitometer, image analyzing systems with cameras or flatbed scanners, TLC-MS interface (Camag) for elution of the separated compounds' zones to the mass spectrometric detector or to the vials for further analysis with other techniques e.g., NMR [13] – enables a lot of possibilities for analysis of dietary supplements. Some applications of TLC-MS in the analysis of dietary supplements will be described later in the section TLC-MS. The instrumental and software possibilities enable also the creation of the spectral libraries separately for the UV/visible, MS, and MS^2 spectra which are used for identification. Such libraries owned by different laboratories are dependent on the applications or research areas of the owners.

Among different developing chambers, twin trough chambers were applied in most of published methods, while horizontal developing chambers and automatic developing chambers were rarely used. Vario chamber (Camag) for simultaneous development of the chromatograms by up to six different developing solvents enables fast method development and is therefore often used. Suitable instrumentation is available also for forced-flow planar chromatographic separations, which can be performed by the application of external pressure (overpressured layer chromatography, OPLC), electric field (electro planar chromatography, EPC), and centrifugal force (rotation planar chromatography, RPC). Available instruments for RPC enable analytical and preparative separations, but depending on the configuration the same instrument can be used also for rotation planar extraction as was described for the extraction, isolation, and analysis of flavan-3-ols [14], which are currently common ingredients of dietary supplements. Problems concerning the instability of the analyte on the sorbent before (on the starting position), during and after the chromatography are usually solved by appropriate impregnation of the sorbent, which is usually performed by predevelopment of the chromatographic plate or by addition of the antioxidant into the developing solvent system, as was shown for analysis of carotenoid in dietary supplements [15]. More details will be given in the section Carotenoids. Additionally, prechromatographic derivatization on the start position before the development can be used to improve the stability of the analyte as well as for improving the separation. The examples of prechromatographic derivatizations are bromination [16,17] or iodation [18] of two triterpenic acids, ursolic and oleanolic acids, on HPTLC silica gel plate that enabled the separation of these positional isomers, which are also used as the bioactive ingredients in dietary supplements. Later it was shown that these two acids can be separated from each other and also from betulinic acid on a C18 RP HPTLC plate without prechromatographic derivatization [17]. Postchromatographic derivatization by different reagents is often used for detection. Derivatization by different reagents including group specific reagents is many times even more informative for the identification of individual sample components. Especially important are selective reagents like reagent with 4-dimethylaminocinnamaldehyde (DMACA) [19], which is much more selective for flavanols than the usually applied anisaldehyde and vanillin detection reagents. It is worth to point out that the optimized derivatization conditions influence not only sensitivity but also the stability [19] of the derivatives, which should remain stable at least so long that a documentation of the developed TLC plate can be performed, and in many cases, also a densitometric scanning as well as the in-situ scanning of the absorption spectra. Different techniques are used for detection of the analytes in TLC. Among them are densitometry [12], image analysis [20], mass spectrometry [21] as well as photoacoustic [22–24] and photothermal [22] techniques. The selection of the detection technique depends on the analyte, the purpose of the analysis as well as the availability of specific instruments, but it usually starts with visual evaluation followed by documentation using image analyzing systems. Currently available commercial image analyzing systems (from Camag, Desaga and Sorbfil) and "homemade" image analyzing system with camera or by flatbed scanner with appropriate software were used also in fingerprinting and quantification analyses of dietary supplements [25–27]. Depending on the performances of these systems, images can be captured at different illumination conditions (white light illumination, illumination at 254 or 366 nm) and different modes of illumination, including reflectance, transmission and a combination of reflectance and transmission at the same time. In addition to capturing the images, these systems enable image processing, providing videodensitograms of the samples and reference standards, profile comparisons based on the videodensitograms and quantification. When image processing is combined with chemometric techniques, data stored in an image of chromatographic plate can become even more informative [28]. However, densitometric scanning by slit-scanning densitometers (from Camag, Desaga) in absorption/reflectance or florescence/reflectance modes is still the most frequently used detection technique for the qualitative [15] and quantitative [29–38] evaluation, but its option of multiwavelength scanning was not often reported for dietary supplements analysis. Comparison of the densitograms (if needed, also multiwavelength scans) and the videodensitograms based track profiles obtained from the images, captured at different illumination conditions, can provide a lot of information. Not only the same R_F values and colors of the zones of the reference standards and determined compounds from the sample test solutions, but also the comparison of in-situ densitometrically scanned spectra of the reference standard and corresponding retention zones of the sample test solutions were often used to support the identification of bioactive ingredients [15] and adulterants [39] in dietary supplements [15]. Additional information for the identification and structural elucidation can be gained from the MS [15,39,40], IR, Raman [41] and NMR spectra. In addition to the MS spectra, the similarity of the fragmentation pattern obtained by TLC-MS/MS [17] should also be examined when needed. Sampling of the entire chromatogram using comprehensive TLC × MS [42] as well as the application of TLC-LC-MS [42–44] or even two-dimensional TLC (2D)-TLC-LC-MS [44] can provide even more data. Although TLC analysis can be performed with minimal sample preparation, it can provide even more reliable results when it is supported by appropriate (analyte and sample-dependent) sample preparation techniques. Two short overviews [45,46] describe the use of TLC as an analytical technique for dietary supplements analysis. For the quality control of dietary supplements, two approaches can be used. Targeted approach concerns only one target compound, while comprehensive approach, also known as chemical fingerprinting for detection of various compounds, provides a more informative view. The use of both approaches in the analysis of the real samples such as raw materials and final dietary supplements products will be described in the following sections.

Images of the developed TLC plates should be available for each method. In case of densitometric screening or densitometric fingerprinting or even at densitometric quantitative determination of the bioactive ingredients in dietary supplements, also densitograms with the full data about the scanning conditions (absorption, fluorescence, reflectance mode, wavelength, slit length and width, and scanning speed) should be described as this can have a big effect on the results [47]. An appropriate range should be selected, and methods providing reasonable applications of standard solutions should be taken into consideration (e.g., 100, 150, 200 ng rather than 104, 153, 216 ng). In the case of densitometric scanning and considering the integration and just visual resolution and not densitometrically obtained baseline separation, it is recommended to work with lower loadings of the sample test solutions and the reference standard solutions on the plate. Additionally, described postchromatographic derivatization and detection conditions should include the recipe for the preparation of the detection reagent and its source (a reference), the way of dipping (time and speed) or spraying, time of drying of the plate thereafter, heating temperature and time on the TLC plate heater, as well as optimal time (after heating) for documentation and evaluation by an image analyzing system or densitometer. Not only secondary chromatography [48,49] but also the other effects – as the residuals of the developing solvents constituents – can effect the background of the chromatogram. For example, at detection of triterpenoids, separated as described in Ref. [50], by anisaldehyde detection reagent, a pink instead of a dark blue background is obtained if the plate is not properly dried. The same background color is obtained also if the reagent is "too old." Additionally, it would be very useful and appropriate to define the stability of the detection reagent and the conditions for its storage (refrigerator, temperature, protected against light, etc.).

21.3.1 TLC-Image Analysis

TLC is the only chromatographic technique that can furnish a picture-like image, which is a big advantage. However, this has only partly been used by analysts and the manufacturers of image analyzing systems since 1985, when quantitative evaluation with image analyzing system was introduced [51]. The comparisons of quantitative evaluation of thin-layer chromatograms by the slit-scanning densitometer and the first commercially available TLC image analyzing system with a CCD video camera [47,52] introduced by Camag in 1996 and its first application for quantification of caffeine based on image analysis [53] were described in 1997. This image analyzing system and the following new systems with CCD cameras or flatbed scanners encouraged the documentation of TLC plate images and quantifications of the analytes by different software possibilities depending on the producer. The availability of improved software for image processing and analysis available online from new vendors has further increased the possibilities for the application of image analysis in TLC. A free trial of Sorbfil TLC Videodensitometer software from Sorbpolymer (Krasnodar, Russia) is available, while ImageJ from the National Institutes of Health (USA) can be used free of charge. A ChromImage densitometer flatbed scanner, which enables scanning of the images at UV 254 nm/visible light and is driven by a Galaxie-TLC software (AR2i Company, Le Plessis Robinson, France) was used for the determination of sterols and fatty acids in prostate health dietary supplements [25]. An office flatbed scanner, HP ScanJet G4010 and the Analtech's TLSee TLC analysis software were applied for the determination of phosphatidylserine in dietary supplement products [26]. A flatbed scanner and Sorbfil TLC

Videodensitometer software were used for the determination of sibutramine hydrochloride adulterated in herbal slimming products, and the obtained results were in good agreement with densitometric results obtained at 225 nm in the absorption/reflectance mode, which confirmed that image analysis can be used as an alternative to densitometry [27].

However, even now, only a small part of the multidimensional information saved on the image of a TLC plate is being used, which means that there is still significant potential for further improvement of these systems. It is expected that certain agreed standardization in the development of the TLC methods would enable the creation of a TLC image library accessible worldwide. Such a library would afford support for the researchers, manufacturers and regulatory authorities working in the field of dietary supplements. It would also be important for the control of commercial trade in dietary supplements in a global market. Furthermore, there is a great potential to develop new systems, which would be able to automatically combine densitometric data with data obtained by image analysis. Such programmed systems should also be able to optimize the coupling of TLC with MS, avoiding manual marking of the bands on the chromatographic plates to perform proper positioning of the elution head of the TLC-MS interface (Camag) for the elution of the compounds from the sorbent.

There is a significant potential for combining chemometric techniques with multivariate image analysis in the analysis of dietary supplements and raw materials for dietary supplements. Recently, HPTLC separation on amino plates, image analysis, and pattern recognition methods (principal component analysis, cluster analysis, and partial least square discriminant analysis) were used for fingerprinting and classification of 52 propolis samples in two major types, orange and blue, supporting the idea of the existence of two types of European propolis [28]. This achievement is valuable for the control of propolis as a raw material often used in dietary supplements.

Unfortunately, several scientific articles describing TLC analytical methods (for different samples, not only for dietary supplements) do not provide the images of the developed chromatographic plates, although the images are much more informative than just the R_F values, densitograms, or video densitograms. The reason why densitograms are mentioned here is that with a selection of a narrow integration window close to the chromatographic peak of interest, commercial software (winCats, Camag) extracts only the peaks in the integration window to be shown in the final densitogram. Of course the entirety of the information is stored in the file with the report, but in most publications, only such "extracted densitograms" are presented (shown as part of the analysis of resveratrol in the section Analysis of bioactive ingredients). Similarly, most of the pharmacopoeial methods do not provide the images, but just the tables, with imprecise descriptions of the chromatograms. However, the first edition of the USP Dietary Supplements Compendium from 2009 [54] includes the section ("Illustrations – Appendix") with botanic and chemical characteristics. This section provides chemical structures of relevant constituents and marker compounds as well as chromatograms for 33 materials used in dietary supplements. Although HPLC methods (28) dominate, TLC methods are also presented for 22 materials, while only 6 examples are given for GC methods (Table 21.2). Color images of TLC chromatograms are expected to assist the performance of the compendial thin-layer chromatographic identification tests. The presented typical TLC or HPTLC chromatograms and details about the plate type, developing solvent, chamber (usually saturated), application volume, detection (even non-USP detection), derivatization reagent and in some cases also saturation time and relative humidity, are really informative. Data are also given for the test solutions and standard solutions, and in addition to the USP, some non-USP thin-layer chromatographic methods are also presented.

21.3.2 TLC-MS

TLC-MS is a very versatile tool for identification and detection of the analytes on all types of TLC, HPTLC, and ultra-thin-layer chromatography (UTLC) sorbents. In spite of its great potential, TLC-MS use in the analysis of dietary supplements was described only in a few publications. Different techniques for the coupling of TLC to MS are available [55,56], and the development of new techniques is still in progress. Compared with indirect sampling TLC-MS (which includes pretreatment by scraping, extraction, filtration and concentration), pretreatment is not necessary for direct sampling of TLC-MS techniques, where currently coupling of TLC with ambient pressure ionization MS is among the most common [56]. The Luftman type elution-head based TLC-MS interface [57], commercialized by Camag in 2009 as the TLC-MS interface, was applied for the identification of the labeled dietary supplements' bioactive ingredients: carotenoids [15], *trans*-resveratrol [40], and rosmarinic acid [37], as well as the identification of adulterant sibutramine [39]. The use of TLC-MS interface requires manual positioning of the analyte's zone on the plate under the elution head of the interface. This can cause difficulties, which depend on the properties of the analytes (intensely colored compounds, compounds fluorescing in visible light, or those that have high absorbance at 254 nm). A TLC darkroom, plates with added fluorescent agent, or densitometric scanning [58] can be helpful in determining and marking the exact positions of the analytes. However, if the analyte's absorbance or quantity is insufficient to show a visible zone on the plate, or if one wants to take advantage of the benefits of the great sensitivity of MS, postchromatographic derivatization or continuous

TABLE 21.2 List of materials for which chemical structures of relevant constituents and marker compounds and chromatographic methods including TLC, HPLC, and GC chromatograms are presented in the USP Dietary Supplements Compendium from 2009 [54]

	Methods		
Material	**TLC**	**HPLC**	**GC**
Ademetionine disulfate tosylate	–	+	–
Powdered bilberry extract	+	+	–
Black cohosh	+	+	–
Capsicum, capsaicin, capsicum oleoresin	–	+	–
Cat's claw	+	+	–
Chamomile	+	+	+
Chaste tree	+	+	–
Cod liver oil	–	–	+
Cranberry liquid preparation	–	+	–
Echinacea angustifolia root, *Echinacea pallida* root, *Echinacea purpurea* root, *Echinacea purpurea* aerial parts	+	+	–
Eleuthero	+	+	–
Feverfew	+	+	–
Fish oil containing omega-3 acids	–	–	+
Garlic	+	+	–
Ginger	+	+	–
Ginko	+	+	–
American ginseng, Asian ginseng	+	+	–
Goldenseal	+	+	–
Powdered decaffeinated green tea extract	+	+	–
Hawthorn leaf with flower	+	+	–
Horse chestnut	+	–	–
Licorice	+	+	–
Lutein	–	+	–
Lycopene, tomato extract containing lycopene	–	+	–
Maritime pine	–	+	–
Milk thistle	+	+	–

TABLE 21.2 List of materials for which chemical structures of relevant constituents and marker compounds and chromatographic methods including TLC, HPLC, and GC chromatograms are presented in the USP Dietary Supplements Compendium from 2009 [54]—Cont'd

	Methods		
Material	**TLC**	**HPLC**	**GC**
Pygeum	−	+	+
Red clover	+	+	−
Saw palmetto	−	−	+
St John's wort	+	+	−
Stinging nettle	+	+	+
Ubidecarenone	−	+	−
Valerian	+	+	−

sampling of the entire chromatogram by means of comprehensive TLC-MS (TLC × MS) [42] can be used. Since TLC-MS analysis requires time, the stability of the analytes on the sorbent during and after development of the plate must be considered during method development. The significant improvement of the stability of carotenoids on the C18 RP HPTLC plates predeveloped by dichloromethane−methanol 1:1 (v/v) and developed by methanol−acetone 1:1 (v/v) + 0.1% 2-tertbutylhydroquinone (TBHQ) in a saturated twin trough chamber [15], further described in the section Carotenoids, was crucial for their analysis in dietary supplements by densitometry and TLC-MS. TLC-MS interface was used for the elution of lutein esters, lycopene, free lutein, and β-carotene zones from the sorbent into the ion trap LCQ system, and an atmospheric-pressure chemical ionization (APCI) source in the positive ion mode was applied for scanning of mass spectra (Figure 21.2 [15]), which additionally confirmed the identity of the analytes. The same interface and MS system were used also in the analysis of *trans*-resveratrol in dietary supplements to confirm the identity of the *trans*-resveratrol zone on the HPTLC silica gel plate developed in a horizontal developing chamber (tank configuration) using *n*-hexane−ethyl acetate−formic acid 20:19:1 (v/v). Electrospray ionization (ESI) source in the positive ion mode was used for scanning mass spectra [40]. The TLC-MS interface was used also for the analysis of rosmarinic acid in dietary supplements [37], where after the separation on the HPTLC LiChrospher silica gel F_{254S} plates developed in Automatic Developing Chamber (ADC2) with

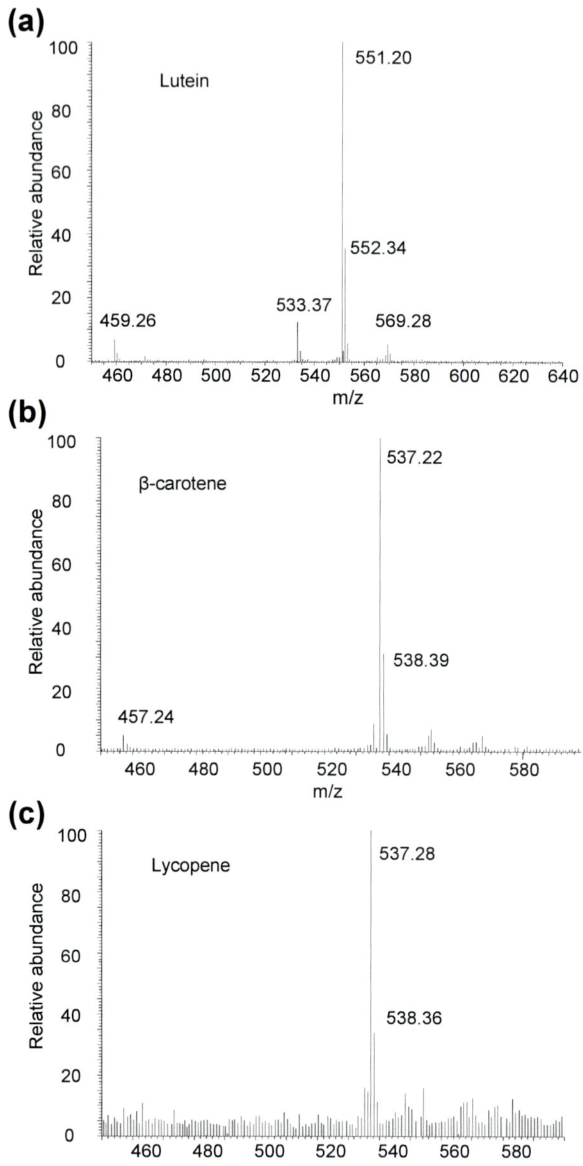

FIGURE 21.2 **Mass spectra of lutein (a), β-carotene (b), lycopene (c), and lutein dipalmitate (d) obtained after elution from C18 RP HPTLC plate developed with methanol–acetone 1:1 (v/v) containing 0.1% 2-tertbutylhydroquinone (TBHQ).** *Adapted from Ref. [15] with permission.*

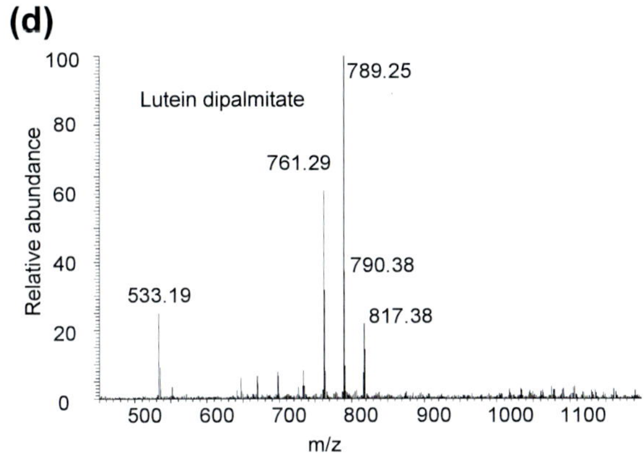

FIGURE 21.2 *Continued*

toluene–ethyl formate–formic acid 6:4:1 (v/v) the identity of eluted rosmarinic acid zone was confirmed by mass spectra obtained by applying ESI source in the negative ion mode.

Another important application area of TLC-MS in the analysis of dietary supplements is in screening of adulterants, in many cases unsafe drugs. An example of this was shown in a recent survey of 50 slimming dietary supplements purchased via the Internet, which confirmed the adulteration with sibutramine in half of the products. These products were analyzed by an HPTLC-UV method on HPTLC silica gel plates, which after the development in saturated ADC2 chamber with toluene–methanol 9:1 (v/v) were scanned by slit-scanning densitometer at 225 nm in the absorption/reflectance mode. Verification of the absence of false positive results was performed by a complementary TLC-MS method using a TLC-MS interface for the extraction of sibutramine retention zone of the samples from the plates and applying the identification in positive ion ESI mode with a Q-trap hybrid mass spectrometer. Visualization of sibutramine standard by postchromatographic derivatization with Dragendorff's reagent was required for TLC-MS analysis, because it enabled proper positioning of the HPTLC plate under the elution head of the TLC-MS interface before the elution of selected zones. No false positive samples were revealed. The obtained results were compared with those of the HPLC-UV and HPLC-MS analyses, proving that the HPTLC-UV method is applicable for routine quality control [39].

Additionally, TLC/desorption electrospray ionization mass spectrometry (DESI-MS) was used for qualitative and quantitative analyses of goldenseal (*Hydrastis canadensis*, a perennial plant from the family *Ranunculaceae*) alkaloids in six dietary supplements in the form of capsules claiming to contain goldenseal root or root extract. The separation on TLC silica gel F_{254} and HPTLC silica gel plates developed in saturated twin trough chamber with ethyl acetate–methanol–formic acid–water 50:10:6:3 (v/v) as a developing solvent. The extended atmospheric sampling capillary of the ion trap mass spectrometer enabled sampling and ionization of the analytes in the alkaloids zones on the chromatographic plates. The computer-controlled surface scanning along the tracks or at fixed R_F value across the plates versus the stationary desorption electrospray emitter was performed by surface positioning software and the appropriate hardware. The limit of detection (LOD) for alkaloids was approximately 5 ng. Qualitative screening using full-scan mass spectra showed a different alkaloid profile in the products and confirmed the presence of at least one additional unlabeled herb in one of the products. The quantities of alkaloids determined in two products by the mass spectral data were in reasonable agreement with the labeled quantities, indicating the quantitative ability of the TLC-MS method. The mass spectral identification and quantification of the analytes even within overlapping zones and the positive identification of even minor species in a mixture were also described [59].

Recently, Brucker introduced the direct coupling of TLC with matrix-assisted laser desorption/ionization time-of-flight mass spectrometry (MALDI-TOF MS) by using an adapter target for TLC silica gel aluminum backed plates and appropriate software (an extension of the standard flexSoftware). Before scanning the MS spectra, the developed TLC plates must be coated with a matrix. Brucker's application describes the analysis of lipids [60] among which was phosphatidylserine, which is also used in dietary supplements as described in Ref. [26]. Unfortunately, the adapter target limits the size of the TLC plates to 5 cm × 7.5 cm, which does not allow the users to develop the plates in the Camag ADC2 chamber. However, it is expected that a new adapter that will overcome the plate size limitation of the ADC2 chamber will soon be introduced.

Soon other more sophisticated ways of using the TLC-MS interface in the analysis of dietary supplements similar to those used in the analysis of the plant extracts (RP-TLC-MS/MS [17], TLC × MS [42], NP-TLC-HILIC-LC [42], NP-TLC-RP-LC-MS [43], 2D NP-TLC-RP-LC-MS [44], 2D RP-TLC-RP-LC-MS [44]) will be introduced.

21.4 ANALYSIS OF BIOACTIVE INGREDIENTS

21.4.1 Carotenoids

Carotenoids are prone to isomerization, oxidation and degradation when exposed to light, heat, oxygen or acids and as such represent a typical example of the analytes that are unstable on the sorbent and can degrade during and immediately after the chromatography. This must be taken into consideration in the preparation and storage of the reference standard stock and working solutions and the sample test solutions regardless of which analytical technique is used for the analysis. In general HPLC is the technique of choice for the analysis of carotenoids, especially if the separation of the isomeric compounds has to be solved. But TLC also has the potential to be used more often, especially for fast screening and depending on the sample complexity also for the quantitative determination of carotenoids in raw materials, extracts, fractions, and final products of dietary supplements. This is needed as the consumption of these supplements is increasing due to the importance of carotenoids in human health (e.g., β-carotene - provitamin A; lack of lutein and zeaxanthin is connected with macular degeneration diseases; lycopene is expected to reduce prostate cancer, etc.). Due to the problems connected with stability on the HPTLC silica gel 60 and HPTLC C18 RP sorbents in most of the published TLC methods relatively high amounts of carotenoid reference standards and sample test solutions were applied on the plates and it was required that documentation by image analysing systems or densitometric scanning of the plates are performed immediately after the development. However, scanning takes time and even more time is required if the absorption spectra are also scanned. Nevertheless, quantitative determination of lycopene in dietary supplements [29] was performed on HPTLC silica gel plates developed in a saturated twin trough chamber using petroleum ether—dichloromethane 9:1 (v/v), followed by densitometric scanning in the absorption/reflectance mode at 416 nm. LOD was 8 ng, limit of quantification (LOQ) was much higher 294 ng, and the calibration curve from 294 to 588 ng was polynomial. For the analysis of lutein in dietary supplements [30], C18 RP HPTLC plates developed in a saturated twin trough chamber by petroleum ether—acetonitrile—methanol 1:2:2 (v/v) and densitometry in the absorption/reflectance mode at 448 nm were applied. LOD and LOQ were approximately 40 ng per band, and the calibration curve from 80 to 320 ng was linear. Lutein and lutein esters were not separated, and the quantification of lutein ester was performed against the free lutein standard. Both methods were validated for sensitivity, linearity, accuracy and precision and were used in the survey of real samples. A high amount of lutein relative to the declared amount was determined (more than 110%, even up to 205%) in all the softgel and hard capsule dietary supplements, which is most probably related with the awareness of the manufacturers of the degradation of lutein during the shelf-life [30]. The situation was different in the case of dietary supplements with lycopene. The determined amounts of lycopene relative to the declared amount were between 90% and 100% only in one sample of softgel capsules. Between 77% and 86% of the declared lycopene amount was determined in the other two

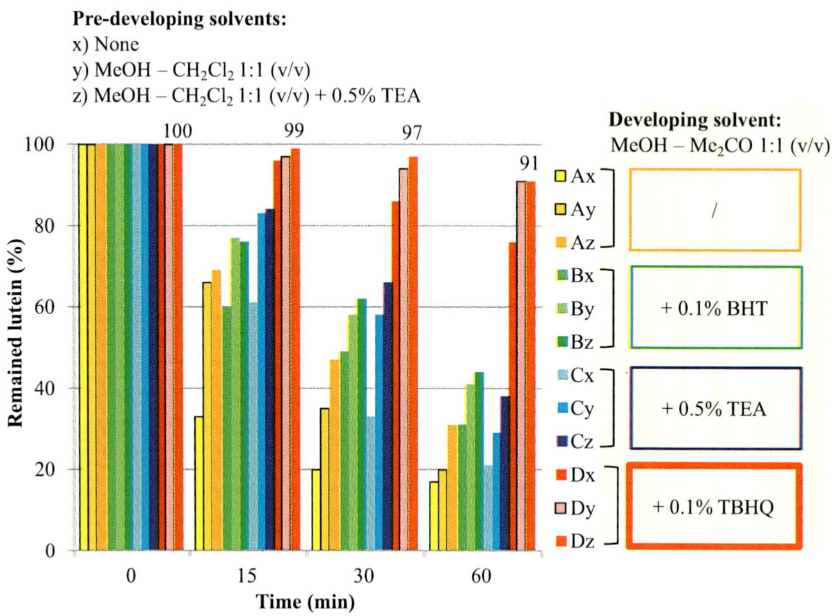

FIGURE 21.3 **Influence of the sorbent predevelopment and addition of antioxidant (triethylamine (TEA), 2,6-di-tert-butyl-4-methylphenol (BHT), or 2-tertbutylhydroquinone (TBHQ)) to the developing solvent methanol–acetone 1:1 (v/v) on stability of lutein standard (15 ng) applied on the C18 RP HPTLC plates after development determined by scanning at 450 nm in the absorption/reflectance mode [15].**

softgel capsule samples, while in the tablets, lycopene was not even detected, which is most probably due to its degradation [29].

In a recent stability study (Figure 21.3) the influence of predevelopment and the addition of antioxidant to the developing solvent methanol–acetone 1:1 (v/v) on the stability of 15 ng of lutein applied on a C18 RP HPTLC plate was demonstrated [15]. After development in a saturated twin trough chamber the plates were scanned in the absorption/reflectance mode at 450 nm. After 15 min only 33% of lutein remained unaltered on the C18 RP HPTLC plate, which was not predeveloped. Predevelopment with methanol–dichloromethane 1:1 (v/v) or methanol–dichloromethane 1:1 (v/v) with addition of 0.5% triethylamine (TEA) significantly improved the stability of lutein: only 34% of lutein degraded [15]. The stability of lutein on the sorbent was further improved by addition of antioxidants 0.5% TEA, 0.1% 2,6-di-tert-butyl-4-methylphenol (BHT), or 0.1% TBHQ to the developing solvent methanol–acetone 1:1 (v/v). Finally, predevelopment of the plate with methanol–dichloromethane 1:1 (v/v) and development of the chromatograms with

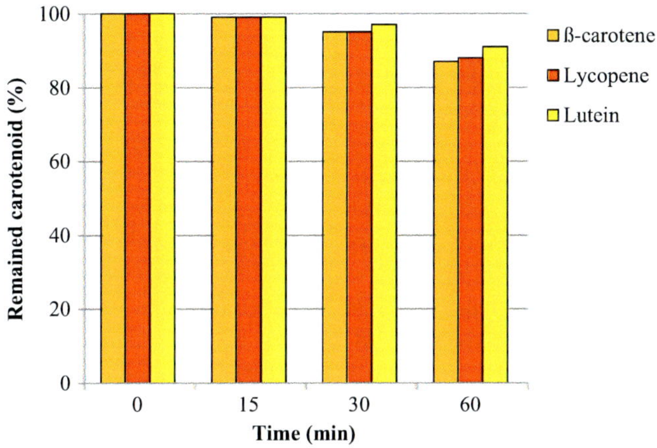

FIGURE 21.4 **Stability of β-carotene (15 ng), lycopene (15 ng), and lutein (15 ng) standards on the C18 RP HPTLC plates predeveloped with methanol–dichloromethane 1:1 (v/v) and developed with methanol–acetone 1:1 (v/v) containing 0.1% 2-tertbutylhydroquinone (TBHQ).** The plates were scanned at 450 nm in the absorption/reflectance mode [15].

methanol–acetone 1:1 (v/v) with 0.1% TBHQ was selected as the best choice, as it preserved lutein almost intact for 15 min and since only 9% of lutein degraded in 1 h after the development (Figure 21.3), which enables quantitative determination of lutein by densitometry and scanning of the absorption and mass spectra. This procedure resulted in by far the lowest LOD being 1.5 ng and LOQ being 5 ng, and enabled quantitative determination of lutein using the polynomial calibration curve from 5 to 30 ng. When the best chromatographic conditions were applied to study stability of other common carotenoids present in dietary supplements, lycopene and β-carotene, the same trend as for lutein was observed within 1 h after the development of the plate (Figure 21.4). This confirms that it is possible to apply C18 RP HPTLC plates also for the analysis of lycopene, which was described in Ref. [29] as critical and therefore not useful on the concentration zone and on the layer itself. The chromatographic conditions enabled a good separation of lutein esters, free lutein, lycopene and β-carotene from dietary supplements and their identification by visible absorption spectra scanned in-situ and by mass spectra [15]. Accuracy, precision and selectivity of the method proved that it can be used in routine quality control of lutein in dietary supplements, food and plant materials (even selection of plant varieties, e.g., *Tagetes spp.*) as well as during the development of new products. The method is also suitable for fast screening of incorrectly labeled dietary supplements products regarding lutein (declared as lutein, but containing lutein esters), which are often found on the market.

The best chromatographic conditions described above were applied to fast screening of the carotenoids lutein, β-carotene and lycopene in dietary supplements of 12 different brands. In the examined brands, carotenoids were declared either as the only bioactive compounds (Figure 21.5(a): tracks 5, 11, 14), as bioactive compounds combined with different plant extracts (Figure 21.5(a): tracks 10, 12), or as compounds combined with several other bioactives (Figure 21.5(a): tracks 4, 6, 7, 8, 9, 13, 15). Due to our previous experience with the analysis of dietary supplements by TLC and HPLC methods, we firstly applied higher amounts of the standards and the sample test solutions (Figure 21.5(a)) to be able to see all labeled carotenoids. As it is evident from the image of the track 7 captured at 366 nm and at white light illumination (Figure 21.5(c)), the chromatographic conditions enable qualitative and quantitative determination of lutein and β-carotene in spite of other compounds from *Calendula officinalis* and blueberry extracts (track 7 captured at 366 nm). Additionally, in the same brand, undeclared lutein esters were detected. In a multi-ingredient dietary supplement only β-carotene was detected, although lycopene was also declared (Figure 21.5(a), track 8). The peaks in the video densitograms (Figure 21.5(b) and (c)) and densitograms (Figure 21.6) are sharp and symmetrical, without tailing or fronting, which is supposed to be a common problem in the TLC analysis of carotenoids. In-situ scanned absorption spectra (Figure 21.6) show characteristic shapes (comparable to those obtained by spectrophotometry) for all examined carotenoid standards and the zones at the same R_F values of the studied sample test solutions.

21.4.2 Caffeine

Caffeine, a central nervous system stimulant, is the world's most widely consumed psychoactive compound, which can be found in many dietary supplements, but only one published TLC method [61] describes its determination in such products. The separation was performed on HPTLC silica gel F_{254} plates in a saturated chamber (not specified) using ethyl acetate–methanol 85:15 (v/v) as a developing solvent and the quantitative evaluation was based on densitometric scanning at 275 nm in the absorption/reflectance mode [61].

We performed qualitative analyses of caffeine in seven dietary supplements of six different brands (Figures 21.7 and 21.8) using the TLC method for alkaloids from green tea with visual evaluation at 254 nm [62]. The separation on HPTLC silica gel F_{254} plates developed up to 7 cm in a horizontal chamber (sandwich configuration) with ethyl acetate–methanol–water 20:2.7:2 (v/v) as a developing solvent was followed by documentation at 254 nm using an image analyzing system (Figure 21.7(a)) and densitometric scanning in the absorption/reflectance

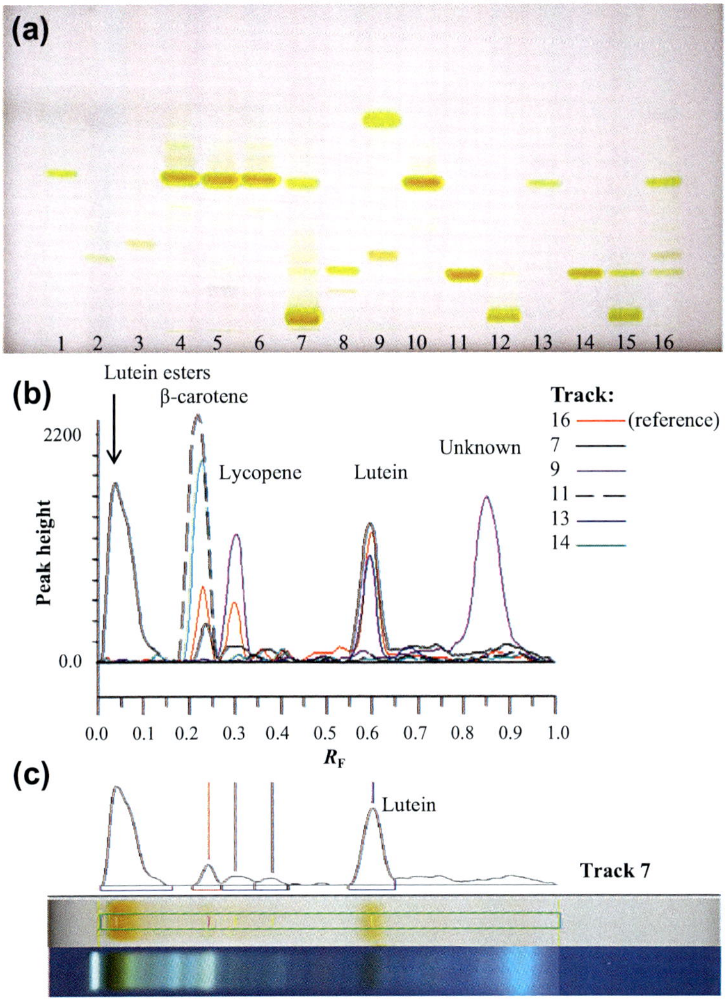

FIGURE 21.5 **Screening of carotenoids in dietary supplements on C18 RP HPTLC plate developed with methanol–acetone 1:1 (v/v) containing 0.1% 2-tertbutylhydroquinone (TBHQ) (a); profile comparison of video densitograms (b); track 7 captured at 366 nm and white light with corresponding video densitogram (c).** Tracks (a): 1: lutein (60 ng); 2: β-carotene (40 ng); 3: lycopene (40 ng); 4: soft capsules with lutein, vitamins, omega-3-fatty acids and *Vitis vinifera* extract with resveratrol; 5: soft capsules with lutein; 6: hard capsules with lutein and vitamins; 7: soft capsules with lutein from *Calendula officinalis*, blueberries extract and vitamins; 8: hard capsules with lycopene, β-carotene from carrot, aloe vera leaves extract and coenzyme Q_{10}; 9: soft capsules with extracts of tomato, curcuma and black pepper containing lycopene, curcumin and piperine; 10: soft capsules with lutein from *Tagetes erecta*; 11: soft capsules with β-carotene; 12: tablets with lutein esters from *Tagetes erecta*; 13: tablets with lutein, vitamins, Pycnogenol®, and extracts of *Calendula officinalis*, blueberries, and grape skin; 14: soft capsules with β-carotene; 15: hard capsules with lutein esters, β-carotene, vitamins and bilberry extract; 16: applications from tracks 1–3.

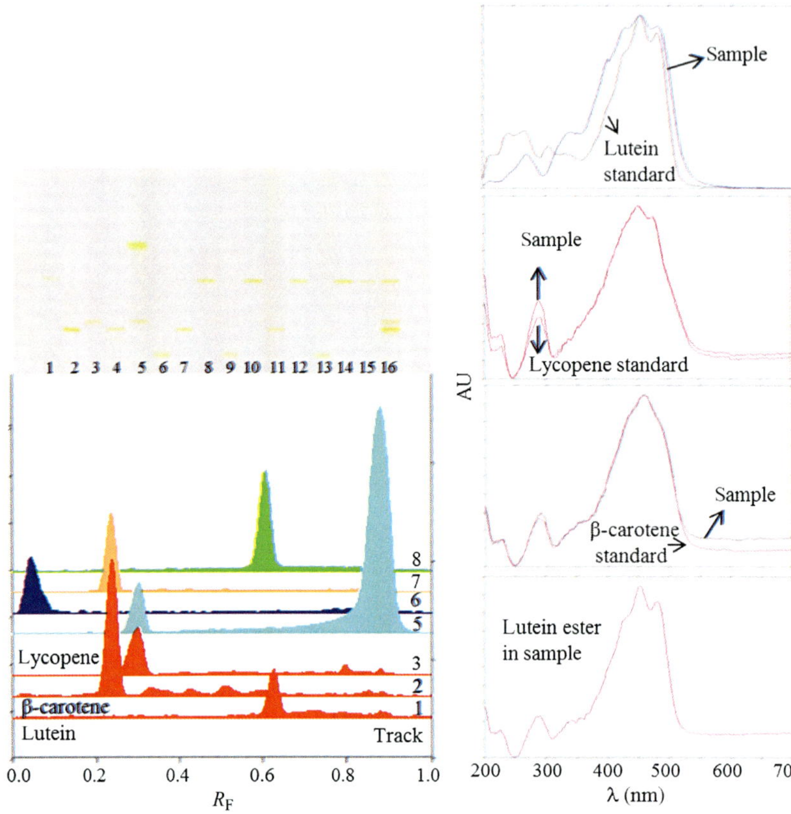

FIGURE 21.6 **The image of a C18 RP HPTLC plate developed with methanol–acetone 1:1 (v/v) containing 0.1% 2-tertbutylhydroquinone (TBHQ) and comparison of the corresponding densitograms (at 450 nm) of carotenoid dietary supplement sample test solutions and reference standards and overlapping of the in-situ absorption spectra of carotenoids' reference standards and the corresponding retention zones of the sample test solutions.** Tracks: 1: lutein (20 ng); 2: β-carotene (40 ng); 3: lycopene (40 ng); 4: hard capsules with lycopene, β-carotene from carrot, aloe vera leaves extract and coenzyme Q_{10}; 5: soft capsules with extracts of tomato, curcuma and black pepper containing lycopene, curcumin and piperine; 6: tablets with lutein esters from *Tagetes erecta*; 7: soft capsules with β-carotene; 8: soft capsules with lutein from *Tagetes erecta*; 9: soft capsules with lutein from *Calendula officinalis*, blueberries extract and vitamins; 10: soft capsules with lutein, vitamins, omega-3-fatty acids and *Vitis vinifera* extract with resveratrol; 11: soft capsules with β-carotene; 12: soft capsules with lutein; 13: hard capsules with lutein esters β-carotene, vitamins and bilberry extract; 14: hard capsules with lutein and vitamins; 15: tablets with lutein, vitamins, Pycnogenol® and extracts of *Calendula officinalis*, blueberries and grape skin; 16: applications from tracks 1–3.

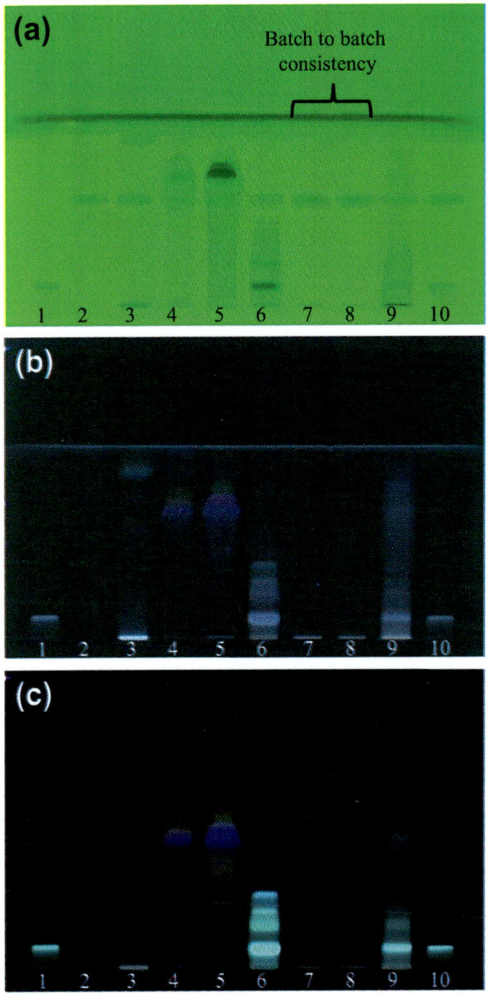

FIGURE 21.7 **Screening of caffeine in dietary supplements on HPTLC silica gel F_{254} plate developed with ethyl acetate–methanol–water 20:2.7:2 (v/v), followed with detection at 254 nm (a), 366 nm before (b), and after derivatization (c) with natural products reagent. Batch-to-batch consistency of hard capsules with guarana seeds containing caffeine on tracks 7 and 8 (a).** Tracks: 1: chlorogenic acid (1.5 μg); 2: caffeine (2 μg); 3: hard capsules with extracts of green tea, black pepper, mango and guarana seeds; 4: soft capsules with green tea and conjugated linolenic acid; 5: hard capsules with caffeine, acai fruit extract and green tea containing polyphenols (epigallocatechin gallate); 6: hard capsules with decaffeinated green coffee and chlorogenic acid; 7,8: hard capsules with guarana seeds containing caffeine; 9: instant black coffee with green coffee extract, caffeine and chlorogenic acid; 10: applications from tracks 1–2.

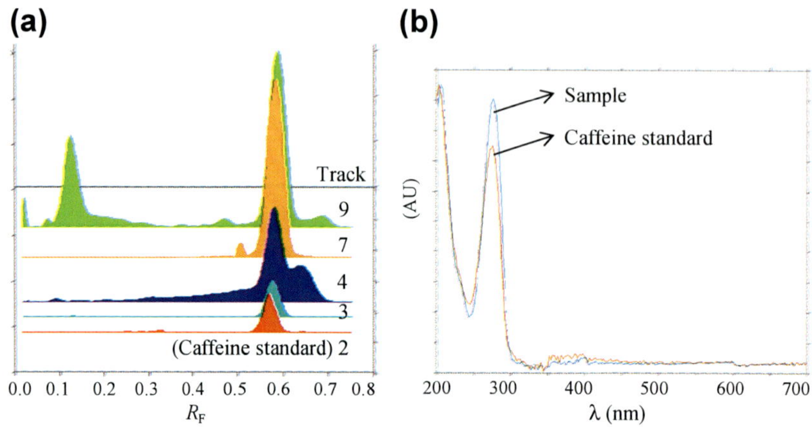

FIGURE 21.8 **Densitograms (a) of selected corresponding tracks from Figure 21.7 scanned at 275 nm in the absorption/reflectance mode and overlapping in-situ absorption spectra (b) of caffeine reference standard and the corresponding retention zone of the sample test solutions.** Tracks: 2: caffeine (0.1 μg); 3: hard capsules with extracts of green tea, black pepper, mango and guarana seeds; 4: soft capsules with green tea and conjugated linolenic acid; 7,8: hard capsules with guarana seeds containing caffeine.

mode at 275 nm (Figure 21.8(a)). Apart from caffeine, this method also enabled simultaneous detection of other bioactive compounds such as chlorogenic acid at 254 nm (Figure 21.7(a), tracks 1 and 10), which was labeled as an ingredient in two analyzed dietary supplement products. Additionally, chlorogenic acid was detected on the same plate at 366 nm before (Figure 21.7(b)) and after derivatization (Figure 21.7(c)) with a natural products reagent [63, p. 277]. The analyzed dietary supplements (Figures 21.7 and 21.8) in capsule formulations (6) and powder (1) were labeled to contain the following: green tea, black pepper, guarana and mango seeds (track 3); green tea and linolenic acid (track 4); caffeine, green tea and acai (track 5); decaffeinated green coffee with chlorogenic acid (track 6); caffeine from guarana seeds (tracks 7, 8); and instant black coffee with green coffee extract and chlorogenic acid (track 9; powder). As can be seen in Figure 21.7, caffeine was detected in all dietary supplements, as expected, although not labeled and also in those with green tea (tracks 3 and 4) and surprisingly even in one containing decaffeinated green coffee (track 6). Analysis of two different batches of the product from the same brand – supplement containing guarana (tracks 7 and 8) – confirmed batch-to-batch consistency. For the densitometric evaluation (Figure 21.8), the applied amount of caffeine reference standard was 20 times lower than for the image evaluation and loadings of the dietary supplements test solutions were lower as well. The identity of caffeine from all analyzed dietary supplements was confirmed by comparison of the in-situ scanned absorption spectra of caffeine zones of the reference standard and from the test solution (Figure 21.8(b)).

21.4.3 *Trans*-Resveratrol

Trans-resveratrol, a well-known stilbene with anticarcinogenic, anti-inflammatory, and also antidiabetic properties, can be found in many dietary supplements. It is primarily isolated from Japanese knotweed (*Polygonum cupsidatum*), but also from red grape skin. Qualitative analysis of resveratrol from dietary supplements was performed on HPTLC silica gel plates developed in a horizontal developing chamber (tank configuration) using *n*-hexane–ethyl acetate–formic acid 20:19:1 (v/v) with UV detection at 303 nm followed by postchromatographic derivatization with anisaldehyde [40]. It was shown that molybdophosphoric acid detection reagent and 5% sulfuric acid in ethanol can also be applied for resveratrol detection. Postchromatographic detection with anisaldehyde reagent at 500 nm is less sensitive than densitometry in the fluorescence ($\lambda = 313$ nm) or absorption ($\lambda = 303$ nm) mode [40], but when dealing with multi-ingredient dietary supplements, derivatization is desirable to avoid potential interferences.

The described procedure using multiple detection (densitometry at 303 nm in the absorption mode following postchromatographic anisaldehyde detection at 500 nm) was proven in our survey of resveratrol in dietary supplements (form of capsules and tablets) of six different brands (Figure 21.9) containing resveratrol: as the only bioactive compound isolated from Japanese knotweed (tracks 3 and 6), with different vitamins and β-carotene (track 4), with different plant extracts (track 5), with different plant extracts, vitamins, lutein and omega-3-fatty acids (track 2), from wine extract, grape seeds' proanthocyanidins and soy phosphatidylcholine (track 7). Due to previous experience with the analysis of dietary supplements by TLC, we first applied higher amounts of the reference standard and test solutions (Figure 21.9) to be able to see the labeled resveratrol in all dietary supplements and to obtain in-situ absorption spectra before (Figure 21.10(a)) and after derivatization (Figure 21.10(b)) with anisaldehyde. But for densitometry at 303 nm (Figure 21.10(a)) and at 500 nm after postchromatographic detection (Figure 21.10(b)), a two-fold lower amounts of the reference standard were applied. As can be seen in Figure 21.9(a) and (b), resveratrol was present in all analyzed products, except in one (track 5), where according to the label, 100 mg of resveratrol per capsule should be present, which is 100 times more than in the product analyzed in track 2 (1 mg per capsule). This confirms that a strict control of the dietary supplements is needed to protect consumers from the misleading labels. As expected, the chromatograms of the sample test solutions of the

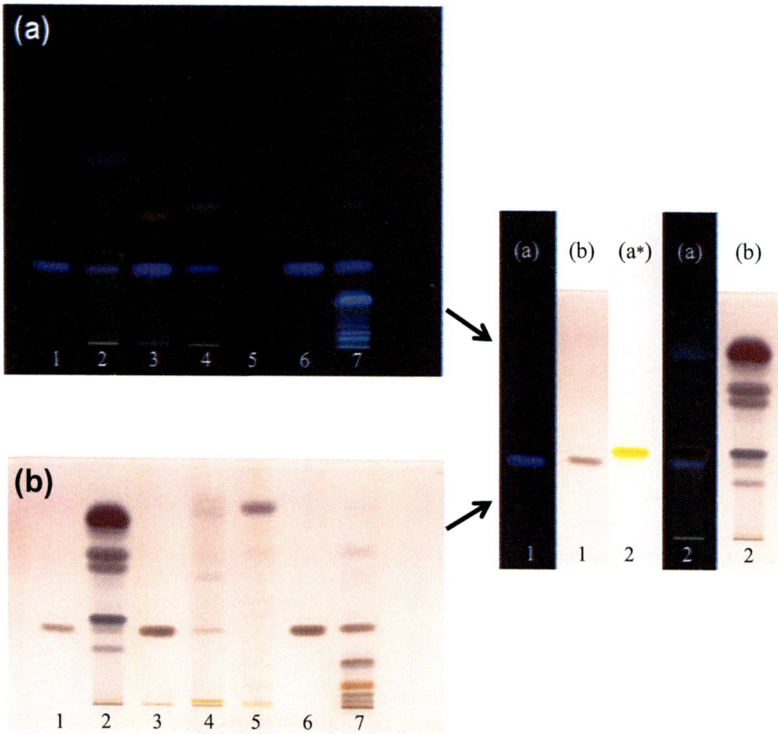

FIGURE 21.9 **Screening of *trans*-resveratrol in dietary supplements on HPTLC silica gel plate developed with *n*-hexane–ethyl acetate–formic acid 20:19:1 (v/v) before (a: 366 nm; a*: white light – detection of lutein) and after (b: white light) derivatization with anisaldehyde detection reagent.** Tracks: 1: *trans*-resveratrol (0.5 μg); 2: soft capsules with *Vitis vinifera* extract containing resveratrol, lutein, omega-3-fatty acids and vitamins; 3: tablets with resveratrol from Japanese knotweed (*Polygonum cupsidatum*); 4: tablets with resveratrol, β-carotene and vitamins; 5: hard capsules with Japanese knotweed containing resveratrol and extracts of goji berries and mangostan; 6: hard capsules with resveratrol from Japanese knotweed; 7: hard capsules with wine extract containing resveratrol, grape seeds' proanthocyanidins and soy phosphatidylcholine.

dietary supplements containing resveratrol as the only bioactive compounds had less additional bands (tracks 3 and 6, Figure 21.9(a) and (b)) than the chromatograms of the other products. Carotenoids (β-carotene or lutein) labeled as additional bioactives in two products (tracks 2 and 4) were detected only in one (track 2) where lutein appeared as a zone with slightly higher R_F value than resveratrol. The same dietary supplement was also analyzed for carotenoids (track 2, Figure 21.4 and track 10, Figure 21.5).

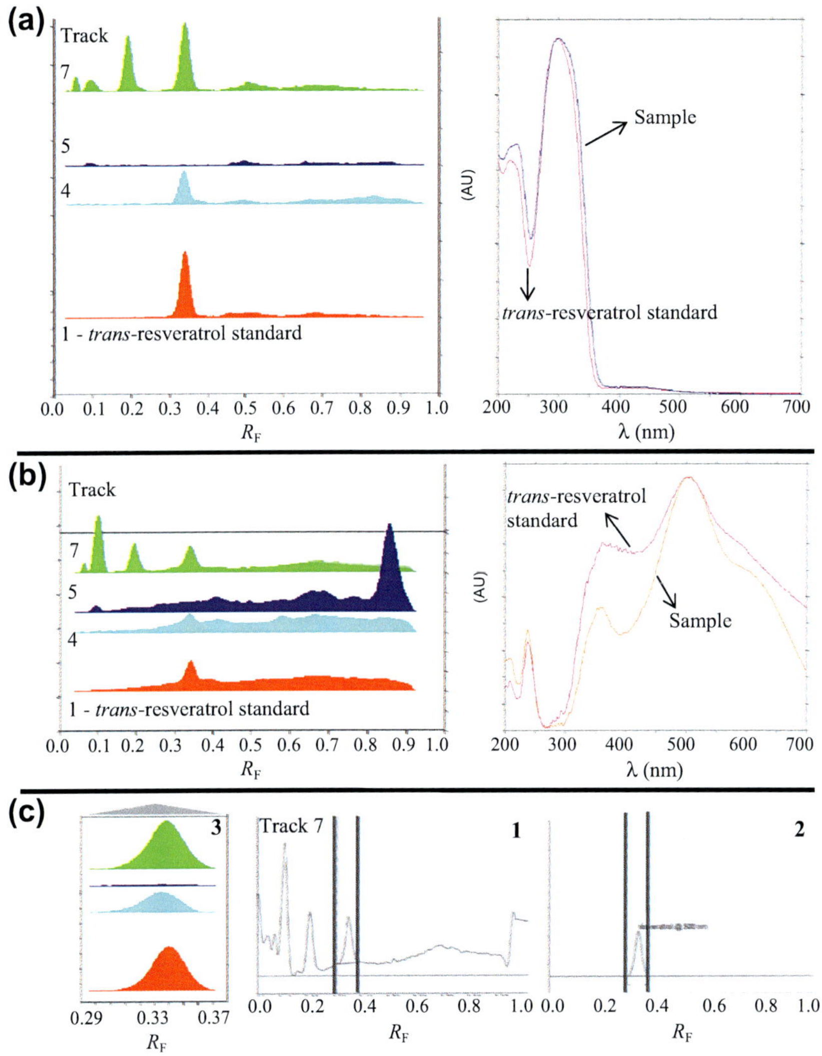

FIGURE 21.10 **Densitograms of selected corresponding tracks from Figure 21.9 scanned before (a: 303 nm) and after (b: 500 nm) derivatization with anisaldehyde detection reagent in the absorption/reflectance mode with overlapping in-situ absorption spectra of *trans*-resveratrol reference standard and the corresponding retention zone of the sample test solution before (a) and after (b) derivatization. Densitograms of track 7: (c1) with marked "integration window"; with "extracted" resveratrol peak from 2D (c2) and from 3D view (c3).** Tracks: 1. *trans*-resveratrol (0.1 μg); 4: tablets with resveratrol, β-carotene, and vitamins; 5: hard capsules with Japanese knotweed containing resveratrol and extracts of goji berries and mangostan; 7: hard capsules with wine extract containing resveratrol, grape seeds' proanthocyanidins and soy phosphatidylcholine.

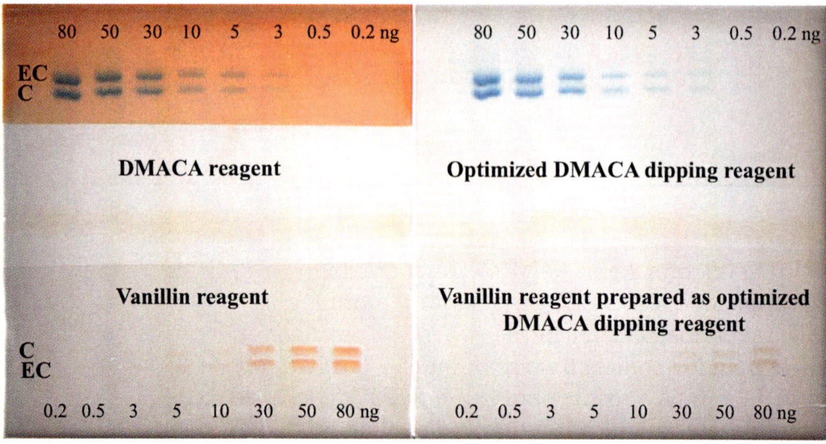

FIGURE 21.11 **Comparison of chromatograms of (−)-epicatechin and (+)-catechin standards on HPTLC cellulose plates after development with water and derivatization.** Left: vanillin reagent [69, p. 440] and 4-dimethylaminocinnamaldehyde (DMACA) prepared in the same way concerning the concentration of aldehyde and acid. Right: optimized DMACA DR [19] and vanillin reagent prepared as optimized DMACA DR. *Adapted from Ref. [19] with permission.*

21.4.4 Flavanols

Flavan-3-ols and proanthocyanidins possess antiinflamatory and vasoprotective properties and are popular ingredients in dietary supplements. Their TLC analyses on HPTLC silica gel and cellulose sorbents enable fast screening and when combined with postchromatographic derivatization and densitometry also quantification. Detection is usually performed by reagents based on the combination of strong acids and aldehydes (vanillin [14,64,65], anisaldehyde [64], DMACA [19, 66–68]), as well as fast blue B salt reagent [54,62]. When the DMACA-HCl reagent [63] (Figure 21.11) is prepared as the most usable vanillin-H_3PO_4 reagent [69, p. 440], concerning the selected concentration of aldehyde and acid, it gives an orange background, which can influence analytes quantification. The study of the influence of the ratio of DMACA reagent ingredients (different acids, alcohols, and DMACA) on the reagent stability and sensitivity finally resulted in an optimized DMACA dipping reagent (DR) prepared by dissolving 60 mg of DMACA in 13 mL of concentrated hydrochloric acid and made up to 200 mL with ethanol [19]. For (+)-catechin and (−)-epicatechin on a cellulose plate derivatized by DMACA DR, LOD and LOQ were 0.2 ng and 0.4 ng, respectively, while by vanillin reagent LOD and LOQ were 0.5 ng and 1 ng, respectively [19]. If vanillin reagent was prepared as the DMACA DR with the same molar concentration of aldehyde and acid, it was significantly less sensitive for the detection of flavan-3-ols (Figure 21.11). The color of the bands derivatized

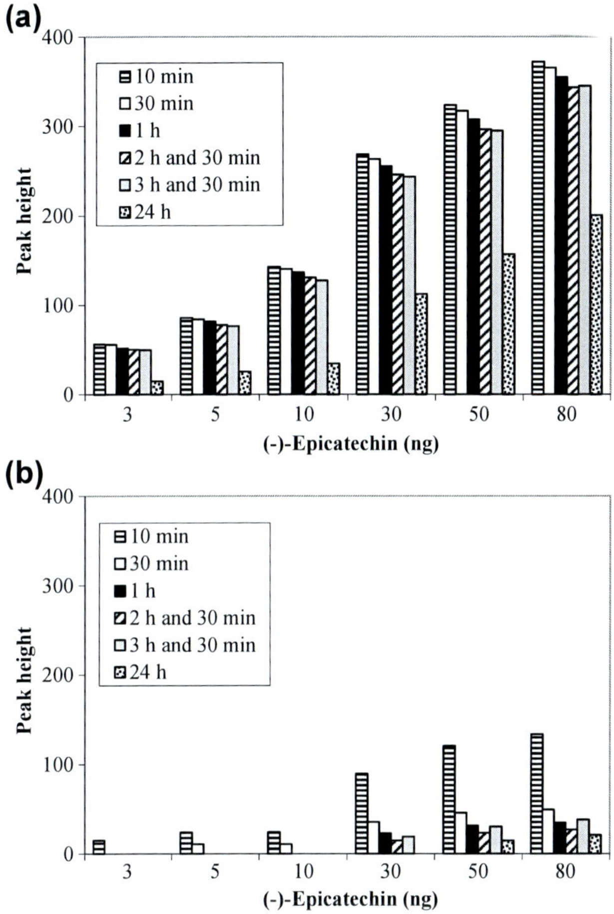

FIGURE 21.12 **Stability of the epicatechin derivatives obtained by the optimized 4-dimethylaminocinnamaldehyde (DMACA) DR (a) and vanillin reagent (b) on HPTLC plate developed with water.** *Adapted from Ref. [19] with permission.*

by the DMACA DR was stable within 1 h then faded slowly but the color of the bands obtained by vanillin reagent was stable for only 10 min, thereafter vanishing rapidly (Figure 21.12). The stability of the derivatives is crucial for image analyses, scanning of the densitograms [70,71], and in-situ absorption spectra, as well as for TLC-MS [42] or TLC × MS analyses [42], where it is helpful in proper positioning of the TLC-MS interface's elution head.

We performed the screening of flavan-3-ols and their oligomeric proanthocyanidins in green tea, grape seeds, pine and cinnamon bark present in multi-ingredient dietary supplements containing also other compounds (resveratrol, linolenic acid, caffeine, etc.) and other plant extracts. All sample test solutions prepared by the extraction with 70% aqueous acetone were analyzed on two types of HPTLC sorbents: silica gel and cellulose, starting with the separation on the silica gel plate using the developing solvent system toluene–acetone–formic acid 3:6:1 (v/v) [68]. In our case, the development in a twin trough chamber was followed by the postchromatographic derivatization with DMACA DR, which resulted in the first indication of the presence of the monomers (the highest R_F value) and other flavanol compounds with higher molar mass (Figure 21.13). Analysis on the HPTLC cellulose plate developed in a horizontal developing chamber (sandwich configuration) with pure water as a developing solvent [14,19,65] enabled fast separation of selected flavan-3-ols and procyanidins in 15 min (data not shown) and also provided data for the optimization of the loading of the sample test solutions. However, due to the matrix-effected deformation of the chromatograms, further sample preparation (e.g. by solid-phase extraction) would be needed. But no deformation was observed when 1-propanol–water–acetic acid 20:80:1 (v/v) mixture [65] was used instead of water as a developing solvent. Therefore, 1-propanol–water–acetic acid 20:80:1 (v/v) mixture was used for further analyses on HPTLC cellulose plates (Figure 21.14). Approximately ten-fold lower loadings were applied on the HPTLC cellulose plates with a sorbent thickness of 100 μm than on the HPTLC silica gel plates with a sorbent thickness of 200 μm to avoid overloading and to obtain the appropriate amount of the analytes for densitometry. As it is evident from the densitograms of the selected reference standards (tracks 2–6) on silica gel (Figure 21.13(b) and (c)) and cellulose (Figure 21.15(a) and (b)) plates presented on the same scale (exported from the 3D view produced by Camag WinCats software), postchromatographic derivatization with DMACA DR (Figures 21.13(c) and 21.15(b)) significantly improved the sensitivity compared with detection before derivatization (Figures 21.13(b) and 21.15(a)). Derivatization by DMACA DR proved to be crucial in achieving the specific detection and thus avoiding interference from other bioactive ingredients absorbing in the same UV range. An example of this is evident by the significant decrease of the peak with the highest R_F value in the densitograms of track 7 (Figure 21.13(b) and (c) and Figure 21.15(a) and (b); dietary supplement test solution containing extracts of mango, green tea and guarana) when comparing scans at 280 and 655 nm (after derivatization). In spite of the multi-ingredient content of the analyzed dietary supplements, the fingerprints (Figures 21.13(a) and 21.14) of those containing green tea (tracks 7, 8 and 9) were equal and showed the same characteristic profile as green tea in Ref. [65], while supplements containing grape seeds (track 10, Figures 21.13(a) and 21.14) and pine bark with grape seeds (track 11, Figures 21.13(a) and 21.14)

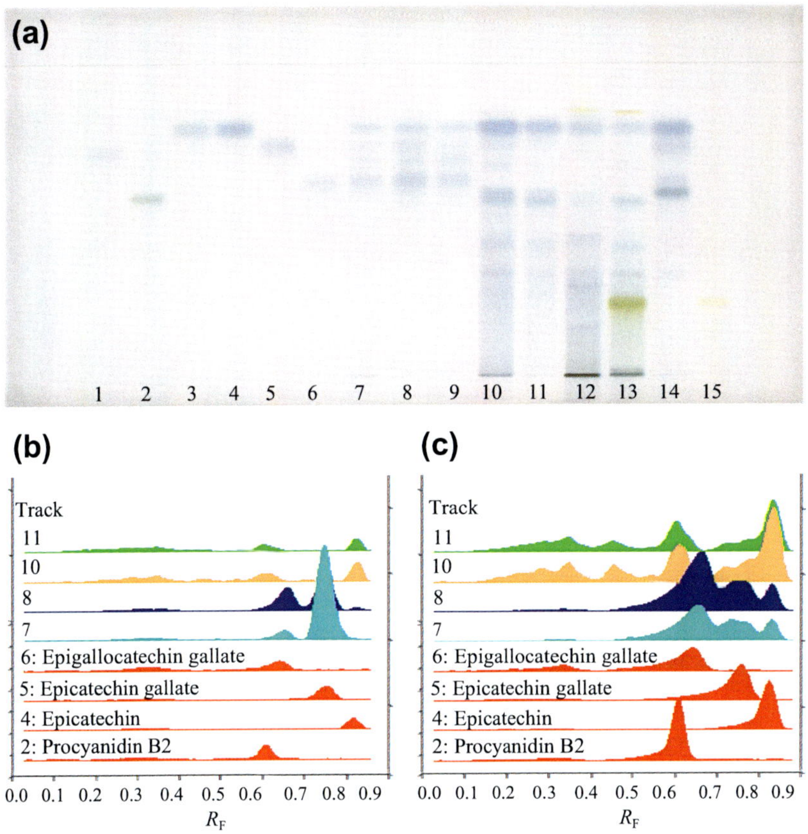

FIGURE 21.13 **Screening of flavanols and procyanidins in dietary supplements on HPTLC silica gel plate developed with toluene–acetone–formic acid 3:6:1 (v/v) after derivatization (a) with 4-dimethylaminocinnamaldehyde (DMACA) DR; densitograms of selected corresponding tracks from (a) scanned in the absorption/reflectance mode before (b: 280 nm) and after derivatization (c: 655 nm) with DMACA DR presented on the same scale.** Tracks: 1: procyanidin A2 (A2) (300 ng); 2: procyanidin B2 (B2) (500 ng); 3: (+)-catechin (C) (300 ng); 4: (−)-epicatechin (EC) (300 ng); 5: (−)-epicatechin gallate (ECg) (300 ng); 6: (−)-epigallocatechin gallate (EGCg) (300 ng); 7: hard capsules with extracts of green tea, black pepper, mango and guarana seeds; 8: soft capsules with green tea and conjugated linolenic acid; 9: hard capsules with green tea containing polyphenols (EGCg), acai fruit extract and caffeine; 10: hard capsules with grape seeds' proanthocyanidins, wine extract containing resveratrol and soy phosphatidylcholine; 11: hard capsules with flavanols and oligomeric proanthocyanidins from French pine bark and grape seeds; 12: hard capsules with extracts of cinnamon bark, *Gymnema sylvestre* leaves, and *Lagerstroemia speciosa* leaves containing procyanidin A2, gymnemic acids and corosolic acid; 13: hard capsules with Pycnogenol®, acerola extracts and rutin; 14: applications from tracks 1–6; 15: rutin (3 μg).

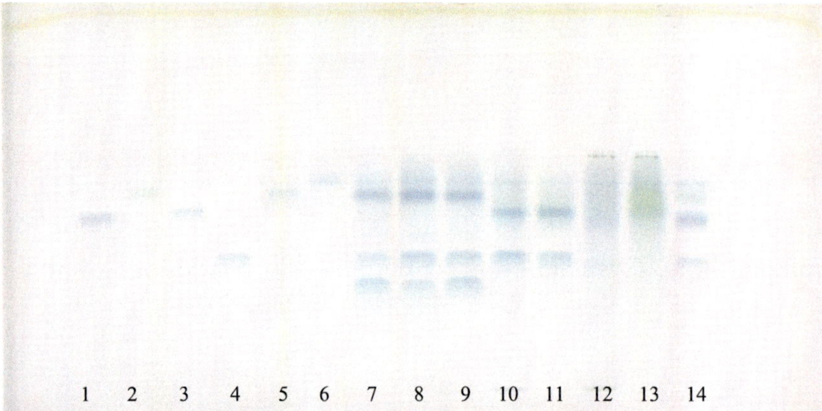

FIGURE 21.14 **Screening of flavanols and procyanidins in dietary supplements on HPTLC cellulose plate developed with 1-propanol–water–acetic acid 20:80:1 (v/v) after derivatization with 4-dimethylaminocinnamaldehyde (DMACA) DR.** Tracks: 1: A2 (50 ng); 2: B2 (50 ng); 3: C (30 ng); 4: EC (30 ng); 5: ECg (40 ng); 6: EGCg (50 ng); 7: hard capsules with extracts of green tea, black pepper, mango and guarana seeds; 8: soft capsules with green tea and conjugated linolenic acid; 9: hard capsules with green tea containing polyphenols (EGCg), acai fruit extract and caffeine; 10: hard capsules with grape seeds' proanthocyanidins, wine extract containing resveratrol and soy phosphatidylcholine; 11: hard capsules with flavanols and oligomeric proanthocyanidins from French pine bark and grape seeds; 12: hard capsules with extracts of cinnamon bark *Gymnema sylvestre* leaves and *Lagerstroemia speciosa* leaves containing procyanidin A2, gymnemic acids and corosolic acid; 13: hard capsules with Pycnogenol®, acerola extracts and rutin; 14: applications from tracks 1–6.

surprisingly have similar fingerprints. Additionally, the fingerprints of the later are remarkably different from the fingerprints of the supplement containing Pycnogenol® (from French maritime pine bark), acerola and rutin (track 13, Figures 21.13(a) and 21.14). This was observed on the silica gel as well as on the cellulose plates, where the difference is even more pronounced. In addition, the interfering rutin zone was yellow on the silica gel layer and greenish on the cellulose layer (track 13). The fingerprint of dietary supplements containing cinnamon bark (track 12, Figures 21.13(a) and 21.14) was different from all other fingerprints. The dietary supplements analyzed on tracks 7, 8 and 9 containing green tea were also tested for the presence of caffeine (tracks 3, 4 and 5; Figure 21.7) despite the missing declaration of the content of caffeine on two of them. The identity of the analyzed flavan-3-ols and oligomeric proanthocyanidins from dietary supplements was also confirmed by comparison in-situ absorption spectra of reference standards and test solutions scanned from the flavan-3-ols and procyanidins zones before (Figure 21.15(a)) and after derivatization (Figure 21.15(b)).

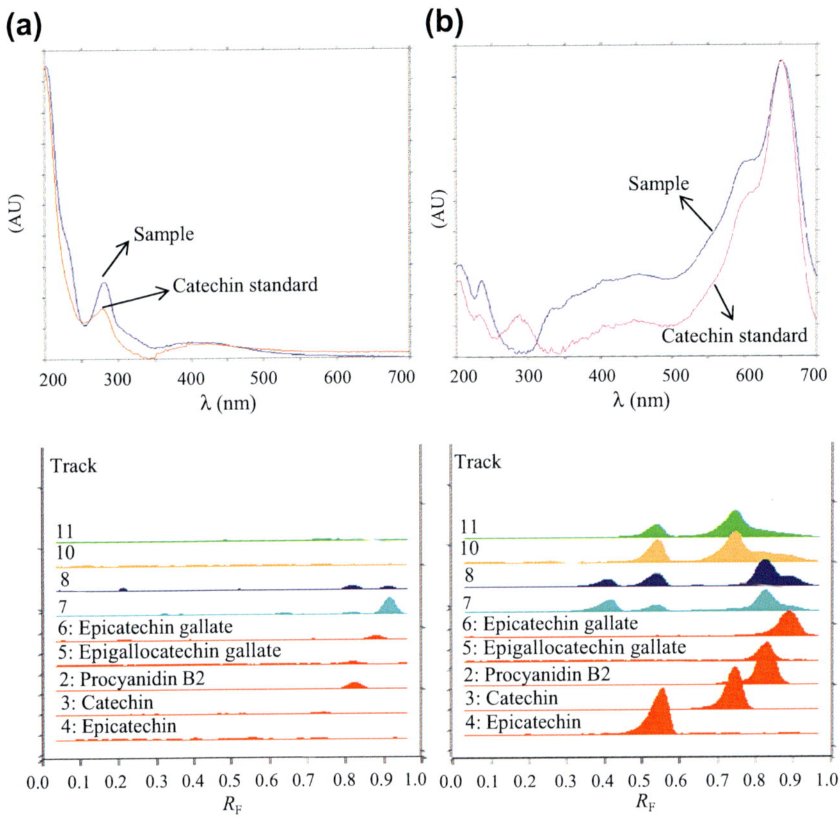

FIGURE 21.15 Densitograms of selected corresponding tracks from Figure 21.14 presented on the same scale after scanning in the absorption/reflectance mode before (a: 280 nm) and after (b: 655 nm) derivatization with 4-dimethylaminocinnamaldehyde (DMACA) DR. Overlapping in-situ absorption spectra of (+)-catechin reference standard and the corresponding retention zone of the sample test solution before (a) and after derivatization (b). Tracks: 2: B2 (50 ng); 3: C (30 ng); 4: EC (30 ng); 5: ECg (40 ng); 6: EGCg (50 ng); 7: hard capsules with extracts of green tea, black pepper, mango and guarana seeds; 8: soft capsules with green tea and conjugated linolenic acid; 10: hard capsules with grape seeds' proanthocyanidins, wine extract containing resveratrol and soy phosphatidylcholine; 11: hard capsules with flavanols and oligomeric proanthocyanidins from French pine bark and grape seeds.

Additional confirmation of identity would also be possible by TLC-MS or a comprehensive TLC × MS procedure (screening the entire chromatogram) after the separation on the cellulose or silica gel sorbents as described in Ref. [42] for the analysis of the same analytes in plant extracts. Although, the sensitivity achieved by using the cellulose sorbent was reported to be superior, the ionization suppression encountered with silica gel sorbent was overcome by a simple online separation on an

HILIC column placed after the elution head of the TLC-MS interface, before the eluent entering the ion source [42].

21.4.5 Glucosamine

Glucosamine, an amino sugar from shell's chitin, is often incorporated in dietary supplements for stimulation of the joint's function and repair. Due to its instability, it is sold in different more stable forms. One of the methods for quantitative determination of glucosamine in dietary supplements is based on the separation on HPTLC silica gel plates developed in a saturated twin trough chamber by 1-butanol–glacial acetic acid–-water 3:1:1 (v/v), followed by postchromatographic derivatization with ninhydrin reagent and densitometry at 580 nm in the absorption/reflectance mode [31]. The method was validated for sensitivity (LOD and LOQ 1 μg), linearity (from 1 to 4 μg), accuracy, precision (repeatability) and ruggedness (intermediate precision). The method was successfully applied to a survey of nine dietary supplement products in the form of capsules and tablets containing different forms of glucosamine (e.g., glucosamine sulfate KCl, glucosamine HCl, and *N*-acetyl D-glucosamine, etc.) and other active (e.g., arginine) and inactive ingredients. The results normalized for free glucosamine enabled comparison between the products in terms of actual free glucosamine content, which cannot be done by consumers when comparing labeled data for different glucosamine forms on different products. *N*-acetyl D-glucosamine form did not react with ninhydrin unless the sample was hydrolyzed before TLC analysis. Determination of glucosamine in a dietary supplement containing dried extracts of the plants traditionally used for rheumatic disorders was performed also on HPTLC silica gel F_{254} plates developed in a saturated twin trough chamber using 2-propanol–ethyl acetate–8% ammonia solution 1:1:1 (v/v) as a developing solvent. Postchromatographic derivatization with modified anisaldehyde detection reagent was followed by densitometric scanning at 415 nm in the absorption/reflectance mode [72]. An additional method for the determination of glucosamine uses the separation on HPTLC amino F_{254} plates developed in a saturated vertical chamber with methanol–ethanol–25% aqueous ammonia 1:1:1 (v/v) as a developing solvent [32], followed by short 10 min heating of the plate at 160 °C to produce a derivative that was detected by absorption (LOD 25 ng) or fluorescence (LOD 15 ng) mode using a diode-array scanner (Tidas TLC 2010 system, J&M Aalen, Germany). This provides linear calibration curves from 25 to 4000 ng for both detection modes. No significant differences were observed between the two detection modes at analyses of glucosamine in different forms (free, sulfate, *N*-acetyl, etc.). The reason is thermal derivatization, which uses the carbonyl group of glucosamine instead of the amino group.

21.4.6 Others

For the quantitative determination of L-arginine hydrochloride in dietary supplements a new method was developed and validated for sensitivity, linearity, accuracy and precision and was applied to a survey of four products (capsules and tablets). After the separation on HPTLC silica gel plates developed with 1-butanol–glacial acetic acid–water 3:1:1 (v/v) in a saturated twin trough chamber followed by postchromatographic derivatization with ninhydrin spray detection reagent, quantitative evaluation was performed densitometrically at 495 nm in the absorption/reflectance mode [33]. LOD and LOQ were approximately 150 ng per band and the calibration curve was linear from 0.2 to 0.8 μg. The selectivity of the ninhydrin detection reagent for compounds with an -NH_2 group enabled determination of arginine even in multi-ingredient formulations containing several additional active ingredients (vitamins A, C, E, B6, B12, thiamine, riboflavin, niacin, folic acid, biotin, pantothenic acid, zinc, selenium, ginseng, and *Gingko biloba*) and several excipients that did not react.

A slightly different ninhydrin detection reagent was applied for the detection of L-carnitine after the separation on HPTLC cellulose plates in a vertical chamber (more details are not provided) using methanol–water–glacial acetic acid 10:2:0.1 (v/v) as a developing solvent [34]. The method for determination of L-carnitine in dietary supplements including drinks, capsules and tablets was validated for sensitivity, linearity, accuracy and precision. LOD and LOQ per band were 2.51 and 7.61 μg, respectively, and the calibration curve was linear from 10 to 40 μg.

A simple and reliable TLC method for quantitative determination of α-lipoic acid after the separation on C18 RP F_{254S} HPTLC plates using 2-propanol–methanol–acetone–water–acetic acid 6:4:2:8:0.2 (v/v) as a developing solvent followed by postchromatographic derivatization with palladium(II) chloride dipping reagent and densitometric scanning at 375 nm in the absorption/reflectance mode [35]. The detection at these conditions was five-fold more sensitive than the detection without derivatization at 332 and 215 nm, because of the poor absorptivity of α-lipoic acid. The method was validated for sensitivity, linearity, accuracy and precision and was used to analyze five dietary supplements products containing α-lipoic acid (capsules and tablets). LOD and LOQ were 0.12 and 0.39 μg, respectively. The calibration curve was linear in the range from 1.0 to 3.0 μg per band, and polynomial from 0.5 to 5.0 μg of α-lipoic acid per band. It was shown that other components such as bilberry extract, alfalfa, and spirulina blen do not interfere.

A TLC method for rapid screening and quantification of the anthocyanins cyanidin-3-O-rutinoside and cyanidin-3-O-glucoside was developed for routine analysis and quality control of raw materials and dietary supplements claiming to be or to contain acai (*Euterpe oleracea* Mart.) berries. Chromatographic separation of the selected anthocyanins was achieved on HPTLC silica gel F_{254} plates developed in a saturated twin trough chamber using ethyl acetate–formic acid–acetic acid–water 10:1.1:1.1:2.6 (v/v) as a developing solvent. Densitometry was performed at 520 nm in the absorption/reflectance mode. Calibration curves from 100 to 500 ng were polynomial for both compounds. The method was validated for precision, accuracy, repeatability and specificity (color and R_F values of the bands and spectral correlation) and was successfully tested in a survey of six dietary supplement products in the form of powders, capsules, and liquids. It was confirmed that the method is specific for the selected markers cyanidin-3-O-rutinoside and cyanidin-3-O-glucoside and any other material such as adulterants failed the test [36].

A new TLC method was developed for the determination of rosmarinic acid in dietary supplements and plant extracts. The separation was performed on HPTLC LiChrospher silica gel F_{254S} plates in an ADC2 chamber (38% relative humidity) with toluene–ethyl formate–formic acid 6:4:1 (v/v) as a developing solvent. Quantitative evaluation was performed by densitometry at 330 nm in the absorption/reflectance mode. The method was validated for accuracy, repeatability, intermediate precision, and linearity (132–660 ng) and was applied to the analysis of two dietary supplements products in the form of capsules [37].

A quantitative method for the determination of sterols and fatty acids in prostate health dietary supplements was developed. The separation on HPTLC silica gel plates developed in a saturated twin trough chamber using petroleum ether–diethyl ether–glacial acetic acid 80:20:1 (v/v) as a developing solvent was followed by postchromatographic derivatization with phosphomolybdic acid reagent (blue zones on a yellow background) and image analysis using ChromImage flatbed scanner densitometer with Galaxie-TLC software (AR2i Company). Sterols were quantified relative to the calibration curve for cholesterol, while fatty acids were quantified relative to the calibration curve for oleic acid. The method is rapid and precise and can be applied for quality control in the manufacturing of such products [25].

TLC with image analysis was used also for the determination of phospholipids (phosphatidylserine, phosphatidylethanolamine and phosphatidylcholine). The separation was carried out on HPTLC silica gel plates developed in a saturated horizontal developing chamber (tank configuration) with chloroform–methanol–25% ammonia solution 65:25:4 (v/v) as a developing solvent, followed by postchromatographic derivatization using a new, modified Dittmer–Lester spray reagent. The applicability of the method was confirmed by the determination of phosphatidylserine in dietary supplement product and by the determination of phosphatidylserine, phosphatidylethanolamine and phosphatidylcholine in biological samples [26].

An important natural anabolic ecdysteroid, 20-hydroxyecdysone, was determined by densitometry in the absorption/reflectance mode at 250 nm after separation on HPTLC silica gel 60 F_{254} plates in a saturated twin trough chamber using chloroform–methanol 8:2 (v/v) as a developing solvent. The method was validated for sensitivity (LOD 60 ng and LOQ 200 ng per band), linearity (from 0.2 to 1.0 μg), accuracy and precision and was applied to a survey of two dietary supplement products in the form of capsules and *Sida rhombifolia* L. methanol extracts. Additionally, postchromatographic derivatization with anisaldehyde/sulfuric acid reagent resulted in unique fingerprints for six different *Sida species* [73].

The indole alkaloid, yohimbine, advertised as a promising compound for treatment of erectile dysfunction, was determined in pharmaceutical preparations and dietary supplements by a simple and sensitive TLC method. TLC silica gel F_{254} plates were developed in a saturated twin trough chamber using chloroform–methanol–ammonia 97:3:0.2 (v/v) as a developing solvent. Quantitative evaluation of yohimbine hydrochloride was performed by densitometry at 280 nm in the absorption/reflectance mode. The method was validated for selectivity, accuracy, sensitivity (LOD 5 ng, LOQ 40 ng per band) and linearity (from 80 to 1000 ng) [38].

21.5 FINGERPRINTING AND DETECTION OF ADULTERANTS

TLC with its multiple detection possibilities is one of the chromatographic techniques used for providing fingerprint profiles of plant materials serving for chemical identification as a part of quality control of raw materials. In addition, TLC has a strong potential for the fast screening concerning safety issues related to intentional and unintentional adulterations (e.g., with plant materials, plant derived materials and drugs), which are unfortunately in many cases unsafe.

The identification of green tea and green tea extract as dietary supplement raw materials was performed on HPTLC silica gel F_{254} plates developed in a twin trough chamber with four different developing solvents providing fingerprints for different bioactives: ethyl formate–toluene–formic acid–water 30:1.5:4:3 (v/v), saturated chamber - for flavonoids and phenolic acids; toluene–acetone–formic acid 9:9:2 (v/v) for flavan-3-ols; ethyl acetate–methanol–water 20:2.7:2 (v/v) for methylxanthines; l-butanol–acetone–acetic acid–water 7:7:2:4 (v/v) for amino acids. The method for flavan-3-ols was validated for specificity, stability, reproducibility, and robustness using botanical reference material. The flavan-3-ols pattern enables the discrimination of green tea from black and other specialty teas [62].

Hoodia gordonii, a Kalahari Desert slow-growing succulent, is traditionally known to have anorexic properties, which are the reason for the increased demand for raw plant material and its products. This results in increased concern for the possibility of adulteration, safety and quality of the products containing *Hoodia*. A simple and rapid chemical fingerprinting of four *Hoodia species*, their seven related genera as well as dietary supplements claiming to contain *Hoodia gordonii* was performed on HPTLC silica gel 60 F_{254} plates developed in a saturated twin trough chamber using dichloromethane–methanol–water 75:17:2.2 (v/v) [74]. Detection was performed by postchromatographic derivatization with anisaldehyde detection reagent. The authentication of *Hoodia gordonii* was achieved by comparison of the colors and R_F values of the bands of the TLC fingerprint with those of 11 marker compounds – pregnane glycosides standards, including the patented oxypregnane glycoside P57. The presence of *Hoodia gordonii* was confirmed only in two of nine analyzed dietary supplement products claiming to contain it. Additionally, some complex mixtures with other botanicals showed that if the sample contains *Hoodia*, it will provide at least a partial match with the fingerprint of a mixture of 11 standard markers.

The roots of a native North American perennial, *Actaea racemosa* L. (black cohosh; syn. *Cimicifuga racemosa* L. Nutt.), are widely used in dietary supplements, primarily for management of menopausal symptoms. Increased consumption increases the possibilities for product adulterations due to insufficient cultivation sources (most come from wildcrafting sources) and misidentification of plant species because of its similar appearance and growing locations as other species such as *Actaea pachypoda* Ell. (white cohosh) and more commonly *Actaea podocarpa* DC (yellow cohosh). As several species of the genus *Actaea* listed as possible adulterants are poisonous, this is a serious health concern, which requires development of inexpensive and effective methodology to evaluate identity and adulteration of black cohosh plant material. Separation on HPTLC Bioluminex silica gel F_{254} plates developed in a saturated twin trough chamber with toluene–ethyl formate–formic acid 5:3:2 (v/v) as a developing solvent was followed by different types of detection documented on images captured before (at 254 and 366 nm) and after postchromatographic derivatization by spraying with 5% H_2SO_4–anisaldehyde detection reagent (at white light). Additionally, Bioluminex detection was performed on the developed plate, coated with buffered luminescent *Vibrio fischeri*. TLC and TLC-bioluminescence methods gave unique profiles for *Actaea racemosa* root and *Actaea podocarpa* root. A marker compound detected in *Actaea pachypoda* samples could be used to distinguish *Actaea pachypoda* from *Actaea racemosa* root. Simultaneous separation of up to 20 complex sample test solutions followed by different types of the detection (UV/visible absorption, bioluminescence) provides multiple fingerprints, which are finally after the evaluation resulting in a unique chemical and biological profile for each analyzed sample test solution. This can be used for the identification of plant material, as well as potentially toxic compounds [75].

A TLC method for rapid screening and quantification of selected markers – anthocyanins cyanidin-3-O-rutinoside and cyanidin-3-O-glucoside described in detail above for routine analysis and quality control of raw materials and dietary supplements claiming to be or to contain acai (*Euterpe oleracea* Mart.) berries is specific for both selected markers and can easily be applied for the detection of adulterants [36].

Recently, a new TLC densitometric method was developed and validated for screening and determination of sibutramine (an approved drug used as appetite suppressant until 2010, withdrawn from European and the U.S. market) in adulterated herbal slimming dietary supplements. The separation on HPTLC silica gel plates developed in a saturated ADC2 chamber (Camag) using toluene–methanol 9:1 (v/v) as a developing solvent was followed by densitometric scanning at 225 nm in the absorption/reflectance mode. A survey of 50 slimming dietary supplements purchased via the Internet confirmed the adulteration with sibutramine in half of the products. Identification of sibutramine was confirmed also by TLC-MS. This proved the need for such simple and rapid analytical methods for detection of sibutramine adulterations in a large number of samples without carryover and without compromising accuracy [39].

A simple TLC method, with quantification based on image analysis and densitometry, was developed for rapid determination of sibutramine hydrochloride in herbal slimming products. The separation achieved on TLC silica gel F_{254} plates developed with toluene–n-hexane–diethylamine 9:1:0.3 (v/v) was followed by postchromatographic derivatization with Dragendorff reagent. Quantification was performed by image analysis of the images captured by a flatbed scanner using Sorbfil TLC Videodensitometer software, as well as by a slit-scanning densitometer at 225 nm in the absorption/reflectance mode. The method was validated for specificity, precision, accuracy and robustness and was applied for determination of sibutramine hydrochloride in five herbal dietary supplement products in the form of capsules as well as in 15 other herbal slimming formulations. LOD and LOQ were 190 and 634 ng, respectively and the calibration curves were polynomial in the range from 1 to 6 μg for image analysis and densitometry. The results obtained by image analysis and densitometry were in good agreement, which confirmed that image analysis can be used as an alternative to densitometry. Sibutramine hydrochloride was determined in 2 of 5 dietary supplements products and in 4 of 15 other herbal slimming formulations [27].

Simultaneous determination of eight phosphodiesterase type 5 inhibitors (PDE5-Is; sildenafil, hongdenafil, homosildenafil, hydroxyhomosildenafil, vardenafil, pseudovardenafil, tadalafil and aminotadalafil) as potential adulterants in herbal dietary supplements for male sexual health was performed by a TLC method [76]. PDE5-Is were separated on HPTLC silica gel plates developed in a saturated twin trough chamber using chloroform–ethyl acetate–methanol–water 40:40:15:11 (v/v) as a developing solvent. Images of the plates captured at UV 254 nm, comparison of the R_F values of the separated bands, and in-situ scanned absorption spectra (also peak purity index) of the reference standards and sample test solutions were used for qualitative analysis in a survey of eight dietary supplement products in the form of tablets, capsules, and liquid. The obtained results were confirmed by an HPLC-PDA-MS method.

A new method was also developed for the detection of antidiabetes drugs (phenformin hydrochloride, metformin hydrochloride, rosiglitazone maleate, and pioglitazone hydrochloride, PIO) used to adulterate botanical dietary supplements to treat diabetes. The separation on TLC and HPTLC silica gel F_{254} plates developed in dichloromethane–methanol–water 8:2:0.2 (v/v) or dichloromethane–cyclohexane–methanol–acetic acid 8:8:0.5:1.5 (v/v) (for PIO detection) was followed by placing silver colloids on the sorbent and qualitative identification of the analytes by surface-enhanced Raman spectroscopy (SERS). The method is robust, selective and sensitive and enables the detection of 0.001% (w/w) of adulteration (1% is usually the smallest amount of adulterant). The TLC-SERS method was used to analyse of 11 dietary supplement products and the results were verified by an LC-MS method [41].

21.6 CONCLUSIONS

Continuous and fast development of dietary supplement analytics is critical for keeping up with the fast appearance of new products with innovative active ingredients (new combinations, various complexes, encapsulated materials, nanomaterials, etc.) and possible adulterations. The importance of analytical methods cannot be stressed enough, because hundreds of known and unknown plant ingredients are included in these products resulting in endless combinations. Instrumental TLC with all of its available detection techniques including image analysis and mass spectrometry combined with chemometrics and worldwide available databases with plates' images and MS (MS^n) spectra will provide an important tool for manufacturers, scientists, and regulatory agencies. TLC when supported by sample preparation techniques can definitely be used in qualitative and quantitative analysis of bioactive ingredients and excipients, depending on the number and chemical diversity of the constituents of the dietary supplement. Apart from being used for chemical fingerprinting, TLC has a tremendous potential (1) in the analysis of adulterants, (2) in supporting the identification and characterization of biomarkers, and (3) in the generation of the standard reference materials.

References

[1] The dietary supplement health and education act of 1994, Public law 103−417. Congress of the United States. 108. Statue 4325. October 25, 1994.

[2] Directive 2002/46/EC of the European parliament and of the council of 10 June 2002 on the approximation of the laws of the Member States relating to food supplements. Off J Eur Union Legis 2002;183:51−7.

[3] Bagchi D, editor. Nutraceutical and functional food regulations in the United States and around the world. 1st ed. Amsterdam: Elsevier Academic Press; 2008.

[4] Bagchi D, editor. Nutraceutical and functional food regulations in the United States and around the world. 2nd ed. Amsterdam: Elsevier Academic Press; 2014.

[5] Guidance for industry: Current good manufacturing practice in manufacturing, packaging, labeling, or holding operations for dietary supplements; small entity compliance guide [Internet]. Silver Spring (MD): U.S. Food and Drug Administration; December 2010 [cited 2014 Feb 15]. Available from, http://www.fda.gov/Food/Guidance Regulation/GuidanceDocumentsRegulatoryInformation/DietarySupplements/ucm238182.htm.

[6] Dietary supplement reference standards. In: Official USP reference standards catalog [Internet], Rovkwille (MD): U.S. Pharmacopeial Convention; 2014. p. 160−6 [cited 2014 May 15]. Available from, http://www.uspcatalog.com/uspcatalog/20140506#pg178.

[7] 110.9 Dietary supplement materials (includes nutraceuticals and herbs) [Internet]. Gaithersburg: National Institute of Standards and Technology; August 15, 2012 [cited 2014 Feb 15]. Available from, https://www-s.nist.gov/srmors/viewTableH.cfm?tableid=79.

[8] F.16 C. HPTLC identification of ginkgo (*Gingko biloba*): Ginkgolic acid [Internet]. Muttenz: CAMAG Chemie-Erzeugnisse & Adsorptionstechnik AG; [date unknown] [cited 2014 March 21]. Available from: http://www.camag.com/en/tlc_hptlc/camag_laboratory/methods.cfm.

[9] F.16 B. HPTLC identification of ginkgo (*Gingko biloba*): Flavonoids [Internet]. Muttenz: CAMAG Chemie-Erzeugnisse & Adsorptionstechnik AG; [date unknown] [cited 2014 March 21]. Available from: http://www.camag.com/en/tlc_hptlc/camag_laboratory/methods.cfm.

[10] PubChem Substance Caffeine [Internet]. Bethesda (MD): U.S. National Library of Medicine; 2014 [cited 2014 Feb 15]. Available from, http://pubchem.ncbi.nlm.nih.gov/summary/summary.cgi?sid=46506408.

[11] Morlock G, Kovar KA. Detection, identification and documentation. In: Sherma J, Fried B, editors. Handbook of thin-layer chromatography. Chromatographic science series. 3rd ed., vol. 89. New York: Marcel Dekker Inc; 2003. p. 207−38. revised and expanded.

[12] Prošek M, Vovk I. Basic principles of optical quantification in TLC. In: Sherma J, Fried B, editors. Handbook of thin-layer chromatography. Chromatographic science series. 3rd ed., vol. 89. New York: Marcel Dekker Inc; 2003. p. 277−306. revised and expanded.

[13] Gössi A, Scherer U, Schlotterbeck G. Thin-layer chromatography-nuclear magnetic resonance spectroscopy – a versatile tool for pharmaceutical and natural products analysis. Chimia 2012;66:347–9.

[14] Vovk I, Simonovska B, Andrenšek S, Vuorela H, Vuorela P. Rotation planar extraction and rotation planar chromatography of oak (*Quercus robur* L.) bark. J Chromatogr A 2003;991:267–74.

[15] Rodić Z, Simonovska B, Albreht A, Vovk I. Determination of lutein by high-performance thin-layer chromatography using densitometry and screening of major dietary carotenoids in food supplements. J Chromatogr A 2012;1231:59–65.

[16] Simonovska B, Vovk I. TLC separation of some triterpenoids. In: Vovk I, Medja A, editors. Proceedings of the international symposium 'Planar chromatography today 2002' held in Novo mesto, Slovenia, October 4–6, 2002. Ljubljana: National institute of Chemistry; 2002. p. 31–5.

[17] Naumoska K, Simonovska B, Albreht A, Vovk I. TLC and TLC-MS screening of ursolic, oleanolic and betulinic acids in plant extracts. J Planar Chromatogr 2013;26:125–31.

[18] Wójciak-Kosior M. Separation and determination of closely related triterpenic acids by high performance thin-layer chromatography after iodine derivatization. J Pharmaceut Biomed Anal 2007;45:337–40.

[19] Glavnik V, Simonovska B, Vovk I. Densitometric determination of (+)-catechin and (-)-epicatechin by 4-dimethylaminocinnamaldehyde reagent. J Chromatogr A 2009; 1216:4485–91.

[20] Vovk I, Prošek M, Kaiser RE. Image analysis. In: Nyiredy S, editor. Planar chromatography: a retrospective view for the third millennium. Budapest: Springer Scientific Publisher, cop; 2001. p. 464–88.

[21] Busch KL. Planar chromatography with mass spectrometric detection. In: Nyiredy S, editor. Planar chromatography: a retrospective view for the third millennium. Budapest: Springer Scientific Publisher, cop; 2001. p. 261–77.

[22] Vovk I, Močnik G. Photoacoustics and photothermy in planar chromatography. In: Nyiredy S, editor. Planar chromatography: a retrospective view for the third millennium. Budapest: Springer Scientific Publisher cop; 2001. p. 312–35.

[23] Gibkes J, Vovk I, Bolte J, Bicanic D, Bein B, Franko M. Photothermal characterization of thin-layer chromatography plates. J Chromatogr A 1997;786:163–70.

[24] Vovk I, Franko M, Gibkes J, Prošek M, Bicanic D. Depth profiling of TLC plates by photoacoustic spectroscopy. J Planar Chromatogr 1997;10:258–62.

[25] Halkina T, Sherma J. Determination of sterols and fatty acids in prostate health dietary supplements by silica gel high performance thin layer chromatography with visible mode densitometry. J Liq Chromatogr Relat Technol 2007;30:2329–35.

[26] Dyńska-Kukulska K, Ciesielski W, Zakrzewski R. The use of a new, modified Dittmer-Lester spray reagent for phospholipid determination by the TLC image analysis technique. Biomed Chromatogr 2013;27:458–65.

[27] Phattanawasin P, Sotanaphun U, Sukwattanasinit T, Akkarawaranthorn J, Kitchaiya S. Quantitative determination of sibutramine in adulterated herbal slimming formulations by TLC-image analysis method. Forensic Sci Int 2012;219:96–100.

[28] Ristivojević P, Andrić FL, Trifković JĐ, Vovk I, Stanisavljević LŽ, Tešić ŽL, et al. Pattern recognition methods and multivariate image analysis in HPTLC fingerprinting of propolis extracts. J Chemom 2014;28:301–10.

[29] Vasta JD, Sherma J. Analysis of lycopene in nutritional supplements by silica gel high-performance thin-layer chromatography with visible-mode densitometry. Acta Chromatogr 2008;20:673–83.

[30] Sechrist J, Pachuski J, Sherma J. Quantification of lutein in dietary supplements by reversed-phase high-performance thin-layer chromatography with visible-mode densitometry. Acta Chromatogr 2002;12:151–8.

[31] Sullivan C, Sherma J. Development and validation of an HPTLC-densitometry method for assay of glucosamine of different forms in dietary supplement tablets and capsules. Acta Chromatogr 2005;15:119–30.
[32] Bleichert M, Eckhardt HS, Klein KF, Spangenberg B. A simple and reliable method for quantification of glucosamine in nutritional supplements. J Planar Chromatogr 2008;21:55–9.
[33] Hess B, Sherma J. Quantification of arginine in dietary supplement tablets and capsules by silica gel high-performance thin-layer chromatography with visible mode densitometry. Acta Chromatogr 2004;14:60–9.
[34] Dąbrowska M, Sieczka E, Starek M. TLC assay of L-carnitine in dietary supplements. J Planar Chromatogr 2012;25:450–5.
[35] Ravanić N, Filipić S, Nikolić K, Popović G, Vovk I, Simonovska B, et al. Analysis of α-lipoic acid in drug formulations and dietary supplement preparations. Acta Chromatogr 2009;21:433–41.
[36] Rumalla CS, Avula B, Wang YH, Smillie TJ, Khan IA. Densitometric-HPTLC method development and analysis of anthocyanins from acai (*Euterpe oleracea* Mart.) berries and commercial products. J Planar Chromatogr 2012;25:409–14.
[37] Coran SA, Mulas S, Mulinacci N. Crucial aspects of high performance thin layer chromatography quantitative validation. The case of determination of rosmarinic acid in different matrices. J Chromatogr A 2012;1220:156–61.
[38] Badr JM. A validated high performance thin layer chromatography method for determination of yohimbine hydrochloride in pharmaceutical preparations. Phcog Mag 2013;9:4–8.
[39] Mathon C, Ankli A, Reich E, Bieri S, Christen P. Screening and determination of sibutramine in adulterated herbal slimming supplements by HPTLC-UV densitometry. Food Addit Contam Part A 2014;31:15–20.
[40] Glavnik V, Simonovska B, Albreht A, Vovk I. TLC and HPLC screening of *p*-coumaric acid, *trans*-resveratrol, and pterostilbene in bacterial cultures, food supplements, and wine. J Planar Chromatogr 2012;25:251–8.
[41] Zhu Q, Cao Y, Cao Y, Chai Y, Lu F. Rapid on-site TLC–SERS detection of four antidiabetes drugs used as adulterants in botanical dietary supplements. Anal Bioanal Chem 2014;406:1877–84.
[42] Smrke S, Vovk I. Comprehensive thin-layer chromatography mass spectrometry of flavanols from *Juniperus communis* L. and *Punica granatum* L. J Chromatogr A 2013;1289:119–26.
[43] Sajewicz M, Staszek D, Natić M, Wojtal Ł, Waksmundzka-Hajnos M, Kowalska T. TLC-MS versus TLC-LC-MS fingerprints of herbal extracts. Part II. Phenolic acids and flavonoids. J Liq Chromatogr Relat Technol 2011;34:864–87.
[44] Sajewicz M, Staszek D, Natić M, Waksmundzka-Hajnos M, Kowalska T. TLC-MS versus TLC-LC-MS fingerprints of herbal extracts. Part III. Application of the reversed-phase liquid chromatography systems with C18 stationary phase. J Chromatogr Sci 2011;49:560–7.
[45] Malet-Martino M, Martino R, Bonnet PA. Nutritional supplements. In: Pico Y, editor. Chemical analysis of food: techniques and applications. 1st ed. Amsterdam: Academic Press; 2012. p. 539–73.
[46] Ankli A, Widmer V, Reich E. Medicines and dietary supplements produced from plants. In: Waksmundzka-Hajnos M, Sherma J, Kowalska T, editors. Thin layer chromatography in phytochemistry. Chromatographic science series, vol. 99. Boca Raton: CRC Press; 2008. p. 37–57.
[47] Vovk I, Prošek M. Reproducibility of densitometric and image analysing quantitative evaluation of thin-layer chromatograms. J Chromatogr A 1997;779:329–36.
[48] Vovk I, Franko M, Gibkes J, Prošek M, Bicanic D. Photoacoustic investigations of secondary chromatographic effects on TLC plates. Anal Sci 1997;13S:191–4.

[49] Vovk I, Prošek M. In-depth distribution in quantitative TLC. In: Wilson ID, Cooke M, Poole CF, editors. Encyclopedia of separation science, vol. 7. London, San Diego: Academic Press cop; 2000. p. 3087–94.
[50] Martelanc M, Vovk I, Simonovska B. Separation and identification of some common isomeric plant triterpenoids by thin-layer chromatography and high-performance liquid chromatography. J Chromatogr A 2009;1216:6662–70.
[51] Prošek M, Medja A, Katic M. Digital analysing system for thin-layer chromatography. In: Kaiser RE, editor. Proceedings of the second international symposium on instrumental high-performance thin-layer chromatography held in Würzburg/West Germany, April 17–19, 1985. Bad Duerkheim: Institute for Chemistry; 1985. p. 367–71.
[52] Vovk I, Prošek M. Quantitative evaluation of chromatograms from totally illuminated thin-layer chromatographic plates. J Chromatogr A 1997;768:329–33.
[53] Vovk I, Golc-Wondra A, Prošek M. Validation of an HPTLC method for determination of caffeine. J Planar Chromatogr 1997;10:416–9.
[54] USP dietary supplements compendium: The authoritative reference. 1st ed. Rockville (MD): U.S. Pharmacopeial Convention; 2009.
[55] Morlock G, Schwack W. Coupling of planar chromatography to mass spectrometry. Trends Anal Chem 2010;29:1157–71.
[56] Cheng SC, Huang MZ, Shiea J. Thin layer chromatography/mass spectrometry. J Chromatogr A 2011;1218:2700–11.
[57] Luftmann H. A simple device for the extraction of TLC spots: direct coupling with an electrospray mass spectrometer. Anal Bioanal Chem 2004;378:964–8.
[58] Vovk I, Popović G, Simonovska B, Albreht A, Agbaba D. Ultra-thin-layer chromatography mass spectrometry and thin-layer chromatography mass spectrometry of single peptides of angiotensin-converting enzyme inhibitors. J Chromatogr A 2011;1218: 3089–94.
[59] Van Berkel GJ, Tomkins BA, Kertesz V. Thin-layer chromatography/desorption electrospray ionization mass spectrometry: Investigation of goldenseal alkaloids. Anal Chem 2007;79:2778–89.
[60] Brucker Daltonics [Internet] Application note # MT-94 Direct read-out of thin layer chromatography (TLC) using MALDI-TOF. Billerica (MA): Brucker Daltonics Inc; January 2009 [cited 2014 Feb 15]. Available from, http://www.bruker.com/fileadmin/user_upload/8-PDF-Docs/Separations_MassSpectrometry/Literature/literature/ApplicationNotes/MT-94-%20MALDI-TLC.pdf.
[61] Abourashed EA, Mossa JS. HPTLC determination of caffeine in stimulant herbal products and power drinks. J Pharm Biomed Anal 2004;36:617–20.
[62] Reich E, Schibli A, Widmer V, Jorns R, Wolfram E, DeBatt A. HPTLC Methods for identification of green tea and green tea extract. J Liq Chromatogr Relat Technol 2006;29: 2141–51.
[63] Jork H, Funk W, Fischer W, Wimmer H. Dünnschicht-Chromatographie: Reagenzien und Nachweismethoden, Bd, Physikalishe und chemische Nachweismethoden: Grundlagen, reagenzien I, vol 1a. Weinheim: VCH Verlagsgesellschaft; 1989.
[64] Vovk I, Simonovska B, Vuorela P, Vuorela H. Optimization of separation of (+)-catechin and (−)-epicatechin on cellulose TLC plates. J Planar Chromatogr 2002;15:433–6.
[65] Vovk I, Simonovska B, Vuorela H. Separation of eight selected flavan-3-ols on cellulose thin-layer chromatographic plates. J Chromatogr A 2005;1077:188–94.
[66] Thies M, Fischer R. New color reaction for microchemical detection and quantitative determination of catechins. Mikrochim Acta 1971;1:9–13.
[67] Schulz JM, Herrmann K. Occurrence of catechins and proanthocyanidins in spices. 5. Phenolics of spices. Z Lebensm-Unters Forsch 1980;171:278–80.
[68] Lea AGH. Phenolics of ciders-oligomeric and polymeric procyanidins. J Sci Food Agric 1978;29:471–7.

[69] Jork H, Funk W, Fischer W, Wimmer H. Thin-Layer Chromatography: Reagents and detection methods. Physical and chemical detection methods: activation reactions, reagent sequences, reagents II, vol 1b. Weinheim: VCH Verlagsgesellschaft; 1994.

[70] Glavnik V, Simonovska B, Vovk I. Comparison of TLC and HPLC methods used for analysis of (−)-epicatechin and its dimer procyanidin B2 in chocolate. J Planar Chromatogr 2010;23:230−2.

[71] Glavnik V, Simonovska B, Vovk I, Mutavdžić Pavlović D, Ašperger D, Babić S. Quantification of (−)-epicatechin and procyanidin B2 in Chocolates. J Planar Chromatogr 2011;24:482−6.

[72] Esters V, Angenot L, Brandt V, Frédérich M, Tits M, Van Nerum C, et al. Validation of a high-performance thin-layer chromatography/densitometry method for the quantitative determination of glucosamine in a herbal dietary supplement. J Chromatogr A 2006;1112:156−64.

[73] Jadhav AN, Rumalla CS, Avula B, Khan IA. HPTLC method for determination of 20-hydroxyecdysone in *Sida rhombifolia* L. and dietary supplements. Chromatographia 2007;66:797−800.

[74] Rumalla CS, Avula B, Shukla YJ, Wang YH, Pawar RS, Smillie TJ, et al. Chemical fingerprint of *Hoodia* species, dietary supplements, and related genera by using HPTLC. J Sep Sci 2008;31:3959−64.

[75] Verbitski SM, Gourdin GT, Ikenouye LM, McChesney JD, Hildreth J. Detection of *Actaea racemosa* adulteration by thin-layer chromatography and combined thin-layer chromatography-bioluminescence. J AOAC Int 2008;91:268−75.

[76] Cai Y, Cai TG, Shi Y, Cheng XL, Ma LY, Ma SC, et al. Simultaneous determination of eight PDE5-is potentially adulterated in herbal dietary supplements with TLC and HPLC-PDA-MS methods. J Liq Chromatogr Relat Technol 2010;33:1287−306.

Index

Note: Page numbers followed by "b", "f" and "t" indicate boxes, figures and tables respectively

C

D

F

M

U

V

Edwards Brothers Malloy
Thorofare, NJ USA
February 19, 2015